JOHN C. PFEIFFER, P.E.
Pfeiffer Engineering Co.
3624 Frankfort Ave.
Louisville, KY 40207

Accident Prevention Manual for Industrial Operations

Engineering and Technology

Occupational Safety and Health Series

The National Safety Council's OCCUPATIONAL SAFETY AND HEALTH SERIES is composed of four volumes written to help readers establish and maintain safety and health programs. The latest information on establishing priorities, collecting and analyzing data to help identify problems, and developing methods and procedures to reduce or eliminate illness and accidents, thus mitigating injury and minimizing economic loss resulting from accidents, is contained in all volumes in the series:

ACCIDENT PREVENTION MANUAL FOR INDUSTRIAL OPERATIONS (2-Volume Set)
  Administration and Programs
  Engineering and Technology
FUNDAMENTALS OF INDUSTRIAL HYGIENE
INTRODUCTION TO OCCUPATIONAL HEALTH AND SAFETY

Other hardcover books published by the Council include:
MOTOR FLEET SAFETY MANUAL
SUPERVISORS GUIDE TO HUMAN RELATIONS
SUPERVISORS SAFETY MANUAL

Accident Prevention Manual for Industrial Operations

Engineering and Technology

Ninth Edition

Project Editor: Patricia M. Laing
Interior Design and Composition: North Coast Associates
Cover Design: Russell Schneck Design
Printed by: R. R. Donnelley & Sons

Library of Congress Cataloging in Publication Data
National Safety Council.
Accident Prevention Manual for Industrial Operations: Administration and Programs
International Standard Book Number 0-87912-136-X
Library of Congress Catalog Card Number: 86-63578
15M1287   Product Number 121.43

Preface

The ninth edition of the *ACCIDENT PREVENTION MANUAL FOR INDUSTRIAL OPER-ATIONS* is published in two volumes. The *Administration and Programs* volume encompasses management techniques, governmental regulations, and programs for safety and health professionals. The *Engineering and Technology* volume covers more technical information vital to the safety and health professional. The National Safety Council's *FUNDAMENTALS OF INDUSTRIAL HYGIENE,* third edition, and *INTRODUCTION TO OCCUPATIONAL HEALTH AND SAFETY* are additional volumes needed to complete readers' safety and health libraries. Contact the National Safety Council for more information.

The ninth edition of the *ACCIDENT PREVENTION MANUAL FOR INDUSTRIAL OPER-ATIONS* is the cumulation of facts and ideas that have become part of the safety movement and should be used to organize and transmit information of value to safety and health professionals—indeed, anyone—committed to preventing accidents and preserving well-being. Covering a broad spectrum of subjects, this *Manual* pinpoints problem areas and directs the reader to the appropriate sources of help.

As used in the *ACCIDENT PREVENTION MANUAL,* the term "accident" means that occurrence in a sequence of events that usually produces unintended injury, illness, death, and/or property damage. Prevention of such occurrences should be the responsibility of employees of every level. A second term, "safety and health professional," is used to mean all those interested in or affected by occupational safety and health.

New material

The ninth edition has expanded and updated material in every chapter. References in each chapter have been revised to reflect current sources and many photos have been updated. Some specific changes are listed below.

Chapter 1—new sections on crane operating rules and on the use of color in industrial facilities for safety and for psychological well-being. Chapter 2—a new section on tower cranes. Chapter 3—new sections on lifting techniques and aerial baskets. Chapter 4—a new section on belt conveyors. Chapter 9—completely revised text with many new photographs. Chapter 13—new sections on hazards; on the effects of toxic gases and particulates and on the health hazards of rays; new material on the changes in permissible exposure limits and exposure standards.

Contributors

The *ACCIDENT PREVENTION MANUAL FOR INDUSTRIAL OPERATIONS* is unique—the compilation of the experience and expertise of contributors from all major occupations and industries. Each of the reviewers and contributors is a practicing expert. To assure uniformity and accuracy, the final version of the text was reviewed by William J. Larson, PE, CSP. The National Safety Council and the editors wish to express their appreciation and gratitude to Mr. Larson and each of the contributors who devoted many hours to updating and checking the accuracy of this publication. These contributors to the *Engineering and Technology* volume include: Gary L. Arthur, Adrian Atkins, William Atkinson, Jr., Lon Ballard, Karl Benson, Roger Brooks, Warren K. Brown, Donald Browning, John Bussema, Charles A. Carlsson, Edward R. Carmody, Barry Cole, Larry L. Delf, William Drake, Walter Garyotis, John Gleichman, George Grady, Joseph W. Hart, G. C. "Red" Haynes, James E. Hodges, William Hooper, Robert D. Jordan, Paul Kuehne, Kevin J. Landkrohn, Al Leffler, Gary E. Lovested, Daniel Mark, Donald R. Maxwell, Floyd Miller, Glenn P. Mills, William M. Montante, George J. Nejmeh, Jr., Cliff Oliver, J. P. O'Donovan, Gary Page, Robert G. Peterson, Richard Peterson, Donald M. Pitsenberger, Charles Popke, Ralph "Cliff" Reel, Donald G. Ryder, Jack P. Short, Charles Simpson, William Slater, George E. Smith, Clarence Soulliard, William Sunter, Anthony Torres, William L. Wachs, Jerry D. Wampner, Ralph W. Wellington, George Williams, and William S. Wood.

The National Safety Council also wishes to thank members of its staff who contributed significantly to this *Manual:* George Benjamin, MD, Thomas Danko, Barbra Jean Dembski, Alan Hoskin, Larry Huey, Joseph Kelbus, Maureen Kerwin, Joseph Koeberl, Ronald Koziol, Joseph Lasek, John Laumer, Robert J. Marecek, Russell E. Marhefka, Robert O'Brien, Carl Piepho, Barbara Plog, Douglas Poncelow, Cynthia Pondel, and Philip E. Schmidt. The National Safety Council expresses special gratitude to Frank E. McElroy, PE, CSP, for his valuable contributions to this edition.

Contents

Industrial Buildings and Plant Layout

DESIGN FOR SAFETY

EFFICIENCY AND SAFETY in industrial operations can be greatly increased by careful planning of the location, design, and layout of a new plant or of an existing one in which major alterations are to be made. Numerous accidents, occupational diseases, explosions, and fires are preventable if suitable measures are taken at the earliest planning stages.

Ideally, safety and health professionals should conduct a hazard and operability study of a proposed plant while the engineering is in the development stage. They could then look for less expensive alternatives without sacrificing standards. They should approach the safety and health evaluation with the philosophy of seeking ways to remove hazards rather than ways to add protective equipment. For example, they might be able to cut inventories of hazardous material or substitute a less hazardous product. They might be able to reduce risks by intensifying a rate of mixing, or storing a gas at a lower temperature and pressure, or just by simplifying the plant design. For more detailed information on these approaches, see the References and Chapter 16 in this volume.

Buildings, processes, and equipment

Major factors determining the size, shape and type of buildings and structures are the nature of the processes and materials, maintenance, mechanical handling equipment, and working conditions. Also to be considered are economic or budgetary restraints.

Catastrophes resulting in large loss of life and heavy property damage often are due to inadequate planning-stage consideration of the physical and chemical properties of dangerous substances and their processing methods.

High hazard processes involving flammable and reactive materials should not be enclosed in buildings. If they are in buildings, ventilation must be adequate to prevent the materials from reaching explosive concentrations.

Personnel facilities, such as lunch rooms and medical, safety, and disaster services, are essential for reasons of health and treatment of injuries. They can be planned and located for maximum effectiveness.

Codes and standards

The policy of companies is to have their safety, health, loss and fire prevention specialists, and their insurance company review plans and specifications for new or remodeled plants. This saves costly alterations and installations that must be made after a plant is in operation in the event of failure to meet local and state requirements covering fire, accident, and health hazards.

Most ordinances and state or provincial laws require that authorized persons approve the means of both normal and emergency egress from buildings. In some states, plans for the installation of emergency lighting, fire alarm, and automatic sprinkler systems must be approved by jurisdictional authorities. Also, in certain states or provinces, exhaust and ventilating installations must also be approved.

Many national and local codes require means of controlling air-polluting industrial contaminants. In the U.S. the Environmental Protection Agency has set specific emission standards for raw wastes disposal and emission standards for industrial by-products. These should be checked in the plant planning stage.

The voluntary safety codes developed by various organizations establish standards for certain structures and equipment. Specifications for the construction of floor and wall openings and railings, for example, are given in American National Standard (ANSI) A12.1, *Safety Code for Floor and Wall Openings, Railings, and Toe Boards.* Proper electric wiring and electrical installations are covered in the *National Electrical Code,* issued by the National Fire Protection Association.

Requirements for fire extinguishing equipment and fire protection standards and codes for flammable liquids and gases, combustible solids, dusts, chemicals, and explosives are provided in the *National Fire Codes,* developed by the National Fire Protection Association. See Chapter 17, Fire Protection. Be sure to check the latest edition of the standards and codes. For the addresses of the appropriate agencies in countries other than the U.S., see Chapter 24, Sources of Help, in the *Administration and Programs* volume.

It must be remembered that "design by the code" is no substitute for intelligent engineering, and that codes only establish a minimum requirement which in many situations must be exceeded.

SITE PLANNING

Selection of a site involves consideration of many factors: possible hazards to the community, the relationship of the new structures to climate and terrain, space requirements, type and size of buildings, necessary facilities, transportation, market, and labor supply. Of the factors listed, those of direct concern to the safety and health professional are discussed next. While the safety and health professional is not directly responsible for specialized plant and process engineering and design, a primary duty is to keep the engineers and architects as safety conscious as possible, and try to have them design safety items into the system before the construction stage begins.

Location, climate, and terrain

Study of the climate of the area is required. The prevailing winds may determine the best location for processing equipment in relation to administration offices. In a hurricane, tornado, earthquake or flood area, plans and specifications should include suitable protective measures for personnel, and safety factors must be designed into the operation. (See the *Administration and Programs* Volume, Chapter 16, Planning for Emergencies.)

Space requirements

The size of a site is determined by both present space requirements and possible future expansion. In the past the growth of business and population has resulted in serious crowding, and the resulting congestion of buildings and other facilities has increased fire, accident, and health hazards. When toxic and flammable materials are handled, a buffer zone should be constructed.

New developments may figure in the size of the plant site. Some companies, anticipating increased use of air transportation, are allowing space for landing fields or heliports as they acquire new property. Plans for such installations should include all necessary safety precautions.

Fire protection codes specify minimum distances between buildings according to their size, type, and occupancy. Laws governing storage of explosives and of highly flammable materials specify minimum distances between manufacturing and storage facilities for various quantities of such materials, as well as minimum distances of the materials from adjoining property.

Ample space for outdoor storage areas is essential. When the area for storing such materials as steel, pipe, and lumber adjacent to shops proves insufficient, space must be provided elsewhere. Storing materials away from the plant necessitates added handling and transportation and increases both costs and accident risk.

Parking lots are best located inside the plant fence for convenience, protection, and safety. Since a considerable area may be necessary to accommodate employee and visitor cars, space requirements are important in site planning.

Well-located disposal areas for solid wastes also must be provided when a site is being laid out. Drainage and waste disposal must be planned in relation to space, terrain, plant needs, and effects upon surrounding municipal systems.

Scale relief models of the site in addition to maps can be very helpful to the planner in predesign, and will also aid in spotting potential safety problems.

These subjects are covered in detail next.

OUTSIDE FACILITIES

In the planning of outside facilities, safety considerations must be kept in mind in order to reduce the accident potential.

Enclosures

A fence around yards and grounds has many advantages—a fence keeps out trespassers who may interfere with work or be injured on the property. Fencing protects people from transformer stations, pits, sumps, stream banks under certain circumstances, and similar dangerous places. A galvanized woven wire fence makes a good enclosure.

A sufficient number of entrances should be provided to accommodate the volume of traffic, with clearance for loaded trucks and for switching personnel riding on the sides of railroad cars.

Because it is unsafe for pedestrians to use the entrances for railroads and motor vehicles, separate gates for pedestrians, convenient to their transportation and to their workplaces, should be provided. If a pedestrian entrance must be located near railroad tracks, part of the right of way should be protected to keep employees from shortcutting along the tracks. Good visibility in all directions is essential at entrances.

If pedestrian entrances must be located on busy thoroughfares or if workers cross railroad tracks on which trains are operated frequently, traffic signals should be installed and subways or pedestrian bridges built (Figure 1-1). Such precautions are especially important where parking lots must be located at a distance from the company.

If site plans call for bridges over streams, ditches, and other hazards, be sure pedestrian traffic is adequately protected by a fence or by handrails 42 in. (1.1 m) high and intermediate rails (Figure 1-1).

Shipping and receiving

Shipping and receiving facilities should fit in with overall material flow within the company or plant, and should aid efficient flow of materials into and out of production areas. Shipping and receiving areas should be designed to keep building heat and cooling losses at a minimum.

Figure 1-1. A bridge provides safe crossover of a freight yard. Be sure construction and personnel protection are adequate. (Printed with permission from Guardian Engineering & Development Company.)

Self-leveling dock boards, truck levelers, and cranes can facilitate loading and unloading.

Railroad siding

This commonly used shipping and receiving facility requires planning, especially if it is advantageous to use bulk raw, process, and maintenance materials. Tank car lots of hazardous materials require proper consideration for pressure piping, breakaway piping, valves, pumps, derails, excess-flow valves, and vapor return lines.

Each sidetrack should be protected from main line and public thoroughfares. Proper clearance between main plants and cars should be observed. More details are given in Chapters 7, Plant Railways and Elevators, and 16, Flammable and Combustible Liquids.

Roadways and walkways

The safety and health professional should always be alert to opportunities to assist the civil engineer in designing for optimum safety. Roadways in plant yards and grounds are sources of frequent accidents unless they are carefully laid out, substantially constructed, well surfaced and drained, and kept in good condition.

Heavy duty truck hauling requires roadways up to 50 ft (15 m) wide for two-way traffic with ample radii at curves. Grades, in general, are limited to a maximum of 8 percent. A slight crown is necessary for drainage, with ditches to carry off water.

Roadways should be located at least 35 ft (11 m) from buildings, especially at entrances. At loading docks an allowance of 1½ truck lengths is desirable to facilitate backing.

The regulation and control of traffic, signs, road layout, and markings should conform to federal and state or provincial practices. Guidance is provided in the *Manual on Uniform Traffic Control Devices for Streets and Highways,* ANSI D6.1.

Traffic signs and signals regulating speed and movements at hazardous locations are essential. Stop signs are specified for railroad crossings and entrances to main thoroughfares, and Sound Your Horn signs are necessary at sharp curves (blind corners) where view is obstructed and at entrances to buildings. Mirrors mounted to afford views around sharp turns or corners of buildings are of great help in preventing accidents if roadways must be built close to buildings. Barricades and Men Working signs are needed for construction and repair work. If roadways are used at night, traffic signs should be made of reflective or luminous materials.

Good walkways between outside facilities prevent injury to employees by helping them avoid stepping on round stones or into holes and ruts in rough ground. As far as possible, walkways should be the shortest distance from one building to another to discourage shortcutting. A walkway that must be next to railroad tracks should be separated from them by a fence or railing. Warning signs should be installed at railroad crossings and other hazardous places. If walkways are located clear of the eaves of buildings, the danger from falling icicles is reduced. In the snow belt, covered walkways are desirable for safety.

Concrete is preferred for sidewalks, especially in principal areas like entrances and between main buildings. Walkways should be kept in good condition, especially where they cross railroad tracks, and should be cleared of ice and snow.

Trestles

If employees are required to perform duties on trestles, a footwalk 5 ft. 1 in. (1.5 m) wide measured from the nearest rail, should be provided on at least one side. The footwalk should have a railing 42 in. (1.1 m) high plus an intermediate rail halfway between the top rail and the floor, and, on the exposed side, toe boards should be 4 in. (10 cm) high.

If employees travel on both sides of the track, crosswalks should be provided at convenient locations.

Metal gratings or screens installed over walkways or passages under trestles furnish protection from falling materials.

Openings for conveyors or hoppers require gratings or a grizzly with bars spaced not more than 12 in. (3 m) apart, to prevent employees from falling into the openings.

Parking lots

To reduce travel in the plant grounds, a desirable location for a parking lot is between an entrance and the locker room. If possible, the location should be such that no one need cross a roadway to go from parking lot to building. The entire parking area should be fenced and separated from other plant areas for security. The surface of the parking lot should be smooth and hard to prevent injuries that might occur from falls on stony or rough ground. Keep lots as level as possible with enough slope for drainage.

The use of white lines 4 to 6 in. wide (10 to 15 cm) to designate stalls reduces confusion and the number of backing accidents. Standard stalls are 9 ft wide and 20 ft long (2.7 m and 6.1 m). The center-to-center distance between parked cars depends upon the method of parking. (See Figure 1-2.) Be sure parking does not encroach on fire hydrant zones, approaches to corners, bus stops, loading zones, and clearance spaces for islands. Driveways should be a minimum of 25 ft (7.5 m) for two-way traffic and there should be no obstructions to viewing.

Angle parking has both advantages and disadvantages. The smaller the angle, the fewer the number of cars that can be parked in the same area. Aisle widths can be narrower but traffic is usually restricted to one way. On the other hand, angle parking is easier for drivers and it does not require a lot of space for sharp turns.

The area allowed per car in parking lots varies from 200 (19 m²) to more than 300 sq ft (28 m²) if aisles are included. Large, economically laid out lots may approach the 200-sq-ft figure; small or poorly configured lots may have a higher percentage of aisle space and may approach 300 sq ft per car. A large, commercial, attended lot is considered efficient if the layout keeps the space requirements to 240 sq ft (22 m²) per car.

Separate entrances for incoming and outgoing cars facilitate orderly traffic movements. Entrances should be designated with suitable signs. Traffic at exits to heavily traveled streets should

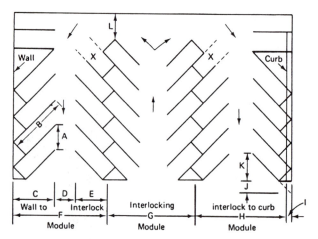

X = Stall not accessible in certain layouts

Parking layout dimensions (in ft) for 9-ft stalls
at various angles

Dimension	On diagram	Angle			
		45°	60°	75°	90°
Stall width, parallel to aisle	A	12.7	10.4	9.3	9.0
Stall length of line	B	25.0	22.0	20.0	18.5
Stall depth to wall	C	17.5	19.0	19.5	18.5
Aisle width between stall lines	D	12.0	16.0	23.0	26.0
Stall depth, interlock	E	15.3	17.5	18.8	18.5
Module, wall to interlock	F	44.8	52.5	61.3	63.0
Module, interlocking	G	42.6	51.0	61.0	63.0
Module, interlock to curb face	H	42.8	50.2	58.8	60.5
Bumper overhang (typical)	I	2.0	2.3	2.5	2.5
Offset	J	6.3	2.7	0.5	0.0
Setback	K	11.0	8.3	5.0	0.0
Cross aisle, one-way	L	14.0	14.0	14.0	14.0
Cross aisle, two-way	—	24.0	24.0	24.0	24.0

Figure 1-2. Parking lots, aisles, and stalls should be designed for maximum safety and efficiency both within the lot and on approaches to adjacent streets. Detailed diagram shows how dimensions vary with the parking angle. Parking angles may be varied within one lot in order to best use the space, if parking stalls are clearly marked and aisle space (D) is adequate for parking angle of largest degree. (Printed from *Parking Principles,* Highway Research Board, Special Report No. 125, 1971, p. 99.)

be controlled either by a traffic light or an acceleration or merging lane.

Speed signs and signs limiting areas to employees or visitors should be installed as circumstances require. These signs should conform to recommended standards, and should be similar to other street and highway signs. Details for signs and marking pavement are given in ANSI D6.1 *Manual on Uniform Traffic Control Devices for Streets and Highways.*

If the lot is used at night, adequate lighting should be provided for safety and for the prevention of theft; about 1 to 5 footcandles per square foot (11 to 54 lux) at a height of 36 in. (90 cm) should be an adequate level.

Equipment for ice and snow removal should be included at the planning stage.

Landscaping

Many companies are landscaping the grounds of both old and new plants. Landscaping should be designed so that trees and shrubbery do not create blind spots at roadway or walkway intersections. Proper maintenance to keep bushes at proper height is important if their growth will otherwise create blind spots. Details on grounds maintenance are given in the next chapter.

Waste disposal

Unsafe methods of waste disposal may cause injury to workers and the public and damage to property. Knowledge of the nature of the wastes is essential in planning suitable disposal methods, which must also conform to applicable municipal and state regulations. Since investigation, treatment, and disposal of wastes require specialized knowledge and training, qualified engineers should be consulted.

If the use of a city or district sewage system is planned, the officials in charge of the facility should be informed of the kind and amount of the wastes. If the wastes have properties that will interfere with operation of the sewage disposal plant, the officials may refuse to accept them.

Under no circumstances should toxic, corrosive, flammable, or volatile materials, or radiation wastes be drained into a public sewage system. It is especially important that plants handling or processing radioactive materials—even in small quantities—conform exactly to local and state regulations for disposal of radioactive wastes.

Many wastes can be disposed of on landfills, if state and local laws permit. Combustible materials, such as wood, scraps, and paper, may be burned in an incinerator. An incinerator should conform to applicable laws, be safely located, and properly attended.

Chemical wastes should be rendered harmless before being dumped; strong acids, for example, should be neutralized. Poisonous materials, magnesium chips, explosives, and similar substances require special procedures for safe disposal.

In some cases, it might be necessary to contract a private scavenger service to dispose of waste materials. If such a service is contracted, they must be told if any of the materials is hazardous. There are special services that will take hazardous substances and dispose of them properly.

Air pollution

Smoke from power plants and other sources and inert dusts from various types of plants may be nuisances or even sources of danger to the public. Provision for complying with smoke control ordinances should be made before construction is begun. Emissions should be checked against federal and local standards to be certain that they are not being violated.

Toxic smoke, fumes, or dusts are a serious problem in some industries. Tall stacks often are used to diffuse gases in the atmosphere. The effectiveness of this method depends on the nature and volume of the gases, the location of the plant, prevailing wind direction, and atmospheric conditions. Rain may absorb harmful gases and cause heavy damage to crops and environment. Conditions causing poor diffusion occasionally have temporarily shut down even those plants that have high stacks.

The possibility of recovering usable or marketable materials from wastes should always be investigated. Dusts and fumes often are recovered by filters, cyclones, electronic precipitators, or similar equipment either because of their intrinsic value or to keep them from polluting the atmosphere. Useful gases and vapors may often be economically recovered by means of spray towers or other recovery equipment.

Confined spaces

Many employees who have entered bins to loosen materials have been buried by slides and suffocated. Stairs provide the safest means of access to the bins, tanks, and similar structures. If fixed ladders are installed, they should conform to the requirements of the ANSI A10 series, *Safety Requirements for Building Construction.* The platform at the top of a tank should have standard railings and toeboards. Standard railings also should be installed along the edges adjoining the platform.

Pits for storage, for access to pipelines, and for other purposes are a source of frequent and serious injury unless precautions are taken to make them safe. Preferably the plans should be changed to eliminate the pit. If plans call for a pit in which employees may be exposed to dangerous gases or vapors, means of ventilating the pit should be provided and adequate air supply should be maintained. Warning signs should also be posted to alert workers to the danger of lack of oxygen or concentrations of dangerous gases in the pit.

When openings require covers, the specifications should require metal covers with hand rings folding flush with the top. Covers should fit snugly and be provided with lugs. Open pits must be safeguarded with fences or rails.

Stairs or retractable or fixed ladders provide safe access to pits. Drains reduce the hazard of falls on muddy, wet, or icy pit floors.

Outside lighting

Outside lighting must serve not only as a production tool and as a safety factor but should also function as part of the plant security system. Maintenance personnel should adjust lighting as daylight hours shorten. Luminaires using many different types of light sources are available for all types of specialized applications. Consideration should be given to lamp life and ease of maintenance of luminaires selected. Lighting equipment used in outdoor locations must withstand exposure to the elements without deterioration. (See a later section in this chapter, Lighting.)

Docks and wharves

The bottom of the sea, lake, or river is an important factor in design and construction of docks and wharves. A soft, deep bottom limits the use of concrete and of heavy fire-resistive materials. Wood piles must be protected if marine borers are present. Flexibility and elasticity are essential where waves force vessels against piers and in tidal waters. Structural consideration should be given to ice build-up from tidal action.

Safety requirements include good illumination for night work, a floor that will withstand heavy trucking, and traffic control equipment. Design of piers should take into account the speed and size of vehicles to be operated on them.

PLANT LAYOUT

Size, shape, location, construction, and layout of buildings and facilities should permit the most efficient utilization of materials, processes, and methods.

Location of buildings and structures

The segregation of raw materials storage, processing buildings, and storage for finished products warrants thorough study in laying out a plant to minimize fire and explosion hazards. Storage of volatile flammable liquids and LP gas in an area apart from processing buildings reduces the fire hazard, and, in the event of fire, control is more easily achieved. Moreover, the cost of separate storage eventually may be less than the investment for storage in a processing building. Ample space should be provided between segregated units and such flame sources as boilers, shops, streets, and adjoining property.

The codes of local and state (or provincial) authorities and of the National Fire Protection Association should be followed in planning the location of the units of a plant.

Federal, state, and local restrictions govern the storage of explosives. Magazines should be located and constructed in accordance with the recommendations of the Institute of Makers of Explosives. The type of retardant required in relation to the horizontal distance between buildings of frame, brick, and fire-resistant construction, and minimum separation distances between buildings are given in NFPA 80A, *Protection from Exposure Fires.*

Many plants use and store flammable liquids having flash points below 100 F (37.8 C). Plans and layouts in such plants should conform strictly to flammable liquids handling and storage specifications developed by the *Flammable and Combustible Liquids Code,* NFPA 30, and to the requirements of local fire prevention authorities. NFPA Standard No. 30 specifies the conditions under which flammable and combustible liquids of various classes can be stored in and around buildings. Also see Chapter 16, Flammable and Combustible Liquids.

Exacting specifications must be met in storing flammable liquids outside of buildings in underground or aboveground tanks. Check current federal and state regulations regarding the amount and kind of liquid that is allowed in an underground tank. Standards for tanks include type and thickness of material, provisions for relieving excessive internal pressure, grounding, insulation, piping, and other appurtenances, and similar details.

The required distance between buildings on adjoining property and the location of an aboveground tank depends upon the content, construction, fire extinguishing equipment, and greatest dimension (diameter or height) of the tank. A secondary containment vault properly lined with monitors may be required. NFPA 30 specifies four different groups of tanks and the minimum distances apart for each.

The material used to construct storage tanks depends upon the type of liquid stored, its corrosive properties, and processing requirements.

Flow sheet

A detailed flow sheet is a useful guide in laying out plants, particularly those using dangerous and harmful materials and complicated processes. The nature of the materials and processes in each manufacturing stage can be studied and provision made to eliminate or control hazards.

An automatic sprinkler system is considered primary protection for hazardous installations. Other safety features include spark-resistant conductive floor surfaces and grounding of equipment and structures. Emergency exit doors should be provided and all tools, trucks, and similar equipment should be made from spark-resistant metal. (See Chapter 15, Electricity, and NFPA 30 for an explanation of equipment designed for use in hazardous locations.)

Layout of equipment

Various methods are used to determine the safest and most effi-cient layout of production machines and equipment. The two-dimensional method consists of templates made to scale and fitted into a plan of the site or floor area. The most effective method is to use three-dimensional models, made to scale and set up on a scaled floor plan. The models can be rearranged until the safest and most efficient layout has been devised.

Layout studies will show the most suitable locations for operations like spray painting, welding, and similar work generally requiring segregation. By using the three-dimensional method, congested areas can be anticipated and avoided. These pile-up areas mean frequent handling of materials and many unnecessary movements that result in bad housekeeping—in itself a prolific source of accidents.

Insufficient headroom at aisles, platforms, pipelines, overhead conveyors, other parts of the structure, and other installations can be discovered in studying the models. A vertical distance of at least 7 ft (2.1 m) generally is specified between passageways and stairways and overhead structures to provide ample clearance. Overhead cranes and conveyors require at least 24 in. (61 cm) of vertical and horizontal clearance.

Trucks and other vehicles require enough room for movement, backing, and turning without endangering workers and equipment. Aisles should be wide enough to permit trucks to pass without colliding. For one-way traffic, aisles should not be less than 3 ft (0.9 m) wider than the widest vehicle. Aisles for two-way traffic should be not less than 3 ft wider than twice the width of the widest vehicle. These minimum widths are exceeded considerably in some new buildings where, because of anticipated heavy traffic, aisles from 12 to 20 ft (3.7 to 6.1 m) wide have been specified. A safe layout of aisles requires an absence of blind corners and an adequate radius to allow for vehicle turns. A 6-ft (1.8 m) radius is enough for small industrial trucks.

Aisles should be clearly defined with approved markings of either traffic paint or striping material. Follow ANSI Z53.1, *Safety Color Code for Marking Physical Hazards.* See the recommendations under Use of Color later in this chapter. Plastic buttons that are glued or fastened with metal fasteners to the floor are used by many because of their durability.

Parking areas for both hand and power trucks should be designated. Large plants often provide garages with room for both storage and maintenance. Battery charging rooms also should be provided. See Chapter 6, Powered Industrial Trucks.

If building plans include ramps for use by both pedestrians and vehicles, a 3-ft wide section should be reserved as a walkway. Sharp turns into aisles at the top and bottom of ramps are hazardous and should be avoided. Provide an abrasive coating where slippery floor conditions may exist.

Integrated computer systems also require an area with controlled temperature, ventilation, and electrical power. Layout studies will reveal the most efficient location for these systems.

Electrical equipment

Complete, metal-enclosed, grounded unit substations have been developed for industrial plants. However, if transformers must be installed in confined areas or near flammable materials, be sure they are noncombustible transformers. Furthermore, if these substations are in enclosed areas, ventilation must be provided to reduce the build-up of gases.

Short-circuit protective devices should have the capacity to carry the load and should be rated to open without any danger from the maximum short-circuit current. Circuit breakers, fuses, and safety switches that fail under short-circuit current may explode and cause serious damage and injuries. Motor controls should be capable of being locked out.

The potential of every grounding system should be measured to determine whether or not the system is capable of conducting the necessary amount of return.

When direct-current voltage is supplied from batteries, isolate the battery room from the work area. A battery room should be well ventilated; smoking should be prohibited.

In an up-to-date industrial electrical system, sections may be deenergized for maintenance and other work without shutting down the entire system. In addition, all electrical installations must conform to the NFPA *National Electrical Code* as well as local ordinances. Also see Chapter 15, Electricity.

Ventilating, heating, and air conditioning

Ventilating, heating, and air conditioning are needed for personal comfort and very often to meet process conditions. Personal comfort is very important because it affects efficiency. Every effort should be made not only to make general office and plant conditions comfortable, but to eliminate — or at least reduce — poor conditions which can contribute to excessive employee fatigue and discomfort. In buildings handling flammable liquids or vapor, ventilation must be adequate to prevent forming explosive concentrations.

Boilers, fans, and air conditioning equipment are usually kept separate from the general work areas because of their noise and vibration. It is important to make certain that boilers receive adequate air and that combustion by-products are exhausted safely. In locating incinerators, too, be sure that a negative pressure differential in a building doesn't cause an incinerator stack to serve as an air source.

Maintenance should be considered. Not only must authorized employees have easy access to the machinery, but there must be sufficient space around the equipment to replace parts as needed. There must be room to pull the tubes if necessary.

The National Safety Council publication, *Fundamentals of Industrial Hygiene*, provides more specific suggestions, as does NFPA 90A, *Air Conditioning and Ventilating Systems (Non-Residential)*.

Inside storage

Sufficient space for raw materials and finished products may be estimated on the basis of maximum production requirements — with allowances for shortages, seasonal shipping, and quantity purchases.

Modern mechanical handling and stacking equipment permits extensive use of vertical space through multiple decking. If this method is anticipated be sure flooring has the capability to accept the maximum projected load.

Plans for storage of supplies, the finished product, and empty or full pallets should allow for easy access. Additional features should include the stability of the piles or stacks as well as a proper functioning fire suppression system. Design density of the system and its area of application are dependent on type of material and the height that it is piled.

Space for storing supplies, tools, flammable liquids, and infrequently used equipment near working areas is seldom included in plans for departmental layouts. The result is that such items often are left in unsafe positions and locations. To discourage employees from leaning large materials or equipment against walls or other places where they can fall, these items should have designated storage places.

Space should be provided for racks, bins, and shelves. Materials that project from such places into the aisle or walkway should be stored above eye level. Metal baskets or special racks provided with drip pans can be used for storing machine parts covered with cutting oils.

Closets should be built for the storage of janitorial supplies such as waxes, soaps, and other cleaning supplies. These closets often contain floor sinks for filling pails and do not require lifting the pail.

Storage of waste material may take considerable room, especially if the waste is bulky or is produced in large quantities. In buildings with basements, chutes to basement storage bins help prevent accumulations in working areas. To dispose of small quantities of sharp-edged waste, boxes with handles that are protected from contact with sharp materials, or similar safe containers may be specified. Chapter 3, Manual Handling and Material Storage, gives additional suggestions.

LIGHTING

Industrial plant lighting may be provided by electric lighting and by daylight.

Daylight

Natural daylight is an uncertain and unpredictable type of illumination. To use it to advantage the following design factors should be taken into account:

- Variations in the amount and direction of incident daylight
- Brightness distribution of clear, cloudy, partly cloudy, and overcast skies
- Variations in sunlight intensity
- Effect of local terrain, landscaping, and nearby buildings on the available light.

The natural light that enters and is made available for use inside a building depends upon the architectural design of the windows and the design and furnishings of the interiors.

Window areas serve at least three useful purposes in industrial buildings: (1) they provide for the admission, control, and distribution of daylight for seeing; (2) they provide a distant focus and thus relax eye muscles; and (3) they eliminate the dissatisfaction some people experience in completely closed-in structures. However, an adequate electric lighting system should always be provided in addition because of the wide variation, with time and weather conditions, in the amount of daylight available.

Control systems, both manual and automatic, are available to vary the output of the electric lighting system and provide only that amount of light necessary to complement the daylight and maintain the recommended task lighting level. For a more comprehensive treatment of this subject see the Illuminating Engineering Society's *Recommended Practice for Daylighting*.

Electric lighting

The prime requirements for industrial lighting are high quality and sufficient quantity of illumination on all work planes. Under these conditions, personnel will be able to observe and effectively control the operations and maintenance of various types of machines and processes.

For most industrial work areas, a sufficient quantity of natural light is often not available, even under optimum daylight conditions. Therefore, electric lighting is required to maintain good seeing conditions. It is essential that the electric lighting system be so designed and installed as to continue the general level of illumination in areas adjacent to the windows or walls, thus ensuring good lighting over the entire working area.

Distribution of light from a luminaire is important. Highly concentrated distributions make high mounting heights economically feasible. Low mounting heights, on the other hand, allow a wide-spread type of distribution.

There are three forms of electric lighting used in industrial areas: general, supplementary, and emergency lighting.

General lighting produces relatively uniform illumination throughout the area involved. Uniform illumination is defined as a distribution of light where the maximum and minimum illumination at any point is not greater than one-sixth above or below the average level in the area. Care must be taken not to exceed the suggested spacing-to-mounting-height ratios for the lighting equipment used.

Supplementary lighting is used to provide higher illumination levels for small or restricted areas where such levels cannot readily or economically be obtained by general lighting methods. Supplementary lighting is also used to furnish a specific brightness, or color, or to permit special aiming or positioning of light sources.

Emergency lighting must be planned for any facility. Such lighting may consist of battery-powered lighting for exits and stairwells or a complete standby generating system.

Quality of illumination

Quality of illumination pertains to the distribution of brightness in the visual environment. Glare, diffusion, direction, uniformity, color, brightness, and brightness ratios all have a significant effect on visibility and the ability to see easily, accurately, and quickly. Installations with poor quality lighting are uncomfortable and possibly hazardous. Moderate deficiencies are not readily detected, although the cumulative effect of even slightly glaring conditions can cause significant loss of seeing efficiency and fatigue.

Quantity of illumination

The desirable quantity of light for any particular installation depends primarily upon the work that is being done. Investigations show that as the illumination of the task is increased, the ease, speed, and accuracy of accomplishing it are also increased. Quantity of illumination is stated in footcandles (1 foot-candle equals approximately 10.8 lux) and is measured with a light meter to give a direct reading of the number of footcandles of light reaching the working plane. Currently maintained levels of illumination for industrial areas as recommended by the Illuminating Engineering Society (IES) are given in the ANSI A11.1 *Practice for Industrial Lighting*. See Table 19-G in Chapter 19 of this volume.

Glare

Glare may be defined as any brightness within the field of vision of such character as to cause discomfort, annoyance, interference with vision or eye fatigue. It reduces the efficiency of the eye and thus may cause discomfort and fatigue. It may reduce the detail of the visual task to such an extent as to seriously impair vision, thus increasing accident hazards. Glare may be direct or reflected.

- *Direct glare* is caused by a source of lighting within the field of view (whether it be daylight or electric). Direct glare may be reduced by: (1) decreasing the brightness of the light source, (2) positioning the light source so that it no longer falls within the normal field of vision, and (3) increasing the brightness of the area surrounding the glare source and against which it is seen.

- *Reflected glare* is caused by high-brightness images or brightness differences reflected from shiny ceilings, walls, desk tops, materials, and machines or other surfaces within the visual field. Reflected glare is frequently more annoying than direct glare because it is so close to the line of vision that the eye cannot avoid it. Furthermore, reductions in contrast caused by veiling reflections often occur and can drastically reduce the task contrast and hence the ability to discern detail. Reduced brightness of light sources will reduce both direct and reflected glare.

Reflected glare may be reduced by: (1) decreasing the brightness of the light source, (2) positioning the light source or the visual tasks so that the reflected image will be directed away from the eye of the observer, (3) increasing the level of illumination by increasing the number of sources in order to reduce the relative brightness of the glare, or (4) in special cases, changing the character of the offending surface to eliminate the specular reflection and the resultant reflected glare.

Soft shadows from general illumination can accent the depth and form of various objects. Avoid harsh shadows, however, since they may obscure hazardous conditions or interfere with visibility at the work area because of undesirable brightness contrasts.

Safety

Because safe working conditions are essential in any industrial plant, the effect of light on safety must be considered. The environment of a plant should be designed to compensate for the limitations of human capability. Any factor that aids seeing increases the probability that a worker will detect a potential cause of an accident and act to avert it.

In most cases where accidents are attributed to poor illumination, the cause is marked down as "very noticeable poor quality of illumination" or "practically no illumination at all." Many less tangible factors associated with poor illumination are, however, important contributing causes of industrial accidents. Some of these are: direct glare, reflected glare from the work, dark shadows—all of which hamper seeing and together cause after-images—and excessive visual fatigue, which alone may lead to accidents.

Seeing tasks tend to become more difficult and vital to profitable operation. Close observations of equipment and instruments and quick physical response to the minute changes indicated will call for more and better lighting to protect major investments in both machines and highly trained personnel.

For extensive lighting installations, a qualified illuminating engineer should always be consulted. The safety and health professional, however, should be familiar generally with the lamps, reflectors, and lighting requirements for industrial environments.

For a more extensive treatment of industrial lighting see Chapter 19, and the following references: ANSI/IES RP7 *Practice for Industrial Lighting,* and the IES *Lighting Handbook.* (See References).

Luminaires

A wide range of types of luminaires permits a choice of designs for industrial applications. When selecting specific types for a proposed installation, consider:

1. Luminance
2. Design of luminaire (1) to avoid objectionable glare under normal seeing conditions and (2) to produce highest initial and sustained light outputs
3. Mechanical construction permitting convenient installation and servicing
4. Environmental suitability for use in normal, classified (hazardous), or special areas, indoor or out.

All interior lighting systems are included in one of five classifications: direct, semi-direct, general diffuse, semi-indirect, and indirect. No *one* system can be recommended to the exclusion of all others—each has characteristics that may or may not match the requirements of a given application. The performance of each should be evaluated to make sure it will efficiently provide the area with lighting both in the quality and quantity required for the task. Luminaire maintenance and the character of the task being performed must be carefully considered.

In many workplaces, a combination of lighting systems can be used. But regardless of the system or combination of systems, as the percentage of indirect lighting increases, the lighting generally becomes softer and more comfortable. However, along with this benefit comes a corresponding disadvantage—that for a given overall level of light, the intensity in the work area is reduced if the light source is pointed away and is reflected back to the worker's area.

This, coupled with more difficult maintenance, generally causes these systems to be less economical. Although all systems may find use to some degree, in most average production areas direct or semi-direct equipment is used.

Hazardous (classified) location

Many areas in industrial plants are classified as hazardous locations by the ANSI/NFPA 70 *National Electrical Code* (NEC). Such areas require the use of specialized lighting equipment (vapor-proof, explosion-proof, and dust-tight luminaires), which can provide the required illumination without introducing hazards to life and property. Because each type of equipment is designed to meet certain requirements, the types are not interchangeable. Article 500 of the NEC should be studied in detail to determine requirements for hazardous (classified) location lighting equipment. In case of question or doubt, the local electrical inspector should be consulted.

Security or protective lighting

Protective lighting is necessary for nighttime policing of outdoor areas to discourage would-be intruders, or to render them visible to plant guards should they attempt entry. It may also reduce fire risk. Illumination for policing, however, is not usually adequate for efficient plant operation; therefore, protective lighting is generally treated as an auxiliary to productive lighting.

Protective lighting is achieved by adequate light upon bordering areas of buildings or, in some cases, by producing glaring light in the eyes of the trespasser with no light on the guard. Lighting should be so arranged that concealing shadows are eliminated.

In general there are four types of lighting units used in protective lighting systems. These are: floodlights, street lights, Fres-

nel lens units, and searchlights. Evacuation of personnel also influences choice. Battery capacity of units and the number of lamps and their wattages should be correlated to provide lighting for at least the brief length of time required for complete evacuation (1–1½ hour if using the recommendations of the *Life Safety* or *National Electrical Code*). Where longer durations of emergency lighting are required, generator sets are used as the power source. These are driven by a prime mover, started automatically upon failure of the normal power supply. Power transfer from the normal to the emergency power supply is made by an automatic transfer switch which reverses the procedure upon reestablishment of the normal power supply.

USE OF COLOR

Color in the workplace

The following section by Linda Trent is used with permission from The Sherwin-Williams Company, Cleveland, Ohio.

At work or at play, consciously or subconsciously, people respond to the colors around them. And, in growing numbers, industrial designers and managers are paying more attention to the interactions between color, lighting, and human behavior.

Managers are becoming more attuned to the industrial psychologists' message that the quality and appearance of work areas can stimulate interest or create boredom. As such, they are receptive to the idea that the proper use of color can generate a positive response to the work environment—favorably affecting workers' housekeeping efforts, safety, and overall productivity. Moreover, since it doesn't cost any more to paint work areas in scientifically-chosen colors than in colors chosen entirely at random, many managers are willing to consult a professional color stylist.

Interaction of color and light inside the industrial plant. The function of a paint, in addition to protecting surfaces, is to absorb or subtract some parts of the spectrum, and to transmit or reflect other parts. The color is determined by which parts of the visible spectrum are reflected when absorption takes place. Consequently, color is visibly modified by different light sources. For that reason, the effects of a plant's lighting system should be a major consideration when selecting colors for the plant's interior.

Typically, a plant's light source is selected on the basis of characteristics such as lumen output per watt consumed, ease of maintenance, ease of shielding and directional control, and overall system economics. The light source's color appearance and color-rendering properties are usually secondary considerations.

Lighting systems currently in use in industrial settings include a range of fluorescent and high-intensity discharge lamps, and combinations of illuminants. The color effects of these various light sources are detailed in Table 1-A.

Interestingly, although light has an effect on color, color also affects the quality of light. When dealing with surface colors, this effect on light is called the light-reflectance value (LRV) of color. It is an important property because the reflections from the painted surfaces—ceilings, walls, machinery, and floors—act as secondary light sources. With proper color styling and recommended reflectances, work area surfaces will maximize use of available light and reduce shadows.

In general, light colors reflect light while dark colors absorb it. White, followed by pastels, has the highest light reflectance

Table 1-A. Color Effects of Light Sources

Lamp Type	Appearance on Neutral Surface	Effect on "Atmosphere"	Colors Strengthened	Colors Grayed	Effect on Complexions	Remarks
FLUORESCENT						
Cool white°	white	neutral to moderately cool	orange, yellow, blue	red	pale pink	Blends with natural daylight; good color acceptance
Deluxe cool white°	white	neutral to moderately cool	all nearly equal	none appreciably	most natural	Best overall color rendition; simulates natural daylight
Warm white†	yellowish white	warm	orange, yellow	red, green	sallow	Blends with incandescent light
Deluxe warm white†	yellowish white	warm	red, orange, yellow, green	blue	ruddy	Good color rendition; simulates incandescent light
INCANDESCENT						
Filament †	yellowish white	warm	red, orange, yellow	blue	ruddiest	Good color rendering
HIGH INTENSITY DISCHARGE LAMPS						
Deluxe white mercury°	purplish white	warm, purplish	red, yellow, blue	green	ruddy	Color acceptance similar to cool-white fluorescent
Metal halide multi-vapor°	greenish to pinkish white	moderately cool, greenish	yellow, green, blue	red	grayed	Color acceptance similar to cool-white fluorescent
High-pressure sodium	golden white	warm, yellowish	yellow, orange, green	red, blue	golden	Color acceptance approaches that of warm-white fluorescent

*Table based on information from The General Electric Co.
°Greater preference at higher levels.
†Greater preference at lower levels.

value; black affords no reflectance. Often, the paint supplier refers to an LRV which is equivalent to the reflectance of a surface of material as defined in the *IES Lighting Handbook,* published by the Illuminating Engineering Society of North America, New York.

In a work environment, the LRV of color can contribute significantly to task visibility. Objects are discerned only in contrast with their surroundings, and the most effective contrast can be obtained by selecting colors for their light-reflectance characteristics. In industrial settings, high-contrast conditions are provided for tasks such as inspection, but deemed unnecessary for tasks such as retrieving goods from seldom-used storage areas.

As a rule-of-thumb, surfaces in industrial plants should be finished to provide light-reflectance values within the ranges listed in Table 1-B. Light reflectance is affected more by color than by type of materials. Wall reflectances should generally be kept within the recommended range, except when unusual conditions make alterations desirable. For example, walls in areas with high, exposed ceilings might have a high-reflectance ceiling carried down to the level of suspended light fixtures, which throw light upward. Using this upper wall surface can increase light reflectance in the room by as much as 10 percent (see K.W. Edmonds, "Selecting Colors for Plant Interiors," *Plant Engineering,* June 23, 1977).

Another exception to the recommended values in Table 1-B involves peripheral areas of a room or work space. If these are not in the direct line of vision of the task, and are restricted to about 10 percent of a worker's visual area, they will usually not affect the efficiency of the lighting system. In fact, deep tones with low reflectance characteristics can be used as accents and focal points, to make the workplace more pleasant.
■ *Geographical location and exposure*—regional color prefer

ences can be employed in a color plan, i.e., desert colors in the Southwest or Colonial colors in the East. In a facility with windows, a particular exposure can be contrasted with color, i.e., warm colors (reds, yellows, oranges) in a northern exposure.

Table 1-B. Reflectance Values Recommended for Plant Surfaces

Surface	Reflectance Values	
	Manufacturing Areas (Percent)	Office Areas (Percent)
Ceilings	80-90	80-90
Walls	50-65	60-70
Floors	15-30	25-40
Machinery	30-50	—
Desk tops	—	40-50

■ *Age of the facility*—newer buildings don't have many windows, so light colors are used to improve overall illumination and morale. In some older buildings, windows no longer access the outdoors, making the area look gloomy and the windowed-wall look cluttered; in this case, windows are painted to increase light-reflectance values and eliminate distractions.

■ *Demographics* (number of female vs. male employees)—color preference studies show that men have a preference for blue, followed by red; women, on the other hand, are more receptive to fashion shades and subtle colors such as peach and mauve.

■ *Noise level*—while no color is going to actually diminish existing noise caused by machinery, the use of blues, greens, and neutrals lessens the workers' psychological response to the noise.

Table 1-C. Summary of OSHA and ANSI Safety Color Code°

Color	Designation
Red	Fire: Protection equipment and apparatus, including fire-alarm boxes, fire-blanket boxes, fire extinguishers, fire-exit signs, fire-hose locations, fire hydrants, and fire pumps. Danger: Safety cans or other portable containers of flammable liquids, lights at barricades and at temporary obstructions, and danger signs. Stop: Stop buttons and emergency stop bars on hazardous machines.
Orange	Dangerous Equipment: Parts of machines and equipment that may cut, crush, shock, or otherwise injure.
Yellow	Caution: Physcial hazards such as stumbling, falling, tripping, striking against, and being caught in between.
Green	Safety: First-aid equipment.
Blue	Warning: Caution limited to warning against starting, using, or moving equipment under repair.
Black on yellow	Radiation: X ray, alpha, beta, gamma, neutron, proton radiation.
Black and white	Boundaries of traffic aisles, stairways (risers, direction, and border limit lines), and directional signs.

°See full text under Section 1910.144 of Occupational Safety and Health standards. For piping colors, see ANSI Standard *Scheme for the Identification of Piping Systems,* A13.1-1981.

- *Type of equipment used and color of material being processed*—most types of equipment are painted in a medium color to reduce glare and provide good contrast for safety colors. Very large machinery, in some instances, can handle a two-tone combination that highlights operating parts and affords visual contrast.
- *Clutter of the environment*—dark, saturated colors make surroundings appear cramped and could make workers depressed; lighter pastels give workers a psychological lift, and create the illusion of spaciousness and calmness. Visually-perceived clutter can also be reduced by painting miscellaneous structural work (pipes, I-beams, etc.) the same color as the adjacent wall or ceiling.
- *Safety*—OSHA and ANSI safety colors are standard throughout industry, and are described later in this article (Table 1-C).
- *Corporate colors*—some companies identify very closely with the colors used in their corporate logo and product packaging, and like to include them in their plant interiors either in graphic application or as subtle accents on doors, etc.

Color as science. When scientists do research, they measure properties and dimensions. And, although Sir Isaac Newton made his great discoveries relating to color and light back in the seventeenth century, the classification of color in terms of dimensions is comparatively recent. The dimensions of color can be described as three related aspects of color—hue, value, and intensity. These terms are defined as follows:

- Hue—the quality giving the color its name (red, yellow, blue, etc.)
- Reflectance value—how light or dark the color is.
- Intensity or saturation—how much hue the color contains in proportion to its greyness.

The first of these dimensions, *hue,* refers to the pure spectral hues produced when sunlight is refracted by a prism. A spectrum consists of light of different wavelengths; each hue can be identified by its corresponding wavelength, or band of wavelengths, in the spectrum. The visible color spectrum spans between red, at about wavelength 750 nanometers (nm), and violet at wavelength 380 nm (1 nm = 10^{-9} m). Any hue can be specified according to the ratio of red, yellow, green, or blue it contains perceptually.

Human response to color. When Newton defined the visual process in *Optiks,* he described the three stages of color perception regarding the physics, physiology, and psychology of the process, in terms that are still valid today:

- The light entering the eye and the factors which determine the spectral composition of light (physics),
- The response to the light in the retina and visual pathways (physiology), and
- The color actually perceived and the psychological response to the color appearance (psychology).

Today, scientists who specialize in studying color are building on Newton's original observations—but using a slightly different vocabulary.

Reflexive responses to color, the most verifiable of the three types, result from the physical structure of the eye. The most apparent reflexive mechanism is the advancing/retreating effect of certain colors caused by the fact that the human eye does not focus all wavelengths of light in the same place. Colors with longer wavelengths—red, yellow, orange—seem to move toward the observer, whereas the cooler, darker colors which have shorter wavelengths are focused such that they appear to move away from the observer. This reflexive response can also be used to help create the illusion of larger or smaller space. Moreover, safety engineers rely on this response, using warm, bright colors to call attention to physical hazards that might cause workers to lose their footing, dangerous machine parts, and fire hazards.

General physiological responses include all the effects of color which cannot be explained as either reflexive or conditioned responses. Of these, the photoreactive effects are the easiest to describe, particularly as used in the treatment of hyperbilirubinemia—a serious condition of newborn babies, in which the blood contains potentially fatal levels of the hemoglobin by-product bilirubin, imparting to the baby a yellow hue (jaundice). Several decades ago, doctors discovered that many cases of this disorder could be treated with blue light, owing to the fact that the bilirubin is photoreactive; that is, it breaks down

Table 1-D. Characteristics and Suggested Uses for Various Colors

Color	Impression	Suggested use for interiors
Warm colors		
Red, orange, yellow	Attract attention, create excitement, promote cheerfulness, stimulate action.	Nonproductive areas, including employee entrances, corridors, lunch rooms, break areas, locker rooms, etc.
Cool colors		
Blue, turquoise, green	Cool, relaxing, refreshing, peaceful, quieting. Encourage concentration.	Production areas, maintenance shops, boiler rooms, etc.
Light colors		
Off-whites and pastel tints	Make objects seem lighter in weight; areas seem more spacious. Will usually give people a psychological lift. Reflect more light than darker tones.	Most production areas, especially small rooms, hallways, and warehouse and storage areas. Poorly illuminated rooms.
Dark colors		
Deep tones, gray, black	Make objects seem heavier; absorb light. Will make rooms appear smaller and surroundings cramped. Long exposure will create monotony and depression.	Not normally recommended for large areas because of light absorption qualities. Use should be confined to small background areas where contrast is needed.
Bright colors		
Notably yellow, as yellow-green, orange, red-orange, red	The purer these colors are, the more compellingly they attract the eye. Make objects appear larger and create excitement.	Complement to basic wall colors; small objects such as doors, columns, graphics, time clocks, time-card racks, bulletin boards, tote boxes, dollies, etc.
White		
	Pure, denotes cleanliness, reflects more light than any other color.	All ceilings and overhead structures, and rooms where maximum light reflection is needed. Can also be used on small objects for greater contrast.

when exposed to certain wavelengths of radiation. Today, the "bililight" is the first treatment used for jaundiced babies (see L. Standwood, "Specifying Color: Is It Really a Science?," *The Construction Specifier*, June 1983, p. 51).

Some other physiological responses studied in color science are the relationships between hue and perceived heat, and the connection between certain hues and metabolic activity. The basic contention of the former is that "warm" colors (reds, yellows, oranges) tend to make people feel warmer, whereas "cool" colors (blues and greens) tend to make people feel cooler. Although early studies suggested that the perceived temperature differences might be as great as 7 F, more carefully-controlled recent studies suggest the differences could be less than 1 F—or completely indistinguishable.

On the basis of these kinds of test results, the researchers have concluded that, in general, cool colors calm people while warm colors excite them.

These hue/activity correlations have practical applications in selecting colors for plant interiors. Cool colors can be used to ease tensions caused by highly detailed work or noisy machinery; warm colors can be used in a lunch room, to dispel boredom and stimulate conversation.

Guidelines for in-plant color. In order to select colors that will enhance plant operations, we have taken into consideration the importance of the lighting system, the psychology of color, and safety standards. Building on this data base, it is possible to establish some common sense suggestions for the use of color in industrial settings. Here are some guidelines:

- Use color schemes to identify and unify work areas that would otherwise be "lost" in large plants which accommodate a variety of production activities.
- Use neutral colors of low light-reflectance values in laboratories where reflected color might prevent accurate observation of materials being tested and analyzed. However, do add interest to the area by using stronger colors on furniture and doors, where visual observation is not so critical.
- Use strong, bright colors in time-clock and locker-room areas. In non-productive areas such as these, color can be used to provide a cheerful atmosphere for the employees at the beginning and end of the workday.
- Use colors with high reflectance values in warehouses, to offset the customary low lighting levels by maximizing the use of available light; color code storage areas to facilitate locating materials.
- Color-style production areas to focus attention on the task, providing contrast to increase visibility.
- Use intense shades of warm colors sparingly, to avoid making the workers confused and anxious.
- Use high-reflectance colors on stairways, and sharp accent colors on rails and doors, to define points of orientation.

More characteristics and suggested uses for various colors are provided in Table 1-D.

Color-coding

Color is used extensively for safety purposes. While never

intended as a substitute for good safety measures and use of mechanical safeguards, standard colors are used to identify specific hazards. Standards have been developed and are given in American National Standard Z53.1, *Safety Color Code for Marking Physical Hazards and the Identification of Certain Equipment.* Be sure to check latest regulations for in-plant use, shipping, or consumer protection. In summary they are as follows:

RED identifies fire protection equipment, danger, and emergency stops on machines.

YELLOW is the standard color for (1) marking hazards that may result in accidents from slipping, falling, striking against, etc.; (2) flammable liquid storage cabinets; (3) a band on red safety cans; (4) materials handling equipment, such as lift trucks and gantry cranes, and (5) radiation hazard areas or containers (Safety Black on Safety Yellow). Black stripes or "checker board" patterns are often used with yellow.

GREEN designates the location of first aid and safety equipment (other than firefighting equipment). (Also see blue, below.)

BLACK AND WHITE and combinations of them in stripes or checks are used for housekeeping and traffic markings. They are also permitted as contrast colors.

ORANGE is the standard color to highlight dangerous parts of machines or energized equipment, such as exposed edges of cutting devices and the inside of (1) movable guards and enclosure doors, and (2) transmission guards.

BLUE is used on information signs and bulletin boards not of a safety nature. (If of a safety nature, use green except in flagging railroad cars. A blue flag is used to mark chocked unloading cars.)

REDDISH-PURPLE is being fazed out as a radiation hazard identification. ANSI Z53.1 states: "The radiation hazard symbol colors shall be Safety Black on Safety Yellow. All present Safety Purple on Safety Yellow or Safety Black on Safety White radiation hazard symbols may be used until replaced."

The piping in a plant many carry harmless, valuable, or dangerous contents, and therefore it is highly desirable to identify different piping systems. ANSI A13.1, *Scheme for Identification of Piping Systems,* specifies standard colors for identifying pipelines and describes methods of applying these colors to the lines. The contents of pipelines are classified:

Classification	Color
Fire protection	Safety Red (7.5R; LVR 12%)
Dangerous	Safety Yellow (5.0Y; LRV 69%)
Safe	Safety Green (7.5G; LRV 6%)
Protective materials (e.g., inerting gases)	Safety Blue (2.5PB; LRV 5%)

(Standard Munsell hue and light reflectance values are given for each safety color.)

The proper color may be applied to the entire length of the pipe or in bands 8 to 10 in. (20-25 cm) wide near valves, pumps, and at repeated intervals along the line. The name of the specific material is stenciled in black at readily visible locations such as valves and pumps. Piping less than ¾ in. diameter is identified by enamel-on-metal tags.

The scheme also recommends highly resistant colored substances for use where acids and other chemicals may affect paints. Other schemes may be equally effective in identifying piping networks.

Accident prevention signs

Accident prevention signs (Figure 1-3) are among the most widely

Figure 1-3. Accident prevention signs should conform to ANSI Z35.1 specifications. Posters and safety bulletins can be of any size or design. (Printed with permission from West Point Manufacturing Company, West Point, GA.)

used safety measures in industry so that uniformity in the color and design of signs is essential. Employees may be unable to speak English or may be color-blind and yet react correctly to standard signs. ANSI Z35.1, *Specifications for Industrial Accident Prevention Signs,* and ANSI Z53.1, *Safety Color for Marking Physical Hazards,* should be consulted.

The following is a digest of requirements:

DANGER—Immediate and grave danger or peril. Red oval in top panel; black or red lettering in lower panel.

CAUTION—Against lesser hazards. Yellow background color; black lettering.

GENERAL SAFETY—Green background on upper panel; black or green lettering on white background on lower panel.

FIRE AND EMERGENCY—White letters on red background. Optional for lower panel: red on white background.

INFORMATION—Blue letters on white background. Used for informational signs, bulletin boards, railroad flags for chocked cars.

IN-PLANT VEHICLE TRAFFIC—Standard highway signs (ANSI D6.1).

EXIT MARKING—See *Life Safety Code,* NFPA 101, section 5-11.

BUILDING STRUCTURES

Stairs, runways, ramps and other access structures are principal sources of injuries. One-fifth of all industrial injuries result from falls; and of those that take place from one level to another, the largest number occurs from stairs and ladders. (See Chapter 2,

Construction and Maintenance of Plant Facilities.)

Many serious injuries can be prevented by careful design and construction. Standards of state and municipal governments and the American National Standards Institute should be followed. A convenient means of access should be provided to most locations more than 4 ft (1.2 m) above the floor.

Runways and ramps

Construction is discussed in Chapter 2, Construction and Maintenance of Plant Facilities. Width should be adequate for anticipated traffic; open sides should be protected with standard railings. Platforms, four or more feet above the floor or ground level, should be guarded by a standard railing, 42 in. (1.1 m) high with intermediate rail and 4 in. (10.2 cm) toeboard. (See Figure 1-4.)

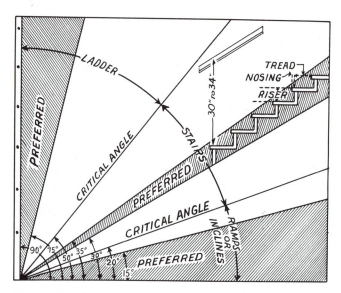

Figure 1-5. Preferred angles for fixed ladders, stairs, and ramps.

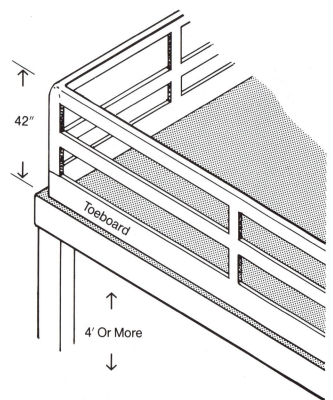

Figure 1-4. Open-sided floors or platforms more than 4 ft (1.2 m) above floor or ground level and scaffolds more than 10 ft (3.1 m) above floor or ground level, should be guarded by a 42 in. high (1.1 m) railing (with midrail). If persons can pass beneath or if there is moving machinery or other equipment from which falling materials could create a hazard, the guardrail should also have a 4 in. high (10 cm) toeboard. Screening can also be added.

Ramps should be of the least slope practicable—some states specify a slope of 1 in 10 (5°43'). Fifteen degrees (a slope of 2.68 in 10) is a recommended maximum, and a slope should never exceed 20 degrees (3.64 in 10). (See Figure 1-5.)

Except where dislodged abrasives would be detrimental to equipment or process materials, ramps should have abrasive coatings or pressure-sensitive adhesive strips to help provide safe footing. Toeboards should be installed where a ramp extends over a workplace or a passageway. Cleats 16 in. (41 cm) apart are needed on steep inclines.

Planks should not overlap and should run the long way of the ramp. Ramps to be used for wheelbarrows should have no cleats on the center plank.

Wire screen enclosures are recommended where materials must be stored on platforms or where persons below will be endangered by falling objects or fragments of materials. Screening can be made of wire netting of No. 16 U.S. gage wire with 1½-in. mesh (1.5 mm diameter wire with 38 mm mesh). Adequate plywood or other fully closed enclosures also may be used.

Stairs and walkways

Circular stairways should be avoided, but if absolutely necessary, should be designed with a minimum variation in tread width. Treads should be covered with a durable slip-resistant material.

The preferred slope for a stairway is between 30 and 35 degrees from the horizontal (Figure 1-5 and Table 1-B). The most suitable slope for a fixed ladder is from 75 to 90 degrees.

A tread width of not less than 9½ in. (19 cm) plus a slip-resistant nosing of 1 in. (2.5 cm) are recommended. Riser height should be not more than 8 in. (20 cm) or less than 5 in. (12.5 cm), and should be constant for each flight.

A flight of stairs having two or more risers should have a standard handrail as specified in American National Standard A12.1, *Safety Requirements for Floor and Wall Openings, Railings, and Toeboards.* Rails should be 30 to 34 in. from the top surface of the stair tread, measured in line with the face of the riser. (Up to 42 in [1.1 m] is permitted on steep-angles.) An intermediate handrail is recommended for stairways more than 88 in. (2.2 m) wide. Consult applicable local and state codes.

Hardwood handrails should be at least 2 in. (5 cm) diameter. If standard black iron pipe is used, the diameter should be at least 1½ in. (3.8 cm) outside diameter. Clearance between the handrail and the wall should be at least 1½ in.

Rails are mounted directly on a wall or partition by means of brackets attached to the lower part of the handrail, to provide a smooth surface along the top and both sides of the handrail. Brackets should be spaced not more than 8 ft (2.5 m) apart, and the mounting should be able to withstand a load of 1600 lb over 8 ft (300 kg/m) applied on the handrail.

Because of space limitations, a permanent stairway sometimes has to be installed at an angle greater than 50 degrees. Such an installation (commonly called an inclined ladder or "ship's ladder") should have handrails on both sides and open risers.

Adequate illumination by lights located so they do not cause glare is important. Outside stairways should be covered to keep off rain, snow and ice.

It is good practice to enclose all inside stairs with partitions of fireproof or fire-resistive material and to install approved fire doors to prevent the spread of smoke or flames from one floor to another. Check codes and insurance representatives about this.

Table 1-E. Slope and Dimensions of Treads and Risers

Angle of Stairway with Horizontal	Riser (inches)	Tread and Nosing (inches)
30°35'	6½	11
32°08'	6¾	10¾
33°41'	7	10½
35°16'	7¼	10¼
36°52'	7½	10
38°29'	7¾	9¾
40°08'	8	9½
41°44'	8¼	9¼
43°22'	8½	9
45°00'	8¾	8¾
46°38'	9	8½
48°16'	9¼	8¼
49°54'	9½	8

Reprinted with permission from OSHA Standards § 1910.24.
(1 in. = 2.54 cm)

Stairs and landings should be able to sustain a live load of not less than 100 pounds per square foot (488.2 kg/m²) with a safety factor of four.

The original building plans should specify elevated walkways and platforms on tanks, bins, big machinery, and other places where workers must go during normal operations or for maintenance purposes. Conveyors should be provided with crossovers having slip-resistant surfaces. Railings and toeboards should also be provided.

Walkways and ramps should be equipped with a standard railing. A standard railing shall consist of a top rail, intermediate rail, and posts, and shall have a vertical height of 42 in. (1.1 m) nominal from upper surface of top rail to floor, platform, runway, or ramp level. A standard toeboard shall be 4 in. (10 cm) nominal in vertical height from its top edge to the level of the floor, platform, runway, or ramp. (See Figure 1-4.)

Exits

Exits must be sufficient both in number and size and located so that in case of fire or emergency the building can be quickly evacuated without loss of life. Plans should be adequate and conform to NFPA, federal, state or provincial, and local requirements, because changing or adding exits after a building is constructed is very costly.

The *Life Safety Code,* NFPA 101, generally requires that two exits be provided on each floor, including basements. One exit from an upper floor may be an inside stairway or smokeproof tower and the other may be a moving stairway or horizontal exit. Some governing authorities require two or more exits that are remote from each other to be provided for all floors of industrial buildings two or more stories high. One exit must be a stair tower made of fire-resistive material, leading directly to the outside at grade (ground) level.

To a considerable extent, the number and width of exits are determined by the building occupancy. In high-hazard occupancy, no part of a building should be farther than 75 ft (23 m) from an exit. For medium- and low-hazard occupancy, 100 to 150 ft (30 to 45 m) is permissible. The *Life Safety Code* also specifies that access to exits provided by aisles, passageways, or corridors shall be convenient to every occupant and that the aggregate width of passageways and aisles shall be at least the required width of the exit. Exit doors should be clearly visible, illuminated, provided with signs, and must open in the direction of exit travel.

Exterior door openings used for hoisting equipment or material on the outside of buildings should be protected by guardrails or gates. Because such doors are a serious hazard when open and not in use, they should be posted with NOT AN EXIT signs.

Flooring materials

Comfort, health, and safety are closely related to the design and specifications for floors. Requirements often vary sharply from department to department, so that a careful study of many factors is essential to determine the best type of floor for a particular location. These factors might include:

Load	Illumination
Durability	Maintenance
Noise	Dustiness
Drainage	Heat conductivity
Resilience	Electrical conductivity
Appearance	Chemical composition

Since a principal cause of floor accidents is slipperiness, the inherent slipping hazard of various types of floor surfaces should be ascertained. (See also the *Administration and Programs* volume, Chapter 21, Nonemployee Accident Prevention.)

Inserts of various materials can be used to reduce slipperiness in specific areas or to combat conditions that cause rapid deterioration of flooring. For example, cast metal inserts are used around woodworking machines. Sheets of soft lead may cover the floor surface where acid is spilled occasionally. Drainage is provided so that the acid can be washed away.

Cast iron plates with checkered or otherwise roughened surfaces laid in cement or asphalt are suitable for some types of rough wear such as that in warehouses. However, it is noisy and highly conductive of heat and electricity. It is relatively non-slippery unless wet or worn smooth. The slipping hazard at the door sills of elevators may be reduced by installing steel grating filled with concrete.

Inserts should be installed flush with the surface of the floor. An insert placed on top of the floor requires a bevel on every side from which it can be approached. The bevel should be at such an angle and low enough that a person will not trip or lose his balance.

Abrasive-coated fabric strips are made for use on metal floor plates, the foot of stairs, stair nosings, and other high-hazard locations. An adhesive binds the strips firmly to the walking surface.

Drainage is essential where wet processes are used. In some instances, floor gratings are installed to reduce slipperiness due to water and other liquids, especially along passageways.

Asphalt is used in various flooring materials; it is dustless, plastic, odorless, and warm to the feet. It makes an especially good floor in some shops because if the aggregate is siliceous, the mixture is resistant to acid. Ordinary grades of asphalt are not recommended for plant roads because they soften in hot weather and cannot accommodate heavy industrial trucking. Asphalt tile, often used in offices, may become slippery when washed or not properly waxed. However, it is quickly being replaced by vinyl-asbestos sheet goods (and other types), which are not only more attractive, but hardly much more trouble to keep clean.

Paving brick, if laid on a solid foundation like concrete, is satisfactory for heavy traffic. Cement mortar joints make a smooth surface. Foundry floors have been made from hard-burned bricks laid face up on a concrete base with sand-filled joints.

Concrete floors are used widely in warehouses and factories. A wood float finish on a mixture of pea gravel, sand, and cement gives a roughened surface that will not crack or "dust." A smooth finish is slippery when wet. Concrete does not withstand acids, and in some types of work, employees consider concrete too hard and too cold. Resilient nonslip mats with low heat conductivity can be placed over the concrete where workers must stand in one position for considerable periods of time.

Cork tile is desirable for its insulation, resiliency, quietness, high nonslip rating, and ability to withstand light traffic for long periods. It is not suitable for wet locations or where there are heavy loadings, and it is expensive.

Asphalt-base or vinyl-base tile or sheet goods are often used in offices, laboratories, and workrooms where cleanliness and good looks are important. The material is easily cleaned, noiseless, and a poor conductor of heat.

Magnesite is suitable for light traffic or where light oils are used. It must be laid on a rigid base and should not be used where there is excessive moisture or hydrostatic pressure, as in basements. A coating of bituminous paint is necessary to protect metals such as pipe from contact, since magnesite corrodes some metals.

Checkered steel plate wears well, but it is noisy and highly conductive of heat and electricity. It is relatively nonslippery unless worn smooth or wet.

Metal grille floors and gratings will not collect dust, dirt, or liquids but, because of noise, are not often used where hand trucking is done regularly. This type of floor material is particularly useful in boiler rooms and over openings.

Parquet is laid on an underfloor and is suitable for offices. If the material is sealed properly, little maintenance is necessary. Parquet wears well but may be noisy.

Rubber flooring is resilient. It also has high dielectric strength, which is undesirable where static electricity is a problem. However, conductive types of rubber flooring are available. Abrasive rubber flooring can be used to overcome slipperiness.

Since terrazzo flooring has no joints in the surface, its use eliminates some of the difficulties that may be encountered with some other types of flooring. It can be made electrically conductive by grounded grilles. It is a conductor of heat. Suitable sealers are necessary to make the floor impervious to most acids. The terrazzo mixture is slippery unless it includes abrasive aggregates.

Plants or laboratories requiring extreme cleanliness and sanitary conditions, such as dairies, use ceramic glazed tile. It should be laid in Trinidad asphalt or in cement with a low lime content. Two or three layers of asphalt roofing felt may be laid beneath the tile.

Wood floors of the proper material and construction are suitable for various services except for the obvious objection of a fire hazard. Plank or board floors of the softer woods are generally unsatisfactory and a source of numerous injuries from slipping and falling as well as from splinters and truck accidents.

A good floor for light manufacturing can be made from matched or jointed hardwood flooring nailed to a subfloor structure or to sleepers in concrete. The thickness of the hardwood flooring will depend on the service to which it will be subjected. It is laid with the grain parallel to the line of truck travel.

Wood blocks meet many of the requirements for a good floor. Properly made, a floor of this type is relatively noiseless and does not become slippery or cause fatigue. If the blocks are laid on a smooth, rigid base, the floor is not likely to crack and will withstand heavy service.

Blocks impregnated with creosote are necessary for floors in contact with liquids or moisture. Expansion joints are required along walls, columns, and similar places. If blocks are laid with a high melting point pitch, hot weather does not create problems. Oils and organic solvents, however, cause trouble because of their solvent effect on bituminous fillers and coatings.

To reduce tracking of mud and dirt into a building, moisture-absorbing mats or runners can be installed at entrances.

In areas where welding is performed regularly and in oven, furnace, boiler and similar areas, noncombustible flooring should be installed.

Openings in floors should be protected by railings or barriers at exposed edges, or otherwise guarded. There should be a top rail, intermediate rail, and toeboard, according to American National Standard A12.1, *Safety Requirements for Floor and Wall Openings, Railings, and Toeboards.* Sheeting or woven wire can be installed around the opening to prevent material from falling in.

Floor loads

Floors should be designed for carrying anticipated loads safely. A registered structural engineer should be consulted. Also check ANSI Standard A58.1, *Minimum Design Loads for Buildings and Other Structures.*

Figure 1–6 presents some floor-loading fundamentals. The ideal load is uniformly distributed over the floor area (as shown in A). The same load concentrated at the center of the span (shown in B) will require twice the structural strength. Conversely stated—if a floor is designed for a given uniform load, only one-half this amount can be concentrated at the center of the span. Figure 1-6C shows the ideal location of aisles and loads.

In estimating, weights of men are figured at 160 lb (73 kg) and women at 138 lb (63 kg). Equipment weights can be obtained from manufacturers; weights of bulk materials, from

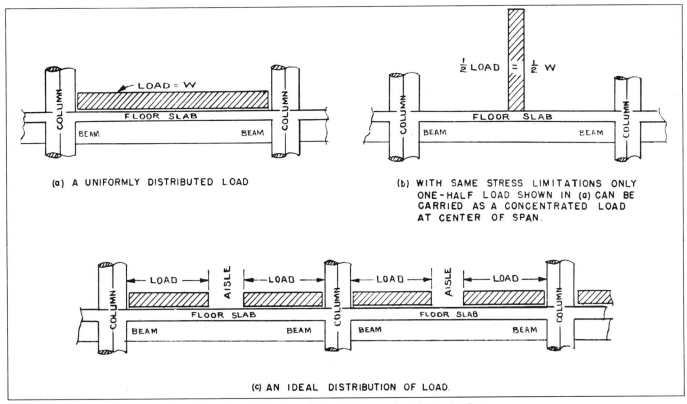

Figure 1-6. Various types of loadings. Changing the location of a load changes the total load a floor can carry.

handbooks. (In piling bulk materials, often air spaces result which reduce the overall density of the material.)

Foundations which distribute loads and vibration over larger areas, or cushion shocks with springs or vibration mounting, help reduce structural reinforcement. Live loads should be figured and future loads anticipated. Industrial trucks when loaded may weigh as much as 60,000 pounds (27.2 metric tons).

REFERENCES

American Conference of Governmental Industrial Hygienists, 6500 Glenway Ave., Bldg. D7, Cincinnati, OH 45211. *Industrial Ventilation.*

American Insurance Assn., 85 John St., New York, N.Y. 10038. *Recommended Good Practice Requirements of the National Board of Fire Underwriters for the Construction and Protection of Piers and Wharves.*

American Society of Heating, Refrigerating, and Air-Conditioning Engineers, 1791 Tullie Circle, N.E., Atlanta, GA 30329. *Guide and Data Book.*

American National Standards Institute, 1430 Broadway, New York, N.Y. 10018.
Manual on Uniform Traffic Control Devices for Streets and Highways, ANSI D6.1.
Minimum Design Loads for Buildings and Other Structures, A58.1.
Life Safety Code, A9.1 (NFPA 101).
Minimum Requirements for Sanitation in Places of Employment, Z4.1.
National Electrical Code, ANSI/NFPA 70.

Practice for Industrial Lighting, ANSI/IES RP7.
Requirements for Fixed Industrial Stairs, A64.1.
Safety Requirements for Construction, A10 Series.
Safety Requirements for Floor and Wall Openings, Railings, and Toeboards, A12.1.
Safety Code for Portable Wood Ladders, A14.1.
Safety Requirements for Window Cleaning, A39.1.
Safety Color Code for Marking Physical Hazards, Z53.1.
Scheme for the Identification of Piping Systems, A13.1.
Specifications for Accident Prevention Signs, Z35.1.

Chowdhury, Jayadev. "Chemical-plant Safety: An International Drawing Card." *Chemical Engineering,* March 16, 1987, pp. 14–17.

Faber Birren and Company, 500 Fifth Ave., New York, N.Y. *Specifications of Illumination and Color in Industry.* Reprinted from *Transactions,* American Academy of Ophthalmology and Otolaryngology.

Factory Mutual System, Engineering Division. *Handbook of Industrial Loss Prevention.* New York, McGraw-Hill Book Co.

Gausch, John P. and Alice Bly Gausch. "Avoiding Catastrophic Losses. A Team Approach to Safety Assurance." *Professional Safety,* Sept. 1985, pp. 26-32.

Illuminating Engineering Society, 345 E. 47th St., New York, N.Y. 10017.
Glare and Lighting Design.
Lighting Handbook.
Recommended Practice for Daylighting.
Practice for Industrial Lighting, ANSI/IES RP7.

Institute of Transportation Engineers, 525 School St., S.W., Suite 410, Washington, D.C. 20024.

Jones, Charles L. *Safety in Lacquer Plants.* Hercules Powder Company, 917 Market St., Wilmington, Del. 19801.

Kletz, T.A. *Make Plants Inherently Safe.* Hydrocarbon Processing, Sept. 1985, pp. 172-80.

Krivan, Steve P. "Avoiding Catastrophic Loss: Technical Safety Audit and Process Safety Review." *Professional Safety,* Feb. 1986, pp. 21-26.

National Fire Protection Association, 470 Atlantic Ave., Boston, Mass. 02210.
  Air Conditioning and Ventilating Systems (Non-Residential), NFPA 90A.
  Construction and Protection of Piers and Wharves, NFPA 87.
  Fire Protection Handbook, latest edition.
  Flammable and Combustible Liquids Code, NFPA 30.
  General Storage, NFPA 231.
  Inhalation Anesthetics, NFPA 56A.
  Life Safety Code, NFPA 101.
  Lighting Protection Code, NFPA 78.
  National Electrical Code, NFPA 70.
  National Fire Codes, Ten volumes.
  Operation of Marine Terminals, NFPA 307.
  Protection from Exposure Fires, NFPA 80A.
  Rack Storage of Materials, NFPA 231C.

National Safety Council, 444 N. Michigan Ave., Chicago, Ill. 60611.
  Industrial Data Sheets
    Firearms for Plant Protection, 413.
    Poison Ivy, Poison Oak, and Poison Sumac, 304.
    Tree Trimming, 244.

Portland Cement Association, 5420 Old Orchard Rd., Skokie, Ill. 60077. (General.)

U.S. Department of Transportation, Federal Highway Administration Bureau of Public Roads. *Manual on Uniform Traffic Control Devices for Streets and Highways.* (ANSI Standard D6.1). Washington, D.C., U.S. Government Printing Office.

U.S. Environmental Protection Agency, 401 M St. SW., Washington, D.C. 20460.

2

Construction and Maintenance of Plant Facilities

CONSTRUCTION ON COMPANY PREMISES

THIS CHAPTER FOCUSES PRIMARILY UPON industrial plants in which incidental construction and demolition operations are carried out. It is for those who need an overview of the controls necessary for construction operations done on their property.

Industrial concerns doing construction work themselves have complete control over their employees and therefore can require adherence to safe practices. When a company hires an outside contractor or subcontractor, the company and the contractor(s) must have a complete understanding of the legal and working relationship between them. The company must insist upon compliance with all provisions of local, state or provincial, and federal safety and health regulations that pertain to the construction work itself. This is spelled out in the contract documents (described later).

Reliance on building codes and construction safety standards is not the answer—the main purpose of building codes is to provide safety to future building occupants, and at best, they represent a minimum requirement for materials and construction. The federal government, many states, and many localities have construction safety standards as well, but these—like building codes—are only as effective as the enforcement behind them. Codes and standards alone cannot guarantee that company or contractor personnel will work safely. This is a result of thorough training and effective supervision.

The best time to "think safety" is during the designing and specifying steps—designing safety into the final product, as well as into its construction. (See Chapter 1.) With this in mind, the safety and health professional should be brought in at the earliest planning stages (as discussed in Chapter 1), and also should discuss the proposed job with the construction department or with outside contractors. (Many safety-conscious contractors try to get a running start on a good safety program and will even insist on a pre-job safety conference.)

Analyzing the construction job

Advance analysis of construction jobs is not new; in fact, some sort of an analysis is made on every construction job. A contractor preparing a bid must make a reasonable, accurate analysis in order to compete with other bidders and avoid a monetary loss. Practical experience in running construction jobs proves one cannot get very far without a set of prints and specifications that clearly defines the work to be done. Accident prevention is a legal requirement and is often included in construction specifications. Unfortunately, many contractors start a job without determining the financial advantages of accident prevention and do little advance planning to eliminate or reduce job exposures.

Most contractors make up their bids under five major divisions: material, labor, plant, overhead, and profit. In the "overhead" is buried an item designated as *insurance*. The amount placed in the bid is usually a percentage of the cost of material, labor, and plant. The amount will vary with the type of job and is affected to some extent by the requirements of owners, architects, and the state or province in which the job is located. This insurance item includes the expense for the performance and payment bonds, worker's compensation, Social Security, unemployment insurance, employer's liability, property damage, automobile insurance, fire insurance, builders' risk insurance, and other types of coverage that the contractor may select. Further, this insurance item is considered one of the fixed charges against the job. Many people believe that, like death and taxes, little can be done about it.

The compensation insurance costs are not fixed but vary with the loss record of the contractor. Some contractors claim that there is no one item in the entire bid which can more easily produce a substantial saving. Some contracts have a "Hold Harmless" clause. According to that clause, any loss is charged against the contractor and is consequently reflected in his insurance costs.

If such costs can be saved on insurance, the next questions are why has it been overlooked and why has there not been more enthusiastic interest by contractors generally. The answers lie with the safety and health professional and may be due to inadequate analysis of the construction job and lack of proper presentation to management.

Safety professionals think and talk mostly in terms of accidents, usually involving personal injury. They use terms like "accident frequencies," usually meaning disabling injury frequency rates. Much of this information is based on hindsight, and, while it has a certain value during the progress of the job, it is unobtainable in practical form at the time the job is bid. Frequency figures cannot easily be translated into the terms of the trade, such as the "cubic feet," the "square yards," or "the dollar," and, therefore, lose much of their value because of lack of understanding. Chapters 5, 6 and 7 of the *Administration and Programs* volume give a detailed discussion of how to understand accident statistics and use them to prevent accidents.

Accident prevention data has its greatest value when it influences construction plans in the following ways:

1. Produces a statement of the contractor's accident record in terms of cost as one item of the bid.
2. Defines the job exposure in specific and measureable terms.
3. Sets up practical and effective safeguards to control these exposures.

Accident prevention is a real factor in the economic success of any construction job. Fortunately, the methods to prevent many accidents are known. Further, there is ample evidence that they are preventable, and it can be very profitable to do so. To the construction industry, it means lower cost and greater efficiency.

Contract documents

The contract documents should spell out that the contractor meet certain minimum safety, health, and equipment requirements, including provisions for protecting company employees (and the public) from construction hazards. As a minimum, the applicable government standards regulations should be noted in the contract. Formally stating these minimum requirements in the contract makes subsequent enforcement easier.

Under no circumstances should a company undertake supervision of the contractor's employees. The usual procedure is to point out matters of safety to the contractor's superintendent or top supervisor who can then exercise the proper authority.

It is important to note that in all construction operations the contractor must comply with applicable codes, laws, and ordinances. Some firms issue a booklet to contractors that spells out safety responsibilities and amplifies the safety requirements spelled out in the general conditions of the contract. Firms should also require copies of contractors' OSHA log for the past 2 or 3 years as well as workers' compensation records.

Contractor safety program

The documents should also spell out that the contractor set up an effective safety program (if one does not already exist). By doing this, a company emphasizes that the human element must be reckoned with—maintaining a good safety record is basically a cooperatively directed accomplishment of the employees.

Take, for example, the way one major contractor makes an active safety program part of everything a worker does on a construction job. The firm spells this out in its comprehensive construction safety program. The safety program on a project is set up by the prime contractor's project manager who maps out a program of safety measures commensurate with the size of the project and the conditions and hazards peculiar to it. This construction safety program involves:

1. A full- or part-time safety and health professional.
2. Adequate first aid facilities and trained personnel.
3. A project safety committee comprised of key personnel (in which all key contractor supervisors are also invited to participate). On some jobs, union representatives and workers are also on this committee, but membership is rotated so that the maximum number of persons will have some opportunity to participate actively in committee work.
4. The committee meets weekly and submits a review copy of the minutes (which includes accident statistics) to the safety department in the home office. The safety committee discusses accidents and near accidents of the previous week. Other hazards are pointed out, housekeeping is evaluated, and plans are made for correcting unsafe conditions and unsafe practices before they lead to an accident. Periodically, a portion of some meetings is dedicated to first aid instruction, resuscitation techniques, firefighting, and other emergency procedures.
5. A safety representative of the week, appointed on a rotating basis at each committee meeting, who acts as the project safety inspector and submits a report to the committee.
6. Crew foremen who hold brief meetings with workers under their supervision at least once a week to discuss the safety operation of their crew, the safety of other workers, and specific problems. The crew meeting should be documented on a "Tailgate Safety Meeting Check List." This list records what was discussed, date, and time. It is signed by all in attendance. Each list is to be kept on file in the construction safety office. This meeting is usually held after the project's safety committee meeting.
7. Safety instructions given to all new workers as part of their first day indoctrination and documented.
8. Special mass safety meetings called by the project engineer. All workers attend these meetings.
9. Safety devices designed to mitigate injury, such as safety hats, eye protection, safety belts, and similar protective equipment provided as necessary.
10. Facilities provided to help prevent accidents (such as suitable roads, lights, barricades, signs, warning devices, guardrails).
11. Safety materials forwarded periodically from headquarters.
12. Periodic visits to projects made by management and the safety supervisor. While there, these people inspect the job and participate in the safety program.

The enthusiasm generated by this program gives a real awareness of job safety as a whole and makes each worker individually safety conscious.

The project manager and staff talk with each contractor's supervisory personnel before work is started in order to explain how the contractor's work will proceed in relation to the work of others. The project safety program and the part of the contractor are discussed. The contractor's key supervisor is invited to serve as a member of the project safety committee.

Prejob safety conference. A preconstruction conference should be held between company (or plant) management (including the safety department) and the contractor's superintendent or other equally authoritative representative. Ways of access (for the contractor's employees, construction materials, and equipment delivery), storage space, and parking areas should be established. Also, the contractor's representative should be acquainted with the plant safety program, first aid facilities, special safety equipment required because of hazards due to plant operations, and how this safety equipment can be obtained.

Prior to commencement of work, a proposed accident-prevention program should be written down. All interested top supervision should then meet in a preconstruction safety conference to discuss the proposed program and to develop mutual understandings relative to the overall aspects of the plan. Provisions should be made to hold periodic staff meetings of top supervision to evaluate and revise the program as required by changing conditions and new problems that may arise. A staff meeting should be held when any new contractor starts the project.

Following is a suggested outline for the preconstruction safety conference.

1. Purpose
   a) Evaluation of proposed program
   b) Discussion of job organization and operating procedures
   c) Preplanning the work and agreement of a means for practical application of standard procedures
2. Notification to all parties
3. Evaluation of proposed program
4. Conference facilities
5. Meeting attendance
6. Conference record (minutes of meetings)
7. Agenda for conference
   a) Orientation
      (1) Explain why we have a program
      (2) Advantages in terms of economy and efficiency
      (3) Prescribed safety standards
      (4) Review of:
         (a) Accident prevention agreements
         (b) General conditions of specifications on safety
         (c) Special conditions of specifications on safety
      (5) Other requirements—local, state, federal
      (6) Supervision
         (a) Organization at project site
         (b) Functions of personnel at the site
         (c) Responsibilities
         (d) Delegated authorities
         (e) Relations regarding enforcement and discipline
   b) Discussion of proposed program
      (1) Plans as to layout of temporary construction, site, buildings, etc.
      (2) Action taken toward planning and coordinating activities between different operations and crafts

   (3) Access to work areas
   (4) Safety indoctrination and safety education
   (5) Delegation of safety responsibilities to supervisors
   (6) Integration of safety into operating methods and procedures
   (7) Housekeeping program
   (8) Safety factors in job built appurtenances
   (9) Traffic control and parking facilities
   (10) Fire protection
   (11) Lighting, ventilation, protective apparel, and medical care
   (12) Safe operating condition of equipment and maintenance
8. General
   a) Methods for meeting objectives
   b) Plans for periodic readjustment of safety objectives
   c) Handling of safety deficiencies
   d) Arrangements for additional meetings and periodic staff meetings
   e) Follow-up of agreements in preconstruction meeting
   f) Three cardinal rules to observe for a workable safety program
      (1) All agreements must be fair
      (2) Paper work should be kept to a minimum
      (3) The program should be simple and deal with facts

Protection of employees and equipment

When construction work is being done in or around an industrial plant, the company employees and equipment should be protected from all construction hazards, including open excavations, falling objects, welding operations, dust, dirt, temporary wiring, and temporary overhead electrical lines. The construction work should be isolated from company operations if at all possible. Barricades, fences, and guardrails should be set up, and appropriate warning signs should be posted. Warning signs are detailed in ANSI Standard Z35.1, *Specifications for Accident Prevention Signs.*

When a construction job is being done in an area that must be kept in operation, a sheeted bulkhead can be erected to keep out dust and dirt and to isolate the operation as much as possible. Flame-retardant materials may be required in some cases.

When construction operations are to be done where flammable vapors and liquids may be present or in other hazardous areas, the contractor's employees should be subjected to the same rigid requirements that apply to the plant employees. Before cutting, burning, or welding is done by employees of the contractor, a permit should be obtained from the plant engineer or safety department to make sure that the necessary fire and safety regulations have been met. The usual requirements for ventilation and health protection also should be observed. (See Chapters 13 and 17 in this volume.)

In an area where no flame is permitted, either the manufacturing process will have to be shut down while welding and cutting are done, or screwed or bolted fittings will have to be used instead of welded connections.

Night lighting (and supplemental daytime lighting) should be provided where necessary, especially in areas where open trenches or ditches create hazards in walkways and roadways. A minimum of 5 footcandles is recommended by the Illuminating Engineering Society for vital exterior locations or structures. (See illumination data in Chapter 19.)

The contractor and plant management should cooperate closely at all times in order to determine what precautions are necessary to prevent accidents. Particularly close liaison should be maintained during tie-in of new piping or equipment to existing lines and equipment to ensure the safety of subsequent operations. The contractor should maintain all work in an orderly, well-kept manner.

Trucking

Contractors working either inside or outside the industrial building or area should take great care to prevent trucks and other mobile equipment from colliding with pipelines, power lines, and other equipment in order to avoid interrupting manufacturing or processing operations.

One method of handling construction deliveries is to have a signalman who serves as the eyes for the truck driver. Standard signals for ready-mix concrete trucks are shown in Figure 2-1. Be sure barricades, guardrails, and warning signs are placed to assure maximum safety.

When trucks, bulldozers, powered wheelbarrows, and other mechanized construction equipment are to be operated within a plant, it is a good idea for the contractor and the plant management to agree upon the traffic flow so that the areas in which

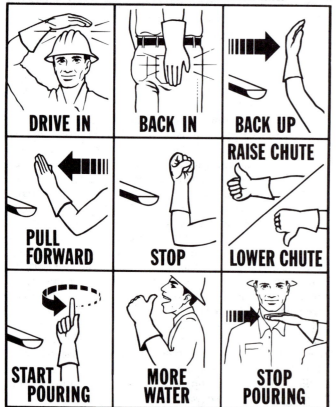

READY MIX TRUCK
SIGNALS
STANDARD SIGNALS FOR MIXER DRIVERS & CONTRACTORS GUIDES

DRIVE IN	BACK IN	BACK UP
PULL FORWARD	STOP	RAISE CHUTE / LOWER CHUTE
START POURING	MORE WATER	STOP POURING

Figure 2-1. Standard signals for mixer drivers and contractor's guides. (Printed with permission from Northern California Ready Mixed Concrete & Materials Association.)

the construction equipment will be operated will be known and isolated where possible. When such agreement is reached, drawings of the areas can be made available to both the contractor's key personnel and the plant personnel affected. To avoid extra handling and vehicle movement, suppliers should be notified of the exact location to make deliveries.

It may be necessary to haul employees in trucks from one location to another within the work area, and, unless controlled, this operation can become a major source of serious injuries. No one should be allowed to stand on the running board or bed of a truck. They should not be permitted to sit at the side or end of a flat-bed truck. Safety belts should be supplied for all passengers who ride in the cabin with the driver. Workers must not be permitted to ride on a loaded truck or other machines not equipped for the purpose. (See Data Sheet 330, *Motor Trucks for Mines, Quarries, and Construction.*)

When people are boarding or descending from a truck, the truck should be standing still. A boarding ladder should be provided. If necesary, a bus should be provided to transport the employees to the work site.

Miscellaneous machinery and equipment

Modern construction requires a variety of machines: tractors and bulldozers for site preparation, power shovels and draglines for excavating, cranes and derricks for placing structural members, concrete mixers or supply trucks, compressors, generators, and many others. No machine or piece of equipment should be placed in operation until it has been inspected by a qualified person and found to be in safe operating condition. Depending on the type and use of the equipment, it may be necessary to inspect the unit daily, monthly, and annually. The inspection reports should be kept on file and necessary action taken to remedy unsafe conditions. (See Chapter 4, Hoisting Apparatus and Conveyors, for more information.)

Guarding, safety devices, platforms, and means of access. Belts, pulleys, sheaves, gears, chains, shafts, clutches, drums, flywheels, and other reciprocating or rotating parts of equipment shall be guarded. No guard or safety appliance or device shall be removed or made ineffective unless immediate repairs or adjustments are required, and then only after the power has been shut off. Guards and devices shall be replaced as soon as repairs and adjustments have been completed.

Current-carrying parts of electrically operated equipment should be properly insulated or guarded. All noncurrent-carrying metal parts should be properly grounded.

High-temperature lines and equipment, located where they endanger employees or create a fire hazard, should be covered with suitable insulating materials.

Exhausts from all equipment powered by steam or internal combustion engines should be properly released and so located that they do not endanger workers or obstruct the view of the operator.

Platforms, footwalks, steps, ladders, hand holds, guardrails, and toe boards should be installed on all equipment where they are needed to provide safe access. Suitable operating floors or platforms, surfaced with slip-resistant material, should be provided for all equipment operators.

Operators of equipment should have protection against the elements, falling objects, swinging loads, and similar hazards.

Windows in cabs or enclosures on equipment shall be made of safety glass, and shall be kept in good repair at all times.

Reverse alarms should be installed on all heavy mobile equipment and trucks unless a signaler is provided.

Positive means should be provided to prevent the starting of equipment by unauthorized persons. This could take the form of key ignition systems or simply blocking the starting apparatus and locking it.

At the end of a work shift, equipment should be set and locked so that it cannot be released, dropped, or activated in any way. The manufacturer's procedure for shut down should be followed. All machine implements, such as bulldozer blades and front-end loader buckets, should be lowered to the ground whenever the machine is stopped.

Accumulations of debris, oil, grease, oily rags, and waste on equipment should not be permitted.

Safe load capacity and operating speeds should be posted on all equipment and should not be exceeded.

Equipment should be placed on an adequate foundation and properly secured.

Before mobile equipment is moved, a survey of the area in which it is located should be made to check for overhead wires, pipelines, excavations, invisible ground conditions, and similar hazards. Timber mats may have to be provided. Equipment with high clearances (such as cranes) should not be moved into or out of, or operated in, any area containing electric power lines until the approval of the superintendent has been obtained. No part, including the load, may reach within 10 ft (3 m) of electric lines, unless power in the lines is shut off. (If local laws specify greater distances, these should prevail.) (See also Chapter 4, Hoisting Apparatus and Conveyors.)

Flammable liquids. Refueling gasoline-operated equipment while the motor is running should be prohibited. Continuously operating equipment should be fueled from properly protected tanks located outside the operating room. Tanks should be adequately grounded and bonded to equipment to prevent static electricity buildup.

Smoking or the use of open flames on or in the immediate vicinity of gasoline-operated equipment while it is being refueled should be prohibited.

No solvent with a flash point below 100 F (37.8 C) should be used for cleaning equipment or parts.

When gasoline and other highly flammable fluids are used, they should be transferred by approved pumps or stored in approved safety cans. Gasoline, fuel oil, and other flammable or combustible liquids should not be stored on equipment except in fuel tanks or approved safety cans with a capacity for only one day's requirements.

Fuel tank filler openings should not be located in such a position that spills or overflows can run down on a hot motor, exhaust pipes, or battery.

A suitable fire extinguisher should be located on or close to each industrial truck. (See Chapters 16 and 17 for more details.)

Repairs. All "out of order" equipment should be shut down for repairs. Suitable signs should be posted and not removed until repairs have been completed. Mobile equipment should, if possible, be removed to a safe location where operations will not interfere with the repair work. Equipment suspended in slings or supported by hoists or jacks for repairs should be blocked or cribbed before anyone is permitted to work underneath it. When

repairs are made remote from the source of power on such equipment as conveyors and cable ways, use chains, blocking, or similar devices to prevent injury in case of accidental starting.

Before repairs on electrically powered equipment are begun, the main switch should be locked in the open (or OFF) position. The key to the switch lock should be retained by the person doing the repairs. If there is more than one repair person, each should lock the main switch with a personal lock and retain the key. Switch boxes should have multiple lock out capability.

Steel erection

Steel erection involves extensive use of cranes, derricks, hoists, ropes, and slings. (See Chapters 4, 5, and 13, Hoisting Apparatus and Conveyors; Ropes, Chains, and Slings; and Welding and Cutting in this volume.)

For lifting heavy loads, wire rope slings are preferable to chains. With either chain or wire rope, the manufacturer's capacity rating should not be exceeded. At points where rope slings pass around sharp corners, padding should be provided.

Eye protection shall be provided to workers who are reaming, drilling, or driving wedges, shims, or pins. Containers should be provided for storing or carrying rivets, bolts, and driftpins. Containers need to be secured to prevent falling.

Air tools. Air flow should be shut off and pressure released before pneumatic hand tools are disconnected. For any adjustments or repairs, never raise or lower the tool by the hose. Always use a handline. Air hose sections should be tied together except when automatic cutoff couplers are used. If air hose must extend across a roadway, it should be protected so vehicles will not damage it.

Bolting. When bolts or driftpins are being knocked out, they should be retrieved so they do not fall on anyone below. Bolts, nuts, washers, and pins should never be thrown; rather, they should be placed in a bolt basket or other good container and raised or lowered by a line.

The use of high-tensile machine bolts or structural rib bolts is popular for field assembly of structural steel. However, contractors must follow the manufacturer's instructions carefully for proper installation, adequate torque application, and prevention of nut back-off.

Impact wrenches should be provided with a locking device to retain the socket.

Welding. Precautions to be taken when securing steel by welding are covered in Chapter 13, Welding and Cutting.

Drilling and reaming machines should be operated by two employees unless the handle is firmly secured to resist the torque created by the machine if the reaming or drilling bit should foul.

Plumbing-up. Hooks or lashings used for plumbing-up should be attached securely before stressing the turnbuckle. Once the turnbuckle is under stress, a device should be used to keep the turnbuckle from unwinding.

Plumbing-up guys should be placed so that the bolters, riveters, or welders can get at the connection points. Guys should not be removed without first getting permission from the job superintendent.

A definite set of directional signals must be established before starting to plumb a structure.

Connecting. When connectors are working together, one

person should give the signals. This person should make sure everyone working on the job is in the clear. All workers should select positions which are clear of all swinging beams.

When connectors are working in pairs, one end of the piece shall be bolted with two bolts before going out to connect the other end; only one connector should go out to fasten the other end. Whenever possible, an employee should straddle the beam instead of walking along the top.

When setting columns before lifting falls are unhitched, either draw down the nuts tightly on the anchor bolts or affix temporary guys. Never release a piece until the required minimum number of bolts have been installed; do not rely on a wrench or driftpin.

Work should be discontinued during rain, high wind, or weather of any sort that might increase the hazard to workers or to others. Proper lighting must be provided at all times.

Erection under plant operation. Steel erection work along with mill or plant operation is especially hazardous as there is usually much congestion from plant personnel and from employees and materials of other contractors. The operating plant supervisor should be responsible for the various phases of the work. The supervisor and others who act as coordinators should be identified to the steel erector and other contractors.

The work area should be clearly defined. The identification and location of gas lines, oxygen lines, electric utilities and electrified rails should be fully established. Responsibility should be established for preliminary work—closing passageways and cleaning grease or other material from crane runways, for example.

All existing mill or plant safety regulations should be observed. Areas should be restricted to operating personnel; others should enter only on permission from the superintendent or first line supervisor.

The supervisor should be responsible for obtaining clearance from the mill or plant supervisor for all phases of the work; this responsibility should not be delegated to a crew member.

Where electric wires are near the work, supervision should determine their voltage and set the necessary clearances. The preferred procedure is to deenergize all power lines. All wires should be assumed to be "hot" until proven otherwise. If they cannot be deenergized, have the power company cover the lines with approved insulated rubber or plastic coverings. When persons must work near hot rails (and current cannot be turned off), they must be provided with adequate insulation and protection.

No one should work on an operating crane runway until the supervisor has been notified and has given permission. When work is being done on or near crane runways, operating crane rail stops should be placed between the worker and the operating crane. If operating conditions do not permit such stops, a safety observer should be in the cab of the crane to protect the worker. Flasher lights or flags are recommended to define the work area and warn the crane operator. Do not permit loose items to remain on cranes or crane runway girders without fastening them.

For additional information, refer to American National Standard *Safety Requirements for Steel Erection,* A10.13. (See References.)

General safe practices. Suggested precautions that should be followed wherever practicable in steel erection work are:
1. Require proper protective equipment to be used.

2. Do not permit employees to ride loads, hooks, or "headache" balls.
3. Do not permit employees to work near electric wires unless the wires are fully insulated.
4. Take precautions to remove from the job any worker who is under the influence of liquor or drugs or who is too sick (in a doctor's opinion) to work.
5. Do not allow employees to work on wet, freshly painted, or slippery steel construction.
6. Have workers wear safety goggles while cutting out rivets, chipping, and doing similar work. Keep adjacent areas clear of personnel or screen such operations.
7. Where it is impractical to provide temporary floors, suspend safety nets below points where employees are working, or have them use fall protection equipment. (See Data Sheet 608. *Safety Nets — Fall Protection for the Construction Industry.*)
8. Where guy cables or braces are used to hold steel during erection, be sure they are guarded to prevent trucks or other equipment from hooking into them and pulling the steel down.

Lateral bracing

Incomplete buildings for which designed lateral support is not yet in place, and all free-standing walls should be adequately braced against the maximum anticipated wind pressures. Exterior masonry walls, whether of load-bearing or nonload-bearing type whatever their height, are subject to wind loads beyond their designed capacity prior to the final set of the mortar or final tie into the structure. These wind loads have caused walls or sections of walls to break off and fall, causing both personal injury (sometimes fatal) and property damage. (See Figure 2-2.)

Masonry walls should follow the erection of permanently installed structural members so that adequate lateral stability is provided. If this is impossible or impractical, temporary brac-

Figure 2-2. Wall bracing must be adequate for anticipated wind loads.

ing should be placed until structural members can be installed.

Usually the architect will include pilasters in the design as well as the requirements for anchors and ties. Also local building codes require that during erection of walls the proper bracing and supporting shall be provided. The question is, what is proper?

During construction of exterior masonry, two external forces should be considered—the weight (vertical) and the wind load (horizontal). Because the vertical load is supported by spandrels and relieving angles, the critical consideration is the horizontal load. Protection or bracing for this load must be provided either by screening or by simple shoring.

Codes and other engineering data indicate the thickness and height of walls which will withstand specific wind loads while unsupported. These specifications should be checked against the local wind conditions and bracing provided when required.

Table 2-A can be used to check bracing that will resist pressures developed by wind at different velocities. The graph shown in Figure 2-3 shows wind velocities that can be withstood by concrete block walls of varying heights.

Table 2-A. Force of Wind for Given Velocities

Miles per hour (V)	Feet per minute	Feet per second	Force in pounds per square foot (0.004V²)	Description
1	88	1.47	0.004	Hardly perceptible
2	176	2.93	0.014	Just perceptible
3	264	4.40	0.036	
4	352	5.87	0.064	Gentle breeze
5	440	7.33	0.1	
10	880	14.67	0.4	Pleasant breeze
15	1,320	22.0	0.9	
20	1,760	26.6	1.6	Brisk gale
25	2,200	29.3	2.5	
30	2,640	44.0	3.6	High wind
35	3,080	51.3	4.9	
40	3,520	58.6	6.4	Very high wind
45	3,960	66.0	8.1	
50	4,400	73.3	10.0	Storm
60	5,280	88.0	14.4	Great storm
70	6,160	102.7	19.6	
80	7,040	117.3	25.6	Hurricane
100	8,800	146.6	40.0	

Printed with permission from Kidder-Parker, *Architects and Builders Handbook.*

Temporary flooring

Where skeleton steel construction in tiered buildings is used, permanent floors should be installed as the erection of the steel progresses. There should not be more than eight stories between the erection floor and the uppermost permanent floor, except where the structural integrity is maintained as a result of the design. At no time, however, should there be more than four floors (or 48 ft [14.6 m]) of unfinished bolting or welding above the foundation or uppermost permanently secured floor.

The derrick (or erection) floor should be solidly planked or decked over its entire surface, except for access openings. Plank-

ing, or decking of equal strength, should be of proper thickness to carry the working load, a minimum of 50 pounds per square foot (psf) (2.4 kPa). Planking should be not less than 2 in. thick (5 cm), full size undressed, and should be laid tight and secured to prevent movement, especially displacement by wind.

There should be a tight and substantial floor or safety nets within two floors (30 ft [9 m], whichever is less) directly under the portion of each tier of beams on which bolting, riveting, or welding is being done, except when gathering and stacking temporary floor planks on a lower floor in preparation for taking these planks to an upper working floor. Bundles to be transferred should not be larger than two planks wide and 15 planks high. Bundles should be choked when hoisted with slings.

Employees should remove such planks successively, working toward the last panel of the floor so that the work is always done from the plank floor. When gathering and stacking floor planks from the last panel, workers should use safety belts and lifelines attached to a catenary line or other substantial anchorage.

During construction in a mill building or other structure where no floors are contemplated and where operation of overhead cranes will not permit temporary flooring, safety belts should be used by the workers.

Once a working floor is provided, a safety line of ⅜ in. wire rope (or equal) should be installed around the perimeter of all temporary planked or decked floors of tier buildings or other multifloored structures. This line should be placed 36 to 42 in. (0.9 to 1.1 m) above the working floor and be attached to all perimeter columns. This line should be left in place until the finished wall is installed.

Metal decking used in place of planks should be of sufficient strength, should be laid tight, and should be tack welded to prevent movement.

Planks should overlap the bearing ends by a minimum of 12 in. (30 cm). Wire mesh or exterior plywood should be used around columns where planks leave an unprotected gap and do not fit tightly. All unused floor openings should be planked over or barricaded until they are needed. Floor planks removed to perform work should be replaced as soon as possible or the open area should be guarded.

Torches and salamanders

Liquid fueled blowtorches and plumbers' furnaces involve the hazards of fire and explosion. Safer equipment should be used, if possible, such as electrically heated soldering irons, paint-remover irons, glue pots, and other devices.

Proper gas or oil space heaters listed by American Gas Association or Underwriters Laboratories are recommended for use in areas containing combustibles. Where flammable or explosive dust or vapors may be present, such torches or furnaces should be used in accordance with local or national laws and regulations.

The storage and handling of gasoline present a hazard in addition to the fire and explosion risks of the torches or furnaces themselves. This hazard can be minimized by careful observance of the requirements for storage and handling of flammable liquids described in NFPA 30, *Flammable and Combustible Liquids Code.*

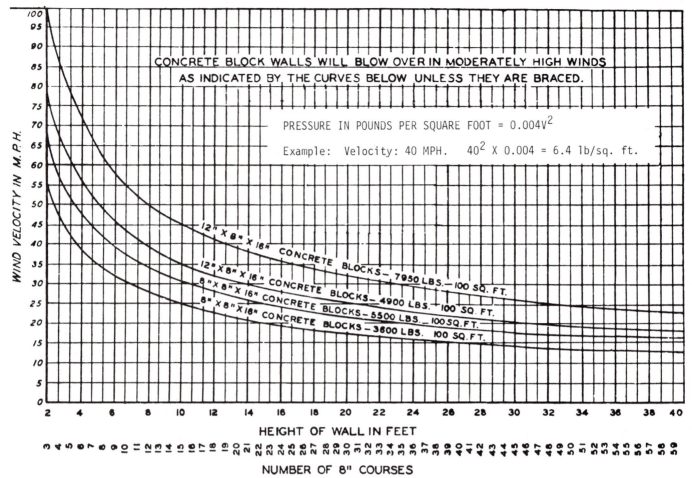

Figure 2-3. Wind velocities that can be withstood by concrete block walls of varying height. (Printed with permission from The Travelers Insurance Company.)

Salamanders and other types of portable heaters are widely used in severe weather to protect masonry, concrete, and plaster from freezing and to provide warmth for the workmen. Gas- or oil-fired, electric, steam, or remotely located heaters with conducted hot air are preferable. These are discussed in detail later.

Solid fuel-burning salamanders are prohibited and should not be used. Liquid fuel-burning salamanders should not be used in confined spaces unless they are vented to the outside. Improperly installed salamanders and other open-flame heaters are particularly dangerous in tool sheds, shanties, and other small enclosed areas because they can give off large amounts of carbon monoxide and consume much oxygen.

If the concentration of carbon monoxide (CO) is greater than 50 ppm at worker breathing zones, the heater should be turned off or additional ventilation provided. Tests for CO should be made about one hour after starting each shift and at least four hours later.

The horizontal clearance between salamanders and combustibles shall be at least 2½ ft (0.76 m); overhead clearance shall be at least 6 ft (1.8 m). Keep tarpaulins and canvas or plastic coverings at least 10 ft (3 m) away. Make sure these are securely fastened to keep them from blowing toward the salamander. Use flameproofed materials whenever possible.

To keep people away from its hot surfaces, a salamander should be surrounded by a noncombustible railing at least 19 in. (45 cm) away. When in use, heaters should be horizontally level, unless otherwise permitted by manufacturer's instructions.

Temporary heating equipment

When using temporary heating equipment, assign a qualified employee to its operation and maintenance. Be sure to follow all instructions of the manufacturer.

Each time the heater is placed in operation, check to make sure it is functioning properly; operation should also be checked periodically when the heater is in use. Heaters shall be equipped with an automatic flame-loss device to stop fuel flow if the flame is extinguished.

Thermostatically controlled heaters should be identified by a warning label advising that the unit may start up at any time.

Fueling. Check and follow manufacturer's instructions as well as applicable regulations. Turn off all flames, including the pilot (if any). Use only the type of fuel specified for the unit. The unit should be cool to the touch. Before and during fueling, check all fuel lines, hoses, and connections for leaks.

Only one day's supply of fuel should be stored inside a building in the vicinity of the heater; this fuel should be stored at least 25 ft (7.6 m) away from a source of ignition. General fuel storage should be outside the structure.

Fan-assisted units. Use only heaters that are designed so that a power failure (or the failure of any electrical components) will not create a fire or electrical shock. Only power supply circuits with three- or four-wire grounding should be used. Grounding continuity must be provided to all parts of the heater unit, including connection to a grounded power supply.

Natural gas-fueled heaters. All piping, tubing, or hose should be leak-tested after assembly and proven free of leaks at normal operating pressure by use of soapsuds or other noncombustible means. Never use a flame.

When placing a unit in operation, make sure it is working properly. Before disconnecting a heater, shut off fuel supply at the source in order to purge the line.

A flexible gas supply line shall be no longer than 25 ft (7.6 m), and shorter if possible. Check all hoses and fittings to make sure they are designed for the pressure and capacity and type of fuel being used. Hoses should have a minimum working pressure of 250 psig (1,730 kPa) and a minimum burst pressure of 1,250 psig (8,600 kPa). All hose connectors should be capable of withstanding a test pressure of 125 psig (860 kPa) without leaking, and a pull test of 400 lb (1,780 N). Hoses should be securely connected to the heater by mechanical means; never use "slip-end" connectors.

Protect hose and fittings from damage and check for deterioration. Hoses shall not be allowed to contact surfaces above 125 F (50 C). Hoses should be placed to minimize any physical damage.

Normal maintenance includes inspection of the hose supply system for cracks, checks, abrasions, and rupture, and leak testing the hose, pipe, and tubing connections. Disconnect electric power supply before repairing heaters.

Liquefied petroleum gas (LP-gas) heaters. Follow the precautions for carbon monoxide concentration, testing operating capability, and leak testing as outlined earlier.

Use only hose labeled "LP-Gas" or "LPG." Minimum working pressure should be 250 psig and minimum burst pressure, 1,250 psig. The hose shall be at least 10 ft (3 m) long, but no longer than 25 ft (7.6 m). All hose should be protected from damage, deterioration, and hot surfaces.

Hose connectors shall be capable of withstanding a test pressure of 500 psig (3,450 kPa) without leaking and a pull test of 400 lb (1,780 N). Hoses should be securely connected to the heater by mechanical means; never use "slip-end" connectors and ring keepers tightened over the hose to give increased force to the metal fitting.

Heaters shall be equipped with an approved regulator in the supply line between the fuel cylinder and the heater unit. Cylinder connectors shall be provided with an excess flow valve to minimize the flow of gas in the event of a fuel line rupture.

For temporary heating, such as in concrete curing, heaters shall be located at least 6 ft (1.8 m) away from LP-gas containers. This does not, however, prohibit the use of heaters specifically designed for attachment to the container or to supporting standard with connecting hose less than 6 ft (1.8 m), provided the design and installation prevent the direct application of radiant heat on to the container. Blower type or radiant heaters shall not be directed toward any LP-gas container within 20 ft (6 m). If two or more heater-container units (of either the integrated or nonintegrated type) are located in an unpartitioned area on the same floor, the container(s) of each unit should be separated by at least 20 ft. The maximum water capacity of individual containers is 245 lb (111 kg—nominal 100 lb gas capac-

ity). The total water capacity of containers manifolded together in an unpartitioned area should not be greater than 735 lb (333 kg—nominal 300 lb LP-gas capacity). These containers should also be separated by at least 20 ft.

On floors on which heaters are not connected together for use, containers may be manifolded together for connection to heaters on another floor, if they meet two requirements. First, the total water capacity of the containers connected to any one manifold is not greater than 2,450 lb (1,111 kg—nominal 1,000 lb LP-gas capacity). Secondly, when more than one manifold having a total water capacity greater than 735 lb (333 kg—nominal 300 lb LP-gas capacity) is located in the same unpartitioned area, the manifolds should be separated by at least 50 ft (15 m).

LP-gas cylinders shall not be refilled inside buildings or structures. Cylinders shall be stored outside of buildings, shall stand on a firm and substantially level surface, and shall be secured in an upright position away from vehicular traffic.

Demolition of structures

Only minor demolition should be done by plant personnel. Specialists in the field should be employed if structures to be removed look as though they will present a problem. Wrecking specialists are familiar with the procedures and precautions necessary to do the work safely, to protect the public and adjacent property, and to comply with applicable federal, state, and municipal codes and regulations.

Following are some fundamental safety procedures and suggestions for minor demolition operations:

1. Make provision to keep the public and unauthorized plant employees at least 15 ft (5 m) away from the structure.
2. Make an engineering survey, by a competent person, of the structure. This is to determine the condition of the framing, floors, and walls and check for any unanticipated conditions. Check for hazardous chemicals, gases, explosives, flammable materials, electrical circuits that may be engaged, asbestos, and hazardous waste, etc.
3. Disconnect utility services (gas, steam, electricity) outside the building. Maintain water lines as long as possible. Maintain or install a temporary water source for fire protection and for wetting down the site to reduce dust.
4. Remove all glass doors and windows throughout the structure.
5. Strip off lath and plaster to eliminate excessive dust during succeeding operations.
6. Remove chimneys and extensions of walls above the roof down to roof level while working from the roof.
7. Remove the roof.
8. Remove walls by picking them apart, using either machine or hand tools. Work from scaffolds supported independently outside the walls.
9. Remove all debris promptly, through chutes or internal holes. To minimize production of dust have a person assigned to wet down the debris.
10. Avoid subjecting walls to lateral pressure from stored material or to lateral impact from falling material.
11. Barricade any area where material is being dumped, and place barricades where necessary to protect workers from flying pieces.
12. Permit no employee to work below others.
13. Require safety hats, goggles, foot protection, respirators, and gloves as needed for all workers. (See Chapter 17,

Personal Protective Equipment, in the *Administration and Programs* volume.)

14. Develop necessary safety procedures for handling hazardous materials.

15. Right-to-know training—regulations for handling hazardous materials must be met.

When the conventional methods for altering or removing concrete installations are unfeasible or undesirable, sometimes "powder cutting," a process that substitutes penetration by intense heat for concussion breakage, can be employed, or demolition by explosives may also be utilized. (For a description of powder cutting, see Chapter 13, Welding and Cutting, in this volume.)

EXCAVATION

To prevent injury and property damage during excavation work, make adequate protective measures part of the job. Study pre-excavation conditions (superimposed loads, soil structure, hydrostatic pressure, and the like) in order to evaluate changes that might occur, to prepare for situations that might develop, and to plan the job ahead.

A major hazard in urban or built-up areas is the presence of underground facilities, such as utility lines (water, electricity, gas, or telephone), tanks, process piping, and sewers. If these are dug into, undercut, or damaged in any way, there may be injury or death to workmen, interruption of service, contamination of water, disruption of processes, and expensive delays. Many states have a "one call" system to locate buried facilities.

Before starting operations, it is important that the company or plant engineer and utility and city or town engineers be consulted. The location of various facilities and their approximate depth below ground must be determined and marked by stakes in the ground or by markings on the floor.

Electronic locators can be especially helpful where an excavation would cross numerous buried obstacles. If the facilities are to be left in place, they must be protected against damage, and sometimes also against freezing.

Contents of buried tanks and piping should be indicated on the location markings. If the contents are flammable or toxic, proper protective equipment should be readily available in case of rupture. The bottom depth of the tank should also be indicated.

No shovel, dragline, or other digging machine should be allowed to excavate close to underground facilities that must be left in place. Establish a proximity limit for machine operations and complete the excavation by hand digging. If personnel are working in a trench deeper than 5 ft (1.5 m), according to OSHA standards, adequate bracing and shoring must be provided or the trench must be sloped. When hand excavation is being done, workers must be warned about driving picks, paving breakers, or other powered tools through buried facilities.

Whenever an excavation must be made within or adjacent to a building and lower than wall or column footings and machinery or equipment foundations, the job should be given to a specialty contractor. The contractor's personnel should make a thorough study to determine the amount and strength of shoring required before work on an open excavation is begun. Such a study will include the nature of the soil, hydrostatic pressure, superimposed loads (both static and live), and other factors. The depth and location of the excavation and the other characteristics will determine the need for sheet piling, shor-

ing, and bracing, which should be designed by an engineer or other person with experience in this type of design.

If underpinning (deeper support under an existing column, wall, or machine) is necessary, it should be done before the open excavation is carried down to final grade.

Excavated material should be placed at least 24 in. (0.6 m) from the edge of the excavation unless toe boards or other effective barricades have been installed to prevent fallback. Tarpaulins, sheeted barricades, or low built-up board barricades can be used to confine the excavated material to the immediate area under construction. Excavated material should not be permitted to accumulate in work areas or aisles, but should be trucked or otherwise removed from the building.

Excavations should be barricaded to prevent employees and others from falling into them. When an excavation must remain open for the duration of the construction work, barricades, fences, horses, and warning signs are necessary. In some cases, watchers and flaggers may be needed. The work area should be guarded by flares, lanterns, or flashing lights at night. (For more information, see Council Data Sheet 482, *General Excavation*.)

Trench excavation

A trench 4 ft (1 m) or more deep should be provided with ladders to facilitate safe entrance and exit. The ladders should be so spaced that no worker in the trench will ever be more than 25 ft (7.6 m) from one of them. The ladders should extend from the bottom of the trench to at least 3 ft (0.9 m) above the surface of the ground.

It is recommended that the side of trenches more than 5 ft (1.5 m) deep be shored unless they are sloped to the angle of repose or unless the trench is in solid rock. Shoring should be adequate to prevent trench wall collapse in whatever soil condition encountered. See Figure 2-4 for four trench-bracing methods.

In hand-excavated trenches, wooden cleats should be spiked or bolted to join the ends of braces to stringers to prevent the braces from being knocked out of place. (For more detail, see Council Data Sheet 254, *Trench Excavation*.)

In a long machine-excavated trench, a sliding trench shield may be used instead of shoring. Sliding trench shields generally are custom made to size for a specific job. They must be designed and fabricated strong enough to withstand the pressures that will be encountered. Also available are metal, portable hydraulic shoring systems.

LADDERS

Construction of all ladders should conform to the provisions of the applicable ladder or safety code of the locality or the state, whichever is more restrictive. Special-use climbing equipment, such as a combination stepladder-work platform, should comply with the applicable codes.

For detailed ladder information, consult the American National Standards Institute publications: *Safety Requirements for Portable Wood Ladders*, A14.1; *Safety Requirements for Portable Metal Ladders*, A14.2; *Safety Requirements for Fixed Ladders*, A14.3; *Safety Requirements for Job-Made Ladders*, A14.4; and *Safety Requirements for Portable Reinforced Plastic Ladders*, A14.5a. Also see Council Data Sheet 568, *Job-Made Ladders*.

The fixed ladder safety standard permits use of safety devices,

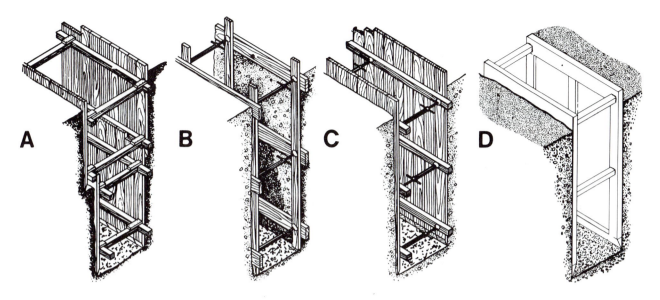

Figure 2-4. Trench bracing: A—bracing used with two lengths of sheet piling; B—bracing with screw jacks in hard soil; C—screw jacks used with complete sheet piling; D—commercially produced trench bracing.

(such as those illustrated in Figure 2-5) in lieu of cage guards on tower, water tank, and chimney ladders longer than 24 ft (7 m) in unbroken length. A landing platform shall be provided at least every 50 ft (15 m) within the length of climb. A rest platform at not over 150 ft (46 m) with a ladder safety device is used.

Ladder safety devices allow a climber to attach a restraint belt to a sliding fixture that travels along a carrier rail or cable anchored to the ladder. The traveling fixture will lock and suspend a person who slips and starts to fall. Many safety professionals consider such devices preferable to cage guards, where they are allowed by the code. A variation is a ladder belt with one or two snap hooks attached. This is not primarily a climbing safety device but provides a means of securing a climber to a ladder and freeing the hands if the climber is to work from the ladder.

The following general requirements apply to fixed ladders (see ANSI A14.3). Fixed ladders must:

- Be designed to withstand a single concentrated load of at least 250 lb (113 kg).
- Have rungs with a minimum diameter of ¾ in. (19 mm) for metal ladders, or 1⅛ in. (28 mm) for wood ladders.
- Have rungs at least 16 in. (40 cm) wide and uniformly spaced no more than 12 in. (30 cm) vertically on center apart.
- Be painted (if metal), or otherwise treated to resist deterioration when location demands.
- Have a preferred pitch of 75–90 degrees for safe use.
- Have 30 in. (75 cm) clearance with minimum 24 in. (61 cm) on the climbing side of the ladder (unless caged).
- Have at least a 7 in. (18 cm) clearance in back of the ladder to provide for adequate toe space.
- Have side rails extend 3½ ft (1 m) above landings.
- Have a clear width of 15 in. (38 cm) on every side of the center line of the ladder (unless with cages or wells).

Nonslip bases and safety tops

It is recommended that all ladders be equipped with slip-resis-

tant bases when there is a hazard of slipping. Slip-resistant bases are not intended as a substitute for care in placing, lashing, or holding a ladder that is being used upon metal, concrete, or slippery surfaces.

Some companies use a rope lashing to tie the top of the ladder to the pipeline or other object being worked on. Other companies replace the top rung on a portable ladder with a chain for work on cylindrical objects like poles and round columns. Such an arrangement will help prevent the ladder from slipping sideways.

Ladder maintenance

Inspection. After receipt, ladders should be inspected promptly for conformity to purchase order specifications and applicable codes noted above.

An inspection program should assure that all ladders are inspected once every three months. A record of each inspection should be kept. A general inspection form is shown in Figure 2-6.

Ladders that are weak, improperly repaired, damaged, have missing rungs, or appear unsafe shall be removed from the job or site for repair or disposal. Before junking a ladder, cut it up so no one can use it again.

Portable ladders must be maintained in good condition at all times, and inspected frequently. Any ladders which have developed defects must be tagged DANGEROUS—DO NOT USE and removed from service for repair or destruction. (See Figure 2-7.)

Coating. The *Safety Requirements for Portable Wood Ladders*, ANSI A14.1, states that "ladders may be coated with a suitable protective material. Non-conductive preservative paint may be used for identification on one side of the rails only.

One large company has this policy:

All ladders upon receipt from vendor should be delivered to the paint shop where, after approval by the purchasing department inspector, they should be given a treat-

CARRIER RAIL

Figure 2-5. Fixed ladder safety devices. (Printed with permission from Safety Tower Ladder Co. for the drawing and Air Space Devices, Norton Company for the photograph.)

ment of water-repellant preservative following the manufacturer's recommendations.

At the discretion of the department using the ladders, they may or may not then be coated with paint, varnish or enamel.

Checks, cracks, splits, and compression failures that may occur subsequently can ordinarily be detected through a transparent coating such as clear varnish, shellac, or other clear preservative.

Markings. Each ladder should be marked with the name of the department to which it belongs. Some companies number

their ladders consecutively so that none will be overlooked during inspection, while others stencil the date on each ladder as it is put into service. Proper identification assists in inspection procedures and also in storage. Warning stickers can also be added (Figure 2-8) to caution users to follow safe practices. (Also see the paragraphs on electrical hazards, under Use of ladders, next.)

Storage. Ladders should be stored where they will not be exposed to the weather and where there is good ventilation. They should not be stored near radiators, stoves, or steam pipes or in other places subjected to excessive heat or dampness.

Ladders can be hung on brackets against a wall horizontally, with more than two supports for long ladders to prevent warping, or placed on edge on racks or on rollers, rather than stored flat. These methods will facilitate removal of ladders.

Ladder storage space should be kept free of obstructions and accessible at all times, so that ladders can be obtained quickly in case of emergency.

Use of ladders

Placement. Workers should observe the following practices when placing ladders:
1. Place a ladder so that the horizontal distance from the base to the vertical plane of the support is approximately one-fourth the ladder length between supports. (For example, place a 12 ft [4 m] ladder so that the bottom is 3 ft [0.9 m] away from the object against which the top is leaning.) See Figure 2-9.
2. Do not use ladders in a horizontal position as runways or as scaffolds. Single and extension ladders are designed for use in a nearly vertical position and cannot be used in a horizontal position or with the base at greater distance from the support than that indicated in the preceding paragraph.
3. Never place a ladder in front of a door that opens toward the ladder unless the door is locked, blocked, or guarded.
4. Do not place a ladder against a window pane or sash. Securely fasten a board (not with nails) across the top of the ladder to give a bearing at each side of the window. Spread attachments are available. On wide windows with metal sash, the bearing may be across the mullions or between window jambs.
5. Place a portable ladder so that both side rails have secure footing. Provide solid footing on soft ground to prevent the ladder from sinking.
6. Place the ladder feet on a substantial and level base, not on movable objects.
7. Never lean a ladder against unsecure backing, such as loose boxes or barrels.
8. When using a ladder for access to high places, securely lash or otherwise fasten the ladder to prevent its slipping.
9. Secure both bottom and top to prevent displacement when using a ladder for access to a scaffold.
10. Extend the ladder side rails at least 3 ft (0.9 m) above the top landing.
11. Do not place a ladder close to electric wiring or against any operational piping (acid, chemical, sprinkler system, etc.) where damage may be done.
12. Ladders are for only one person at a time.
13. Do not overload a ladder. Do not hit it.

LADDER INSPECTION CHECKLIST

General Item To Be Checked	Needs Repair	Condition O.K.
Loose steps or rungs (considered loose if they can be moved at all with the hand)	☐	☐
Loose nails, screws, bolts, or other metal parts	☐	☐
Cracked, split, or broken uprights, braces, steps, or rungs	☐	☐
Slivers on uprights, rungs, or steps	☐	☐
Damaged or worn nonslip bases	☐	☐
Rusted or corroded spots	☐	☐

Stepladders

	Needs Repair	Condition O.K.
Wobbly (from side strain)	☐	☐
Loose or bent hinge spreaders	☐	☐
Stop on hinge spreaders broken	☐	☐
Broken, split, or worn steps	☐	☐
Loose hinges	☐	☐

Extension Ladders

	Needs Repair	Condition O.K.
Loose, broken, or missing extension locks	☐	☐
Defective locks that do not seat properly when the ladder is extended	☐	☐
Deterioration of rope, from exposure to acid or other destructive agents	☐	☐

Trolley Ladders

	Needs Repair	Condition O.K.
Worn or missing tires	☐	☐
Wheels that bind	☐	☐
Floor wheel brackets broken or loose	☐	☐
Floor wheels and brackets missing	☐	☐
Ladders binding in guides	☐	☐
Ladder and rail stops broken, loose, or missing	☐	☐
Rail supports broken or section of rail missing	☐	☐
Trolley wheels out of adjustment	☐	☐

Trestle Ladders

	Needs Repair	Condition O.K.
Loose hinges	☐	☐
Wobbly	☐	☐
Loose or bent hinge spreaders	☐	☐
Stop on hinge spreader broken	☐	☐
Center section guide for extension out of alignment	☐	☐
Defective locks for extension	☐	☐

Sectional Ladders

	Needs Repair	Condition O.K.
Worn or loose metal parts	☐	☐
Wobbly	☐	☐

Fixed Ladders

	Needs Repair	Condition O.K.
Loose, worn, or damaged rungs or side rails	☐	☐
Damaged or corroded parts of cage	☐	☐
Corroded bolts and rivet heads on inside of metal stacks	☐	☐
Damaged or corroded handrails or brackets on platforms	☐	☐
Weakened or damaged rungs on brick or concrete slabs	☐	☐
Base of ladder obstructed	☐	☐

Fire Ladders

	Needs Repair	Condition O.K.
Markings illegible	☐	☐
Improperly stored	☐	☐
Storage obstructed	☐	☐

Figure 2-6.

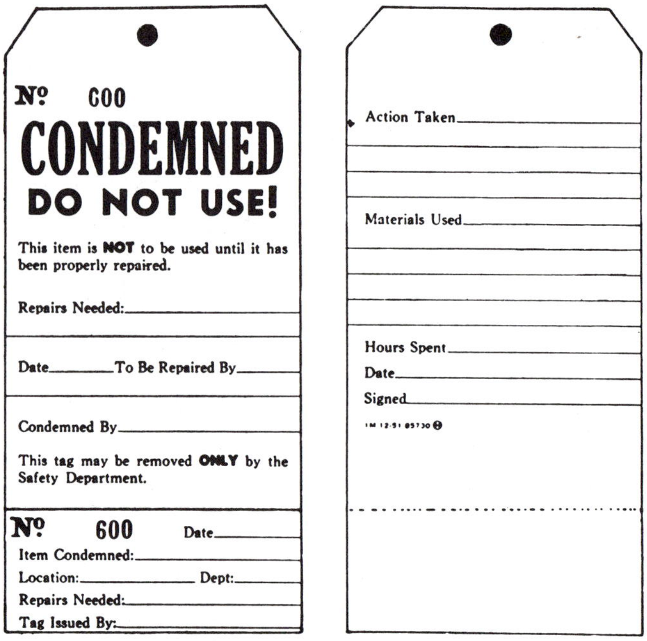

Figure 2-7. "Condemmed—Do Not Use" tag shows that a ladder or other piece of equipment should not be used until repaired.

Figure 2-8. Warning stickers added to ladders alert users to follow safe practices. (Printed with permission from Patent Scaffolding Co.)

Ascending or descending ladders. Workers should observe the following practices when ascending or descending ladders:

1. Hold on with both hands when going up or down. If material must be handled, raise or lower it with a rope either before going down or after climbing to the desired level.
2. Always face the ladder when ascending or descending.
3. Never slide down a ladder.
4. Be sure your shoes are not greasy, muddy, or slippery before you climb.
5. Do not climb higher than the third rung from the top on straight or extension ladders or the second tread from the top on stepladders.
6. Tools may be carried on a tool belt.

Other recommended practices. When using ladders:

1. Do not use makeshift ladders, such as cleats fastened across a single rail.
2. Be sure that a stepladder is fully open and the metal spreader locked before you start to climb it.
3. Before using a ladder, inspect it for defects. (See details under Ladder maintenance earlier in this section.)
4. Never use a defective ladder. Tag or mark it so that it will be repaired or destroyed (Figure 2-7).
5. Do not splice or lash short ladders together. They are designed for use in their original lengths and are not strong enough for use in greater lengths. Also, most splicing methods, particularly "on-the-job methods," are not recommended.
6. Keep ladders clean and free from dirt and grease, which might conceal defects.
7. Do not use ladders during a strong wind except in emergency, and then only when they are securely tied.
8. Do not leave placed ladders unattended. Remember that children may be attracted to them.

9. Ladders shall not be used as guys, braces, or skids, or for other than their intended purposes.
10. Adjustment of extension ladders should only be made by the user when standing at the base of the ladder, so that the user may observe when the locks are properly engaged. Never attempt adjustment while user is standing on the ladder.
11. The length of a straight portable ladder is 30 ft (9 m) or less. On two-section extension ladders, the minimum overlap is specified by ANSI A14.1–1982.

Electrical hazards and metal ladders. Since metal ladders are electrical conductors, they should not be used around energized electrical circuits or equipment or in places where they may come in contact with electrical circuits. The importance of these electrical hazards cannot be overemphasized, and those using metal ladders should be warned of the danger. Many construction projects forbid metal ladders.

In addition to this warning, metal ladders should be marked with signs or decals reading CAUTION—DO NOT USE NEAR ELECTRICAL EQUIPMENT. These decals may be placed on the inside of the side rails at about eye-level from the bottom of the ladder. (See Figure 2-10.) Glass fiber ladders, as well as wood, should be considered for use near electrical hazards.

SCAFFOLDS

A scaffold is an elevated working platform for supporting both personnel and materials. It is a temporary structure, its main use being in construction and/or maintenance work. Scaffolds should be designed to support at least four times the anticipated weight of workers and materials which will use them.

Scaffolding is the structure (made of wood or metal) that supports the working platform.

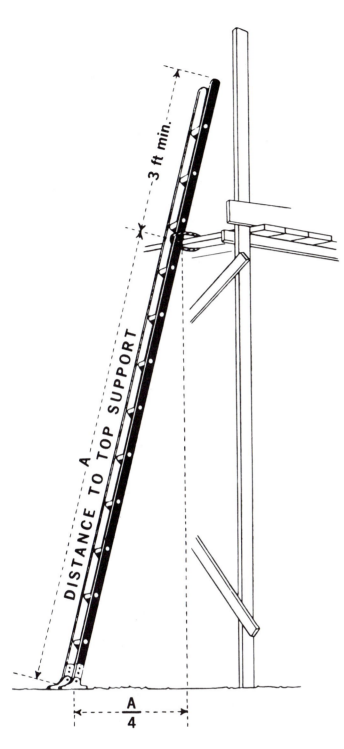

Figure 2-9. Safe procedure in setting up a ladder. The base should be one-fourth the ladder length from the vertical plane of the top support. Where the rails extend above the top landing, ladder length to the top support only is considered.

Figure 2-10. Place decals (like this one) on the side rails of metal ladders.

than 9 ft (3 m) above the working platform and should be planking or other strong suitable material.

Means of access

A safe and convenient means must be provided to gain access to the working platform level. Means of access may be a portable ladder, fixed ladder, ramp or runway, or stairway.

Ladders used for access to scaffolds should conform to the requirements of the applicable ladder code. (See the previous section. Also see Chapter 1 for information concerning runways and ramps.)

Types of scaffolds

A number of types of scaffolds are available. The major types are wooden pole scaffolds, tube and coupler scaffolds, and tubular welded frame scaffolds. Specialized scaffolds are not discussed in this Manual but are covered in OSHA regulations and the NIOSH "Health and Safety Guide for Masonry, Stonework, Tilesetting, Insulation, and Plastering Contractors." (See References.) Mobile (rolling) scaffolds, swinging (suspended) scaffolds, and boatswain's chairs are discussed later in this section.

General requirements for scaffolds must be followed, if applicable, unless more specific requirements for a particular type of scaffold are more applicable:

- Every time scaffolding is leased or purchased, safety instructions for erections and use should accompany the equipment. Never interchange the scaffolding components of different manufacturers.
- The footing or anchorage for scaffolds shall be level, sound, rigid, and capable of carrying the maximum intended load without settling or displacement. Unstable objects such as barrels, boxes, loose bricks, or concrete blocks are not to be used to support scaffolds or planks.
- No scaffold shall be erected, moved, dismantled, or altered except under the supervision of competent personnel.
- Guardrails, midrails, and toe boards shall be installed on all open sides and ends of platforms more than 10 feet (3 m) above the working surface (floor).
- Guardrails shall be 2 × 4 in. (5 × 10 cm) or the equivalent, a minimum of 36 in. (0.9 m) and a maximum of 42 in. high (1 m), with a midrail when required. Supports shall be at intervals not to exceed 8 ft (2.5 m). Toeboards shall be a

Overhead protection

Whenever work is being done over personnel working below on a scaffold, overhead protection should be provided on the scaffold for those personnel. This protection should be not more

- minimum of 4 in. (10 cm) in height. (Check individual state requirements.)
- Where persons are required to work or pass under the scaffold, scaffolds shall be provided with a screen between the toeboard and the guardrail, extending along the entire opening, and consisting of No. 18 U.S. gage (1.25 mm) ½-inch wire mesh or the equivalent.
- Any scaffold or component of a scaffold that is weakened or damaged must be repaired or replaced immediately.
- All load-carrying timber members of scaffolds shall be a minimum of 1,500 fiber (stress grade) construction grade lumber.
    All planking shall be scaffold grades, or equivalent. The maximum permissible spans for 2 × 10 in. (5 × 25 cm) or wider planks shall be as shown in the following:

	Full Thickness Undressed Lumber			Nominal Thickness Lumber	
Working Load (pounds per square foot)	25	50	75	25	50
Permissible span (ft)	10	8	6	8	6

- Lumber sizes, except where otherwise stated, are nominal sizes.
- The maximum permissible span for 1¼ × 9-in. or wider plank of full thickness shall be 4 ft (1.2 m) with medium duty loading of 50 psf (240 kg/m²).
- All planking of platforms shall be overlapped a minimum of 12 in. (30 cm) or secured from moving.
- Scaffold planks shall extend over their end supports between 6 (if cleated) to 12 in. (15 to 30 cm) only.
- The poles, legs, or uprights of scaffolds shall be plumb, and securely and rigidly braced to prevent swaying and displacement.
- Slippery conditions on scaffolds must be eliminated as soon as they occur.
- Wire, synthetic, or fiber rope used with scaffolds must be capable of supporting at least six times the rated load and should be inspected before each use.

Wooden pole scaffolds. The following regulations are applicable where wooden pole scaffolds are used.
- Scaffold poles (uprights) must bear on a foundation of sufficient size and strength to spread the load from the pole over a sufficient area which will prevent settlement. All poles must be set plumb.
- Where wood poles are spliced, the ends must be squared and the upper section shall rest squarely on the lower section (square butt joints). Wood splice plates (scabs) at least 4 ft (1.2 m) in length must overlap the wooden poles and cannot be less than the cross-sectional width of the wooden pole.
- Independent pole scaffolds shall be set as near to the wall of the building as practicable.
- All pole scaffolds shall be securely guyed or tied to the structure or building. Where the height or length of the scaffold exceeds 25 ft (7.6 m), it must be secured at intervals not greater than 25 ft vertically or horizontally.

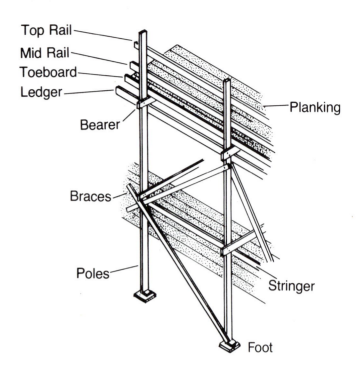

Figure 2-11. Nomenclature of a wooden pole scaffold. (Based on a NIOSH illustration.)

- Ledgers (sometimes called ribbons) must be long enough to extend over two pole spaces and cannot be spliced between the poles. Ledgers must be reinforced by bearing blocks securely nailed to the side of the pole to form a support for the ledger (Figure 2-11).
- Putlogs or bearers must be set in place with their greater dimension vertical and long enough to project over the ledgers of the inner and outer rows of poles at least 3 in. (7.5 cm) for proper support.
- Every wooden putlog on single pole scaffolds must be reinforced with a ³⁄₁₆ × 2-in. steel strip, or equivalent, and secured to its lower edge throughout its entire length.
- Diagonal bracing shall be provided to prevent the scaffold from moving, swaying, or buckling. Diagonal bracing shall be fitted in both directions on the exterior scaffolding face, cover the whole (tying in each level and pole) in one or more units, and not exceed 45 degrees to the horizontal.
- Cross bracing shall be provided between inner and outer sets of poles in independent pole scaffolds.
- Platform planks must be laid with their edges close enough together so that tools or materials cannot fall through.
- Where planking is lapped, each plank shall lap its end supports at least 12 in. (30 cm). Where the ends abut, forming flush surfaces, the butt joint shall be at the centerline of a pole. The abutted ends shall rest on separate putlogs. Intermediate beams must be used to prevent dislodgement of planks where necessary and ends shall be secured to prevent dislodgment.
- When a scaffold materially changes direction, the platform planks shall be laid to prevent tipping. The planks that meet the corner putlog at an angle shall be laid first, extending over the diagonally placed putlog far enough to have a good, safe bearing, but not far enough to involve any danger from

tipping. The planking running in the opposite direction at an angle shall be laid so as to extend over the rest on the first layer of planking.

- Guardrails made of lumber not less than 2 × 4 inches (or other material providing equivalent protection), approximately 42 in. (1 m) high, with a midrail of 1 × 6-in. lumber or equivalent material, and toeboards at least 4 in. (10 cm) high shall be installed on all open sides and ends on all scaffolds more than 10 ft (3 m) above a working surface. Where scaffolds are erected above walks or work areas, the space between toeboard and railing should be screened.
- Wood pole scaffolds greater than 60 ft (18 m) in height shall be designed by a qualified engineer competent in this field, and it shall be constructed and erected in accordance with such design. All wood pole scaffolds 60 ft or less in height shall be constructed and erected in accordance with federal occupational safety and health regulations, Title 29, *Code of Federal Regulations,* paragraph 1019.28(b).

Tubular metal scaffolds. Because tubular metal scaffolding is readily available, versatile, adaptable to all scaffolding problems, and economical to use, it is generally used. Most tubular metal scaffolding manufacturers and suppliers provide engineering service to help in the design of adequate scaffolding for any situation. Many suppliers also furnish erection and dismantling service.

Following are some common-sense rules for erecting, disassembling, and using metal scaffolding, recommended by the Scaffolding, Shoring and Forming Institute:

1. Post scaffolding safety rules in a conspicuous place and make sure people follow them.
2. Abide by all state, local, and federal codes, ordinances, and regulations.
3. Inspect all equipment before use. Never use equipment that is damaged or deteriorated in any way.
4. Keep equipment in good repair. Avoid using rusted equipment; its strength is not known.
5. Inspect erected scaffolds regularly to be sure they are maintained in safe condition.
6. Consult the scaffolding supplier when in doubt. Never take chances.

Tube and clamp (coupler) scaffolds are an assembly consisting of tubing that serves as posts, bearers, braces, ties, and runners, a base supporting the posts and special couplers that serve to connect the uprights and to join the various members (Figure 2-12).

A tube and coupler scaffold shall have all posts, bearers, runners, and bracing of nominal 2-in. O.D. steel tubing. All tube and coupler scaffolds shall be constructed and erected to support four times the maximum loads expected.

When other structural metals are used for these scaffolds, they must be designed to carry an equivalent load. No dissimilar metals can be used together.

- Runners (ribbons) shall be erected along the length of the scaffold, located on both the inside and outside posts at even height. Runners shall be interlocked to form continuous lengths and coupled to each post. The bottom runners shall be located as close to the base as possible. Runners must not be placed more than 6 ft-6 in. (2 m) on centers.

Figure 2-12. The Statue of Liberty was enclosed in scaffolding during its restoration. (Printed with permission from the National Park Service, Statue of Liberty National Monument.)

- Bearers (putlogs) shall be installed transversely between posts and shall be securely coupled to the posts bearing on the runner coupler. When coupled directly to the runners, the coupler must be kept as close to the posts as possible.
- Bearers (putlogs) shall be at least 4 in. (10 cm) but not more than 12 in. (30 cm) longer than the post spacing or runner spacing.
- Cross bracing shall be installed across the width of the scaffold at least every third set of posts horizontally and every

fourth runner vertically. Such bracing shall extend diagonally from the inner and outer runners upward to the next outer and inner runners.

- Longitudinal diagonal bracing on the inner and outer rows of poles shall be installed at approximately a 45-degree angle from near the base of the first outer post upward to the extreme top of the scaffold. Where the longitudinal length of the scaffold permits, such bracing shall be duplicated beginning at every fifth post. In a similar manner, longitudinal diagonal bracing shall also be installed from the last post extending back and upward toward the first post. Where conditions preclude the attachment of this bracing to the posts, it may be attached to the runners.
- The entire scaffold shall be tied to and securely braced against the building at intervals not to exceed 30 ft (9 m) horizontally and 26 ft (8 m) vertically.
- Guardrails are required as set forth in the Wooden pole scaffold section.

Tubular welded frame scaffolds are an assembly consisting of factory-welded frames and attachable metal crossbrace members, leveling screws, jacks and/or baseplates, and other accessories that are available to form a complete system. (See Figure 2-13.)

- Metal tubular frame scaffolds, including accessories such as braces, brackets, trusses, screw legs, ladders, etc., shall be designed, constructed, and erected to safely support four times the maximum rated load.
- Spacing of panels or frames shall be consistent with the loads imposed.
- Scaffolds shall be properly braced by cross or diagonal braces for securing vertical members together laterally, and the cross braces shall be of such length to automatically square and align vertical members so that the erected scaffold is always pumb, square, and rigid.
- The frames shall be placed one on top of the other with coupling or stacking pins to provide proper vertical alignment of the legs.
- Panels shall be locked together vertically by pins or other equivalent suitable means.
- Drawings and specifications for all frame scaffolds over 125 ft (38 m) high above the base plates shall be designed by a registered professional engineer.
- Guardrails are required as set forth in the Wooden pole scaffold section.

Mobile (rolling) scaffolds are caster-mounted sections of tubular metal scaffolding, or are made of components specifically made for the purpose. (See Figure 2-14.)

1. Do not ride rolling scaffolds.
2. Secure or remove all material and equipment from platform before moving scaffold.
3. Apply caster brakes at all times when a scaffold is not being moved.
4. Attach casters with plain stems to the panel or adjustment screw by pins or other suitable means.
5. Do not try to move a rolling scaffold without sufficient help. Watch out for holes in the floor and for overhead obstructions.
6. Do not extend adjusting screws more than 12 in. (30 cm).
7. Horizontal bracing should be used on a scaffold, starting with the base, at 20 ft (6 m) intervals.

8. Do not use brackets on rolling scaffolds.
9. Do not let the working platform height exceed four times the smallest base dimension, unless guyed or otherwise stabilized (Figure 2-14).

Testing scaffold planks. Scaffold planks should not be proof-tested because this may result in concealed or unrecognized damage that may subsequently cause failure. To check scaffold planks, the following procedure is recommended:

1. Examine the plank for large knots, excessive grain slope, shakes, decay, and other defects that may render it unfit. Discard the plank upon visible or audible evidence of failure or if it has an obvious deflection.
2. Determine the safe load for a plank of its size and species.

Swinging (suspended) scaffolds

Swinging (two-point suspension) scaffolds (stages) are usually factory built. The four types of platforms in common use are ladder, plank, beam, and metal. They are designed for light duty, primarily for workers using hand tools or washing windows. Swinging scaffolds should be suspended by wire or synthetic fiber ropes capable of supporting at least six times the maximum intended load.

Swinging scaffolds must be hung securely from parapets or roof, or other reliable supports, with properly placed hooks or outrigger beams of sufficient strength to provide a minimum safety factor of four (Figure 2-15). Whenever using a beam support with counterweights attached, never extend the beam more than 30 in. (0.75 m) beyond the face of the building and make certain that the beam is tied off in the rear as an added precaution.

Swinging scaffold supports should be tied off with a secondary rope secured to a stable roof-mounted structure.

Anchorages should be carefully inspected before the hooks or beams are placed and be tied off to a separate support from that from which the scaffold is suspended.

Other safe practices to be followed in the use of swinging scaffolds are:

1. Test the installation before using it following the manufacturer's procedures.
2. Inspect the raising and lowering mechanism frequently.
3. Be sure there are at least three turns of wire rope on drums at all times. Do not lower the scaffold below this point.
4. Equip the stage with toeboards, properly supported guardrail, and wire mesh when required between rail and toeboard.
5. Supply each person who works on the swinging scaffold with a safety harness and a lanyard that is securely attached to a separate line extending from roof to ground and attached to a separate point on the ground. Each lanyard should be tied to the lifeline with triple sliding hitch or with a mechanical rope grab.

Boatswain's chairs

Boatswain's chairs should be erected with care by those who will use them. The seat should not be less than 2 ft (60 cm) long by 1 ft (30 cm) wide. In wooden seats cleats should be nailed to the underside of each end of the chair to prevent the board from splitting. The chair should be supported by a sling attached to a suspension rope.

Where blow torches, cutting torches, or open flames are used,

2—Construction and Maintenance of Plant Facilities

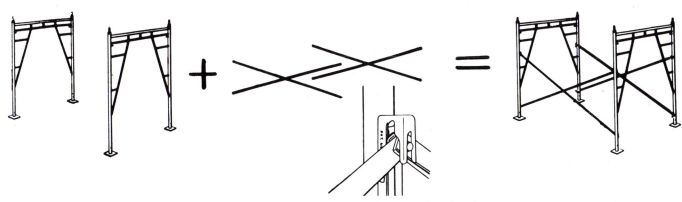

Figure 2-13. Tubular welded frame scaffold is quicker and easier to set up and is consistently safer than a wooden scaffold; it can be used over and over again for different applications. This type scaffold can be used in conjunction with a tube and clamp scaffold to provide design flexibility. (Printed with permission from Patent Scaffolding Co.)

slings should be made of wire rope. A safety belt, lanyard, and separate lifeline attached to a separate substantial portion of the structure should be provided. The slings should be at least 5/8 in. (15 mm) abaca (manila) rope (or its equivalent in strength if synthetic fiber or wire rope is used), doubled and passed through holes bored in the seat.

For information on other types of scaffolds, see applicable portions of federal/state OSHA regulations.

HOISTS USED AT CONSTRUCTION SITES

Material hoists and personnel hoists are made of tubular steel structural members. Be sure to consult with tubular steel manufacturers or suppliers for current technical data. The employer must comply with the manufacturer's specifications and limitations.

Hoists may be erected in hoistways inside the building or in outside towers. Personnel hoists are used just for the transport of people. Personnel must never be permitted to ride on a material hoist.

Work in or on the hoistway while the hoist is in operation should not be permitted. For additional information, be sure to check the latest ANSI Standards (A10.4, *Safety Requirements for Personnel Hoists;* and A10.5, *Safety Requirements for Material Hoists.*)

Rated load capacities, recommended operating speeds, and special hazard warnings or instructions must be posted on cars and platforms.

Inside material hoistways

If the material hoist is installed inside the building, the hoistway should be enclosed. Solid enclosure is preferable, heavy wire screening (½-inch mesh, No. 18 U.S. gage wire—1.25 mm) is often substituted. Adjacent hoistways should be partitioned.

Entrances should be protected by solid or slatted wood gates at least 5½ ft. (1.7 m) high and within 4 in. (10 cm) of the hoistway. Gates should be counterweighted and have latching or locking mechanisms.

Protective covering of heavy planking should be provided below the cathead of all hoists to prevent objects from falling down the hoistway.

More details are in Chapter 7, Plant Railways and Elevators.

Outside material hoistways

Hoisting towers are usually of tubular steel and are used on construction sites. Design should be based on a safety factor of at least 5. The tower should be on a level and solid foundation and should be well guyed or fastened to the building.

The tower should be enclosed with heavy wire screening and equipped with a fixed ladder extending the full height of the tower. Runways connecting the tower to the building should have standard railings and toe boards.

Material hoist platforms

Material hoist platforms should be built with a safety factor of at least 5. Flooring should be of not less than 2-in. (5 cm) timber, and sides not used for loading should be enclosed with heavy wire screening and should have 6-in. (15 cm) toe boards. There should be an overhead plank covering at the crosshead. The covering can be built in hinged sections to permit handling of long material. Where wheelbarrows are handled, stop cleats should be attached to the floor. The car should be equipped with a broken-rope type of safety device. If car cable breaks, the car safety clamps or dogs are thrown into position on the guide rails to stop the car.

Signal systems

A good signal system is necessary for safe operation. Electrically operated lights or bells, bells operated by pull cords, a combination of bells and lights, or telephone system can be used. Standard signals should be adopted and posted at each entrance and at the engine.

Personnel hoists

Any hoist used for carrying passengers should be made to conform to the safety requirements of ANSI A10.4, *Safety Requirements for Personnel Hoists.*

Figure 2-14. Mobile scaffolds. *Left:* Light-duty aluminum ladder scaffold used for general maintenance work. Note outboard safety supports used when scaffold is more than four times higher than narrowest base dimensions. *Right:* Sectional rolling steel frame scaffold has horizontal bracing at base. Outriggers are not required because the height (15 ft, 4.5 m) is less than four times the narrowest base dimension (5 ft, 1.5 m). (Printed with permission from Patent Scaffolding Company.)

Figure 2-15a. Suspended scaffolding on the face of a building under construction. Note that scaffolding is constructed to fit around protuberances of the building. (See closeup photo, Figure 2-15b.) (Printed with permission from Patent Scaffolding Company.)

Rack and pinion hoists can be used to carry personnel or material (but never both) according to strict manufacturing specifications and safety rules. All safety devices, including an over-speed governor, normal limit, and final limit switches, should be included, and the hoist should be regularly and thoroughly inspected and maintained.

Temporary use of permanent elevators. Permanent passenger or freight elevators in buildings under construction, modification, or demolition may be used for carrying workers or materials or both, provided they are approved for such use and a temporary permit is issued for the class of service. Elevators should conform to *Elevators, Escalators, and Moving Walks,* ANSI/ASME. A17.1-1984.

Towers, masts and hoistway enclosures. The tower or mast

Figure 2-15b.

construction forming the supports for the machinery and guide members should be designed and installed to support the load and forces specified.

For hoists located outside of a structure, the enclosures (except the one at the lowest landing) may be omitted on the side where there is no floor or scaffold adjacent to the hoistway. Enclosures on the building side of the hoistway should be full height or

a minimum of 10 ft (3 m) at each floor landing. Enclosure at the pit should be not less than 8 ft (2 m) on all sides.

For hoists located inside a structure, the hoistway should be enclosed throughout its height.

Hoistway enclosures should be so constructed that when they are subjected to a horizontal pressure of 100 lb (445 N), (1) they cannot deflect more than 1 in. (2.5 cm), and (2) the running clearance between the car and the hoistway enclosure is not reduced below ¾ in. (2 cm), except on the sides used for loading and unloading. If on openwork, hoistway enclosures should be provided on all sides within the building or structure with an unperforated kickplate extending not less than 12 in. (30 cm) above the level of each floor above the lowest.

Foundations of hoists should distribute the transmitted load so as not to exceed the safe load-bearing capacity of the ground upon which it is set. Hoist structures should be anchored to the building (or other structure) at vertical intervals not exceeding 25 ft (8 m). If tie-ins cannot be made, the hoist structure should be guyed to adequate anchorages to ensure stability. When wire rope is used for guys, it should be at least ½-in. (2 cm) diameter. Tie-ins should conform to, or be equal to, the manufacturer's specifications and should remain in place until the tower or mast is dismantled.

Where multiple hoistways are used and one or more of the cars are designed solely as a material car (in accordance with ANSI A10.5), personnel cars are prohibited. Each personnel hoist should be independently powered and operated. Chicago booms should never be used on a hoist structure.

Doors (or gates) for hoistways should not be less than 6 ft-6 in. (2 m) high. If a solid door is used, it should have a vision panel, not wider than 6 in. (15 cm) and not larger than 80 sq in. (0.05 m²), covered with expanded metal. All hoistway entrances must be protected by substantial gates or bars.

Landing doors should lock mechanically so they cannot be operated from the landing side. At the landings other than the lowest one, these locks should be of a type that can be released only by a person in the car. If the door at the lowest terminal landing is locked automatically when closed with the car at the landing, it should be provided with a means to unlock it from the landing side to permit access to the car. A hook-and-eye should never be used as a door-locking device.

Car platforms. Each car should have a platform or protective covering extending over the entire area of the car enclosure. It should be nonperforated and fire retardant, and supported by the car frame. Both the frame and the floor should be designed to handle the anticipated loads.

Car enclosures. Car enclosures and linings should be made of metal or fire-retardant wood. Personnel hoist cars should be permanently enclosed on the top and all sides, except the entrance and exit. This enclosure should be securely fastened to the car platform and so supported that it cannot become loosened or displaced when the car safety or buffer is engaged. The enclosure walls should be strong enough so that their running clearance is reduced by no more than ¾ in. (2 cm) when a force of 100 lb (445 N) is applied horizontally to the walls of the enclosure.

An emergency exit with an outward-opening cover should be provided in the top of all cars. The opening should be not less than 400 sq in. (0.025 m²) area, with a minimum dimension of 16 in. (40 cm). It should provide a clear passageway unobstructed by fixed hoist equipment on or in the car.

Do not locate a working platform, or place equipment that is not required for the operation of the hoist or its appliances, on the top of the hoist car, unless specifically provided in ANSI A10.4.

Wire glass (or the equivalent) should be used for vision panels. Plain glass should be used only for car operating appliances.

Wire ropes and sheaves

Hoisting ropes not less than ½ in. (1.2 cm) diameter shall be used except on such equipment as small winches, like those on gin poles. In any case, they should provide a factor of safety conforming to the requirements of applicable elevator codes. Ropes shall be inspected frequently and kept lubricated. They shall be replaced when inspection discloses that wear, breakage, or corrosion has reduced their strength below the permissible limit of safety.

Where clip fastenings are used, there shall be at least three clips, with the U-side on the dead end of the rope. Ropes shall be guarded at points where persons may come in contact with them and where they may strike or rub against objects. Sheaves should be well aligned, and bearings should be kept lubricated. In general, sheave diameter should be at least twenty times the rope diameter.

For additional details, see Chapter 4, Hoisting Apparatus and Conveyors, and Chapter 5, Ropes, Chains, and Slings.

Hoisting engines

Hoisting engines should not be located in public streets. Where they must be so located, they should be enclosed with barricades for protection of the public. In any case, a roof to protect equipment and operator from the elements is advisable.

Engines should have brakes capable of stopping and holding 150 percent of the rated safe load; in addition, there should be a pawl for holding suspended loads.

Exposed gears, shafting, and couplings should be enclosed. Exposed steam pipes should be covered, and exhaust pipes should be placed where steam cannot strike nearby persons.

Where electric hoists are used, switches should be the enclosed safety type, and all current-carrying parts should be enclosed or guarded to prevent personal contact. Installation should be grounded.

PLANT MAINTENANCE

A sound, efficient maintenance program is essential in any industrial establishment. Such a program will keep the physical plant in good condition and will reflect favorably in the safety record.

Maintenance should include proper, long-term care of the building and grounds, as well as of the equipment. It should include routine care to maintain service and appearance, as well as repair work required to restore or improve service and appearance.

Too often, mainenance is thought to mean only repair. In this discussion, considerable emphasis will be placed on *preventive* maintenance and on the type of inspection that will discover conditions pertaining to the building or equipment which, if uncorrected, might result in accidents.

The maintenance program may be under the supervision of the plant engineer or maintenance superintendent. However, because the safety of employees is closely tied in with the condition of buildings and equipment and because there is a close relationship between maintenance and safety, the safety professionals will find that the items covered have an important bearing on the safety program as well. They should not hesitate to point out—to the proper authorities—items that need correcting or replacement.

Foundations

Footings and columns. It is rather difficult to detect structural irregularities of footings, but it is possible to check unusual settlement of building columns and footings. Level marks of known elevation, placed about 5 ft (1.5 m) above the basement floor, can be checked periodically for settlement.

Excessive settlement may threaten the stability of a building, as well as the effective working of the machines and equipment in it. Excessive settlement should be reported at once to a structural engineer for immediate action.

Dry rot around the bottom of wood columns at the basement floor level results if the basement floor is damp, subject to water seepage, or for other reasons alternately dry and wet. Rust at the base of steel columns should be scraped away and the columns given a coating of paint. Also see Structural members described in the next section.

Foundation walls. The inside of foundation walls should be inspected for cracks, which may result from settlement of the building and shrinkage of concrete. Since these cracks are below grade, they may admit water to the basement area. If enough water comes in through large cracks, it may cause settlement of the backfilled earth around the outside of the foundation walls. Sidewalks and adjacent roadways may be damaged.

Minor or small cracks can be repaired from the interior by application of a water-proofing material. If the cracks are relatively large, it may be necessary to dig down outside the building to the bottom of the wall, clean off all earth and other foreign material, and apply a waterproofing compound. A membrane covering should then be applied directly on the compound and covered with another coating of compound. When unusual settlement is noted, a check should be made and settlement readings taken like those for footings and columns.

Pits. Inspection should also include pits, with cracks noted and repaired. No debris or rubbish should be allowed to collect in pits. Guardrails or covering should be supplied where needed.

Structural members

Joists, beams, and girders. In many instances, joists, beams, and girders are covered by suspended or sealed ceilings and are relatively inaccessible. In such cases, excessive deflection may be indicated only by a badly sagged floor. At least once a year, the entire floor system on each floor level should be examined.

Joists, beams, and girders should be checked, and deflection, twisting, tipping, or other unusual conditions corrected. If major repairs are necessary, the floor should be cleared immediately of stored materials. Before repairs are made, a qualified engineer should be consulted.

Columns. Building columns should be examined for unusual distortion or buckling. Excessive eccentric column loadings should be avoided, and a check should be made for holes cut in or through columns. If holes are found, they should be called to the attention of a qualified engineer.

Steel. Steel I-beams, channels, columns, angles, girders, and other steel structural members should be checked for rust once a year, more often in corrosive atmospheres. Where rust is noted, the steel member should be scraped and a paint applied.

Concrete. Floor slabs, beams, girders, and columns should be checked regularly for cracks, spalling, and sections of concrete chipped away from the reinforcing steel. Because rust may form on exposed reinforcing rods, repairs should be made at once. Guniting will provide a protective coating for the exposed reinforcing. Any damage may be evidence of more serious problems and a competent engineer should investigate.

Wood. A wood floor system should be inspected for shakes, checks, and splits in joists, planks, beams, stringers, posts, and columns. Decay or dry rot in wood columns, joists, and other members should be looked for. It is important that beams, joists, and girders have full bearing, and every evidence of movement or slippage should be fully investigated.

Walls

Exterior walls of brick, concrete, terra cotta, stone, cement or cinder block, and stucco need inspection for cracks or joint separation. Cracks may be caused by expansion, contraction, vibration, or settlement. When cracks are found, they should be filled immediately because water may freeze in them and cause additional damage.

Windows. Masonry buildings require periodic minor repairs of walls and windows. Mortar joints will loosen and disintegrate from settling of the building and from weathering. Such joints should be raked and pointed. If such joints are not repaired, moisture eventually will get into interior wall surfaces and cause progressive damage.

A similar precaution applies to window openings. Because of settlement of the building, drying out of the wood, or improper setting of metal frames, caulked joints will crack open, allowing moisture to enter.

Before recaulking is done, all loose material should be removed and all cracks cut out so that the new compound will be well bonded. A suitable quality of calking compound should be applied with a gun that will force the compound well into the openings (and not just cover the surface joints).

Parapet repairs include the maintenance and repair of masonry, metal, and wood parapets. Serious repairs can be avoided if masonry walls are checked carefully once a year for cracks or spalling.

Brick parapets and walls above grade should never be coated on either side with any material such as pitch, roof paper, or asphalt roof coating that will not allow the wall to "breathe." Such coating, which is commonly misapplied to brick parapets, causes spalling of the brick and disintegration of the mortar joints, particularly in areas of the country where freezing temperatures occur.

Sometimes brick walls are painted in the outside, either for decoration or for a sign. If there are high humidities inside a building, excessive moisture will penetrate the brick and condense under the impervious paint film, again causing spalling and joint disintegration during the winter.

Stone caps and other stone work on brick walls should be checked for cracks at all mortar joints. Cracks might allow moisture to enter and eventually loosen the stone. They should be filled with a cement grout or mastic filler. If these are on high buildings, their falling off could cause injuries and damage.

Interior walls and ceilings. The same rigid inspection should be given to partitions, cross walls, interior sides of main walls, and ceilings. Inspectors should look for such defects as cracks in interior walls, holes, loose mortar in joints, broken or missing brick, and spalled or worn areas where power trucks may have frequently scraped tile or brick walls. To prevent damage from trucks, standard 3-in. (8 cm) or 4-in. (10 cm) iron pipe railings can be put near the floor level as barriers.

Ceilings may need painting, cleaning, or repair. Unusual sag should be investigated immediately and corrected. If sag is found in a suspended ceiling, hangers and fastenings should be checked. However, the sag may be the result of excessive loading of the floor above, a situaiton which should be corrected.

Floors

Accidents due to inadequate maintenanance of floors are a major source of injuries in many plants. Slippery conditions, particularly, account for numerous falls by workers. Holes and other irregularities in wood and concrete floors, both inside and outside plant buildings, result in frequent injuries from stumbling and falling and, in addition, cause many truck accidents.

Repair procedures. Wood flooring should be inspected for rot, wear, and unusual stress as indicated by sag. In many cases, a section of wood flooring may need to be replaced. New flooring should be installed flush with the existing flooring. Badly worn or loose wood block flooring should be replaced by anchored wood blocks.

To anchor loose wood finish flooring, holes are drilled at an angle through the finished floor into the subfloor, and flooring nails larger than the holes then are driven into the subfloor.

For concrete floors, the damaged area should be chipped out, cleaned thoroughly and wet down. Cement mortar (1 part Portland cement and 3 parts sand) is troweled in. Minimum patch thickness is 1 in. (2.5 cm). Patches 2 in. (5 cm) thick or more may require wire mesh or reinforcing steel. For finishing, a wood float gives a less slippery surface than does a steel trowel.

For proper curing, traffic should be kept off the patch for at least three days, unless a quick-setting cement has been used.

If mixed and applied properly, epoxy resin repair materials give excellent results. A thickness of as little as $\frac{1}{8}$ to $\frac{3}{16}$ in. (2-3 mm) will give an extremely tough wearing surface. Be sure to take adequate precautions when using these mixes.

Maintenance and housekeeping procedures. Use of the wrong cleaning materials, methods, and surfacing often causes even the most suitable types of flooring to deteriorate and become slippery. Alkaline cleaners should not be used on terrazzo, but mild alkaline cleaners may be used on asphalt tile. Oils are unsuitable for rubber tile and, when applied to wood floors, increase the fire hazard. To keep floors clean, safe, and sanitary, the recommendations of the flooring manufacturer should be followed. Procedures should be standardized and spelled out in detail. (See Figure 2-19.)

In general, the routine maintenance procedure for linoleum, marble, terrazzo, asphalt tile, and other types of flooring used in offices, institutions, and similar buildings is to clean the floors with a soft floor brush or vacuum cleaner and, when necessary, mop them with a mop dampened with clean, cold water.

One section of floor is cleaned at a time, and, if traffic in the area is heavy, that section is roped off. When soap is used, and soapy film must be removed by thorough rinsing, to avoid a slippery condition.

Ordinary wax for polishing wood, tile, and similar floor surfaces is unsuitable because of its inherently slippery nature. However, according to the manufacturers, floor oils and waxes can be used to surface various types of floors without adding unduly to their slipperiness, provided that instructions regarding their application are followed. Soft floors (such as asphalt, vinyl, and linoleum) are often refinished four times a year. Hard floors (such as concrete and terrazzo) are cleaned and sealed once a year.

Oils and greases, water, paper, sawdust, and numerous other foreign materials on floors create slipping hazards. Leaks of oil from machines and of water and other liquids from pipelines and spillage from processing equipment often can be eliminated by good maintenance of the equipment, such as promptly tightening loose connections.

Pans and absorbent materials can be used when leakage cannot be readily eliminated at the source. However, a study of such sources often reveals ways of keeping slippery materials from getting on the floors, such as the installation of splash guards on machines using cutting oils.

Slippery materials spilled on floors should be cleaned up promptly. To remove grease and oils, the area can be covered with slaked lime to a depth of about $\frac{1}{4}$ in. (5 m). After two or three hours, the lime then is removed with a scraper or stiff brush. Sand and various commercial cleaners also can be used.

Even as innocuous a substance as coffee can cause an accident if spilled in a high traffic area. In some company offices, cups containing beverages must either be covered, or be on a tray if carried to an employee's desk.

Aisles should be kept clear of machinery, equipment, raw and manufactured materials. In many cases, the allowable floor loading has been figured on clear aisle space with no allowance made for power trucks using aisles. It is advisable for efficient, safe operation to determine whether floors and aisles are capable of sustaining loads of power trucks. Lines to indicate aisle width should be well maintained.

Overloading of floors, which may be caused by the installation of new heavy equipment, excessive weight and unequal distribution of stored raw and finished materials, and heavy truck transportation, is particularly hazardous. Signs stating allowable floor loads and horizontal lines showing the maximum height to which materials may be piled should be painted on the walls. (See Figure 2-16.)

Strength and load determinations. Accurate information on floor load capacity is a prime requisite for a physical plant survey. If these data are not already known or readily obtainable from building plans, a structural analysis by a qualified engineer may be necessary. Rough estimates based upon experience or conclusions arrived at by a casual glance at a handbook are dangerous. Use accurate weight data.

Load distribution is often a complex problem. Most buildings are designed to carry uniform loads. A concentrated load places twice as much stress on supporting members as a uniform load of equal weight. For this reason, most heavily concentrated loads, such as machines, are placed directly over beams or girders rather than on slabs or joists.

As in the case with floor load capacity, the safety professional should consult a structural engineering specialist for an accurate determination of concentrated loads.

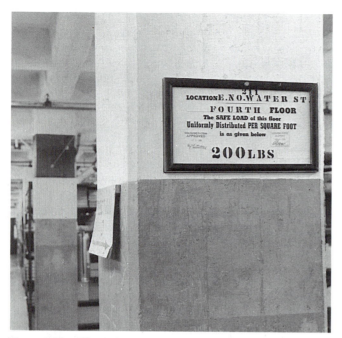

Figure 2-16. Where the same type of material is stored regularly, a line can be marked on the wall to indicate the height to which material can be piled without exceeding the allowable floor load. In addition to safe floor load limit signs, a warning sign reading DO NOT PILE ABOVE LINE can be placed on the wall.

Overloading determinations. The relationship between design load capacity and actual floor loading will indicate accurately whether a floor is safely loaded or is overloaded.

Although visual evidence of overloading is not always to be found, the inspector should look for it. Deflection is the most common evidence of overloading in wood and steel beams. A sag or deflection greater than 1/360th of the span length is a warning that the floor may be overloaded.

To measure the deflection, a cord is stretched between two columns of the span 5 ft (1.5 m) down from the underside of the beams or girders at ceiling level. The distance from the center of the taut cord to the bottom of the beam or girder is measured, and the difference between this measurement and 5 ft is the deflection.

Overloaded wood beams will check and crack. In reinforced concrete beams and girders, concrete will spall and fall away from the tensile side. Wood floors will be punctured, and flat concrete slabs will crack and spall. Timber columns will split and crack. Concrete columns will spall, the concrete falling away from the reinforcing. On steel columns, flanges will be twisted.

Bearing walls of masonry will show extensive cracking, disintegration of the bricks, bulges, and pulling away of floor joists.

Whenever there is any doubt as to the amount of deterioration of floor supporting members and therefore the load-carrying capacity of the floor, inspection and determination should be done by a qualified structural engineer.

Roofs

Abnormal loading. Roofs are usually designed to carry the maximum snow load expected for the locale. Roof configura-tion and wind frequently combine to deposit heavy drifts over portions of a roof. Where practical, these accumulations should be removed as quickly as possible to prevent roof collapse or lesser damage. Ice is a serious problem and can also cause overloading.

Inspection. All roofs should be inspected periodically, perhaps once every six months. Roof flashings should be checked for cracks at the parapet wall. Roof gutters and drain connections should be checked for cracks at the roof line. Areas around dormers, chimneys, and valleys, where metal is used, must be checked to see that the metal is tight with the roof as well as with the drain. Roof drains and overflow gutters through parapet walls should be kept open; leaks will occur if water should rise above the flashings.

It is often difficult to find the source of a leak, since the fault may be distant from the place where the leak shows inside the building. On sloping roofs, a check should be made above and to the sides of the place where the leak shows. Flat decks, laid with concrete, are difficult to check because water frequently follows a crack and shows up at a distant point.

Gutters should be kept clean, and gutter areas checked for ice damage in the early spring months.

Repairs. A leaky roof should be patched as soon as possible after the leak is first apparent.

A new roof is indicated when maintenance becomes excessive and leaks occur after each rain. Cold mastic applications usually are good for about 5 years, whereas an application of hot pitch covered with pea gravel may last from 5 to 10 years. Reputable roofing manufacturers will provide a guaranteed-performance bond for a number of years commensurate with the quality of the job specified.

On a repair job, a roof can be punctured by tools, boards with nails, stones, or other sharp objects. Runways and protective runboard covering should be used. If hot pitch or roofing compound is used, workers should wear gloves, goggles, and leather or fire-resistant duck leggings.

Roof anchorage is a very important point to check during inspection. The lifting of unanchored roofs accounts for a large percentage of the total wind damage loss to American industry.

The roofs of all buildings should be securely anchored. Anchors can be readily installed, and the cost is reasonably low.

Roof-mounted structures. Roof inspection should include penthouses, stacks, vents, and supports for water tanks where these structures are flashed at the main roof level. The roofs of penthouses should be checked, since most penthouses contain elevator machinery which may be damaged by water seepage.

Penthouse windows, skylights, and monitor sash should have necessary reglazing, reputtying, frame caulking, or painting. When operating mechanisms are being repaired or repainted, workers should work on safe scaffolds and should have safety lines tied off to lifelines. They should never work on or over unprotected skylights.

Tanks and towers

If a tank is more than 20 ft (6 m) above the ground or building roof, a wood or steel balcony should be placed around the base of the tank, designed to support a weight of at least 100 pounds per square foot (486 kg/m²). For a tank not more than 15½ ft

(4.7 m) diameter, the width of the balcony should be at least 18 in. (45 cm); for tank diameters greater than 15½ ft, the balcony width should be 24 in. (60 cm) with railings. See ANSI A12.1, *Safety Requirements for Floor and Wall Openings, Railings, and Toeboards.*

Maintenance of tanks and towers is important, not only for fire protection, but also because structural failures may cause serious accidents.

When tanks are to be cleaned or painted on the inside, all precautions relating to tank entry should be strictly observed. (For a detailed discussion of the required precautions, see Cleaning Tanks in Chapter 16, Flammable and Combustible Liquids.)

Stacks

Stacks should be inspected at least once every six months. They are subject to deterioration, both inside and out, from weathering, high winds, lightning, foundation settlement, and the action of corrosive flue gases.

A brick or concrete chimney can be protected by lightning rod conductors of low resistance and ample carrying capacity if they are installed from the top of the chimney to a good electrical ground. Lightning rods can be readily installed on chimneys when repairs necessitate the erection of scaffolding. Underground water pipes or buried copper plates afford good ground connections.

Stack ground wires should be checked for effective grounding.

Platforms and loading docks

Mechanized traffic on platforms and docks often causes damage to platform surfaces. With no edge protection, concrete may become spalled or chipped, and ruts in the concrete finish may cause power or hand trucks to swerve and run off a dock or into employees or material. For these reasons, angle iron or channel iron protection should be provided at the edge of the platform, and kept well maintained. Badly rutted platforms may be resurfaced with concrete or epoxy cement.

Wood platforms should be checked for decay or dry rot, loose or uneven planking, weakened or broken supporting members, and repairs should be made immediately.

Canopies

A canopy roof should receive the same careful inspection as that given to the roof of the main building. Evidence of pulling away from the building should be noted and corrected. Drainage is important, and downspouts and gutters should be kept open and in good repair.

Supporting members, if wood or steel, should be scraped and painted periodically.

Sidewalks and drives

Concrete sidewalks and drives should be repaired as soon as spalling or cracking of the concrete creates a hazardous condition. In cold climate areas, inspections should be made in the spring after the ground has thoroughly thawed. Sections of sidewalks can be relaid, but bituminous driveways will need patching with hot tar or similar material.

Drainage of a driveway is important, whether it is made of concrete, asphalt, or gravel. Repairs should be made as soon as possible so as to keep damage to a minimum. To keep asphalt drives in good condition, they should be recoated periodically by someone who understands paving techniques.

Colored plastic disks can be used to mark parking areas, to eliminate the need for frequent repainting of stripes.

Underground utilities

Sewers. At least two persons should work on a sewer maintenance job, whether or not dangerous gases are suspected. Before anyone goes into a manhole structure or sewer, tests should be made for oxygen deficiency, methane, hydrogen sulfide, carbon monoxide, carbon dioxide, and nitrogen. If any of these gases are found, the necessary precautions must be strictly observed.

Proper respiratory equipment (see Chapter 23, Respiratory Protective Equipment, in National Safety Council's *Fundamentals of Industrial Hygiene*, 3rd ed.) should be used, or complete ventilation should be provided with either blowers or suction fans. Blowers are preferable since their source of supply is known. Suction fans may draw poisonous gases from unseen pockets or crevices into the area.

Utility trenches and tunnels. Where workers must be in trenches over 5 ft (1.5 m) deep, shoring is required. If the trench is adjacent to machinery foundations, or other superimposed loading, a thorough design of the bracing and shoring to be installed should be made before work is started.

It is important to check atmospheric conditions in utility trenches, tunnel trenches, tunnels, and manholes before work is started. Proper ventilation is necessary for the safety of the workers, who should work in pairs. Manholes, tunnels, and trenches known to be contaminated should be tagged or otherwise identified for the work crews.

Smoking should not be permitted in or near manholes, tunnels or trenches.

The use of open-flame devices, such as solder pot furnaces and welding equipment, should not be permitted in or near manholes, tunnels, or trenches, in which tests indicate the presence of flammable gas.

Waste disposal facilities. When on-site waste disposal is planned, a permit must be obtained from the U.S. EPA. The U.S. EPA specifies all requirements.

Other underground pipelines. Before repairs are begun, pipelines should be completely drained, the connecting systems blanked off, and the valves closed and locked. Whenever workers open a line or valve, they should watch for back pressure in the line.

Workers should be cautioned to open valves slowly and to equalize pressure slowly. Sudden changes in pressure can wreck equipment and endanger lives.

After a steam valve has been opened completely, it should be backed off at least one-half turn so that thermal expansion will not lock the valve in the open position.

When people are to work on an underground pipeline, it is important that its approximate location be known before digging is begun. Workers cutting into or working on a pipeline should know what the line is carrying and the dangers involved.

When pipe or trenches must remain open overnight, they must be protected by barricades, signs, and lanterns.

Lighting systems

All lighting systems—fluorescent, incandescent, or mercury

vapor—require regular maintenance to maintain maximum light output.

To reduce the number of persons who might be exposed to flying glass and dust if lamps are broken, fixtures can be relamped on weekends or at other times when personnel are not in the area. Gloves should be worn by those who handle the lamps.

To dispose of a few lamps, put them in special containers for regular rubbish removal or wrap in several thicknesses of newspaper or wrapping paper, place outdoors in a disposal container, and crush with a heavy stick or shovel. Large numbers of tubes should be broken in commercially available machines and in a ventilated enclosure equipped with an exhaust and a dust collector. Captured dust should be wet down and removed.

Lamp burn-outs and depreciation, dirt accumulation, voltage drops, and light absorption by dirty walls and ceilings will cause a loss of light. Because dark or dirty surfaces absorb as much as 80 percent of the light that strikes them, a regular cleaning schedule is important. For dirty areas, this may be every two weeks.

A light meter can be used to check conditions, and the readings then can be checked with the minimum standards of illumination for industrial interiors, ANSI/IES RP7-1983, *Practice for Industrial Lighting*. These standards, recommended by the Illuminating Engineering Society, are summarized in Table 19-G, entitled Levels of Illumination, in Chapter 19, Safety Engineering Tables. These values can be used as a guide to formulate a regular maintenance program. If cleaning lamps and replacing burnouts fail to increase the illumination to standard levels, then a qualified illuminating engineer should be called upon to make a complete survey of the present system.

Fluorescent fixtures should be of the type that locks the tubes in place. Older fixtures should be equipped with shields or grids beneath the tubes to prevent the tubes from falling if vibration should loosen them.

Where reflectors or glassware cannot be taken down, the current should be shut off and a cleaner that requires no rinsing should be used, followed by wiping with a cloth.

Maintenance problems can be simplified by laborsaving devices. When possible, disconnecting-type reflectors should be used. They permit cleaning and relamping from the floor rather than at the outlet. For relatively low mounting heights, stepladders should be used because of their convenience and portability. However, they should be of sufficient height so that the top two steps of the ladder will not be used to stand on during maintenance activities. Clips and hooks which hold spare lamps and cleaning rags enable a person to do an entire cleaning and relamping job with one trip up the ladder.

Where the entire installation is cleaned at frequent intervals, special cleaning trucks can be used which have separate compartments for cleaning solutions, for warm rinse water, and for clean rags.

Portable maintenance platforms can be used where there are a great many luminaries at a given mounting height. These platforms are made of lightweight material and are equipped with casters so that one person can handle them easily. This type of equipment permits a person to reach several units safely and not reposition the platform.

Manufacturers have thought of various means to reach all types and styles of lamps. One of the simplest devices is the clamp grip mounted on the end of a pole and formed to fit fluorescent tubes. In many industrial plants with installations having open bottoms and exposed lamps, these pole-changing devices are used between periods of regular maintenance for emergency lamp replacement. They are well suited for recessed reflector lamps.

Stairs and exits

The following items should be noted when the condition of stairways and exits is checked:
Lack of handrails—handrails too low—handrails rough
Improper illumination
Poor housekeeping
Improper or inadequate design, construction, or location
Wet, slippery, or damaged surfaces
Faulty treads or mats on stairs
Lack of curbing on ramps

Whenever any one of these items is detected in the maintenance inspection, it should be repaired or corrected immediately. Various types of nonslip surface materials that can be applied directly to stairways are on the market.

Passageways should not serve as storage areas. The floor surface should be smooth and the area well lighted. Hazardous conditions should be corrected as soon as possible after inspection.

Operation of exit doors should be checked. There should be no obstruction to their free movement, nor should the paths to them be blocked in any way.

Exit signs and lights should be kept in good repair. Exit signs and emergency lighting designed to operate in the dark in case of failure of the lighting systems should be tested in darkness.

For details, see National Fire Protection Association No. 101, *Life Safety Code*.

GROUNDS MAINTENANCE

Prevention of accidents and injuries from tools and machines used in grounds maintenance requires that equipment be chosen for its specific purpose and that it be used and maintained properly. Fuel and chemicals should be stored and used properly. Workers should be thoroughly trained and should wear proper clothing and use protective equipment, as required.

It is important that maintenance workers be able to recognize poisonous vines, shrubs, fruits, and insects. They should avoid contact with poison oak and ivy, and destroy permanently all poisonous growths. They should guard against insects and infections, and scrub hands thoroughly after working outdoors. Treat all cuts and scratches received outdoors with proper antiseptic covering. Remove all foreign matter, such as glass, metal, and wire, from the grounds to be maintained. Use gloves and wear safety shoes and appropriate garments for protection at all times.

Hand tools

The same rules apply for using tools safely outside of the plant as well as inside. The primary rules include the following: (1) Use battery-powered tools outside, if possible. (2) Use double insulated tools. (3) Use the correct tool for the task at hand. (4) Keep landscaping tools in good condition. (5) Use tools in the way they were intended to be used. (6) Store tools in a safe place. (7) Keep cutting tools sharp and clean. (8) Keep tool handles smooth and strong. (9) Shovels, spades, and other digging tools should have points that are smooth and properly shaped.

Electric landscaping tools. Read the operator's manual carefully before switching on the tool and putting it to work. If a tool is equipped with a three-prong plug, it should be in a three-hole grounded receptacle and a three-wire extension cord should be used. Use a GFCI outside. All visitors should be kept a safe distance from work areas, and the area should be clear of debris, animals, etc. Don't force the tool—it will do the job better and safer at the rate for which it was designed. Keep both hands on the tool if a second handle is provided.

Never use electric power tools in the rain, or when grass or shrubs are wet. Don't abuse the electric cord. Never carry a tool by its cord or yank it to disconnect it from the receptacle. Always use cord with proper wire size to carry the current. Be sure to avoid cutting the cord with the tool or equipment.

Always wear eye protection when using landscaping tools. Hold tools in position ready for use before switching on current. Use both hands to hold and guide a tool. Always shut off current when resting arms, picking off cuttings, or before changing direction of cut. If a tool becomes jammed or fails to start, always turn off the switch before attempting to free the jam or before looking for the trouble. When leaving the work scene, even for a few minutes, always shut off the current and disconnect the plug. Always keep other people well away while tools are in operation.

Electrical hedge trimmers. The usual injuries from electric hedge trimmers are amputated fingers, serious cuts on fingers and hands, and cuts on knees and legs when lowering the trimmer to rest the arms. Hedge trimmer injuries result from five types of actions—changing hand position with the trimmer running, holding branches away from the cutting bar, removing debris from the trimmer, holding the trimmer with only one hand, and failing to wait for the blades to stop after turning the trimmer off.

Hedge trimmers should be chosen with the following features in mind: Select one that is not too heavy to hold comfortably for a long time, and one that has a large support handle for the second hand to hold. As before, a switch that requires continuous finger pressure to maintain power is safer. Use battery powered double insulated tools. Look for the UL label.

When using the trimmer, workers should get into a comfortable position, use both hands, avoid cramped spaces, take their time and not "force" the work. They should not overreach or lean off a stepladder. When they leave, even for a coffee break, they should take the trimmers with them or follow company regulations on securing tools if there is any chance that some unauthorized person might try to use them.

Lawn trimmers are useful for cutting the grass around tree trunks or along fences. Along with edgers, which cut borders along the edges of sidewalks, driveways, or gardens, such trimmers injure several thousand people each year. Edgers and trimmers should be treated with as much caution as mowers because they, too, have a metal cutting blade that can throw debris or cut a finger.

Keep guards in place and in working order. Keep the blades sharp. Do not put hands near the working area unless the machine is turned off.

Nylon-cord weed trimmers cannot hurt as seriously as metal-blade trimmer-edgers, but getting hit by the line can sting. The operator must disconnect the power cord when adjusting the cutter cord length or changing the reel, applying the same precautions as would be taken with any electrical appliance. Be careful of wet areas and periodically check the cord for cracks or breaks in the insulation.

Gasoline-powered equipment

Observe the following safety rules when handling gasoline:

- Never use gasoline for cleaning floors, tools, clothes, or hands. Gasoline is only to be used in engines as a source of energy.
- Always store gasoline in an approved closed container.
- Pouring gasoline from one container to another may generate a charge of static electricity. A metal-to-metal contact must be maintained.
- Gasoline spills should be cleaned up immediately to prevent accumulation of vapors. Do not allow electrical switches to be turned on until the gasoline vapors have dispersed.
- If gasoline is spilled on a person, remove the saturated clothing immediately and keep the person and clothing away from sources of ignition. Wash the affected area of the skin with soap and water to avoid a skin rash or irritation. If the eyes are involved, flush with water and get the person to medical attention.
- Gasoline tanks or equipment parts that are likely to contain gasoline should be drained or dismantled only out-of-doors or in a well-ventilated area, free from sources of ignition.
- Never smoke in fueling areas, fuel system servicing areas, maintenance areas, bulk fuel delivery areas, and similar fuel present areas.
- Never dispense gasoline into the fuel tank while the engine is running, or a motor is hot.
- Never store equipment with fuel in the tank inside a building where vapors could reach an open flame or spark. Allow the engine to cool before storing in any enclosure.
- Never run an engine indoors.

Gasoline-powered mowers and tractors should meet the standards of *Safety Specifications for Turf Care Equipment—Power Lawn Mowers, and Lawn and Garden Tractors*, ANSI/OPEI B71.1-1986. Snow throwers should meet ANSI B71.3-1984, *Safety Specifications for Snow Throwers*.

Power lawn mowers, especially the rotary type, have proven to be a mixed blessing. Although they save time and effort and leave a lawn neatly manicured, they also take lives and have caused more than 100,000 personal injuries a year. These injuries range in severity from minor cuts to amputations.

Before starting, make sure the operator is well trained in using the mower. If it is the first time in the season that the mower is being used, review the instruction manual. Before starting to mow, the operator should pick up rocks, glass, tree branches and twigs, and any other objects that could become lethal missiles if thrown out by the mower blade. The operator also should observe the location of fixed objects, such as pipes, lawn sprinkler heads, and curbs, that could damage the mower or break off and become missiles. Make any wheel height adjustment before starting the mower. Disconnect the spark plug wire when cleaning, repairing, or inspecting the mower. Unauthorized persons are not to be in the mowing area. The operator should make a quick inspection for loose nuts and bolts, check the engine oil level, and fill the fuel tank before starting. Use a vented can with a flex spout. The operator should wear safety footwear and safety glasses. A brimmed hat and full-length trousers and shirt will protect against sunburn.

The operator should be instructed to mow in daylight or good artificial light and to push the mower forward as much as possible because feet can be injured when pulling a mower backward. When mowing on a slope or terrace, make a series of horizontal passes along the incline. If the operator pushes up the incline, the mower could drift back onto a foot. If it is pushed down, footing can be lost and the operator can fall into the mower.

Don't use the mower when the grass is wet and slippery. If the grass is damp or high, cut it at a slower speed (if possible) and set the cutting height higher than for dry grass, otherwise the discharge chute may clog up.

Rotary blades can also pick up stones, pieces of wire, nails, or other objects hiding in the grass and throw them out of the discharge chute at terrific speeds. Newer models should have a guard over the discharge chute that deflects objects downward, but the guard must be removed if the grass catcher is used. Some mowers have guards that automatically snap back into place when the grass bag is taken off. Others require that the guard be bolted in place any time the catcher is not used.

The operator must shut off the engine and be sure the blade has stopped completely before taking off the grass catcher to empty it, or attempting to free obstructions from the discharge chute, adjusting the cutting height, or performing any operation requiring the person to have hands or feet near the blade.

Reel mowers are like the "old fashioned" kind with several blades that shear the grass against a horizontal stationary edge. The reel mower does not need the speed that a rotary mower does, so it is safer. However, it is not as good for cutting tall grass or weeds.

Riding mowers. If large areas around the plant must be mowed, a riding mower would be most efficient. Suggested safe practices include the following:

- The operator must be fully instructed and know the controls and how to stop quickly. The owner's manual should be read at the beginning of each mowing season.
- The work area must be cleared of objects that might be picked up and thrown. Fixed objects that might damage the mower should be identified. Realize that all areas cannot be reached by a riding mower and that some corners or sharp slopes will have to be mowed by a power mower. When planning landscaping, leave enough space around new plantings for easy mower access, and allow for future growth.
- Disengage all attachment clutches and shift into neutral before attempting to start the engine (motor). Disengage power to attachment(s) and stop the engine before making any repairs or adjustments. Disengage the power to the attachments when transporting them or when they are not in use. Take all possible precautions when leaving the vehicle unattended, such as disengaging the power takeoff, lowering the attachment(s), shifting into neutral, setting the parking

brake, stopping the engine, and removing the key from the ignition.

- When mowing, stay alert for holes in the terrain and for other hidden hazards. Do not start or stop suddenly when going uphill or downhill. Avoid steep slopes. Do not mow steep slopes when wet from rain or morning dew. Do not mow between large trees that the rear wheels will not pass through. Large mowers are known to turn over backwards as a result of the extreme power in the rear wheels. Mow up and down the face of steep slopes, never across them, because the wheelbase is longer than the tread, so the unit is more stable that way. Reduce speed on slopes and in sharp turns to prevent tipping or loss of control. Use extreme caution when changing direction or turning around, especially on slopes. Do not back up without looking to make certain it is safe to do so. Watch out for traffic when crossing or near roadways. When using attachments, be sure to direct discharge of material away from anything that could be hurt or damaged by it.
- Maintain the vehicle and its attachments in good operating condition and keep safety devices in place. Keep all nuts, bolts, and screws tight and make sure the equipment is in safe working condition; check especially blade mounting bolts. If the vehicle or its attachment(s) strike a solid object, stop and inspect for damage; the damage should be repaired before restarting and operating the equipment. Do not change the engine governor settings or overspeed the engine.

Garden tractors. Some of the accident patterns associated with these tractors are:

- *Overturning*—This can occur when driving over uneven terrain, steep slopes, or embankments. The rider can come in contact with the tractor when it overturns or sustain injuries during the fall. Garden tractors may also overturn if they are used to pull heavier vehicles out of mud or from a ditch. The front end of the garden tractor can rise and turn over on the victim.
- *Garden tractor running over the victim*—This can occur when a garden tractor goes into reverse. The victims are often young children who were not seen by the operator of the tractor.
- *Ignition of flammable liquids*—The use of gasoline around a garden tractor can be hazardous if the gasoline spills and is ignited by a spark or heat source.

The U.S. Consumer Product Safety Commission offers the following suggestions for the purchase, safer use, and maintenance of garden tractors.

Purchase. Specify that garden tractors including mower attachments have safeguards for all moving parts to reduce the hazard of contacting belts, chains, pulleys, and gears.

- Buy a garden tractor with throttle, gears, and brakes that are accessible and can be operated smoothly and with minimum effort.
- Be sure that safety instructions are provided with the garden tractor. There should be warning labels on the machine itself.

Use. The operator should read the owner's manual and pay attention to its recommendations before each use of the garden tractor.

- Never allow children or unauthorized persons to operate the tractor, and keep them away from the areas during operation.

- The operator should wear sturdy, rough-soled work shoes and close-fitting slacks and shirts to avoid entanglement in the moving parts. Never operate a garden tractor in bare feet, sandals, or sneakers.
- Always turn off the machine and disconnect the spark plug wire when the machine is to be adjusted.
- Drive up and down the slopes—rather than across—for greater stability when using a garden tractor on a hill. (This instruction is different from that for power lawnmowers.)
- Start the garden tractor outdoors, not in a garage where carbon monoxide gas can collect.
- Don't smoke near the garden tractor or gasoline storage can. The gasoline vapors can easily ignite.
- Replace or tighten all loose or broken parts, especially blades.
- Get expert servicing regularly—it may prevent serious injuries.

Snow throwers. All snow throwers are potentially dangerous. Their large, exposed mechanism, which is designed to dig into the snow, is difficult to guard. However, with proper handling, snow blowers offer a service that is safer than the back-breaking, heart-straining shoveling method. Safer snow blowers have guards on the drive chains, pulleys, and belts.

Without a doubt the auger, at the front of the blower, presents the greatest hazard. Injuries usually occur when the operator attempts to clear off debris while the motor is running. Some models have automatic stop devices that take effect when the handle is released. Some also have an additional auger for extra throwing power. These, along with moving gears, drive chains, and belts can be a source of danger to anyone tampering with a blower when it is running.

Although snow blowers can handle dry, powdered snow without too much difficulty, their performance in wet, sticky snow is not as effective. Wet snow tends to clog the blades and vanes, and often jams and sticks in the chute. Throwers are also capable of picking up and throwing ice, stones, and other hard objects.

There is a strong tendency for persons to reach in and clear the blower without first stopping the engine. Even cleaning the machine with a stick can be extremely dangerous if the motor is left on. The stick can be pulled from the operator's hand by the spinning blades and tossed back at the person with great force if the clutch lever is also held on.

The following safety suggestions for operators are recommended by the Outdoor Power Equipment Institute of Washington, D.C. Protect yourself and others by following these safety tips.
- Know the controls and how to stop quickly. Read the operator's manual. Do not allow children to operate the machine nor allow adults to operate it without proper instructions.
- Disengage all clutches and shift into neutral before starting the motor. Keep hands, feet, and clothing away from power-driven parts.
- Know the controls and how to stop the engine or how to throw the unit out of gear quickly.
- Disengage power and stop the motor before cleaning discharge, removing obstacles, making adjustments, or when leaving the operating position.
- Never put a hand or foot near any moving part, and never place your hand inside the discharge chute or even near its outside edge with the engine running.

- Keep personnel a safe distance away.
- Adjust height to clear gravel or crushed rock surface.
- Exercise caution to avoid slipping or falling, especially when operating in reverse.
- Do not operate the machine on slopes or ground where you risk a slip or fall.
- Never direct discharge at by-standers nor allow anyone in front of the machine; debris may be hidden in the snow. (See Figure 2-17.)
- Keep the machine in good operating condition and keep safety devices in place.

Snow shoveling. If the area to be cleared of snow is small, or if there is no snow thrower, maintenance people will have to shovel by hand. This should only be done by someone who is in good physical condition and good general health who has been getting regular exercise.

First, this person should mentally divide the area into sections and clean one part; then rest before going on to the next section. Whenever the snow begins to feel especially heavy, he should take a rest break. Keep in mind:
- Wet snow is much heavier than dry snow. Govern the rate of shoveling accordingly.
- Push or sweep as much of the snow as possible.
- If an icy crust has formed on top of several inches of snow, shovel the snow in layers.
- Make use of small quantities of rock salt or other ice-melting materials to make the job as easy as possible.
- Dress warmly while shoveling snow because cold itself can pose a strain on the circulation. Don't bundle up so heavily, however, that movement is difficult.
- Don't shovel snow right after eating, and don't smoke right before, during, or right after shoveling snow.
- If you experience chest pain, weakness, or other signs of physical stress, stop shoveling at once and seek medical attention.

Pesticides

Insecticides, herbicides, fungicides, disinfectants, rodenticides, and animal repellants are all pesticides. The safe use of pesticides is everyone's responsibility. The user, however, has the major responsibility, which begins the day a pesticide is selected and purchased and continues until the empty container has been disposed of properly. The U.S. Department of Agriculture County Extension Agent can help you choose the proper pesticide poison for the pest you wish to control, and also help you decide when to use the poison. All pesticides sold in the United States must carry an EPA registration number on the label. This means the Environmental Protection Agency has reviewed the product and found it safe and effective when used according to directions. All labels must include a list of what the product will control, directions on how to apply the pesticide, a warning of potential hazards, and safety measures to follow.

Before using any pesticide poison, be sure to read the label carefully. The label gives you some idea of the hazards involved and antidotes and first aid instructions. Those poisons that have DANGER-POISON on the label are highly toxic. If you breathe or eat them, and frequently if you simply allow them to remain on your skin, they could kill you. Poisons that have WARNING are moderately toxic and can be quite hazardous. Poisons that have CAUTION on the label have low toxicity but could harm you if the poison is eaten or grossly misused. Follow the

Figure 2-17. Snow throwers are an efficient way to remove large amounts of snow from a plant area. In this model of snow thrower, should any missile-like debris be scooped up with the snow it will be deflected away from the driver. (Printed with permission from International Harvester Company.)

label instructions for mixing, handling, and applying. Be sure—don't guess.

Application. Any restricted-use pesticide used around the plant has to be applied by a certified handler according to Public Law 92.516.

Use the least toxic pesticide for the job in order to reduce hazards. Manufacturers have formulated different compounds to control the same pest, so use the one that is easiest on the growing plants you want to keep. A severe infestation may require the use of a phosphate ester (organophosphate) insecticide. These require severe precautions (Figure 2-18); be sure to consult the supplier or your county agricultural extension agent.

Try to purchase just enough pesticide to last one season. This should cut down on storage and disposal problems. The following precautions should be observed:

- Use pesticide poisons only for the purposes given on the label.
- Keep pesticide poisons in the original labeled container. Check for leaks or container damage.

- Mix pesticide poisons carefully, outdoors, if possible, *keep off your skin,* and avoid breathing dust or vapors. Use protective clothing and equipment, including respirators for toxic chemicals.
- Set aside a special set of mixing-tools—measuring spoons and a graduated measuring cup—for use with sprays and dusts only. Keep them with your chemicals.
- Avoid spilling. Set aside a level shelf or bench in a well-ventilated area preferably outside, for mixing chemicals. A level, uncluttered surface helps avoid spills. If chemicals do spill, wash hands at once with soap and water. Then hose down the mixing area.
- Never smoke or eat while spraying or dusting. Cover food and water containers when spraying around watchdog areas.
- Someone should always keep an eye on those who apply dangerous pesticides.
- During application, stay out of the spray drift. Avoid outside application on a windy day.

Figure 2-18. When applying insecticides, be sure personnel is well trained and wears proper equipment. Phosphate ester (organophosphate) insecticides require extreme precautions to be used. Always choose the least toxic pesticide that will do the job.

- Avoid spraying near lakes, streams, rivers, and make every effort to see that toxic residues do not enter waterways.
- Canister respirators that serve the proper respiratory protection are recommended.
- If pesticide poison gets on skin or clothing, immediately remove the clothing and take an all-over bath or shower (be sure to shampoo); use plenty of soap and water. Wash clothing before reuse.
- When finished, wash immediately with soap and water—*Do not smoke, eat, or drink without washing first.*
- Never allow unauthorized personnel around treated areas or pesticide poison mixing, storage, and disposal area.

Safe storage of pesticides. Store all pesticide poisons in a well-ventilated, locked area or building.

Poisons should be kept in tightly closed, original containers so the label can give information needed in case of accidents.

Do not store clothing, respirators, lunches, cigarettes, or drinks with pesticide poisons. They may pick up poisonous fumes or dusts or soak up spilled poisons.

Keep soap and plenty of water handy. Seconds count when washing poisons from your skin. Their labels give treatment information.

Disposal of pesticides. Dispose of all pesticides according to label instructions.

Emergency information. If an emergency does occur, additional advice and information on antidotes for specific pesticides may be obtained from the following list of agencies that maintain a current information file on all compounds, their constituents, and recommended treatment in case of poisoning:
The local Poison Control Center;
The state department of health;
The country agricultural extension agent;
The regional office of the U.S. Environmental Protection Agency.

MAINTENANCE CREWS

Maintenance employees should be selected for their experience, alertness, and mechanical ability, and for their capacity to learn the essential safety principles of machines or operations for which they may be made responsible.

Training

Maintenance employees should have more thorough training in accident prevention than regular workers. Safety for them involves not a set pattern of activity, but a complex and constantly changing set of problems. Furthermore, they must know how to use not only ordinary tools but also ladders, various kinds of protective equipment, chains, slings, ropes, and many others.

Maintenance crews must be made aware of job hazards and be receptive to proper training in order to protect themselves and others working close by. Their training program should include first aid and life-saving techniques. In industries where irritating, toxic, or corrosive dusts, gases, vapors, or fluids are present, maintenance employees should be given special training to make sure that they are familiar with the properties of these substances and with the methods of controlling the hazards.

Some companies have arrangements for the purchasing department to notify crews when new chemicals are purchased so that the necessary precautions can be planned.

The crew should be called together at the start of a nonroutine job to discuss the hazards involved and the method of doing it safely. The crew should check the equipment and call the supervisor before work that looks unsafe is started. For especially complicated or hazardous jobs, the safety professional may be called upon to help in the job planning. Scale models can be constructed to determine clearances, the best methods of moving, and sequences of action. After several trials have been made and the crew has agreed on a safe procedure, the various steps should be recorded as a guide for each worker (Figure 2-19).

The tools and tackle required to do such special jobs should be inspected for wear and defects before they are used. When special tools can be devised to make a job safer, the engineering department can provide design and construction specifications.

In the course of its daily work, the maintenance crew travels throughout the plant, becoming familiar with every machine and process. If properly selected and trained, each crew member can do much to locate and correct unsafe conditions in both plant and equipment.

In smaller companies, maintenance personnel may also be responsible for inspection and care of portable power tools, extension cords, and the like. If so, special procedures and training are suggested.

Predictive maintenance—by computer

Predictive maintenance (PM) can reduce employee exposure to

ABBOTT LABORATORIES—CORPORATE ENGINEERING—DIVISION BASIC OPERATING PROCEDURES			
SUBJECT Sanitizing Sterile Room #7026—R-1B 7th. Floor			
WRITTEN BY			**NO.**
ORIGINAL RELEASE DATE	**PRIOR EVALUATION DATE**	**CURRENT REVISION DATE**	Page 1 of 5

1. PURPOSE: Cleaning and disinfecting of Room 7026

2. EQUIPMENT: 1 - 9" Sponge Mop
     1 - Cotton Wet Mop
     1 - 32 qt. Bucket (stainless steel)
     1 - Mop Handle
     1 - Mop Wringer (stainless steel)
     1 - 10 qt. Bucket (stainless steel)
     1 - 22 oz. Spray Bottle Complete
       Window Solution
       Disinfectant
       Shop Towels
     1 - Window Squeegee

3. RECORD: Abbott Housekeeping Manual

4. PROCEDURES:

STEPS	KEY POINTS
4.1 Take Sanitizing Equipment to Sterile Area.	4.1a From room 7028 pick up 8 liters of sterile water. From room 7030 pick up disinfectant and window cleaning solution (16% in spray bottle). Bring to Change Room.
	4.1b Obtain coveralls, hood and fast cover from locker inside room 7027.
4.2 Procedures for entrance to Sterile Area.	4.2a On N.E. exterior wall of outer air-lock room cut off ultraviolet light, cut on fluorescent lights.
	4.2b From clothes hooks mounted on wall, take eye goggles, put on and enter change room.

Figure 2-19. Basic operating procedures are spelled out in detail so that each job is completed efficiently, thoroughly, and safely. Shown here is the first page of a detailed procedure for sanitizing a sterile room. (Printed with permission from Abbott Laboratories.)

hazards, decrease equipment downtime, and optimize the effectiveness of maintenance expenditures. By reducing costs due to equipment being out-of-service and repair costs, an effective PM program reduces total controllable maintenance costs to a minimum.

These savings come from the following:

- Reduced employee exposure to malfunctioning equipment
- Reduced production lost time
- Fewer emergency failures of equipment
- Scheduled equipment outages
- Lower repair frequency and cost
- Improved and safer labor utilization
- Extended equipment life

By systematically monitoring servicing and equipment before trouble starts, the PM program becomes the "ounce of prevention" needed to avert the costs incurred from unexpected breakdowns. But the benefits of a PM program typically extend beyond the maintenance operation to encompass other areas of the plant, including the safety function.

An effective PM program, for example, not only alerts the proper personnel to potentially hazardous conditions—such as equipment failures—it also facilitates the recordkeeping required to conform to state and federal safety regulations. Here's how a typical PM program is established.

Systems development. Exisiting PM systems, records, and forms are revised as needed to provide a record and control system capable of supporting and monitoring the PM operation.

Computerized assistance. The most effective and economical method to plan, schedule, and track PM tasks is to schedule them from the last service date—a process that can be accomplished effectively only with computerized assistance. A computer software program can be designed to be consistent with concepts and techniques on which maintenance systems consulting services can be based.

Principle features of the program can be as follows:

- PM task definition and identification of the cycle on which the tasks are to be performed
- PM specifications and procedures that define the craft, standard hours, work procedures, and measurements to be performed for each task
- Automated scheduling, based on last service date and predetermined task cycles
- PM work order preparation for all crafts according to available labor
- Reports, including compliance, performance, forecast, and budget planning reports
- Equipment history records for each PM task.

For example, in an electrical group, baseline data can be gathered on every major piece of electrical equipment at the time of installation. A vibration analysis can be done before a motor is coupled to its load, providing a record on the no-load speed condition of the motor when new. Then, when a vibration reading is taken during routine maintenance and compared with the original reading, it can indicate impending problems. Because maintenance costs can account for as much as 60 percent of total controllable plant operating costs, an effective computerized predictive maintenance program directly affects the bottom line (Figure 2-20).

Scheduling preventive maintenance

Because the function of plant maintenance is to keep physical equipment in top operating condition, a good maintenance system must catch breakdowns before they happen. That is, it must be *preventive* maintenance.

A preventive maintenance plan can be set up at least for critical equipment and for machinery that might seriously affect the safety of workers. A good preventive maintenance program starts with a list of buildings, machinery, and equipment that require periodic inspection, adjustment, cleaning, and lubrication, as well as adjustment of guards or changes in the types of guards.

Detailed engineering drawings and specifications should be on file for each machine or structure. For machinery, these give dimensions, weights, sizes, and locations of utility service connections, lubrication requirements, details on bearings, and power transmission or motor drive data. For buildings, they include data on general layout, services available, floor load capacities, ceiling clearances, column spacing, and many other features.

After management has made an analysis of its maintenance responsibilities, an organization chart should be formulated for the selection and safe assignment of workers. A chart gives the supervisor a comprehensive picture of his force, pointing out the status of trained key-persons and reserves. Outside specialists who can be called in for unusual or very hazardous jobs should be listed. In emergencies when competent people are needed in a hurry, this information is quite valuable.

Equipment inspections

An inspection schedule based on actual trouble points listed on maintenance records is essential to preventive maintenance. The schedule can be determined by the number of inspection reports turned in by regular full-time inspectors or by maintenance crew personnel.

An inspector must know what to look for, being something of an expert on, or at least thoroughly familiar with, the equipment being inspected. This knowledge is particularly important for electrical equipment inspectors. Electrical equipment usually gives few obvious indications of impending trouble. Mechanical equipment often warns of deterioration by such easily recognized signs as noise, unusual appearances, or substandard output.

It is important that inspectors be equipped with suitable instruments for making their observations. While there is no difficulty in recognizing burned contacts on a motor starter or in hearing the pounding of a worn gear, checking insulation resistance of a motor or measuring the wear on a shaft requires the proper instruments.

Personal protective equipment

Maintenance crews should dress properly for their specific job. Garments should be snug-fitting, with a few small pockets. Breast pockets are often sewn closed or removed to prevent items from dropping into machinery or inaccessible places when the wearer leans over. Rings, wristwatches, and other jewelry should not be worn.

Neckties and loose clothing should not be worn by maintenance crews. Loose rags must be kept clear of moving machinery.

If a person carries so few tools that he does not need a tool bag, he should wear a special belt fitted with tool carriers. To prevent back and spine injuries in case the worker should fall, tools should be carried at the side instead of in the back portion of the belt. (See the discussion in Chapter 14, Hand and Portable Power Tools.)

Gloves or hand leathers should be worn by a person handling rough or sharp objects. Welding gloves, rubber gloves for electrical insulation, and chemical-resistant gloves for handling acids should be used as needed, but never worn around moving machinery.

Goggles should be in every maintenance tool kit. Moreover, unless the repair person knows the exact conditions to be faced, equipment should include necessary types of goggles to provide protection against flying objects and molten metal, injurious heat and light rays, dust and wind, and acid splashes.

When a person needs to work in high places or to enter a manhole, bin, or tank, a life belt fall protection shall be worn, with the lifeline attached to some permanent support. In all cases, on jobs involving such risks, an extra person should be stationed close by.

Every repair person's kit should include an explosion-proof flashlight, since they are often needed for work in dark or gaseous places.

See Chapter 17 in the *Administration and Programs* volume for details on personal protective equipment.

Lockout of mechanical and electrical equipment

Maintenance of power presses, various other types of machinery, electrical equipment, and boilers is dealt with in detail in the respective chapters of this volume. Zero mechanical state procedures and instructions are described in Chapter 8, Principles of Guarding, and electrical lockouts are discussed in Chapter 15, Electricity. Safe tank-entry procedures are listed in Chapter 18, Boilers and Unfired Pressure Vessels.

Piping

Proper identification. Accidents have occurred because of improper identification of lines. In factories and power plants where high and low pressure steam lines run adjacent to compressed air, sprinkler system, and sanitary lines, maintenance crews must be especially protected against opening wrong valves or disconnecting wrong pipes.

Piping must be identified. A system of color schemes, tags, and stencils can be worked out so line contents can be identified at a glance. Identification is particularly important when outsiders are called in for service or when emergencies occur. Standardized identification does not present a new learning problem to newcomers. To prevent confusion, use a consistent piping color code, such as ANSI A13.1, *Scheme for the Idenitification of Piping Systems*. Also see the section Use of Color in the preeceding chapter.

Preventive and predictive maintenance (PPM) can reduce employee exposure to hazards, decrease equipment downtime, and optimize the effectiveness of maintenance expenditures.

As shown in Figure 1, by reducing costs due to equipment being out-of-service and repair costs, an effective PPM program reduces total controllable maintenance costs to a minimum.

These savings come from:

• Reduced employee exposure to malfunctioning equipment;

• Reduced production lost time;

• Fewer emergency failures of equipment;

• Scheduled equipment outages;

• Lower repair frequency and cost;

• Improved and safer labor utilization;

• Extended equipment life.

By allowing systematic monitoring of equipment and servicing it before trouble starts, a PPM program is the "ounce of prevention" needed to avert the costs incurred from unexpected breakdowns.

But the benefits of a PPM program typically extend beyond the maintenance operation to encompass other areas of the plant, including the safety function.

An effective PPM program, for example, not only alerts the proper authorities to potentially hazardous conditions—such as equipment failures—it also facilitates the record-keeping required to conform to state and federal safety regulations.

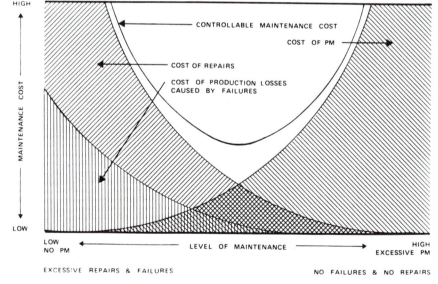

Figure 1

Figure 2-20. Controlling maintenance costs using computerized predictive maintenance.

Proper isolation and handling. Before work is done on a pipeline, the line must be shut off, valves locked and tagged, the section of the line relieved of pressure and drained. When hazardous materials are encountered, the supervisor should ascertain the need for special protective equipment, such as chemical-type goggles, protective suits, rubber gloves, or respiratory protective equipment.

To prevent hands from slipping, maintenance personnel should carry a piece of waste or a rag in their pockets to wipe off excessive oil on pipes and fittings. Gloves should be worn when handling pipes and fittings, especially when ends are threaded. Pipes should be checked for burrs and these should be filed off immediately.

Shutoff valves are usually of the gate type, but if globe valves are used, the pressure-side should be under the valve seat so that the packing will not have to hold the pressure when the valve is closed.

If pipelines that carry chemicals must run overhead, they should be isolated or covered so they will not drip on workers or materials underneath. Emergency showers should be provided with plainly marked locations. The emergency showers should be tested periodically and be properly maintained. Full instructions should be given to maintenance crews (as well as operating personnel) in their location and use.

There are many special safety precautions which should be taken in maintenance of pipelines, valves, and bolted flanges—especially when hazardous materials are involved. The section on Problems with Hazardous Materials in Chapter 3, Manual Handling and Material Storage, contains pertinent recommendations for both operating and maintenance persons.

Industrial gas lines. When industrial gases such as propane and butane are encountered, mechanics should understand their behavior, storage characteristics, pipeline arrangement, and have a copy of the piping layout which shows equipment location and safety features, such as soft heads and backfire preventers. Sectional and main shutoff valves throughout the entire plant should be located so in case of mishap, maintenance personnel can find these valves quickly.

Handling pipe. When maintenance involves a considerable amount of pipe work, lengths of pipe, valves, and fittings to be used should be placed so that the floor is not overloaded. If possible, the material should be moved to strategic points as the job progresses, to eliminate spot accumulation and reduce the amount of handling.

When lengths of pipe are transported by mechanical means, red warning flags should be attached to material that extends beyond the conveyance. If the load is allowed to stand near passageways, the flags should remain and additional warning signs or barricades should be used to prevent other workers from bumping into the load.

Long lengths of pipe being run overhead and continuously joined should be pulled up from the floor with overhead rigging to the proper location and secured with tie ropes, wires, or fixture straps. Use of overhead rigging will prevent back strains and possible falls from ladders and will keep the material itself from dropping.

Crane runways

Before repair personnel work on or near an overhead crane runway, they should notify the crane operator and make sure that the operator understands what is being done and what part the operator is to take in the work. Repair personnel should also provide a temporary rail stop between the crane and the point where the work is to be done. See Chapter 4, Hoisting Apparatus and Conveyors.

When work must be done on the crane itself, the person in charge should be held responsible for locking out safety switches, placing warning signals to indicate that people are working above, and seeing that a careful inspection is made after the job is completed. This inspection should include a check for parts or tools left on the crane, which later might drop into the mechanism or fall on persons.

A handline should be used to eliminate the need for carrying material up a ladder. The line should be tied to a crane part, not attached to the worker's body because the employee might become entangled or pulled off.

Final check—tools and guards

After a machine has been repaired, it should, if possible, be turned over first by hand. Mislaid tools or materials thus may be discovered in time to prevent their wrecking the machinery and perhaps causing injury. Guards should be replaced and securely fastened, tools picked up, and the work area left as nearly in its original condition as possible. Repair personnel can aid greatly in preserving good housekeeping.

Too much emphasis cannot be given to the fact that guards are a part of a machine and must be replaced before ending a job.

To prevent tools from being left on or near machinery after a repair job has been completed, a tool box with a special rack for each personal tool should be used so that repair personnel can quickly see when a piece is missing. Also, a tool check system will help them keep track of special tools used for the job.

Lubrication

As a rule, lubrication is handled by a separate schedule. Nevertheless, it is an important part of a preventive maintenance program. Improper lubrication sets up special problems; lack of lubrication can cause overheating of bearings, shutdowns, and fires.

Over-oiling of motor bearings causes oil to drop or to be thrown onto the insulation of the electrical windings. The oil deteriorates the insulation, exposing live conductors that will arc and cause fires or electrically charged ungrounded surfaces. If the maintenance personnel work on this equipment, they may accidentally complete the circuit to ground with their body and be fatally shocked.

If the windings become sticky from oil and dirt accumulates on them, defects may be covered up and go unnoticed until serious trouble or a complete breakdown occurs. The oil and dirt accumulations would result in heat buildup that could cause a deterioration of the electrical windings and overheating of the bearings resulting in shutdown or fire.

A complete survey should be made of the plant to determine lubrication requirements, and the information should be entered on the machinery records. The machinery then should be checked for missing fittings, missing oil cups, and plugged oil holes.

If possible, improvements should be made at this time. Automatic feed oil cups, mechanical force-feed lubrication systems, and special fixtures that will enable the oiler to reach parts in remote locations without danger from moving parts should be installed.

The oiler should be provided with a simple diagram of each machine with all the parts that require lubrication plainly marked. The diagram should include a table that tells the oiler the kind of lubricant to use and how often it is to be applied. Some companies devise a color code that indicates the various lubricants and the frequency of application. The point of lubrication, the oil cups, oil feeds, and oil holes are painted according to the color code, and the oiler is supplied with a color code chart.

Each oiler should have a supply of necessary oils, greases, guns, fittings, and wiping cloths before starting out for the day's work. In large plants, a special cart should be provided to hold all the equipment and supplies needed for the day.

Special precautions or instructions to the oiler should be transferred from the master sheet and entered on his personal schedule. The notations would concern certain types of machines that must be stopped to assure the oiler's safety or those that involve exposure to electrical elements and must have power lines shut off.

Since it is sometimes necessary to remove a guard from the lubrication point, workers should be trained to replace the guard in its fittings and holding devices properly so that it will clear all moving parts. If the guard or fittings are worn badly or damaged, workers should make a report immediately, requesting repairs or replacement.

If possible, shafting should be oiled while the machinery is at rest. Special pump oil cans with long spouts can be used to reach overhead hangers or out-of-the-way bearings. Some of these oiling devices can be used without a ladder.

Ladders should not straddle machines in operation, nor should the oiler attempt to reach several countershafts from one position. Makeshifts, such as chairs and boxes, should never be used instead of ladders.

Shop equipment maintenance

Repair facilities should be provided for maintenance crews. The stock of spare parts and units should be large enough to meet probable demand. Machine tools should be modern and arranged to assist the maintenance crew in easy flow of work. Hoists should be provided to handle heavy machinery, and power or hand trucks should be available to transport material from one department to another.

The maintenance shop should have a special area that is well ventilated, especially if welding, paint spraying, or cleaning of metal parts of machinery is done in the shop. Fire extinguishers should be located in this area, and the crew should be trained in the types to use on grease, oil, electrical, and other types of fires.

Special tools

Spark-resistant tools of nonferrous materials sometimes are advised for use in areas where flammable gases, highly volatile liquids, and other explosive substances are stored or handled.

Nonferrous tools need inspection before each use to be certain that they have not picked up steel particles, which could produce friction sparks.

Keeping up-to-date

The maintenance department should not overlook the poten-

tialities of new products, cleaners, lubricants, paints, wood preservatives, insulation, floor repair materials, protective coatings, alloys, and other developments. It should also keep up with the new applications of existing products. New products or applications usually lead to better methods and safer practices.

Increased mechanization requires careful assessment of new potential hazards, especially in high-speed equipment and processes. Use of color throughout the plant, as specified in the previous chapter will contribute to accident prevention and help develop high standards of operation.

Sometimes mechanical means can help offset potential hazards. Centralized lubrication, floor cleaning machines, centralized spray painting equipment, steam cleaners, and other devices will help make a maintenance program safer. For frequently performed maintenance jobs, permanent accessories, such as hoists, fixed ladders, and catwalks, should be provided.

Maintenance procedures should be reexamined periodically for safer ways to do the job. A special suggestion system should be set up for maintenance crews so that they will be able to present new ideas or corrective measures.

Engineering books and service manuals should be a part of the maintenance resource materials. Supervisors and workers should be familiar with them and use them when necessary.

REFERENCES

American Gas Association, 1515 Wilson Blvd., Arlington, Va. 22209.

American Institute of Architects, 1735 New York Ave., NW., Washington, D.C. 20006. *Standard Form of Agreement Between Owner and Contractor.*

American National Standards Institute, 1430 Broadway, New York, N.Y. 10018.

Elevators and Escalators (Inspector's Manual), ANSI/ASME, A17.2.

Elevators, Escalators, and Moving Walks, ANSI/ASME, A17.1.

Lifting Devices, B30 Series.

Safety Requirements for Portable Wood Ladders, A14.1.

Safety Color Code for Marking Physical Hazards, Z53.1.

Safety in Welding and Cutting, Z49.1.

Safety Requirements for Demolition, A10.6.

Safety Requirements for Fixed Ladders, A14.3.

Safety Requirements for Floor and Wall Openings, Railings, and Toeboards, A12.1.

Safety Requirements for Job-Made Ladders, A14.4.

Safety Requirements for Personnel Hoists, A10.4.

Safety Requirements for Portable Metal Ladders, A14.2.

Safety Requirements for Scaffolding, A10.8.

Safety Requirements for Steel Erection, A10.13.

Safety Requirements for Temporary and Portable Space Heating Devices and Equipment Used in the Construction Industry, A10.10.

Scheme for the Identification of Piping Systems, A13.1.

Specifications for Accident Prevention Signs, Z35.1.

Associated General Contractors of America, Inc., 1957 E St., NW., Washington, D.C. 20006. *Manual of Accident Prevention in Construction.*

Building Officials & Code Administrators International, Inc., 4051 Flossmoor Rd., Country Club Hills, Ill. 60477.

The BOCA Basic Material Building Code (issued every three years).

The BOCA Basic Fire Prevention Code (Issued every three years).

Building Operations Manual. Champaign, Ill., University of Illinois Press, 1976.

"Health and Safety Guide for Masonry, Stonework, Tilesetting, Insulation, and Plastering Contractors," DHEW (NIOSH) Publication No. 78-208. Cincinnati, Ohio, National Institute for Occupational Safety and Health, Div. of Technical Services, 1978.

Huntington, Whitney Clark. *Building Construction: Materials and Types of Construction,* 5th ed. New York, N.Y., John Wiley & Sons, Inc., 1981.

Insecticide, Fungicide, and Rodenticide Act, Public Law 92-516, Oct. 21, 1972. Regulations published in *Code of Federal Regulations* Title 40—Protection of Environment, Chapter 1—Environmental Protection Agency, Part 165.

Laborers' International Union, 905 16th St., NW., Washington, D.C. 20006. (General.)

National Fire Protection Association, Batterymarch Park, Quincy, Mass. 02269.

Flammable and Combustible Liquids Code, NFPA 30.

Life Safety Code, NFPA 101.

National Safety Council, 444 North Michigan Ave., Chicago, Ill 60611.

Industrial Data Sheets.

Acetylene, 494.

Aerial Baskets, 572.

Asphalt, 215.

Atmospheres in Subsurface Structures and Sewers, 550.

Barricades and Warning Devices for Highway Construction Work, 239.

Blowtorches and Plumbers' Furnaces, 470.

Chains (Alloy) for Overhead Lifting, 478.

Conveyors, Belt (for Bulk Material) Equipment, 569.

Conveyors, Belt (for Bulk Material) Operation, 570.

Conveyors, Roller, 528.

Cutting and Clearing Vegetation, 575.

Diving in Construction Operations, 555.

Electromagnets Used with Crane Hoists, 359.

Excavation, General, 482.

Flexible Insulating Protective Equipment for Electrical Workers, 598.

Load-Haul-Dump Machines in Underground Mines, 576.

Hoists, Construction Material, 511.

Job-Made Ladders, 568.

Motor Graders, Bulldozers, and Scrapers, 256.

Motor Trucks for Mines, Quarries, and Construction, 330.

Paving with Portland Cement Concrete, 541.

Pavement Marking of Streets, Roads, and Highways, 643.

Power Lawn Mowers, 464.

Ready-Mixed Concrete Trucks, 617.

Safety Hats, 561.

Safety Nets—Fall Protection for the Construction Industry, 608.

Sidewalk Sheds, 368.

Silicon Diodes Grounding Devices, 581.

Steel Plates, Handling for Fabrication, 565.

Temporary Electric Wiring for Construction Sites, 515.

Tilt-Up Concrete Construction, 513.

Tools, Live Line, 498.

Tractor Operation and Roll-Over Protective Structures, 622.

Trench Excavation, 254.

Wire Rope Slings, Recommended Loads on, 380.

Claire, Frank. "Preventive and Predictive Maintenance—by Computer." *National Safety News,* Aug., 1984.

Outdoor Power Equipment Institute, 1901 L St., NW., Washington, D.C. 20036. (General.).

Sack, Thomas F. *A Complete Guide to Building, and Plant Maintenance,* 2nd edition, New York, N.Y., McGraw-Hill-Book Company, 1971.

Scaffolding and Shoring Institute, 1230 Keith Bldg., Cleveland, Ohio 44115, "Steel Scaffolding Safety Rules."

Tucker, Georgina, and Madelin Schneider, *The Professional Housekeeper,* 2nd edition, Boston, Mass., Cahners Books, 1982.

U.S. Army, Corps of Engineers, Washington, D.C. *General Safety Requirements,* Manual EM385-1-1.

Manual Handling and Material Storage

MATERIAL IS HANDLED BETWEEN OPERATIONS in every department, division, or plant of a company. It is a job that almost every worker in industry performs—either as a sole duty or as part of the regular work, either by hand or with mechanical help.

Mechanized material handling equipment is being used with increasing frequency. In many industries, material could not be processed at low cost without efficient mechanical handling. Although mechanical handling creates a new set of hazards, the net result (entirely aside from increased efficiency) is fewer injuries.

The problems and safe handling techniques involved in the manual and mechanical handling of material are discussed in this and the next four chapters.

The amount and extent of manual lifting involved in a particular employment position should be considered in job selection and placement. Preemployment physical examinations can often identify those employees who are most likely to incur serious back injuries or hernias.

Areas that should be considered when reviewing material handling are the work environment, the need for specific training, and proper material handling engineering.

PREVENTING COMMON INJURIES

Material handling problems

Handling of material accounts for 20 to 25 percent of all occupational injuries. These injuries occur in every part of an operation, not just the stockroom or warehouse. On an average, industry moves about 50 tons of material for each ton of product produced. Some industries move 180 tons for each ton of product.

Strains, sprains, fractures, and contusions are the common injuries. They are caused, primarily, by unsafe work practices, such as lifting or carrying too heavy a load, incorrect gripping, failing to observe proper foot or hand clearances, and failing to wear personal protective equipment.

To gain insight into the material handling injury problem, the safety professional should consider the following:
1. Can the job be engineered to eliminate or reduce manual handling?
2. Can the material be conveyed or moved mechanically?
3. In what ways do the materials being handled (such as chemicals, dusts, rough and sharp objects) cause injury?
4. Can employees be given handling aids, such as properly sized boxes, adequate trucks, or hooks, that will make their jobs safer?
5. Will protective clothing, or other personal equipment, help prevent injuries?

These are not the only questions that might be asked, but they serve as a start toward overall appraisal and detailed inquiry. The largest number of injuries occur to fingers and hands. People need instructions if they are to learn the proper way to lift objects and to set them down. Training in safe work habits, breakdown and study of even the simplest job operations, and adequate supervision can help minimize these accidents.

General pointers that can be given to those who handle materials include:
1. Inspect materials for slivers, jagged or sharp edges, burrs, rough or slippery surfaces.
2. Grasp the object with a firm grip.

3. Keep fingers away from pinch and shear points, especially when setting down materials.
4. When handling lumber, pipe, or other long objects, keep hands away from the ends to prevent them from being pinched.
5. Wipe off greasy, wet, slippery, or dirty objects before trying to handle them.
6. Keep hands free of oil and grease.

In most cases, gloves, hand leathers, or other hand protectors should be used to prevent hand injuries. Extra caution must be used when working near moving or revolving machinery. (See details in the *Administration and Programs* volume, Chapter 17, Personal Protective Equipment.)

In other cases, handles or holders can be attached to objects themselves, such as handles for moving auto batteries, tongs for feeding material to metal-forming machinery, or wicker baskets for carrying control-laboratory samples.

Feet and legs sustain a major portion of material handling injuries—the greater percentage occurring to the feet. Workers should be instructed to wear foot protection, such as safety shoes equipped with metatarsal guards.

A worker's eyes, head, and trunk can also be injured. When opening a wire-bound or metal-bound bale or box, a person should wear eye protection equipped with side shields as well as stout gloves, and take special care to prevent the ends of the binding from flying loose and striking the face or body. The same precaution applies to handling coils or wire strapping, or cable. In many cases, special tools are available to safely cut bands, strapping, and the like. Workers should always read the labels on packages for special instructions.

If material is dusty or toxic, the person handling it should wear a respirator or other suitable personal protective equipment.

Manual handling of materials increases the possibility of injury and adds to the cost of a product. To reduce the number of material handling injuries and to increase efficiency, manual material movement should be minimized by combining or eliminating operations and by introducing ergonomic principles to job design. Mechanical movement of materials should be used as much as possible. For those jobs that cannot be mechanized, here are some suggestions.

Manual lifting

Physical differences make it impractical to establish safe lifting limits applicable to all workers. A person's height and weight, although important, do not necessarily indicate lifting capability, because some small, thin individuals can handle heavier loads than some tall, heavy persons. (See Kroemer in References.) A job safety analysis (described in the *Administration and Programs* volume) and medical recommendations should also be used when establishing lifting standards.

When a worker lifts a heavy or bulky object and carries it to another location, the route over which the object is moved should be inspected beforehand to make sure that there are no obstructions or spills that could cause slipping or tripping injuries. If clearance is not adequate for handling the load, then an alternate route should be taken.

The object should be inspected next to decide how it should be grasped and how to avoid sharp edges, slivers, or other things that might cause injury. The object may have to be turned over before attempting to lift it. Also, if the object is wet or greasy,

it should be wiped dry so that it will not slip. If this is not practical, the worker should use a rope sling or other device that will give a positive grip.

Most lower back overexertion injuries come from lifting tasks, although they may also be caused by lowering, pushing, pulling, carrying, twisting, and other activities.

Several techniques can be used to manually lift objects with reasonable safety. However, each of these lifting techniques has limitations. All three main factors in manual lifting—load location, task repetition, and load weight—must be considered in determining what is safe or unsafe to lift.

Load location refers to the distance horizontally from the body and vertically from the floor that the load is located at the beginning of the lift. The greater the distance a load is horizontally from the body, the more it weighs to the lower back. The closer a load is to the floor, the more a worker must bend to lift it.

Repetition refers to fatigue. The more often a lift must be performed in a time period, the greater the potential for fatigue. Fatigued muscles are prone to injury.

The third factor to consider is load weight. The more a load weighs, the harder it is to lift. However, load weight cannot be considered alone. A 25 lb (13 kg) object lifted at arm's length weighs more to the lower back than the same weight cradled against the stomach.

If the loads lifted are beyond the physical capabilities of most workers, then worker training and the specific lifting technique used will not result in reduced lower back injuries. (See *Administration and Programs* volume, Chapter 10, Ergonomics: Human Factors Engineering, for discussion on lifting evaluations.)

Two main types of two-hand lifting are used: the squat lift ("bend your knees") (Figure 3-1) and the stoop lift (Figure 3-2). In addition, there are several variations of these two lifting techniques. The two-hand squat lift is often the preferable lift technique, assuming that the load lifted can be brought between the knees to reduce the horizontal load distance and that the task is not performed repetitively.

When the squat lift cannot be used (e.g., lifting objects out of large containers such as wood box pallets, steel tubs, etc.), the worker must use the two-hand stoop lift. In the stoop lift, the worker bends over horizontally from the waist to reach into the container. Since the majority of the body weight of the lifter is in the upper body, the farther a worker bends over, the more stress to the lower back regardless of the load weights lifted.

An alternative lifting technique for lifting objects out of large containers is the assisted one-hand lift. This technique features using one hand to lift the object while the other hand pushes down on the top edge of the container. Using this method, the worker provides support for the upper body as it is being raised out of the container, thus reducing stress to the lower back.

The correct application of the two-hand squat lift and assisted one-hand lift are discussed below. One caution must be noted, however. While using correct lifting techniques may prevent or reduce some injuries, the greatest reduction of lifting injuries will be achieved by eliminating lifts from ground level. The safety and health professional should study and modify work activities to avoid this maximum lift injury potential.

Two-hand squat lift

Six basic factors apply in the two-hand squat lift technique. Their order of application is shown in Figure 3-1.

Squat Lifting Techniques

Lifting is so much a part of many everyday jobs that most of us don't think about it. But it is often done wrong, with bad results: pulled muscles, disk lesions, or painful hernia. Here are six steps to good squat lifting techniques:

1. Keep feet parted—one alongside, one behind the object.
2. Keep back straight, nearly vertical.
3. Tuck elbows and arms in, and hold load close to body.
4. Grasp the object with the whole hand.
5. Tuck your chin in.
6. Keep body weight directly over feet.

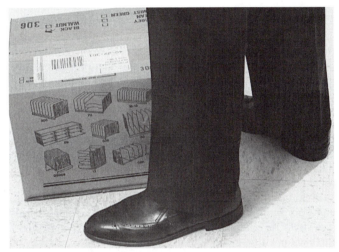

Figure 3-1a. Feet should be parted, with one foot alongside the object to be lifted and one behind. Feet, comfortably spread, give greater stability; the rear foot is in position for the upward thrust of the lift.

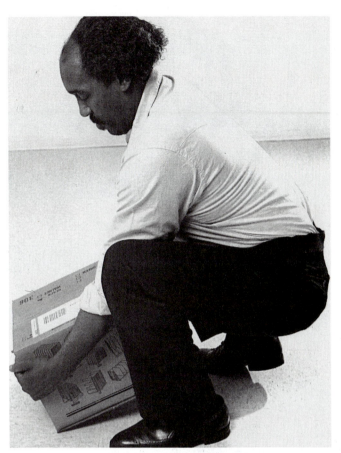

Figure 3-1b. Back. Use the sit-down position and keep the back straight—but remember that "straight" does not mean "vertical." A straight back keeps the spine, back muscles, and organs of the body in correct alignment. It minimizes the compression of the abdomen that can cause hernia.

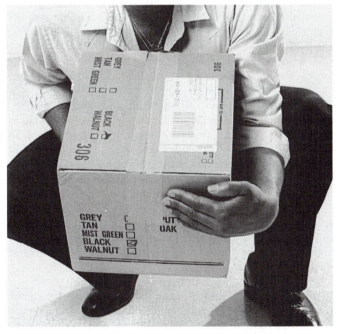

Figure 3-1c. Arms and Elbows. The load should be drawn close, and the arms and elbows should be tucked into the side of the body. When the arms are held away from the body, they lose much of their strength and power. Keeping the arms tucked in also helps keep body weight centered.

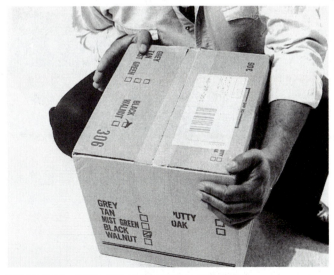

Figure 3-1d. Palm. The palmar grip is one of the most important elements of correct lifting. The fingers and the hand are extended around the object you're going to lift. Use the full palm; fingers alone have very little power. Pull load in between knees and as close to body as possible. (The glove has been removed to show the finger positions better.)

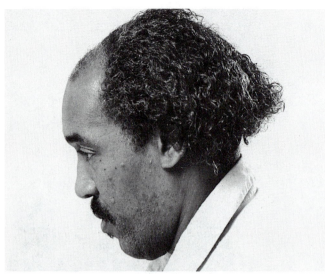

Figure 3-1e. Chin. Tuck in the chin so your neck and head continue the straight back line and keep your spine straight and firm.

Figure 3-1f. Body weight. Position body so its weight is centered over the feet. This provides a more powerful line of thrust and ensures better balance. Start the lift with a thrust of the rear foot.

1. Correct position of feet
2. Straight back and bent knees
3. Load held close to the body (for lifting and carrying)
4. Correct grasp
5. Chin in
6. Use of body weight
   Each will be discussed in turn.

Correct position of feet. One of the causes of muscle injury, particularly to the back, is the loss of balance due to working with the feet too close together. Lifting off the ground, pushing and pulling, or reaching (and in many instances overreaching) may cause an off-balance body condition. A common reaction to this condition is a stiffening of muscles in the lower limbs and back.

In the kinetic method, however, the feet are correctly positioned with one placed in the proposed direction of movement and the other in a position where it can give thrust to the body. The forward foot can be either left or right.

Straight back and bent knees. A straight back is not necessarily a vertical back. In the kinetic method, the back is often inclined, particularly when lifting weights from the ground, but the inclination should be from the hips so that the normal curvatures are maintained. This normally curved spine is termed a "straight back."

With "straight back lifting" the spine is fairly rigid (nonmedically speaking) and the pressure on the lumbar intervertebral disks is evenly distributed. When lifting with the back bent, the spine forms an arc, with the result that the lower back muscles are subject to strain, causing an uneven pressure on the disks.

In addition to the risk of intervertebral disk lesions, lifting an object with the back bent and the legs straight imposes excessive stress on the muscles of the back for two reasons. First, the back must be inclined at a greater angle to the vertical for the hands to reach the object. Since the "effective weight" (the object plus the upper part of the worker's body) increases rapidly as this angle is increased, a much greater effort is required to raise the back to its vertical position. Second, muscular effort is required to straighten the spine.

When a weight is properly lifted from the ground, with maximum effective use of the legs, the back is straight but inclined forwards. As the worker begins to rise by extending the knees, the back returns to the vertical position.

The position of the feet and the flexion of the knees are the key factors for maintaining a straight back.

Load close to the body. While lifting and carrying weights, the worker should keep the load close to the body. The closer the load is to the body, the less it weighs to the lower back. To do this, the arms should be close to the body and remain straight whenever possible. Flexing the elbows and raising the shoulders imposes unnecessary strain on the upper arm and chest muscles.

Carrying involves a static posture of the arms, and, in the case of long distance carrying, any assistance given by the body in supporting the weight will lessen the tension in the muscles. Carrying with the arms lowered assures that the weight will rest against the body.

Correct grasp. An insecure grip may be caused by taking the load on the finger tips, thus creating undue pressure at the ends of the digits and strain to certain muscles and tendons of the arm. Because greasy surfaces often prevent a secure hold, wipe surfaces clean before grasping. Use suitable, properly fitted gloves, if necessary.

A full palm grip will reduce local muscle stress in the arms and decrease the possibility that the load will slip.

Chin in. Raising the top of the head and tucking the chin in straightens the whole spine, not merely the neck. This position automatically raises the chest and the shoulders for more efficient arm action. See illustrations on previous page.

Tuck the chin in immediately before lifting and hold it in throughout the procedure. The worker will be looking down at the early stages of the lift, which may conflict with the desire to raise the head to see where he is going. However, as the worker returns to the upright position, the head will automatically be raised at the same time.

Use of body weight. Employing the correct positioning of the feet and the flexion and extension of the knees, the weight of the body can be effectively used to push and pull objects and to initiate a forward movement, such as placing an object on a shelf or walking.

When lifting an object from the ground, the thrust from the back foot, combined with the extension of the knee joints, will move the body forward and upward, and for a brief period it will be off balance. However, this position is immediately countered by bringing the back leg forward, as in walking, and by this time the lift is completed. This forward movement of the body results in a smooth transition from lifting to carrying. Do not "jerk-lift" load as this multiplies the stress to the lower back.

Here are some techniques for specific situations.
1. If the object is too bulky or too heavy to be handled by one person, help should be sought.
2. Before lifting the load to be carried, the worker should consider the distance to be traveled and the length of time that the grip will have to be maintained. The worker should select a place to set the load down and rest in order to allow for the loss of gripping power during a long-distance carry. Pausing to rest is especially important when negotiating stairs and ramps.
3. *To place an object on a bench or table,* the worker should first set it on the edge and push it far enough onto the support to be sure it will not fall. The object should be released gradually as it is set down. It should be moved in place by pushing with the hands and body from in front of the object. This method prevents pinched fingers.
4. It is especially important that an object, placed on a bench or other support, be securely positioned so that it will not fall, tip over, or roll off. Supports should be correctly placed and strong enough to carry the load. Heavy objects, like lathe chucks, dies, and other jigs and fixtures, should be stored at approximately waist height.
5. *To raise an object above shoulder height,* the worker should lift it first to waist height, then rest the edge of the object on a ledge, stand, or hip, and shift hand position, so the object can be boosted after the knees are bent. The knees should be straightened out as the object is lifted or shifted to the shoulders. As in the case of lifts originating at ground level, the upper extreme of muscle stress begins at shoulder height. Good job design should be applied to eliminate or minimize above shoulder lifting.
6. *To change direction,* the worker should lift the object to the carrying position and turn the entire body, including the feet. Avoid twisting the body. In repetitive work, the person and the material should both be positioned so that twisting the body when moving the material is unnecessary.

7. To deposit an object manually in a tight space, it is safest to slide it into place with the hands in the clear.

Assisted one-hand lift

When it can be performed, the assisted one-hand lift is a viable alternative lifting technique to the two-hand stoop lift. In this method, the worker rests one hand on top of the container, bends over to grasp an object in the container, and then pushes down with the nonlifting hand resting on top of the container to force the upper body back to a vertical position. This method features using the large arm and shoulder muscles to perform the lift instead of the vulnerable muscles of the lower back. By using the nonlifting hand to raise the upper body, stress is distributed across the shoulders and arm muscles and reduced in the lower back (Figure 3-2).

Figure 3-2. The assisted one-hand lift method. The lift is made by pushing down on the top of the container to push the upper body up into a vertical position. This shares stress across the shoulders and arm and reduces stress to the lower back.

For a good assisted one-hand lift, follow these steps:
1. Place the nonlifting hand on the container top, bend over container, and assume lift position.
2. While bending over, kick the foot on the same side as the nonlifting hand rearward to provide forward body balance (optional).
3. Reach and grasp object to be lifted.
4. Push down with the nonlifting hand on the container top, raising the upper body to a vertical position. (Be sure to let the nonlifting hand, not the back, do the work!)

As with any lifting technique, there are limitations to its use. The application of the assisted one-hand lift assumes that the object to be lifted can be grasped by one hand, that it is not too long that it becomes awkward to handle and that the weight is not excessive. Studies show that a load weight of 15-20 lbs (6.8-9.1 kg) would not be excessive for most workers to lift with one hand.

The assisted one-hand lift technique also may be used to lift objects off the floor or pallet flats by placing the nonlifting hand on a spot above the knee and pushing down on it after grabbing an object with the other hand.

Team lifting and carrying

When two or more people carry one object, they should adjust the load so that it rides level and so that each person carries

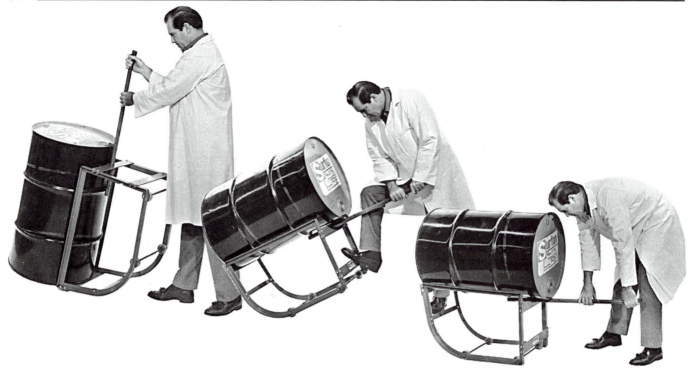

Figure 3-3. A drum tilter is used to minimize strain during upending and lowering. (Printed with permission from Equipment Company of America.)

an equal part of the load. Test lifts should be made before proceeding.

When two people carry long sections of pipe or lumber, they should carry them on the same shoulder or walk in step. Shoulder pads will prevent cutting of the shoulders and reduce fatigue.

When a gang of workers carries a heavy object such as a rail, the supervisor should make sure that proper tools are used. Work direction should be provided; frequently, a whistle or direct command can signal "lift," "walk," and "set down." The key to safe gang carrying is to make every movement in unison.

Handling specific shapes

Boxes, cartons, and sacks. The best way to handle boxes and cartons is to grasp the alternate top and bottom corners and to draw a corner between the legs.

Sacked materials are also grasped at opposite corners. Upon reaching an erect position, the worker should let the sack rest against the hip and belly and then swing the sack to one shoulder.

As the sack reaches the shoulder, the worker should stoop slightly and put a hand on the hip, so that the sack rests partly on the shoulder and partly on the arm and back. The other hand should be holding the sack at the front corner. When the sack is to be put down, it should be swung slowly from the shoulder until it rests against the hip and belly. While the sack is being lowered, the legs should be flexed, and the back kept straight.

Barrels and drums. Those who handle heavy barrels and drums require special training. A barrel is generally less hazardous to handle than a drum because the shape of the barrel aids in upending it. Since the weight and contents of a barrel or drum may vary greatly, special attention should be given to these factors.

Frequently, only one person is available to handle a drum, in which case it is better to wait for help or use mechanical assistance. A commercially available drum tilter equipped with wheels is commonly used. An extension handle provides control and leverage during the tilting operation. The wheels allow for easy transport of the tilted drum over short distances (Figure 3-3). Another commercial device that makes tilting and transporting easier is a two-wheeled dolly equipped with large rubber tires.

If it is necessary to roll a barrel or drum, the worker should push against the sides with the hands. To change direction of the roll, the worker should grip the chime rather than kick the drum. A clamp device for carrying a drum is available.

To lower a drum or barrel down a skid, the drum should be turned and slid endwise. Rolling a drum or barrel up a skid takes two persons, who should stand outside the skid, not inside the rails nor below the drum or barrel being raised or lowered.

If drums or barrels are to be handled on an incline or skid, ropes or other tackle should be used to control their motion. The drum or barrel should be snubbed with a rope, one end of which is securely fastened to the platform from which the drum or barrel is to be lowered. The rope should then be passed around the barrel or drum, and the operator, keeping a firm grip on the free end, can gradually lower the load.

Sheet metal usually has sharp edges and corners and should be handled with leather gloves, hand leathers, or gloves with metal inserts. Gauntlet-type gloves or wristlets will give added protection to wrist and forearm. Bundles of sheet metal should be handled with power equipment. A sheet metal "grab" can be purchased.

Flat glass should be handled by persons equipped with gloves or hand laps; wrists and arms should be protected with leather cuffs and safety sleeves. The worker should wear a leather apron,

Figure 3-4. Service truck holds panes of glass between fixed and hinged panels to ensure safe transport. (Printed with permission from Globe Glass Co., Chicago, Ill.)

leggings, and safety shoes with metatarsal guards. Unless the glass plates are small, the worker should carry only one at a time and walk with care. The plate should be lifted carefully and carried with its bottom edge resting in the palm (turned outward) and the other hand holding the top edge to steady it. Glass plate should never be carried under the arm because a fall might break the glass and sever an artery.

To transport larger glass plates over any distances, handling equipment should be used (Figure 3-4). Plate glass must not be carried in such a manner that it could bend.

Larger flat glass should be handled by equipment specifically designed for that purpose. Equipment such as cranes equipped with vacuum frames, C-frames or spreader bars, and special wagons or dollies normally is used to transport heavy glass. If large plates must be transported by hand, two workers wearing safety hats, safety sleeves, cuffs, gloves, and safety shoes should be assigned to the job.

For details on safe glass handling, see Oresick article in References.

Long objects. Long pieces of pipe, bar stock, or lumber should be carried on the shoulder. A second worker should guide when going around corners. Workers should wear shoulder pads for this operation.

Irregular objects present special problems. Often the object must be turned over or up on end, so that the best possible grip can be secured. If the worker feels unable to handle the object, because of either its weight or shape, assistance should be requested.

Miscellaneous objects. Figure 3-5 shows two safe methods of securing miscellaneous material and small objects on overhead platforms or when raising or lowering them to different elevations.

Scrap metals. In a scrap storage area, the best possible housekeeping practices should be observed. Irregularly shaped, jagged pieces may be tangled in such a way that strips or pieces may fly when a piece is removed from a pile. Workers, therefore, should be provided with goggles, leather gloves or mittens, safety shoes, safety hats, and protection for the legs and body. Workers should be cautioned against stepping on objects that may roll or slide.

Heavy, round, flat objects (such as railroad car wheels or tank covers) present considerable danger even to skilled personnel if rolled by hand. The operation requires careful training and exacting precautions. It is preferable to use a hand truck or power equipment designed for the purpose. Heavy rolls can be safely secured and handled by specially designed devices (Figures 6-3 and 6-25 in Chapter 6).

Machines and other heavy objects

Manual movement of heavy machinery and equipment requires special skill and knowledge. Sometimes machines or castings weighing 100 tons or more must be moved from freight cars to ground level and into permanent position without use of heavy-duty cranes or similar equipment.

Only general safety principles for such jobs can be suggested. Each one presents its own problems and requires careful study and thorough planning. Some companies build scale models of the machines and the blocking, jacks, rollers, and other equipment to be used and then work out the procedure in miniature.

In all cases, the safe floor load limits for areas over which the machine or part will move, as well as for the place in which it is to be installed or stored, should be determined.

Blocking and timbers should be selected with great care. They should be of hardwood, preferably oak, and of the proper sizes to allow the machine to be safely blocked or cribbed as it is raised or lowered. Do not use wood that has round corners or shows signs of dry rot.

For sufficient strength, cribbing should have a safety factor of at least four. Be cautious about the natural tendency to underestimate the load. Cribbing must be placed on a foundation in such a manner that it can be removed readily as the machine is lowered.

ACCESSORIES FOR MANUAL HANDLING

In handling materials, a variety of hand-operated accessories can be used. Each tool, jig, or other device should be kept in good repair and used only for the job for which it is designed.

 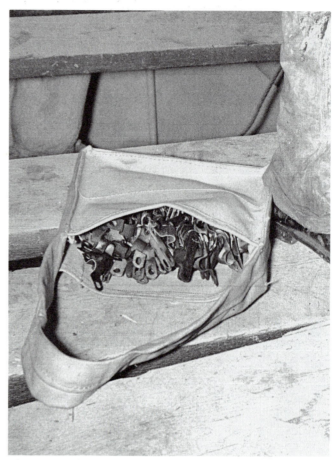

Figure 3-5. Special slings, buckets, and bags are used to handle various materials around platforms and staging. The barrel hitch (left) prevents bail from pulling out when heavily loaded; material in buckets is not to be handled over employees. On the right, small material should be carried and laid on staging in canvas bags. (Printed with permission from Newport News Shipbuilding and Dry Dock Co.)

Hand tools

Hooks. The worker should be trained to use hand or packing hooks in such a manner that they will not glance off hard objects with possible injury to the worker. If a hook is to be carried in the belt, the point should be covered.

Hook handles should be of hardwood and maintained in good condition. Hooks for handling logs, lumber, crates, boxes, and barrels should be kept sharp and inspected daily and before each use.

Bars. The principal hazard in the use of a crowbar is slippage. A dull, broken crowbar is more likely to cause injury than a sharp one. The point or edge should have a good "bite." The worker should be positioned to avoid falls or pinched hands if the bar slips or the object moves suddenly—never work astride an object. Hands and gloves should be dry and free of grease or oil.

Crowbars not in use should be stored so that they will not fall or cause a tripping hazard.

Ordinarily crowbars should not be used to move cars on steel rails. Car movers that do not readily slip are available. When two persons are needed to move a car, two car movers should be used. Each person should operate a mover.

Rollers. Heavy, bulky objects must often be moved by means of rollers. The principal hazard is that the fingers or toes may be pinched or crushed between a roller and the floor. Rollers should extend beyond the load to be moved and be sufficiently strong. Rollers under a load should be moved with a sledge or bar, not with the hand or foot.

Jacks

When a jack is used, the capacity plate or other marking on the jack should be checked to make sure the jack can support the load. If the identifying plate is missing, the maximum capacity of the jack should be determined and painted on the side. If a properly rated jack is used, it should not collapse under the load.

Jacks should be inspected before and after each use. Any sign of hydraulic fluid leakage is sufficient reason to remove the jack from use.

A heavy jack is best moved from one location to another on a dolly or special hand truck. If it has to be manually transported, it should have carrying handles, and at least two workers should form a team to move it. The operating handle should never be left in the socket while a jack is being carried, because it may strike other workers.

Workers should make certain that jacks are well lubricated,

but only at points where lubrication is specified, and should inspect them for broken teeth or faulty holding fixtures. A jack should never be thrown or dropped upon the floor; such treatment may crack or distort the metal, causing the jack to break when a load is lifted.

The floor or ground surface upon which the jack is placed must be level and clean and the safe limit of floor loading capacity must not be exceeded. If the surface is earth, the jack base should be set on substantial hardwood blocking (at least twice the size of the jack), so that it will not turn over, shift, or sink. If the surface is not perfectly level, the jack may be set on blocking leveled by substantial shims or wedges placed so securely that they cannot be crushed or forced out of place.

To prevent the load from slipping, no metal-to-metal contact should be permitted between the jack head and the load. A hardwood shim, longer and wider than the face of the jack head, should be placed between the jack head and the contact surface of the load. Two-inch wood stock is suitable for this purpose.

"Extenders" of wood or metal should never be used. Instead, either a larger jack should be obtained or higher blocking that is correspondingly wider and longer should be placed under the jack.

All lifts should be vertical, with the jack correctly centered for the lift, the base on a perfectly level surface, and the head with its shim bearing against a perfectly level meeting surface. When an emergency requires that the lifting force be applied at an angle, extra precautions must be taken, including:

1. A base of blocking, securely fastened together and to the ground, to make an immovable surface at right angles to the lift for the jack base to sit on,
2. Cleats on the blocking to prevent shifting of the jack base,
3. A meeting surface at right angles to the direction of the lift for the jack head with its shim to bear against,
4. Props or guys to the load to prevent its swinging sidewise when lifting begins.

When a jack handle is placed in the socket, the worker should make sure that the area is clear and that there is ample room for an unobstructed swing of the handle before appling pressure. A faulty movement in the load may cause the handle to pop up and strike another worker. The person operating the handle should stand to one side so that, if the handle kicks, it will not strike any body or facial areas. When releasing a jack, the worker should keep all parts of the body clear of the movement of the handle.

After the load is raised, metal or heavy wooden horses or blocking should be placed under it for support in case the jack should let go. A raised load should never be allowed to remain supported only by jacks. The handles of the jacks should be removed immediately and placed out of the way to prevent workers from tripping over them.

Hydraulic jacks may settle after raising a load. It is, therefore, especially important to place blocking under a load that has been raised by such jacks.

Screw jacks have a tendency to twist when a heavy load causes the floating head of the jack to bind. The base of a screw jack should be anchored as securely as possible, so that the jack base will not twist and slip out from under the load when force is applied on the bars to raise the screw.

To raise a large piece of equipment with screw jacks, two or more jacks should be used. The load should be equally distributed on each jack. Each jack should be raised a little at a time to keep the load level and the strain equal on each screw jack head. Special signals are required to verify that all jacks are rising uniformly.

Workers using jacks should wear safety shoes and instep-guard protection, metatarsal guards, because jack handles may slip and fall or parts of machinery or equipment may become loose and drop while the load is being lifted or shifted. Wiping material should be furnished to jack operators to remove oil from their hands and from the jack handles, so that they will always have a firm grip.

Oil that has collected in the bases of equipment or machines to be jacked should be removed before the operation is begun, to prevent spillage when the equipment or machines are tilted. Spillage of residual oil should be wiped up immediately.

When a jack begins to leak, malfunction, or show any signs or defect, it should be removed from service, repaired, and tested under load before being returned to service.

Handtrucks, carts, dollies, and wheelbarrows

Many special types of handtrucks and dollies are available, such as two-wheeled, flat, platform, refrigerator, appliance, box, special racks or dollies, and lift trucks. Trucks can be purchased or designed for objects of various sizes and kinds. Operators should wear gloves and safety shoes.

Two-wheeled trucks and wheelbarrows should be equipped with knuckle guards to protect hands from being jammed against door frames or other obstructions.

Some two-wheeled trucks have brakes so that the worker need not hold the truck with a foot on the wheel or axle. Most commonly used trucks do not have brakes (Figure 3-6).

To reduce the hazard to toes and feet, wheels should be as far under the truck as possible. Wheel guards can be installed on many types of trucks.

Tongues of flat trucks should be provided with counterweights, springs, or hooks to hold them vertical when not in use. If this is not possible, workers should be trained to leave handles in such a position that they will not be a tripping hazard.

Equipment should be inspected daily and kept in good repair. Repair and maintenance records should be kept to describe the condition of each piece of equipment. Axles should be kept well greased.

The type of truck most suitable for the work at hand should be used. No one truck is suitable for handling all types of material.

Two-wheeled trucks look deceptively easy to handle, but there are safe procedures that must be followed.

1. Tip the load to be lifted forward slightly, so that the tongue of the truck goes under the load.
2. Push the truck all the way under the load to be moved.
3. Keep the center of gravity of the load as low as possible. Place heavy objects below lighter objects. When loading trucks, truckers and loaders should keep their feet clear of the wheels.
4. Place the load well forward so the weight will be carried by the axle, not by the handles.
5. Place the load so it will not slip, shift, or fall. Load only to a height that will allow a clear view ahead.
6. When a two-wheeled truck or wheelbarrow is loaded in a

Figure 3-6. Lightweight, two-wheel trucks are commonly used. Large semipneumatic tires move loads more easily over rough surfaces.

Figure 3-7. With this design, it is more natural for the operator to grasp the handle away from the outside framework.

horizontal position, raise it to traveling position by lifting with the leg muscles and keeping the back straight. Observe the same principle in setting a loaded truck or wheelbarrow down—the leg muscles should do the work.

Figure 3-8. On older design hand trucks, workers' hands can be protected by sturdy gloves and knuckle guards on the handles. Such guards can be made and installed by maintenance personnel.

7. Let the truck carry the load. The operator should only balance and push.
8. Never walk backwards with a hand truck.
9. For extremely bulky items or pressurized items, such as gas cylinders, strap or chain the item to the truck.
10. When going down an incline, keep the truck ahead so that it can be observed at all times.
11. Move trucks at a safe speed. Do not run. Keep the truck constantly under control.

Four-wheel truck or cart operation follows rules similar to those for two-wheeled trucks. To avoid lower back injuries, trucks or carts should not require excessive physical force to push or pull. Extra emphasis should be placed on proper loading, however. Four-wheeled trucks should be evenly loaded to prevent their tipping. Four-wheeled trucks should be pushed rather than pulled because there is less stress on the lower back while pushing an object than while pulling on it.

Trucks should not be loaded so high that operators cannot see where they are going. If there are high racks on the truck, two workers should move the vehicle—one to guide the front, the other to guide the back. Handles should be placed at protected places on the racks or truck body, so that passing traffic, walls, or other objects will not crush or scrape the operator's hands.

Truck contents should be arranged so that they will not fall or be damaged in case the truck or the load is bumped.

General precautions. Handlers of two- and four-wheeled trucks should be made aware of three main hazards: (1) running wheels off bridge plates or platforms, (2) colliding with other trucks or obstructions, and (3) jamming their hands between the truck and other objects (Figures 3-7 and 8).

Figure 3-9. Various methods of stocking and piling materials are used. Palletized materials are speedily and efficiently handled by lift trucks; however, pallets must be securely and evenly loaded. (Printed with permission from Towmotor Corporation.)

Figure 3-10. Barrels and kegs stacked on pallets must be adequately supported by planking or channels between layers, or else stored on end. (Printed with permission from Towmotor Corporation.)

Workers should operate trucks at a safe speed and keep them constantly under control. Special care is required at blind corners and doorways. Properly placed mirrors can aid visibility at these places.

When not in use, trucks should be stored in a designated area, not parked in aisles or other places where they would be a tripping hazard or traffic obstruction. Trucks with drawbar handles should be parked with handles up and out of the way. Two-wheeled trucks should be stored on the chisel with handles leaning against a wall or the next truck.

Powered hand trucks and power trucks are discussed in Chapter 6.

Pallets are constructed of paper, wood, glass fiber, and plastic. They should always be kept in good condition and repair, preferably through systematic inspection, repair and replacement.

Pallets not in use should not be left scattered around because of the tripping hazard that they present. When not in use, pallets should be stacked, but never on end.

For more information on pallets, see Chapter 6.

Conveyors, cranes, and hoists. See Chapter 4, Hoisting Apparatus and Conveyors, for a description of many types of lifting and transporting devices used to reduce physical strain and facilitate material handling.

STORAGE OF SPECIFIC MATERIALS

Temporary and permanent storage of materials should be neat and orderly to eliminate hazards and conserve space. Materials piled haphazardly or strewn about increase the possibility of accidents to employees and of damage to materials.

The warehouse supervisor must direct the storage of raw material (and sometimes processed stock) kept in quantity lots for extended periods of time. The production supervisor is usually responsible for storage of limited amounts of material and stock for short periods of time near the processing operations. Planned materials storage reduces the handling necessary to bring materials into production, as well as to remove finished products from production to shipping.

When planning materials storage, allow adequate ceiling clearance under the sprinklers. Be sure that automatic sprinkler system controls and electrical panel boxes are free and clear. Unobstructed access to fire hoses and extinguishers must be maintained. All the exits and aisles must be kept clear at all times (Figures 3-9 and 10). The amount of clearance between storage and automatic sprinkler heads will vary with the commodity being stored and the height of storage. (See NFPA 13, *Installation of Sprinkler Systems.*)

If materials are to be handled from the aisles, allow for the turning radius of any power truck to be used. Employees should keep materials out of the aisles and out of the loading and unloading areas. These areas should be marked with painted or taped lines.

Use bins and racks to facilitate storage and reduce hazards (Figure 3-11). Material stored on racks, pallets, or skids can be moved easily and quickly from one work station to another with less material damage and fewer employee injuries. When possible, material piled on skids or pallets should be cross-tied.

Storage racks should be secured to floor, wall, and to each other. If flue spaces are provided, stock should never block the flue. If automatic sprinklers or fire prevention devices are pro-

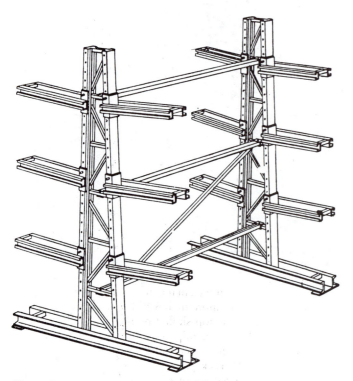

Figure 3-11. Heavy-duty cantilever rack allows good visibility and easy access to materials. (Printed with permission from The Paltier Corporation.)

Figure 3-12. Clearly marked pickup and drop stations promote good housekeeping, reduce hazards, and conserve space.

vided in the racks, exercise special care to avoid damage to them. Racks, when damaged, should be repaired. Never allow employees to climb racks.

In an area where the same type of material is stored continuously, it is a good idea to paint a horizontal line on the wall to indicate the maximum height to which material may be piled. This will help keep the floor load within the proper limits and the sprinkler heads in the clear.

Pickup and drop stations should be clearly marked (Figure 3-12).

High rack storage. The modern trend in storing uniform-sized containers or stock is high bay storage. Some European storage facilities now have approximately 100-ft (30.5 m) high automated storage, and more of these are being constructed in the U.S.

High rack facilities require unique, specially designed, high-lift material handling equipment. Some, up to 30 ft (9 m), are operated manually. Others are operated under computer automation or control. Applicable standards include the ANSI B56 series, "Powered Industrial Trucks," and the Crawler Cranes section of ANSI B30.5, *Safety Code for Crawler, Locomotive, and Truck Cranes.* Not only do these represent unique material handling problems, but also special fire protection problems. (See NFPA 231C, *Rack Storage of Materials.*)

The protection of personnel who operate and maintain such facilities must include disaster and emergency planning, physical protection at the operating position, and visual and audible warnings on moving equipment.

Special procedures and equipment are also required for maintenance and physical inventory.

Rigid containers

Large metal containers and box pallets. There are three general types of large containers commonly used in industry: wire mesh or expanded metal containers, solid-sided metal tubs, or skids and box pallets (Figure 3-13). There are also many variations on these types.

Safe stacking of these large containers requires a level and stable stacking surface. These containers must be in good condition and have the ability to nest or interlock with the container below. Containers of different types (e.g., metal tubs and wood box pallets) should not be intermixed in a stack unless designed for each other and full safety is assured.

A rule of thumb for stacking heights is three times the minimum base diameter of the container. For example, a container with a 2 ft (61 cm) × 3 ft (91 cm) base may be stacked 6 feet high if the other stacking conditions have been met. Do not exceed container weight capacities and weight bearing capacities of the stacked containers. For visibility, containers should not be stacked near the corners of working (non-storage) aisles.

Fiberboard/cardboard cartons. Loaded cartons should be stored on platforms as protection against moisture since even low piles, when wet, will collapse. Preferably, cartons should be stored on pallets or racks. Any sign that the lower cartons are being crushed requires a restacking of the pile.

Since the height of piles of cartons is regulated by the kind of materials in the cartons, it is not possible to establish a standard height. An important factor to consider is that the sides of the cartons will not support much load. Sheets of heavy wrapping paper placed between layers of cartons help prevent the shifting of the pile. Interlocking the cartons increases the stability of the piles.

Certain bulky materials, such as skids of paper, should not be stacked to maximum allowable heights in the rows bordering on aisles, especially those which carry hand- or power-truck traffic. A good rule is to make the first row one item high unless the material can be tightly interlocked.

Uncrated stock

Lumber. Except for the amount for immediate need, lumber is best stored outdoors or in a building separate from the general warehouse. Lumber should be sorted by size and length and stored in separate piles. It is important for outdoor lumber piling that firm ground be selected. The area should be well drained to remove surface water and prevent softening of the ground. A periodic check should be made to determine if any material has shifted position.

For long-term storage, substantial bearings or dunnage is recommended. Concrete with spread footing extending below the frost line is a good method.

For temporary piling, heavy timbers may be used to support the crosspieces. This type of support should be inspected periodically for deterioration, which may cause the pile to list dangerously.

If the lumber must be removed by hand, it should be stored in low piles or in racks, with galleries provided, if necessary, to permit workers to reach the top of the piles.

If lumber must be handled manually to or from a higher pile, the pile should be not more than 6 ft (1.8 m) high, and a safe means of access to the top should be provided.

Twenty feet (6 m) is generally considered the maximum safe height for lumber piles when lumber is piled and removed mechanically by lift trucks.

Tie pieces are needed not only to stabilize the pile, but also to provide air circulation. Tie pieces should not extend into walkways, but should be cut flush with the pile. Green lumber should have tie pieces on every layer, whether stored indoors or outdoors.

Lumber stored outdoors should be covered to prevent checking or twisting. Lumber stored indoors should be in a well-ventilated building.

Bagged material should be cross-tied with the mouths of the bags toward the inside of the pile. The bags should be neatly stacked; avoid overhangs that could be ripped, thus spilling the contents. This precaution also applies to stacking on pallets and stacking several tiers high by lift truck.

Pipe and bar stock places a heavy load on the floor. Therefore, the floor area must be selected with bearing strength in mind. Because of the hazard to passersby while stock is being withdrawn from the racks, fronts of pipe and bar stock racks should not be located on main aisles.

Pipe and other round materials should be piled in layers with strips of wood or iron between the layers. Either the strip should have blocks at one end, or the end should be turned up.

Material such as lumber or pipe is particularly dangerous because of its tendency to roll or slide. Dropping instead of placing such objects on a pile frequently causes them to slide or bounce. Employees are likely to be injured if they attempt to stop rolling or sliding objects with their hands or feet.

Bar steel stock in the larger sizes should be stored in racks that are designed to rest the bars on rollers. The center distance between rollers will be governed by the sizes of the bars and should permit their easy withdrawal. Rollers with multiple sections make withdrawal easier.

Racks should incline toward the back so that bars cannot roll out. Light bar stock may be stored vertically in special racks. Special A-frame racks of metal can hold a variety of pipes

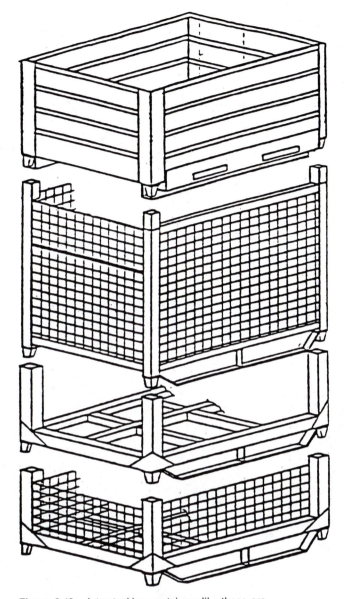

Figure 3-13. Interstacking containers like these are found in many industrial plants. To stack safely, be sure surface is level and stable.

Barrels and kegs. Piles of barrels should be symmetrical and stable, preferably in the shape of a pyramid. The first or bottom row should be blocked to prevent the barrels from rolling. If barrels or kegs are piled on end, place planks between rows.

When barrels or kegs are piled on their sides other than in pyramid form, they should be laid on specially constructed racks. Otherwise, planks should be laid between rows and the ends of the rows should be blocked.

Drums are discussed below under Portable Containers in the Liquids section.

Rolled paper and reels. Clamp-type trucks stack these items three or four high. Extreme care is required to assure even stacking. If there is any physical damage to lower rolls or reels the material must be restacked.

Compressed gas cylinders. For their safe handling and storage refer to Chapter 13, Welding and Cutting.

and bars safely in quantity, if loaded evenly and supported properly.

Sheet metal. Racks, similar to those for bar stock, may be provided for plate and sheet stock, except that rollers are not always applicable. Oiled sheets require additional caution in handling.

Sheet metal usually has sharp edges. It should be handled with hand leathers or leather gloves, or gloves with metal inserts. Large quantities should be handled in bundles by power equipment. These should be separated by strips of wood to facilitate handling, when the material is needed for production, and to lessen chances of shifting or sliding of the piles of material.

Tin plate strip stock is heavy and razor sharp. Should a load or partial load fall, it could badly injure anyone in the way. Two measures can be taken to prevent spillage and injuries: (1) band the stock after shearing, and (2) use wooden or metal stakes around the stock tables and pallets that hold the loads. It is the responsibility of the supervisor and all who handle the bundles to make sure the load is banded properly and that the stakes are in place when the load is on the table. (Special cutting tools are required to remove bands.)

Burlap sacking. When burlap sacking is stored in high stacks, heat is generated by the weight, creating a spontaneous ignition hazard. One way to reduce the hazard is to cut the size of the stack by breaking it up into smaller stacks. This can be done either by making smaller stacks (which would increase the number of stacks and take more space) or by placing blocks at intervals in the stack, so that, in effect, there would be a number of small piles, one atop the other. Additional protection should be provided by constructing the storage room of fire-resistant materials and having sprinklers and dust-tight lights.

Straw, excelsior, and other packing materials are usually received baled and should preferably be stored that way, either in a separate building or in a fire-resistant room provided with sprinklers and dust-proof electric equipment.

Because these materials are a fire hazard, only the amount necessary for immediate use should be taken into the packing room. For storing enough for immediate use, the best bins are made entirely of metal, or of wood lined with metal, and are provided with covers that are normally closed.

Large bins may have several compartments with counterweighted covers. The counterweight ropes should have fusible links to ensure automatic closing of covers in case of fire. Counterweights should be boxed in to prevent injury if the ropes break.

No Smoking signs must be posted.

PROBLEMS WITH HAZARDOUS MATERIALS

Storage and handling of specific hazardous materials are discussed in other chapters: gases in Chapter 13, Welding and Cutting; flammable liquids (including refrigerants) and tank car and tank truck loading and unloading, in Chapter 16, Flammable and Combustible Liquids; NFPA hazard symbols in Chapter 17, Fire Protection; and incompatible materials in Chapter 19, Table 19-N.

Liquids

Drums. Filled drums containing volatile liquids should be stored in a protected area out of the sun. Heat from any source causes liquids to expand and the resulting buildup of pressure could cause leaks with subsequent fire or explosion. It is recommended that an approved drum vent be installed in the bung opening as soon as a drum is received (Figure 3-14).

Reuse of drums also causes problems if drums are not thoroughly cleaned beforehand. Drums should be carefully purged and cleaned out with water and steam; a chemical expert should be consulted in order to set up safe procedures. Often the top of a drum is removed so the drum can be used as a receptacle. *Never* burn the top of a drum out with a torch because some liquid or vapor left behind in the drum could cause an explosion. It is much safer to follow the drum-purging methods described in Chapter 13, Welding and Cutting, and then remove the top with a mechanical opener.

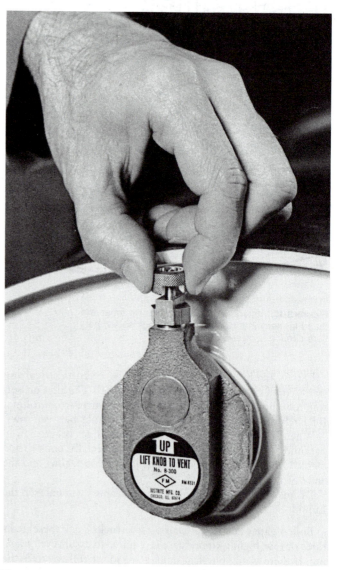

Figure 3-14. This simple device automatically relieves drum pressure and can be lifted by hand to relieve vacuum. (Printed with permission from Justrite Manufacturing Co.)

More details about handling and storing drums are given in the section on Portable Containers below.

Tanks. The structure of a new building should be designed with an ample safety factor to permit supporting the weight of storage tanks. However, inspection by a competent structural engineer should be made before a tank is installed in an old building. Storage tanks for hazardous liquids are preferably installed outdoors.

There are many advantages in underground installation of outdoor storage tanks. However, the danger of undetected leaks in underground tanks containing corrosive or toxic materials probably outweighs the advantage of freedom from drips and sprays. When an outdoor tank is located in a pit, the pit should be large enough to permit easy access to all parts of the tank. A permanent ladder and an access door that can be fastened shut should be provided.

No one should be permitted to enter a pit without an approved type of supplied air mask or hood or oxygen breathing apparatus, unless tests have established the presence of enough oxygen and the absence of dangerous amounts of toxic vapors. In any case, a worker who enters a pit should wear a safety belt with attached lifeline and a similarly equipped observer should be stationed to watch.

Process tanks that will contain volatile or corrosive liquids should be installed only at or above grade, in areas having adequate drainage, and should be separated from the processing area by construction having a fire-resistive rating of at least two hours.

Tanks should be installed where traffic cannot pass under them. If people must walk beneath them, drip pans should be installed and provided with drainage to a safe disposal or recovery location.

Tanks should be provided with permanent stairs or ladders and walkways with standard guardrails and toeboards. Tanks should be emptied, cleaned, and inspected for structural weaknesses at regular intervals, and records kept of each inspection.

Tanks for holding volatile materials should be bonded, grounded, and provided with emergency venting devices. Venting should be in accordance with provisions of the *Flammable and Combustible Liquids Code,* NFPA 30. If tanks are inside buildings, vents should discharge outside the building at a location free from any ignition source, and one that will not cause contact with personnel. Be sure to consider the effects of corrosion on venting devices for tanks that will contain corrosive liquids.

Connections for filling and emptying tanks are preferably made through the top, to minimize liquid loss and the possibility of injury from a broken fitting. Fill lines should be plainly labeled and equipped with a drain. Use of compressed air is not permitted for the transfer of flammable and combustible liquids, according to NFPA 30, paragraph 5–2.4.5.

Cleaning tanks can be an exceedingly dangerous operation. An exact and specific procedure should be established, preferably in written form, and strictly followed. Specifications for tank-cleaning procedure are set forth in NFPA 327, and API publication RP 2015. (See References.)

The procedure should be modified for toxic compounds only to the extent that more complete protective equipment may be required. Liquid aromatic nitro compounds and amines are solvents for rubber and are absorbed through it gradually as well as through the skin. For such exposures, personal protective equipment should be made of one of the inert synthetic rubbers or plastics. (This includes gloves, aprons, boots, respiratory- and eye-protective equipment).

The handling of phenolic compounds, such as carbolic acid, cresylic acid, and the cresols, requires precautions against skin absorption, although these compounds are not such effective rubber solvents as the nitro and amino compounds.

To prevent pumps from being primed with dangerous liquids where connection is made through the top of a tank, self-priming pumps or pumps that generate enough suction to lift the liquid from the bottom of the tank should be used.

Pipelines for carrying chemicals are preferably installed in trenches or tunnels. If they must be installed overhead, they should be isolated so that they will not drip on anyone working underneath. Pipelines for carrying flammables should not be installed in tunnels. All pipelines must be identified as to content.

There are three major sources of chemical injury in pipeline work:

- FAILURE OF PACKING IN VALVE STEMS OR OF GASKETS IN BOLTED FLANGES. To minimize injuries from valve packing failure, the valve stem can be surrounded by a sheet metal box or hood which will deflect spray away from the person operating the valve. So far as possible, packing should be renewed without pressure on the valve.

- OPENING THE WRONG VALVE. To prevent injuries and accidents from this source, pipelines and valves should be indentified by tags, lettered markings, and distinctive colors. (See the next section, Identification of Piping.) It is also desirable to have valves well separated and the immediate area well lit to assure quick and easy identification.

- FAILURE TO CHECK THAT VALVES ARE CLOSED AND LOCKED AND THE LINES DRAINED BEFORE TENSION IS RELEASED ON FLANGE BOLTS. The opening between the faces of the flange may be temporarily covered with a piece of sheet lead while the flange bolts are being loosened and the faces separated.

 The bolts farthest away should be loosened first so that drainage will tend to go away from the worker. Blinds should be inserted in the flanges as soon as they are opened. For lines that are opened often, blinds permanently pivoted on a flange bolt, with one end acting as a gasket and the other as a blind, can be used.

At the conclusion of a job on a pipeline containing corrosive chemicals, tools and personal protective equipment should be thoroughly washed with a reagent to neutralize or remove the corrosive material and then rinsed in clean water before the equipment is removed and the tools stored.

Identification of piping. Distinctive colors for identifying piping have been standardized in American National Standard A13.1, *Scheme for the Identification of Piping Systems,* and are described in Chapter 1, Industrial Buildings and Plant Layout.

Specific identification of piping should be provided by a lettered legend which names the material being piped, summarizes the hazards involved, and gives directions for safe use. Legends should be moisture-resistant and contain pigments that are colorfast. Stencils or decals may be used to apply legends.

Portable containers. Where liquid chemicals are used in quan-

Figure 3-15. Special racks facilitate safe multiple tiering of drums. The rounded corners minimize the hazard of punctures. Bungs must always be tightened to prevent leaks.

Floors in the storage area of corrosive liquids should be made of cinders, concrete treated to decrease its solubility, or other resistant material. Concrete is also satisfactory for flammable liquids. Good drainage will permit easy cleaning in case a container in the storage area leaks or breaks. Pollution control regulations at local, state, and federal levels usually prohibit draining these spills into sanitary or storm sewer systems.

The storage area should be well ventilated, as per OSHA regulations. Whenever it can be used, natural ventilation is preferable to mechanical, because it involves no operating problems.

Full drums should be stacked in racks, preferably with a separate rack for each material. These racks should be arranged to permit easy access for moving the drums in and out and for ready inspection of stock (Figure 3-15).

Barrels may be stacked vertically with dunnage between the tiers, but are more conveniently handled if they are kept in racks similar to those used for drums. The safest way to handle drums is to use mechanical-powered lift equipment with drum-lifting clamps. Transporting drums on pallets is a common practice in many companies. Workers handling these drums should use bung nut wrenches (Figure 3-16) to tighten the bung nuts and prevent leaks during storage and transportation.

Be sure grounding is adequate if flammable liquids are involved. See Chapter 16, Flammable and Combustible Liquids, for details.

Different materials should be stored in different designated areas, separated by wide aisles. Boxed carboys should generally be stacked not higher than two tiers and never higher than three. Not more than two tiers should be used for carboys of strong oxidizing agents, such as concentrated nitric acid or concentrated hydrogen peroxide.

Before acid carboy boxes are handled, they should be inspected for corrosion of nails or weakening of the wood by acid.

Before empty carboys are piled, they should be thoroughly drained and the stoppers replaced.

Special equipment should include a long-handled truck (Figure 3-17) that picks up the boxed carboys under the handling cleats or between the bottom cleats provided on all standard 12- and 13-gallon (45 and 49 liter) boxed carboys. These trucks have handles long enough to keep persons handling them away from splashes, in case carboys are dropped.

There is also less danger of dropping a carboy with this kind of truck than with the standard two-wheeled truck. This truck becomes a much safer device for handling drums and barrels if it has a bed curved to fit the drum and a hook to catch the chime.

The safest way to empty a carboy is to move the liquid by suction from a vacuum pump or aspirator or start a siphon with a rubber bulb or ejector. The carboy incliner, if it holds the carboy firmly by the top, as well as the sides, and automatically returns to the neutral position on being released, is satisfactory.

Compressed air, even from a hand pump, should not be used on a carboy unless the carboy is enclosed by another container so that the pressures inside and outside remain the same (Figure 3-18). *Pouring by hand or starting pipettes or siphons by mouth suction should never be permitted.* Mechanical pumps are not desirable unless they are self-priming or have sufficient suction force to start themselves.

A corrosive or poisonous liquid sometimes requires a specially identified container. The simplest is a glass or plastic jug or jar, with a good closure, placed into a metal can. A container

tity, it is generally better to install pipelines and outside storage tanks than to use portable containers. Spillage is reduced and localized, thereby making it easier to handle.

Portable containers, such as drums, barrels, and carboys, should be properly stored. A minimum amount of liquid should be kept at the point of operation; only enough for use on one shift is a common rule. The main stock should be stored in a safe, isolated place.

If the liquid is corrosive or highly toxic, the storage area should be isolated from the rest of the plant by impervious walls and floor. A provisional plan for safe cleanup of spillage and disposal of contaminated materials should be posted in the storage area. Otherwise, a separate building should be used.

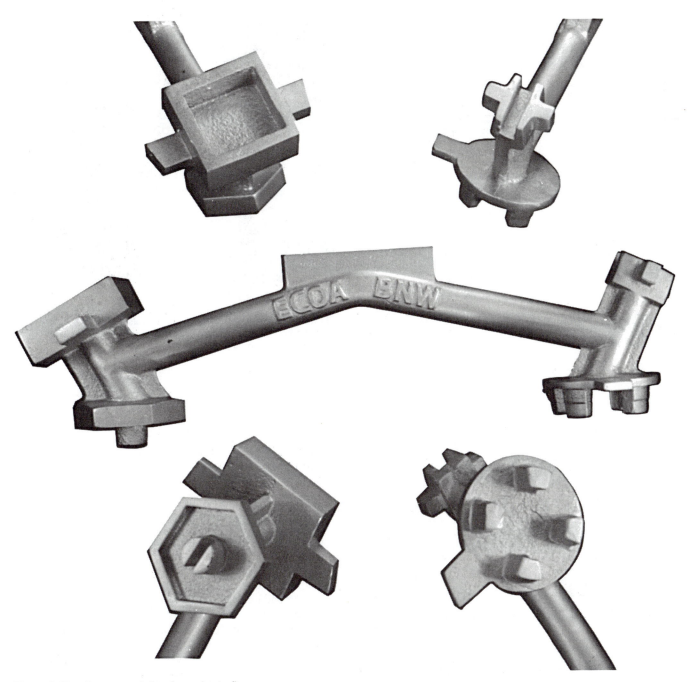

Figure 3-16. Bung wrench has four grips to fit various bung nut configurations.

resistant to shock can be improvised by placing a jug into a metal pail and filling the space between the pail and the jug with pitch or foamed plastic. A container may be made to fit the jug with only a thin layer of padding, such as a layer of gasket rubber. Containers are also available commercially.

Highly toxic substances, such as cyanides and soluble oxalates, should be kept in containers of a distinctive shape if they must be handled manually. All containers must be tagged or labeled, identifying the contents and specifying the appropriate hazard warning as required by OSHA-CFR 1910.1200. Materials should be locked up at all times and dispensed only by authorized personnel.

Where caustics or acids are stored, handled, or used, emergency showers or eyewash fountains must be available. Workers should be provided with chemical goggles, rubber aprons, boots, gloves, and other protective equipment necessary to handle the particular liquid. (See *Administration and Programs* volume, Chapter 17, Personal Protective Equipment.)

Tank cars. Isolate tank cars on sidings by derails and by blue stop flags or blue lights, chock car wheels (Figure 3-19), and set hand brakes before the cars are loaded or unloaded. Before the car is opened, it should be bonded to the loading line. The track and the loading or unloading rack should be grounded, and all connections checked regularly. (Also see Chapter 16, Flamma-

Figure 3-17. Carboy truck is safe and convenient.

Figure 3-19. Tank cars should have wheels chocked and be marked by an approved visible sign, flags, or blue lights. A sign stating that the tank car is connected is another good safety precaution. (Printed with permission from T & S Equipment Company.)

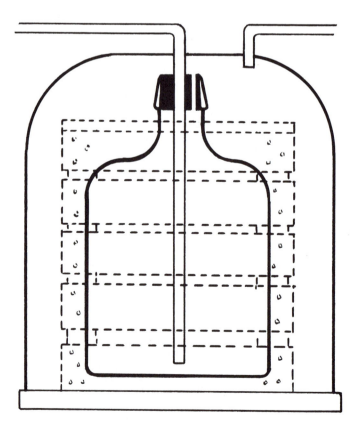

Figure 3-18. Air pressure admitted through the short pipe forces liquid from the carboy through the long pipe. Pressure is exerted on the bell cover, not the carboy.

ble and Combustible Liquids, for a detailed discussion of loading and unloading tank cars.)

Chemical tank cars should be unloaded through the dome rather than through the bottom connection. If the contents are nonflammable, air pressure not to exceed 25 psi (172 kPa) may be used for unloading. The connections should be equipped with a safety valve and gage, so that the pressure on the tank can be determined at any time (Figure 3-20).

Before the car is opened, the cap of the unloading pipe should be gently backed off without being completely removed. Pressure in the tank car should be allowed to escape gradually before the cap is entirely removed.

The unloading dock should be equipped with a walkway at the height of the tank car domes and with drawbridges that can be lowered to make a firm walkway directly to the domes of the cars. Standard handrails and non-slip surfaces are required. If corrosive materials are handled, emergency showers and eyewash stations should be provided along the walkway.

Some materials normally shipped in tank cars solidify at temperatures reached during shipment. Tank cars used for such materials are ordinarily equipped with steam lines for melting the contents. The eduction lines and valves should be thoroughly thawed and clear before unloading is begun.

The plant line to the unloading dock must be completely drained after unloading. To facilitate this procedure, the line should be installed with a slope toward the storage tank so it drains by gravity. Should vessel entry be necessary for maintenance or cleaning, check with your supervisor regarding special procedures at your facility.

Solids

Bins. When new bins are to be installed in an old structure or when new materials are to be stored in old bins, the mechan-

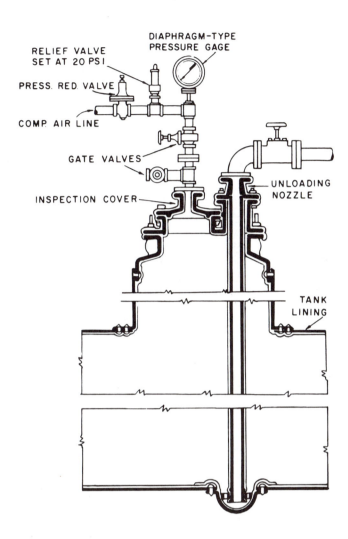

Figure 3-20. Cross section of one type of tank car designed for unloading under air pressure. Note that air pressure should not be used for discharging flammable liquids.

ical strength of the structure should be checked. Solids vary in unit weight, and the more dense materials produce high unit loads.

A fundamental factor in the design of equipment for the handling of bulk solids is sufficient slope in the cone bottom of a tank or bin to permit the solid to run freely and to prevent its arching over. Where arching takes place, there should be a method to restart the flow without a worker entering the bin either above or below the solid material.

A vibrator to shake the bottom of a small metal bin, or an agitator bottom, is a simple means to start the flow. Both are standard equipment, and one or the other can be applied to bins of almost any size or shape.

It is sometimes possible to work either from the bottom or from the top of a bin. If a person can break up the arch from the top with long tools without entering the bin, the job is reasonably safe; it is dangerous when attempted from the bottom. There is also an ever-present temptation to step inside and work with a little more convenience until the material starts to

flow. However, falls into open storage bins often result in injuries.

Bin openings at floor level or within 2 ft (61 cm) of it should be surrounded with standard guardrails and toe boards. If guardrails are impracticable, or if the opening is not easily accessible, like the fill opening of a high bin, the opening can be covered with a grating, which will not materially obstruct the opening but will prevent a worker from falling in.

Many bin openings can be covered with a 2-in. (5.1 cm) mesh, and most of them can be covered by a 6-in. (15.2 cm) grating or by parallel bars on 6-in. centers.

Before repair and maintenance work is started, a test should be made for oxygen content of the air and for the presence of toxic materials, particularly carbon monoxide. If the bin has contained an organic material, there may have been a reaction that dangerously depleted the oxygen content of the air without noticeable rise in temperature or other warning signs.

The lockout procedure described in Chapter 8, Principles of Guarding, and Chapter 15, Electricity, should be followed before a worker enters a bin. The filling equipment should be made inoperative so that it cannot be started again, except by the worker after leaving the bin or by the immediate supervisor after checking the worker out of the bin.

The worker entering the bin should be provided with a safety belt and lifeline, and a similarly equipped worker should be stationed with the sole duty of tending the line and observing the person inside the bin. Where bins are filled and emptied by continuous conveyors, the control of dust is likely to be a serious problem. In filling a bin, material is generally dropped from a belt conveyor. If the material is dropped through a chute from an elevator conveyor, even more dust may be produced.

Escape of dust into the rest of the plant can generally be prevented by enclosing the bin, except for the fill opening, with a skirt of either metal or fabric, and by providing an exhaust up through the filling chute.

If the material is scraped from a belt conveyor, it is usually enough to cover the conveyor at the point of discharge with an exhaust hood and provide a closed chute from the discharge point to the bin.

Since the dust in these cases is released at a low velocity, it is sufficient to provide an inward air velocity of about 50 fpm (0.25 m/s) through all the openings, provided that there are no seriously disturbing air currents to blow dust out of the openings and that enough velocity is provided through the rest of the system.

The same general principles apply to the discharge of bins onto conveyors. There is seldom a serious dust problem except at the loading and discharge points of the conveyors. These points must generally be ventilated and covered with hoods, because it is seldom feasible to provide them with dust-tight enclosures or to reduce the dust by wetting. Approved respirators should be worn by those who are exposed to these conditions.

Combustible solids. Where combustible materials are handled, the dust content of the air must be kept below the lower explosive limit. In addition to tight enclosures and dust collection systems, good plant housekeeping will go far toward preventing disaster.

Dust explosions commonly occur as a series, not as a single shock. The first explosion uses up the dust in the air, but the shock stirs up more dust from the building members, which is in turn set off. If the building is kept clean, this sequence cannot occur and damage will be minimized.

All sources of ignition should be excluded from the area of a potentially explosive dust. Wiring, lights, and switches should be in compliance with the NFPA series 61A–D and NFPA 70, *National Electrical Code,* for hazardous locations. Electric motors should be of the totally enclosed, explosion-proof type or should be in a tight enclosure, which is independently ventilated from a nonhazardous area and kept under positive pressure.

Bearings should be large and well protected; a hot bearing may ignite many types of dust. Heating systems should be of the indirect type only, and radiators should be constructed to permit easy cleaning. The no smoking rule must be rigidly enforced.

Static electricity is the source of ignition in many fires. It can be prevented from accumulating on most surfaces, if relative humidity is maintained at 60 to 70 percent. If this cannot be maintained, use a ground to minimize static buildup.

Static electricity can be removed from moving parts, such as conveyor belts and shafts, by methods that use static collectors, grounding brushes, and conductive V-belts and leather belt dressings.

Metal bins should be grounded to the conveyor frame. Electrical interconnection at loading and discharge points of conveyors should be checked especially. Static voltmeters will quantitatively measure the effectiveness of the grounding system, or a rough estimate can be made with a simple electroscope.

Automatic sprinkler protection should be installed inside bins and processing equipment containing combustible materials. Where water is undesirable, either because of reaction with or damage to the material, protection can be provided by inert gas extinguishers.

If a particularly hazardous material like metal powder is being handled, the apparatus may be completely enclosed and flooded with an inert gas like carbon dioxide, nitrogen, or helium to remove or reduce the oxygen content and thus prevent ignition.

The area should be explosion-vented, preferably by windows and skylights that swing out on friction catches.

Portable containers. The same general rules that apply to the storage and handling of containers of liquids apply also to the storage and handling of containers of solid materials. The most popular containers for solids are 50- and 100-pound (23 and 45 kg) paper bags. These bags are free from sifting or leaking, but they must be handled with some care to prevent mechanical damage. A few slipover bags should be available to cover the occasional broken or leaking container in order to save material and to prevent skin contact with the dust.

Full bags should be stacked on pallets or staging to prevent water damage. Interlocked stacking on pallets generally leads to better piling and less mechanical hazard in moving material. Bags should be protected from the weather, although some of the laminated bags are remarkably weather resistant.

Large quantities of solid chemicals are handled and shipped in bulk, cloth bags, barrels, and barrels with paper liners.

The filling of bags or barrels with solids is always a potentially dusty operation and, if the material being handled is finely divided or dangerous, the health and fire hazards may be seri-

ous. The simplest solution to the problem is to moisten the material, so that it does not produce fine dust. This solution, however, cannot be used with some materials, and other methods of handling or using hoods and exhaust ventilation must be provided.

The common way to open both cloth and paper bags is to slit one side crosswise and fold back the top and bottom. The hazard of knife cuts can be reduced by keeping knives in scabbards when not in use and by using knives with hilts or guards.

The emptying of bags and barrels involves health hazards similar to those of filling. However, prevention of dust and skin contact is somewhat more difficult when such containers are emptied than when they are filled because of the tendency to dump them suddenly, with consequent rapid dispersion of dust that may not be trapped by the collecting systems.

Exhaust hoods that are larger than the containers and careful supervision can help solve this problem. In some instances, toxic materials can be handled by providing complete enclosures. The head of the barrel is broken in after the enclosure is sealed, and all dust is removed before the next barrel is put in.

Explosives must be stored in magazines of approved fireproof and bulletproof construction, located at a safe distance from railroads and other buildings. Federal, state, and local codes regarding storage of explosives must be consulted and followed closely. NFPA Standard 495 gives detailed specifications for handling and storage, as does the Institute of Makers of Explosives and the duPont *Blasters Handbook* (see References).

Explosives must be stored under lock and key, and records maintained of all explosives issued. Storage should be arranged so that the oldest explosives are used first.

No matches, flammable materials, metal, or metal tools should be brought into an explosives magazine. Floors must be kept clean and free from loose explosives. The floors, which are usually of wood, should be blind-nailed; no nail or bolt-head should be exposed.

Magazines should be clean, dry, and well ventilated. Ventilation openings should not exceed 110 sq in (710 cm²) in area and should be screened to prevent the entrance of sparks and rodents.

Only portable lights, approved for such use, are to be permitted in a magazine. Fire or sparks should not be allowed near a magazine, and the surrounding ground should be kept clear of brush, leaves, grass, debris, and other flammable material. Explosives should not be exposed to the direct rays of the sun.

Ammonium nitrate requires special precautions including stacking limitations, air space, and ventilation. No oils or hydrocarbons are to be permitted near ammonium nitrate.

Packages of explosives should always be opened at least 50 ft (15 m) from the magazine. Only wood wedges and wood, fiber, rawhide, zinc, babbitt metal, or rubber mallets should be used to open cases of explosives.

Blasting caps or detonators of any kind should never be kept in the same magazine with other explosives.

Explosives and blasting supplies should always be kept in the magazines in a place where access to them by animals, unauthorized persons, or children is impossible. Many children have been killed or crippled because they have obtained detonators from unwatched or unguarded sources.

STORAGE AND HANDLING OF CRYOGENIC LIQUIDS

Most gases used in plants are also available as cryogenic liquids. Among the most common are oxygen, nitrogen, argon, helium, and hydrogen.

Liquid oxygen is frequently delivered to a plant—and even to a construction site—and then vaporized for use in flame cutting, welding, metallizing, or heating. Other uses include oxygen injection into a foundry cupola and oxygen-based processes such as paperpulp bleaching and steelmaking.

Liquid nitrogen is also very common. A variety of processes have been developed that use the liquid primarily because of its high refrigeration values. Examples include freezing food, stripping scrap rubber from tires and cables, and removing parting lines and risers from plastic injection-molded parts. It is even used as a super-cold quencher for high-alloy steels to transform retained austenite.

The availability of large volumes of liquid helium has made possible the rapid development of superconductivity. And these examples are only a few from only some of the major industrial gases.

The key to expanding use of cryogenic liquids is economics. The cost of delivering and storing the liquid is often lower than buying the gas in compressed-gas cylinders. At room temperature (70 F or 20 C) and atmospheric pressure, nitrogen occupies 700 times as much space as the same amount of nitrogen in liquid form. The reduction in cost for containers, demurrage, shipping, and storage is enormous. (See Figure 3-21.) However, handling liquified gases that are stored and used at very low temperatures requires some special knowledge and special precautions. To use these gases safely, the plant engineer and his employees must know the specific properties of each liquified gas and its compatibility with other materials, and must follow some common-sense procedures.

Characteristics of cryogenic liquids

A cryogenic liquid has a normal boiling point below -238 F (-150 C). (The term is defined in depth in NBS Handbook 44. See References.) The most commonly used industrial gases that are transported, handled, and stored at cryogenic temperatures are oxygen, nitrogen, argon, hydrogen, and helium. Three rare atmospheric gases—neon, krypton, and xenon—are used in the liquid state. Natural gas, liquefied natural gas (LNG) or liquid methane, and carbon monoxide also are handled as cryogenic liquids, although they are not usually classified as industrial gases. Liquefied ethylene, carbon dioxide, and nitrous oxide are transported and stored as liquids, but are not classified as cryogenic.

Handling cryogenic liquids in large volumes is not new. Liquid oxygen was first shipped by tank truck in 1932, and today it is common to see portable liquid containers, cryogenic trailers and trucks, and railroad tank cars hauling large quantities of liquefied gases across the country. Cryogenic tanker ships transport LNG overseas, and aircraft move other liquefied gases, especially liquid helium, from one place to another.

Many safety precautions that must be taken with compressed gases (see Chapter 13, Welding and Cutting) also apply to

Figure 3-21. Liquified gases occupy much less space than their gaseous counterparts. One person can handle this container (using a handcart), but an entire crew would be needed to move the equivalent amount of compressed oxygen stored in cylinders.

liquefied gases. However, some additional precautions are necessary because of the special properties exhibited by fluids at cryogenic temperatures.

Both the liquid and its boil-off vapor can rapidly freeze human tissue and can cause many common materials such as carbon steel, plastic, and rubber to become brittle or fracture under stress. Liquids in containers and piping at temperatures at or below the boiling point of liquefied air (-318 F or -194 C) can cause the surrounding air to condense to a liquid.

Extremely cold liquefied gases (helium, hydrogen, and neon) can even solidify air or other gases to which they are directly exposed. In some cases, even plugs of ice or foreign material will develop in cryogenic container vents and openings and cause the vessel to rupture. Following the supplier's operating procedures can help prevent plugging. If a plug should form, contact the supplier immediately. Do not attempt to remove the plug; move the vessel to a remote location.

All cryogenic liquids produce large volumes of gas when they vaporize. For example, 1 volume of saturated liquid nitrogen at 1 atmosphere vaporizes to 696.5 volumes of nitrogen gas at room temperature at 1 atmosphere. The volume expansion ratio of oxygen is 860.6 to 1. Liquid neon has the highest expansion ratio—1445 to 1—of any industrial gas.

Vaporized in a sealed container, these liquids produce enormous pressures. For example, when 1 volume of liquid helium at 1 atmosphere is vaporized and warmed to room temperature in a totally enclosed container, it has the potential to generate pressure of more than 14,500 psig (100,000 kPa). Because of

Table 3-A. Safety Properties of Cryogenic Fluids

	Xenon (Xe)	Krypton (Kr)	Methane (CH$_4$)	Oxygen (O$_2$)	Argon (Ar)	Carbon Monoxide (CO)	Nitrogen (N$_2$)	Neon (Ne)	Hydrogen (H$_2$)	Helium (He)
Boiling point, 1 atm										
F	−163	−244	−259	−297	−303	−313	−321	−411	−423	−452
C	−108	−153	−161	−183	−186	−192	−196	−246	−253	−268
Melting point, 1 atm										
F	−169	−251	−296	−362	−309	−341	−346	−416	−435	−•
C	−112	−157	−182	−219	−189	−207	−210	−249	−259	−
Density, boiling point, 1 atm lb/cu ft	191	151	26	71	87	49	50	75	4.4	7.8
Heat of vaporization, boiling point Btu/lb	41	46	219	92	70	98	85	37	193	10
Volume expansion ratio, liquid at 1 atm boiling point to gas at 60 F, 1 atm	559	693	625	861	841	−	697	1445	850	754
Flammable	No	No	Yes	No †	No	Yes	No	No	Yes	No

• Helium does not solidify at 1 atmosphere pressure.

† Oxygen does not burn, but will support combustion. However, high oxygen atmospheres substantially increase combustion rates of other materials and may form explosive mixtures with other combustibles. Flame temperatures in oxygen are higher than in air.

this high pressure, cryogenic containers usually are protected with two pressure-relief devices: a pressure-relief valve and a frangible disk.

Relief devices should function only during abnormal operation and emergencies. If they are triggered, the system should be checked for loss of insulating vacuum or for leaks. Do not tamper with the safety valve settings. Report leaking or improperly set relief valves to the gas supplier and have them replaced or reset by qualified personnel. Similarly, all safety valves with broken seals or with any frost, ice formation, or excessive corrosion should be reported.

Most cryogenic liquids are odorless, colorless, and tasteless when vaporized to a gas. As liquids, most have no color; liquid oxygen is light blue. However, whenever the cold liquid and vapor are exposed to the atmosphere, a warning appears. As the cold boil-off gases condense moisture in the air, a fog that extends over an area larger than the vaporizing gas forms.

General safety practices

The properties of cryogenic liquids affect their safe handling and use (Table 3-A). The table presents data on flammability limits in air and oxygen, spontaneous ignition temperature in air at 1 atmosphere, and other information to help determine safe handling procedures. None of the gases listed is corrosive at ambient temperatures, and only carbon monoxide is toxic.

The liquids are listed by decreasing boiling point. Although xenon boils above -238 F (-150 C), it also has been included. Natural gas is not listed because it is a mixture of methane and

other hydrocarbons; its boiling point depends on its composition. However, natural gas is primarily methane and methane data are included.

- Always handle cryogenic liquids carefully. They can cause frostbite on skin and exposed eye tissue. When spilled, they tend to spread, covering a surface completely and cooling a large area. The vapors emitted by these liquids are also extremely cold and can damage delicate tissues.
- Stand clear of boiling or splashing liquid and its vapors. Boiling and splashing always occur when a warm container is charged or when warm objects are inserted into a liquid. These operations should always be performed slowly to minimize boiling and splashing. If cold liquid or vapor comes in contact with the skin or eyes, first aid should be given immediately. (See Figure 3-22, Treating Cold-Contact Burns.)
- Never allow any unprotected part of the body to touch uninsulated pipes or vessels that contain cryogenic fluids. The extremely cold metal will cause the flesh to stick fast to the surface and tear when withdrawn. Touching even nonmetallic materials at low temperatures is dangerous.

Tongs should be used to withdraw objects immersed in a cryogenic liquid. Objects that are soft and pliable at room temperature become hard and brittle at extremely low temperatures and will break easily.

Workers handling cryogenic liquids should use eye and hand protection to protect against splashing and cold-contact burns. Safety glasses are also recommended. If severe spraying or splashing is likely, a face shield or chemical goggles should be

---

TREATING COLD-CONTACT BURNS

Workers will rarely come in contact with a cryogenic liquid if proper handling procedures are used. In the unlikely event of contact with a liquid or cold gas, a cold-contact "burn" may occur. Actually, the skin or tissue freezes.

Medical assistance should be obtained as soon as possible. In the interim, the following emergency measures are recommended:

- Remove any clothing that may restrict circulation to the frozen area. Do not rub frozen parts, as tissue damage may result.

- As soon as is practical, immerse the affected part in warm water (not less than 105 F or more than 115 F, or 40 C to 46 C). Never use dry heat. The victim should be in a warm room, if possible.

- If the exposure has been massive and the general body temperature is depressed, the patient should be totally immersed in a warm-water bath. Supportive treatment for shock should be provided.

- Frozen tissues are painless and appear waxy and yellow. They will swell and be painful and prone to infection when thawed. Do not rewarm rapidly. Thawing may require 15 to 60 minutes and should continue until the pale blue tint of the skin turns pink or red. Morphine or tranquilizers may be required to control the pain during thawing and should be administered under professional medical supervision.

- If the frozen part of the body thaws before the doctor arrives, cover the area with dry sterile dressings and a large, bulky protective covering.

- Alcoholic beverages and smoking decrease blood flow to the frozen tissues and should be prohibited. Warm drinks and food may be administered.

Figure 3-22.

worn. Protective gloves should always be worn when anything that comes in contact with cold liquids and their vapors is being handled. Gloves should be loose fitting so that they can be removed quickly if liquids are spilled into them. Trousers should remain outside of boots or work shoes.

Special precautions

Some liquefied gases require special precautions. For example, when oxygen is handled, all combustible materials, especially oil or gases, should be kept away. Smoking or open flames should never be permitted where liquid oxygen is stored or handled. No SMOKING signs should be posted conspicuously in such areas.

Oxygen will vigorously accelerate and support combustion. Because the upper flammable limit for a flammable gas in air is higher in an oxygen-enriched air atmosphere, fire or explosion is possible over a wider range of gas mixtures.

Liquid oxygen or oxygen-rich air atmospheres should not come in contact with organic materials or flammable substances. Some organic materials—oil, grease, asphalt, kerosene, cloth, tar, or dirt containing oil or grease—react violently with oxygen and may be ignited by a hot spark. If liquid oxygen spills on asphalt or on another surface contaminated with combustibles (for example, oil-soaked concrete or gravel), no one should walk on, and no equipment should pass over, the area for at least 30 minutes after all frost or fog has disappeared.

Any clothing that has been splashed or soaked with liquid oxygen, or exposed to a high gaseous-oxygen atmosphere, should be changed immediately. The contaminated systems should be aired for at least an hour until they are completely free of excess oxygen. Workers exposed to high-oxygen atmospheres should leave the area and avoid all sources of ignition until the clothing and the exposed area have been completely ventilated. Clothing saturated with oxygen is readily ignitable and will burn vigorously.

Finally, oxygen valves should be operated slowly. Abruptly starting and stopping oxygen flow may ignite contaminants in the system.

Inert gas precautions

The primary hazards of inert gas systems are rupture of containers, pipelines, or systems, and asphyxiation. A cryogen cannot be indefinitely maintained as a liquid even in a well-insulated container. Any liquid, or even cold vapor trapped between valves, has the potential for causing enough pressure buildup to cause violent rupture of the container or piping. The use of reliable pressure-relief devices is mandatory.

Loss of vacuum in vacuum-jacketed tanks will increase evaporation in the system, causing the relief devices to function and vent the product. The vented gases should be routed to a safe outdoor location. If the gases are not vented outdoors, adequate ventilation must be maintained; instruments should be used to monitor the area.

Flammable gas precautions

Smoking or open flames should not be permitted where flammable fluids are stored or handled. Clothes that minimize ignition sources should be worn in atmospheres that may contain concentrations of flammable gases.

All major stationary equipment should be properly grounded. Ground connections should be provided between stationary and mobile equipment before any flammable gas is loaded or unloaded. All electrical equipment used in or near flammable-gas loading and unloading areas, or in atmospheres that might contain explosive mixtures, should conform to National Fire Protection Association (NFPA) standards 50B, *Liquefied Hydrogen Systems at Consumer Sites,* and 59A, *Production, Storage, and Handling of Liquefied Natural Gas,* or to Article 500 of the *National Electrical Code (NEC),* NFPA 70. When flammable cryogenic liquids and gases are handled inside, adequate positive mechanical ventilation is necessary. Electrical equipment and wiring must conform to Article 501 of the NEC. (See details in Chapter 15, Electricity.

Flash-off gas from closed liquid-hydrogen containers used or stored inside should be piped through a laboratory hood to the outside, or should be vented by other means to a safe location. If hydrogen is vented into duct-work, the ventilation system should be independent of other systems, and sources of ignition must be eliminated at the exit.

Asphyxiation

All gases except oxygen will cause asphyxiation by displacing breathable air in an enclosed workplace. They should be used and stored in well-ventilated areas. Only oxygen will support life. The presence of these gases cannot be detected without instrumentation. Asphyxiation can be sudden or may occur slowly without the worker being aware that he is in trouble. (Refer to Safety Bulletin SB-2 from the Compressed Gas Association for additional information.)

Unless large quantities of inert gas are present, the problem is easily prevented by using proper ventilation at all times. Nitrogen should be vented outside to safe areas. Analyzers with alarms should be installed to alert workers to oxygen-deficient atmospheres. Constant monitoring, sniffers, and other precautions should be used to survey the atmosphere when personnel enter enclosed areas or vessels. When it is necessary to enter an area where the oxygen content may be below 19 percent, self-contained breathing apparatus or a hose mask connected to a breathing-air source must be used. A conventional gas mask will not prevent asphyxiation.

Most personnel working in or around oxygen-deficient atmospheres rely on the buddy system for protection. But, unless equipped with a portable air supply, a coworker may also be asphyxiated when entering the area to rescue an unconscious partner. The best protection is to provide both workers with a portable supply of respirable air. Lifelines are acceptable only if the area is free of obstructions and one worker is capable of lifting the other rapidly and easily.

Training

The best single investment in safety is trained personnel. Some workers will need detailed training in a particular type of equipment, cryogen, or repair operation. Others will require broader training in safe handling practices for a variety of cryogenic liquids. The following subjects should be familiar to everyone involved in using, handling, storing, or transferring cryogens:

- Nature and properties of the cryogen in both its liquid and gaseous states
- Operation of the equipment
- Approved, compatible materials
- Use and care of protective equipment and clothing
- First aid and self-aid techniques to employ when medical treatment is not immediately available
- Handling emergency situations such as fires, leaks, spills
- Good housekeeping practices.

Good housekeeping is essential to safety. Few cryogens are spontaneously hazardous, but each liquefied gas poses another hazard.

Liquid oxygen may form mixtures that are shock sensitive with fuels, oils, or grease. Porous solids, such as asphalt or wood, can become saturated with oxygen and also become shock sensitive. Ignition is more likely with weaker sparks and lower temperatures than would be required in air.

Flammable gases such as hydrogen and methane are lighter than air. At normal temperatures, they will rise. But at the first temperatures that exist just after evaporation from the liquid state, the saturated vapor is heavier than air and tends to fall. Wind or forced ventilation will affect the direction of the released gases and must be considered during disposal of any leaking fluid.

The location and maintenance of safety and firefighting equipment are important. Outside personnel also should be informed of all necessary safeguards before entering a potentially hazardous area. In general, good housekeeping rules and demanding a high level of worker conduct everywhere in the plant will minimize negligence.

Safe storage recommendations

Cryogenic liquids are stored and transported in a wide range of containers from small Dewar flasks to railroad tank cars and tank trucks (Figures 3-23 and -24). Only equipment and containers designed for the intended product, service pressure, and temperature should be used. If any questions arise about correct handling or transport procedures, or about the compatibility of materials with a given cryogen, the gas supplier should be consulted.

Cryogenic liquids ordinarily should not be handled in open containers unless they are specifically designed for that purpose and for the product. Cryogenic containers should be clean and made from materials, such as austenitic stainless steels, copper, and certain aluminum alloys, suitable for cryogenic temperatures.

Cryogens should be transferred into warm lines or containers slowly to prevent thermal shock to the piping and container and to eliminate possible excessive pressure buildup in the system. When liquids are transferred from one container to another, the receiving container should be cooled gradually to prevent shock and reduce flashing. High concentrations of escaping gases should be vented so that they do not collect in an enclosed area.

- *Do not drop warm solids or liquids into cryogenic liquids.* Violent boiling will result and liquid can splash onto personnel and equipment.
- *Avoid breathing vapor from any cryogenic liquid source* except for liquid-oxygen equipment designed to supply warm breathing oxygen. When cryogenic liquids are being discharged from drain valves or blowdown lines, open the valves slowly to prevent splashing. Smoking should never be permitted.

Two types of portable liquid-storage vessels are generally used to hold and dispense cryogenic liquids—nonpressurized Dewar containers and pressurized liquid cylinders (Figures 3-23 and -24).

Dewar containers for liquids are open-mouthed, nonpressurized, vacuum-jacketed vessels usually used to hold liquid argon, nitrogen, oxygen, or helium. Some of these containers are designed for lightweight liquids such as helium and for maximum holding times; their internal support system cannot hold some of the heavier cryogens, such as argon. When they are

Figure 3-23. Cryogenic liquids are stored and transported in a wide range of containers. A typical example is this flask of liquid helium, an open-mouthed, unpressured, vacuum-jacketed vessel.

used, be sure that no ice accumulates in the neck or on the cover and causes a blockage and subsequent pressure buildup. Laboratory Dewar flasks with wide-mouthed openings have no cover to protect the liquid. Most are made of metal, but some smaller units are of glass.

Liquid cylinders are pressurized containers, usually vertical vessels, designed and fabricated according to U.S. Department of Transportation specifications. There are three major types: for dispensing liquid or gas, for gas withdrawal only, and for liquid withdrawal only. Each type of liquid cylinder has appropriate valves for filling and dispensing and is adequately protected with a pressure-control valve and a frangible disk.

Some liquid cylinders can be handled manually, but it is preferable to move them using portable handcarts. A strap should be used to secure the cylinder to the handcart and keep it from slipping off.

An unusually cold outside jacket on a cryogenic vessel indicates some loss of insulating vacuum. Frost spots may appear. A vessel in this condition should be drained, removed from service, and set aside for repair. Such repairs should be handled by the manufacturer or qualified company.

Cryogenic containers must be handled very carefully. They should not be dropped or tipped on their sides.

Transfer lines

Many types of filling or transfer lines are used to handle the flow of cryogenic fluids from one point to another—small, uninsulated copper or stainless steel lines; large-diameter rigid lines; or flexible hose systems, vacuum-jacketed lines or other insulated systems.

Liquid can be transferred by three methods—the simplest is gravity. In this case, the height of the stored liquid serves as the transfer medium. Pressurized transfer uses the vapor pressure of the product, or pressure from an external source, to move the liquid to the lower-pressure receiving container.

Various types of cryogenic pumps are also available. Flow rates may vary from less than one to several hundred gallons per minute. The product should be in liquid form in the transfer lines. Any vaporization of liquid within the system may cause excessive pressure drop and two-phase (liquid and gas) flow and cavitation that is detrimental to the operation of cryogenic pumps.

Short transfer lines used for intermittent service are normally not insulated, but lines used for continuous transfer of cryogens usually are. All liquid transfer hoses should have dust caps.

Vacuum-jacketed lines are required to transfer liquid hydrogen and liquid helium because of their extremely cold temperatures and low heats of vaporization. Vacuum-jacketed lines occasionally are used for in-plant transfer of atmospheric cryogenic fluids to reduce costly line and flash-off losses.

SHIPPING AND RECEIVING

The supervisor of the shipping and receiving room must be aware of Department of Transportation (DOT) regulations and labels. Hazardous or flammable materials must be properly labeled and identified. Bills of lading or shipping must identify the item.

Figure 3-24. Large volumes of cryogenic liquids can be handled easily and safely. Here, more than 50,000 gallons of liquid helium at -452 F are lifted by container carrier for loading aboard ship. Loss of helium from vaporization during a two-week voyage is so small that it is nearly undetectable. The liquid helium is surrounded by a liquid nitrogen chamber and contained in a specially insulated outer jacket.

Floors, ramps, and aisles

Floors in warehouses, storerooms, and shipping rooms must be level. Unevenness of floors may lead to the toppling of piles of stored materials.

Safe floor load capacities and maximum heights to which specific materials may be piled should be posted conspicuously.

Where bulk material, boxes, or cartons of the same weight are regularly stored, it is good practice to paint a horizontal line on the wall indicating the maximum height to which the material may be piled.

The strength of floors should be checked before the use of power trucks is adopted. Floor load capacity should be determined by a structural expert from architectural data, the age and condition of the floor members, the type of floor, and other pertinent factors.

Wherever materials are stored or transported, the surface of floors, platforms, and ramps should be kept in good condition. Damaged structures should be repaired immediately; the area around doorways and elevator entrances should particularly be watched.

Ramps should have nonskid surfaces. When ramps are used for hand trucking, a nonskid foot strip may be laid in the center, or in the center of each lane for two-way traffic. Ramps should have handrails and, where there is heavy trucking, substantial curbs. A separate pedestrian lane, divided from the truck lane by a handrail, is a good idea for ramps used by both pedestrians and trucks.

Good housekeeping contributes to safety in manual or mechanical transport of material. A fall by a person carrying an object might result in more serious injury than it would if his hands were free.

Mirrors, placed at blind intersections, help to prevent collisions (Figure 3-25). Warning signs and signals at such locations also serve as useful reminders, particularly to operators of power equipment. Doorways and entrances to tunnels and elevators may be similarly protected.

Mobile equipment used in storage areas should be equipped with backup warning devices.

Aisles should be wide enough to enable employees to move about freely while handling material or removing it from bins, racks, or piles, and to allow safe passage of loaded equipment. Aisleways and unloading areas should be clearly marked with white paint or black and white stripes. (See American National Standard Z53.1.) Trucks not in use, material, and other objects should not be allowed to stand in or extend into aisles.

Aisles leading to sprinkler valves and fire extinguishing equipment should be kept clear. Materials should not be piled closer than 18 in. to sprinkler heads. Closer spacing may reduce the effectiveness of the heads in the event of fire. For overly large, closely packed piles of combustible cases, bales, cartons, and similar stock, up to 36 in. of clearance should be provided. (See *Installation of Sprinkler Systems,* NFPA 13.)

Lighting

General illumination of warehouses and storage rooms should follow American National Standard *Practice for Industrial Lighting,* ANSI/IES RP7, published by the Illuminating Engineering Society. (See Table 19G in Chapter 19. Also see the *IES Handbook.*)

Special lighting should be provided for operations requiring greater illumination. All lighting fixtures and wiring should meet the requirements of the *National Electrical Code,* NFPA 70.

Stock picking

The movement of full pallet loads is most readily accomplished with the modern powered industrial truck. However, many industrial operations today require sequence picking of componenet parts for final assembly and fabrication.

With the variety of smaller finished products manufactured in industrial plants, shipments are typically made up by truck and/or rail carloads of mixed lots.

In both operations, such stock is usually found in racks or bins and, with the increase in high bay storage, special order picking equipment is required. Such operations lend new efficiency

Figure 3-25. Blind corner hazards can be minimized by using convex mirrors that increase vision around obstructions.

to material movement as well as new hazards in the areas of traffic, personal injury, and fire protection. The worker operating from a mobile order-picking truck is exposed to falls from a height as well as falling objects and material handling accidents.

Often, a worker is required to climb a ladder to get small parts or stock. Only heavy-duty material handling ladders should be used. These may be on rollers, equipped with a braking mechanism with rubber feet that contact the surface as the worker's weight is imposed on the ladder. A working platform is provided, and standard guardrails protect the worker from falls as he reaches for stock or parts.

Under no circumstances should employees be allowed to use ordinary stepladders (particularly the short 2- or 3-step stools). Heavy-duty material handling equipment is required. Employees must never climb racks or shelves.

Dockboards (bridge plates)

Dockboards, used in trailer and rail car loading and unloading, should be designed to carry four times the heaviest expected load and be wide enough to permit easy maneuvering of hand or power trucks. Dockboards are also known as bridge plates, dock plates, gangplanks, and bridge ramps. (Information given here applies to both hand and power truck operations.)

Many modern facilities today use automatic dock levelers and fixed-position hydraulic dockboards on both truck and rail docks. Dock shelters, usually found in inclement weather zones, effectively keep moisture from the dockboard and give the shipping and receiving department personnel a safer working environment.

These units require regularly scheduled maintenance. Most operate hydraulically and will provide a solid working and walking surface for heavy industrial trucks as well as hand truck operations. The design must be considered. All shear points must be guarded. The edges of movable sections should be painted yellow to denote a possible tripping hazard.

Trailers being loaded or unloaded by trucks must have wheel chocks at each wheel, and the nose provided with a special jack, when the equipment operates in the forward portion of the trailer.

Dockboards should be designed and maintained so that when they are in a secure position the ends will have substantial contact with the dock (or loading platform) and the carrier to prevent the boards from rocking or sliding when they are being used. The sides of dockboards should be turned up at right angles, or otherwise designed, to prevent trucks from running over the edge.

Dockboards should have a nonskid surface to prevent employees and trucks from slipping. They should be kept clean and free of oil, grease, water, ice, and snow.

Handholds, or similar devices, should be provided on dockboards to facilitate safe handling. Where practicable, fork loops or lugs should be fitted to the plates, so that they can be handled by fork trucks (Figure 3-26). Another method for lifting steel dockboards uses a low-voltage magnet, which is hung from the forks of a forklift truck and powered by the truck's battery.

When dockboards are handled manually, they should be lowered or slipped into place and not dropped. Enough workers should be assigned to the job to permit safe and easy handling. Extra care should be taken to prevent foot injury.

Dockboards must be secured in position. They should be either anchored down or equipped with devices to prevent slippage from the platform or the car threshold.

When not in use, dockboards should be stored in a safe place provided for that purpose. In some cases, heavy boards may be stacked horizontally by power trucks.

Protective devices should be used to prevent engines or car pullers from moving railroad cars while dockboards are in position and workers are on them. Standard blue flags for daytime and blue lights for night work should be used to warn train crews. Local railroad authorities should be consulted on the specific warning device.

A regular program of dockboard inspection and maintenance should provide for necessary repairs or replacement.

Machines and tools

Machines that may be used in receiving and shipping, such as shears, saws, and nailing machines, should be guarded. Workers should wear the necessary protective clothing—goggles, for instance—when operating a nailing or banding machine.

Tools should be of high quality and kept in good condition. Edged tools should be sharp, and holsters provided for workers to carry them. Files should be provided that have a handle on the tang.

When the worker uses a drawknife instead of a scraper to remove markings from cases, boxes, and barrels, he should never brace the work with his knees. Should the drawknife slip, injury would almost certainly occur.

Steel and plastic strapping and sacking

Steel and plastic strapping, which may be flat or round, require that the worker be trained in both application and removal. In all cases, the worker should wear safety goggles and leather-palm gloves (Figure 3-27). Steel-studded gloves may be required for heavy strap.

Equipment designed for applying and removing strapping should be used. When operating a strapping tool, the worker should face in the direction of pull, one foot ahead of the other. Then, if the strap breaks or the tool slips, the worker is in a position for self-protection.

When final tension is being attained, the operator should get out of direct line of the strap so that, in case of breakage, ends of strap will not strike the face or body. Excess strap beyond the tension-holding seal should be broken or cut off before a bound shipment is considered safe for further handling or shipment.

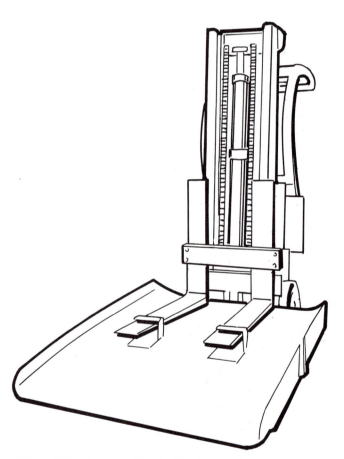

Figure 3-26. Lugs are fitted to this dockboard so it can be transported by a forklift. Most dockboards are designed to properly fit a particular use.

Before attempting to move bound merchandise or material, the worker should examine it for broken bands or loose ends. Broken bands should be removed and safely disposed of and, if possible, replaced to keep the shipment from coming apart.

A box, carton, or package should never be handled by the steel bands, either manually or with a lift truck or other mechanical handling device. Stored packages, boxes, cartons, or other

Figure 3-27. Note gloves and eye protection used by packers of electronic crossbar frame. Packers stand slightly to one side, so that they are out of direct line of the strap in case it breaks. Spectacles with sideshields or goggles are preferred protection. (Printed with permission from Western Electric Co., Omaha Works.)

material should be checked for loose or protruding banding ends that might cut or otherwise injure passersby.

To remove strapping from bound containers, a cutter designed for the work should be used. Workers should never break the steel strap by applying leverage with a claw hammer, chisel, crowbar, or other tool. Before cutting strapping, workers should make sure that no one is standing close enough to be hit by loose ends of strap.

To cut bands safely, the worker should place one gloved hand on the nearest portion of the strap. Then, if the strap springs, it will be held to one side and fly away from the worker's face. In addition, the face should be held out of direct line of the strap. Eye protection should be worn.

Strap should be cut square, never at an angle. Strap cut at an angle has much sharper ends. Containers should be available to dispose of plastic and steel bands separately.

Burlap and sacking are often received baled. Opening these bales is a job requiring some skill and experience. The supervisor of the department should thoroughly instruct employees in the exact procedure to be used. Burred ends of wire used to tie sacks may cause many cuts. The employee should hold down one end of the wire when making the cut, and should stand clear of the free end.

Glass and nails

Broken glass, often found in containers being unpacked, is a serious hazard in the shipping room. When unpacking glass or crockery, the employee should assume that broken material may be present. If possible, gloves should be worn.

Companies operating large shipping rooms report many injuries from flying nails. When driving a nail, the worker should start it with a few light taps, so that it will take a good hold and will not fly. Eye and face protection should be worn.

Workers should be instructed to pull out nails that have been started at the wrong angle and drive them in properly. Nails should be driven flush so that no part of the nail projects above the surface. Not only fellow workers, but also employees of the customer or carrier, are likely to be injured by projecting nails as they handle the cases. Poorly driven nails may cause merchandise loss if the package should come apart while being shipped.

Loose nails should not be left on the floors. Loaded trucks passing over the nails may drive them into the floor with the points up. In this position, they are a serious hazard since they can easily be driven through workers' shoes and can also puncture pneumatic tires. The best practice is to have nails picked up from the floor at regular intervals during the day.

Employees who open packing cases should be instructed to bend nails over or remove them from the boards. If the box wood is to be used to make other or smaller boxes, removal of nails is important.

More often than not, crating lumber is discarded. Where quantities of such lumber are found, special carts should be provided for prompt removal.

If nails are used directly from kegs, supervisors should make certain that the nails holding the keg head in place are pulled out. Kegs used for storage of nails at the work station should be placed in an inclined rack so that the nails will feed out of the kegs.

Pitch and glue

To protect export shipments, parts are wrapped first with paper, plastic, or cheesecloth and often covered with burlap. Pitch or other material is used for sealing. Hot pitch, however, can severely burn the skin not only because it is hot, but also because it is difficult to remove.

This work should be given to skilled personnel who have been instructed in the hazards and know how to avoid them. Goggles, face masks, gloves with sleeves rolled over the gauntlets, and aprons should be worn. Wherever possible, it is better to use cold mastic asphalt instead of hot pitch.

Labeling glue, which often contains silicate of soda, causes discomfort when it splashes into the eyes. Workers should wear eye protection and use the glue brush carefully.

Barrels and kegs

Projecting nails, jagged hoops and metal bands, ends of wire, splinters and slivers cause many barrel-handling injuries, some of which lead to infection. Before handling barrels and kegs, employees should inspect them and take precautions against these hazards.

One method of opening a barrel is to use a lather's hatchet or a crate-opening tool to remove the nails. Then, when the top hoop is removed and the second hoop loosened, the head can easily be removed intact. Loosening nails on a barrel with wood hoops is simple if the hoop is struck sharply with a hammer or hatchet near the point where the nail is inserted. The nail can then easily be pulled. This method of opening barrels not only preserves the barrel for future use, but also prevents contamination.

The opening of single-trip drums with a hammer and chisel is a frequent source of cuts and scratches. A commercial drum opener will open these drums without hazard and leave a smooth rolled edge. Drum handling, storage, and opening are discussed above in the section on Problems with Hazardous Materials.

Boxes and cartons

Employees who open boxes or cartons may incur wire punctures, nail punctures, or cuts from the device used for opening pasteboard cartons. When wirebound boxes or nailed boxes are opened, wires should be bent back and nails should be turned over or removed. Eye protection should be worn.

The boxes and covers should be piled neatly out of passageways. Safety carton openers, made of a protected sharpened blade, are available. The blade is slid along the edge of the carton.

These tools are useful only, of course, where the carton will not be reused. Cartons that are to be reused should be pried open with a flat steel pry bar, so that the flaps will not be damaged.

Corners of boxes and crates receive more blows than other parts and, therefore, should be constructed strongly. The interlocking-corner crate (Figure 3-28) is stronger than any other type, requires less lumber, and should be used wherever possible.

Figure 3-28. The interlocking corner method (right) of crate construction is stronger than the other method (left).

Diagonal braces help greatly in making crate sections rigid. One diagonal brace in the section will give more stiffening than several parallel slats. The diagonal should extend from corner to corner and should be placed so that it does not project beyond the other members of the crate.

Skids constructed as an intergral part of a box should be made of sound lumber, free from knots, and of sufficient size to support the box without breaking. The skids should be firmly attached to withstand dragging across the floor.

The principal reason for failure of boxes and crates is poor nailing. Correct nailing is essential to safe shipment. Where trouble is experienced with the nails pulling out, cement-coated nails, which hold better than uncoated nails, should be used.

Broken or damaged containers of consumables. Handlers should not attempt to sample or distribute food or other commodities that are available when containers are damaged or broken. These commodities could have become contaminated, tainted, or otherwise unsafe while en route. In one documented case that occurred in Columbia, South America, contents of insecticide and food bags became mixed during rough shipment

and the consequent use of the contaminated flour caused the deaths of many people. Careless handling of samples could also endanger health.

Car loading

Heavy machinery shipped on skids should be braced inside the car to prevent shifting. Lag screws should not be used to fasten a skid to the car floor, because a worker may drive them in with a hammer rather than use a wrench. Using a hammer damages the wood, thus reducing the holding power of the lag screws.

Skids with large knots are hazardous when used on shipments of heavy machinery. When rollers are used to move the object, the skid is likely to break when the roller comes under a knot. (Refer to Association of American Railroads Pamphlet No. 21 in References.)

Before rail cars are opened, the doors should be carefully inspected. If damaged, runners should be repaired and special precautions taken. Door openers should be used and employees should stand clear, in case an improperly loaded car is received.

A lift truck's fork should not be used to open doors because the angle of the fork can lift the door off its track and risk injury to employees and damage to equipment.

Workers who are opening and closing car doors may catch their hands between the doors and the car doorposts. Workers should be instructed never to grasp the leading end of the door, which might cause their fingers to catch between the car door and the side of the car. Likewise, they should keep their hands and fingers away from the doorpost when they are closing the door.

To avoid leaving hazards for railroad or other employees, as well as to avoid contamination of future lading or damage to it, consignees should clean cars after they have been unloaded.

PERSONAL PROTECTION

Certain items of protective equipment are desirable for the prevention of various types of materials handling injuries. Since toe and finger injuries are among the most common types, handlers should wear safety shoes and stout gloves, preferably with leather faces. Other special protective clothing, such as eye protection, aprons, and leggings, should be required for the handling of certain types of materials.

Gloves should be dry and free of grease and oil. Hand protection should be worn when wooden crates are handled to prevent injury from splinters. Clean leather-palm gloves give better holding power on smooth metal objects than do cotton or other types. However, it may be unsafe for workers to wear gloves near conveyors, or whenever the risk of catching exists. Care should be taken not to bruise or squeeze the hands at doorways or other points where clearance is close. During inclement weather, special weather apparel may be necessary.

Where toxic or irritating solids are handled, workers should take daily showers to remove the material from their persons before they leave the plant. Even though the exposure does not necessitate showers, workers should be encouraged to wash thoroughly at the end of their shifts. Cleansing materials, shower stalls, and wash basins should be provided.

Washable suits of tightly woven fabric, preferably full-length coveralls, and washable caps should be worn. Suits, caps, socks and underwear should be laundered daily at the plant. Clothing should be laundered less frequently only at the express direction of the plant medical department.

Skin contact with chemicals should be avoided.

REFERENCES

American Petroleum Institute, 1801 K St., N.W., Washington, D.C. 20006. *Cleaning Petroleum Storage Tanks*, RP 2015.

American National Standards Institute, 1430 Broadway, New York, N.Y. 10018.
Gray Finishes for Industrial Apparatus and Equipment, Z55.1.
Practice for Industrial Lighting, ANSI/IES RP7.
Precautionary Labeling of Hazardous Industrial Chemicals, Z129.1.
Mobile and Locomotive Truck Cranes, ANSI/ASME B30.5.
Safety Requirements for Fixed Ladders, A14.3.
Safety Standards for Powered Industrial Trucks, B56.1-7.
Safety Color Code for Marking Physical Hazards, Z53.1.
Scheme for the Identification of Piping Systems, A13.1.

Association of American Railroads, Operating Transportation Div., 1920 L St., N.W., Washington, D.C. 20006. *Recommended Methods for Loading, Bracing and Blocking Carload Shipments of Machinery in Closed Cars*, Pamphlet No. 21.

Braver-Mann, S., *A Comparison of the Dynamic Balanced One Hand Lift and the Two Hand Stoop Lift*, Unpublished Master's Thesis, University of Iowa, Iowa City, Ia., 1985.

Chemical Manufacturers Association, 1825 Connecticut Ave., N.W., Washington, D.C. 20009 (General)

Compressed Gas Association, Inc., 500 Fifth Ave., New York, N.Y. 10036. *Oxygen-Deficient Atmospheres*, Bul. SB-2.

E. I. du Pont de Nemours & Co. (Inc.), Explosives Dept., Wilmington, Del. 19898. *Blaster's Handbook*.

Illuminating Engineering Society, 345 East 47th St., New York, N.Y. 10017. *IES Handbook*.

Institute of Makers of Explosives, 420 Lexington Ave., New York, N.Y. 10017. *Safety in the Handling and Use of Explosives*.

Kroemer, K. H. E. *Material Handling: Loss Control Through Ergonomics*. Alliance of American Insurers, 2nd Edition, Schaumburg, Ill. 60172. 1983.

Lovested, G. E., "Materials Handling Safety in Industry," *Materials Handling Handbook*, John Wiley & Sons, New York, N.Y., 1985. "The One-Hand Lift as a Lifting Method Alternative," *National Safety News*, Chicago, Ill., June 1981.

National Bureau of Standards, Washington, D.C. 20234. *Specifications, Tolerances, and Regulations for Commercial Weighing and Measuring Devices*, Handbook 44.

National Fire Protection Association, Batterymarch Park, Quincy, Mass. 02269.
Cleaning or Safeguarding Small Tanks and Containers, 327.
Manufacture, Transportation, Storage, and Use of Explosive Materials, 495.
Explosion Venting Guide, 68.
Flammable and Combustible Liquids Code, 30.
Installation of Sprinkler Systems, 13.
Liquefied Hydrogen Systems at Consumer Sites, 50B.
National Electrical Code, 70.
National Fire Codes, Vol. 4, "Flammable Liquids and Gases."
Powered Industrial Trucks, NFPA 505.
Production, Storage, and Handling of Liquefied Natural Gas, 59A.
Rack Storage of Materials, 231C.
Series on Prevention of Fires and Dust Explosions, 61A–D.

National Insititute of Occupational Safety and Health. *Work Practices Guide for Manual Lifting*. DHEW (NIOSH) Publication 81-122, Cincinnati, Ohio, 1981.

National Safety Council, 444 North Michigan Ave., Chicago, Ill. 60611.
Industrial Data Sheets
Automotive Hoisting Equipment, 437.
Chains, (Alloy Steel) for Overhead Lifting, 478.
Confined Space Entry Control System,—R & D Operations, 704.
Conveyors, Belt (Equipment), 569.
Conveyors, Belt (Operations), 570.
Conveyors, Roller, 528.
Conveyors, Underground Belt, 725.
Cranes, Pendant-Operated and Radio-Controlled, 558A.
Dock Plates and Gangplanks, 318.
Electromagnetics Used with Crane Hoists, 359.
Emergency Showers & Eyewash Fountains, 686.
Entry into Grain Bins and Food Tanks, 663.
Flammable and Combustible Liquids in Small Containers, 532.
Forging Industry, Handling Materials in, 551.
Freight Cars, Unloading Bulk Grain from, 521.
Front-End Loaders, 589.
Fusees and Torpedoes—Used in Railroad Operations, Handling and Storage of, 639.
Hand Trucks, Powered, 317.
Handling Bottles and Glassware in Food Processing and Food Service, 355.
Hoisting Equipment, Automotive, 437.
Hoists, Construction Material, 511.
Liquefied Petroleum Gases for Industrial Trucks, 479.
Load-Haul-Dump Machines in Underground Mines, 576.
Lumber Handling and Piling, 345A.
Motor Trucks for Mines, Quarries, and Construction, 330.
Oil Field Pipe, Handling Large-Diameter, 463A.
Pipes & Fittings—Water Transportation, Handling & Storage, 623.
Power Presses, Removing Piece Parts from Dies in Mechanical, 534.
Powered Industrial Trucks, 653.
Pulpwood, Unloading at the Mill, 274.
Railroad Flatcars, Loading & Unloading of Trailers on, 672.
Scrap Ballers, 611.
Sheet Metal, Handling and Storage of, 434.
Steel Plate, Handling for Fabrication, 565.
Stores, Fire Prevention in, 549.
Sulfur, Handling and Storage of Solid, 612.
Sulfur, Handling Liquid, 592.
Tractor Operation and Roll-over Protective Structures, 622.
Winches, Truck Mounted Power, 441.
Wire Rope Slings, Recommended Loads for, 380.

Oresick, Andrew. "Safety Techniques in Glass Handling," *ASSE Journal*, May 1973, pp. 22-29.

"Pre-employment Strength Testing," DHEW (NIOSH) Publication No. 77-163, Cincinnati, Ohio, National Institute for Occupational Safety and Health, Div. of Technical Services, 1977.

"R. M. Graziano's Tariff," Bureau of Explosives, 1920 L St., N.W., Washington, D.C. 20036. *Hazardous Materials Regulations of the Department of Transportation.*

Szymanski, Edward. "Safe Storage and Handling of Cryogenic Liquids," *Plant Engineering* Magazine (June 14, 1979).

U.S. Department of Transportation, Office of Hazardous Materials, Washington, D.C. 20590. "An Index to the Hazardous Materials Regulations." Available from Superintendent of Documents, U.S. Government Printing Office, Washington, D.C. 20402.

U.S. Department of the Treasury, Internal Revenue Service, Washington, D.C. *Published Ordinances: Explosives — State Laws and Local Ordinances Relevant to Title 18 U.S.C., Chapter 40.* Publication 740.

Hoisting Apparatus and Conveyors

HOISTING APPARATUS

To raise, lower, and transport heavy loads for limited distances, hoisting apparatus has been in use for centuries. Many thousands of hoists (electric, air, and hand powered) are in use in industry. Typically, they range from ¼ to 10 tons (2200 to 88,000 newtons) in capacity, but greater capacities are not unusual. Today there are traveling cranes in steel mills, power plants and naval shipyards capable of lifting hundreds of tons. (See also the discussion of material and passenger hoists in Chapter 2, Construction and Maintenance of Plant Facilities.)

The safe load capacity of each hoist shall be shown in conspicuous figures on the hoist body of the machine. In addition, all hoists must have affixed to the hoist, hook block, or controls, a label or labels in a readable position covering safe operating procedures.

All hoists should be securely attached to their supports (fixed member or trolley) with shackles. Hoists can be either rigid suspended or hook suspended. If the hoist is hook suspended, the support hooks should be moused or have hook latches. Latches are recommended also for load hooks. Hoist supports should also have an adequate design factor for the maximum loads to be imposed. Overhead hoists operating on rails, tracks, or trolleys should have positive stops or limiting devices on the equipment, rails, tracks, or trolleys to prevent the overrunning of safe limits. Flanges on hoist drums with single-layer spiral grooves should be free of projections that could damage a rope.

Figure 4-1. Company safety rule states: "Be watchful of loads suspended in air; keep out from under them." Fortunately everyone was in the clear when this 150-lb (70 kg) crane hook failed. (Printed with permission from Colorado Fuel & Iron Corp.)

A load should be picked up only when it is directly under the hoist. Otherwise, stresses for which the hoist was not designed may be imposed upon it. If the load is not properly centered, it may swing (upon being hoisted), and injury could result. Everyone must stay out from under raised loads (Figure 4-1). When operating an overhead hoist, care must be taken to avoid injury to the operator or any other worker nearby.

Caution: Hoists or cranes must not be used to lift, support, or otherwise transport people. The standard commercial hoist or crane does not provide a secondary means of supporting the load should the wire rope or other suspension element fail. (See section on aerial baskets later in this chapter.)

Hoists should be examined for evidence of wear, malfunction, damage, and proper operation of devices such as load hooks, ropes, brakes, clutches, and limit switches. Deficiencies shall be carefully examined and, if determined hazardous, corrected immediately (Figure 4-2).

Electric hoists

Rope-operated electric hoists should have nonconducting control cords unless they are grounded. Control cords should have handles of distinctly different contours so that, even without looking, the operator will know which is the hoisting and which is the lowering handle.

Figure 4-2. To check the rope of this 1000-lb (4450 N) electric hoist, inspect 2- to 3-ft (61 to 91 cm) increments after the rope has come to complete stop. To check for frays, wipe a rag up and down rope. (Printed with permission from Acco Hoist and Crane Division.)

Each control handle should be clearly marked "hoist" or "lower." Some companies attach an arrow to each control cord, pointing in the direction in which the load will move when the rope is pulled. Also, it may be advisable to pass the control cords through the spreader to keep them from becoming tangled. The spreader can be a 1- by 3-in. (2.5 cm by 7.5 cm) board or other nonconductive material with equally spaced holes, resting upon the pull handles or lowest position. Control cords, usually made of fiber or light wire rope, should be inspected periodically for wear and other defects.

Means for effecting automatic return to the OFF position should be provided on the control, so that a constant pull on the control rope or push on the control button must be maintained to raise or lower the load (Figure 4-3). The pendent station must be supported in a manner that will protect the electrical conductors against strain. The station must also be grounded in case a ground fault occurs. Pushbutton control circuits must be limited to 150 Vac and 300 Vdc.

A limit switch should be installed on the hoist motion. The minimum of two turns of rope shall remain on the drum when the load block is on the floor, except when a geared lower limit switch is used. One turn is permitted. If a load block can enter a pit or hatchway in the floor, the installation of a lower travel limit-switch is recommended and if that switch trips, the turns of the rope remaining on the drum can be reduced to one when the block has been stopped.

Air hoists

After a piston air hoist has been in operation for a time, the locknut that holds the piston on its rod may become loose so that the rod will pull out of the piston, allowing the load to drop. It is recommended that the locknut be secured to the piston rod by a castellated nut and cotter pin. Whenever an air hoist is overhauled, a check should be made to see that the piston is well secured to the rod.

On a cylinder load balancer or hoist, if an ordinary hook is used to hang the hoist from its support, the cylinder may come unhooked if the piston rod comes in contact with an obstruction when lowering. A clevis or other device should be used to prevent the hook from being detached from the hoist support.

To prevent the hoist from rising or lowering too rapidly, a choke (available from the hoist supplier) can be placed in the air line coupling.

It is recommended that a rope drum air hoist be provided with a closing loadline guide.

Hand-operated chain hoists

Chain hoists may be portable, but can also be either permanently hooked onto a monorail trolley, or built into the trolley as an integral part. They are suitable for many operations on which a block and tackle, fitted with fiber rope, is used because the chain hoists are stronger, more dependable, and more durable than fiber-rope tackle.

There are three general types of chain hoists: spur-geared, differential, and screw-geared (or worm drive). The spur-geared type is the most efficient as it will pick up a load with the least effort on the part of a person. The differential type is the least efficient.

Screw-geared and differential hoists are self-locking and will automatically hold a load in position. Since the spur-geared type is free running, it tends to allow the load to run itself down.

Figure 4-3. Pendent-controlled electric hoists should be designed so that a constant push on the control button is needed to raise or lower the load. Note that the pendent station is supported by a chain in order to protect the electric cable against strain. (Printed with permission from Acco Hoist and Crane Division).

Therefore, an automatic mechanical load brake, similar to that on a crane, is provided to control the rate of descent of the load.

Hoist chain inspection is described in Chapter 5, Ropes, Chains, and Slings.

CRANES

Design and construction

In the U.S. OSHA requires that all new top-running, multiple-girder cranes, constructed and installed on or after August 31, 1971, meet the design specifications of the American National Standard (ANSI) B30.2, *Overhead and Gantry Cranes (Top Running Bridge, Single or Multiple Girder, Top Running Trolley Hoist)*. OSHA requirements do not cover single girder cranes.

Single girder underhung cranes are covered in ANSI B30.11; single girder top-running cranes, in ANSI B30.17; overhead hoists used on single girder cranes, in ANSI B30.16. Warning: Check for reverse phasing when installing a hoist or crane.

All parts of every crane (Figure 4-4), especially those subject to impact, wear, and rough usage, should be of adequate strength for its rated service. Journals and shafts should be of sufficient strength and size to bring the bearing pressure to within safe limits.

Open hooks shall not be used to support loads that pass over workers or loads where there is danger of relieving the tension on the hook due to the load or hook catching or fouling.

Outdoor storage gantry cranes shall be provided with remotely operated rail clamps or other equivalent devices. Parking brakes may be considered as minimum compliance with this rule. Rail clamps should only be applied when the crane is not in motion. When rails are used for anchorages, they shall be secured to withstand the resultant forces applied by the rail clamps. If the clamps act on the rail, any projection or obstruction in the clamping area must be avoided. A wind-indicating device shall be provided which will give a visible and audible alarm to the crane operator at a predetermined wind velocity.

Each independent hoisting unit of a crane shall have brakes complying with the requirements of ANSI B30 Series, "Safety Requirements for Cranes, Derricks, Hoists, Hooks, Jacks and Slings."

The rated load of the crane shall be plainly marked on each side of the crane. If the crane has more than one hoisting unit, each hoist shall have its rated load marked on it or its load block. The marking should be clearly legible from the ground or floor. The crane shall not be loaded beyond its rated capacity, except for testing.

The general arrangement of the cab and the location of control and protective equipment should be such that all operating handles are within convenient reach of the operator when facing the area to be served by the load hook or when facing the direction of travel of the cab. The arrangement should allow the operator a full view of the load hook in all positions.

Each controller and operating lever should be marked with the action it controls and the direction. These levers should have spring returns, so that if the operator releases a lever, it will automatically move into the OFF position. ANSI Standard B30.2 requires that "cranes not equipped with spring-return controllers, spring-return master switches or momentary contact pushbuttons shall be provided with a device which will disconnect all motors from the line, in the event of a power failure, and will not permit any motor to be restarted until the controller handle is brought to the OFF position, or a reset switch or button is operated."

Other factors that should be considered:
1. Platforms, footwalks, steps, handholds, guardrails, and toe boards shall be provided for safe footing and accessways.
2. All machinery, equipment, and material hoists operating on rails, tracks, or trolleys shall have positive stops or limiting devices on the equipment, rails, tracks, or trolleys to prevent overrunning the safe limits.
3. All points requiring lubrication during operation shall have fittings so located or guarded as to be accessible without hazardous exposure.

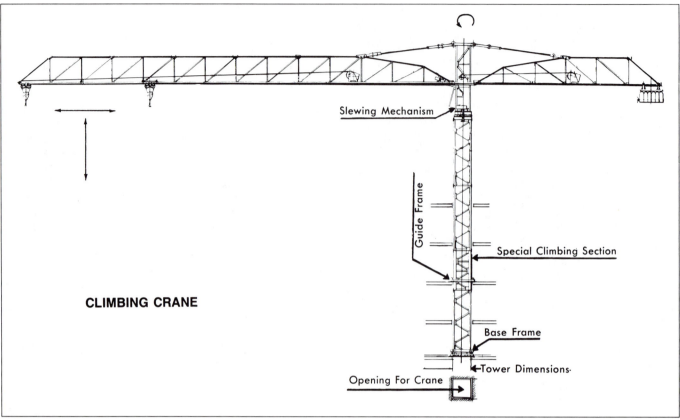

CLIMBING CRANE

Slewing Mechanism

Guide Frame

Special Climbing Section

Base Frame

Tower Dimensions·

Opening For Crane

Figure 4-4a.

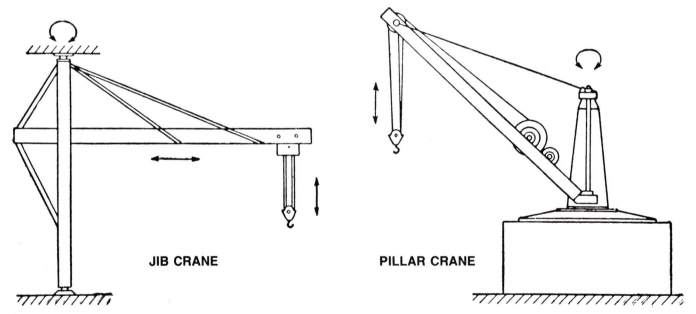

JIB CRANE

PILLAR CRANE

Figure 4-4b. Diagrammatic sketches of various types of cranes. (Printed with permission from ANSI Standard Series B30, "Safety Standards for Cableways, Cranes, Derricks, Hoists, Hooks, Jacks and Slings," except for climbing crane).

4. Platforms, footwalks, steps, handholds, guardrails with intermediate rails, and toeboards shall be provided on machinery and equipment to provide safe footing and accessways. Platforms and steps should be of slip-resistant material.

5. Access to the crane cab and/or bridge walkway shall be pro-

vided by conveniently placed fixed ladders, stairs, or platforms whose steps leave no gap exceeding 12 in. (30.5 cm). Fixed ladders should be in conformance with the ANSI A14.3, *Safety Requirements for Fixed Ladders.*

6. A dry chemical or equivalent fire extinguisher shall be kept in the crane cab.

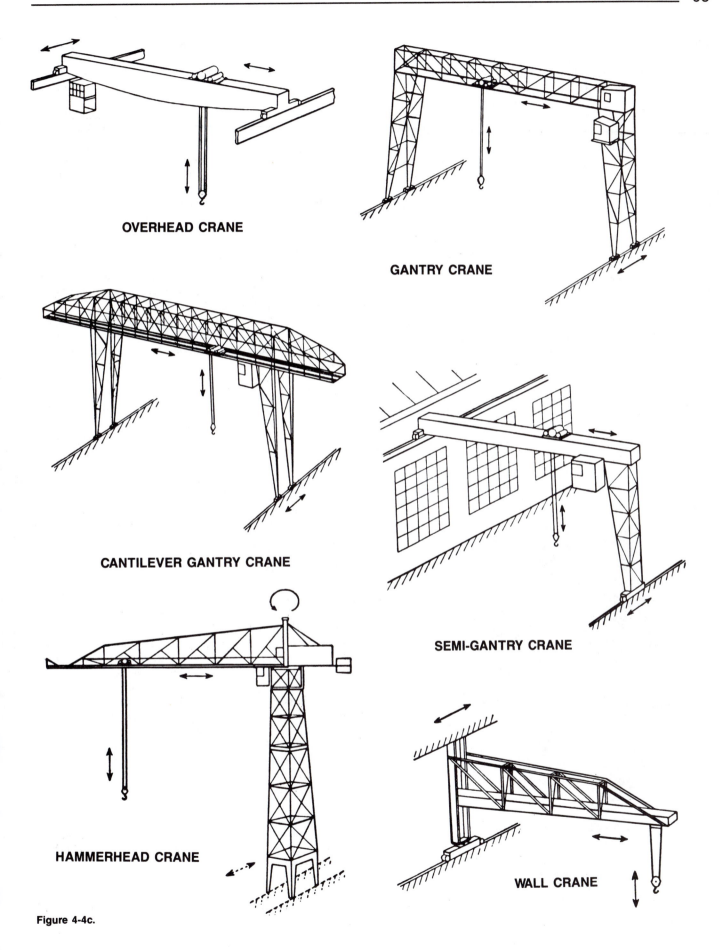

OVERHEAD CRANE

GANTRY CRANE

CANTILEVER GANTRY CRANE

SEMI-GANTRY CRANE

HAMMERHEAD CRANE

WALL CRANE

Figure 4-4c.

Guards and limit devices

If contact can be made with gears and other moving parts during normal operating conditions, they shall be totally enclosed, covered by screen guards, or placed out of reach.

No overhung gears shall be used, unless means are provided to prevent their falling if they should break or work loose.

The bolts in shaft couplings should be recessed, so that the tops of the nuts do not project.

To prevent crushed fingers, large load hooks on cranes shall have handles, so that a person can hold or guide the hooks when slings are being placed on them. Also, on small cranes, the pinch points where cables pass over sheaves in the load block shall be guarded (Figure 4-5).

The hoist motion of every crane, with the exception of boom-type cranes and derricks, must have an overtravel limit switch in the hoisting direction to stop the hoisting motion. Lower travel limit switches are recommended for all hoists where the load block enters pits or hatchways in the floor.

Limit devices always shall operate on a normally closed circuit and should be tested at the beginning of each shift. To make this test, the unloaded block shall be carefully run up so as to actuate the device. Here is a suggested testing procedure:
1. Inch block into limit switch.
2. Lower load approximately 10 ft (3 m), stop and operate at full speed into limit switch.
3. The test should always be conducted away from equipment and employees.

Figure 4-5. Load blocks are safer when guarded.

Hoist limit switches are operational safety devices to prevent accidental overtravel of the load block and are not intended for constant-duty service. Questions about requirements for hoists with constant-duty limit switches should be referred to a crane or hoist manufacturer.

The hook block should be designed so that it will lift vertically without the wire ropes of the cable twisting around each other. The hook should be of solid forged steel or built-up steel plates. Bronze and stainless steel hooks are frequently used for fire protection. Large hooks should swivel on roller or ball bearings.

Electric equipment wiring shall be installed in accordance with Article 610, "Cranes and Hoists," of the *National Electrical Safety Code* (ANSI C2 and NFPA 70).

On electric-power operated cranes, the power supply to the runway conductors should be controlled by a switch or circuit breaker, which is located on a fixed structure accessible from the floor and arranged to be locked in the OFF position.

Other precautions include:
1. No guard, safety appliance, or device shall be removed or otherwise be made ineffective in machinery or equipment, except for the purpose of making immediate repairs, lubrications, or adjustments and, then, only after the power has been turned off, except when power is necessary for making adjustments.
2. All guards and devices shall be replaced immediately after completion of repairs and adjustments.
3. Traveling cranes shall be equipped with a warning device that can be sounded continuously while the crane is traveling.

Before starting maintenance or repair work on a crane, workers shall apply their personal padlocks to the main power switch while it is in OFF position. When working on a multiple-crane runway, positive provision should be made to prevent other cranes from running into the crane being repaired.

Ropes and sheaves

Hoisting ropes should be of a recommended construction for crane or hoist service. The crane or rope manufacturer shall be consulted whenever a change is contemplated. The rated load divided by the number of parts of rope should not exceed 20 percent of the nominal breaking strength of the rope.

Sheaves and drums should be inspected for wear. If the grooves become enlarged by wear or corrugated from excessive rope pressure, they should be replaced. These conditions will cause rapid wear and loss of rope strength. In cases where considerable material must be removed to regroove the drum or sheave, the strength of these parts may be impaired. In such cases, it is recommended that the hoist or crane manufacturer be consulted prior to regrooving. (For further information, refer to ANSI/ASME B30.2, *Overhead and Gantry Cranes [Top Running Bridge, Single or Multiple Girder, Top Running Trolley Hoist].*)

To reduce the strain on the hoist rope where it enters the socket or anchorage on the drum, the minimum of at least two wraps shall remain on the drum when the load block is at the lowest elevation. However, one turn is permitted when a geared lower limit switch is used. The drum end of the rope should be anchored to the drum by a socket arrangement approved by the crane or rope manufacturer or both. As with electric hoists, if the load block can enter a pit or hatchway in the floor, a lower

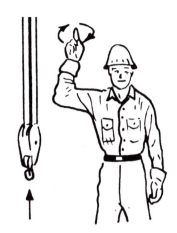

HOIST. With forearm vertical, forefinger pointing up, move hand in small horizontal circle.

LOWER. With arm extended downward, forefinger pointing down, move hand in small horizontal circles.

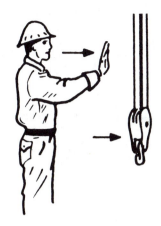

BRIDGE TRAVEL. Arm extended forward, hand open and slightly raised, make pushing motion in direction of travel.

TROLLEY TRAVEL. Palm up, fingers closed, thumb pointing in direction of motion, jerk hand horizontally.

STOP. Arm extended, palm down, move arm back and forth.

EMERGENCY STOP. Both arms extended, palms down, move arms back and forth.

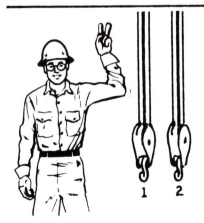

MULTIPLE TROLLEYS. Hold up one finger for block marked "1" and two fingers for block marked "2". Regular signals follow.

MOVE SLOWLY. Use one hand to give any motion signal and place other hand motionless in front of hand giving the motion signal. (Hoist slowly shown as example.)

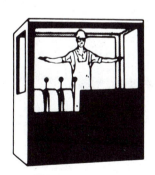

MAGNET IS DISCONNECTED. Crane operator spreads both hands apart palms up.

Figure 4-6. Standard hand signals for controlling operation of overhead and gantry cranes. (Printed with permission from ANSI Standard Series B30, "Safety Standards for Cableways, Cranes, Derricks, Hoists, Hooks, Jacks, and Slings," The American Society of Mechanical Engineers, New York).

98

HOIST. With forearm vertical, forefinger pointing up, move hand in small horizontal circle.

LOWER. With arm extended downward, forefinger pointing down, move hand in small horizontal circles.

USE MAIN HOIST. Tap fist on head; then use regular signals.

USE WHIP LINE. (Auxiliary Hoist) Tap elbow with one hand; then use regular signals.

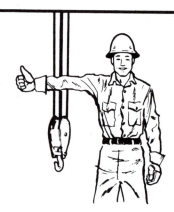

RAISE BOOM. Arm extended, fingers closed, thumb pointing upward.

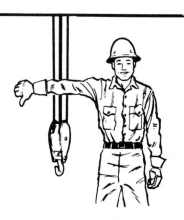

LOWER BOOM. Arm extended fingers closed, thumb pointing downward.

MOVE SLOWLY. Use one hand to give any motion signal and place other hand motionless in front of hand giving the motion signal. (Hoist Slowly shown as example)

RAISE THE BOOM AND LOWER THE LOAD. With arm extended thumb pointing up, flex fingers in and out as long as load movement is desired.

LOWER THE BOOM AND RAISE THE LOAD. With arm extended, thumb pointing down, flex fingers in and out as long as load movement is desired.

Figure 4-7. Standard hand signals suitable for crawler, locomotive, and truck boom cranes. One-hand signals for extending or retracting boom (not shown above): Extend boom—one fist in front of chest with thumb tapping chest. Retract boom—one fist in front of chest, thumb pointing outward and heel of fist tapping chest. (Printed with permission from ANSI Standard Series B30, "Safety Standards for Cableways, Cranes, Derricks, Hoists, Hooks, Jacks, and Slings," The American Society of Mechanical Engineers, New York.)

SWING. Arm extended point with finger in direction of swing of boom.

STOP. Arm extended, palm down, hold position rigidly.

EMERGENCY STOP. Arm extended, palm down, move hand rapidly right and left.

TRAVEL. Arm extended forward, hand open and slightly raised, make pushing motion in direction of travel.

DOG EVERYTHING. Clasp hands in front of body.

TRAVEL. (Both Tracks) Use both fists, in front of body, making a circular motion, about each other, indicating direction of travel; forward or backward. (For crawler cranes only)

TRAVEL. (One Track) Lock the track on side indicated by raised fist. Travel opposite track in direction indicated by circular motion of other fist, rotated vertically in front of body. (For crawler cranes only)

EXTEND BOOM. (Telescoping Booms) Both fists in front of body with thumbs pointing outward.

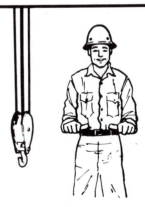

RETRACT BOOM. (Telescoping Booms) Both fists in front of body with thumbs pointing toward each other.

Figure 4-7 (continued).

travel limit switch is recommended; and if that switch trips, the turns of the rope remaining on the drum can be reduced to one when the block has been stopped.

Crane and hoist signals

Crane movements shall always be governed by a standard code of signals, transmitted to the crane operator by the crane director (signaler). Signals may be given by any mutually understood and officially adopted method, but preferably by motion of the hand. Signals must be discernible or audible at all times. No response shall be made unless signals are clearly understood. There should be only one designated person who is qualified to give crane signals to the operator.

The operator should move the hoisting apparatus only on signals from the proper person, but a stop signal should be obeyed regardless of who gives it.

Unless obedience would result in an accident, the operator should be governed absolutely by the signal. However, if an accident seems unavoidable by obeying the signal, the signaler should be notified at once, so that corrective measures can be taken immediately.

If signalers are changed frequently, they shall be provided with one (and only one) conspicuous armband, hat, glove, or other badge of authority, which must be worn by the person currently in charge.

A simple code of one-hand signals is appropriate for an overhead crane or bridge crane. The ANSI set of signals, adopted by many companies, is shown in Figure 4-6 (ANSI Series B30). A set of one- and two-hand signals for a locomotive or crawler crane, or any other boom rig, is shown in Figure 4-7.

Where visual or audible signals are inadequate, telephone or portable radio communication should be used. A remote radio-control system for overhead cranes, which eliminates the need for hand signals, is available (Figure 4-8). The operator controls all movements of the bridge, trolley, and hoist from the plant floor. Circuits are so designed that failure of a system component causes all crane motions to stop.

Employees who work near cranes or assist in hooking on or arranging loads should be instructed to *keep out from under loads.* Supervisors shall see that this rule is strictly enforced. One manufacturing company publishes this warning: "From a safety standpoint, one factor is paramount: Conduct all lifting operations in such a manner that, if there were an equipment failure, no personnel would be injured. This means *keep out from under raised loads!*" (Figure 4-1.)

Selection and training of operators

Cranes shall be operated by employees who are physically qualified, particularly with regard to acuity of vision, depth and color perception, hearing, muscular coordination, and reaction time. A physical examination should be required prior to assignment and annually thereafter. Employees must be of legal age to operate this type of equipment, and be capable of reading and understanding instructions as well as speaking the dominant language of the workers.

Operators should undergo a course of training to operate cranes. In the U.S. they should be completely familiar with applicable ANSI standards and current safety requirements of the Occupational Safety and Health Act (OSHA). Equally important are training and authorization of personnel who actually will be rigging the load. Some companies require both

Figure 4-8. Operator controls overhead crane by means of hand-held remote radio (lower right).

operators and local rigging personnel to have permits, renewable at intervals of a year or two upon reexamination.

Inspection

Overhead and gantry cranes must be inspected in accordance with the procedure noted in ANSI, B30.2.

Hooks must be inspected in accordance with the procedure in ANSI/ASME B30.10:

Hooks having any of the following deficiencies shall be removed from service unless a qualified person approves their continued use and initiates corrective action. Hooks approved for continued use shall be subjected to periodic inspection.

1. Crack(s),
2. Wear exceeding 10 percent (or as recommended by the manufacturer) of the original sectional dimension,
3. A bend or twist exceeding 10 degrees from the plane of the unbent hook,
4. For hooks without latches, an increase in throat opening exceeding 15 percent (or as recommended by the manufac-

turer); for hooks with latches, an increase of the dimension between a fully-opened latch and the tip section of the hook exceeding 8 percent (or as recommended by the manufacturer),

5. If a latch which is provided becomes inoperative because of wear or deformation, and is required for the service involved, it shall be replaced or repaired before the hook is put back into service. If the latch fails to fully close the throat opening, the hook shall be removed from service or moused until repairs are made.

A crane operator shall not attempt to make repairs. Any condition that might make the crane unsafe to operate should be reported to the supervisor. Certain faults may be so dangerous that the crane should be shut down at once and not operated until the faults are corrected.

A list of unsafe conditions to be checked mainly by operators of overhead traveling cranes follows (also see next section).

Bearings: loose, worn

Brakes: shoe wear

Bridge: alignment out of true (indicated by screeching or squealing wheels)

Bumpers on bridge: loose, missing, improper placement

Collector shoes or bars: worn, pitted, loose, broken

Controllers: faulty operation because of electrical or mechanical defects

Couplings: loose, worn

Drum: rough edges on cable grooves

End stops on trolley: loose, missing, improper placement

Footwalk: condition

Gears: lack of lubrication or foreign material in gear teeth (indicated by grinding or squealing)

Guards: bent, broken, lost

Hoisting cable: broken wires

Hook block: chipped sheave wheels

Hooks: straightening

Lights (warning or signal): burned out, broken

Limit switch: functioning improperly

Lubrication: overflowing on rails, dirty cups

Mechanical parts (rivets, covers, etc.): loose

Overload relay: frequent tripping of power

Rails (trolley or runway): broken, chipped, cracked

Wheels: worn (indicated by bumpy riding)

Many companies believe in performance tests for all hoisting equipment. They make sure that all hoisting equipment satisfactorily completes a performance (operating) test when placed in service on a project. The test should be repeated prior to unusual or critical lifts of load bearing or load controlling parts; after alteration, modification, or reassembly; and at least every year. The test results should be recorded and kept available for review.

Operating rules

The following operating rules for crane operators are from the Crane Manufacturers Association of America, Inc.:

One measure of a good crane operator is the smoothness of operation of the crane. Jumpy and jerky operation, flying starts, quick reversals and sudden stops are the "trademarks" of the careless operator. The good operator knows and follows these tried and tested rules for safe, efficient crane handling.

1. Crane controls should be moved smoothly and gradually to avoid abrupt, jerky movements of the load. Slack must be

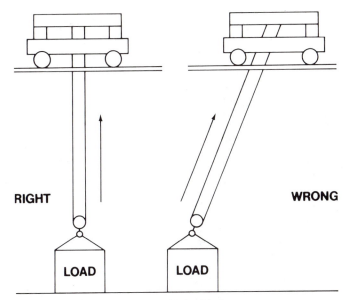

Figure 4-9. Center crane over the load before starting the hoist. When depositing a load, do not swing it to reach a location not under the crane.

removed from the sling and hoisting ropes before the load is lifted.

2. Center the crane over the load before starting the hoist to avoid swinging the load as the lift is started. Loads should not be swung by the crane to reach areas not under the crane (Figure 4-9).

3. Crane hoisting ropes should be kept vertical. Cranes shall not be used for side pulls.

4. Never lower the block below the point where less than two full wraps of rope remain on the hoisting drum. Should all the rope be unwound from the drum, be sure it is rewound in the correct direction and seated properly in the drum grooves or otherwise the rope will be damaged and the hoist limit switch will not operate to stop the hoist in the high position.

5. Be sure everyone in the immediate area is clear of the load and aware that a load is being moved. Sound the warning device (if provided) when raising, lowering or moving loads wherever people are working to make them aware that a load is being moved (Figure 4-10).

6. Do not make lifts beyond the rated load capacity of the crane, sling chains, rope slings, etc.

7. Do not operate the crane if limit switches are out of order or if ropes show defects or wear.

8. Make certain that before moving the load, load slings, load chains, or other load lifting devices are fully seated in the saddle of the hook.

9. When a duplex hook (double saddle hook) is used, a double sling or choker should be used to assure that the load is equally divided over both saddles of the hook.

10. On all capacity or near capacity loads, the hoist brakes should be tested by returning the master switch or pushbutton to the OFF position after raising the load a few inches off the floor. If the hoist brakes do not hold, set the load on the floor and do not operate the crane. Report the defect immediately to the Supervisor.

Figure 4-10. Caution—Overhead Cranes sign with warning lights and audible signal is used in a high-traffic area to alert personnel when crane will be passing by.

11. Check to be sure that the load is lifted high enough to clear all obstructions and personnel when moving bridge or trolley.

12. At no time should a load be left suspended from the crane unless the operator is at the master switches or pushbutton with the power on, and under this condition keep the load as close as possible to the floor to minimize the possibility of an injury if the load should drop. When the crane is holding a load, the crane operator should remain at the master switch or pushbutton.

13. When a hitcher is used, it is the joint responsibility of the crane operator and the hitcher to see that hitches are secure and that all loose material has been removed from the load before starting a lift.

14. Do not lift loads with any sling hooks hanging loose. (If all sling hooks are not needed, they should be properly stored or use a different sling.)

15. All slings or cables should be removed from the crane hooks when not in use. (Dangling cables or hooks hung in sling rings can inadvertently snag other objects when the crane is moving.)

16. Crane operators should not use limit switches to stop the hoist under normal operating conditions. (These are emergency devices and are not to be used as operating controls.)

17. Do not block, adjust or disconnect limit switches in order to go higher than the switch will allow.

18. Upper limit switches (and lower limit switches, when provided) should be tested in stopping the hoist at the beginning of each shift, or as frequently as otherwise directed.

19. Never move loads carried by magnets or vacuum devices over anyone. Loads, or parts of loads, held magnetically may drop. Failure of power to magnets or vacuum devices will result in dropping the load unless a backup power supply is furnished.

20. Molten metal shall never be carried over people.

21. If the electric power goes off, place your controllers in the OFF position and keep them there until power is again available.

22. Before closing main or emergency switches, be sure that all controllers are in OFF position so that the crane will not start unexpectedly.

23. If plugging protection is not provided, always stop the controllers momentarily in the OFF position before reversing—except to avoid accidents. (The slight pause is necessary to give the braking mechanism time to operate.)

24. Whenever the operator leaves the crane, this procedure should be followed:
    (a) Raise all hooks to an intermediate position.
    (b) Spot the crane at an approved designated location.
    (c) Place all controls in the OFF position.
    (d) Open the main switch to the OFF position.
    (e) Make visual check before leaving the crane.
    Note: On yard cranes (cranes on outside runways), operators should set the brake and anchor securely so the crane will not be moved by the wind.

25. When two or more cranes are used in making one lift, it is very important that the crane operators take signals from only one designated person.

26. Never attempt to close a switch that has an OUT OF ORDER or DO NOT OPERATE card on it. Even when a crane operator has placed the card, it is necessary to make a careful check to determine that no one else is working on the crane, before removing the card.

27. In case of emergency or during inspection, repairing, cleaning or lubricating, a warning sign or signal should be displayed and the main switch should be locked in the OFF position. This should be done whether the work is being done by the crane operator or by others. On cab operated cranes when others are doing the work, the crane operator should remain in the crane cab unless otherwise instructed by the supervisor.

28. Never move or bump another crane that has a warning sign or signal displayed. Contacts with runway stops or other cranes shall be made with extreme caution. The operator shall do so with particular care for the safety of persons on or below the crane, and only after making certain that any persons on the other cranes are aware of what is being done.

29. Do not change fuse sizes. Do not attempt to repair electrical apparatus or to make other major repairs on the crane unless specific authorization has been received.

30. Never bypass any electrical limit switches or warning devices.

31. Load limit or overload devices shall not be used to measure loads being lifted. This is an emergency device and is not to be used as a production operating control.

General maintenance safety rules

The following maintenance safety rules are taken from ANSI/ASME B30.2, *Overhead and Gantry Cranes (Top Running Bridge, Single or Multiple Girder, Top Running Trolley Hoist)*:

1. To be repaired, a crane must be moved to a location where there will be minimum interference with other cranes and operations in the area.

2. All controllers should be in the OFF position.

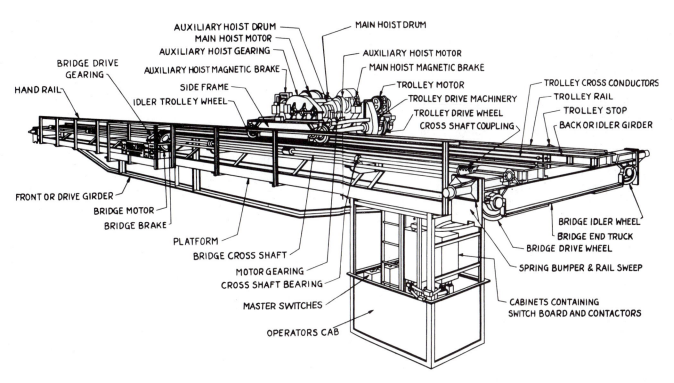

Figure 4-11. Essential parts of a typical cab-controlled overhead traveling crane. (Printed with permission from Shaw Box Crane and Hoist Division of Manning, Maxwell & Moore, Inc.)

3. The main power source should be disconnected, deenergized and locked, tagged, or flagged in the deenergized position.
4. WARNING or OUT OF ORDER signs should be placed on the crane, on the floor beneath, or on the hook where they are visible from the floor.
5. If other cranes are in operation on the same runway, rail stops or other suitable devices shall be provided to protect the idle crane.
6. Where rail stops or other devices are not available or practical, a person should be located where he can warn the operator of reaching the limit of safe distance from the idle crane.
7. Where there are adjacent craneways and the repair area is not protected by wire mesh or other suitable protection, or if any hazard from adjacent operations exists, the adjacent runway must also be restricted. A signaler shall be provided when cranes on the adjacent runway pass the work area. Cranes shall come to a full stop and may proceed through the area on being given a signal from the designated person.
8. Trained personnel shall be provided to work on energized equipment when adjustments and tests are required.
9. After all repairs have been completed, guards shall be reinstalled, safety devices reactivated, and maintenance equipment removed before restoring crane to service.

Overhead cranes

An overhead crane (Figure 4-11) may be operated either from a cab or from the floor. In the latter case, control devices may be either pendent pushbuttons or pull ropes. (In some cases they are radio controlled.) The control handles shall be clearly identified by signs and by shape or position so that the operator, while maintaining visual contact with the signaler, can identify each control by touch. Controls on all floor-operated over-

head traveling cranes should be identified. Likewise in cab-operated cranes the controls should be identified.

If there are several cranes on the same runway or in the same building, all should have the controls in identical positions so that a substitute operator will not be confused.

Safe means shall be provided for the operator to pass from the cab to the footwalk. Stairs are preferable to ladders. If ladders are used, they should meet the appropriate safety standards. Furthermore, the space that the operator must step across in going from the landing or the runway girder to the crane should not exceed 12 in. (30 cm). Safe access also shall be provided to the bridge motor and brake and to the equipment on the crane trolley—OSHA requires that a clearance of not less than 3 in. (7.6 cm) overhead and 2 in. (5.08 cm) laterally be provided and maintained between the crane and obstructions. Finally, stanchions or grab irons shall be installed to enable a person to climb onto the trolley in safety.

In case of an emergency, the operator must be able to escape from the crane regardless of its location on the runway. The operator would be in particular danger if a fire were to occur while the crane is attached to a load and could not travel at once to the access landing. Unless a means of escape via the bridge and runway is provided, an emergency means of escape should be installed in the cab. A CO_2 dry chemical or equivalent fire extinguisher should also be installed in the cab.

Strict precautions shall be taken to restrict personnel from servicing or riding the crane while it is in motion. This will prevent a person from being brushed off the crane by low beams or trusses. Service personnel must stay on the footwalk. Also, the movement of the crane must *never* jeopardize the safety of personnel. This is one reason the safety and health professional should be active during the building design stage.

Footwalks and platforms should be substantial, rigidly braced, and protected on open sides with standard railings and toeboards. The footwalk should be reached by one or more fixed ladders not less than 16 in. (40 cm) wide. The outside edge of the walk should be not less than 30 in. (76 cm) from the nearest part of the trolley. The bridge walkway should have a 42-in. (1 m) high handrail, an intermediate rail, and a 4-in. (10 cm) toeboard, and the space at the squaring shaft should be guarded so that a person cannot fall between the walkway and the crane girder.

A footwalk shall be provided also, if headroom permits, along the entire length of the bridge of any crane having the trolley running on the tops of the girders. This footwalk shall be on the drive side of the bridge and should have toeboards and metal handrails. Safe access to the opposite side of the trolley also shall be provided.

Floors of walks should be neatly fitted so as to leave no openings and shall have a nonslip surface. Vertical clearance between the floor of the walk and overhead trusses, structural parts, or other permanent fixtures shall be at least 6.5 ft (2.0 m). Where such clearance is structurally impossible, built-in members should be distinctively painted or striped and padded where necessary. Toeboards shall be provided at the edges of flooring on the trolley to prevent tools from falling to the floor below.

To guard against electric shock, a heavy rubber mat shall be provided at the control panel in the cab. The operator should have an unobstructed view of the load hook in any possible position.

The bridge truck wheels and the trolley wheels shall have sweeps to push away a person's foot or hand. To prevent a serious impact on the crane, if a bridge wheel should fail, the end frames or trucks should have safety lugs not more than 1 in. (2.5 cm) above the top of the rails.

Runway rails, after years of service, may become distorted or the span between them may be altered because of settling of the column footings, so that wear on the flanges of the bridge wheels results. Rail alignment, therefore, should be checked every few years. If the tread of bridge truck wheels is tapered, the crane will run square with the runway constantly.

Rail stops or bumpers must be so located that, when contacted, the crane bridge is square with the runway.

Operation. When not in use, a crane should be parked with the load hook (and the slings if they remain on the hook) raised high enough to clear the heads of anyone at work on the floor below, and the operator should throw all controls into OFF position and open the main switch.

The operator should center the trolley over the load, when about to make a lift, and should accelerate and brake the lifting and lowering motions slowly to minimize stresses on the crane.

A check should frequently be made to see that the crane is square with the runway rails, by carefully contacting the rail stops. If the crane is out of square, the operator should apply power for travel motion (without a load on the crane and with the trolley at the end of the bridge where the wheel is away from the rail stop). The drive wheel at one end of the crane then will slip on the rail while the other drive wheel rolls to the rail stop.

If a warning device is furnished, it shall be activated each time before traveling and intermittently when approaching workers (ANSI Standard B30.2).

Figure 4-12. An electromagnet is used to load scrap metal into railroad cars.

Two or more cranes operated in the same bay or runway should be kept at least 30 ft (9 m) apart where possible. If they operate closer, operators and anyone else in the area should take special care to avoid collision.

When repairs to one crane are necessary, every precaution must be taken to prevent other cranes from colliding with it while repairs are underway. Safety stops shall be installed.

Electromagnets are often used with electric overhead cranes, as well as with gantry cranes and several other types, to handle ferrous scrap and hot or cold ingots and to move iron and steel products (Figure 4-12).

Magnets shall not be used either close to steel machines or parts or near ferrous materials in process, to prevent those parts from becoming magnetized. Watches and other delicate instruments should be kept out of the electromagnetic field.

Switches or switchboxes controlling power to the magnet should be labeled DANGER—DO NOT OPEN SWITCH—POWER TO ELECTROMAGNET. The magnet switch must have a means of discharging the inductive load of the magnet.

The metal body of an electromagnet shall be grounded. The magnet's power supply circuit should have a battery backup system. Even with a backup system, however, a load suspended from a magnet *never* should be moved over personnel. The switchboard, wiring, and all other electrical equipment should comply with the American National Standard C2 *National Electrical Safety Code* (NFPA 70). For further information, see National Safety Council Data Sheet 359, *Electromagnets Used with Crane Hoists.*

Special hook-on devices can be designed and made for handling special shapes. A custom-made clamping device for lifting rolls of paper is shown in Figure 4-13. A grab for positioning steel coils is shown in Figure 4-14.

Figure 4-13. Custom-made clamping device for lifting rolls of paper. As chain is pulled upward, clamp (center) tightens on roll.

Storage bridge and gantry cranes

Storage bridge and gantry cranes are similar to traveling cranes, but travel on rails at ground level instead of on elevated runway girders. Gantry cranes have relatively short spans, whereas storage bridge cranes may have a span up to 300 ft (92 m) or more, sometimes with a cantilever on one or both ends. Storage bridge cranes usually are used for handling coal or ore. (See the Conveyors section later in this chapter.) Ordinarily, a caged ladder on one of the legs of the crane provides access to the cab.

Track riding cranes should be provided with substantial rail scrapers or track clearers at each end of the trucks. Rail scrapers should be effective in both directions of travel.

Because there may be a serious shearing or crushing hazard in the area between cab-operated cranes and adjacent structures or stored material, a gong or other warning device should sound intermittently from the time the travel controller handle is first moved from OFF position until it is returned to OFF position.

The wheel truck of gantry cranes should have adequate side clearance.

Bumpers, made of cast steel plates and angles, should be provided on storage bridge and gantry cranes and should be at least one-half the diameter of the truck wheels in height. Both truck wheel and trolley bumpers should be fastened to the girder and not to the rail.

Spring bumpers usually are provided where bridge axles have antifriction bearings. If compression springs are used, they should be at least 5 in. (12.7 cm) in diameter at the point of contact and so arranged that, if a spring or guide pin breaks, no part can fall on the crane.

To prevent the crane from being moved down the track by a strong wind, the operator should apply rail clamps before leaving the cab, even for a short time. The holding power of the clamps should be sufficient to withstand wind pressure of 30 lb/ft² (1.4 kPa) of projected area of the crane.

Figure 4-14. Mounted on a 25-ton (22,680 kg) capacity crane, this grab has full access to the outdoor coil storage area. The grab can rotate 360 degrees for precise positioning of the coils. (Printed with permission from Acco Hoist and Crane Division).

The electric contact rails or wires shall be so located, or so guarded that persons normally could not come into contact with them.

So that a bridge crane can be squared to the track, the squaring shaft should have a clutch. One end of the bridge can then be moved, while the other end remains stationary, to bring the crane into proper position.

All outside cranes should have the following features:
1. Floors of the footwalk constructed to provide drainage.
2. The operator's cab constructed of fire-resistant material, weatherproof, with provision for heating and ventilation, with ample space for control equipment, and with the operator located so that signals can be clearly seen.
3. The floor of the cab extended to an entrance landing and equipped with a handrail and toeboard of standard construction.
4. A rope ladder or other means of emergency escape from the cab.
5. Locking ratchets on wheel locks, rail clamps, and brakes so that the crane will not move in a high wind.

6. Skew switches to prevent excessive distortion of the bridge.
7. A screen or other barrier (preferably nonconductive) between the contact bars and the bridge walkways to prevent accidental contact with the current conductor.

The main line switch should be so constructed that it can be locked in OPEN position and should be mounted above the cab where it can be reached conveniently from the footwalk.

Monorails

A monorail system consists of one or more independent trolleys, supported from or within an overhead track, from which hoists are suspended. All applicable safety features should conform to ANSI standard B30.11, *Monorails and Underhung Cranes*.

Monorail hoists are used extensively in many industries to raise, lower, and transport materials. They can be classified in three major groups: hand-operated, semihand-operated, and power-operated. On the hand-operated monorail, the material is raised with a hand-powered hoist, and the trolley is propelled by hand. The semihand-operated monorail has a power hoist and is moved horizontally by hand. The power-operated monorail is fully power actuated for both vertical and horizontal movements.

No attempt should be made with a monorail hoist to lift or otherwise move an object by a side pull, unless the hoist has been designed for such use. Monorail hoists, operated in swivels, should have one or more safety catches or lugs that will support the load should a suspension pin fail. All trolleys for monorail systems and underhung cranes should be designed to accommodate the load imposed and the service requirements established.

Monorail track supports and track shall be designed to carry safely the intended loads and erected according to good engineering practice. Both the track and its support should be inspected frequently for signs of weakening and wear, and necessary repairs should be made as soon as possible.

Stops shall also be provided at open ends of tracks such as interlocked cranes, track openers, and track switches.

If an electric monorail carrier or crane is operated from a cab, a fixed access platform or ladder shall be provided to give the operator access to the cab. A means must also be provided for the emergency escape of the operator. The electric contact wires and the current collectors shall be so located that an operator cannot inadvertently make contact with them upon entering or leaving the cab or while in the normal operating position inside the cab.

Jib cranes

A jib crane is a crane capable of lifting, lowering, and rotating the load within the circular arc covered by a rotating arm or jib. The jib and the trolley running on it usually are supported from a building wall or column or from a pillar. A hoist (chain, air, or electric), with which the loads are lifted, is usually suspended from a trolley that travels on the jib boom.

Before a jib crane is mounted to a building wall or column of the building, the strength of the structure shall be checked by a qualified engineer to determine whether or not the column or wall to be used is strong enough to support the jib, hoist, and load. Free-standing jibs must also have good foundation support. The jib shall be braced or guyed if necessary to withstand the loads it is expected to carry.

A stop plate or angle iron shall be installed at the outboard end of the jib to prevent the load trolley from running off the beam. The end stop requires frequent inspection to see that it is not becoming loose or rusting off.

Derricks

The principal types of derricks are the A-frame derrick, the stiff-leg derrick, and the guy derrick. There are other types, but these are the most common. All derricks must have every part firmly anchored.

The A-frame derrick (Figure 4-15), as the name implies, has a frame of steel or timber shaped like the letter "A" and erected in a vertical plane, with a brace or leg extending from the top of the A at a 45 degree angle to the ground. The sills tie this brace to the bottom of the A-frame. The boom is hinged at the horizontal member of the A. The base of the A-frame and the rear brace must be firmly weighted down.

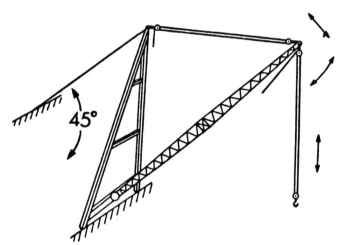

Figure 4-15. A-frame derrick. Brace is set at 45 degrees to the ground.

The stiff-leg derrick (Figure 4-16) has a mast with two braces at a 90 degree angle to each other and at a 45 degree angle to the ground. Usually steel or timber sills tie the mast and the braces together at the ground. To withstand the uplift caused by a heavy load on the boom, sandbags, cast iron weights, or concrete blocks are used to hold down the stiff legs.

The hoist engines for both stiff-leg and A-frame derricks usually are bolted to the sills or ground members. On smaller derricks, the suspended loads may be slewed by being pushed manually. The boom of a large derrick may be swung by a "bull-wheel" to which cables from another drum on the hoist engine are attached.

Since a loaded cable may whip considerably and cause severe injury, the horizontal cables between the hoist engine and the boom hinge shall be barricaded, and workmen should be prohibited from crossing over or under them.

The guy derrick (Figure 4-17) is used largely for erection of the structural steel of tall buildings, especially those over 10 stories high that cannot be reached by the boom of a crawler crane operating on the ground. Such derricks usually are of latticed steel and have an odd number of equally spaced wire rope guys, each equipped with a turnbuckle and attached to the steel beams or columns on the erection floor.

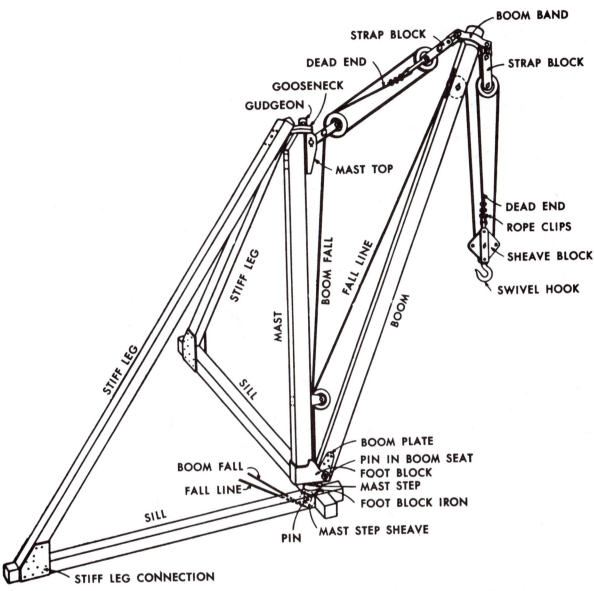

Figure 4-16. Diagram of stiff-leg derrick shows names of various parts. (Printed with permission from Travelers Insurance Co.)

If the derrick is erected on the ground, the guys should be secured to heavy steel anchors buried deep in the ground, with additional weights placed on the anchorages.

Steel beams or heavy timbers, 12 by 12 in. or 12 by 16 in. (30 × 30 cm or 30 × 40 cm), should be placed on the floor beams to support the foot of the mast. These foot blocks must be braced against the stubs of the building columns to prevent their being "kicked" out of position when a heavy load is picked up with the boom at a low angle. Wire rope and turnbuckles may be used in place of 8 by 8 in. (20 × 20 cm) timbers to secure the base of the mast.

The hoist engine, whether on the same level with the derrick base or on the ground many floors below, should be securely anchored by steel cables or shoring timbers to prevent it from being pulled towards the base of the mast by the tension on the cables.

A unique feature of the lattice-type steel guy derrick is its ability to lift itself from one erection floor to the next. First,

the hinge pin is removed to disconnect the boom from the mast. The boom hoist cable (or topping lift) then lifts the boom in a vertical position and sets it on the foot blocks close to the mast. The boom is rotated 180 degrees, the normally unused guys are secured, and the boom then stands as a guyed gin pole.

The load hoist of the boom then is used to pick up the mast, the mast guys being slackened off a few at a time and reattached at the upper level. When the mast is secured at the new erection floor, the boom is raised and again connected at the hinge.

To swing guy derricks, the workers push either on the suspended load or on a pipe "bull stick" attached near the base of the mast. A bullwheel can also be used as a mechanical means to swing the derrick.

Other types of derricks are the gin pole and the breast derrick. The gin pole is merely a mast slightly out of plumb, with a hoisting tackle suspended from its upper end. The gin pole is supported by a number of guys, most of which are on the

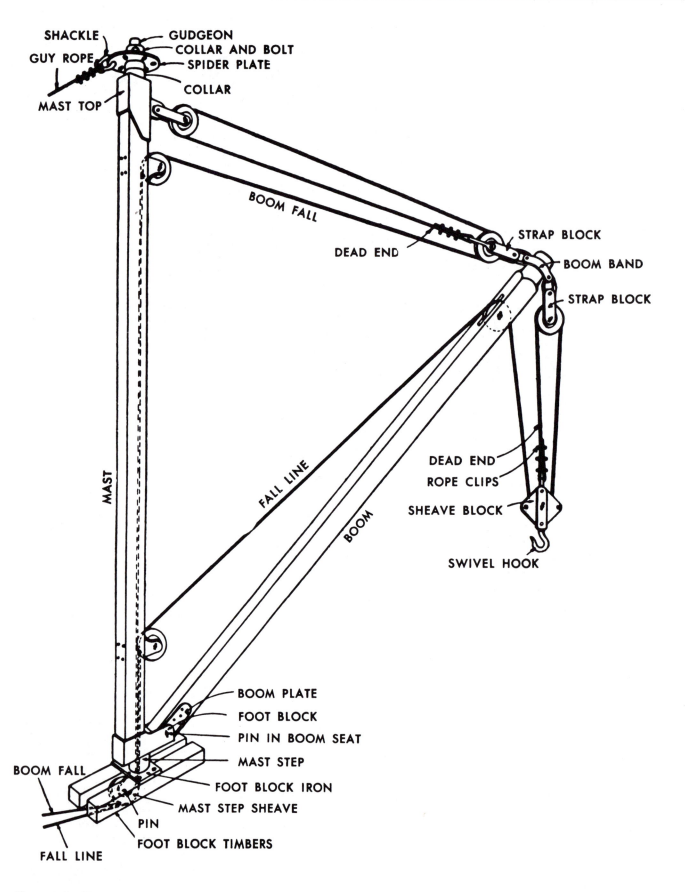

Figure 4-17. Diagram of guy derrick. (For simplicity, only one guy is shown.) (Printed with permission from Travelers Insurance Co.)

side away from the load. This rig is used for raising and lowering a load that needs to be moved only a few feet horizontally.

The breast derrick is a small portable A-frame with a winch attached to it. Like the gin pole, it is erected in a nearly vertical position, with one or two guys to support it. Care must be exercised to prevent the base from slipping and causing the load to fall. The men should be warned to watch their fingers when they operate the winch.

Tower cranes

There are many design variations for the tower crane (Figure 4-18), depending upon the manufacturer and the intended use. Tower cranes can be erected on a minimum of ground area or within a building; for example, within an elevator shaft or other floor opening. To increase their range and versatility, some tower cranes are mounted on undercarriages running on rails, rather than on a fixed-base; there is also a truck-mounted type.

The following procedures have been established as being significant among the more common causes of tower crane failure:

- Improper erection of the crane.
- Lifting loads above the rated capacity of the crane, or lifting eccentric loads.
- Improper bracing of the crane.
- Bracing, or attachment to material or structural members that are insecure or unable to provide the needed support.
- Erection within a building, the design of which has not provided the necessary allowances for the crane weight, or support at the application point of the crane weight.
- Operators not being fully cognizant of the limitations or operating characteristics of tower cranes.

Figure 4-18. A tower crane is used during construction of this multi-storied building.

- Tampering with limit switches or other safety devices.
- Failure to have instructions given clearly in English or the native language of the operator.
- Failures resulting from use during high winds.

General accident prevention information for tower cranes includes:

- Stresses for steel used for fabrication and construction conforming to American Institute of Steel Construction specifications. (If special materials such as high-tensile steel or aluminum alloys have been used in the crane structure, the crane should bear a notice to this effect.) All parts of the crane and supports should be designed and constructed to withstand maximum stresses resulting from intended use. (The design and construction should provide the safety factors specified by the authority having jurisdiction.)
- A secure attachment of counterweights; and safety ropes, rods, or chains to hold the counterweights, in addition to the basic attachment.
- The strength of the system used to anchor the rope on a winding drum with an ample safety factor exceeding the normal working load of the rope.
- Flanges of winding drums projected well above the height of the highest layer of rope wound on the drum in normal operation.
- Nonrotating hoist rope (except on receiving systems which do not require it.)
- Guarding of all moving parts including pulley block and sheave guards.

Operation. Only personnel of recognized ability should operate a tower crane. They should be mature in attitude, have quick responses, and be in good health. Their backgrounds should include both training and experience in the operation of this type of equipment.

Operators should possess a general knowledge of the crane's construction and the necessary knowledge of electricity, hydraulics, trade terms, parts identification, and of maintenance needs for this work. They also should have a knowledge of safety codes and standards applicable to crane operation, and of any special safety recommendations of the crane manufacturer.

Operators also should be knowledgeable concerning procedures involving visibility and signaling, lifting and lowering, swinging, and shutting down.

The crane operator should never stand on, or climb upon, the framework outside the cab while the crane is in operation.

Climbing to the end of the jib should be prohibited except when necessary, for which prescribed special precautions and equipment should be used.

Safety belts and other necessary personal protective equipment should be available, and used when necessary.

Mobile cranes

Mobile cranes include locomotive cranes, crawler cranes, wheel-mounted cranes (Figure 4-19), and industrial truck cranes. The first three types are well standardized in design, but there is a great variety of industrial truck cranes.

All mobile cranes have booms with load hoists and boom hoists. In most instances, the crane swings or rotates on a turntable, which rests on a railroad car, crawler, or wheel chassis. Power is provided by electric motor, steam, gasoline, or diesel engine.

Figure 4-19. Wheel-mounted hydraulic crane has telescoping boom. (Printed with permission from Link-Belt Division of FMC).

Electric powered equipment should be grounded as specified in the ANSI *National Electrical Safety Code, C2.* Repairs or adjustments shall be made only by qualified personnel. Power shall be disconnected before repairs are made. Trailing cables should be kept off the ground whenever possible and should be handled only with insulated hooks.

Gasoline-operated cranes require protection against the hazards of fire and explosion. Engines shall not be refueled while running. If refueling is done by hose connection from a tank truck or from drums by means of pumps, metallic bonding connection between the hose nozzle and fill pipe should be maintained.

If fuel is transported to the crane by hand, safety cans should be used. Open lights, flames, and sparks should be eliminated, and lights on the equipment should be of an approved explosion-proof type. A fire-extinguisher of 5BC rating shall be kept in the cab of the rig.

To prevent the boom from being dropped accidentally, the boom hoist mechanism should be operated by gearing or chain and the lowering speed controlled by the speed of the engine. A self-setting brake and locking pawl or other positive locking means should be provided. Furthermore, with the exception of some hydraulic cranes, boom stops—preferably of the shock-absorbing type—should be installed on all cranes.

It is desirable, if practicable, to install a hoist line, two-blocking limiting device. This device could be controlled in conjunction with the load-line hoist clutches and brakes. The operator must have full control of all crane functions at all times.

No attempt should be made to lift the boom by means of the load hoist cable. A load should not be lifted by the boom hoist line unless the crane is designed for this purpose.

Every crane should have on it a capacity plate or a sign plainly legible to the crane operator, signaler, and rigger, stating the safe

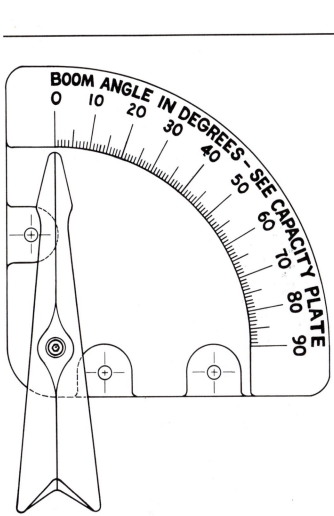

Figure 4-20. Indicator has freely suspended pointer that tells the boom angle. Each crane has a maximum angle that for given loads must not be exceeded, and the pick must be within the limits prescribed by the load chart.

loads at various radii from the centerpin of the turntable. A boom angle indicator can be mounted on the side of the boom near the hinge, with a pointer actuated by gravity suspended freely in front of it (Figure 4-20).

A capacity chart for the operators should indicate boom length, boom angle, and capacity. A load indicator device will better enable the operator to handle the load. For locomotives, using outriggers fully extended, the maximum load rating is 80 percent of tipping load. Where structural competence governs lifting performance, load ratings are reduced and the rating chart should so indicate. When loads which are limited by structural competence rather than by stability are to be handled, the person responsible for the job shall ascertain that the weight of the load has been determined within plus or minus 10 percent before it is lifted.

The operator must have safe access to and egress from the cab or seat. To operate the crane safely, the operator must have an unobstructed view of the load hook and the point of operation at all times or rely on someone giving signals. The operator must also be able to see ahead of the crane when it is traveling on the ground, whether the chassis is moving foward or backward. On some cranes, visibility to the rear is obstructed, and the operator must use extreme caution, and rely on a signaler.

A crane operated after dark should have clearance lights. Floodlights should illuminate the area beneath the boom, and lights mounted on the underside of the boom are recommended.

A warning bell or horn and an automatic backup alarm are necessary equipment for a wheel-mounted crane. A signaling warning bell or horn should be provided on crawler cranes.

Workers should *always* be kept beyond the range of the cab swing and out from under the boom and the load.

Operation. Extended outriggers are considered a part of the counterweight on new crane load charts. There are separate charts for stating crane capacity for a traveling load and for lifting a load without using outriggers. Whether traveling or stationary, the crane's turntable shall remain level.

A boom must never be swung too rapidly; if it is, the suspended load will be swung outward by centrifugal force, so that the crane may rock or even be upset and the load may swing and strike a person or object.

Operating a crane on soft or sloping ground or close to the sides of trenches or excavations is dangerous. A crane should always be level before it is put into operation. Outriggers can be relied upon to give stability only when used on solid ground. Heavy timber mats should be used whenever there is doubt as to the stability of the soil on which a crane is to be operated.

The use of any makeshift methods to increase the capacity of a crane, such as using timbers with blocking or adding counterweight, should not be permitted.

If the crane tips when hoisting or lowering a load, the operator should lower the load as quickly as possible by snubbing it lightly with the brakes. Workers, therefore, should never be allowed to ride a load that is being hoisted, swung, or transported.

When operating a crane with the boom at a high angle, the operator should take care that the suspended load does not strike the boom and bend the steel lattice bars on its underside. Bending these members will weaken the boom; and, when it picks up the next heavy load, it may collapse. Likewise, if the main members of the boom are bent even slightly, the strength of the boom may be materially reduced.

When an extended boom is used on a crane, as for structural steel erection, the operator must use extreme care in lowering it to the ground at the completion of the job. An extended boom never should be lowered to one side of the chassis or crawler, for the stability of the crane is noticeably reduced in that position and the crane may upset.

When using a boom tip extension or jib, the allowable load on the jib is limited. Its capacity must be known by the operator. The operator must refer to the crane's capacity chart in order not to exceed load limits. The crane operator must use care when swinging with a load, especially when the jib is lowered at an angle to the main boom.

The hook must be centered over the load to keep it from swinging while it is being lifted. Employees should keep their hands out of pinch points, when holding the hook or slings in place, while the slack is taken up. A hook, or even a small piece of board, may be used for the purpose. If a person must use his hand, it should be placed flat against the sling to hold it. The hook-on person, the rigger, and everyone else must be in the clear before the load is lifted. A tag line should be used for guiding loads.

A heavy load should never be removed from a truck by hooking a crane to the load and then having the truck pull out from under it. If the load should prove too heavy for the crane, the crane will upset before the operator can lower the load to the ground. The load should be lifted clear of the truck body, and the operator should make sure that the crane can handle it safely before the truck is moved out from under it.

A crane should never be used to jerk piling. If piling cannot be pulled by a straight steady pull limited to rated capacity, a pile extractor should be used. When a pile extractor is used, the boom angle should be kept at or less than 60 degrees above horizontal.

Before leaving the crane at the end of the work day, the operator should lower the load block or bucket onto the ground in such a manner that it cannot be upset.

The manufacturer of the equipment should be consulted before any modifications are made. Such changes shall maintain at least the same factor of safety as the original designed equipment. Maintenance and repair work should be performed only by trained and qualified personnel. The operator, however, is responsible for keeping the unit clean (Figure 4-21).

Travel of cranes. Except for very short distances, a crane should not travel with a load suspended from the boom. When a crawler crane must travel on a public thoroughfare, the boom should point forward and someone with a flag should walk ahead of it.

A wheel-mounted crane and a crawler crane on a semitrailer should be transported with the boom pointing toward the rear

Figure 4-21a. Ways that the crane operator can contribute to all-around safety. Because a dirty or cluttered deck invites falls, the operator should sweep the deck and wipe up any oil or grease. (Printed with permission from Power Crane and Shovel Association).

Figure 4-21b. Even if the crane is in perfect working order, a dirty windshield can obstruct the view. Spray-on window cleaner and paper towels make cleaning easy.

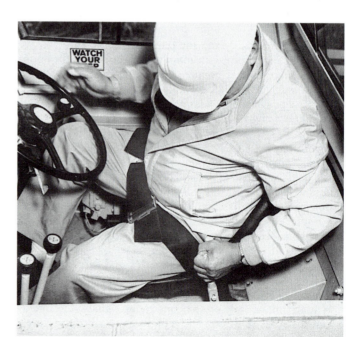

Figure 4-21c. Operator must set the handbrake of his crane whenever he is not operating the unit, regardless of the terrain. Wearing a safety belt is an important precaution.

and high enough to clear an automobile. Wheel-mounted cranes with short booms may travel with the boom forward in the boom rest. One of the work crew should follow in a car or truck to keep other vehicles from traveling beneath the boom. Otherwise, if the carrier wheels should roll into a low spot in the pavement, the boom might suddenly crash through a car roof.

If not disassembled, a crane being transported should have the crane engine running and the operator in the cab swinging the boom, when necessary, to avoid fouling trees, poles, or buildings when the vehicle turns corners.

Before heavy, slow-moving equipment or heavy equipment on a low-slung trailer is moved over any public or private railroad grade crossing, a responsible respresentative of the railroad company should be notified. The railroad then can provide flag protection to guard against a train's striking the equipment, while it is moving over the crossing or if the equipment becomes stalled or "hung up" on the crossing. This is an important precaution for the benefit of both the equipment owner and the railroad company.

Electric wires. Any overhead wire should be considered an energized line until either the person who owns the line or the electric utility authorities indicate that it is not energized. Compliance with recommended practices, and not reliance on other devices, shall be followed in determining proximity of the crane and its protuberances, including load, to electric power lines. A qualified signaler shall be assigned to observe the clearances and give warning before approaching the stated limits. The boom load line and cables of a crane shall be kept away from all electric wires, regardless of their voltage. In the United States, OSHA requires that, except where the electrical lines have been deenergized and visibly grounded at the point of work, or where insulating barriers, not a part of or an attachment to the crane, have been erected to prevent physical contact with the lines, cranes shall be operated near power lines only in accordance with the following:

- For lines rated 50 kv or below, minimum clearance between the lines and any part of the crane must be 10 feet (3 m).
- For lines rated over 50 kv, minimum clearance between the lines and any part of the crane must be either 10 feet plus 0.4 inch (3 m plus 10 mm) for each 1 kv over 50 kv, or twice the length of the line insulator but never less than 10 feet.
- In transit and with no load and boom lowered, the clearance should be a minimum of 4 feet (1.2 m).

If cage-type boom guards, insulating links, or proximity warning devices are used on cranes, such devices must not be a substitute for the requirements of a specifically assigned signal person, even if such devices are required by law or regulation. In view of the complex, invisible and lethal nature of the electrical hazard involved, and to lessen the potential of false security, devices shall be used and tested in the manner and at intervals prescribed by the device manufacturer.

If the boom load line or cables accidentally come into contact with a wire, the operator should swing the crane to get clear. If the wire has been broken and the boom cannot be cleared from it, the operator shall stay on the crane and remain calm.

A crawler crane, if the ground is wet or damp, will be electrically grounded and, when the boom touches a power line, the wire will, in turn, be grounded and the power company circuit breaker will open. Some arcing may occur. After a few sec-

onds, however, the circuit breaker will automatically close and reenergize the wire. Again the circuit breaker may open, and again it will close. Thus the wire may be "dead" at one instant, but live a few seconds later.

If the boom of a wheel-mounted crane on rubber tires should become tangled with a "hot" electric wire, the entire crane may be energized because the rubber tires may or may not insulate the crane and chassis from the ground. Depending upon the voltage and the soil conditions, the tires on the crane may burn and melt, and thus lose any insulating qualities. Hence, the circuit breaker may not open, and the wire *and* the crane may remain energized.

Stepping from the crane to the ground is often fatal, for one hand and one foot may be in contact with the crane when the other foot touches the ground. Therefore, the operator should remain on the crane until the emergency crew from the electric company arrives and frees the crane from the live wire. However, if the gasoline tank should become ignited, or if for any other reason it is impossible to remain on the crane, the operator should *jump,* making sure that all body parts are clear of the crane before touching the ground.

Crawler and wheel-mounted cranes are frequently used in structural steel erection for tall buildings. Extension sections may be inserted to lengthen the booms of these cranes. When the boom is assembled, check all parts of the structure for damaged and missing members.

In one documented case, a lattice bar was accidentally omitted. Later, when a heavy load was picked up, the boom buckled at this point and several men were killed.

Locomotive cranes. In the cab of a locomotive crane (Figure 4-22), a clear passageway should be provided from the operator's platform to an exit door on the nearer side of the cab. Doors should be hinged at the rear edge and should open outward. Sliding doors should slide to the rear to open.

The motor and all power transmission apparatus should be guarded.

Enough light should be provided in the cab to permit the operator to work safely and to see the gages and indicators plainly.

Steps and handholds should be installed for safe access to the cab. Some state laws require that footboards and handholds be provided at each end of the truck bed, or that the truck have a pilot or fender. No one except necessary operating personnel should be permitted on a rig while it is operating.

An on-track crane should have standard automatic couplers and uncoupling levers, and air brakes as well as hand brakes.

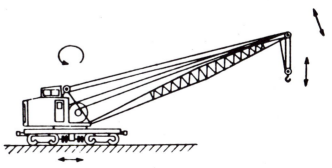

Figure 4-22. Locomotive crane.

It should have rail clamps at each corner to hold it in a stored position. A guard should be provided at the end of the boom to prevent the thimble on the cable from coming into contact with the sheave.

An on-track crane should be moved only on signal from an authorized signaler or switchman, who should walk ahead of the crane to warn others and to see that switches are properly set and the track is free of obstructions. When no signaler or switchman is employed, the operator should move the crane only on order from the supervisor of the department in which the crane is working.

The crane should not be swung across another track until the craneman and signaler have made sure that cars are not on that track and will not be moved to it.

When moving the crane about the yard or working space, the operator should keep the crane and boom parallel to the track to avoid striking buildings or other structures and should carry the boom low enough to clear overhead wires. Buckets and magnets should not be carried on the boom when the crane is going from one location to another.

Aerial baskets

Aerial lift equipment is now commonly used for working above ground. These boom-mounted buckets, baskets, or platforms are used extensively in construction and maintenance of electric and communication lines. Because of their capabilities, however, their use is increasing in harbor and port work, the aircraft industry, highway sign and lighting construction, and in maintenance, painting, sandblasting, and firefighting work.

Hazards. The most frequent causes of accidents while using mobile aerial baskets include:
1. Failure to observe proper precautions against electrical hazards to personnel both aloft and on the ground.
2. Improper positioning of vehicle or outriggers, lack of sufficient blocking under outriggers, or overloading the boom. Any one of these can cause the apparatus to overturn or fail.
3. Over-reaching from basket or other improper working procedures.
4. Failure to use proper personal protective equipment.
5. Improper moving of the truck while the boom is raised, or moving it where there is inadequate clearance for the boom.
6. Failures of structural or mechanical parts, or failure or jamming of controls.
7. Swinging the boom or basket against overhead obstructions or energized equipment.
8. Moving the boom into positions that interfere with traffic.
9. Failure to adequately train personnel.

Operation. The operating and maintenance instruction manuals issued by the manufacturer shall be followed. Lift controls shall be tested each day prior to use to determine that such controls are in safe working condition. Aerial baskets shall be inspected daily for defects. Mechanical equipment shall be inspected each day before using to make sure it is safe to operate. Additional safe operating procedures include:
1. Load limits of the boom and basket shall be posted and shall not be exceeded.
2. A warm-up period and test of the hydraulic system is required.
3. Basket equipment approved for use on energized equipment

shall be dielectrically tested periodically.
4. When working near energized lines in aerial basket trucks and aerial ladder trucks, the trucks shall be grounded. Where grounds are not permitted by the client, barricading shall be required.
5. The insulated portion of an aerial lift boom and basket shall not be altered in any manner that might reduce its insulating value.
6. Drivers of trucks with mounted aerial equipment shall be constantly alert to the fact that the vehicle has exposed equipment above the elevation of the truck cab, and they shall provide necessary traveling clearance.
7. The truck shall not be moved unless the boom is lowered and the basket or ladder is cradled.
8. Riding in the basket while truck is traveling shall not be permitted. (Employees may ride in the basket at the work location for short moves if the basket is returned to the cradled position for each move.)
9. Available footing for the truck wheels and outriggers shall be examined carefully to be assured of a stable setup. Hand brakes, chocks, and/or cribbing, when needed, shall also be used to insure stability. The truck shall sit approximately level when viewed from the rear.
10. Before lowering stabilizers, outriggers, or hydraulic jacks, the operator shall be certain there is no one in an unsafe position.
11. When the boom must be maneuvered over a street or highway, necessary precautions shall be taken to avoid accidents with traffic and pedestrians.
12. The operator shall always face the direction in which the basket is moving and ascertain that the path of the boom or basket is clear when it is being moved.
13. An employee shall not stand or sit on the top or the edge of the basket. Ladders shall not be used in the basket. While in the basket, the employee's feet shall always be on the floor of the basket.
14. Employees shall not belt to an adjacent pole or structure. Employees shall always belt to the boom or ladder. Belting to the basket equipment shall be done upon entering the basket.
15. When working with rubber protective equipment on energized circuits or apparatus above 300 volts, the following minimum conditions shall be met in addition to all other rules governing the use of protective equipment:
   a. Rubber gloves with leather protectors and rubber sleeves shall be worn.
   b. An employee shall make physical contact with protective devices installed on energized primary conductors with only rubber gloves and rubber sleeves.
   c. Employees shall be isolated from all grounds by using approved protective equipment or other approved devices.
   d. In no case, when employees are working on the same pole, substation structure, or from a bucket truck, shall they work simultaneously on energized wires or equipment of different phase of polarities.
16. An employee shall not enter or leave the basket by walking the boom.
17. Employees shall not transfer between the basket and a pole. On dual basket trucks, employees shall not transfer between the baskets.

18. Climbers shall not be worn by employees in baskets.
19. When two workers are in the basket or baskets, one of them shall be designated to operate the controls. One employee shall give all signals and make sure these signals are thoroughly understood by all persons concerned.
20. Baskets shall be located under or to the side of conductors or equipment being worked. Raising the basket directly above energized primary conductors or equipment shall be kept to a minimum.
21. Only non-conductive attachments shall be allowed on baskets.
22. The operator shall be sure that hoses or lines attached to tools cannot become entangled with the levers that operate the boom.
23. Air or hydraulic operated tools shall be disconnected from the source when not in use.
24. In no case, when the employees are working on the same pole, substation structure, or from a bucket truck, shall they work simultaneously on energized wires or equipment of different phase of polarities.
25. When employees are working from the basket, extreme care shall be taken to avoid contacting poles, crossarms, or other grounded or live equipment.

Inspection. An effective daily inspection should cover the following:
1. Visual inspeciton of all attachment welds between actuating cylinders and booms or pedestals.
2. Visual inspection of all pivot pins for security of their locking devices.
3. Visual inspection of all exposed cables, sheaves, and leveling devices for both wear and security of attachment.
4. Visual inspection of hydraulic system for leaks and wear.
5. Check of lubrication and fluid levels.
6. Visual inspection of boom and basket for cracks or abrasions.
7. Operation of boom from ground controls through one complete cycle. Look for unusual noises and deviations from normal operation.

Defects found should be reported and corrected before they develop into dangerous conditions.

Basket safeguards. Aerial baskets should be equipped with (1) safety belts and lanyards to be worn by all persons working from the baskets, and (2) a means for attaching the lanyard to the equipment. In general, it is more satisfactory to anchor the lanyard to the boom. But, if this will interfere with the controls or if other considerations are involved, then anchor it to the basket. Lanyards should only be long enough to allow movement within the basket and should prevent climbing onto the rim. Such lanyards limit free fall in an accident, but do not restrict work or entangle workers' feet.

Traveling. Each basket operator and driver should be thoroughly trained in the use of the equipment before running it on a job. They should not only know the particular equipment involved but also the type of work it will do in the field.

Proper barricading should be established if the public is exposed to possible contact with the vehicle because a boom contact with energized equipment might electrically charge the truck.

Good solid footing for the wheels and outriggers is essential, since many accidents have been caused by inadequate foot-

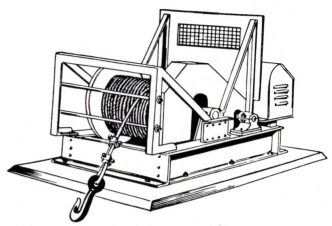

Figure 4-23. Operator of this car puller winch stands behind the shield, which protects the employee if the rope breaks.

ing. Snow, ice, mud, and soft ground call for extra caution because additional firm footing under the outriggers may be necessary.

Crabs and winches

Crabs and winches (Figure 4-23) may be either hand operated or electrically driven. Some form of brake or safety lowering device should be installed, and portable units should be anchored securely against the pull of the hoisting rope or chain.

Barricade guards should be installed to protect the operator against flying strands of wire and the recoil of broken ropes. Also, be sure gears are fully guarded. Power-driven crabs and winches should have their moving parts encased and should be electrically grounded.

The locking pawl on the ratchet of a winch frequently presents a serious finger hazard, particularly when the operator attempts to disengage the pawl. To reduce this hazard, a small lever can be welded to the pawl so that it can be grasped safely.

A major danger with hand-operated equipment (that has a crank handle instead of a hand wheel) is that the operator may be struck by the revolving crank handle if he loses control while lowering a load. A dog should be provided to lock the gears. A pin through the end of a crank will keep it in the socket during hoisting operations.

To lower loads rapidly, a strap brake is practicable. Before using the brake, the crank should be removed, or other steps taken, to prevent the crank handle from flying around, such as replacing a spur gear and dog with a worm gear.

Blocks and tackles

A factor of safety of 10 is recommended for determining the safe working load of manila rope falls in a block-and-tackle assembly. The purpose of this large safety factor is to allow for error in estimating the weight of the load, for vibration or shock in handling the load on the tackle, for loss of strength at knots and bends, and for deterioration of the rope due to wear or other causes.

The governing value usually is the safe working load limit of the blocks, rather than of the falls (rope). The reason is that multiplying the number of sheaves and rope parts multiplies the weight of the load that can be handled by the rope but does not correspondingly increase the strength of the blocks. Calculation will show that, in most instances, using a safety factor of 10 for the rope will automatically keep the load on blocks corresponding to the rope size within safe working load limits.

Blocks should be plainly marked with their safe working loads, as specified by their manufacturers, and the total weight on the tackle never should exceed this safe load. Breaking strengths of fiber ropes are shown in Chapter 5, Ropes, Chains, and Slings. Safe working loads for rope used in block and tackle assemblies are conversely 1/10 of the block's breaking strength, based on a safety factor of 10.

To find the required breaking strength for new rope, proceed as follows: For each sheave 3 in. (7.6 cm) in diameter or larger, add 10 percent to the weight of the load to compensate for friction loss. Divide this figure by the number of ropes or parts running from the movable block, and multiply the resultant figure by a safety factor of 10.

For example, a load to be lifted weighs 2,000 (900 kg) and the tackle consists of two double blocks—four sheaves, four rope parts at the movable block. Friction loss (10 percent for each sheave) = 40 percent or 800 lb, 2,000 + 800 = 2,800 lb, which divided by 4 (the number of parts at the movable block) = 700. Applying the safety factor of 10 (10 × 700) gives 7,000 lb (3,200 kg), the required breaking strength of the rope. New manila rope of ⅞ in. (22 cm) has a breaking strength of 7,700 lb (3,500 kg) and, therefore, is the proper size for the load. (Synthetic fibers would have greater tensile strength. See Table 5-A in Chapter 5.)

The safe working load for two double blocks made for rope of ⅞ diameter as given by a prominent manufacturer for one series of its standard blocks (regular mortise, inside iron strapped blocks, with loose side hooks, intended for use with manila rope), is 2,000 lb (900 kg)—the equivalent of the total load in the example.

The rope should be attached to the block with a thimble and a proper eye splice. A mousing of yarn or small rope should be placed on the upper hook of a set of falls as a precaution against its accidental detachment.

Blocks should be inspected thoroughly and frequently, with particular attention to all parts subject to wear.

Figure 4-24 shows how tackle blocks should be reeved.

If the sheave holes in blocks are too small to permit sufficient clearance, excessive surface wear of the rope will occur. Likewise, excessive internal frictional wear on the fibers will occur if the diameter of the sheave is too small for the rope.

When using block and tackle in confined spaces, the pulley block must be guarded so that a person's hands cannot be caught between the pulley and the rope.

When blocks and falls are used to lift heavy materials or to keep heavy loads in suspension, as on heavy duty scaffolds, wire rope is more serviceable than fiber rope.

Portable floor cranes or hoists

Portable floor cranes or hoists are hoists mounted on wheels. They can be moved from place to place, either by hand power or under their own power. They can raise and lower loads in a vertical line, but cannot rotate around a fixed point.

Portable floor hoists are useful where overhead construction, belting, or shafting prevents the use of overhead hoists or cranes and where more expensive equipment is not justified because of

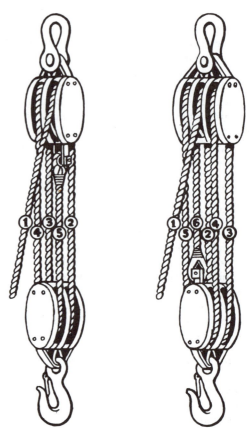

Double and Double   Triple and Double

Figure 4-24. Reeving tackle blocks. Lead line and becket line should come off a middle sheave when blocks contain more than two sheaves. The upper and lower blocks then will be at right angles to each other. This eliminates tipping and the accompanying loss in efficiency.

infrequent use. These hoists are handy for placing work on machines, loading heavy material on trucks, and transporting material from one location in the shop to another.

Portable floor hoists usually are operated by hand or by electric power. The lifting mechanism ordinarily is either a winch with wire rope and block or a chain hoist.

Hoists operated by electric power should be effectively grounded to prevent shock in case of short circuit. A special ground wire can be used. One end is fixed permanently to the frame of the hoist, and the other end is equipped with a device that can be attached to a grounded building column, water line, or other direct-to-ground connection.

Where conditions permit, it may be advisable to install sweep guards on the truck wheels to prevent foot injuries when the hoist is moved.

Truck handles on hoists should be designed to stand upright when not in use. If permitted to project horizontally from the hoist or to lie on the floor, these handles present a serious tripping hazard.

Tiering hoists and stackers

The tiering hoist—sometimes called a stacking elevator, portable elevator, tiering machine, or platform hoist—is designed to raise material in a vertical line on a moving platform. This hoist is portable and is used extensively in warehouses for piling and storing materials. It is operated either manually or electrically. The large capacity machines usually are power driven.

Tiering hoists should have a braking device to permit safe lowering of the platform, and a rachet lock (or dog) to lock the platform in position for loading and unloading operations. Workers should not be permitted to ride the platform because they may be crushed should the platform meet an obstruction.

Tiering machines, especially the revolving type, should be operated so that they will not tip over. Essential precautions include having the machine solidly on the floor, making sure that its safe capacity is not exceeded, and placing the load properly on the platform.

Before the machine is used, the casters should be lifted off the floor. One type of hand-operated hoist is arranged so that the platform cannot be moved unless the machine stands solidly on the floor—on the frame and not on the casters.

Material in the form of rolls and other round objects should be blocked to prevent their rolling off the platform.

Power-operated tiering hoists should be grounded, and the wiring preferably should be safeguarded by armored cable. On two-section machines, locks should be provided to prevent release of the upper section. Gears and channel iron guides should be guarded to prevent shearing of fingers.

Some tiering machines have practically all the safeguards that a freight elevator has, including limit stops for top and bottom travel on the hoisting cable drum as well as for the shipper rope, if one is provided.

CONVEYORS

Because there has been much confusion as to what is a conveyor and what is not, and as to correct names for different types of conveyors, the general definition given in the American National Standard B.20.1, *Safety Standards for Conveyors and Related Equipment,* here quoted in full, including exceptions:

■ CONVEYOR. A horizontal, inclined, or vertical device for moving or transporting bulk material, packages, or objects, in a path predetermined by the design of the device, and having points of loading and discharge, fixed or selective. Included are skip hoists, and vertical reciprocating and inclined reciprocating conveyors. Typical exceptions are those devices known as industrial trucks, tractors, and trailers; tiering machines; cranes, hoists, power shovels; power scoops, bucket drag lines; platform elevators designated to carry passengers or the operator; manlifts, moving walks; moving stairways; highway or rail vehicles; cableways; or tramways, pneumatic conveyors, or integral transfer devices.

By industry agreement, nomenclature and definitions have been standardized and published as *Terms and Conveyor Definitions,* ANSI/CEMA 102. Terms and definitions in this Manual and in ANSI/CEMA 102 follow this standard.

General precautions

The most common conveyors are of the belt, slat, apron, chain, screw, bucket, pneumatic, aerial, portable, gravity, live roll, en masse, flight, mobile, and vertical types. All should be designed and constructed to conform with applicable codes and regulations. Figure 4-25 shows some suggested guards which can be used along the conveyor's entire length.

Manually loaded conveyors, traveling partially or entirely in a vertical path, should have a conspicuous sign at each loading point showing the safe load that can be raised or lowered.

Gears, sprockets, sheaves, and other moving parts must be protected either by standard guards or must be positioned in such a way to ensure against personal injuries.

The entire conveyor mechanism should be inspected periodically, and any part showing signs of excessive wear should be replaced immediately. Particular attention should be paid to brakes, backstops, antirunaway devices, overload releases, and other safety devices to ensure that all are operative and in good repair.

All machine parts should be lubricated according to the manufacturer's instructions. If grease nipples are installed on long tubes or pipes that permit oilers to keep a safe distance from moving parts, then the conveyor need not be shut down for greasing.

All conveyors within 6 ft-8 in. (2 m) of a floor or walkway surface must have alternate passageways that comply with *Life Safety Code* NFPA 101 requirements if the walkway is a means of egress. Frequently, a work platform on a movable conveyor tripper can be used as a crossover, if properly railed.

Underpasses should have sheet metal ceilings. Where overhead conveyors dip down at work stations, guards or handrails should be provided. Guards should be provided below all conveyors passing over roads, walkways, and work areas.

Conveyors running in tunnels, pits, and similar enclosures should be provided with adequate drainage, lighting, ventilation, guards, and escapeways wherever it is necessary for persons to work in or enter such areas. Sufficient side clearance should be provided to allow safe accessway and operating space for essential inspection, lubrication, repair, and maintenance operations.

Where conveyors pass through building floors, the openings should be guarded by standard handrails and toeboards. As a fire precaution, each opening should be protected against the passage of flame or super-heated gases, from one floor to the next, by doors that close automatically or by fog-type automatic sprinklers, so placed as to provide a curtain of waterfog across the opening. Where a conveyor passes through a fire wall, similar protection should be provided. Conveyor tunnels under stock piles of materials should be open at both ends.

If the top of a loading hopper is at or near the level of a floor or platform, the hopper should be protected by a bar guard with openings not greater than 2 in. (5 cm) in one dimension such as 2 × 12, 14 × 2, 2 × 16 in. (5.1 × 30.5 cm, 35.5 × 5.1 cm, 5.1 × 40.6 cm) or by standard railings and toeboards.

Elevated conveyors should have access platforms or walkways on one or both sides (Figures 4-25a and 4-26a and b). Handrails should be 42 in. (1 m) high with an intermediate rail, and platforms should have 4 in. (10 cm) toeboards. Flooring should be of checkered plate or other nonslip surface, particularly on sloping walkways.

Both sideboards along edges and at corners and turns of overhead conveyors and screen guards underneath high runs will protect workers from falling material.

Crossovers or underpasses with proper safeguards should be provided for passage over or under all conveyors. Crossing over or under conveyors, except where safe passageways are provided,

SUGGESTED SAFEGUARDING OF CONVEYORS

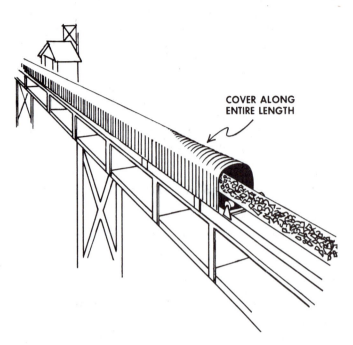

Figure 4-25a. Covered conveyor.

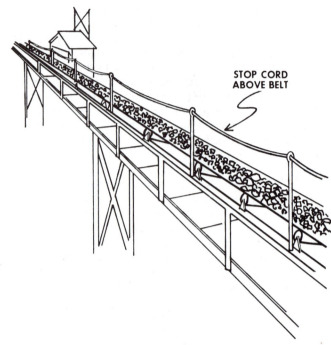

Figure 4-25b. Emergency stop cord along entire length of conveyor.

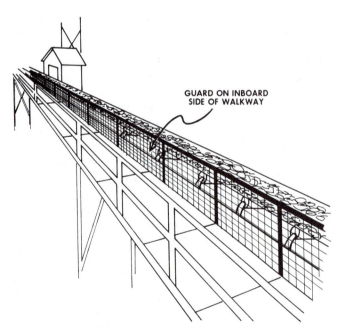

Figure 4-25c. Expanded metal along entire length.

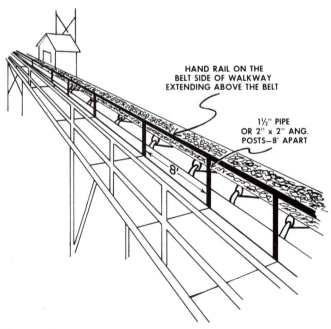

Figure 4-25d. Detail of guardrail.

should be prohibited. Riding on a conveyor should be absolutely forbidden.

Operating precautions. The starter button or switch for a conveyor should be located so that the operator can see as much of the conveyor as possible. If the conveyor passes through a floor or wall, then each area should be equipped with starting and stopping devices, and the simultaneous operation of all starting buttons or switches should be required to start the conveyor. These start-stop devices should be marked clearly, and the area about them must be kept free of obstructions so they can be seen and reached easily. All personnel working on or about the conveyor should be instructed in the location and operation of all stopping devices.

Electrical or mechanical interlocking devices (or both types) should be provided which will automatically stop a conveyor when the unit it feeds (another conveyor, bin, hopper, or chute) has been stopped or is blocked, so that it cannot receive additional loads.

If two or more conveyors operate in series, controls should be designed so that, if one conveyor is stopped, all conveyors feeding it are stopped also.

Emergency stop devices should be located not more than 75 ft (23 m) apart along walkways by the conveyor. For some installations, a good solution is to have a lever-operated emergency stop device at the tail end of the conveyor, with a strong cord or wire strung on each side of the conveyor for its entire length. A pull on the cord or wire will stop the conveyor (Figure 4-25b).

On conveyors, where there is a possibility of reversing or running away, antirunaway and backstop devices can be provided or the conveyor track should be designed so that the load (or conveyor parts) cannot slide or fall in event of mechanical or electrical failure. If such design is not practicable, guards capable of withstanding the impact shock and of holding the falling load should be installed.

Electric machines, with brakes that are mechanically applied and released by the movement of operating devices, must be so designed that, if the power is interrupted with the brakes in the Off position, the load can descend only at a controlled speed.

In addition to the overload protection customarily provided for electric motors, there should be an overload device designed to protect the conveyor and mechanical drive parts. In the event of an overload, the device must shut off the electric power quickly, disconnect the conveyor or drive parts from the motive power, or limit the applied torque. Shear pins and slip or fluid couplings are examples of overload protective devices.

When a conveyor has stopped because of an overload, all starting devices should be locked out and the cause of the overload removed. The entire conveyor should be inspected before it is restarted.

The loading and discharge points of a conveyor carrying material in fine or powdered form should be covered with exhaust hoods and should have good general ventilation, to prevent the formation of dust clouds. Dust removal must be in accordance with government regulations.

If the material is combustible, the concentration of dust must be kept below the lower explosive limit. Only approved explosion-proof electrical fixtures should be used, and all sources of ignition should be excluded from the dusty area. The conveyor should be grounded and its parts bonded to prevent differences in electric potential. The container into or from which the material is conveyed also should be grounded and its parts bonded.

Persons working near or on conveyors should wear close-fitting clothing that cannot become caught in moving parts. Safety shoes are recommended. If the conveyor galleries are dusty, goggles and, if necessary, respirators should be worn.

Maintenance precautions. Before maintenance personnel commence working on a conveyor, they should lock out the main power control in the Off position. The maintenance supervisor should carry the only key to this lock. If two gangs work

on one conveyor, the supervisor of each gang should lock out the master switch.

Areas should be readily accessible so that maintenance personnel can change the position of pulleys, sprockets, or sheaves to compensate for normal working conveyor stretch and wear.

The on-running belt should be guarded for at least 18 in. (55 cm) from the points of tangency between the belt and the head and tail pulleys and between the belt and the tripper and hump pulleys. If the hazard is out of reach (more than 8 ft (2.4 m) above the floor or platform) or close to a wall or other obstruction, workers in the normal course of their duty should not be exposed to it.

The fact that the skirtboards of the loading boot are close to the upper surface of the belt means that, if a person's arm should be caught, the belt could not be raised sufficiently to allow the arm to ride over an idler pulley under the belt where the arm undoubtedly would be badly mangled. Therefore, guards should be installed at the sides of the conveyor at the loading boot.

The points of contact where the wheels of movable trippers roll on the rails also must be guarded. Where the operator travels on the tripper, a platform should be provided. It should be so located and constructed as to protect the operator from slipping and falling and to prevent contact between the operator or his clothing and moving parts. To help prevent the operator falling into the hopper, the platform should have handholds and railings.

At the conveyor floor above the coal bunkers in power plants, the slot through which coal from the tripper chute is discharged into the bunker is protected by bars placed across it about 12 in. (30 cm) apart at floor level. A piece of discarded belting of the required width and length can be placed to cover the slot, and its ends are securely anchored. At the tripper platform are four pulleys that raise this belt vertically so that it passes over the access to the tripper platform. This device not only provides safety, but also seals the slot and thus keeps the dust in the bunker.

Whenever a person might fall onto a conveyor, a gate or paddle can be suspended as low as practicable above the belt near the head pulley, so that a person's body riding on the belt would automatically pull a stop rope and quickly stop the conveyor.

On belt conveyors at floor level or on balconies or galleries, a shield guard or housing should completely enclose each end. Guardrails and toeboards should extend the length of the conveyor.

To help remove static from belt conveyors, tinsel or needle-point static collectors can be placed close to the outrunning sides of the drive pulleys and idlers, which, along with the shafting, can be grounded through carbon or bronze brushes running on the shaft.

A belt that does not move too fast can be grounded to the drive gear by a continuous strip of copper foil on the pulley side. Other belt conveyors can be grounded by being treated with conductive belt dressings.

One of the dangers of belt conveyors is that workers are tempted to clean off material that sticks to the tail drums or pulleys while they are in motion. Fixed scrapers and revolving brushes eliminate the need for a hand operation.

Barrier guards can be placed directly in front of the pinch points of the belts and drums to protect workers should they attempt to clean or dress the belts while they are moving. The

Figure 4-26a.

Figure 4-27. On this positive drive carton conveyor note the underpass and triangular pinch-point guards.

Figure 4-26b. Note the service walkway and handrails on this belt conveyor system. (Printed with permission from Vulcan Materials Co.)

belt and drum should be guarded on the side also at a sufficient distance from the drum to prevent contact.

Belt conveyors

A belt conveyor is an arrangement of mechanical components that supports and propels a conveyor belt, which in turn carries bulk material. The five principal components of a typical belt conveyor are: (1) the belt, which forms the moving and supporting surface on which the conveyed material rides; (2) the idlers, which form the supports for carrying the belt; (3) the pulleys, which support and move the belt and control its tension; (4) the drive, which imparts power to one or more pulleys to move the belt and its load; (5) the structure, which supports and maintains the alignment of the idlers and pulleys and supports the driving machinery.

Like all moving machinery, belt conveyors present hazards to workers and must be safeguarded. Operators must be trained in safe work procedures around belt conveyors. These procedures should stress that workers stand clear from moving conveyors. They should be trained to avoid pinch points and other areas where their fingers or hands may be caught during operations. All pinch points should be guarded (Figure 4-27).

Other frequent causes of conveyor accidents are improper cleanup of conveyors and shoveling of spillage back on the belt.

Severe accidents also arise from (1) attempting repairs or maintenance on moving conveyors, (2) attempting to cross moving belts where no crossover exists, and (3) attempting to ride a moving belt. Moreover, workers standing or working on conveyors, particularly maintenance personnel, can be injured by falls or crushed against stationary objects, particularly if the conveyors are accidentally started.

The following injuries and hazards may be reduced by proper guards and environmental controls on bulk handling conveyors:

- Injuries from pinch points. Usually employees are pulled in when their hands, clothing, or tools are caught in the pinch point.
- Injuries from pieces of material falling from moving conveyors.
- Injuries and deaths to workers crushed against stationary objects by moving conveyors.
- Injuries to workers who fall from moving conveyors while riding on them or from conveyors that are started while they are standing or working on them.
- Injuries to workers who reach in around guards at head and tail pullies that are running.
- Injuries to workers who service moving conveyor belts.
- Injuries to workers who fall while trying to cross over conveyors where there is no crosswalk.
- Fires from friction, overheating, or static or other electrical sources.
- Explosions of dust raised by combustible materials at transfer points, where belts are loaded or discharged.
- Hazards to health and vision from irritating or toxic dusts.
- Electrical shock from ungrounded or improperly installed controls or conductors.

Transmission equipment and other power-driven parts should be guarded in accordance with the *Safety Standard for Mechanical Power-Transmission Apparatus,* ANSI B15.1, and with any state (or provincial) or federal regulation governing the safety, health, and welfare of employees. The ANSI standard stipulates how pulleys, chains, sprockets, belts, couplings, and other parts of conveyor drives should be enclosed.

Suitable sweeps should be provided for shuttle conveyors, movable trippers, traveling plows, and hoppers and stackers to push objects ahead of the moving pinch points between the wheels and the rails, in order to guard against nips. Again, it is imperative to comply with all state (or provincial) and federal regulations.

Pinch points at the head, tail, and take-up pulleys should be completely guarded.

An idler pulley becomes a hazard when skirt plates and chute skirts are so positioned as to force the belt against the idler and create a pinch point.

Mechanical belt cleaners, such as fixed or tension scrapers, revolving brushes, or rubber disks, sometimes eliminate manual cleaning. They also eliminate a major reason for working around moving pulleys.

In order to lubricate a conveyor in continuous operation, extension grease lines should be installed so that a worker cannot be caught by rollers and bearings when working around them. All grease fittings inside a guard enclosure (except those that move with the part they serve) should be fitted with extension pipes to make them accessible from outside the guard.

Slat and apron conveyors

The slat conveyor has one or two endless chains operating on sprockets and usually runs horizontally or at a slight incline (Figure 4-28). Attached to the chain or chains are nonoverlapping, noninterlocking slats that are spaced closely.

Apron conveyors have overlapping or interlocking plates that form a continuous moving bed. They vary greatly in size: one may be part of a bottling machine, and another may handle billets and castings in a steel mill.

Pinch points between slats or plates, and between them and the chain, sprockets, and guides, should be guarded. Where slats are spaced farther apart than 1 in. (2.5 cm), a serious shearing hazard exists between the slats and the conveyor substructure. When a slat conveyor is located at floor level or in working areas, the space under the top run of the slats should be filled in solid.

When designed for handling heavy material, slat and apron conveyors usually are installed flush with the floor to facilitate loading and unloading. When so located, conveyors should be guarded by handrails (except at loading stations) so that workers will not step onto the moving conveyor. Openings in the floor at the loading platforms should be guarded or covered.

Chain conveyors

Chain conveyors take many forms, but they all carry, pull, push, haul, or tow the load either directly by the chain or by means of attachments, pushers, cars, or similar devices.

Types in which the chain itself directly moves the load are drag, rolling, and sliding chain conveyors.

Those consisting of an endless chain that is supported by trolleys from an overhead track or that runs in a track above, flush with, or under the floor, with attachments for towing trucks, dollies, or cars, are known as tow conveyors. Tow conveyors use 4-wheel carts that are held in place by a pin through the chain. These pins should be equipped with an automatic disengaging mechanism. Many have a horizontal bumper or kick mechanism that will disengage with 3 to 5 lb (1.4 to 2.3 kg) of pressure. Workers are consequently protected from being caught between carts.

Trolley conveyors. Where an endless chain propels a series of trolleys (Figure 4-29) supported from or within an overhead track, with the loads suspended from the trolleys, the assembly is known as a trolley conveyor.

The return portion of the chain should be well guarded. Whenever possible, guardrails should be installed along both sides of the trough to prevent a person from stepping or falling into it. The path of travel of such overhead chain conveyors should be clearly designated. Emergency stop switches should be provided each 75 ft (23 m).

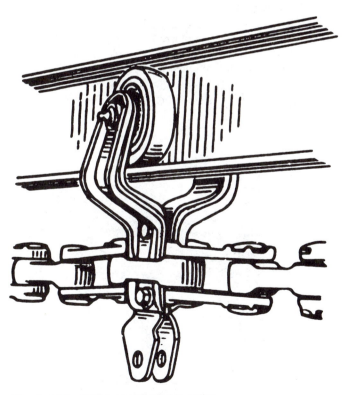

Figure 4-29. Trolley can be used in trolley conveyors, overhead tow conveyors, overhead chain conveyors, and on jibs and monorails to support hoists. (Printed with permission from Conveyor Equipment Manufacturer's Association.)

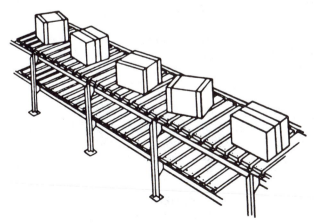

Figure 4-28. Slat conveyor.

Shackle conveyors

A shackle conveyor consists normally of chain-type conveyors, with suspended shackles evenly spaced along the line for conveying poultry or other meat products through a processing plant. In the case of a poultry processing plant, the shackle conveyor or conveyors carry the poultry from the beginning of the line where live poultry is hung on the line, from each shackle, head down, through all processing to the packing department where the finished product is removed for packaging and shipment. In some plants, one continuous conveyor travels through all departments and processes, while in other plants two or more similar conveyors carrying suspended shackles are used.

The speed of such conveyors varies from plant to plant dependent on the number of workers on any given conveyor line, with the speed generally increased with more workers on the line. Poultry is normally processed by hanging head down and, after cutting and bleeding, it is then repositioned feet down before going through scalders and beaters and then to viscerating and cleaning.

The unique hazard involves workers accidentally placing a thumb or finger in one of the shackles and being dragged through a beater or scalder or to the end of a platform where a worker would be suspended by a finger resulting in loss of life, a finger, or a thumb. The hazard should be controlled by placing emergency stop switches, actuated by pull wires positioned below and perpendicular to the line of travel, at critical points such as the end of a platform or before entry into an area hazardous to workers.

Screw conveyors

Screw conveyors convey bulk materials horizontally, on inclines, and vertically. They are employed in many industries and handle a variety of materials. A screw conveyor essentially consists of a continuous helix mounted on a pipe or shaft. The screw rotates in a stationary trough, generally U-shaped. Material introduced into the trough is conveyed by the screw rotation (Figure 4-30).

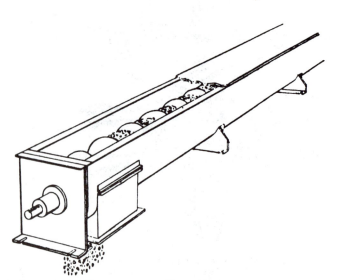

Figure 4-30. Screw conveyor. (Cover has been cut away to show interior.) (Printed with permission from Conveyor Equipment Manufacturer's Association.)

A screw conveyor which is not enclosed presents the hazard of entrapment, as feet, hands, or other portions of the body may be caught between the rotating screw and stationary trough. Such accidents can occur, and may result in serious injury or fatality. (See Figures 4-31 and 32 for warning label and cover fastening methods.)

These accidents can be avoided by implementation and enforcement of an in-plant safety program, and by training employees to follow certain basic safety rules:

- Covers, gratings, and guards shall be securely fastened before operating the conveyor. Conveyors should be provided with solid covers, securely fastened. When an exposed feed opening is required, it should be covered with a securely fastened grating. If an open housing is functionally necessary, the entire conveyor should be guarded by a fence or railing. Moving drive components shall be guarded.
- Never step or walk on covers, gratings, or guards.
- Lock out power before removing covers, gratings, or guards by padlocking the main disconnect in the OFF position.

More detailed information on screw conveyor operation and safety is available in Conveyor Equipment Manufacturers Assn. book No. 350, *Screw Conveyors;* and ANSI Standards B20.1 and Z244.1. (See References.)

Bucket conveyors

The three general types of bucket conveyors all have an endless belt, chain, or chains that carry elevator buckets, which are either fixed or pivoted (Figure 4-33).

- A BUCKET ELEVATOR carries fixed buckets in a vertical or inclined path. The buckets discharge by gravity as they pass over the head sheave or drum.
- A GRAVITY-DISCHARGE conveyor-elevator has fixed buckets and operates in vertical, inclined, and horizontal paths. The buckets act as flights while carrying material along a trough in the horizontal plane to a point of gravity-discharge.
- A PIVOTED-BUCKET conveyor also operates in horizontal, inclined, and vertical paths. The buckets remain in carrying position until they are tipped or inverted to discharge.

To protect operating personnel, bucket conveyors should be totally enclosed in a housing. No attempt should be made to take samples when the conveyor is in motion.

The pivoted-bucket conveyor has tripping devices for emptying the buckets. Frequently these tripping devices are movable so materials can be distributed evenly in the storage bins. If the control lever mechanism for shifting and locking the trippers can be operated by remote control, workers will not have to go on the conveyorway to change the position of the trippers.

For the safety and convenience of repair personnel and operators, a permanent footwalk alongside a conveyor that hoists material and carries it over stokers or bins is advisable. It should be well lit and equipped with standard handrails and toeboards.

Pneumatic conveyors

A pneumatic conveyor is an arrangement of tubes or ducts through which solid objects, such as cash, mail, and other small items or grain, dust, and similar bulk materials, are moved by using compressed air or a vacuum.

Solid objects are placed inside a cylindrical cartridge having packing on its outer surface to seal it inside the smooth tubing. Compressed air injected into the tubing system pushes the cylinder forward at a relatively high velocity.

Figure 4-31. This warning label is used by screw conveyor manufacturers to illustrate nature of the hazard, the severity of an accident and precautions to avoid it. Serious injury or death may result due to failure of personnel to follow basic safety rules.

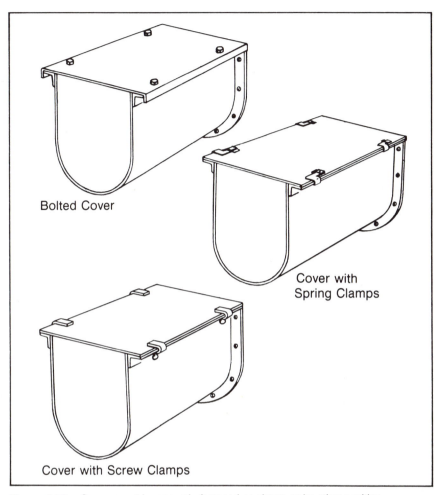

Bolted Cover

Cover with Spring Clamps

Cover with Screw Clamps

Figure 4-32. Covers must be securely fastened as shown or by other positive-locking methods. Interlocking safety switch devices may fail, become inoperative, and lull the operator into a false sense of security. They may also tempt the operator to ignore the basic safety rules. If interlocking safety devices are incorporated, they must be considered "secondary protection;" improper use or lack of understanding of their function can actually cause the devices to become a hazard.

To convey bulk material, compressed air is injected into the piping though a nozzle below the loading chute. The air mixes with the material making it flow rather like a fluid through the piping.

A constant volume, variable pressure (positive type) blower should have a relief valve on or adjacent to it. Doors of blower conveyors should be interlocked, so they cannot be opened if there is positive internal pressure. To keep conveyed material from being thrown against workers or into the working area if a gasket leaks, gaskets holding the line pressures should be shielded.

Where suction lines are large enough to draw in a person, bar guards or screening are needed over the intake and employees must be instructed to stay at a safe distance.

Receivers and storage bins should have full-bin indicators or controls to prevent overfilling.

A pneumatic conveyor serving an area containing contaminated air must be arranged so that no contaminated air enters the conveyor tube and is carried to other areas.

When transporting material that has a dust explosion hazard, the air velocity should exceed the critical velocity for flame propagation of the material to prevent propagation of an explosion from one point to another within the system. Equipment should be bonded electrically and grounded to prevent static electricity being an ignition source. Details are given in the General precautions section, earlier in this chapter.

Aerial conveyors

There are two types of aerial cable systems:
- A CABLEWAY is a wire rope-supported system in which the materials handling carriers are not detached from the operating span and the travel is wholly within the span;
- A TRAMWAY is a wire rope-supported system in which the travel of the materials handling carriers is continuous over the supports of one or more spans.

Aerial conveyors are used frequently in industrial plants, and particularly on large construction sites to carry material from point to point. They also move coal and ore. Workers are gen-

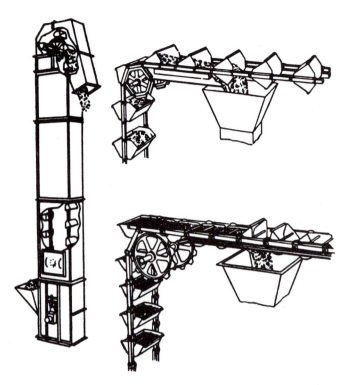

Figure 4-33. Three types of bucket conveyors. Left: bucket elevator. Upper right: gravity-discharge conveyor-elevator. Lower right: pivoted-bucket conveyor. (Printed with permission from Conveyor Equipment Manufacturer's Association.)

erally injured by falling material, or when inspecting and oiling the cables and carriages, and if signals are misunderstood.

No one should work directly under the conveyor except, of course, at the loading and unloading stations where exceptional precautions should be taken. Wherever workers must pass under the conveyor, a covered passageway should be provided. Heavy wire screens suspended under the conveyor from pole to pole that are wide enough to catch the material also provide protection from material falling on roadways or working spaces.

The equipment should be inspected regularly with special attention to the sheave wheels and bearings, the rope fastening, the bucket latch and trunnions, and all load-sustaining parts.

Keep all ropes well oiled, for operation efficiency and protection against the weather. Hauling rope should be continuously lubricated by means of a controlled drop feed from an oil reservoir at the point (one or both ends of the line) where the rope leaves the drive sheave and passes over a support sheave.

In one plant, a worker had to ride a trolley carriage to oil a tramway rope used for conveying coal. An automatic lubrication system was devised to eliminate this dangerous practice.

Suitable lighting should be provided at critical points for night operation and repairs.

Every aerial cableway must have a suitable signaling system. A telephone or electric pushbutton system is advisable. On large construction jobs, a portable telephone system allows a signaler to direct the operation of raising and lowering loads that are out of the operator's sight.

For a tramway, at least three control systems are recommended:
1. Pushbutton stations and a bell signal code that indicates stop, start, slow speed, high speed, and reverse.
2. An all-metallic, aerial, wire-circuit telephone with instruments at certain points along the line in addition to terminal sets.
3. A second telephone circuit, which may be grounded if desired.

The U.S. Army Corps of Engineers suggests these precautions be taken:
1. The control console compartment should be kept locked when the cableway is in use. No one should be permitted in the compartment with the operator while the cableway is in operation.
2. At least two communication and control systems should be maintained continuously between the signaler and the cableway operator. This dual system should include voice communication by telephone and radio. Lights or bells may be included with or substituted for one of the voice systems.
3. Only authorized inspection and maintenance personnel should be permitted to ride cableway carriages.
4. The riding of cableway load blocks should not be permitted.

Portable conveyors

Belt, flight, apron, extendable, and fixed-bucket conveyors are made as inclined portable units setting on a pair of large wheels. They are used to load railroad cars and trucks with bulk materials and to raise construction material from one elevation to another (Figure 4-34).

Portable conveyors should be provided with guards as specified for the corresponding types of fixed conveyors.

Electrical equipment on portable conveyers should be weatherproof. With three-phase power, the flexible cord that is connected to the power outlet should be a four-conductor, the fourth wire being grounded in all plugs and receptacles. Arrange the cable so it cannot be run over by trucks or other machines. If it must cross a driveway, hang it on poles at a minimum height of 14 ft (4.3 m). If two or more sections of cable are required, the connectors should be kept above the ground.

Skirtboards or sideboards (not less than 10 in. [25 cm] high)

Figure 4-34. Portable conveyor.

keep heavy material from falling over the sides and light or loose material from blowing off. This safety precaution applies to belt conveyors, as well as to the other types, because troughing of the belt gives insufficient protection against such spillage.

The conveyor should be stable and fitted with a locking device to hold the conveying unit at various fixed elevations.

The mechanism that raises and lowers the boom on all types of portable conveyors should be checked closely periodically—it is the major hazard area of such machinery. While any positive type may be used, the self-walking worm or jack-screw type is preferable. Gear drives should be completely housed and run in oil. All chains within easy reach of workers should be thoroughly guarded. A system for conveniently oiling chains is needed. In warehouses, in order to work in small spaces portable conveyors usually have booms consisting of two sections. These conveyors should be designed so material will not roll back from one section to the other at the transfer point.

When loading or unloading railroad cars with a portable conveyor, a suitable safety device is needed to prevent the car from shifting during the operation. Do not remove the device until the crew is sure no one is in the car. A red banner or a standard blue warning sign at each end of the car is recommended as a warning to the switching crew. (See the section, Plant Railways, in Chapter 7.)

When portable conveyors are used to raise or lower construction materials, ample stairs or ladders are needed in the immediate vicinity of the conveyor. Walking on idle or moving conveyors must be prohibited.

Gravity conveyors

Because gravity conveyors (Figure 4-35) depend wholly upon the natural pull of gravity, their operation often leads to the disregard of some necessary safe practices. If employees climb on a conveyor to release a blockage, they may slip on the rollers or be knocked down should the jam suddenly be released. Therefore, workers should be prohibited from climbing onto conveyors. In addition, the installation of steel or wood plates between rolls helps assure that neither a worker's body nor limb can fit between the rolls.

Chute conveyors. Polished metal sheets or bars are used to lower packing cases, cartons, and crates from one floor to

another and from the sidewalk to the basement of a building through a sidewalk elevator shaft or other opening.

The inclined chute may be straight, or have a vertical curve of large radius to deliver packages onto the lower floor without impact or damage.

Some gravity-chute conveyors are built like a spiral around a vertical pipe with a slope at the outer edge between 18 and 30 degrees.

In removing packages from the delivery end of spiral chutes, workers frequently injure their hands when they are caught by descending packages or are mashed against others on the delivery table. Where the chute is enclosed, a warning sign should be placed over the delivery end. A simple mechanical or electrical device that will give warning when a package is about to be delivered from the chute is advisable, especially where the packages cannot be plainly seen in descent.

Spiral chutes present a serious fire hazard because they form flues from lower floors to upper floors through which fire will spread quickly. Two methods of eliminating this hazard are to enclose the chute in a tower made of fire-resistant material, such as steel, concrete, or masonry; or to provide automatic fire doors (draft checks) where the chute passes through floors.

The enclosed tower has doors at each charging station and a door at the delivery end. The charging station doors should be kept closed, except when charging is being done. The door at the delivery end should close automatically in case of fire.

Shutoff doors (draft checks) are of two kinds—the vertical sliding and the shutter type. Both types should have fusible links so that they will close automatically in case of fire.

Where an open chute is used, a guardrail and toeboard should be provided at each floor. The charging stations should be guarded either by a movable railing or by a hinged door or gate.

Roller or wheel conveyors are similar to chute conveyors, except that the angle of slope is much less (2 to 4 percent). The conveyors (Figures 4-36 and 37), therefore, can be used to convey packages for considerable distances on one floor. If the rollers or wheels are placed radially instead of parallel, the course of travel can be changed from a straight line to a curve.

The principal hazards in the use of roller conveyors are that material may run off the edge of the rollway and fall to the floor, and that loads may run away. A guard railing often is provided on each side of the roller conveyorway, to guide the material and prevent it from running off. Such guardrails are especially advisable at corners and turns and on elevated conveyors under which workers must work or pass.

When heavy loads are conveyed, retarders, brakes, or similar means help prevent the loads from running away. A power conveyor on which speed can be controlled also could be substituted.

A vertically swinging hinged section of a roller or wheel conveyor should be hinged to that end of a stationary section of the conveyor from which the material is flowing to help block the oncoming material. The open end of the conveyor line should have a stop that automatically projects above the level of the rollers or wheels when the hinged section is opened and automatically retracts when the hinged section is closed.

Where a horizontally swinging hinged section occurs in a conveyor, the open ends of stationary sections (the two ends adjacent to the hinged section) should be equipped with retractable stops to prevent loads from dropping off when the hinged section is open.

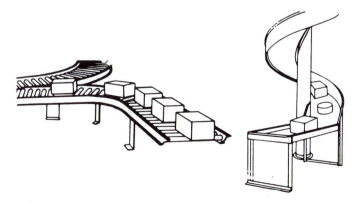

Figure 4-35. Types of gravity conveyors. Left: roller conveyor. Right: spiral chute. (Printed with permission from Conveyor Equipment Manufacturer's Association.)

Figure 4-36. Note the catwalk (black area) and footing surfaces between rollers on this built-in floor receiving conveyor with load on plywood slab.

Figure 4-37. An access stile was built for this conveyor that is over 20 in. (50.8 cm) high.

Live roll conveyors

A live roll (or roller) conveyor consists of a series of rolls over which objects are moved by power applied to all or some of the rolls through belts or chains.

Where installed at floor level or used in working areas, live roll conveyors should be designed to eliminate hazards from pinch points and moving parts, unless other provisions are made to prevent personnel from coming in contact with or crossing the conveyor.

Vertical conveyors

Vertical conveyors handle packages or other objects in a vertical or substantially vertical direction. In some cases, a hinged section, interlocked to the power in the main system, will be provided for access to the work station. The interlock should shut down the entire system until the section is restored to its position. There are three basic types:

- VERTICAL RECIPROCATING CONVEYOR: a power- or gravity-actuated unit that receives objects on a carrier or car bed, usually constructed of a power or roller conveyor, and elevates or lowers them to other locations.
- SUSPENDED TRAY CONVEYOR: a vertical conveyor having one or more endless chains with pendant trays, cars, or carriers that receive objects at one or more elevations and deliver them to another or several elevations. Protection of the underside is critical to prevent materials from falling on people.
- VERTICAL CHAIN CONVEYOR (opposed shelf type): two or more vertical elevating conveying units opposed to each other, each unit consisting of one or more endless chains whose adjacent facing runs operate in parallel paths. Thus, each pair of opposing shelves or brackets receives objects (usually dish trays) and delivers them to any number of elevations.

Where vertical conveyors are loaded and unloaded automatically, guards should be provided to protect personnel from contact with moving parts. Where they are loaded and unloaded manually, guards and safety devices, such as lintel and sill switches and deflectors, should be installed.

Carriages of vertical reciprocating conveyors designed to reg-ister at a floor, balcony, gallery, or mezzanine level never should have a solid bed. This type of conveyor is not intended to carry passengers or operators, or to have its car or carriage called to a station by a manually operated pushbutton.

REFERENCES

American National Standards Institute, 1430 Broadway, New York, N.Y. 10018.
Terms and Conveyor Definitions, ANSI/CEMA 102.
National Electrical Safety Code, C2.
"Safety Requirements for Cranes, Derricks, Hoists, Jacks, and Slings," B30 Series.
Safety Requirements for Floor and Wall Openings, Railings, and Toeboards, A12.1.
Safety Requirements for Personnel Hoists, A10.4.
Safety Requirements for the Lock Out/Tag Out of Energy Sources, No. Z244.1.
Safety Standards for Conveyors and Related Equipment, ANSI/ASME B20.1.
Safety Standard for Mechanical Power-Transmission Apparatus, ANSI/ASME B15.1.
Conveyor Equipment Manufacturers Association, 152 Rollins Ave., Rockville, MD 20852.
Screw Conveyors, Book No. 350.
Electric Controller and Manufacturing Co., 2704 East 79th St., Cleveland, Ohio 44104. *How to Operate a Crane,* Booklet 920.
International Union of Operating Engineers, 1125 17th St., NW., Washington, D.C. 20036.
National Fire Protection Association, Batterymarch Park, Quincy, Mass. 02269.
Life Safety Code, NFPA 101.
National Electrical Code, NFPA 70.
National Safety Council, 444 N. Michigan Ave., Chicago, Ill. 60611.
Industrial Data Sheets
Aerial Baskets, 572.
Automotive Hoisting Equipment, 437.

5

Ropes, Chains, and Slings

SPECIAL SAFETY PRECAUTIONS APPLY to using and storing ropes, rope slings, wire rope, chains, and chain slings. The safety and health professional should know the properties of the various types used and the precautions both in use and maintenance.

FIBER ROPE

Fiber rope is used extensively in handling and moving materials. The rope is generally made from manila (abaca), sisal, henequen, or nylon. Manila or nylon ropes give the best uniform strength and service. Other types of rope on the market today include those made from polyester and polypropylene.

Natural fibers

The properties of abaca (commonly known as manila) fiber make it the best-suited natural fiber for cordage. Manila rope is often recommended for capstan work because of its ability to render or pay out evenly when so used. High-grade manila rope, when new, is firm but pliant, ivory to light yellow in color, and with considerable luster. Its good reputation in fresh and salt water has been established for many years.

The properties of the several agave fibers (commonly known as sisals) do not give these ropes the high general acceptance of manila. The sisals are confined mostly to the smaller size ropes. The sisals are not as satisfactory for general use partly because their breaking strengths are generally lower than those of manila. Sisal rope varies in color from white to yellowish white and lacks the gloss of high-grade manila. The fibers are stiff and harsh and tend to splinter. This makes the ropes uncomfortable to handle.

Both sisal and manila fibers deteriorate when in contact with acids and caustics (and their mists or vapors). This deterioration is accelerated by hot, humid conditions: both fibers lose 50 percent of their strength at 180 F (80 C) and burn at over 300 F (150 C).

Sisal and henequen are not as satisfactory as manila because their strength varies in different grades. Sisal rope is about 80 percent as strong as manila; henequen is about 50 percent as strong, but is more resistant to atmospheric deterioration.

Other natural fibers are also used in ropes but to a lesser or negligible degree or for very special reasons. These fibers include: cotton, flax, coir, straw, asbestos, istle, jute, kenaf, silk, rawhide, and sanseveria.

Synthetic fibers

The popularity of synthetic fiber ropes is greater than that of natural fiber ropes. There are several reasons.

- Greater knowledge of the properties of various synthetics is an important reason for their increased use. Successful use of synthetics depends largely on the selection of the synthetic with the physical properties and characteristics that most closely match the requirements of the job.
- Splices can be made readily and can develop nearly the full strength of the rope. Tapered splices are highly recommended for rope sizes with a 1-in. (2.5 cm) diameter and larger.

Nylon rope has over 2½ times the breaking strength of manila rope and about 4 times its working elasticity. It is, therefore, particularly suitable for shock loading, such as required for restraint lines. Its resistance to abrasion is remarkably high in comparison to other ropes. While nylon rope is wet or frozen, its breaking strength is reduced by 10 to 15 percent.

It also is highly resistant to organisms that cause mildew and rotting and to attack by marine borers in sea water. Atmospheric exposure produces little loss of strength over a considerable period. Wet nylon rope runs through blocks as easily as dry nylon rope, since there is no swelling. Although resistant to petroleum oils and most common solvents and chemicals, nylon strength is affected by drying oils, such as linseed oil or the phenols, and is quickly vulnerable to strong mineral acids, phenalic compounds, and heat.

Nylon loses some of its strength at 300 F (150 C) and all of it at 482 F (250 C), its melting point. Short of melting, most strength is regained on cooling to normal temperature. Nylon of higher melting point can be secured.

Nylon, more than any other rope material, will absorb and store energy in the same manner as a spring. This energy, released at break, will make the moving ends as dangerous as a projectile. Caution must therefore be exercised when working lines around corners, capstans, timber heads, and the like.

Polyester makes probably the best general-purpose rope available, especially for critical uses. Polyesters stretch about half as much as nylon, so energy absorption is also about half. It is not weakened by rot, mildew, or prolonged exposure to sea water. It retains its full strength when wet and shows little deterioration from long exposure to sunlight. It has good resistance to abrasive wear. Polyester is somewhat vulnerable to alkalis, and its ultraviolet resistance is good to excellent. It burns at about 480 F (250 C) and loses strength over 390 F (200 C).

Polyolefin ropes, in general, are strong and inexpensive; they float and are unaffected by water. Polyolefins, like the polyesters, are not hygroscopic; therefore, they do not shrink or swell with water. The movement of crossed ropes, as well as other types of abrasion, must be avoided because of a very rapid friction sawing which even modest loads will cause. They are unaffected by rot, mildew, and fungus growth and are highly resistant to a wide variety of acids (except nitric) and alkalis, as well as to alcohol-type solvents and bleaching solutions. They swell and soften with hydrocarbons, particularly at temperatures above 150 F (66 C). There are two types:

- POLYPROPYLENE, with a specific gravity of 0.91 and a softening point of 300 F (150 C), is made in several different size filaments and from film with or without longitudinal fracturing. Polypropylene rope is about 50 percent stronger than manila, size for size. Pure polypropylene ropes have relatively poor rendering properties. It burns at 330 F (166 C) and loses some strength at 150 F.
- POLYETHYLENE, with a specific gravity of 0.95 and a softening point of 250 F (120 C), is characteristically slippery and has very little springiness. Stretch is low and strength is good. It has a comparatively low softening point and low coefficient of friction.

Composite ropes (combining several types of synthetic fibers or synthetic and natural fibers) are available. They result from attempts to give the surface of the rope or strand more wear resistance, internal tensile strength, or structural strength to resist deformation. Many of these ropes are made to match the requirements of specific jobs. These ropes can be expected to perform to the ends their makers declare.

Other types of rope are available, but enjoy only a small

percentage of the market for reasons of cost, limited use, or short supply. These ropes include in various modifications: paper, glass, acrylic, rayon, polyvinyl chloride, fluorocarbon, rubber, cellulose acetate, and polyurethane.

Figure 5-1 gives a checklist of factors that may be important to specify when obtaining rope for a specific application.

ROPE CHECKLIST	
Strength	Friction melting
Stretch with load	Combustibility
Impact load	Sunlight resistance
Permanent stretch	Latitude and altitude
Recovery from stretch	Color and type
Length	Diameter and construction of
Size	rope
Yardage	Frequency of use
Floatability	Storage methods
Flexibility	Marine growth resistance
Twist direction and torque	Rot resistance
Flex life in bending	Chemical resistance
Slipperiness	Color
Texture	Aging
Water repellency	Contamination
Hygroscopicity	Uniformity
Ruggedness in shape	Service cost
Temperature resistance	Toughness against wear

Figure 5-1. Factors that may be of significance when specifying rope for a specific use.

Strengths and working loads

Tables 5-A and 5-B list linear density, new rope tensile strength, safety factors, and working load limit for manila and synthetic ropes.

Safety factors. Because safety factors are not the same for all ropes and are based upon static loading, one must be cautious in using them.

Working load. One must also be cautious in using the working load figures. Because of the wide range of rope use, rope condition, exposure to the several factors affecting rope behavior, and the degree of risk to life and property involved, it is impossible to make blanket recommendations as to working loads. However, to provide guidelines, working loads are tabulated for rope in good condition, with appropriate splices in noncritical applications, under normal service conditions, and with very modest dynamic loads included.

A higher working load may be selected only with expert knowledge of conditions and professional estimate of risk. Factors to consider include (1) whether the rope has been subject to dynamic loading or other excessive use, (2) whether it has been inspected and found to be in good condition, (3) whether it is to be used in the recommended manner, and (4) whether the application involves elevated temperatures, extended periods under load, or obvious dynamic loading (see the following explanation) such as sudden drops, snubs or pickups. For all such applications and for applications involving more severe exposure conditions, or for recommendations on special applications, consult the manufacturer.

Many uses of rope involve serious risk of injury to personnel or damage to valuable property. This danger is often obvious, as when a heavy load is supported above one or more workers.

Table 5-A. Specifications for Synthetic and Natural Fiber Rope

Three-Strand Laid and Eight-Strand Plaited—Standard Construction

Nominal Size		Manila				Polypropylene				Polyester			
Diameter	Circumference	Linear Density[1] (lbs/100 ft)	New Rope Tensile Strength[2] (lbs)	Safety Factor	Working Load[3] (lbs)	Linear Density[1] (lbs/100 ft)	New Rope Tensile Strength[2] (lbs)	Safety Factor	Working Load[3] (lbs)	Linear Density[1] (lbs/100 ft)	New Rope Tensile Strength[2] (lbs)	Safety Factor	Working Load[3] (lbs)
3/16	5/8	1.50	406	10	41	.70	720	10	72	1.20	900	10	90
1/4	3/4	2.00	540	10	54	1.20	1,130	10	113	2.00	1,490	10	149
5/16	1	2.90	900	10	90	1.80	1,710	10	171	3.10	2,300	10	230
3/8	1 1/8	4.10	1,220	10	122	2.80	2,440	10	244	4.50	3,340	10	334
7/16	1 1/4	5.25	1,580	9	176	3.80	3,160	9	352	6.20	4,500	9	500
1/2	1 1/2	7.50	2,380	9	264	4.70	3,780	9	420	8.00	5,750	9	640
9/16	1 3/4	10.4	3,100	8	388	6.10	4,600	8	575	10.2	7,200	8	900
5/8	2	13.3	3,960	8	496	7.50	5,600	8	700	13.0	9,000	8	1,130
3/4	2 1/4	16.7	4,860	7	695	10.7	7,650	7	1,090	17.5	11,300	7	1,610
13/16	2 1/2	19.5	5,850	7	835	12.7	8,900	7	1,270	21.0	14,000	7	2,000
7/8	2 3/4	22.4	6,950	7	995	15.0	10,400	7	1,490	25.0	16,200	7	2,320
1	3	27.0	8,100	7	1,160	18.0	12,600	7	1,800	30.4	19,800	7	2,820
1 1/16	3 1/4	31.2	9,450	7	1,350	20.4	14,400	7	2,060	34.4	23,000	7	3,280
1 1/8	3 1/2	36.0	10,800	7	1,540	23.8	16,500	7	2,360	40.0	26,600	7	3,800
1 1/4	3 3/4	41.6	12,200	7	1,740	27.0	18,900	7	2,700	46.2	29,800	7	4,260
1 5/16	4	47.8	13,500	7	1,930	30.4	21,200	7	3,020	52.5	33,800	7	4,820
1 1/2	4 1/2	60.0	16,700	7	2,380	38.4	26,800	7	3,820	67.0	42,200	7	6,050
1 5/8	5	74.5	20,200	7	2,880	47.6	32,400	7	4,620	82.0	51,500	7	7,350
1 3/4	5 1/2	89.5	23,800	7	3,400	59.0	38,800	7	5,550	98.0	61,000	7	8,700
2	6	108	28,000	7	4,000	69.0	46,800	7	6,700	118	72,000	7	10,300
2 1/8	6 1/2	125	32,400	7	4,620	80.0	55,000	7	7,850	135	83,000	7	11,900
2 1/4	7	146	37,000	7	5,300	92.0	62,000	7	8,850	157	96,500	7	13,800
2 1/2	7 1/2	167	41,800	7	5,950	107	72,000	7	10,300	181	110,000	7	15,700
2 5/8	8	191	46,800	7	6,700	120	81,000	7	11,600	204	123,000	7	17,600
2 7/8	8 1/2	215	52,000	7	7,450	137	91,000	7	13,000	230	139,000	7	19,900
3	9	242	57,500	7	8,200	153	103,000	7	14,700	258	157,000	7	22,400
3 1/4	10	298	69,500	7	9,950	190	123,000	7	17,600	318	189,000	7	27,000
3 1/2	11	366	82,000	7	11,700	232	146,000	7	20,800	384	228,000	7	32,600
4	12	434	94,500	7	43,500	276	171,000	7	24,400	454	270,000	7	38,600

An equally dangerous situation occurs if personnel are in line with a rope under excessive tension. Should the rope fail, it may recoil with considerable force—especially if the rope is made of nylon. Persons should be warned against standing in line with the rope. In all cases where any such risks are present, or there is any question about the loads involved or the conditions of use, the working load should be substantially reduced and the rope be properly inspected. Manufacturers should be consulted regarding their recommendations on working loads.

Dynamic loading voids working loads. Working loads are not applicable when rope is subject to significant dynamic loading. Whenever a load is picked up, stopped, moved, or swung, there is an increased force due to dynamic loading. The more rapidly or suddenly such actions occur, the greater this increase will be. In extreme cases, the force put on the rope may be two, three, or even more times the normal load involved, such as when picking up a tow on a slack line or using a rope to stop a falling object. Therefore, in all such applications as towing lines, lifelines, safety lines, climbing ropes, etc., working loads as given do not apply.

Users should be aware that dynamic effects are greater on a low elongation rope such as manila than on a high elongation rope such as nylon, and greater on a shorter rope than on a longer one. The working load listed contains provision for very modest dynamic loads. This means, however, that when this working load has been used to select a rope, the load must be handled slowly and smoothly to minimize dynamic effects and avoid exceeding the provision for them.

Inspection

Before being placed in service, new rope should be thoroughly inspected along its entire length to determine that no part of it is damaged or defective. Any irregularity in the uniformity of appearance is evidence of possible degradation.

There is no agreement on what determines when a rope should be removed from service. Synthetic rope damage is not always visible.

In-service rope should be inspected every 30 days under ordinary conditions and much oftener if used in critical applications, such as to support scaffolding on which employees work. Inspection consists of an examination of the entire length of the rope, inch by inch, for wear, abrasions, powdered fiber between strands, broken or cut fibers, displacement of yarns or strands, variation in size or roundness of strands, discoloration, and rotting. To inspect the inner fibers, the rope should be untwisted in several places to see whether the inner yarns are bright, clear, and unspotted. If exposed to acids, natural fiber ropes, such as manila, should be scrapped or retired from critical operations, as visual inspection will not always

Table 5-A (continued).

Nominal Diameter (in.)	Composite[4]				Nylon				Sisal				Nominal Diameter (in.)
	Linear Density[1] (lbs/100 ft)	New Rope Tensile Strength[2] (lbs)	Safety Factor	Working Load[3] (lbs)	Linear Density[1] (lbs/100 ft)	New Rope Tensile Strength[2] (lbs)	Safety Factor	Working Load[3] (lbs)	Linear Density[1] (lbs/100 ft)	New Rope Tensile Strength[2] (lbs)	Safety Factor	Working Load[3] (lbs)	
3/16	.94	720	10	72	1.00	900	12	75	1.50	360	10	36	3/16
1/4	1.61	1,130	10	113	1.50	1,490	12	124	2.00	480	10	48	1/4
5/16	2.48	1,710	10	171	2.50	2,300	12	192	2.90	800	10	80	5/16
3/8	3.60	2,440	10	244	3.50	3,340	12	278	4.10	1,080	10	108	3/8
7/16	5.00	3,160	9	352	5.00	4,500	11	410	5.26	1,400	9	156	7/16
1/2	6.50	3,960	9	440	6.50	5,750	11	525	7.52	2,120	9	236	1/2
9/16	8.00	4,860	8	610	8.15	7,200	10	720	10.4	2,760	8	345	9/16
5/8	9.50	5,760	8	720	10.5	9,350	10	935	13.3	3,520	8	440	5/8
3/4	12.5	7,560	7	1,080	14.5	12,800	9	1,420	16.7	4,320	7	617	3/4
13/16	15.2	9,180	7	1,310	17.0	15,300	9	1,700	19.5	5,200	7	743	13/16
7/8	18.0	10,800	7	1,540	20.0	18,000	9	2,000	22.5	6,160	7	880	7/8
1	21.8	13,100	7	1,870	26.4	22,600	9	2,520	27.0	7,200	7	1,030	1
1 1/16	25.6	15,200	7	2,170	29.0	26,000	9	2,880	31.3	8,400	7	1,200	1 1/16
1 1/8	29.0	17,400	7	2,490	34.0	29,800	9	3,320	36.0	9,600	7	1,370	1 1/8
1 1/4	33.4	19,800	7	2,830	40.0	33,800	9	3,760	41.7	10,800	7	1,540	1 1/4
1 5/16	35.6	21,200	7	3,020	45.0	38,800	9	4,320	47.8	12,000	7	1,710	1 5/16
1 1/2	45.0	26,800	7	3,820	55.0	47,800	9	5,320	59.9	14,800	7	2,110	1 1/2
1 5/8	55.5	32,400	7	4,620	66.5	58,500	9	6,500	74.6	18,000	7	2,570	1 5/8
1 3/4	66.5	38,800	7	5,550	83.0	70,000	9	7,800	89.3	21,200	7	3,030	1 3/4
2	78.0	46,800	7	6,700	95.0	83,000	9	9,200	108	24,800	7	3,540	2
2 1/8	92.0	55,000	7	7,850	109	95,500	9	10,600	—	—	7	—	2 1/8
2 1/4	105	62,000	8	8,850	120	113,000	9	12,600	146	32,800	7	4,690	2 1/4
2 1/2	122	72,000	7	10,300	149	126,000	9	14,000	—	—	7	—	2 1/2
2 5/8	138	81,000	7	11,600	168	146,000	9	16,200	191	41,600	7	5,940	2 5/8
2 7/8	155	91,000	7	13,000	189	162,000	9	18,000	—	—	7	—	2 7/8
3	174	103,000	7	14,700	210	180,000	9	20,000	242	51,200	7	7,300	3
3 1/4	210	123,000	7	17,600	264	226,000	9	25,200	299	61,600	7	8,800	3 1/4
3 1/2	256	146,000	7	20,800	312	270,000	9	30,000	—	—	7	—	3 1/2
4	300	171,000	7	24,400	380	324,000	9	36,000	435	84,000	7	12,000	4

[1]Linear density (lbs/100 ft) shown is "average." Maximum is 5% higher.

[2]New rope tensile strengths are based on tests of new and unused rope of standard construction in accordance with Cordage Institute Standard Test Methods.

[3]Working loads are for rope in good condition with appropriate splices, in non-critical applications, and under normal service conditions. Working loads should be exceeded only with expert knowledge of conditions and professional estimates of risk. Working loads should be reduced where life, limb, or valuable property are involved, or for exceptional service conditions such as shock, loads, sustained loads, etc.

[4]Composite rope. Materials and construction of this polyester/polypropylene composite rope conform to MIL-R-43942 and MIL-R-43952. For other composite ropes, consult the manufacturer.

(Printed from the Cordage Institute.)

reveal acid damage. A rope, like a chain, "is only as strong as its weakest link" (or with rope, its cross section). If there is a visible core or core damage, the rope must be replaced or spliced out.

Natural fiber rope loaded to over 50 percent of its breaking strength will be permanently damaged; synthetics loaded to over 65 percent may be damaged. Damage from this cause may be detected by examining the inside fibers. These will be broken into short lengths in proportion to the degree of overload. A good estimate of the strength of fibers can be made by scratching the fibers with a fingernail—fibers of poor strength will readily part. This "fingernail test" is a quick test for chemical damage.

If the diameter of a rope is more than 5 percent worn away, replace the rope. In small ropes (up to 3/4 in., 19 mm, diameter), surface wear that has progressed to the center of the twisted element (yarn) may account for more than an 80 percent loss of rope strength. In ropes with a 3/4 in. diameter or more, surface wear may destroy the strength of the cover yarns, yet not affect the original strength of the core yarns; the remaining strength of the rope will be in the proportion that the core yarns are to the original total of yarns.

If sample fibers can be secured from the rope, an approximation of rope strength can be made. Estimate the fiber strength by manually breaking fiber samples, and estimate the distribution of fibers in a cross section, quartered to allow for twist configuration.

Due to the motion of slippage on a supporting surface when under high tension, synthetics sometimes melt on the surface and form a skin. This skin may be evidence of degradation.

Ropes, using multifilament synthetic fiber on the surface, will often "fuzz," which is due to the minute fiber breakage. If it is very fuzzy, replace the rope and look for the source of abrasion.

Care of rope in use

Safety variations in different types of rope result from factors such as chaffing, cutting, elasticity, diameter-strength ratio, and

Table 5-B. Specifications for Double Braided Nylon Rope

Nominal Size		Double Braided Nylon Nylon Cover-Nylon Core			
Diameter (in.)	Circ. (in.)	Linear Density[1] (lbs/100 ft)	New Rope Min. Strength Strength[2] (lbs)	Safety Factor	Working Load[3] (lbs)
¼	¾	1.56	1,650	11	150
⁵⁄₁₆	1	2.44	2,570	11	234
⅜	1⅛	3.52	3,700	11	336
⁷⁄₁₆	1⁵⁄₁₆	4.79	5,020	10	502
½	1½	6.25	6,550	10	655
⁹⁄₁₆	1¾	7.91	8,270	9	919
⅝	2	9.77	10,200	9	1,130
¾	2¼	14.1	14,700	8	1,840
¹³⁄₁₆	2½	16.5	17,200	8	2,150
⅞	2¾	19.1	19,900	8	2,490
1	3	25.0	26,000	8	3,250
1¹⁄₁₆	3¼	28.2	29,300	8	3,660
1⅛	3½	31.6	32,800	8	4,100
1¼	3¾	39.1	40,600	8	5,080
1⁵⁄₁₆	4	43.1	44,700	8	5,590
1⅜	4¼	47.3	49,000	8	6,130
1½	4½	56.3	58,300	8	7,290
1⅝	5	66.0	68,300	8	8,540
1¾	5½	76.6	79,200	8	9,900
2	6	100	103,000	8	12,900
2⅛	6½	113	117,000	8	14,600
2¼	7	127	131,000	7	18,700
2½	7½	156	161,000	7	23,000
2⅝	8	172	177,000	7	25,300
3	9	225	231,000	7	33,000
3¼	10	264	271,000	7	38,700
3½	11	329	338,000	7	48,300
4	12	400	410,000	7	58,600

[1]Linear density (LD) shown is average and is determined from the equation LD=25×(Nom. diameter).[2] Tolerance is ±5%.
[2]Minimum tensile strength (MTS) is based on a large number of tests by various manufacturers and represents a value 2 standard deviations below the mean. Minimum strength is determined by the formula MTS=1057 LD[995].
[3]Working loads are for rope in good condition with appropriate splices, in non-critical applications, and under normal service conditions. Working loads should be exceeded only with expert knowledge of conditions and professional estimates of risk. Working loads should be reduced where life, limb, or valuable property are involved, or for exceptional service conditions such as shock loads, sustained loads, etc.
(Printed from the Cordage Institute.)

general anticipated mishandling. If possible, a rope should not be dragged as this abrades the outer fibers. If the rope picks up dirt and sand, abrasion within the lay of the rope will rapidly wear it out.

Precautions should be taken to keep rope in good condition. Kinking, for example, strains the rope and may overstress the fibers. It may be difficult to detect a weak spot made by a kink. To prevent a new rope from kinking while it is being uncoiled, lay the rope coil on the floor with the bottom end down. Pull the bottom end up through the coil, and unwind the rope counterclockwise. If it uncoils in the other direction, turn the coil of rope over, and pull the end out on the other side.

Twisted rope should be handled so as to retain the amount of twist (called balance) that the rope seeks when free and relaxed. If rotating loads and improper coiling and uncoiling change the balance, it can be restored by proper twisting of either end. Severe unbalance can cause permanent damage; localized overtwisting causes kinking or hocking.

Sharp bends over an unyielding surface cause extreme tension on the fibers. To make a rope fast, an object with a smooth round surface of sufficient diameter should be selected. If the object does have sharp corners, pads should be used. To avoid excessive bending, sheaves or surface curvatures should be of

suitable size for the diameter of rope, as shown in Table 5-C.

Table 5-C. Sheave Sizes for Fiber Ropes of Varying Thickness

Diameter of Rope (in.)	Diameter of Sheave (in.)
¾	6
⅞	7
1	8
1¼	10
1⅜	11

Conversion factor: 1 in.=2.54 cm

When lengths of rope must be joined, they should be spliced and not knotted. A well made splice will retain up to 100 percent of the strength of the rope, but a knot retains only 50 percent (Table 5-D).

Rope must be thoroughly dried out after it becomes wet, otherwise it will deteriorate quickly. A wet rope should be hung up, or laid in a loose coil in a dry place until thoroughly dry. Rope will deteriorate more rapidly if it is alternately wet and dry than if it remains wet. Wet rope should not be allowed to freeze.

Table 5-D. Efficiency of Manila Fiber Rope With Splices, Hitches, and Knots

Jointure	Percent Efficiency
Full strength of dry rope	100
Eye splice over metal thimble	90
Short splice in rope	80
Timber hitch, round turn, half hitch	65
Bowline, slip knot, clove hitch	60
Square knot, weaver's knot, sheet bend	50
Flemish eye, overhand knot	45

Use of wet rope, or of rope reinforced with metallic strands, near power lines and other electrical equipment is extremely dangerous.

Care of rope in storage

To maintain the existing strength of any rope that is properly prepared, it should be stored safe from deleterious fumes, heat, chemicals, moisture, sunlight, rodents, and biological attack.

Rope should be stored in a dry place where air circulates freely about it. Air should not be extremely dry, however. Small ropes can be hung up and larger ropes can be laid on gratings, so air can get underneath and around them.

Rope should not be stored or used in an atmosphere containing acid or acid fumes, as it will quickly deteriorate. Signs of deterioration from this cause are dark brown or black spots on the rope.

Rope should not be stored unless it has been cleaned. Dirty rope can be hung in loops over a bar or beam and then sprayed with water to remove the dirt. The spray should not be so powerful that it forces the dirt into the fibers. After washing, the rope should be allowed to dry and then shaken to remove the rest of the dirt.

WIRE ROPE

Wire rope is used widely instead of fiber rope because:
- It has greater strength and durability under severe working conditions.
- Its physical characteristics do not change when used in varying environments.
- It has controlled and predictable stretch characteristics.

Construction

Wire rope is composed of steel wires, strands, and core. The individual wires are cold drawn to predetermined size and breaking loads according to required grades. Grades include iron, tractor, mild plow steel, plow steel, improved plow steel, and extra improved plow steel. The wires are then laid together in various geometrical arrangements according to strand construction requirements and wire rope classification (6×19, 6×37, etc.). In making a strand, carefully selected pitch or lay lengths are used which have a definite ratio to the length of lay or pitch used in forming the finished wire rope. After the individual strands are made, the required number are helically laid or formed around the core, which supports the load carrying strands. The core may be sisal or synthetic fiber, a metallic strand core or independent wire rope core (IWRC).

The size, number and arrangement of wires, the number of strands, the lay, and the type of core in a rope are determined by the service for which the rope is to be used.

The most popular and generally used wire rope constructions fall within two classifications, the 6×19 and 6×37. The 6×19 classification contains a variety of wire rope constructions ranging in number of wires per strand, 15 to 26. Typical constructions are 6×19 Seale (Figure 5-2), 6×25 filler wire, 6×19 Warrington, etc.

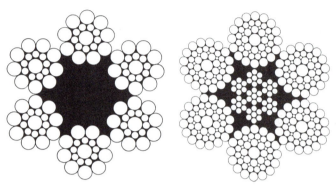

Figure 5-2. Wire rope is made from a number of individual wires grouped in strands, then laid together over a core member (fiber core, IWRC, or strand core). The number of wires per strand and the number of strands per rope depend upon the expected working conditions and the amount of flexibility required. The cross section at the left above shows one construction of the 6×19 fiber core (FC) classification containing 114 wires. The cross section at the right above shows one construction of the 6×37 classification containing 343 wires, including those of the independent wire rope core (IWRC).

The 6×37 classification also covers a large number of wire rope designs and constructions with the number of wires per strand, 27 through 49. Typical constructions of this classification are 6×41 filler wire (Figure 5-2), 6×37, 6×36 Warrington Seale, 6×49 filler wire, etc. Generally speaking, the greater the number of wires per strand, the more flexible the wire rope. However, the fewer number of wires per strand, the more abrasion and crush resistant is the rope. In the large rope diameters, 2½ in. (6.4 cm) and larger, practically all wire rope is produced in the 6×37 or 6×61 class. Therefore, because of the larger number of possible rope constructions available, the user must exercise care in making the proper selection (see Table 5-E). For more details, see the American Iron and Steel Institute (AISI) *Wire Rope Users Manual* (References).

Depending upon service requirements and conditions, six-strand ropes may have a fiber core (FC), a wire strand core (WSC), or an independent wire rope core (IWRC). They may be either regular lay (wires in the strands are laid in the opposite direction from that of the strands in the rope) or lang lay (wires in the strands are laid in the same direction as those of the strands in the rope).

Where maximum flexibility is required eight-strand hoisting ropes are also used. They are usually of the 19-wire strand classification with regular lay and a fiber or IWR.

Flexibility is not a requirement for guy wires, highway guards, and similar services. Therefore, wire rope of six-by-seven construction (six strands with seven wires per strand) is suitable. When rope for a particular service is to be selected, it is recom-

Table 5-E. Nominal Breaking Strength in Tons of Improved Plow Steel (IPS) and Extra Improved Plow Steel (EIPS) Ropes

Diameter (in.)	6x19 & 6x37 IPS[1] FC[3]	6x19 & 6x37 IPS[1] IWRC[4]	6x19 & 6x37 EIPS[2] IWRC	8x19 IPS FC	8x19 IPS IWRC	8x19 EIPS IWRC	19x7 IPS FC	19x7 EIPS IWRC
3/8	6.10	6.56	7.55	5.24	5.76	6.63	5.59	6.15
7/16	8.27	8.89	10.2	7.09	7.80	8.97	7.58	8.33
1/2	10.7	11.5	13.3	9.23	10.1	11.6	9.85	10.8
9/16	13.5	14.5	16.8	11.6	12.8	14.7	12.4	13.6
5/8	16.7	17.9	20.6	14.3	15.7	18.1	15.3	16.8
3/4	23.8	25.6	29.4	20.5	22.5	25.9	21.8	24.0
7/8	32.2	34.6	39.8	27.7	30.5	35.0	29.5	32.5
1	41.8	44.9	51.7	36.0	39.6	45.5	38.3	42.2
1 1/8	52.6	56.5	65.0	45.3	49.8	57.3	48.2	53.1
1 1/4	64.6	69.4	79.9	55.7	61.3	70.5	59.2	65.1
1 3/8	77.7	83.5	96.0	67.1	73.8	84.9	71.3	78.4
1 1/2	92.0	98.9	114	79.4	87.3	100	84.4	92.8
1 5/8	107	115	132					
1 3/4	124	133	153					
1 7/8	141	152	174					
2	160	172	198					
2 1/8	179	192	221					
2 1/4	200	215	247					

[1]IPS=Improved plow steel.
[2]EIPS=Extra improved plow steel.
[3]FC=Fiber core.
[4]IWRC=Independent wire rope core.

mended that engineers of reliable wire rope manufacturers be consulted for the most suitable type.

Some service conditions require rope with special qualities. Fiber cores are affected by temperatures above 250 F (120 C). Under such conditions, a metallic core provides greater efficiency and safety. A zinc-coated or stainless steel wire rope effectively resists some types of corrosion. Specific corrosion problems should be referred to a rope manufacturer.

Since preformed wire rope does not unravel, it has advantages for certain services, such as for slings and on construction and similar heavy equipment. Preformed rope is less likely to set or kink, and broken wires are less likely to protrude and create a hazard to workers.

It is recommended that hoisting rope have at least the strength of the "improved plow steel" grade. For many applications the "extra improved plow steel" grade, which is the greatest in strength, should be used to provide an adequate design factor and better service.

Design factor for hoisting ropes

The operating or design factors for hoisting ropes are calculated by dividing the nominal catalog strength of the rope by the sum of the maximum loads to be hoisted. It is normal practice to base this on static loads.

In the case of mine hoisting rope, the maximum weights to be hoisted include the weight of the skip or car or cage, plus the weight of the material, plus the weight of the suspended rope when the skip or cage are at the lowest point in the shaft. In some cases acceleration stresses are also considered.

The minimum design factors for hoisting rope depends upon the type of service and the federal, state or provincial, or local codes covering the particular hoisting operation. Many of these codes describe exactly how the design and operating factors are to be figured. Therefore, it is prudent to check what codes are in force before making a final determination or selection. The advice of a reliable wire rope manufacturer should also be obtained.

Causes of deterioration

Deterioration of wire ropes is due largely to a number of factors, which vary considerably in importance, depending on the conditions of service (Figure 5-3). For example, corrosion often is the principal cause of deterioration of mine hoisting rope in wet mine shafts because of moisture and the presence of acid in the water. Among other factors are:

- *Wear,* particularly on the crown or outside wires, from contact with sheaves and drums.
- *Corrosion,* particularly of the interior wires, indicated by pitting. This condition is difficult to detect and highly dangerous. Wear is accelerated by corrosion.
- *Kinks,* acquired in improper installation of a new rope, hoisting with slack in the rope, and so forth. A kink cannot be removed without creating a weak place.
- *Fatigue,* indicated by a square type of wire end fracture can be caused by bending stresses from sheaves and drums with small radii, whipping, vibration, pounding and torsional stresses.
- *Drying out of lubrication,* often hastened by heat and operating pressure.
- *Overloading,* including dynamic overloading, if acceleration and deceleration are factors of importance.
- *Overwinding* results when rope length is greater than drum can accommodate in a single layer. This can cause heavy abrasion and excessive wear at crossover points. However, successful overwinding can be achieved by the use of a special engineered drum grooving.
- *Mechanical abuse,* caused by careless handling, running over rope with equipment and permitting equipment obstructions to remain in the rope travel path. It is rather common that more wire rope is thrown away because of abuse rather than use.

WIRE ROPE WEAR AND DAMAGE

The evidence in these illustrations will aid the inspector in determining the actual cause of wear or damage in any wire rope.

A wire rope which has been kinked. A kink is caused by pulling down a loop in a slack line during improper handling, installation, or operation. Note the distortion of the strands and individual wires. Early rope failure will undoubtedly occur at this point.

Localized wear over an equalizing sheave. The danger of this type wear is that it is not visible during operation of the rope. This emphasizes the need of regular inspection of this portion of an operating rope.

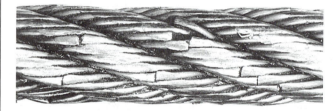

A typical failure of a rotary drill line with a poor cut-off practice. These wires have been subjected to excessive peening causing fatigue type failures. A predetermined, regularly scheduled, cut-off practice will go far toward eliminating this type of break.

A single strand removed from a wire rope subjected to "strand nicking." This condition is the result of adjacent strands rubbing against one another and is usually caused by core failure due to continued operation of a rope under high tensile load. The ultimate result will be individual wire breaks in the valleys of the strands.

A "bird cage." Caused by sudden release of tension and resultant rebound of rope from overloaded condition. These strands and wires will not return to their original positions.

An example of a wire rope with a high strand—a condition in which one or two strands are worn before adjoining strands. This is caused by improper socketing or seizing, kinks or dog legs. Picture A is a close-up of the concentration of wear and B shows how it recurs in every sixth strand (in a six-strand rope).

Figure 5-3. Typical characteristics and causes of broken wires in wire ropes. (Printed with permission from Wire Rope Corp. of America, Inc.)

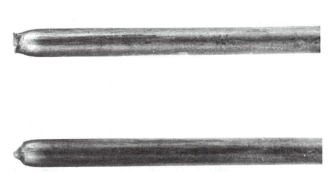

An illustration of a wire which has broken under tensile load in excess of its strength. It is typically recognized by the "cup and cone" appearance at the point of fracture. The necking down of the wire at the point of failure to form the cup and cone indicates that failure occurred while the wire retained its ductility.

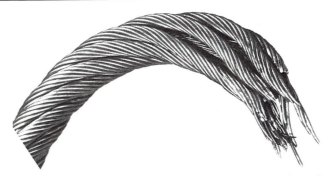

A wire rope which has jumped a sheave. The rope itself is deformed into a "curl" as if bent around a round shaft. Close examination of the wires show two types of breaks—normal tensile "cup and cone" breaks and shear breaks which give the appearance of having been cut on an angle with a cold chisel.

A wire rope which has been subjected to repeated bending over sheaves under normal loads. This results in "fatigue" breaks in individual wires—these breaks are square and usually in the crown of the strands.

An illustration of a wire which shows a fatigue break. It is recognized by the squared off ends perpendicular to the wire. This break was produced by a torsion machine, which is used to measure the ductility. This break is similar to wire failures in the field caused by excessive bending.

An example of "fatigue" failure of a wire rope which has been subjected to heavy loads over small sheaves. The usual crown breaks are accompanied by breaks in the valleys of the strands—these breaks are caused by "strand nicking" resulting from the heavy loads.

An example of a wire rope that has provided maximum service and is ready for replacement.

A close-up of a rope subjected to drum crushing. Note the distortion of the individual wires and displacement from their normal position. This is usually caused by the rope scrubbing on itself.

A fatigue break in a cable tool drill line caused by a tight kink developed in the rope during operation.

Figure 5-3 (continued).

Sheaves and drums

Fatigue resulting from bending stresses is dependent upon drum and sheave diameters and loading—the larger the diameter of the drums and sheaves, the more favorable will be the rope life. However, in some designs a sacrifice must be made because of other equipment designs and considerations. Table 5-F should be considered as a design base only since many types of equipment are operating successfully with smaller drum-sheave rope ratios and others use much larger ratios. A case in point is the drum and sheave requirements as contained in most mining codes, elevator codes, and shovel, hoist, and crane codes.

Table 5-F. Recommended Tread Diameters of Sheaves and Drums for Wire Rope

Rope Classification	Average Recommended (times rope diameter)	Minimum
6 × 7	72	42
6 × 19	45	30
6 × 37	27	18
8 × 19	31	21

Printed from *ANSI/ASME* A17.1, *Elevators, Escalators, and Moving Walks.*

The safety and service life of hoisting rope installations can be greatly increased by the use of sheaves and drums of suitable size and design, proper lubrication, and good maintenance of the rope and the hoisting equipment.

It is essential that head, idler, knuckle, curved sheaves, and grooved drums have grooves which support the rope properly. Before a new rope is installed, the grooves should be inspected and, where necessary, machined to proper contour and groove diameter, which should exceed the nominal rope diameter by the amount shown in Table 5-G. For recommended grooving for drums and sheaves, consult a wire rope manufacturer's handbook.

Table 5-G. Groove Diameter in Relation to Wire Rope Diameter

Rope Size (inches)	Amount that Groove Diameter Should Be Larger than Nominal Rope Diameter (inches)	
	Used	New
¼ and ⁵⁄₁₆	¹⁄₁₂₈	¹⁄₆₄
³⁄₈ to ¾ incl.	¹⁄₆₄	¹⁄₃₂
¹³⁄₁₆ to 1⅛ incl.	³⁄₁₂₈	³⁄₆₄
1³⁄₁₆ to 1½ incl.	¹⁄₃₂	¹⁄₁₆
1⁹⁄₁₆ to 2¼ incl.	³⁄₆₄	³⁄₃₂
2⁵⁄₁₆ and larger	¹⁄₁₆	⅛

So far as the service life of wire rope is concerned, the condition and contour of sheave grooves is important. Sheave grooves should be checked periodically (Figure 5-4) and should not be allowed to wear to a smaller diameter than those shown for used grooves in Table 5-G. If they become worn more than this, a reduction in rope service life can be expected. Reconditioned sheave grooves should conform to the tolerance, shown in Table 5-G, for new (or remachined) grooves.

On all new sheaves, the grooves should be made for the size of rope specified. The bottom of the groove should have a

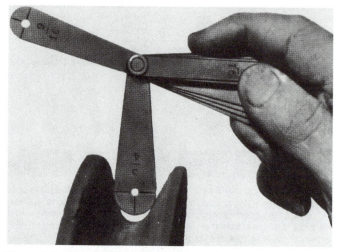

Figure 5-4. Check sheave grooves with a gage designed with one-half the allowable oversize. If light is seen between the gage and sheave, replace the sheave. (Printed with permission from Armco Steel Corporation.)

150-degree arc of support, and the sides of the groove should be tangent to the ends of the bottom arc. The depth of the groove should be 1¾ times the nominal diameter of the rope, and the radius of the arc should be one-half the nominal rope diameter plus one-half the value shown in Table 5-G, for new (or remachined) grooves.

Multiple layer winding on drums should be avoided, if possible, because the rope will wear and its life will be shortened materially, particularly at the point where the rope rises to the next layer. Where practicable, the drum should be of such diameter and length as to take all the rope in a single layer.

Crushing and excessive wear of wire rope are minimized by use of helically grooved drums and capacity for one layer of rope. In any case, the number of layers should be limited to three. Rope lifters at the flanges are recommended when two or more layers are wound on drums. To distribute wear at crossover points uniformly, 1¼ wraps can be cut off every six months or three or four times during the life of the rope. In no case should there be fewer than two full wraps on a drum (OSHA requirement) and three is preferred.

In general, wire rope reverse bending—first in one direction and then in the opposite—over sheaves or drums wears out rope faster and should be avoided.

The fleet angle is the included angle between the rope winding on the drum and a line perpendicular to the drum shaft and running through the head or lead sheave. (See Figure 5-5.)

It is desirable that the fleet angle not exceed 1 degree 30 minutes to reduce any tendency for the rope to open-wind. Also, a minimum angle of 0 degree 30 minutes for smooth drums and 2 degrees for grooved drums to assure the rope's starting back on the next layer, has been found desirable. Adherence to these specifications will help achieve uniform winding on smooth-faced drums and will also increase the winding efficiency of grooved drums. For smooth-faced drums, proper direction of lay of rope for specified winding conditions further helps achieve uniform winding.

Installation of a wire rope on a plain (smooth) face drum requires a great deal of care. The starting position should be

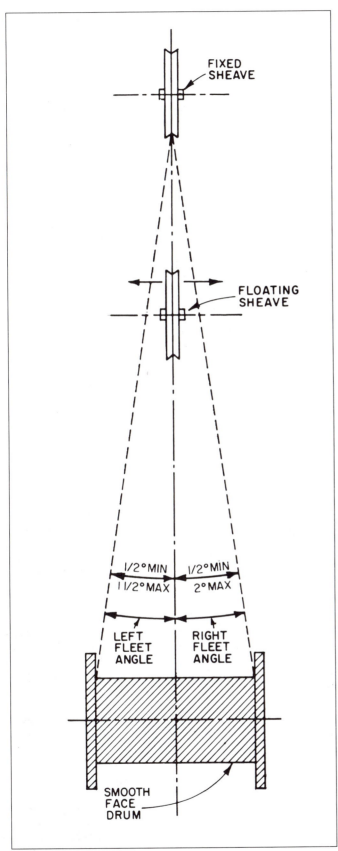

Figure 5-5. The fleet angle is graphically defined in this illustration of wire rope running from a fixed sheave, over a floating sheave, and then on to a smooth drum.

at the drum end so that each turn of the rope will wind tightly against the preceding turn (Figure 5-6). Here too, close supervision should be maintained all during installation. This will help make certain that:

1. The rope is properly attached to the drum
2. Appropriate tension on the rope is maintained as it is wound on the drum
3. Each turn is guided as close to the preceding turn as possible, so that there are no gaps between turns
4. There are at least two dead turns on the drum when the rope is fully unwound during normal operating cycles.

Loose and uneven winding on a plain- or smooth-faced drum, can and usually does create excessive wear, crushing, and distortion of the rope. The results of such abuse are lower operating performance and a reduction in the rope's effective strength. Also, for an operation that is sensitive in terms of moving and spotting a load, the operator will encounter control difficulties because the rope will pile up, pull into the pile, and fall from the pile to the drum surface. The ensuing shock can break or otherwise damage the rope.

Sheaves should be checked for proper alignment when installed. During rope changes the sheaves should be checked for worn bearings, broken flanges and proper groove size, smoothness, and contour. Heavily worn or damaged sheaves should either be reconditioned or replaced.

Sheave groove bearing pressures can become very high depending upon operating conditions and rope loadings. This can cause excessive sheave wear and shorten wire rope life. It therefore is necessary to consider this factor and to make sure proper sheave materials and liners are selected at the time of installation. Information on this subject may be found in most wire rope manufacturers' handbooks and the American Iron and Steel Institute's *Wire Rope Users Manual* (see References).

Lubrication

When possible, a wire rope should be cleaned before lubrication. Regular application of a suitable lubricant to hoisting wire rope prevents corrosion, wear from friction, and drying out of the core. Good lubricants are free from acids and alkali and have adhesive strength to stick. They also have the ability to penetrate the strands. The lubricant should be insoluble under the prevailing conditions. Ropes should be dry, when lubricant is applied, so that moisture will not be entrapped by the lubricant. Thin lubricants can be applied by hand, but the best arrangement is to provide some means of dripping them on the rope, or using a spray device to apply the proper quantity automatically.

Periodic monthly cleaning, such as done in mine shafts, for removal of dirt, abrasive particles and corrosion producing moisture is recommended. Cleaning fluids should not be used because of their detrimental effect on the core lubricant. Light oils are sometimes used to loosen the coating of lubricant and harmful materials. A compressed air or steam jet (or other mechanical method) cleans a rope effectively and thoroughly.

Overloading either the wire rope or the equipment is not recommended since damage to either may occur but not become known until some time after the overload.

Rope fittings

Methods of attaching wire rope to fittings are important for

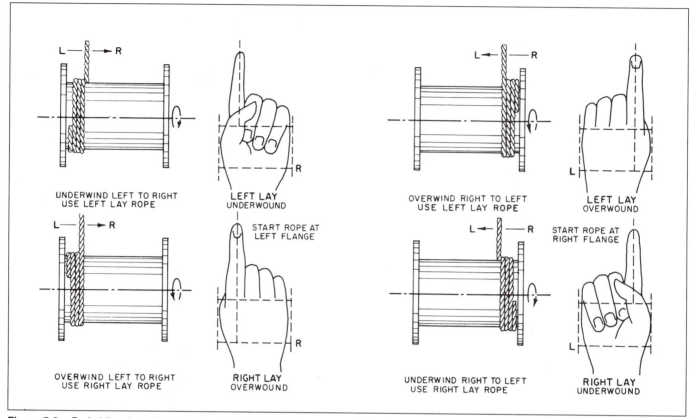

Figure 5-6. By holding the left or right hand with the index finger extended, palm up or palm down, the proper procedure for installing left- or right-lay rope on a smooth drum can be easily determined. (Printed with permission from American Iron and Steel Institute.)

safety because they develop from 75 to 100 percent of the breaking strength of the rope. Manufacturers also specify fittings of suitable size and design for ropes of different sizes. The strength of an attachment is attained only when the connection is made exactly according to the manufacturer's instructions (Figure 5-7).

Some types of attachments, such as pressed fittings or mechanical sleeve splices which are used in making slings, must be made at either a wire rope manufacturer's plant or a properly equipped commercial sling shop.

Efficiencies of properly made hand tucked splices vary according to the ability of the splicer and the rope diameter but can be up to 90 percent (Figure 5-7). The efficiency of mechanical sleeve splices varies from 90 to 95 percent when IWRC type wire rope is used.

Rope often is connected to the fittings of conveyances by means of clips and clamps. They are rated to develop 75 to 80 percent of the breaking strength of the rope. Figure 5-7 shows how the clips should be attached, and Table 5-H gives the number of clips and the spacing required for ropes of different sizes. It is important to retighten the nuts on all clips after the initial load-carrying use of the rope, as well as at all subsequent regular inspection periods.

Socketing with zinc and a thermostate plastic resin will develop 100 percent of the breaking strength of the rope. Since there is no ready way to detect flaws in the finished job, the recommended procedure must be followed exactly. In high-speed hoisting, fatigue is especially likely to develop with this type of attachment. For this reason, the section adjacent to the conveyance should be cut off and discarded at frequent intervals.

The required interval in some state mining laws is every 6 months. Figure 5-8 shows zinc-poured and swaged sockets.

Square and other types of knots have low and unpredictable efficiencies 40 percent or less. Their use is likely to result in failure of a rope assembly and under certain conditions result in a serious accident. In the United States, OSHA and other industrial and constructional codes prohibit the use of knots in wire rope.

Inspection and replacement

The frequency of inspections and replacement of a rope depends considerably on service conditions. OSHA requires wire rope or cable to be inspected when installed and weekly during use. In any case, at regular intervals, a specially trained inspector should examine the ropes on which human life depends. Some plants and mines, for instance, make a daily inspection for readily observable defects, such as kinking and loose wires, and a thorough inspection weekly. For the latter inspection, the rope speed is generally less than 60 fpm.

The inspector specifically checks for wear of the crown wires, kinking, high strands, corrosion, loose wires, nicking, and lubrication (Figure 5-3). Rope calipers (Figure 5-9) and micrometers are used to determine changes in the cross section of rope at various locations. In most cases sudden changes in rope length and/or diameter is a warning that the wire rope is nearing the end of its useful life and should be removed from service. The reason for this change is general deterioration of the interior rope structure such as corrosion of uninspectable wires and general deterioration of the wire rope core. Reduction in diameter

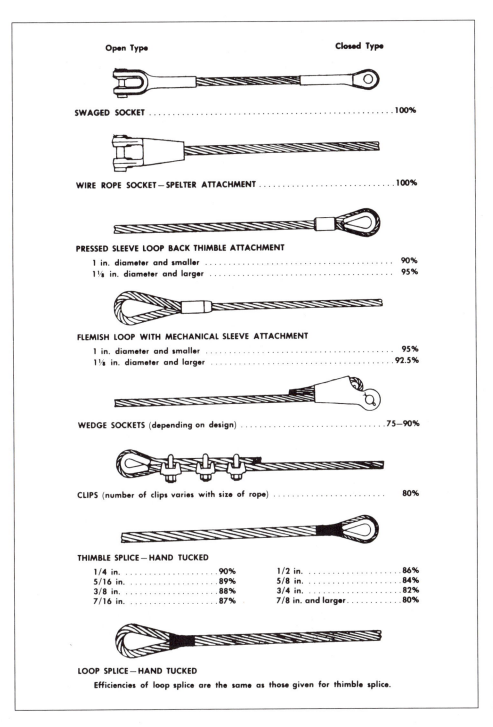

Figure 5-7. Typical efficiencies of attaching wire rope to fittings in percentages of strength of rope.

which may also occur is difficult in many cases to determine.

The number of broken wires per lay is one of the principal bases for judging the condition of a rope. If most of the broken wires in a lay are concentrated in several strands, that section of the rope is weaker than it would be if the broken wires were uniformly distributed throughout all strands and along the length of the rope. If, however, the number of broken wires along the length of a rope increases rapidly between inspections, the rope is becoming fatigued and nearing the end of its useful life.

Inspection codes may vary from state to state with regard to rope inspection and allowable degrees of deterioration. Usually inspections are based on numbers of broken wires per strand in one rope lay or number of broken wires per rope lay in all strands. For specific information within an operating area, consult state or provincial codes and the specific OSHA regulations covering the type of operation. Electronic inspection devices are available for determination of loss of strength due to corrosion, loss of metallic area, and broken wires.

Experience and judgment of all factors, combined with the length of time in service, the tonnage hoisted, or other unit for judging the work done by the rope determine when it should be discarded and replaced. At intervals throughout the life of the rope, a short section should be cut off at the socket end. This practice has two purposes: (1) to remove wires damaged

Table 5-H. Number of Spacing of Clips for Ropes of Various Sizes

Diameter of Rope (inches)	Minimum Number of Clips	Length of Rope Turned Back Exclusive of Eye (inches)	Torque (ft-lb)
⅛	2	3¼	—
¼	2	3¼	—
½	3	11½	65
⅝	3	12	95
¾	4	18	130
⅞	4	19	225
1	5	26	225
1⅛	6	34	225
1¼	7	44	360
1⅜	7	44	360
1½	8	54	360
1⅝	8	58	430
1¾	8	61	590
2	8	71	750
2¼	8	73	750

1 in. = 2.54 cm.
1 ft-lb = 1.36 newton-meter.

The number of clips shown is based upon using right regular or Lang lay wire rope, 6×19 class or 6×37 class, fiber core or IWRC, IPS, or XIPS. If Seale construction or similar large outer wire type construction in the 6×19 class is to be used for sizes 1 in. (2.5 cm) and larger, add one additional clip.

The number of clips shown also applies to right regular lay wire rope, 8×19 class, fiber core, IPS, nominal sizes 1½ in. and smaller; and right regular lay wire rope, 18×7 class, fiber core, IPS or XIPS, nominal sizes 1¾ in. and smaller.

For other classes of wire rope not mentioned above, it may be necessary to add additional clips to the number shown.

Turn back the specified amount of rope from the thimble. Apply the first clip one base width from the dead end of the wire rope (U-bolt over dead end—live end rests in clip saddle). Tighten nuts evenly to recommended torque.

Apply the next clip as near the loop as possible. Turn on nuts firm but do not tighten.

Space additional clips if required equally between the first two. Turn on nuts—take up rope slack—tighten all nuts evenly on all clips to recommended torque.

NOTE: Apply the initial load and retighten nuts to the recommended torque. The rope will stretch and shrink in diameter when loads are applied. Inspect periodically and retighten.

The efficiency rating of a properly prepared termination for clip sizes ⅛ to ⅞ in. is approximately 80 percent and for sizes 1 to 3 in. is approximately 90 percent. This rating is based upon the catalog breaking strength of wire rope. If a pulley is used in place of a thimble for turning back the rope, add one additional clip.

Figure 5-8. End fittings should be of best possible type for specific use. Zinc-poured sockets (bottom) are efficient in straight tension but are not as fatigue-resistant as swaged sockets (above).

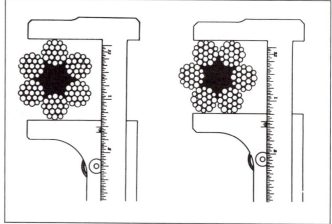

Figure 5-9. Correct way (left) and wrong way (right) to measure wire rope. Always read the widest diameter. (Printed with permission from Armco Steel Corporation.)

by vibration dampened at the socket, and (2) to change the positions of critical wear points throughout the system.

FIBER AND WIRE ROPE SLINGS

The safety of a sling assembly for special or ordinary uses is determined particularly by these factors: use of rope or chain and fittings of suitable strength for the load; the method of

Figure 5-10. Braided slings are resistant to kinking. Be sure loads are hoisted uniformly and that all slings have a minimum safety factor of five.

fastening the rope or chain to the fittings; the type of sling, such as single or three-leg; the kind of hitch; and regular inspection and maintenance.

Materials

Since the strength of fiber rope is affected by chemicals, freezing, high temperatures, and sharp bends, these factors should be considered when such rope is used for slings. Fiber rope is particularly suitable for the handling of loads that might be damaged by contact with metal slings.

Wire rope slings are usually made of extra improved plow steel grade rope. In some cases where the extra improved plow steel grade is not available, the improved plow grade of rope is used. The difference between the two grades is 15 percent. Normally wire rope of IWRC construction is used in extra improved plow and improved plow slings where mechanical type loop endings are employed or where swaged or pressed on terminations are used. In the smaller wire rope diameters up to and including 1⅛ in., the 6×19 classification wire rope is used. For rope diameters larger than 1⅛ in. the 6×37 classification rope is used. The most popular type of sling in use is the "strand laid" made from the above two-wire rope constructions.

Other popular types of wire rope slings are known as "cable laid" made from a multiple of individual wire rope laid up into one rope structure and used where its greater flexible characteristics are of value. "Braided slings" (Figure 5-10) made with a number of ropes braided into a single unit are used where flexibility, high strength, and rotation resistance are prime requisites. Because braided slings are braided in an open manner, they are fairly easy to inspect.

Methods of attachment

All hooks and rings used as sling connections should develop the full rated capacity of the wire rope sling. Sockets, when properly attached and compression fittings properly attached, should develop 100 percent of the rated wire rope strength. Swaged sleeve sling endings should develop 92–95 percent of the nominal wire rope strength. Compression and swaged sleeve fittings can be secured from the wire rope manufacturer or any properly equipped sling shop.

Hand-tucked splices develop about 90 percent of the rope strength in diameters less than ½ in. and 80 percent for larger diameters. Fittings used with hand-tucked slings should develop the same strength efficiencies as those used with mechanical type slings. The recommended load rating for a sling assembly is usually based on one-fifth the calculated strength of the assembly, although there may be cases where engineered lifts are made that do not meet this value.

Table 5-I. Rated Capacity Limits (in Tons) of Wire Rope Slings, Using Preformed Improved Plow Steel Wire Rope
(Depending on method of attaching the rope to the fittings)

| Rope Diameter (in.) | Single Leg | | | | | | Two-Leg Bridle or Basket Hitch | | | | | | | | | | | |
| | Vertical | | | Choker | | | Vertical* | | | 30 Degrees Vertical | | | 45 Degrees | | | 60 Degrees Vertical | | |
	S	MS	HT	S	MS	HT	S	MS	HT	S	MS	HT	S	MS	HT	S	MS	HT
6×19 Classification Construction																		
3/8	1.3	1.2	1.2	.92	.92	.92	2.6	2.4	2.4	2.3	2.1	2.0	1.8	1.7	1.7	1.3	1.2	1.2
1/2	2.3	2.2	2.0	1.6	1.6	1.6	4.6	4.4	4.0	4.0	3.8	3.5	3.3	3.1	2.8	2.3	2.2	2.0
5/8	3.6	3.4	3.0	2.5	2.5	2.5	7.2	6.8	6.0	6.2	5.9	5.2	5.1	4.8	4.2	3.6	3.4	3.0
3/4	5.1	4.9	4.2	3.6	3.6	3.6	10.0	9.8	8.4	8.7	8.5	7.3	7.1	6.9	5.9	5.0	4.9	4.2
7/8	6.9	6.6	5.5	4.8	4.8	4.8	14.0	13.0	11.0	12.0	11.0	9.5	9.9	9.3	7.8	7.0	6.6	5.5
1	9.0	8.5	7.2	6.3	6.3	6.3	18.0	17.0	14.0	16.0	15.0	12.0	13.0	12.0	10.0	9.0	8.5	7.2
1 1/8	11.0	10.0	9.0	7.9	7.9	7.9	22.0	20.0	18.0	19.0	17.0	16.0	16.0	14.0	13.0	11.0	10.0	9.0
6 × 37 Classification Construction																		
1 1/4	14	13	11	9.7	9.7	9.7	28	26	22	24	23	19	20	18	16	14	13	11
1 3/8	17	15	13	12	12	12	34	30	26	29	26	23	24	21	18	17	15	13
1 1/2	20	18	16	14	14	14	40	36	32	35	31	28	28	25	23	20	18	16
1 3/4	27	25	21	19	19	19	54	50	42	47	43	36	38	35	30	27	25	21
2	34	32	28	24	24	24	68	64	56	59	55	48	48	45	40	34	32	28
2 1/4	43	40	34	30	30	30	86	80	68	74	69	59	61	57	48	43	40	34

*If slings are used to handle loads with sharp corners, pads or saddles should be used to protect the rope. The radius of bend should not be smaller than five times the diameter of the rope. If the radius of bend is smaller, a choker hitch rating should be used.

S = Socket or swaged terminal attachment.
MS = Mechanical sleeve attachment.
HT = Hand-tucked splice attachment.

Note 1. Table is based on a design factor of 5, sling angles formed by one leg and a vertical line through the crane hook, and uniform loading.
Note 2. For 3-leg bridle slings, multiply safe load limits for 2-leg bridle slings by 1.5, and for 4-leg bridle slings, multiply by 2.0.

Table 5-J. Rated Capacity Limits (in Tons) of Wire Rope Slings, Using Preformed Extra Improved Plow Steel
(Depending on method of attaching the rope to the fittings)

| Rope Diameter (in.) | Single Leg | | | | | | Two-Leg Bridle or Basket Hitch | | | | | | | | | | | |
| | Vertical | | | Choker | | | Vertical* | | | 30 Degrees Vertical | | | 45 Degrees | | | 60 Degrees Vertical | | |
	S	MS	HT	S	MS	HT	S	MS	HT	S	MS	HT	S	MS	HT	S	MS	HT
6×19 Classification Construction																		
3/8	1.5	1.4	1.3	1.1	1.1	1.1	3.0	2.8	2.6	2.6	2.4	2.3	2.1	1.0	1.8	1.5	1.4	1.3
1/2	2.7	2.5	2.3	1.9	1.9	1.9	5.4	5.0	4.6	4.7	4.3	4.0	3.8	3.5	3.3	2.7	2.5	2.3
5/8	4.1	3.9	3.5	2.9	2.9	2.9	8.2	7.8	7.0	7.1	6.8	6.1	5.8	5.5	4.9	4.1	3.9	3.5
3/4	5.9	5.6	4.8	4.1	4.1	4.1	12.0	11.0	9.6	10.0	9.7	8.3	8.3	7.9	6.8	5.9	5.6	4.8
7/8	8.0	7.6	6.4	5.6	5.6	5.6	16.0	15.0	13.0	14.0	13.0	11.0	11.0	11.0	9.0	8.0	7.6	6.4
1	10.0	9.8	8.3	7.2	7.2	7.2	20.0	20.0	17.0	17.0	17.0	14.0	14.0	14.0	12.0	10.0	9.8	8.3
1 1/8	13.0	12.0	10.0	9.1	9.1	9.1	26.0	24.0	20.0	23.0	21.0	17.0	18.0	17.0	14.0	13.0	12.0	10.0
6 × 37 Classification Construction																		
1 1/4	16	15	13	11	11	11	32	30	26	28	26	23	23	21	18	16	15	13
1 3/8	19	18	15	13	13	13	38	36	30	33	31	26	27	25	21	19	18	15
1 1/2	23	21	18	16	16	16	46	42	36	40	36	31	33	30	25	23	21	18
1 3/4	31	28	24	21	21	21	62	56	48	54	49	42	44	40	35	31	28	24
2	40	37	32	28	28	28	80	74	64	69	64	55	57	52	45	40	37	32
2 1/4	49	46	40	35	35	35	98	92	80	85	80	69	69	65	57	49	46	40

*If slings are used to handle loads with sharp corners, pads or saddles should be used to protect the rope. The radius of bend should not be smaller than five times the diameter of the rope. If the radius of bend is smaller, a choker hitch rating should be used.

S = Socket or swaged terminal attachment.
MS = Mechanical sleeve attachment.
HT = Hand-tucked splice attachment.

Note 1. Table is based on a design factor of 5, sling angles formed by one leg and a vertical line through the crane hook, and uniform loading.
Note 2. For 3-leg bridle slings, multiply safe load limits for 2-leg bridle slings by 1.5, and for 4-leg bridle slings, multiply by 2.0.

ANGLE STRENGTH LOSS FROM RATED CAPACITY

ANGLE	FACTOR
70°	.3420
60°	.5000
50°	.6428
40°	.7660
30°	.8660
20°	.9397
10°	.9848
0°	1.0000

FACTOR	ANGLE
.2588	75°
.4226	65°
.5736	55°
.7071	45°
.8192	35°
.9063	25°
.9659	15°
.9962	5°

The increased angle of the sling leg reduces its capacity. See chart for loss factor. Determine the angle between the sling leg and the vertical plane. Then multiply the sling rating by the appropriate loss factor from the chart. This will determine the slings reduced rating.

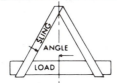

EXAMPLE:
Assume sling capacity
  2,000#
If angle = 50° then loss
  factor = .6428
Multiply: 2,000# x .6428
  1,286# = rated capacity
  of sling at 50°

Figure 5-11. Increasing the angle between the sling leg and vertical increases the stress on each leg of the sling and reduces its capacity (Printed with permission from Web Sling Association.)

Rated capacities

The rated capacities as given in the various sling catalogs and tables are based on newly manufactured slings. As the sling is used, factors such as abrasion, nicking, distortion, corrosion, and other such factors of use, for example bending around small radii, will affect the load rating and must be given consideration when lifts are contemplated.

Because slings can be used at various angles and since the rope stress increases rapidly with the angle of lift, it is essential that all slings be ordered with this consideration in mind. Fortunately, most wire rope sling catalogs have tables worked out with the load ratings given for the most used and critical angles of lift. These should be consulted for safe rigging practices.

When the rope is made into a sling and placed in position on a load, the angle formed by the ropes and the horizontal should be determined and given careful consideration. The rated load capacity of the sling decreases sharply as the angle formed by the sling leg and the horizontal becomes smaller. When this angle is 45 degrees, the rated capacity has decreased to 71 percent of the load that can be lifted when the legs are vertical. As this angle decreases the rated capacity continues to decrease. (See Tables 5-I and -J.)

Figure 5-11 shows how the tension on a leg of a sling increases as the angle decreases from the vertical. When the angle formed by a leg and the vertical is 30 degrees the rated capacity is only 87 percent of that of both legs vertical. For an angle of 60 degrees, the rated capacity is only 50 percent of that for both legs vertical. These losses are proportional to the cosine of the sling angle with the vertical; the actual stress is equal to the amount of the load that a leg must support, divided by the cosine of the angle that the leg is away from being vertical. (See Table 19-N, Natural Trigonometric Functions.) Excessive angles may be avoided by the use of longer slings, if head room permits.

Tables showing rated sling capacities should be posted conspicuously around the shop. Also, each sling should bear a tag indicating its rated load capacity (as shown in Figure 5-12).

Allowances in rated capacity are also required for different types of hitches. For example a decrease of at least 25 percent in rated capacity for a single leg vertical sling is made when a choker hitch is used. The suitable load for a basket hitch is based on the angle of the legs.

Special clamps are used in handling of steel plate, flanged castings and similar shaped products. The slings using the horizontal- and vertical-type clamps have the same rated capacity as other bridle-type slings provided the strength of the clamp is equal to the other sling components.

If loads that have sharp edges or sharp corners must be lifted, pads or saddles should be used to protect the ropes or chains.

For strand-laid wire rope slings in a basket hitch the minimum diameter of curvature of the sling in contact with the load should be not less than 20 times the sling rope diameter. For cable-laid slings the minimum diameter of curvature should not be less than 10 times and the braided slings the minimum diameter of curvature should not be less than 20 times the diameter of the component rope.

Thimbles spliced in the ends of slings will materially reduce wear.

Inspection

Employees should be trained to check slings daily and whenever they may suspect damage after a lift, to report promptly any questionable conditions of the equipment or assembly. In the United States, OSHA inspection requirements for industrial slings should be used as a guide in evaluating the sling's condition. A thorough inspection by a trained person should be made at least every 6 months.

Slings that do not pass inspection requirements should be promptly withdrawn from service and made unusable by burning or cutting before being discarded.

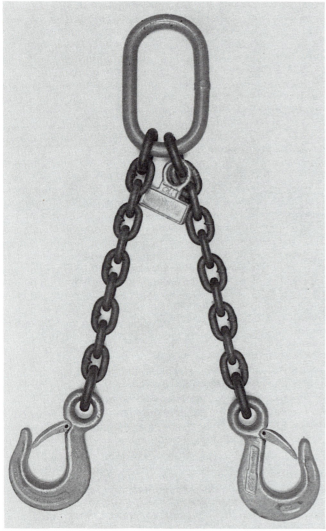

Figure 5-12. Typical double-chain sling. All components, such as the oblong master link, the body chain, and the hook, are carefully matched for compatibility. Note permanent identification tag. (Printed with permission from Columbus McKinnon Corporation.)

CHAINS AND CHAIN SLINGS

Materials

Alloy steel has become the standard material for chain slings. Alloy steel chain has high resistance to abrasion and is practically immune to failure from the cold-working of the metal. More than one chain can be used (Figure 5-13).

Special-purpose alloy chains are made from stainless steel, monel metal, bronze, and other materials. They are designed for use where resistance to corrosive substances is required, or where other special properties are desirable.

Proof coil chain (also known as common or hardware chain) is used for miscellaneous purposes where failure of the chain would not endanger human life or result in serious damage to property or equipment. Proof coil chain should never be used for slings.

Figure 5-13. Double, triple, or quad chain slings, as in this application, can be rigged to handle loads of virtually any shape or size. (Printed with permission from Campbell Chain Company.)

Properties of alloy chain

Alloy chain is produced from heat-treatable alloy steel in conformance with ASTM specifications A391–1975. Chain after heat treatment has mechanical properties as follows:

Tensile Strength	115,000 psi minimum
Elongation	15 percent minimum

Table 5-K shows the recommended working load limits, proof test loads, and the minimum breaking strengths of alloy steel chain.

The working load limit (safe load strength) is arrived at by dividing the breaking strength (ultimate strength) by a specified safety factor.

The values, shown in Table 5-K, represent the maximum loads that should ever be applied in direct tension to a length of alloy chain. All alloy chain is tested in direct tension under the proof test loads shown, prior to final inspection and shipment. The data for other types of chain may be obtained from the National Association of Chain Manufacturers' Specification No. 3001.

The tensile strength of alloy steel chain increases in proportion to its hardness (produced by heat treating). Resistance to abrasion also increases proportionately with hardness.

Impact conditions caused by faulty hitches, bumpy crane

Table 5-K. Working Load Limits, Proof Test Loads and Minimum Breaking Loads for Alloy Steel Chain

Nominal Size of Chain (in.)	Working Load Limit (lb)	Proof Test (lb)	Minimum Break (lb)
¼	3,250	6,500	10,000
⅜	6,600	13,200	19,000
½	11,250	22,500	32,500
⅝	16,500	33,000	50,000
¾	23,000	46,000	69,500
⅞	28,750	57,500	93,500
1	38,750	77,500	122,000
1⅛	44,500	89,000	143,000
1¼	57,500	115,000	180,000
1⅜	67,000	134,000	207,000
1½	80,000	160,000	244,000
1¾	100,000	200,000	325,000

Reprinted from *Specification for Alloy Chain*, American Society for Testing and Materials, A391-1975. *Alloy Steel Chain Specifications*, No. 3001, National Association of Chain Manufacturers.

tracks, and slipping hookups can materially add to the stress in the chain. If severe impact loading may be encountered, a lower working load limit should be used, regardless of the chain type.

Alloy steel chains are suitable for high temperature operations. However, continuous operation at a temperature of 800 F (425 C) (the highest temperature for which continuous operation is recommended) requires a reduction of 30 percent in the regular working load limit. These chains may be used at temperatures up to 1000 F (540 C) at 50 percent of the regular working load limit, but only for intermittent service. The chain working load limit must be permanently reduced by 15 percent after having become subjected to this temperature.

The general strength and working load limits of alloy steel chain are not altered appreciably by low atmospheric temperatures.

Chain slings should preferably be purchased complete from the manufacturer and, whenever repairs are required, should be sent back to him.

Hooks and attachments

As a general rule, hooks, rings, oblong links, pear-shaped links, coupling links, and other attachments should be made of the same, or equivalent, heat-treatable alloy steel as the chain itself. In most cases, attachments will be installed on the chain by the chain manufacturer, who will then heat-treat and proof-test the assembly. (See Figure 5-12.)

If emergency conditions make it necessary for the user to replace an attachment, the grade and size should be selected with extreme care. High-strength, heat-treatable alloy connecting links of the same analysis type as that used by the chain manufacturer should be employed.

Unalloyed carbon-steel hooks, repair links, rings, pear-shaped links, and other attachments should not be used. Homemade or makeshift bolts, rods, shackles, hooks without safety catches, or other attachments should never be used.

Standard items produced from alloy steel include sling hooks, grab hooks, foundry hooks, grab links, rings, oblong links, pear-shaped links, and repair links. All such attachments used with the recommended chain size provide a safety factor equal

to or greater than that of alloy chain itself. Dimensional specifications for these attachments will vary somewhat with the individual manufacturer.

Many injuries have resulted when employees have caught their fingers between the hook attachment and the load. To prevent such injuries, handles can be attached to the assembly hook or end attachment. To increase operating efficiency, handles are also frequently used on large hooks, master links, and other attachments.

Inspection

Most of the causes of chain failures can be detected before failure occurs if the proper inspection procedure is followed.

A good inspection plan provides for two inspections; one daily by the personnel using the chain, and the other biannually or oftener by a person qualified by experience or training. The former should be able to detect those links and hooks which have become visibly unsafe because of overloading, faulty rigging, or other unsafe practices. The latter should be an experienced person who has the authority to remove damaged assemblies from service for reconditioning or replacement.

A link-by-link inspection should be made to detect the following:
1. Bent links.
2. Cracks in weld areas, in shoulders, or in any other section of link.
3. Tranverse nicks and gouges.
4. Corrosion pits.
5. Elongation resulting from stretching can only be caused by overloading. The best way to detect such elongation and also link wear is by a visual link-by-link inspection. Overall sling length measurements and even measurements of 1- to 3-ft lengths are inadequate because not all links are affected uniformly. Likewise caliper readings only over several links can miss such a condition (Table 5-L and Figure 5-14).

Table 5-L. Maximum Allowable Wear at Any Point of Link

Chain Size (in.)	Maximum Allowable Wear (in.)
¼	³⁄₆₄
⅜	⁵⁄₆₄
½	⁷⁄₆₄
⅝	⁹⁄₆₄
¾	⁵⁄₃₂
⅞	¹¹⁄₆₄
1	³⁄₁₆
1⅛	⁷⁄₃₂
1¼	¼
1⅜	⁹⁄₃₂
1½	⁵⁄₁₆
1¾	¹¹⁄₃₂

When inspecting the hook, measurement must be made between the shank and the narrowest point of the hook opening. Whenever the throat opening exceeds 15 percent of the normal opening, the hook should be replaced. Special attention should be given to slings to which hooks have been added, in order to make sure that the hooks are secure.

Unless full and adequate facilities for repair are available, chain showing faults by inspection should be returned to the manufacturer for reconditioning.

148

Figure 5-14. Extreme wear at bearing surfaces. Links are turned to detect wear.

Alloy steel chains and hooks should never be annealed or normalized because these processes reduce the hardness and, therefore, greatly reduce the strength.

Safe practices

In the use of chains, recognized safe practices will do much to prevent failures:

1. Never splice a chain by inserting a bolt between two links.
2. Never put a strain on a kinked chain. Workers should be trained to take up the slack slowly and see that every link in the chain seats properly.
3. Do not use a hammer to force a hook over a chain link.
4. Permanent identification tags are usually attached to chain slings by the manufacturers. Tags should never be removed.
5. Remember that decreasing the angle between the legs of a chain sling and the horizontal increases the load in the legs.
6. Use chain attachments (rings, shackles, couplings, and end links) designed for use with the chain to which they are fastened.
7. See that the load is always properly set in the bowl of the hook. Loading on or toward the point (except in the case

of grab hooks or others especially designed for the purpose) overloads the hook and leads to spreading and possible failure.

8. Chains not in use should be stored in a suitable rack. Do not let them lie on the ground or floor where they can be damaged by lift trucks or other vehicles.
9. Secure "out-of-balance" loads properly (see Figure 5-15).

Figure 5-15. This "out-of-balance" load is secured by a double sling with chain leg adjusters (the two short chains with grab hooks attached to the master link). The chain leg passing through the bore of this casting has been protected from damage by adequate padding. (Printed with permission from American Chain Division of Acco.)

SYNTHETIC WEB AND METAL MESH SLINGS

Two widely used slings today are the synthetic web and metal mesh types. They are strong and dependable slings, if selected to meet the needs of the particular operation and if used in the proper manner.

Synthetic web slings

Web slings are useful for lifting loads that need their surface protected by the soft suppliant web sling surface. This usage is well suited for tubular, nonferrous, ceramic, painted, polished, highly machined, and other products with a fine

Figure 5-16.   Web sling protects roller's polished surface. (Printed with permission from The Wear-Flex Corp.)

or delicate surface (Figure 5-16). To this end, web sling service life is secondary to load protection; and the user should be aware that web slings are relatively easily cut and have small abrasion resistance compared with that of chain or wire rope. (The types of web slings are shown in Figure 5-17.)

The web fibers most used are nylon and polyester. Each has specific advantages and disadvantages of elongation, strength, and chemical resistance.

Design factor. According to ASTM standard B309–1984, minimum design factor for synthetic web slings shall be 5 to 1; more simply expressed, the minimum breaking strength shall be 5 times the rated capacity (see Tables 5-M and -N).

Identification. Every synthetic web sling shall bear following legible identification information—name or identification of manufacturer; sling code number, rated capacities for usable types of hitches, and type of material (polyester, nylon, etc.).

Effect of sling angles. When two slings (or one sling in bas-

ket hitch) are used to lift a load from one crane hook, sling capacity is reduced. The load-carrying capacity of the sling is determined by applying the appropriate factor × hitch capacity.

Types of inspection.
1. Initial inspection—both new and repaired slings shall be inspected prior to use to determine that the sling meets the requirements of the purchase order, that it is the correct type and has the proper rated capacity for the application, and that it has not been damaged in shipment, unpacking, or storage.
2. Frequent inspection—the sling shall be inspected by the person handling it each time it is used.
3. Periodic inspection—this inspection shall be made by designated personnel experienced in the inspection of synthetic web slings. Frequency of periodic inspection shall be based on the following factors—frequency of use, severity of service conditions, and experience gained relative to service

Type I. Triangle and choker end fittings usable in vertical, choker, and basket hitches.

Type II. Triangle fittings each end usable in vertical and basket hitches only.

Type III. Flat eye ends usable in vertical, choker, and basket hitches.

Type IV. Twisted eye ends usable in vertical, choker, and basket hitches.

Type V. Endless (or grommet) usable in vertical, choker, and basket hitches.

Type VI. Reversed (or return) eye. Essentially an endless sling, butted on the sides with wear pad(s) on body. Usable in vertical, choker, and basket hitches.

Figure 5-17. Basic synthetic web sling types.

Table 5-M. Rated Capacity in Pounds for 1600 lb/in. Web Slings

Single-Ply Capacities for Various Type Slings in Vertical, Choker, and Vertical Basket Hitches

Web Width in In.	Types 1, 2, 3, & 4 Hitches			Type 5 Endless Hitches			Type 6 Reversed Eye Hitches		
	Vertical	Choker	Basket	Vertical	Choker	Basket	Vertical	Choker	Basket
1	—	—	—	2,600	2,100	5,200	—	—	—
2	3,200	2,400	6,400	5,100	4,100	10,200	4,500	3,600	9,000
3	4,800	3,600	9,600	7,700	6,200	15,400	—	—	—
4	6,400	4,800	12,800	10,200	8,200	20,400	7,700	6,200	15,400
5	8,000	6,000	16,000	12,800	10,200	25,600	—	—	—
6	9,600	7,200	19,200	15,400	12,300	30,800	11,000	8,800	22,000

Rated capacities shall never be exceeded.
See manufacturer's rated capacities for multiple-ply slings.

Table 5-N. Rated Capacity in Pounds for 1200 lb/in. Web Slings

Single-Ply Capacities for Various Type Slings in Vertical, Choker, and Vertical Basket Hitches

Web Width in In.	Types 1, 2, 3, & 4 Hitches			Type 5 Endless Hitches			Type 6 Reversed Eye Hitches		
	Vertical	Choker	Basket	Vertical	Choker	Basket	Vertical	Choker	Basket
1	—	—	—	1,900	1,500	3,800	—	—	—
2	2,400	1,800	4,800	3,800	3,000	7,600	3,500	2,800	7,000
3	3,600	2,700	7,200	5,800	4,600	11,600	5,000	4,000	10,000
4	4,800	3,600	9,600	7,700	6,200	15,400	6,800	5,400	13,600
5	6,000	4,500	12,000	9,600	7,700	19,200	—	—	—
6	7,200	5,400	14,400	11,500	9,200	23,000	8,000	6,400	16,000

Rated capacities shall never be exceeded.
See manufacturer's rated capacities for multiple-ply slings.

life of slings used in this and similar conditions. Slings should be inspected at least annually.

Inspection parameters for synthetic web slings.
1. Excessive abrasive wear on webbing and any fittings
2. Cuts, tears, snags, punctures, holes, and crushed fabric
3. Worn or broken stitches, particularly that of the laps
4. Burns, charring, melting damage, or weld spatter damage
5. Acid, caustic, or other chemical damage
6. Broken, distorted, or excessively worn fittings
7. Other defects, such as knots, which cause doubt as to safety of the sling
8. Ultraviolet degradation can be judged only from prior experience. Consult the manufacturer.

Inspection records may be of considerable help and, if used, should identify the sling, and give dates of inspection; they should record the condition at the time of each inspection.

Repair of web slings. Although generally not repaired, web slings may be repaired by any manufacturer or qualified person who shall identify the work and certify the rated capacity. All repaired slings shall be proof tested to two times their newly rated capacity. No temporary repairs should be made.

Operating practices. The ANSI/ASME standard B30.9-1984 recommends the following practices for slings:

1. Slings having suitable characteristics for the type of load, hitch, and environment shall be selected in accordance with appropriate tables (see 9-5.3 and 9-5.5 in ANSI/ASME B30.9-1984).
2. The weight of load shall be within the rated load (rated capacity) of the sling.
3. Slings shall not be shortened or lengthened by knotting or other methods not approved by the sling manufacturer.
4. Slings that appear to be damaged shall not be used unless inspected and accepted as usable under 9-5.6.
5. Sling shall be hitched in a manner providing control of the load.
6. Sharp corners in contact with the sling should be padded with material of sufficient strength to minimize damage to the sling.
7. Portions of the human body should be kept from between the sling and the load, and from between the sling and the crane hook or hoist hook.
8. Personnel should stand clear of the suspended load.
9. Personnel shall not ride the sling.
10. Shock loading should be avoided.
11. Slings should not be pulled from under a load when the load is resting on the sling.
12. Slings should be stored in a cool, dry, and dark place to prevent environmental damage.
13. Twisting and kinking the legs (branches) shall be avoided.
14. Load applied to the hook should be centered in the base (bowl) of hook to prevent point loading on the hook.
15. During lifting, with or without load, personnel shall be alert for possible snagging.
16. In a basket hitch, the load should be balanced to prevent slippage.
17. The sling's legs (branches) should contain or support the load from the sides above the center of gravity when using a basket hitch.
18. Slings should be long enough so that the rated load (rated capacity) is adequate when the angle of the legs (branches) is taken into consideration.
19. Slings should not be dragged on the floor or over an abrasive surface.
20. In a choker hitch, slings shall be long enough so the choker fitting chokes on the webbing and never on the other fitting.
21. Nylon and polyester slings shall not be used at temperatures in excess of 194 F (90 C).
22. When extensive exposure to sunlight or ultraviolet light is experienced by nylon or polyester web slings, the sling manufacturer should be consulted for recommended inspection procedure because of loss in strength.

Metal mesh slings

The safe use of metal mesh slings is primarily determined by two factors: (1) use of the right sling for the right load; and (2) the construction of the sling (see Figure 5-17). Figure 5-18 shows the structure and nomenclature of a metal mesh sling. All metal slings are properly identified with their safe working load limit stamped in their vertical basket hitch and choker hitch (metal handle). They can be used efficiently with all weights that fall below that limit. They are classified as either heavy-duty, medium-duty, or light-duty. The design factor of metal mesh slings is five to one, or five times the amount stated on the sling. All metal mesh slings are proof tested to a minimum of 200 percent of their rated capacity. This removes all permanent elongation when used at its rated capacity.

With metal mesh slings, the handling of sharp-edged metals, concrete in its many prestressed forms, and high-temperature materials up to 500 F (260 C) is a safe operation, except elactomer coated slings (usually up to 200 F [93 C]). Any danger in the use of metal mesh slings stems mainly from improper use. Damaging the slings at the edges by faulty loading or dragging a sling out from under a load may eventually cause wear to the spirals that comprise the mesh, thus reducing the wire diameter and requiring the sling to be taken out of service. Figure 5-19 shows a metal mesh sling made into a four-legged basket hitch.

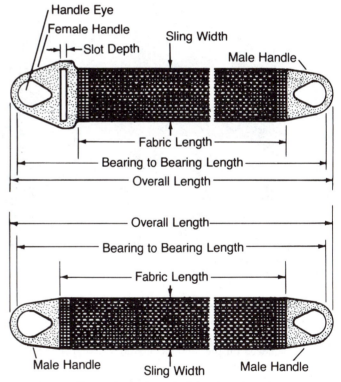

Figure 5-18. Nomenclature for typical metal mesh slings.

Metal mesh slings should never be shortened by using knots, bolts, or other unapproved methods. Consult the sling manufacturer on how to do this if shortening becomes necessary. Tampering with the surface of any sling can weaken it and make it highly dangerous. Precautions should be taken never to twist or kink the legs of a metal mesh sling or use one when the spirals are locked. A sudden jolt may break the spirals, cause the load to shift, and create havoc on the floor below.

With metal mesh as well as all other slings, adhering to

Figure 5-19. A four-legged, basket-hitch sling of steel mesh can take the sharp edges of lumber without failure and without damaging the wood.

rules for certain hitches also assures safe handling of a load.

One of the most important precautions in the use of slings is regular inspection by a qualified person. Inspection frequency should be based on the amount of sling use and the severity of conditions. The user should look at the sling each time it is used and thoroughly at least once a year. It would be wise to keep written inspection records with identification of each sling reported as it is provided on the sling by the manufacturer. Metal mesh slings should be removed from service if a broken weld or brazed joint is discovered along the sling edge. Other breakdowns in construction to guard against are broken wires in any part of the mesh, a reduction of 25 percent in wire diameter due to abrasion, a lack of sling flexibility, cracked end fitting, and visible distortion. Any one of these conditions or a combination of them, if ignored, could eventually result in sling breakdown. See the preceding section for operating practices applicable to all slings.

To sum it up, it would be wise to return to the primary fac-

tors of proper sling selection, construction, and maintenance. Keeping these three items in mind, and applying the information concerning environmental conditions, type of load, and inspection, should keep the use of metal slings within the boundaries of safe operation.

REFERENCES

American Iron and Steel Institute, Committee of Wire Rope Producers, 1000 16th Street, NW., Washington, D.C. 20036. *Wire Rope Users Manual,* 1979.

American National Standards Institute, 1430 Broadway, New York, N.Y. 10018.
  "Safety Code for Cableways, Cranes, Derricks, Hoists, Hooks, Jacks, and Slings," B30 Series.
  Elevators, Escalators, and Moving Walks, A17.1.
  Safety Color Code for Marking Physical Hazards, Z53.1.
  Specifications for and Use of Wire Ropes for Mines, M11.1.

American Petroleum Institute, 1220 L St. N.W., Washington, D.C. 20005, *Recommended Practice on Application, Care, and Use of Wire Rope for Oil-Field Service,* Code No. API-RP-9B.

American Society for Testing Materials, 1916 Race St., Philadelphia, Pa. 19103. *Specification for Alloy Steel Chain,* A391–1975.

Broderick & Bascom Rope Co., Rt. 3, Oak Grove Industrial Park, P.O. Box 844, Sedalia, MO 65301.
Rigger's Handbook, 1986.
Wire Rope Handbook.

Cordage Institute, 314 Lincoln St., Suite 568, Hingham, MA 02043.

Electric Controller and Manufacturing Company, 2704 East 79th St., Cleveland, Ohio 44104. *How to Operate a Crane,* Booklet 920.

Hess, Owen, "Metal Mesh and Nylon Slings." *National Safety News,* 103: 78–79 (June 1971).

National Association of Chain Manufacturers, 20 N. Wacker Dr., Suite 3318, Chicago, IL 60606. *Alloy Steel Chain Specifications,* No. 3001.

National Fire Protection Association, Batterymarch Park, Quincy, MA 02269.
Flammable and Combustible Liquids Code, NFPA 30.
National Electrical Code, NFPA 70.

National Safety Council, 444 North Michigan Ave., Chicago, Ill. 60611. Industrial Data Sheets.
Alloy Chains for Overhead Lifting, 478.
Recommended Loads for Wire Rope Slings, 380.

U.S. Department of the Interior, Bureau of Mines, Washington, D.C. 20240. *Recommended Procedures for Mine Hoists and Shaft Installation, Inspection, and Maintenance,* Information Circular 8031.

Powered Industrial Trucks

FACTORIES, WAREHOUSES, DOCKS, AND TRANSPORTATION TERMINALS use powered industrial trucks to carry, push, pull, lift, stack, and tier material. Each of these trucks requires guarding for the operator's protection and for the safety of other workers. The establishment of safe practices for the operation, maintenance, and inspection of powered industrial trucks is also essential.

This chapter does not discuss powered industrial trucks employed in airport and air terminal areas. For discussion of trucks and operators in these areas, see *General Aviation Ground Operations Safety Handbook,* published by the National Safety Council. See References.

TYPES OF TRUCKS

Powered industrial trucks may be classified by power source, operator position, and means of engaging the load. Power sources include electric motors energized by storage batteries; engines using gasoline, LP-gas, or diesel fuel; or trucks using a combination of gas or diesel and electricity. Provisions for safe operation, maintenance, and design of powered industrial trucks are given in ANSI *Safety Standard for Powered Industrial Trucks,* B56.1. Users are not permitted to modify trucks without the written approval of the manufacturer.

▪ One class of powered industrial truck is designed to be controlled by a riding operator (Figure 6-1). The widely used lift truck, with its cantilever load-engaging means, vertical uprights, and elevating mechanism, is usually a rider truck. Some rider trucks use a platform to engage the load. Both types may be either high-lift—with an elevating mechanism that permits the tiering of one load on another (Figure 6-2)—or low-lift—with a mechanism that raises the load only enough to permit horizontal movement.

The development of attachments has increased the versatility of the lift truck. Clamps, rotators, shifters, stabilizers, pushers, pullers, up-enders, bottom dumpers, top lifters, rams, cranes, scoops, as well as other modifications, have been developed to meet specific needs. Two or more motions have been built into one attachment; for example, it could clamp and rotate, or side-shift and push and pull (Figure 6-3). It is possible to interchange attachments so that one truck can be used for various type loads.

Straddle carriers (Figure 6-11) carry long material, such as pipe or lumber, under the truck body, which rides on four high legs. Powered industrial tractors draw trailers, nonpowered trucks, and other mobile loads. In warehouse operations, order-picker trucks (Figure 6-5) are used to raise the operator to the desired height.

▪ A second category of powered industrial trucks is the motorized handtruck controlled by a walking operator (Figure 6-6). They also have a platform or lifting forks to engage the load and may be either high-lift for tiering or low-lift to raise the load only enough for horizontal movement.

A unique powered industrial truck is an electronically controlled vehicle without an operator (Figure 6-7). It travels a prearranged route, outlined on or under the floor, and is controlled by frequency sensors, a light beam, or induction tape.

SAFEGUARDING

A powered industrial truck (with a riding operator) capable of lifting loads higher than the operator's head or operated in areas

Figure 6-1. Lift trucks are available in various capacities and may be powered by electricity, gasoline, or LP-gas. Note load backrest extension and backward tilt of mast. (Printed with permission from Eaton Corporation, Industrial Truck Division.)

where there is a hazard from falling objects must be equipped with an overhead guard (Figures 6-1, -2, -3, -4). This guard should not interfere with good visibility. Openings in the guard should be small enough to protect the operator from being struck by objects or material falling from an overhead load or stack, and should conform to American National Standard B56.1. A load backrest extension (Figure 6-1) should always be used when the type of load presents a hazard to the operator. The top of a load should not exceed the height of the backrest.

Requirements

Trucks should be equipped with platforms extending beyond the operator's position, strong enough to withstand a compression load equal to the weight of the loaded vehicle applied along the longitudinal axis of the truck with the outermost projection of the platform axis of the truck against a flat vertical surface. Note platform in Figure 6-8.

The work environment must be considered when purchasing or modifying powered industrial trucks. New approaches to add to operator safety on these vehicles include operator restraint systems that limit the operator to within the vehicle itself.

Exposed tires should have guards that will stop particles from being thrown at the operator. Hazardous moving parts, such as chain and sprocket drives and exposed gears, should be guarded to protect the operator when in a normal operating position.

Although many lift trucks may come with steering wheel knobs, their use is prohibited by many companies. If knobs are used, they should be of the mushroom type to engage the palm of the operator's hand in the horizontal position, and should be mounted within the periphery of the wheel. The steering mechanism should minimize transmission of road shock to the steering wheel.

Every powered industrial truck should carry a name plate showing the weight of the truck and its rated capacity as specified by the ANSI B56.1, *Safety Standard for Powered Industrial*

Figure 6-2. High-lift fork truck for pallet storage on racks proceeds down this aisle. (Printed with permission from Allstate Insurance Company, Training Division.)

Trucks, previously mentioned (Figure 6-9). Other vehicle apparatus may have limitations that also should be observed by the operator.

Powered industrial trucks should have horns or other warning devices that make a distinctive sound loud enough to be heard clearly above other local noises. The warning device should be under the operator's control; a backup alarm, however, works whenever the truck backs up. Where excessive noise could cause confusion, flashing lights mounted on the overhead guard can warn employees of approaching trucks.

Specifications of steering, braking, and control arrange-

Figure 6-3. A special clamp affixed to this lift truck permits handling of paper rolls without damage. (Printed with permission from Hyster Co.)

Figure 6-5. The operator of this order-picker truck can control the truck's movement and the mast from the platform. Note the handholds at the top of the backrest and the safety strap attached to the employee.

Figure 6-4. The screen covering the top of this lift truck prevents smaller objects from striking the operator. This is particularly advantageous when a truck is used in areas where rocks fall.

ments should conform in detail to ANSI B56.1.

All steering controls, except for motorized handtrucks with a steering handle, should be confined within the plain view outline of the truck or guarded to prevent injury to the operator when passing obstacles (Figure 6-10).

Hazardous location classification

Powered industrial trucks should also be constructed and equipped to comply with Underwriters Laboratories "Standards for Safety," UL 558 and UL 583, as specified in NFPA 505, *Type Designations, Areas of Use, Maintenance and Operation of Powered Industrial Trucks.*

The definitions of hazardous locations are given in NFPA 70, *National Electrical Code.* Class I locations are those in which flammable gases or vapors are or may be present; Class II, combustible dust; and Class III, easily ignitible fibers or flyings. The B56.1 *Safety Standard for Powered Industrial Trucks* stipulates that trucks, electric- or gasoline-powered, shall not be used in certain hazardous locations unless they either comply with NFPA requirements or are specifically approved by the inspection authority for the location involved. Definitions of hazardous locations will be found in NFPA 505, *Powered Industrial Trucks.*

Figure 6-6. The walking operator controls this electric pallet-lift (walkie) truck. (Printed with permission from Big Joe Manufacturing Co.)

Powered industrial trucks (whether or not they are designed for hazardous locations) should be equipped with an appropriate fire extinguisher that all operators are trained to use. An important incidental advantage is that the truck-mounted fire extinguisher and trained operator are quickly available to combat other fires on the premises.

Lift trucks

If an overhead guard is attached to the rear of the truck body, it also should be attached to the body in front and not the mast (Figure 6-1). An exception is made for those trucks that have the tilt cylinder as an integral part of the overhead guard construction. The overhead guard shall be designed so as to prevent injury to the operator should the mast-tilting mechanism fail.

Forks should be locked to the carriage, and the fork extension, if used, should be designed to prevent unintentional lifting of the toe or displacement of the fork extension.

Lift trucks should have means or be equipped with mechanical hoist and tilt mechanisms to prevent overtravel of hoist and tilt motions. If the lifting systems are hydraulically driven, a relief valve should be installed in the system and suitable stops provided to prevent overtravel.

Straddle trucks

Straddle trucks (Figure 6-11) should be provided with horns or other warning devices, and with headlights and tail lamps for night operation. Safe access ladders, wheel guards, and chain drive guards should be provided. Certain types of operation may require a rigid overhead guard for the operator.

Overhead clearances can be determined by telltales installed in advance of overhead obstructions across railways and other passageways where straddle trucks operate. Gage rods may be mounted on the truck at front and rear.

Straddle trucks present a special problem for operators. Because they sit so high off the ground, their angle of sight is reduced for objects immediately to the front or rear.

While precautions always must be taken to avoid striking pedestrians, this especially is true when carrying long loads. Red flags may be attached to the ends of such loads or signal persons may be stationed in congested areas. Particular care must be taken if the truck is used after dark.

Crane trucks

Although some have three, crane trucks usually have four wheels (Figure 6-12). One model is designed so the operator sits behind a small pillar-type jib crane mounted on a chassis, while in another, the operator stands on a platform and operates a fully or partly rotating crane. Still another type has a fixed boom (which cannot be swung), so that to make side motions the entire rig must be moved from one position to another.

A crane carrying a load should be driven at the lowest possible speed, and the load should be carried as low as possible. When it is traveling without a load, the hook should be fastened to the lower end of the boom to prevent the hook block from swinging.

The operator should have a helper to hook on the load and to give signals. When a long load is being carried, the helper should walk alongside, and keep the load from swinging and striking against objects along the way by means of a tag line.

Tractors and trailers

Tractors and trailers should incorporate the necessary safety features in the coupling used to make up the tractor-trailer train. The type of coupling used depends on the construction of the trailer, the loads carried, and whether or not the route traveled includes sharp curves, ramps, or inclines. The coupling must be one that will not come unhitched on grades nor permit the trailer to whip or cut in on curves. Loads on trailers should be secured to the trailer to avoid widespread scattring of material shoudl the load shift or the trailer roll over.

Motorized handtrucks

A powered handtruck should be equipped so that its brakes will be applied when the handle is in either the fully raised or the fully lowered position (Figure 6-13).

In operating a motorized handtruck, theprinciple hazards are the operator being pinned between the truck and a fixed object, and the truck running up on the operator's heels. Operators should walk ahead of such trucks, leading them from either side of the handle and facing the direction of travel. When a truck must be driven close to a wall or other obstruction, down an incline, or onto an elevator, the operator should put the truck in reverse and walk behind it facing the direction of travel. Some motorized trucks are designed to be ridden over longer distances as well as guided while walking (Figure 6-14).

Steering handles should have a guard to prevent the operator's hand or the controls from coming into contact with obstacles when the truck must be maneuvered in close quarters (Figure 6-10).

The wheels of many powered handtrucks can be considered guarded by their position under the frame or lift platform. However, where the wheels might injure the operator or others, guards should be installed. High-lift platform rollers and chain sprockets should be guarded to protect operators in their normal operating positions.

Figure 6-7. This electronically controlled industrial tractor is designed to travel a prearranged route outlined along the floor. (Printed with permission from Clark Equipment Co., Industrial Truck Division).

Trucks with a platform upon which the operator rides in a standing position must be designed so the platform extends beyond the operator's position and is strong enough to meet the requirements of the B56.1 standard (Figure 6-8). If an operator enclosure similar to the one in Figure 6-8 is provided, it shall provide easy ingress to and egress from the platform. To discourage operators from sitting on the truck during operation, a prism-shaped cover can be installed over the battery-box.

Automated Guided Vehicle (AGV)

Because trucks guided by remote control (automated guided vehicles or AGVs) operate without an operator on the scene, they must be provided with some means of completely stopping if someone steps in front of them (Figure 6-7 and -15). Such trucks should be equipped with a lightweight, flexible bumper that, when contacted, shuts off the power and applies the brakes. Sufficient clearance between the bumper and the front of the truck is needed so the truck can come to a full stop without contacting anything in its path.

The use of such "robot" trucks requires that aisles where the truck operates should be clearly marked and clear of material. Employees should be forbidden to jump on or off, or ride these trucks. No attempt should be made to load or unload an AGV that is in motion.

Transfer conveyors should be designed so they can be moved out of the way except when transferring loads to and from the AGVs. This avoids having a pinch point between the vehicle and the conveyor. The same logic should apply when laying out the route next to machine and columns.

OPERATING PRINCIPLES

Powered industrial truck traffic accidents can be prevented by using the same safe practices that apply to highway traffic. For example, excessive speed can lead to accidents both in the plant and on the road. Safe speed is the rate of travel which will permit the truck to stop well within the clear distance ahead, or make a turn without overturning. Wet or slippery floors require a slower-than-ordinary speed.

Depending on the application and operating conditions, specific speed limits for trucks should be established. A speed limit of 6 mph (9.6 km/hr) for the warehouse may be too high in congested areas. Many companies have installed governors, that lock in some instances.

Collisions between trucks and stationary objects often occur while trucks are backing, usually when they are turning and maneuvering. In such cases, the operator may be so intent on handling the load that where the rear of the truck is going is forgotten. Because backing accidents usually result from a failure to look, operators should be required to look in the direction of travel, maintaining a clear view of it (Figure 6-16). Some trucks permit the operator to sit sideways to make looking backward and forward more easy. Stunt driving and horseplay must not be permitted.

The operator should stop at blind corners and before passing through doorways, and go ahead only when it can be seen that the way is clear. Many companies have installed large convex mirrors at blind corners (Figure 6-25), so that operators and pedestrians can see each other approaching. To be effective, these mirrors have to be kept clean and properly adjusted.

Operators should keep trucks a safe distance apart during operation—some companies specify three truck lengths. Operators must not pass other trucks traveling in the same direction at intersections, blind spots, or other dangerous locations. They must keep to the right, if aisle width permits it without passing dangerously close to machine operators and others. Where aisles are not wide enough for continuous two-way traffic, vehicles should run in the middle of the aisle except when in a passing situation where the vehicle then moves to the side and passes with caution.

Operators should avoid making quick starts, jerky stops, or quick turns at excessive speed. They should use extreme caution when operating on turns, ramps, grades, or inclines. On

Figure 6-8. The enclosure on this stand-up truck protects the operator from crushing injury if the truck collides with other objects. (Printed with permission from The Raymond Corp.)

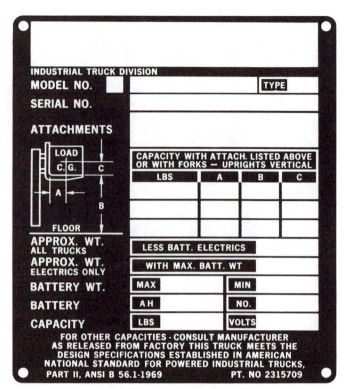

Figure 6-9. A name plate must be attached to every lift truck by the manufacturer. Pertinent identification and reference information must include the weight of the truck, its rated capacity, and model and serial numbers.

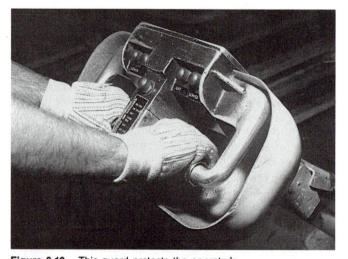

Figure 6-10. This guard protects the operator's hands from coming in contact with obstacles when the truck is being maneuvered in close quarters.

descending grades, trucks should be kept under control, so that they can be brought to an emergency stop in the clear space in front of them (Figure 6-17). The reverse control should never be used for braking.

No powered industrial truck should be used for any purpose other than the one for which it is designed. Common dangerous misuses include bumping skids, pushing piles of material out of the way, makeshift connections to move heavy objects, using the forks as a hoist, and moving other trucks. Disabled trucks should not be pushed or carried by another lift truck. They should be moved by towing with tow bar and safety chain. Powered industrial trucks should never be used to tow or push freight cars, nor used to open or close freight car doors unless the truck has a device specifically designed for this purpose.

Trucks should not be driven onto an elevator unless the operator has been authorized to do so. The truck should approach the elevator slowly and at right angles to the door. Enter the elevator only after the car is properly leveled, and after checking to make sure the weight of the truck, load, and driver do not exceed the capacity of the elevator. After making sure that the load is lowered to the floor, the brakes are set, the power is shut off, and the controls are in neutral, the operator should get off the truck. It is advisable for all other personnel to leave the elevator before a truck enters or leaves.

Powered industrial trucks should be driven carefully and slowly over bridge plates that are properly secured (Figure 6-18). Trucks should cross railroad tracks diagonally wherever possible and park at least 6 ft (1.8 m) from the nearest rail.

Highway trucks, trailers, and railroad cars should have their brakes set and their wheels securely blocked while they are being loaded or unloaded by powered industrial trucks. Before entering a trailer, fork truck drivers should determine that wheel chocks are squarely placed in front of the rearmost tires on dual

Figure 6-11. Straddle trucks are designed to carry loads of pipe, lumber, and other long materials. They should have many safety features—drive chain guards, an access ladder and handholds, horn, lights, and weather protection for the operator. (Printed with permission from Towmotor Corp.)

Figure 6-12. This truck-mounted hydraulic crane is capable of swinging a full 360 degrees. Outriggers provide stability. The enclosed cab (right) offers the operator protection plus clear visibility. (Printed with permission from Master Craft Engineering, Inc.)

axle trailers. On tri-axle or quad-axle trailers two additional chocks should be squarely placed in front of the foremost tires which are still on the ground. When a trailer is next to a ramp or wall, place at least two wheel chocks squarely in front of the outside rear and foremost tires. When trailers are parked without the tractor, some companies also place jacks under the front of the trailer. A sequence for proper use of jacks and flags is shown in Figure 6-19. Trailer restraint systems also are available (Figure 6-20).

Feet and legs should be kept inside the guard or the operating station of the truck. Driving with a foot or leg outside is unsafe, as is placing a hand, arm, or leg on or between the uprights of a truck. When in close quarters, operators should keep their hands where they cannot be pinched between the steering or control levers and projecting stationary objects. Steering handles on motorized hand trucks should have guards to protect against injuries of this kind.

The operator should leave a truck unattended only after the controls have been put in neutral, the power shut off, the brakes set, the key removed, or the connector plug pulled, and the load-engaging means placed in a lowered and inoperative position. (OSHA defines an "unattended truck" as one where the operator is more than 25 ft (7.6 m) from it or cannot see it.) Although it is not usually good practice to leave a truck on an incline, when such action is necessary, the wheels should be blocked as an added precaution.

It is the operator's responsibility never to park a truck in an aisle or doorway, nor to obstruct material or equipment to which another worker may need access. Accidents often happen when a truck is blocking a passageway, and an unauthorized employee tries to move it.

Looking out for pedestrians should also be the truck operator's responsibility, and the horn should be sounded when approaching pedestrians. Excessive hornblowing, however, is to be discouraged. Having sounded the warning, the operator

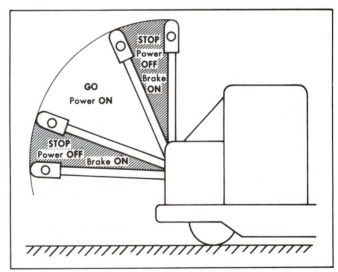

Figure 6-13. The hazard of contacting the frame or wheels of a walkie lift truck can be minimized if the "power on" area is limited to approximately the area shown.

should proceed with caution, passing only when the pedestrians are aware of the truck's presence and are in the clear. The operator should not use the horn to "blast" a way through. Trucks should never be driven directly toward anyone who is standing in front of a bench or other fixed object.

Operators must be especially careful when turning because

Figure 6-14. This truck enables the operator to ride the vehicle while transporting a load over longer distances. (Printed with permission from Clark Equipment Company.)

Figure 6-15. Two automatically guided, driverless tractors electronically decide which one will have the right of way, thus eliminating potential collisions. (Printed with permission from Jervis B. Webb Co.)

the rear (steering) wheels project beyond the truck enclosure, presenting a hazard.

Pedestrians also have a responsibility to watch out for trucks and to get out of the way with reasonable promptness. Ill feeling and accidents can result if pedestrians refuse to move out of the way when a truck approaches—consideration by both sides is needed.

Passengers must never be permitted to ride on a truck, form, coupling, or trailer. It is the operator's responsibility to keep riders off.

Loads, whether on trucks, trailers, skids, or pallets, should be stable. Objects should be neatly piled and crosstied, if the shapes permit (Figure 6-21). Irregularly shaped objects should be loaded so that they cannot roll or fall off. Heavy, odd-shaped objects should be placed with the weight as low as possible. Round objects, like pipe or shafting, should be blocked and, if necessary, tied so that they cannot roll (Figure 6-22). Loading to an excessive height not only blocks the view ahead, but makes it likely that part of the load may fall.

Operators should not permit gasoline engines to idle for long periods in enclosed or semi-enclosed areas because of the accumulation of exhaust vapors and combustion gases.

Figure 6-16. An operator must keep alert when backing and face in the direction of travel. (Printed with permission from Clark Equipment Co., Industrial Truck Division).

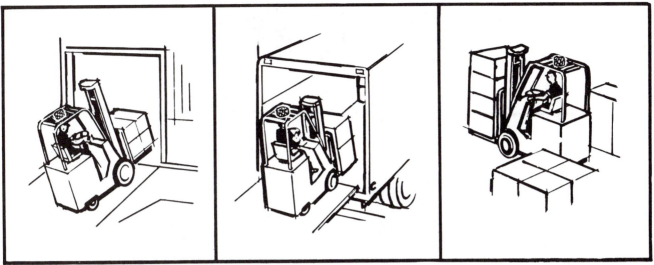

Figure 6-17. Driving with the load pointing upgrade permits better control. When forward vision is obscured by big bulky loads, driving in reverse (right) permits better visibility.

Concentration of carbon monoxide (CO) gas in areas where powered industrial trucks are operated should not exceed the levels established by OSHA or specified by local, state, or municipal codes. Sampling should be done by a qualified industrial hygienist.

Catalytic exhaust purifiers are available that considerably reduce the level of carbon monoxide and other noxious gases in the engine exhaust. Even though exhaust purifiers are installed on lift trucks, in order to maintain a clean atmosphere the user must provide adequate ventilation for enclosed areas.

Proper maintenance is essential to avoid a feeling of false security. Batteries should not be operated beyond their rated capacity.

Lift trucks

Because the operation of a lift truck has some basic differences from that of an automobile or a highway truck, the prospective operator should be impressed with the fact that a lift truck:

Figure 6-18. Dock plate is placed in a freight car door by a lift truck. After positioning, the dock plate is anchored by pins (connected to chain at sides) which drop into holes in the edge of the dock and the car door.

1. Is generally steered by the rear wheels,
2. Steers more easily loaded than empty,
3. Is driven in the reverse direction as often as in the forward,
4. Is often steered with one hand—the other hand being used to operate the controls.

Because a lift truck is generally steered by the rear wheels, the swing of the rear of the truck always must be carefully watched (Figure 6-23). Beginners usually try to turn too sharply. Some lift trucks when traveling forward have a peculiarity known as "free turning," that is, once the turn is started, the truck tends to turn more and more sharply in smaller and smaller circles. To counteract this tendency and to slow down the sharpness of the turn, the operator must apply force on the steering wheel in the opposite direction. When such a truck is traveling in reverse, the opposite holds true—the operator must apply force in the direction of the turn.

Operators should learn to judge the correct aisle width for the truck size and load. They should also observe the general operating safety rules, given earlier in this chapter, as well as the specific rules for lift trucks that are discussed next.

All starts and stops should be easy and gradual to prevent the load from shifting. Turns should be made smoothly and gradually at a safe speed.

The operator should be particularly careful, while either traveling or maneuvering, to avoid striking overhead structures and nearby objects, such as sprinkler piping, electrical conduit, or fixed structures. Critical equipment and materials, such as electrical panels, fire equipment, fire doors, and load supporting structures, should be protected by strong barriers, posts, or curbing to prevent damage.

Loads should not be raised or lowered en route. Loaded or empty, the forks should be carried as low as possible, but high enough not to strike any raised or uneven surface. Tilting back the upright keeps the load steady and secure.

If a bulky load is to be carried that cannot be lowered enough to prevent its obstructing the view, the truck should be driven backward so the operator can see where the truck is going. Some companies have a policy that trucks are to be driven in reverse any time they are carrying a load regardless of size.

RED FLAGS

DURING LOADING & UNLOADING - Position the flag on the dock, next to the outside left edge of the trailer. Make certain the semi-driver can see the words "Do Not Move Trailer."

AFTER LOADING OR UNLOADING IS COMPLETE - Position the flag in the center of the dockplate as a warning to dock personnel to stay off the trailer.

JACK-STANDS

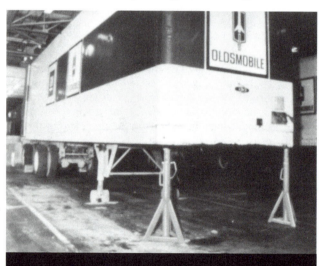

ALL TRAILERS - Two jack-stands must be placed beneath the nose of the trailer so that the frame of the trailer is as close as possible to the top of the jack-stands.

TRAILERS WHOSE AXLES ARE MOUNTED 3 OR MORE FEET FORWARD OF THE TAILGATE - Place two additional jack-stands beneath the rear of the trailer so that the frame of the trailer is as close as possible to the top of the jack-stands.

Figure 6-19. This poster alerts operators to place jack stands and red flags properly. (Printed with permission from Oldsmobile Division of General Motors Corp.)

Figure 6-20. This low-profile vehicle restraint shows a hooking mechanism. (Printed with permission from Rite-Hite Corporation).

Figure 6-22. This diesel powered lift truck moves these electrodes which range in size from 1 ft × 6 ft (0.3 m × 1.8 m) to 4 ft × 10 ft (1.2 m × 3 m). (Printed with permission from Hyster Company.)

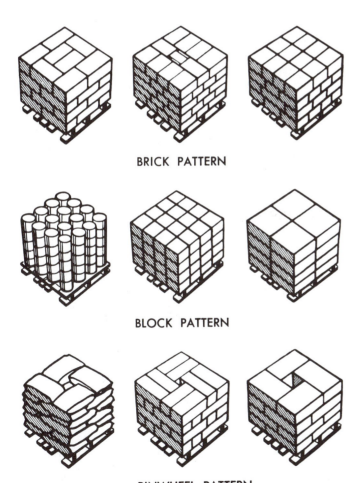

BRICK PATTERN

BLOCK PATTERN

PINWHEEL PATTERN

Figure 6-21. Typical pallet-loading patterns. (Printed with permission from The Industrial Truck Assn.)

Trucks should ascend or descend grades slowly. When ascending or descending grades in excess of 10 percent, loaded trucks should be driven with the load upgrade (Figure 6-17). Unloaded trucks should be operated on all grades with the load-engaging means downgrade.

Keep in mind that high-lift order-picker trucks are not designed for steep grade operation. Consult the manufacturer's operating instructions for recommended operating procedures.

On all grades, the load, and load-engaging means should be tilted back, if applicable, and raised only as far as necessary to clear the road surface. Low gear or the slowest speed should be used when the truck is descending a grade. The operator should keep clear of the edge of loading docks and ramps and never make a turn on a ramp.

Before driving over them, it is important that the operator check bridge plates, to make sure they are properly secured, and the floors of freightcars and trucks, to be sure they are in good condition and will bear the weight of the truck and load. Also check the truck or trailer to see that it is properly chocked. Failure to do this may result in the bridge plate shifting or the trailer or truck moving away from the dock (Figure 6-24).

Lift trucks have a rated capacity, which is stated in pounds at a 24 in. (60 cm) load center distance from the vertical face of the forks. Thus, the rated capacity may be "2000 pounds (908 kg) at 24 in. (60 cm)." A wide, flat load of 2000 pounds with the center of gravity 30 in. (75 cm) from the fork is, therefore, a considerable overload. Every operator should be familiar with the maximum load limits of the truck being operated and may be required to observe them.

Placing extra weight on the rear of a lift truck to counterbalance an overload shall not be permitted as it may strain chains, forks, tires, axles, and the motor and also may cause accidents.

The stability of lift trucks is covered in the ANSI B56.1. Most lift truck manufacturers use the stability values in the standard as criteria for design and to determine the rated capacity for various truck models.

Operators should never, under any circumstances, attempt to operate a truck with an overload. Such a load is dangerous

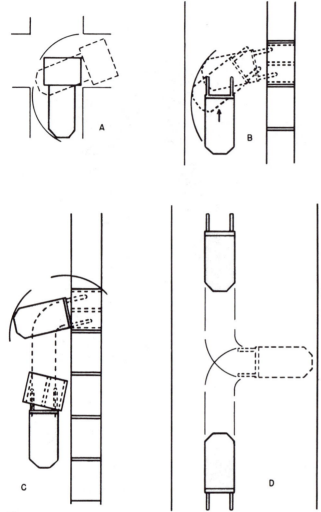

Figure 6-23. Lift truck maneuvers. A—Turning a sharp corner. B—Turning across an aisle. C—Turning in an exceptionally narrow aisle. D—Turning around in a narrow passage. Driver should allow ample space for rear-end swing and make the turns carefully.

because it removes weight from the steering wheels, which affects the steering. Standing on a truck or adding counterweights to compensate for an overload should never be permitted.

Side stability is a critical factor in making turns at speed or on a slope or ramp. Back tilt of uprights reduces side stability on high lifts, and allowance should be made for this factor.

Particular care should be taken not to exceed floor load limits. The force exerted by a truck on a floor varies with the speed, load, and total weight distribution. It also is affected by the number of wheels, wheelbase, and other factors. All questions about floor capacities should be referred to a qualified building architect or structural engineer.

When standard forks are used to pick up round objects, such as rolls or drums, care must be taken to see that tips do not damage the load or push it against workers. The uprights should first be tilted, so that the tips of the forks touch the floor, and then moved forward so that the forks can slide under the object. Tilting the uprights backward will then cause the load to roll

back against the vertical face of the forks and/or carriage and the load backrest extension—a secure carrying position (Figure 6-25). A block or wedge should be against the drum or roll.

To unload a large case or similar object without a pallet, the operator should first drive into position for stacking. The load should then be lowered onto a base having a block near the edge and the forks withdrawn so that only their tips hold up the end of the load. Next withdraw the block, tilt the uprights forward, and back away (Figure 6-26).

In attempting to pick up a palletized load, the forks should be fully and squarely seated in the pallet, an equal distance from the center stringers and well out toward the sides. Forks to be inserted in a pallet should be level, not tilted forward or back. If the forks are placed close together, the pallet tends to drop at the sides and seesaw, causing strain and instability (Figure 6-27).

When raising or lowering loads (Figure 6-28) while standing still, the operator should not leave the truck in gear with the clutch depressed; the shift should be returned to neutral and the clutch disengaged.

Operators of lift trucks should be told to refuse improperly loaded skids or pallets, broken pallets, or loads too heavy for the truck. All chemicals should be properly identified before relocation, observing material safety guidelines and regulations. The proper storage of materials can be accomplished when there is knowledge of material properties.

When a lift truck is parked, the forks should be placed flat on the floor. No one should be allowed to stand or walk under elevated forks.

Using a lift truck as an elevator for employees (for example, to service light fixtures) should be done only if a work platform that is securely seated on the forks, fastened to the vertical face, and provided with handrails and toeboards is used (Figure 6-29). The truck should also have an overhead guard for the operator's protection. The operator must not leave the controls while the truck is used as a lift. Special trucks are built for this purpose.

OPERATORS

Selection

Trainees should have valid vehicle operator's driver's licenses, good driving records and few, if any, traffic violation tickets. They should have good attitudes toward the responsibility of operating expensive and heavy industrial equipment under new and difficult situations. For this reason it is advisable to verify the trainee's previous experience, both off and on the job, wherever possible.

The trainees must appreciate the importance of the program. Their attitude is important to good performance even if some of the finer points of the training are forgotten and if supervision is not as close as it should be.

Trainees should meet certain physical standards and should be examined by a qualified physician familiar with the job requirements. Minimum requirements would be 20/40 vision, corrected if necessary, good reaction time, depth perception of no less than 90 percent of normal and good hearing preferably without the benefit of a hearing aid. It is desirable to give drivers a physical examination every two years and also to check their driving record off the job as well as on the job.

Figure 6-24. Never assume that a trailer is properly chocked and jacked up. Always make certain. Here is what happened when a trailer dipped under the weight of a lift truck and moved away from the dock.

Good judgment together with respect for the safety of personnel and property is required. It should be made clear to operators that they carry considerable responsibility and that reckless or careless operation will not be tolerated. They should be checked on the job from time to time for observance of this rule.

Each plant should adopt a set of rules governing the operation of powered industrial trucks. Because plant conditions and equipment vary widely, rules that cover specific conditions should be set up.

For ready identification, many companies issue badges to authorized truck operators. Badges also tend to remind operators of their responsibilities and give them pride in their job.

Training

OSHA regulations state that "only trained and authorized operators shall be permitted to operate a powered industrial truck." A truck operated incompetently may cause severe injury or substantial property damage; no company can afford even one improperly trained truck operator.

To be effective, a training program should center around the company's policies, operating conditions, and type of trucks used. All new operators, regardless of claimed previous experience, should receive a training course. Operators also should be given a refresher course every two years. Information on training programs may be obtained from the National Safety Council, truck manufacturers, and trade associations.

As with any program, management support is essential if the truck operator's training is to be effective and lasting.

Before starting a training program, the company should determine the facts regarding problems with industrial trucks such as, numbers and types of accidents, extent of economic losses, and operating habits of operators. This information can then be worked into the training program.

Management must also understand that any training program will cost money—materials and equipment, payment for nonproduction time, lost production, possible damage to materials, and other costs—and accept the fact. Practical training can be conducted on obstacle courses (Figure 6-30).

Check standards and codes for permissible floor loads, clearances for sprinklers, travel routes of the trucks, adequacy of lighting and ventilation, and noise control. Whenever possible, correct unsafe or hazardous conditions before starting the training. If there are conditions that cannot be corrected, be sure that instructors know about them and why the conditions cannot be changed. Operators will not be impressed with the need for safe job performance if they feel that management permits other unsafe conditions to exist.

It is important that the supervisors fully understand the need for the program so they will adjust their schedules, if necessary, to make people available for training either during or after working hours.

Maintenance personnel should be involved in the program from the start. Not only are mechanics possible instructors because of their knowledge of the truck and its operation, but this will avoid possible conflict when the truck operators start to request additional repairs and adjustments because they have become "safety conscious."

Other factors to consider in setting up a training program are:
1. Who will be in charge?
2. Qualifications for both the instructor(s) and the trainees.
3. Number and length of sessions for both classroom instruction and practice driving.
4. Location of the sessions for both classroom and practice driving.
5. Number of trainees in a class.

6. Should experienced operators who are taking a refresher course be in the same class with new trainees?
7. Informing all employees of the program and making sure they understand the importance of it.
8. The establishment and maintenance of a record system to be sure the program stays current and for use by OSHA compliance officers reviewing the program.

An effective training program does not end with the presentation of a certificate. Management has the responsibility to continue to maintain safe operating conditions and insist on safe performance by all employees after the operators have completed the course. A record of each operator's performance should be kept (Figure 6-31).

INSPECTION AND MAINTENANCE

Powered industrial trucks should be thoroughly inspected by maintenance personnel on a regular schedule and a complete overhaul given after regular periods of operation. Operators should make daily inspections of controls, brakes, tires, and other moving parts. This should be done at the start of each shift in multishift operations. Checklists (Figure 6-32) should be used to record conditions requiring correction and a detailed schedule of inspection and repair records kept for each vehicle (Figure 6-33). Forks should be magnafluxed on a regularly scheduled basis determined by use. Defective brakes, controls, tires, lights, power supply, load-engaging mechanism, lift system, steering mechanism, and signal equipment should be repaired before trucks are allowed to go back into service. Operators should be prohibited from making truck repairs.

Before repairs are made to any part of a powered industrial truck, the operating mechanism should be locked "off."

Electric trucks

Handling and charging storage batteries for electric trucks introduce several hazards. By wearing chemical goggles, rubber gloves, aprons, and rubber boots, operators of charging equipment can be protected against acid burns during refilling or handling operations. Wood slat mats, rubber mats, or clean floorboards will help prevent slips and falls and will protect against electric shock from the charging equipment. See *Powered Industrial Trucks*, NFPA 505, and National Safety Council Data Sheet No. 635, *Lead-Acid Storage Batteries*.

Battery changing and charging operations must be performed only by trained and authorized personnel. Truck operators may or may not be so authorized depending upon individual plant setups.

Battery charging installations must be located in areas designated for that purpose. Facilities must also be provided for flushing and neutralizing spilled electrolyte, for fire protection, for protecting charging apparatus from damage by trucks, and for adequate ventilation to disperse flammable hydrogen gases and

Figure 6-25. To pick up round objects, forks should be tilted (left), not used flat (right).

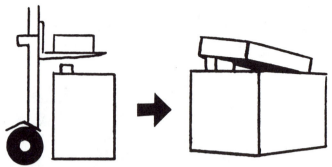

Figure 6-26. Unloading a large case or similar object without a pallet requires a base and a movable block.

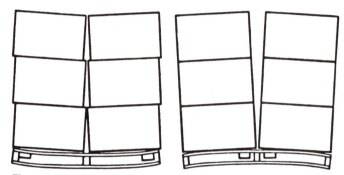

Figure 6-27. With forks spread wide (left), load is well distributed and tends to bind itself together. The effort of placing the forks too close is shown at the right.

Figure 6-28. This fork lift is used to lift cargo into an airline hold. Note the lights for outdoor night operation.

Figure 6-29. Work platform attaches to lift truck for use in overhead maintenance work; it is equipped with guardrails and toe boards. (Printed with permission from Ford Motor Company.)

fumes from batteries. Eye safety includes having eyewash facilities and employees wearing eye protection.

When racks are used for support of batteries, they should be made of materials nonconductive to spark generation or be coated or covered to achieve this objective.

When charging batteries, acid should be poured into water; never the reverse. A carboy tilter or siphon should be provided for handling electrolyte. If acid or electrolyte is spilled on the worker's skin or clothing, it should be washed off immediatley with plenty of water.

Trucks should be properly positioned and the brake applied before attempting to change or charge batteries. Reinstalled batteries should be properly positioned and secured in the truck.

When charging batteries, the vent caps should be kept in place to avoid electrolyte spray. Care should be taken to assure that vent caps are functioning. The battery (or compartment) cover(s) should be open to dissipate heat. Manufacturer's recommendations should be referred to in performing charging and maintenance procedures.

Sulfuric acid used for refilling should not be allowed to run into ordinary cast iron, lead, steel, or brass drains.

Precautions should be taken to prevent open flames, sparks, or electric arcs in battery charging areas. Electrical installations should conform to the local codes and the *National Electric Code.* Smoking should be prohibited in the charging area.

Battery terminals should be clean, connections tight, and the battery securely locked in place in the trunk. Tools or metal parts should never be laid on a battery.

To prevent operator strains from manual handling of heavy or awkward loads, a roller conveyor, an overhead hoist, or equivalent material handling equipment should be provided. Chains, hooks, yokes, and other parts of the hoisting mechanism should be insulated or nonconductive to prevent short-circuiting. Care must be given to the points where cables contact reels and suspension attachments to prevent insulation wear which may produce arcing.

Gasoline-operated trucks

Gasoline for trucks should be handled and stored in accordance with the provisions of the *Flammable and Combustible Liquids Code,* NFPA 30.

Fuel tanks on gasoline-operated trucks and tractors should be filled at designated locations, preferably in the open air, with the filling hose and equipment properly grounded and bonded. Locations outside main buildings should be selected to minimize the chances of involving combustible material in a fire.

Safety cans used for fuel handling should be tested and

Figure 6-30. The driving ability of this lift truck operator trainee is tested by maneuvering through a mockup of a crowded aisle. All training programs should contain a driving test as well as a written examination. (Printed with permission from Clark Equipment Co., Industrial Truck Division.)

approved by Factory Mutual or listed by Underwriters Laboratories. (See Chapter 16, Flammable and Combustible Liquids.)

Engines must be stopped and operators must be off trucks before they are refueled. No smoking should be permitted during refueling. Gasoline tanks should be drained at safe locations into grounded self-closing cans.

Workers should avoid spilling gasoline or overflowing the gasoline tank during refueling. Before an attempt is made to start the engine, the gasoline tank cap should be replaced and spilled fuel flushed down or allowed to vaporize.

Liquefied petroleum gas trucks

The use of liquefied petroleum gas (LP-gas) as a fuel for industrial power trucks is increasing, largely through conversion of gasoline-powered units.

A properly adjusted engine burning LP-gas will generally produce a substantially lower concentration of carbon monoxide in the exhaust than a similar engine that uses gasoline as fuel. Only air sampling, however, can prove whether or not the CO concentration in an area is below the maximum allowable. The 1979 ACGIH TLV listing gives 50 ppm for an eight-hour exposure as the maximum exposure allowable.

Fittings not listed by a nationally recognized agency or incor-

rectly installed, or connections not properly tightened before refueling is begun may fail and release combustible gas into the air. Only conversion units and fittings listed by an agency, such as Underwriters Laboratories or Factory Mutual, should be used. They should include all safety features that are incorporated in LP-gas-fueled trucks listed by the testing agency. The units and fittings should be installed in strict conformity with requirements specified in *Storage and Handling of Liquefied Petroleum Gases,* NFPA 58, and Underwriters Laboratories Standards for Safety No. 558, *Internal Combustion Engine-Powered Industrial Trucks,* to provide maximum protection against damage to the system by vibration, shock, or objects striking against it, and against failure from other causes.

The manufacturer of a truck which is to be converted to LP-gas operation should be consulted. The manufacturer may be able to supply listed conversion units and assign a qualified representative to supervise the installation. Conversions should be attempted only by qualified mechanics who are familiar with handling LP-gas equipment and who are using listed parts.

Only listed fuel containers, designed in accordance with DOT or ASME standards, should be used (Figure 6-34). Permanently mounted and removable fuel containers should be filled outdoors and storage facilities and filling equipment should meet standards of *Storage and Handling of Liquefied Petroleum Gases,* NFPA 58.

Refueling of LP-gas trucks with permanently mounted containers shall be done out-of-doors. Although the exchange of removable fuel containers preferably should be done outdoors, it may be done indoors if one of the two methods specified in the NFPA Standard 58 is followed.

A special building or outside storage is recommended for the storage of fuel containers. Where the cylinders must be stored inside of the building, they must be kept in a special room or designated safe area in accordance with NFPA Standard 58. NFPA standards permit no more than two containers for LP-gas motor fuel on each industrial truck and require that storage inside a building not frequented by the public be limited to a total gas capacity of 300 lb (135 kg). Containers should be enclosed in a separate room that is well ventilated and of ample size. The walls, floor, and ceiling of the room are required to be of specified fire-resistive construction. Openings to other parts of the building should be protected by specified fire doors.

Proper container filling is of the utmost importance. The person filling the containers must be trained in the safe handling of LP-gas. Container filling from bulk storage must be done at least 10 ft (3 m) from the nearest important masonry-walled building and at least 25 ft (7.5 m) from important nonmasonry buildings and openings in masonry and non-masonry buildings (Figure 6-35). The filling plant must conform to NFPA Standard 58 and applicable state or local insurance regulations.

Trucks themselves must comply with NFPA Standard 505. LP-gas-fueled trucks should be garaged in a well-ventilated area. Ventilation should be provided at floor level because fuel gas is heavier than air. The trucks should not be garaged in the same room with stored cylinders.

A rigid and thorough inspection and maintenance procedure for LP-gas-fueled trucks should be followed. LP-gas trucks can be stored or serviced inside garages provided the fuel system is leak-free and the container is not filled beyond the limit

FORKLIFT TRUCK OPERATOR PERFORMANCE TEST

OPERATOR'S NAME_____ DEPT. _____ DATE _____

1. Operator's ability to perform check-sheet inspection for safe operation of truck prior to use.

    Uses check-sheet satisfactorily ☐   Failed to check safety items ☐

2. Proper use of controls (understands proper technique and proper direction of movement of control to get desired direction of movement).

    (a) Clutch operation.   Yes ☐   No ☐

    (b) Inching control (auto transmissions).   Yes ☐   No ☐

    (c) Tilt control.   Yes ☐   No ☐

    (d) Lift control.   Yes ☐   No ☐

    (e) Attachment controls.   Yes ☐   No ☐

    (f) Steering techniques for type of machine being used by operator.   Yes ☐   No ☐

    (g) Proper positioning of all controls, switches, parking brakes when leaving machine unattended.   Yes ☐   No ☐

    (h) Service brake.   Yes ☐   No ☐

    (i) Parking brake.   Yes ☐   No ☐

3. Maneuvering skills.

    (a) Smooth starting and stopping.
    Acceptable ☐   Needs practice ☐   Poor control ☐

    (b) Sharp turns forward and reverse.
    Proper speed.   Yes ☐   No ☐
    Looks in direction of travel.   Yes ☐   No ☐
    Carries forks low.   Yes ☐   No ☐
    Clears obstacles by safe distance.   Yes ☐   No ☐

4. Selecting loads.

    (a) Proper capacity for truck used.   Yes ☐   No ☐

    (b) Proper size load for visibility and safety of handling.   Yes ☐   No ☐

    (c) Load tilted back against back rest.   Yes ☐   No ☐

    (d) Carries load low (just high enough to clear floor obstacles).   Yes ☐   No ☐

Figure 6-31a. Typical form used to keep record of operator's performance.

5. Driving with load.

    (a) Smooth starting and stopping.   Yes ☐   No ☐

    (b) Proper speed.   Yes ☐   No ☐

    (c) Sounds horn at intersections and corners.   Yes ☐   No ☐

    (d) Keeps to the right in aisles used for two-way traffic.   Yes ☐   No ☐

    (e) Travels at least three lengths behind other vehicles.   Yes ☐   No ☐

    (f) Handles load in manner to prevent product damage.   Yes ☐   No ☐

6. Stacking.

    (a) Approaches loads squarely.   Yes ☐   No ☐

    (b) Stacks straight and squarely.   Yes ☐   No ☐

    (c) Does not tier too high.   Yes ☐   No ☐

    (d) Deposits load safely; does not use excessive tilt action.   Yes ☐   No ☐

    (e) When selecting top load for pickup, uses proper form spread for load.
    Yes ☐   No ☐

    (f) Removes load and lowers to safe level before making turn to proceed in direction of desired travel.   Yes ☐   No ☐

7. Dock Safety.

    (a) Checks bridge plates (dock boards) before crossing.   Yes ☐   No ☐

    (b) Checks trailers for proper wheel chocking before entering and proper jack installation of trailers where required.   Yes ☐   No ☐

    (c) Checks rail freight cars for proper positioning and safe loading conditions.   Yes ☐   No ☐

ADD ANYTHING THAT MAY PERTAIN TO YOUR PARTICULAR OPERATION IN MATERIAL HANDLING ON WHICH YOU MAY WANT TO TEST YOUR OPERATORS' PERFORMANCE.

2C97567   Printed in U.S.A.   Stock No. 199.78

Figure 6-31b.

OPERATOR'S DAILY REPORT

Battery-Powered Lift Trucks

Truck No._____ Make _____ Date _____ Shift _____

Hour Meter Reading: Start _____ End_____ Hours for Shift _____

CHECK EACH ITEM If OK write OK	SHIFT			Explain below if not OK or any other action taken
	Start	During	End	
1. Battery plug connection				
2. Battery charge				
3. Battery load test				
4. Brakes — service and seat brake				
5. Lights — head, tail and warning				
6. Horn				
7. Hour meter				
8. Steering				
9. Tires				
10. Hydraulic controls				
11. Other conditions				

Remarks and additional explanation or suggestions _____

_____

_____

Operator's Signature _____

2C97567   Printed in U.S.A.   Stock No. 199.77

Figure 6-32. Operators can use a checklist to make a daily inspection of their industrial trucks.

NATIONAL SAFETY COUNCIL
FORKLIFT TRUCK OPERATORS TRAINING COURSE

INSPECTION AND MAINTENANCE LOG

Truck No.	Make	Work Done		Work Description or Remarks	Cost
		Date	Hour Meter		

2C97567   Printed in U.S.A.   Stock No. 199.74

Figure 6-33. A detailed inspection and repair record should be kept for each truck.

Figure 6-34. Example of LP-Gas fuel storage container.

specified in Chapter 4 of NFPA Standard 58, the container shut-off valve is closed except when the engine is operated, and the truck is not parked near sources of heat, open flames, other sources of ignition or near inadequately vented pits.

If the fuel container is permanently mounted, major repairs should be made outdoors or in a well-ventilated, fire-resistive area provided for this purpose.

Subparts of the OSHA regulations that apply to this chapter must be strictly adhered to.

PALLETS

Most companies buy ready-made pallets (Figure 6-36). No matter who builds them, however, procedures should be established to inspect pallets, both before they are put into service and at regular intervals, to be sure that they are in a safe condition. Top deckboards should be sound and securely fastened to runners. Splintered, broken, or loose parts should be repaired or replaced. Loose nails or chunks of wood can cause injury to workers and damage to trucks.

A safe place, out of the way of traffic and work areas, should be provided for pallet storage. They should be neatly stacked

and limited in height so that they are stable and secure against sliding or collapse. Also, they should not be left standing on edge or in a leaning position from which they might topple onto people or other objects. Storage of large blocks of pallets within a building should be avoided to prevent overpowering the sprinklers in case of a fire. Large stacks in outside storage should be kept well away from buildings.

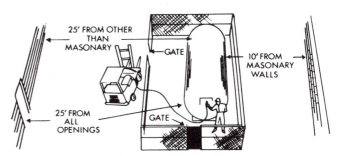

Figure 6-35. Schematic shows minimum distances allowed for industrial truck refueling operations. (Printed with permission from the LP-Gas Assn.)

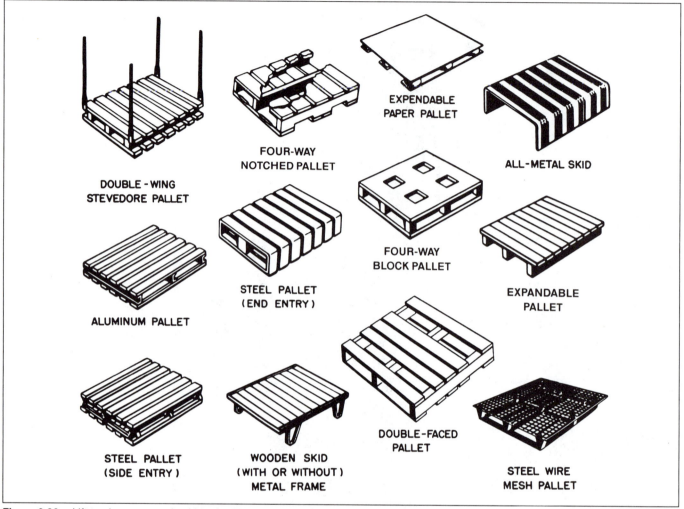

Figure 6-36. Lift truck operators should be familiar with pallet and skid types. (Printed with permission from Industrial Truck Assn.)

REFERENCES

American Conference of Governmental Industrial Hygienists, 6500 Glenway Ave., Bldg. D-7, Cincinnati, Ohio 45211. "Threshold Limit Values" (latest edition).

American Insurance Association, 85 John St., New York, N.Y. 10038. *Safe Handling and Use of LP-Gas.*

American National Standards Institute, 1430 Broadway, New York, N.Y. 10018.
Fork and Fork Carriers for Powered Industrial Fork Lift Trucks, Hook Type, ANSI/ASME, B56.11.4.
Safety Standard for Powered Industrial Trucks—Low Lift and High Lift Trucks, ANSI/ASME, B56.1.
Safety Color Code for Marking Physical Hazards, Z53.1.

National Fire Protection Association, Batterymarch Park, Quincy, Mass. 02269.
Flammable and Combustible Liquids Code, NFPA 30.
Liquefied Petroleum Gases, NFPA 58.
National Electric Code, NFPA 70.
Self-Service Gasoline Stations, NFPA 30E.
Type Designations, Areas of Use, Maintenance and Operation of Powered Industrial Trucks, NFPA 505.

National LP-Gas Association, 1301 W. 22nd St., Oak Brook, Ill. 60521.

National Safety Council, 444 North Michigan Ave., Chicago, Ill. 60611.
General Aviation Ground Operations Safety Handbook.
"Fork Lift Truck Operators Training Course."
Industrial Data Sheets
Carbon Monoxide, 415.
Lead-Acid Storage Batteries, 635.
Liquefied Petroleum Gases for Industrial Trucks, 479.
Powered Hand Trucks, 317.
Powered Industrial Lift Trucks, 653.

Underwriters Laboratories, Inc., 333 Pfingsten Road, Northbrook, Ill. 60062.
"Industrial Trucks, Electric-Battery-Powered," Standard for Safety, No. 583.
"Industrial Trucks, Internal Combustion Engine-Powered," Standard for Safety, No. 558.

U.S. Department of Health, Education, and Welfare, National Institute for Occupational Safety and Health, Division of Technical Services, Cincinnati, Ohio 45226. *Outline for Training of Powered Industrial Truck Operators,* DHEW (NIOSH) Publication 78-199.

Plant Railways and Elevators

PLANT RAILWAYS

Clearances and warning methods

INSOFAR AS POSSIBLE, plant railway hazards should be eliminated in the design of a new plant. Horizontal and vertical track clearances are primary considerations.

It is recommended that in this and all other phases of track construction, American Railway Engineering Association (AREA) standards be adopted, provided municipal, state or federal regulations do not conflict. Some state regulations are even more stringent than the AREA recommendations for overhead clearances.

Where platforms, building entrances, or structures are located along curved track, additional clearance must be allowed on both sides of the curve for lateral movement of cars and loaded flatcars at center and at ends. Additional track clearance must be allowed when awnings or jalousie windows are installed in adjacent buildings.

Close clearance warning signs should be placed at points where buildings or other obstructions are so close to a track that they will not clear a person riding on the side of a car. In addition, a red light may be installed and allowed to remain lit at all times. Side or overhead telltales sometimes are installed to give positive warning of approach to structures with close clearance.

Standard clearances may not be sufficient, especially where tracks pass doorways or corners of buildings or other places where workers may walk directly onto tracks in front of moving cars. These locations may be safeguarded with fixed railings that force pedestrians to detour a short distance before stepping onto the track. If a barrier railing is impractical, hinged bars or gates swinging horizontally through an angle of not more than 90 degrees are effective.

Still another means is a convex mirror located at the intersection of the passageway and the track at a 45-degree angle, in which approaching railway equipment can be seen before the intersection is reached. This would be similar to that shown in Figure 3-25.

Various methods are used to warn pedestrians at railway crossings inside plants. Automatic wigwags, blinking lights, and gongs are more effective than signs. Heavily traveled roadways can be safeguarded effectively by gates equipped with red lights and by crossing guards, who should be provided with a whistle and a shielded red lantern.

The visibility of gates is increased by painting them with alternate red and white stripes at an angle of about 45 degrees. Another means of protection is the warning system shown in Figure 7-1.

Plant yards where plant or railroad personnel may be required to switch cars or perform other work at night should be adequately illuminated by high-intensity high-pole lights, arranged so as to cast as few shadows as possible.

Track

All track, fittings, and structures should be installed in accordance with AREA specifications for the service required.

In addition to meeting standard Federal Railroad Administration (FRA) specifications, rails and fittings should be inspected periodically and repaired as required. Serious accidents may result if defects are permitted to go unremedied.

Tracks should be level at loading points. Workers have been killed when cars roll down slight slopes and trap them. If it is

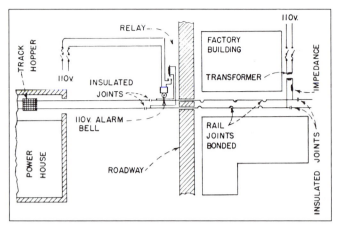

Figure 7-1. Layout of bell warning system for blind level crossings inside the plant.

necessary to spot cars on grades, brakes should be set tightly and car blockers, rail clamps, or track skates used.

Derails are needed at the bottom of steep slopes and at the connection of sloping switch tracks to main lines. Derails also should be installed on the approach to permanent shipping and receiving areas, whether located within or outside a building. Where such areas are on tracks open at both ends, derails should be placed beyond both ends of the shipping and receiving area. Derails should not be located in hard-paved areas, as this type of surface will defeat the purpose of the derail.

Trestles should have a footwalk not less than 5 ft-1 in. (1.5 m) wide, measured from the nearest rail. Railings should be 42 in. (1.0 m) high and toeboards 6 in. high (15.2 cm) and meet the standard side clearance of the railroad or the controlling state requirements. If footwalks are necessary on both sides of the track, crosswalks should connect them. Trestles must be designed and built to carry anticipated loads and withstand vibration and shock.

Openings at ground level for conveyors and similar equipment used to unload cars should have covers that are kept in place when equipment is not in use, to prevent persons from falling into them.

Hoppers or trackside bins into which material is dumped should also be covered. Heavy steel bars should be spaced so as to prevent a person from falling or being carried through the openings. It may also be necessary to cover walkways that are under trestles, in order to protect those who use them from being hit by falling materials. Where tracks are at a dead end, a standard bumping post or earth mound should be installed.

A switch with a handle swinging parallel to the rails is safest. Targets should have rounded corners to lessen the possibility of cuts, scratches, and torn clothing. Switch lamps or reflectorized targets are needed if tracks are to be used at night. Blocking should be installed in switch points and frogs used to prevent employees from getting their feet caught.

Loading and unloading

Tracks at loading and unloading areas warrant special attention because numerous accidents occur when loads are trucked into and out of cars. Excessive clearance requires heavy, large bridge plates or dockboards. If the dock level is considerably above or below the level of car door openings, trucking is more hazardous.

The dock should be wide enough to provide a temporary storage area without interference with truck movements. A narrow dock may force trucks to turn onto dockboards at a dangerous angle. The hazard may be reduced somewhat by making dockboards wider at the dock side, with flanges on the plates to turn wheels away from the edge. Portable dockboards must be securely anchored and be strong enough to carry the load imposed on them. Handholds or other means must be provided to permit safe handling.

Cars spotted for loading or unloading should be protected against being moved by switching crews. Standard railroad blue flags for daytime, and blue lights for night use, furnish warnings to train crews. Signals should be placed between the rails at both ends of a car that is accessible from either direction. Train crews should be strictly prohibited from coupling engines or cars to any cars so protected. The blue signals should be removed only by the employees, engaged in the loading or unloading operations, and only when they are ready to release the cars.

Bells and oscillating warning lights can be installed along the tracks in working areas to warn personnel that switching operations are going on. The warning lights and bells should be turned on by plant supervision before switching operations are begun.

It is the responsibility of the local plant management to clear employees from cars before releasing the track to railroad employees for switching. Therefore, before derails and blue flags are removed, plant supervision must make sure that:

- Building railroad doors are opened and other obstructions cleared to provide standard side clearance from the track area,
- All plant personnel are cleared from railroad cars and track area,
- All overhead building cranes in the area being switched have stopped operations and are clear of the tracks,
- All dockboards (bridge plates) are removed from cars,
- All counterweighted, retractable, service platforms are retracted and secured,
- All car moving equipment (cables, hooks, etc.) is removed from cars, and
- All car doors, hopper doors, etc. are closed and properly secured.

It is preferable that tracks for loading or unloading flammable liquids or other dangerous materials are used only for those purposes. For additional protection, switches must be provided with locks. Specific recommendations in regard to tank cars, grounding, rail bonding, etc., are given in Chapter 16, Flammable and Combustible Liquids.

After unloading the car, the consignee of a carload shipment of material or merchandise should clean it before releasing it. In addition, it is the consignee's responsibility to make sure that the doors are properly closed and secured. Plug doors on box cars create a distinct hazard when the doors are not properly secured (Figure 7-2). Most rail carriers have issued instructions to switching crews to not move such cars until the doors have been secured. When pieces of crating and dunnage, nails, and strapping are left loose in the car, they become serious hazards to railroad employees and others who may have to enter it later. So proper safety precautions may be taken, when equipment being loaded or unloaded is damaged or in need of repair the rail carrier or switching crew should be notified.

Figure 7-2. Plug doors should be locked before a car is moved. (Printed with permission from Inland Steel Co.)

Overhead crane runways

A very serious hazard exists at points where an overhead crane runway crosses above a railroad track, either inside or outside. The hazard lies in the possibility of crane loads or hook blocks striking locomotives or cars while switching is going on.

A system of interlocked signal lights, with one set visible to the crane operator and the other to the switch crew, can be installed to guard against movement of the crane near the track while it is occupied. All personnel involved, especially crane operators, must be trained to respect the signals without exception. The signals can be actuated manually by a key switch, whose key is kept by the area supervisor. The signals can also be interlocked with a derail.

Another method is to provide a zone power cutoff for the crane runways in the vicinity of the track, with the cutoff being actuated by a key switch under the control of the area supervisor. Other methods that assure positive control of the crane movements may be used.

Switch crews should be required to get clearance from the area supervisor before moving into the area, and the supervisor should be responsible for keeping cranes clear until the switch engine and cars move out.

Overhead cranes are also discussed in Chapter 4, Hoisting Apparatus and Conveyors.

Types of motive power

All plant locomotives (and cars as well) should be equipped with all safety appliances and with standard automatic couplers and air brakes, as required for common carrier railroads by federal law. All equipment and appliances should be maintained in sound, safe operating condition.

The type of motive power for a plant railway is important from the standpoint of accident and fire prevention. Explosive gases are easily ignited by flames or sparks from fuel-fired locomotives. Where such gases may be present, electric, compressed air, or storage battery locomotives should be used.

Where ventilation is insufficient to keep the concentration of noxious and even toxic exhaust gases at a safe level, as in mines, the use of fuel-fired locomotives should be prohibited. How-

ever, since diesel engines can be equipped with devices to eliminate toxic gases from the exhaust, this type of locomotive is being used in some adequately ventilated mines (other than coal).

A major hazard with electric locomotives, in addition to that of sparking in explosive atmospheres, is the possibility of employee contact with the trolley. The trolley should be guarded at all points where employees may pass under it and should be high enough to prevent contact anywhere along it.

Boilers of steam locomotives are constructed in accordance with the American Society of Mechanical Engineers (ASME) *Boiler and Pressure Vessel Code* (see References) and inspected in accordance with Federal Railroad Administration *Laws, Rules and Instructions for Inspection and Testing in Steam Locomotives and Tenders and Their Appurtenances* (see References). Although the regulations of the administration are not mandatory for most industrial plant railways, the standards are an excellent guide.

Diesel locomotives, which operate more quietly than steam locomotives, should be equipped with bells. Some companies paint the front and rear of diesel locomotives in contrasting colors, such as yellow and black stripes. Deck walks should have handrails around the outside. Each locomotive should carry an extinguisher for oil fires.

The tanks of compressed air locomotives should be constructed in accordance with the ASME standard *Boiler and Pressure Vessel Code* (see References). Each air receiver requires an air pressure gage, a pop safety valve, and drainpipe with valve at bottom.

A battery locomotive should have a deadman control, so that the operating lever will return automatically to the "off" position when released, or when the operator leaves the controls and the circuit is broken. Safety provisions for battery charging rooms are in Chapter 6, Powered Industrial Trucks.

Tools and appliances

Devices, tools, and appliances are important for the safe operation of a plant railway.

Automatic couplers eliminate the serious hazard involved when workers go between either standard or narrow gage cars to insert or withdraw pins manually.

Hopper bottom cars should be opened only with standard car wrenches. A special tool for closing the latches on bulk cars eliminates the need for a person to go on top of the car and risk falling. Rerailers should be provided for narrow gage cars; other means of rerailing cars generally are dangerous.

Moving cars

Moving cars using hand methods often results in accidents. The safe procedure is to use a switch engine.

If the ordinary hand car mover is used, there should be a shield around the bar, so that employees will not strike their hands or otherwise injure themselves, if the tool should slip. Crowbars and other makeshift tools should not be used to move cars.

When a car is on a grade, the handbrake should be tested to make sure that it takes hold and that excess slack in the brake chain is taken up. A worker should remain on the brake platform to stop the car with the brake at the required point.

Although winch-type car pullers are used extensively, if the rope breaks, the operator may be killed. Because the operator

Figure 7-3. Self-propelled, rider-operated car mover. (Printed with permission from Whiting Corp.)

is in line with the rope while operating the equipment, a shield of steel plate or expanded metal should be installed for protection. A forged steel hook should fasten the rope to the car.

A self-propelled, rider-operated car mover is used in many industries (Figure 7-3). This vehicle has rubber-tired wheels and steel rail wheels, both retractable, so that it can run either on plant grounds or on rails. It is fitted with standard couplers and, depending upon the model, can develop a drawbar pull of from 8,400 to 18,000 pounds (38 to 80 kN). Its use eliminates the need to use hand car movers, winches, capstans, or power equipment, which are not designed for the purpose and which are often hazardous.

It also should be understood that use of a push-pole to move cars by a locomotive is dangerous under any circumstances and is not recommended.

Safe practices

It is vital that transportation personnel—as well as all other employees—observe safe practices. They can be guided by the same safety rules observed by all large railroad systems. When plant railway safety problems arise, help and suggestions can be sought from safety professionals and operating officers of the plant's connecting railway line.

Safety meetings and other educational activities should be used to inform plant personnel about railway hazards.

To include all the safe practices required for the safe operation of railway equipment would require a book in itself. However, a few of the more important ones are included here:

1. Stop and look both ways before crossing any track.
2. Expect trains or cars to move at any time, on any track, in either direction.
3. *Step over* rails when crossing tracks. Never step, walk, or sit on any rail.
4. Never go between moving cars (or cars that may move) for the purpose of adjusting couplers or for any other purpose. (The once common practice of kicking couplers to align them is particularly dangerous.)

5. Give a hand or lamp stop signal, and receive an acknowledgment of the signal, before going between standing engines or cars.
6. To close a boxcar door, place one hand on the door handle and the other on the back end of the door.
7. Step down from cars—do not jump.
8. *Never* attempt to ride the leading footboard of an engine. (This rule applies to *all* employees, including switchmen.)

POWER ELEVATORS

Information essential to the safe operation and maintenance of elevators is provided. This discussion is generally limited to electric-drive passenger and freight elevators and hydraulic elevators. It is not intended as a guide for design or specification of new equipment, nor is it a manual for skilled elevator inspectors and mechanics.

Use of codes

Before new elevators are specified or major alterations scheduled for existing installations, reference should be made to the latest edition of the American National Standard A17.1, *Safety Code for Elevators, Dumbwaiters, Escalators, and Moving Walks.* For the rest of this section, this will be referred to as the Elevator Code. This standard is the basis for much of the material presented in this section.

To make sure operations are safe, the Elevator Code requires inspections. These should be made as recommended by the latest edition of the *Practice for the Inspection of Elevators, Escalators, and Moving Walks (Inspectors' Manual),* ANSI A17.2, including addenda. This is discussed later under Inspection and Maintenance.

New equipment should be specified so that it will conform to Elevator Code requirements, unless federal, state, or local codes are more stringent, in which case those should be observed as the minimum requirements. Usually, the Elevator Code will be more strict and, in this case, it should be referred to in the specifications. Two commonly used wordings are:

"The elevator and associated equipment shall meet the requirements of American National Standard A17.1, *Safety Code for Elevators, Dumbwaiters, Escalators, and Moving Walks,* latest edition, except as hereinafter specifically exempted or modified."

The second typical paragraph is: "Except as changed or modified herein, the elevator and associated equipment shall meet the requirements of American National Standard A17.1, *Safety Code for Elevators, Dumbwaiters, Escalators, and Moving Walks,* latest edition, and also shall comply with all applicable local laws and/or ordinances."

Such paragraphs will generally take care of items not specifically spelled out in the specifications and drawings.

Types of drives

To provide the safety professional with a working knowledge of various elevator types and their limitations, this section has been divided into two major topics, each covering a type of elevator drive: (1) electric elevators and (2) hydraulic elevators. (Belt-drive and chain-drive machines have practically disappeared; their installation is now prohibited by the Elevator Code. However, many are still in use and repairs to existing installations are permitted by the Elevator Code.)

Electric elevators. The two general types of drives for electric elevators are traction drive and winding drum. The winding drum type, however, is now almost obsolete and is presently being used only in new dumbwaiter construction and freight elevators with restrictions as specified by the Elevator Code.

The Elevator Code requires that all driving machines be of the traction type, except that it permits the use of winding-drum machines on freight elevators which travel not more than 40 ft (12.2 m) at speeds not exceeding 50 ft per minute (0.25 m/s) and which are not provided with counterweights.

In the winding-drum type, the hoisting rope is anchored in and winds on a spirally grooved drum. This is a positive drive. If the machine is not stopped at the limits of travel, the car may be pulled into the overhead structure and, if the motor is powerful enough, the ropes may be pulled from their anchorage.

In the traction-drive type (Figure 7-4), the hoisting rope is not attached to the machine. The elevator is moved by the traction (friction) of the ropes in grooves on the drive sheave. The grooves may be semicircular (U-groove) or undercut U-groove. Some V-groove sheaves are still in use.

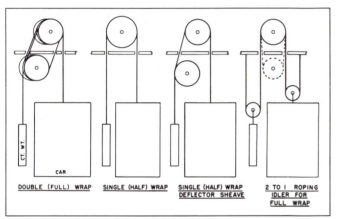

Figure 7-4. Common elevator traction drives showing double (full) and single (half) wraps.

With the U-groove, the ropes run through one set of grooves on the drive sheave, over an idler sheave, and then through a second set of grooves on the drive sheave—effecting a nearly double (full) wrap.

A simple rule is that when the drive sheave has twice as many grooves as there are hoisting ropes, the machine is a double (full) wrap. The total angle of contact of each rope with the drive sheave generally is between 300 and 360 degrees, and the traction relation changes very little with wear.

Most companies use some form of undercut U-groove for their single-wrap machines. The friction is higher than an ordinary U-groove because of the pinching action. The width of the undercut is varied to suit traction needs. The traction remains substantially constant until the groove has worn to near the bottom of the undercut. The groove should be checked periodically to make sure that it is not worn so much that the rope bottoms.

The V-groove likewise has higher friction when new but, as the V-groove wears, the rope seats deeper and deeper, so that the area of contact of rope with groove is increased and the unit pressure is decreased. When the rope reaches the bottom of the groove, much of the driving traction probably will be lost,

and a loaded (or empty) car may "break traction" and slide with the brake locked. For this reason, such grooves should be checked frequently for wear and should be machined before the rope wears the groove down to the bottom. A minimum grooving bottom diameter must be observed.

Generally, the V-groove or undercut U-groove is used on geared machines and the U-groove is used on gearless machines, but there are exceptions.

Traction machines have an inherent safety feature: when the descending member (counterweight or car) bottoms, driving traction is lost and, in most cases, the ascending member cannot be pulled into the overhead. With extremely high rises, the weight of rope hanging on the downrun side may be sufficient to maintain traction after the car or counterweight has landed but, since the compensative sheave is tied down, there is no danger.

For safe and successful operation, rope tension on the car side must bear a definite relation to rope tension on the counterweight side. Normally, this relation ranges from 2-to-1 to 1-to-2. See discussion of overloads in Operation, later in this chapter.

Hydraulic elevators are being installed in many new buildings averaging up to six stories in height. New installations are usually the electrohydraulic type. The problems associated with older hydraulic elevators, generally lower efficiency, the difficulty of keeping the valves and stuffing boxes tight, etc., have been eliminated by the use of modern technology and materials and with a proper maintenance program. Hydraulic elevators are power elevators where the energy is applied by means of a liquid under pressure in a cylinder equipped with a plunger or piston.

The Elevator Code requires that new installations be equipped with anti-creep leveling devices and with hoistway-door locking devices, electric car-door or gate contacts, hoistway access, and parking devices the same as those required for electric-drive elevators.

Because most electrohydraulic elevators do not have a counterweight, the motor must supply pressure to lift the entire weight of the car and the load, and it must be more powerful than the motor of a traction-drive electric machine, on which the weight of the car and part of the load is compensated for by the counterweight, to maintain the same speed. As built, the electrohydraulic type has all the electrical protective devices (interlocks, car gate contacts, limit switches, and similar devices) found on an electric elevator.

No car safety is required on electrohydraulic elevators, since they can come down no faster than the fluid can be forced out of the cylinder by the descending plunger. If counterweights and/or car safeties are on the elevator, they must comply with the Elevator Code.

Belted machines. The Elevator Code prohibits the installation of belt-driven and chain-driven machines. Existing installations should be provided with electrically released brakes, terminal stopping, and safety devices as required for electric elevators. Some jurisdictions have outlawed this type of installation while others allow its continued use.

New elevator equipment

Requirements. When new elevators are under consideration, care should be taken to be sure that all requirements of the latest edition of the Elevator Code are met. A few examples of the requirements to be checked are: safe and convenient access to the machine room and the pit, adequate lighting in the machine room and overhead spaces, and convenient electric outlets on the crosshead and in the pit.

An inspection station (with slow-speed UP and DOWN operating buttons and an emergency STOP switch) is required by the Elevator Code on the top of the car for the use of maintenance personnel and inspectors.

The elevator is required by the Elevator Code to have normal and final limit stops, interlocks on all hoistway doors, a contact on the car door, and a car emergency exit or exits. If a single elevator hoistway exists, emergency access doors must be provided to the blind portions of such a hoistway. Unless the elevator is the hydraulic-plunger type, it should have a governor-actuated safety that meets the latest code requirements (city, state or provincial, federal, and/or ANSI standards).

Buffers are required to absorb the energy of descending cars and counterweights at the limits of travel. The two types in use are spring buffers and oil buffers.

Spring buffers are permissible under the Elevator Code only for cars where the rated speed is not in excess of 200 fpm (1.0 m/s), but a few states and cities permit them for higher speeds. A spring at best is a poor absorber of energy. A good spring will return approximately 95 percent of the stored energy, so that a spring buffer stop affords only a series of decreasing surges.

Capacity ratings. The size, capacity, and speed of new freight elevators will depend upon the purposes for which they are to be used.

When preparing specifications for new freight elevators, it is important to anticipate future load requirements as well as present ones. The safety professional should, therefore, find out from production personnel what size units (both freight "package" and freight carrier) will probably be in use 20 years hence.

The Elevator Code classifies freight elevators as follows:

- Class A—Elevators loaded by hand or by handtrucks and the load is distributed. Here, the weight of any single piece of freight or of any single handtruck and its load is limited to a maximum of one-fourth the rated load capacity.
- Class B—Elevators used solely to carry automobile trucks and passenger automobiles up to rated capacity of the elevator.
- Class C1—Industrial truck loading where the truck is carried by the elevator.
- Class C2—Industrial truck loading where the truck is *not* carried by the elevator but used only for loading and unloading.
- Class C3—Other loading with heavy concentrations where a truck is *not* used.

For Class C1, Class C2, and Class C3 loadings, the Elevator Code requires that the rated load of the elevator be not less than the load (including any truck) to be carried, and that the elevator be provided with a two-way automatic leveling device.

With palletized loads increasing in size year by year and with lift trucks being made larger to handle them, old elevators that are overloaded can start downward when the last loaded lift truck is run onto the car. In some cases, the brakes do not hold. In others, the traction relation is broken, and the hoisting ropes slip through the drive sheave.

Adherence to the Elevator Code should prevent such cases in new installations. Where old elevators are used to handle heavy palletized loads, however, safe load limits and safe oper-

ating procedures must be determined and strictly enforced, if serious accidents are to be prevented.

The rated capacity of the passenger elevators should conform to the Elevator Code.

Hoistways and machine rooms

Hoistways. Most building elevator codes require that new elevators be installed in two-hour fire-resistant hoistways, with one and one-half hour fire doors that fill the entire opening, in order to prevent the rapid spread of fire from floor to floor.

Hoistways, guide rails, and all other appurtenances should be thoroughly cleaned of grease and dirt accumulations, at frequent intervals, to eliminate fire hazards. Projections in the hoistways should be bevelled at an angle of not less than 75 degrees with the horizontal as required by the Elevator Code. Windows in the walls of the hoistway enclosures are prohibited.

It is particularly important that no pipe conveying gas or liquids, which—if discharged into the hoistway—would endanger lives, be installed in or under any elevator or counterweight hoistway. However, low-pressure steam (5 psig or less) or hot water pipes only for heating the hoistway and the machine room (or penthouse) are permitted, if certain conditions are met. All electrical equipment and wiring shall conform to the *National Electrical Code* (NFPA 70; ANSI C1). Only such electrical wiring and equipment used directly in connection with the elevator may be installed in the hoistway.

Pits. To protect persons working in elevator pits, a minimum clearance of 2 ft (60 cm) is necessary between the lowest projection on the underside of the car platform (not including guide shoes and aprons attached to the sill) and any obstruction in the pit (exclusive of compensating devices, buffers, buffer supports, and similar devices). Measurements should be made when the car is resting on fully compressed buffers.

Counterweight runways should be enclosed from a point not more than 1 ft (30 cm) above the pit floor to a point at least 7 ft (2 m) above the pit floor and adjacent pit floors, except where compensating chains or cables are used.

Screen partitions, at least 7 ft high between adjacent pits, will protect persons in one pit from cars and counterweights in adjacent pits and will protect employees from hazards, when adjacent pits are at different levels.

An elevator pit should never be used as a thoroughfare or storage space. It should be fully enclosed, and the entrances kept locked. To remove water, a sump pump should be provided. Drains connected directly to sewers are not allowed by the Elevator Code.

Pits should be kept clean and free of debris. Rubbish never should be swept into pits. A vertical ladder should be provided for easy access.

Lighting of at least five foot-candles (54 lux) at the pit floor level should be provided. A light switch, accessible from the pit access door, is required by the Elevator Code. An emergency stop switch shall be installed in every pit and shall be accessible from the pit access door.

Machine rooms should have safe and convenient access, as specified in the Elevator Code. So that persons repairing or inspecting elevator hoisting machinery have sufficient room and are safe, there should be at least 7 ft of headroom between the machinery platform and the machine room roof.

Like pits, elevator machine rooms should never be used as

thoroughfares. The one exception would be if the elevator equipment were in a separate locked enclosure. Rooms should be well-ventilated and lighted with not less than 10 foot-candles (108 lux) at floor level. Doors should be kept locked, to prevent entry of unauthorized persons, with a sign affixed to the door stating such.

Machine rooms should be kept clean and should not be used for storage. Small quantities of ordinary maintenance supplies should be placed in a wall cabinet. A portable Class C fire extinguisher should be kept within reach of someone standing at the door. (See Chapter 17, Fire Protection.)

Overhead protection. If the elevator machine is located in the penthouse, a substantial grating or floor of fire-resistant construction should be provided under the machine. For detailed information concerning construction, the Elevator Code should be consulted.

On all overhead machine installations there should be a cradle below the secondary sheaves if they extend below the floor or grating.

No elevator machinery, except the idler or deflecting sheaves, should be hung underneath the supporting beams at the top of the hoistway. When the governor or other devices (other than terminal stopping switches) must be installed below the machine floor, they should be set on a substantial secondary floor.

For winding-drum machines, a substantial beam or bar should be placed at the top of the counterweight guide rails and beneath the counterweight sheaves, to prevent the counterweights from being drawn into the sheaves.

Landings and doors

The Elevator Code specifies that hoistway landing openings of all elevators must be provided with hoistway doors that guard the full height and width of the openings.

Hoistway doors (passenger elevators). Records show that most elevator accidents occur at the hoistway door. Most are "tripping" accidents. Some are the "caught and crushed" or "fell down hoistway" type. These are on old elevators, because the interlocks required by the Elevator Code, if properly installed and maintained, practically eliminate these two types of accidents.

Many serious accidents and fatalities have occurred on older elevators with doors that necessitated using a special key to open on the corridor side, regardless of the relative position of the elevator. Accidents occurred when the victim erroneously assumed the elevator to be at the landing, opened the door, and stepped into the shaft.

The Elevator Code permits the use of unlocking devices and other means for emergency access to elevators equipped with doors that are unlocked, when the car is at the landing or can be unlocked from the corridor side without tools if the car is in the landing zone.

All emergency keys should be kept where they are accessible only to qualified personnel who will guard and use them carefully.

To help prevent hoistway accidents in older elevators, conduit can be installed in each hoistway so an electric light can be installed opposite each opening. If the door is opened with an emergency key and the car is not at the landing, the light, which burns continuously, will help a person see that the shaft is empty in time to prevent him or her from stepping into it (Figure 7-5).

Figure 7-5. View from pit looking up hoistway illumination (arrows). (Printed with permission from Architect U.S. Capitol.)

These requirements can be met in either of these two ways:
1. Provide only two means of access to the hoistway—one at an upper landing to permit access to the top of the car, and one at the lowest landing if this landing is the normal point of access to the pit.
2. Where elevators operate in a single hoistway, provide hoistway doors that can be unlocked when closed with the car at the floor, or locked but possible to open from the landing only when the car is in the landing zone. This requirement may vary depending upon the jurisdiction; it should be checked with local authorities.

In general, three types of doors at landings are used: vertically or horizontally sliding doors, combination sliding and swinging doors, and swinging doors. All three types should have direct-acting interlocks. Nothing less than direct-acting mechanical interlocks should be used.

Power-operated vertical-slide doors and gates must operate in sequence if the elevator is used for passengers and, in any case, if the doors close automatically.

For new installations, according to the Elevator Code, the distance between the hoistway side of the hoistway door opposite the car opening and the hoistway edge of the landing threshold should be not more than ½ in. (13 mm) for swinging doors and 2¼ in (57 mm) for sliding doors. The face of the hoistway door should not project into the hoistway beyond the edge of the landing sill. On existing installations, if this distance exceeds 1½ in. (38 mm) for swinging doors and 2½ in. (64 mm) for sliding doors, it is advisable to fill in the excess space.

No automatic fire door, whose functioning depends on the action of heat, should be designed to lock any landing opening in the hoistway of any elevator or any exit leading to the outside of the building.

The loading platform for at least 2 ft (60 cm) back from the door should be so constructed and maintained that persons will not readily slip. Many members of the National Safety Council use rubber mats, firmly secured, or adhesive abrasive strips, or abrasive-surfaced concrete directly in front of all hoistway doors. (See Council Data Sheet 595, *Floor Mats and Runners.*) To eliminate tripping hazards, such surfaces should be made flush with the surrounding floor.

Hoistway doors and gates (freight elevators). Like those for passengers, most freight elevators are now installed in two-hour fire-resistant hoistways with one and one-half hour fire doors that fill the entire opening, as required by practically all codes. However, there are some older elevators—mostly freight—which have hoistways enclosed only to a 6- or 7-ft (1.8 or 2.1 m) height and hoistway gates of hardwood slats which are 5 or 6 ft (1.5 or 1.8 m) high and have a clearance under them of as much as 8 to 10 in. (20 to 25 cm). Many of these elevators are shipper-rope operated, so it is necessary to reach into the hoistway to bring the car to a landing—a procedure which has been prohibited in new installations by the Elevator Code for many years.

Hoistway doors (Figure 7-6) should comply with applicable state and municipal requirements for fire resistance. Doors closed by hand should be so arranged that it is not necessary to reach back of any panel, jamb, or sash to operate them.

To facilitate movement of trucks, a door sill is recommended to fill the gap between the landing and the car.

Hoistway gates in older installations should be replaced with doors as soon as practicable. Where gates are used, special attention must be given to maintain them at the best possible standard of safety. The openings in gates made of grille, lattice, or other open-work should reject a ball 2 in. (5 cm) in diameter.

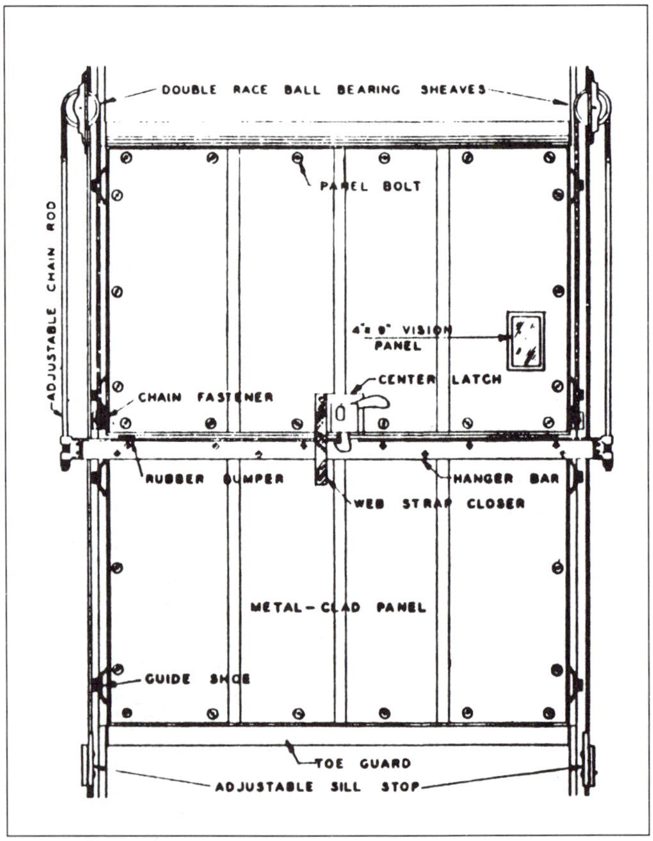

The bottom of the gate should come down within 1 in. (2.5 cm) of the threshold to prevent objects from sliding under the gate into the hoistway. Where lack of headroom precludes a standard gate at the lowest landing, the gate should be made in two sections.

Gates should have convenient handles or straps for manual operation. However, an attachment for closing the gate by power, controlled from the landing buttons, saves time and labor and decreases the possibility of leaving or propping the gate open.

Power-operated doors and gates are recommended. For freight elevators, these are usually of the continuous-pressure-operation type.

The growing weight and speed of forklift trucks requires that freight elevator hoistway doors and gates be protected. A 5-ton (4,500 kg) load moving only 2 mph (2.93 fps) has an impact of 2,690 foot-pounds (F = ½ MV²). If it is assumed that the door or gate can deflect 1 in., a force of 32,300 lb (144 kN) results—certainly more than any hoistway door can stand.

One solution to this problem is installing a heavy wire rope across the opening. This must be lowered by the truck operator before the truck can be run onto the car platform.

For savings in wear and tear, as well as the elimination of unsafe practices, placing one person in charge of operating the elevator and responsible for its operation is also a solution.

Interlocks and electric contacts (passenger and freight elevators). Unless the hoistway door is locked in the closed position, to prevent a car from moving away from the landing, hoistway doors should be equipped with interlocks that comply with the Elevator Code (Figure 7-7). Interlocks should be direct-acting mechanical-activated devices. All interlocking devices should be of a type that cannot be plugged or made inoperative in any way.

In addition to preventing movement of the car when the hoistway door is unlocked, the interlock also prevents opening of

Figure 7-6. Design of a typical vertical sliding biparting steel hoistway door for freight elevators, as seen from inside the car. (Printed with permission from The Peelle Co.)

Figure 7-7. An electromechanical interlock designed for use on biparting, manually operated steel doors on pushbutton-controlled freight elevators.

the hoistway door from the landing side, except by emergency key, unless the car is within the landing zone and is either stopped or being stopped.

Locks and contacts (not interlocks) are permitted in a few cases. Contacts are required for the car door or gate, which is considered closed if within 2 in. (5 cm) of the nearest face of the jamb.

The Elevator Code defines the "closed position" of hoistway doors as being ⅜ in. (9.5 mm) from the jamb or between panels of center-opening doors, except that, for vertical-slide biparting doors, the dimension is ¾ in. (19 mm). Under certain conditions, the elevator may be started when the doors are 4 in. (10 cm) from full closure.

The Elevator Code requires that all interlocks be so designed that the door must be locked in the closed position, as defined here, before the car can be operated.

Like any safety device, an elevator interlock or electric contact will be useless if it easily gets out of order. It is advisable, therefore, to use only those devices that comply with the Elevator Code and that have been either tested and listed by competent, designated testing laboratories, or approved by the city or state authorities that have jurisdiction. Also, it is essential that interlocking devices be inspected frequently and maintained in proper working order.

The car-leveling device automatically brings the car to a stop when the platform is level with the desired landing. A tripping hazard from uneven surfaces thus is eliminated. Moreover, in the case of freight elevators, the level surface automatically provided for the passage of trucks saves wear and damage to sills, and reduces the possibility of material being jarred off trucks.

Cars

Enclosure. All elevator cars are required by the Elevator Code to be enclosed on the sides and top, except for the side or sides used for exit and entrance. Openings for ventilation, emergency exit, signal, operating, or communication equipment are allowed.

The sides and top of every passenger elevator car should be made of metal or of an approved, fire-retardant treated material.

Although wood and openwork have been used in the past for freight elevator cars, the Elevator Code now specifies a solid metal enclosure to a height of at least 6 ft (1.8 m) for enclosures. However, they may have perforations above the 6-ft level, except in the area where the counterweight passes the car. That portion of the enclosure, for 6 in. (15 cm) on each side of the counterweight and up to the crosshead or car top, should be solid.

The enclosure should be fastened securely to the floor and to the suspension sling. If the enclosure is cut away at the front to provide access to the hand rope, the opening should be low enough to prevent injury to the operator's hand.

Enclosures should be kept in good condition to prevent injuries to persons loading and unloading the car. Broken wooden wainscoting or torn sheet metal may cause severe lacerations. One way to prevent such damage is to install bumper strips at the level of truck platforms and truck push bars. They should be made of oak, ash, hickory, or similar tough, resilient wood and should be at least 1½ in. (38 mm) thick (2 in. [50 mm] thick where heavy trucks are used). Steel channels are also effective.

Strips should be wide enough to match all heights of platforms and push bars of trucks that are transported in the elevator.

Painting the inside of the car a flat black, from the floor to the top of the upper guard strip, will make bumps and scrapes less conspicuous. If the sides above the top strip are painted aluminum, the high reflectance value of the paint will help overall illumination.

Top covers. Freight elevator cars may have either solid or openwork top covers. If openwork is used, the Elevator Code requires that the openings be small enough to reject a ball 1½ in. in diameter; some companies, however, recommend ¾-in. openings.

Wire mesh is recommended for openwork covers, because it combines maximum strength with minimum weight and offers little interference with light and air. Not less than No. 10 gage (2.5 mm dia.) steel wire should be used.

The cover should be strong enough to sustain a load of 300 pounds (135 kg) on any square area, 2 ft (61 cm) on a side, or of 100 pounds (45 kg) applied at any one point.

An emergency exit of at least 400 sq in. (0.258 m²) in area and not less than 16 in. (40.6 cm) on any one side should be in the top of every elevator car to provide an unobstructed passageway for easy escape if the car becomes stalled between floors or some other emergency arises.

The exit cover should open outward and be so hinged or attached that it can be opened easily from the top of the car only. A heavy cover should be counterweighted or divided into several hinged sections, which should be kept down except when being used.

The Elevator Code now requires an unobstructed refuge space of not less than 650 sq in. (0.4 m²) on top of the car enclosure. It shall not measure less than 16 in. on any side. The top emergency exit may open into this space provided that there is an area adjacent to the opening available for standing when the emergency exit cover is open. The minimum vertical distance in the refuge area between the top of the car enclosure and the overhead structure or other obstruction shall not be less than 3.5 ft (1 m) when the car has reached its maximum upward movement.

Doors or gates. To conform to Elevator Code requirements for new installations, a car door or gate must be provided at each entrance to the car. If the elevator is electrically operated, the car door or gate should have electric contacts that will prevent movement of the car unless the door or gate is within 2 in. (5 cm) of the fully closed position.

It is recommended that existing elevator cars having more than one opening be equipped with a car gate on each entrance. The gate should be provided with a contact to prevent its being opened when the car is in motion. When closed, the gate should guard the full width and height of the opening (except for vertically sliding gates, which should extend from a point not more than 1 in. [2.5 cm] above the car floor to a height of at least 6 ft [1.8 m]). Collapsible gates, when fully expanded, should reject a ball 3 in. (7.6 cm) in diameter. They should have convenient handles with guards that protect the operator's fingers.

Floors of cars should be kept in good condition. Ordinary metal sheets should not be used for surfacing or repairing floors because they soon become smooth and slick. In time, even sheet metal with raised surface markings may wear smooth and then should be replaced.

Neither the edge of the car platform nor the edge of the landing should be permitted to become slippery or badly worn. Cast iron and steel sills at such points are slipping hazards unless provided with antislip surfaces. Abrasive strips or welded beading may be applied to eliminate slipperiness.

Loads. Even though the rated capacity (or safe working load) of an elevator includes a safety factor, the rated capacity should not be exceeded and should be indicated by a conspicuous sign inside the car. Metal signs with stamped, etched, or raised letters and figures, not less than ¼-in. (6 mm) high, are satisfactory for passenger elevators and a minimum of 1-in. (25 mm) high letters and figures for freight elevators.

Lighting. The Elevator Code requires that cars and landings be well-lit at all times when in use. Code requirements are minimums and should be exceeded if necessary.

A car should have at least two lights that provide a minimum illumination of 5 foot-candles (5.4 lux) for passenger elevators and 2½ foot-candles (27 lux) for freight elevators at the landing edge of the car platform, when the car and loading doors are open.

Car lights may be omitted on freight elevators with perforated car tops that travel no more than 15 ft (4.57 m), if at least two electric lights are installed at the top of the hoistway and furnish the minimum illumination specified.

Illumination on landing thresholds should be at least 5 foot-candles (5.4 lux). The Code requires that new installations, in addition to the regular lighting, have emergency lighting that goes on automatically 10 seconds after the regular lighting fails.

Hoisting ropes

Car and counterweight ropes must be of iron (low-carbon steel) or steel having the commercial classification "Elevator Wire Rope" or rope specifically constructed for elevator use. Ropes less than ⅜ in. (9.5 mm) in diameter may not be used on passenger elevators. Traction elevators should have not fewer than three hoisting ropes; winding-drum elevators should have not fewer than two hoisting ropes and two ropes for each counterweight used.

The Elevator Code requires that the diameter of sheaves or drums for hoisting or counterweight ropes be not less than 40 times the diameter of the ropes. In practical application, however, this ratio is usually higher than 40.

On winding-drum machines, rope should be long enough so there will not be less than one full turn of rope on the drum, when the car is at the extreme limit of its overtravel. Drum ends of ropes are usually secured by tapered babbitted sockets or by clamps on the inside of the drum.

Rope fastenings. Ends of car and counterweight suspension ropes are usually fastened by individual babbitt-filled tapered sockets (Figure 7-8). The Elevator Code gives detailed specifications on proper application.

Rope sockets must develop at least 80 percent of the breaking strength of the rope used. Shackle rods, eyebolts, and other means used to connect sockets to the car or counterweight must have a breaking strength at least equal to that of the rope.

Governor ropes should be of iron, steel, Monel metal, phosphor bronze, or stainless steel and be of regular lay construc-

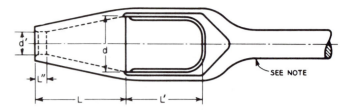

NOTE—ROPE SOCKET AND SHACKLE ROD MAY BE IN ONE PIECE, AS SHOWN, (UNIT CONSTRUCTION) OR THE SOCKET AND ROD MAY BE SEPARATE. (See Elevator Code Rule 212.9c)

Rope Diameter (in., nom.)	Maximum Diameter of Hole
⅜ to ⁷⁄₁₆ (inc.)	³⁄₃₂ in. larger than nominal rope diameter
½ to ¾ (inc.)	⅛ in.
⅞ to 1⅛ (inc.)	⁵⁄₃₂ in. larger than nominal rope diameter
1¼ to 1½ (inc.)	³⁄₁₆ in.

1 in. = 2.5 cm

Figure 7-8. Tapered babbitted rope sockets. Diameter (d) of the hole at the small end of the tapered portion of the socket shall not be more than the chart shows. (Printed with permission from the American National Standards Institute.)

tion not less than ⅜ in. (9.5 mm) in diameter. Tiller rope is prohibited by the Elevator Code. Rope used as a connection from the safety to the governor rope including rope wound on the safety drum must be corrosion-resistant.

Operating controls

Stopping devices. Winding-drum elevators are required by the Elevator Code to have a direct-driven adjustable, automatic, machine-limit stop mechanism, which will stop the car if it overruns the highest and lowest landings. In addition, limit switches should be placed either in the hoistway or on the car.

Electric elevators with traction machines are required to have stopping switches in the hoistway, on the car, or in the machine room if the switches are operated by the motion of the car.

Drum-type elevator machines are required to have a slack rope device to shut off power to the machine and brake in case the rope becomes slack. This device must be such that it won't reset automatically when slack in the rope is removed.

Every electric elevator must have an emergency stop switch in the car, adjacent to the operating device, which will cut off the power. The button should be clearly identified and be red in color. Contacts of such switches should be directly opened mechanically and should not depend solely upon springs for opening contacts. Some jurisdictions have outlawed the emergency stop switch; local authority should be consulted.

Grounding. The motor frame (and the operating cable, if insulated from the motor frame) should be grounded. All switches and wiring should conform to NFPA 70, *National Electrical Code* (ANSI C1).

Safety devices. Every elevator car suspended by wire ropes is required by the Elevator Code to have one or more car safety devices that will catch and stop the car in the event of overspeed or failure of the hoisting ropes. The safeties must be attached to the car frame, and one safety must be located within or below the lower members of the car frame (safety plank).

The Elevator Code classifies three types of safeties:

- Type A (instantaneous) safeties rapidly increase their pressure on the guide rails to give a very short stopping distance. They are permissible on elevators having a rated speed of not more than 150 fpm (0.76 m/s).
- Type B safeties apply limited pressure to the guide rails, and the retarding forces are reasonably uniform after full application. They are permissible on elevators of any speed and may be used in multiple.
- Type C safeties (Type A with oil buffers between the safety plant and elevator car) are permissible on elevators with a rated speed of not more than 500 fpm (2.5 m/s).

Counterweight safeties should conform to the car safety requirements.

Car switches. The handle of the car switch (operating control) should be designed to return to the "stop" position and lock, when the operator's hand is removed.

Signal system. Every elevator (except automatic-operation and continuous-pressure-operation elevators) should have a signal system that can be operated from any landing to signal the car when it is wanted at that landing.

Inspection and maintenance

Inspections and tests must conform to requirements and regulations set forth by the particular municipality or regulatory agency. Much of the discussion in this section is based on ANSI/ASME A17.2 *Inspectors' Manual for Elevators and Escalators,* published by the American Society of Mechanical Engineers and also available from the American National Standards Institute. It is important that anyone concerned with maintenance and inspection should have and use a copy of the Elevator Codes and the Inspectors' Manual.

A guide for inspection is given in the Inspectors' Manual. It contains these requirements:
1. Acceptance inspections and tests of all new installations and alterations should be made by an inspector employed by the enforcing authority.
2. Routine inspections and tests of all installations should be made by a person qualified to perform such services, in the presence of an inspector employed or authorized by the enforcing authorities. It is recommended that periodic inspections and tests be made at intervals not longer than:
   - Six months for power passenger elevators and escalators.
   - Twelve months for power freight elevators, hand elevators, and power and hand dumbwaiters.
3. Periodic inspections and tests shall be made by a person qualified to perform such service in the presence of an inspector employed or authorized by the enforcing authority. Car and counterweight safeties, governors, and oil buffers, should be given periodic inspections and tests at least every year with no load and every five years with a full load.

Inspection program. Careful maintenance is essential to the conservation and safe operation of elevators and their appurtenances. It also reduces the need for repairs. Frequent and thorough inspections by qualified personnel are the first requisite of an efficient maintenance program. However, minor day-to-day inspection and maintenance can be performed by qualified plant personnel.

Items of regular maintenance that require special consideration include: inspection of hoisting and counterweight wire ropes and lubrication of oil buffers; cleaning and lubrication of guide rails, controller contactors and relays, and car safety mechanisms; cleaning of hoistways and pits, machine rooms, and tops of cars.

Hoistways and landings. Accidents occurring or originating at landings usually are due to (1) tripping or slipping at the car entrance or landing, (2) being caught inside the car, (3) falling down the hoistway, and (4) being caught by the doors.

To help prevent tripping and slipping accidents at hoistway landings, the following points should receive attention:
1. The leveling of the car should be checked. Improper leveling may be due to careless operation or, where cars are leveled automatically, to improper adjustment. Elevator operation should be observed and necessary instructions given to the operator. With an automatic car-leveling device, the operator has no control over the final stop. The automatic car leveler should be adjusted by a competent mechanic, preferably one trained or employed by the elevator manufacturer.
2. The condition of landing sills and floors should be watched. Landing sills should be of nonslip material; if they are not, they should be replaced or roughened if worn smooth. Broken sills, holes in flooring, worn floor coverings, and other conditions that create tripping hazards should be repaired. The finished surface should be flush with the surrounding floor.
3. Illumination of landings should be checked, with particular attention to those near building entrances where the difference between outdoor and indoor light intensity is noticeable. Globes and reflectors should be clean, and lamps should be of adequate size.

To provide the required protection at landings, interlocks and contacts must be well maintained and correctly adjusted.

Track grooves for hoistway doors should be clean, so that the doors will move freely. There should be no excessive play. Vision panels should be clean and unbroken. Counterweights should operate freely, and sheaves should be properly aligned.

Because the entire load is transferred to the guide rails when the safety operates, the guide rails must be properly aligned at the joints and securely attached to the brackets, and the bolts or welds with which the brackets are attached to the building structure must remain tight. Alignment of rails can be checked easily by sighting along the faces; bracket bolts must be tested individually.

Except where roller guide shoes, which run on dry rails, are provided, guide rails should be kept properly lubricated to help reduce wear between the rails and the guide shoes. The lubricant used should be that specified by the elevator manufacturer. Roller guide shoes should be clean and, if necessary, adjusted for pressure against the rails.

Elevator cars should be checked for structural defects, such as loose bolts and other fastenings, excessive play in guide shoes, and worn or damaged flooring.

Each emergency exit should be tested by being opened. Each exit panel should be checked to assure that it is securely fastened in the closed position. If panels are held in place by locks, the key should be kept at the building in a location not accessible to the public, but available for emergency use; and if the panels are held by thumbscrews, they should be removable without having to use pliers.

Car doors or gates should be subjected to the same examination as hoistway doors, and the same standards of maintenance should be followed. Contacts should be checked for adjustment in the same manner as hoistway door contacts.

The car-operating switch, when provided, should be tested to see that it returns to the neutral position, when released by the operator, and locks there. On a cable-operated car, the cable lock should be checked to make sure that, when it is locked, the cable cannot be operated. All switch contacts in the car-operating device should be examined, and the glass cover should be in place over the emergency release switch. Emergency release switches, however, are prohibited on automatic elevators by the Elevator Code.

Car lights should be checked for operation and for loose or missing screws and broken or cracked glassware. Glassware or bulbs should be clean. The emergency lighting should be checked periodically.

Safety devices. Maintenance of safety devices often is neglected because they do not affect normal operation of the car. However, in case of emergency, the safety of passengers depends entirely upon proper performance of these devices. It is of utmost importance, therefore, that they be maintained in proper working condition.

At frequent intervals, all the safety devices and equipment should be cleaned and lubricated, inspected for worn, cracked, broken, or loose parts; and tested to determine their ability to stop and hold the car. Safety devices should be tested and adjusted only by a person qualified to perform such services, in the presence of an inspector employed or authorized by the enforcing authorities. (Test intervals are specified in the Elevator Code.)

Limit switches. The inspector should check the switches for proper alignment and the mounting of switches and cams for rigidity.

Buffers. Since oil buffers lose some oil during normal usage, the oil levels should be checked at least once every month and each time the buffer is known to have been compressed.

For refilling, an oil of the type specified by the manufacturer must be used. Buffers that have been submerged by floods or pit leakage should be emptied, cleaned, and refilled with fresh oil. The alignment and the tightness of bolts in the anchorage should be checked.

Spring buffers should be checked for alignment and for proper seating in the cups or mountings. Springs should be examined for deformation and permanent set.

Hoisting machines should be examined carefully at each inspection. The machine base should be checked for misalignment and cracks, and defects should be repaired immediately. The inspection should also include the following details:
1. The oil level in the motor (if used),
2. Brake operation,
3. The oil level in the gear housing (if provided), and
4. Sheaves and drums for cracks and wear in the grooves.

Belted machines. At frequent intervals, belts should be examined for proper tension, wear, burns, condition of splices, and cuts and breaks in the surface. Chains should be checked for excessive wear. Machine fastening bolts, belt guards, and the fastenings of platforms under any ceiling machinery also should be checked regularly.

Hoisting ropes. Records show that it is relatively unusual for an elevator accident to be caused by the parting of the hoisting ropes. When it does happen, though, it is more likely to occur with winding-drum machines than with traction machines. Periodic resocketing intervals for drum elevators are specified in the Elevator Code. Although ropes usually are installed to give the high safety factors specified in the Elevator Code, they should be inspected closely to avoid the possibility of accident. (Refer to *Inspectors' Manual,* ANSI 17.2, for wire rope inspection.)

Rope life is shortened by improper brake action. Unduly sudden stops may result from brake defects, such as heavy spring pressure or brakeshoe wear.

Unnecessary starting and stopping also shorten rope life. Inching for landings should be avoided. In some cases, it can be eliminated by proper adjustment of the stopping devices; in most cases, it is due to faulty operation and the operator should be reinstructed.

Inspection routine. Persons making elevator inspections should take all necessary precautions for their own safety. American National Standard A17.2, *Inspectors' Manual for Elevators and Escalators,* contains not only detailed instructions for the conduct of all tests, but also comprehensive information on personal safety for inspectors working in machine rooms, in and on top of cars, and in pits.

Close-fitting clothing, preferably one-piece overalls with all buttons fastened and without cuffs, is recommended. Gloves should not be worn, except when checking wire rope.

Inspectors must pay close attention to moving objects, such as counterweights, to hoistway projections, and to limited overhead and pit clearance. Before electrical parts are inspected, main line disconnect switches should be locked in the open position, so that current-carrying parts cannot be energized. When a worker is on top of a moving car, one hand should be kept free to hold onto the crosshead or another part of the car frame on top. Hoisting ropes should not be held for support. Safety belts should be attached to a fixed structure of the car or frame, not to the ropes.

If controls are not provided on top of the car for the inspector's use, specific instructions should be given to another person who operates the elevator car, while the inspector is inspecting the top of the car and other appurtenances.

The order in which the various parts are inspected depends upon the type of installation and the preference of the inspector. However, inspection time and interference with elevator operation can be reduced by planning—by determining which parts of the job can be done from each location.

A written report of each inspection should be prepared and kept on file. Such a record is particularly valuable for checking the progress of defects in ropes. Each report, therefore, should give definite details concerning rope condition—diameter, number of broken wires per unit length, and estimated percentage of wear. Most insurance carriers can supply the most recent elevator inspection reports.

A report of each service interruption should be made by the mechanic who corrects the trouble. The report should be entered on the log-sheet for the particular elevator, so that the maintenance engineer can spot defective equipment that causes repeated interruptions and correct the basic fault.

Operation

Training and placing operators. Many common causes of elevator accidents can be eliminated by safe operating practices. Best results are obtained where a properly instructed operator can be assigned to full-time duty. In any case, only specified employees who have been properly instructed should be permitted to act as elevator operators.

Elevator operators should be selected with extreme care because of the risks involved. In many cases, the lives of others depend upon the operator's efficiency. Moreover, an incompetent or poorly trained operator may damage valuable and indispensable equipment.

Faulty practices, such as starting a car before the doors and gates are closed, blocking gates open, and permitting crowding on cars not provided with gates, have caused many serious injuries. Actions, such as improper loading, unnecessary starting and stopping, and reversing of the controller, have caused damage to equipment.

Persons selected to operate elevators should be mentally alert, not easily excited, and capable of carrying out instructions and of insisting upon compliance with rules.

Some companies have found that elevator operators work more safely if they are given cards which state that they have completed the training course and are authorized to operate the equipment. In some cases, company rules require that operators have such cards.

Operating rules. Definite operating rules should be adopted. In the case of industrial freight elevators, the rules should be posted in the car, and persons serving as operators should be required to know and observe them. Many concerns that employ a number of elevator operators prepare pocket-size rule books for their guidance. In plants with a number of elevators that are used under variable conditions, specific rules for specific elevators or groups of elevators are prepared.

Accident investigations. Careful investigation of minor elevator accidents often will disclose conditions or practices that, if left uncorrected, may later cause a serious accident. Every accident, therefore, should be carefully investigated to determine the responsible conditions or practices, and prompt corrective action should be taken.

Overloads. Overloading elevators may result in injury to personnel, mechanical failure of the machine or the car, or both. Many machines, still in service, were built with much lower safety factors than are now required, and with them the hazard is particularly great. However, even though machines meet current safety factor requirements, the danger of overloads cannot be overemphasized, as the consideration of a few engineering factors will show.

In a traction elevator, the ratio between rope tension on the driving sheave (car side) and rope tension on the counterweight side must be kept within certain limits. The counterweight normally is equal to the weight of the car plus 40 percent of the rated capacity. The motor torque and the brake are designed to handle this difference in weight.

In the case of a freight traction elevator with a car weighing 8,000 lb (3,600 kg) and a rated capacity of 10,000 lb (4,500 kg) the counterweight would equal 12,000 lb (5,450 kg)—the weight of the car, or 8,000 lb, plus 40 percent of the rated capacity, or 4,000 lb (1,800 kg). The motor would be designed to lift, and the brake to hold, a 6,000-lb (2,700 kg) load.

If the elevator is overloaded by 50 percent of its rated capacity (for this example 5,000 lb or 2,250 kg), the total platform load will be 15,000 lb (6,800 kg). If 4,000 lb (overbalance of the counterweight) are subtracted from the 15,000 lb on the platform, a net weight of 11,000 lb (5,000 kg) must be handled by a motor and brake designed to handle only 6,000 lb. This is an 83 percent overload.

In such a case, the traction relation may be broken, and the motor may not even pick up the load. As the brake is lifted, the car may start to move downward, probably a fuse will blow, and other mechanical failures, as well as injury to personnel, may result.

In no case should an elevator be overloaded unless the manufacturer, familiar with the design of the particular equipment, has checked the entire installation for its ability to handle the load. Failure to do so may result in serious injury or death to employees and in serious damage to valuable equipment.

If an overload or a one-piece load (a transformer, for instance) as heavy as or heavier than that rated capacity of the car must be handled, the company that installed the elevator should be called in to see that:
1. The machine is strong enough to handle the load,
2. The elevator structure, including the car frame (sling), platform, and undercar safeties, is adequate, and
3. The traction relation will not be exceeded.

If the machine is otherwise strong enough for the overload, the elevator company's crew may increase the counterweight to maintain the traction relation while the overload is being lifted.

Emergency procedures

It is recommended that an emergency procedure be implemented, similar to the one shown in Figure 7-9 for the safe removal of persons from elevators stalled between floors. (ANSI/ASME A17.4 [*Emergency Evacuation of Passengers from Elevators*] is available from the American Society of Mechanical Engineers and from ANSI.) As is illustrated, emergency instructions for persons involved in such incidents are spelled out in two printed self-adhesive stickers. It is suggested that these stickers be mounted behind a protective transparent material to preserve legibility. In addition, when deemed appropriate, stickers in languages other than English should be posted.

In gaseous or toxic environments, tests should be taken to determine accident/injury potential and, if necessary, other emergency procedures should be implemented.

A telephone or other communication means is recommended for existing cars, but the Elevator Code requires such equipment on new installations. Having it may prevent occupant panic and may assist coordination of emergency procedures. Elevator emergency operation and signal devices are discussed in the Elevator Code.

Requirements for the handicapped

To help further increase the mobility of handicapped individuals, it is advisable to incorporate the requirements found in the National Elevator Industry, Inc. booklet, *Suggested Minimum Passenger Elevator Requirements for the Handicapped*, for all new installations, and, where practical, on existing installations.

ELEVATOR EMERGENCY INSTRUCTIONS

1. When the alarm sounds, the elevator has stopped between floors.

2. Notify the Fire Department or Rescue Team, at once, by calling

   _____
   (Telephone Number)

3. Notify the elevator maintenance company at once by calling

   _____
   (Telephone Number)

4. Do not attempt to rescue until trained, authorized maintenance personnel arrive. Assure the stuck passengers that help is on the way and that they are safe.

Figure 7-9. This sign should be posted in each elevator car.

SIDEWALK ELEVATORS

A sidewalk elevator is defined in the Elevator Code as an elevator of the freight type for carrying material, exclusive of automobiles, and operating between a landing in a sidewalk (or other area exterior to a building) and the floors below the sidewalk or grade level.

Because sidewalk elevators present hazards not easily eliminated, it is always preferable to locate them inside the building line or in an area not accessible to the public. Sidewalk elevators should conform to the requirements of the Elevator Code for sidewalk elevators.

Except by permission of the administrative authority, the maximum dimensions of sidewalk openings should be 5 ft (1.5 m) at right angles to and 7 ft (2.1 m) parallel with the building line, and the side of the opening nearest the building should be not more than 4 in. (10 cm) from the building wall.

Where hinged doors or vertically lifting covers are provided at the sidewalk or at other areas exterior to the building, bow-irons or stanchions shall be provided on the car to operate such doors or covers. A loud audible signal should sound when the car is ascending.

A sidewalk elevator with winding-drum machinery should have a normal terminal stopping device on the machine and one either in the hoistway or on the operating device.

Operation

The Elevator Code requires that a sidewalk elevator be raised and lowered only from the sidewalk or other exterior area. Either a key-operated, continuous-pressure up-and-down switch, or continuous-pressure up-and-down buttons on the free end of a detachable flexible cord, 5 ft (1.5 m) or less in length, should be used.

Hatch covers

Automatic hatch covers, when closed, should be capable of sus-taining not less than 300 lb (1,136 kg) applied on any area 2 ft (610 mm) on a side and not less than 150 lb (68 kg) applied at any point.

Hatch covers are required by the Elevator Code to be self-closing. Fastening or holding them open when the car is away from the top landing is forbidden by the Code.

The covers should be made of metal and should be vertically lifting or hinged with the line of the hinges at right angles to the building wall. When the covers are fully open, there should be minimum clearance of 18 in. (46 cm) between them and any obstruction.

Hatch covers should have a coefficient of friction not less than that of the surrounding walkway surface. No hinges, locks, or flanges should project above the closed covers to constitute tripping hazards to passersby.

HAND ELEVATORS

Hand-powered elevators once were used extensively, but today they are usually only found in storage rooms, warehouses, and private residences. Few companies currently manufacture hand elevators. In no case is it advisable to install a hand elevator where it is to be used constantly during the working day. Instead, some form of power equipment should be installed wherever an elevator is a basic part of the manufacturing process or service function.

Mechanical power never should be applied to hand elevators by means of rope-grip or similar attachments. If power is to be used, the entire installation should be changed, and all the protective devices called for by the Elevator Code for power elevators are required.

A hand elevator having a travel of more than 15 ft (4.5 m) is required by the Elevator Code to have a safety attached to the underside of the car frame which is capable of stopping and sustaining the car and its rated capacity.

The construction of hand elevators is specified in the Elevator Code, Part II. No hand elevator car upon which persons are permitted to ride should have more than one compartment or be arranged to counterbalance another car.

With regard to hoistways, hoistway openings, pits, machinery spaces, supports, and foundations, hand elevators should conform to the requirements of the Elevator Code for power elevators.

Hoistway doors

Every hoistway door should have conspicuously displayed on the landing side, in letters not less than 2 in. (5 cm) high: DANGER—ELEVATOR—KEEP CLOSED.

Hoistway openings may have self-closing doors that extend to the floor; doors made in two parts, one above the other, and so arranged that the lower part can be opened only after the upper part has been opened; or doors equipped with two spring locks or latches, one located 6 ft (1.8 m) above the floor.

Safety devices and brakes

Hand elevators should have hand or automatic brakes that operate in either direction of motion and, when applied, remain locked in the ON position until released.

Where the travel exceeds 40 ft (12 m), a driving machine with a hand-operated brake also should have an automatic speed retarder.

DUMBWAITERS

In the Elevator Code, Part VII, a dumbwaiter is defined as a hoisting and lowering mechanism equipped with a car that moves in guides and has a floor area not exceeding 9 sq ft (0.8 m²), a compartment height not exceeding 4 ft (1.2 m), a rated capacity not greater than 500 lb (225 kg), and which is used exclusively for carrying materials. It may be hand- or power-operated.

Hoistways and openings

The Elevator Code requirements for dumbwaiter hoistways are substantially the same as those for elevator hoistways, with certain exceptions. The requirements for dumbwaiter landing openings and doors are designed to protect persons from falling into the hoistways and from being struck by the car as it rises or descends.

The Elevator Code specifies that the hoistway landing openings of power-driven dumbwaiters must be provided with hoistway doors that guard the full height and width of the openings.

With certain specified exceptions, the Code requires that power dumbwaiter doors be equipped with hoistway-unit-system hoistway-door interlocks that will prevent machine operation if any hoistway door or gate is open.

Every hoistway door of hand dumbwaiters should have conspicuously displayed on the landing side, in letters not less than 2 in. (5.1 cm) high, the words, DANGER—DUMBWAITER—KEEP CLOSED.

Safety devices and brakes

The Elevator Code specifies that power dumbwaiters (except hydraulic dumbwaiters) have brakes that are automatically applied when the power is cut off or fails, and that they also have an automatic means to stop the car within the limit of overtravel at each terminal. Hand-driven machines should have hand-operated or automatic brakes capable of sustaining the weight of the car and its load.

A power dumbwaiter having winding-drum machinery, a travel greater than 30 ft (9 m), and a capacity in excess of 100 lb (45 kg) requires a slack-rope device that will cut off the power from the motor and stop the car, if it is obstructed in its descent.

ESCALATORS

An escalator is defined as a power-driven, inclined, continuous stairway for raising or lowering passengers. The Elevator Code, Part VIII, sets escalator standards.

Safety devices and brakes

Emergency STOP buttons (or other type of manually operated switches having red buttons or handles) shall be accessibly located in the right-hand newel base on new installations, at or near the top and bottom landings of each escalator and shall be protected against accidental operation. An escalator STOP button with an unlocked cover over it, which can readily be lifted or pushed aside, shall be considered accessible. The operation of either of these buttons or switches shall interrupt the power to the driving machine. It shall not be possible to start the driving machine by these buttons or switches.

Buttons or switches used for starting the units should be key operated and located within sight of the escalator steps. Means

should be provided to cut the power in case an ascending escalator accidentally reverses its travel.

Each escalator should have a speed governor that will interrupt the power if the speed exceeds a predetermined value (not more than 40 percent greater than the rated speed). The overspeed governor is not required where a low-slip, alternating current, squirrel cage induction motor is used and the motor is directly connected to the driving machine.

If a tread chain breaks, a broken-chain sensing device should cut the power. Where an escalator has tightening devices that are operated by tension weights, provision should be made to retain these weights in the escalator truss if they should fall.

Each escalator should have an electrically released and mechanically applied brake capable of stopping the fully loaded escalator when it is traveling either up or down. The brake should stop the escalator automatically if any safety devices function.

Machinery

Every escalator machine room should have a light that can be lit without having to pass over or reach over any part of the machinery. Reasonable access to the interior of the escalator should be provided for inspection and maintenance. For the protection of maintenance personnel, all chains in escalator machinery compartments should be guarded.

While bearings on modern installations are sealed and require no oiling, the chains do require lubrication. Care should be taken not to overlubricate. An oil pan should be provided in the bottom of the truss to catch oil or grease that may drip from moving parts as well as dust and dirt dropping between the treads.

The oil pan should be cleaned periodically to eliminate any fire hazard from the accumulated oil-soaked dust and dirt. A brush available for this purpose can be attached to a step axle and drawn down over the drip pan to brush all the foreign matter to the lower end of the truss. The sweepings then can readily be removed. Reversing the unit returns the brush to the top, where it can be disconnected.

The moving handrail, if of the common canvas duck and rubber construction, will show some stretch over a period of time. It should be checked at intervals, and the handrail drive adjustment should be used to take up the slack.

All parts of escalators and their driving machinery should be inspected at regular intervals and be well maintained. All safety devices should be tested for proper functioning.

Protection of riders

Most escalators are installed in public places, and their principal hazards arise from misuse by the public. Escalators have been developed to a high degree of safety, and their accident record, in general, is good.

Most escalator accidents occur when shoe heels, fingers, or toes are caught between the surface grooves or slots on the treads and the combplate. The Elevator Code limits the width of each slot to not more than ¼ in. (6.3 mm) and the depth to not less than ⅜ in. (9.5 mm), with a center-to-center spacing of not more than ⅜ in. between adjoining slots.

In some accidents, edges of shoe soles have been caught between the step and the vertical side member (skirt guard). The Elevator Code limits this clearance to 3/16 in. (4.8 mm) on each side.

The Elevator Code requires that each balustrade have a handrail moving in the same direction and at substantially the same speed as the steps. Each moving handrail is to extend at normal handrail height (not less than 12 in. [305 mm] beyond the points of the combplate teeth at the upper and lower landings). Hand or finger guards are to be provided at the point where the handrail enters the balustrade.

In some localities, barefoot passengers have been injured, and signs warning barefoot persons not to ride the escalator have been posted. Umbrella tips frequently are caught between the grooves and the combplate, a type of accident that occasionally results in minor injuries and in damage to equipment.

Many accidents have resulted from attempts to handle baggage on escalators. Suitcases and handbags should not be placed on the steps and, if carried by hand, always should be held parallel to the run of the escalator.

Escalator treads and landings should be of incombustible material affording secure foothold.

Some riders, through inexperience or infirmity, have trouble seeing the parting point where treads rise or descend and the point where they level off. To aid them especially, the Elevator Code calls for illumination on all tread surfaces. Additional warnings in green demarcation lights are required by the Elevator Code to be mounted inside the truss at top and bottom to shine through the treads where they break away and come together—the trouble points.

Some manufacturers mark the edges of the steps to emphasize the lines between adjacent steps. One manufacturer, for example, adds a distinctive color strip to the edge of the step adjacent to the riser of the next step.

Signs reading PLEASE HOLD HANDRAIL are often posted at the top and bottom and are found to be helpful. It is important that no overly conspicuous or distracting signs (such as advertising) be placed near these critical points (Figure 7-10).

At the various floor levels, directional signs and floor number markings are recommended in order to improve traffic flow from the escalators.

When traffic on escalators is extremely heavy, as in department stores during the holiday shopping season, proprietors often station uniformed employees at each flight to repeat warnings and, if necessary, help riders on and off.

It is extremely important that no object or construction of any kind obstruct the free flow of passengers from the area at the exit of an escalator. (This area is not a part of the escalator.) Serious accidents have occurred where the flow of traffic from the exit was restricted by a fence or barrier placed at some distance from the escalator.

Fire protection

Protection of escalator floor openings against the passage of fire and smoke is required by the applicable building code. One of the best safeguards against burnout is to divide buildings of fire-resistive construction into limited areas in which fire can be readily controlled. Vertical openings need to be protected against the passage of fire from story to story. This principle often is disregarded where escalators are installed.

Escalators accredited as a required means of egress must be fully enclosed in accordance with the requirements of local laws and ordinances.

Escalators not accredited as a required means of egress must have the floor openings protected by one of the following methods, in accordance with national and local standards and regulations:

Figure 7-10. Signs should be prominently displayed at the top and bottom of an escalator.

1. Full enclosures,
2. Kiosks,
3. Automatic rolling shutters,
4. Sprinklers so spaced as to protect the exposed sides of the opening, or
5. Spray nozzles (only where the building area is fully protected by a supervised automatic sprinkler system).

For detailed specifications, see *Life Safety Code*, NFPA 101.

MOVING WALKS

A moving passenger-carrying device, on which persons stand or walk and in which the passenger-carrying surface remains parallel to its direction of motion and is uninterrupted, is called a moving walk.

Criteria for the design, construction, installation, operation, inspection and testing of moving walks installed for the purpose of transporting passengers are set forth in the Elevator Code, Part IX.

Moving walks may operate in a horizontal plane or in a slope up to a maximum of 12 degrees except the slope shall not exceed 3 degrees within 3 feet of the entrance or exit. Operating speed and treadway width are governed by the slope.

Comments as to protection of passengers, found in the previous Escalators section of this chapter, apply also to moving walks.

MANLIFTS

The following are the principal hazards in the use of manlifts:
1. The rider may be carried over the top.
2. The rider may be unable to make an emergency stop.
3. The rider may jump on or off after the step has passed the floor.
4. His or her head or shoulders may strike the edges of floor openings if there is not a conical hood as required by OSHA.
5. The rider may be unable to reach the landing because of power failure and belt stoppage.
6. Parts of the manlift may fail or operate unsafely.

Construction

Manlifts should be constructed, maintained, and operated in strict compliance with the recommendations of ANSI/ASME, 90.1, *Belt Manlifts*. A safety factor of 6 (based on a 200-lb [90 kg] load, on each step, on both the up and down runs) should be used, and all equipment should be braced securely at top, bottom, and intermediate landings. The manlift rails must be secured to prevent their spreading apart, vibrating, and becoming misaligned. Steps should be nonslip. The entire manlift should be suspended from the top to prevent bending or buckling of the rails.

Handholds (either open or closed type) should be painted a conspicuous color, such as orange or yellow. They should be not less than 48 in. (1.2 m) nor more than 56 in. (1.4 m) above each step, and steps should not be less than 16 ft (5 m) apart.

Floor landings or emergency landings should be provided for each 25 ft (7.6 m) of manlift travel. Clearance between the surface of any landing and the lower edge of the conical guard suspended from the ceiling should be at least 7½ ft (2.3 m). The minimum clearance between the center of the head pulley and the roof or other obstruction should be 5 ft (1.5 m).

The bottom landing on the up side should have steps to a platform level with the manlift step as it rises to a horizontal position.

Floor openings and conical guard openings. At floor landings, standard 42-in. (1.06 m) guard rails and 4-in. (10 cm) toeboards should be provided around floor openings in such a way as to permit a landing space at least 2 ft (0.6 m) wide. Rails or guards should have maze or staggered openings or self-closing gates that open away from the manlift (Figure 7-11). At each floor opening on the up side, funnel-shaped (conical) guards must be installed (Figures 7-12 and -13).

Brakes, safety devices, and ladders

A brake, designed to be applied automatically when the power is shut off, should be installed on the motor shaft for direct-connected units and on the input shaft for belt-driven units. The brake should get its power or force from an external source. The brake must be capable of quickly stopping the manlift and holding it when the down side is loaded with 250 lb (110 kg) on each step. It should be electrically released.

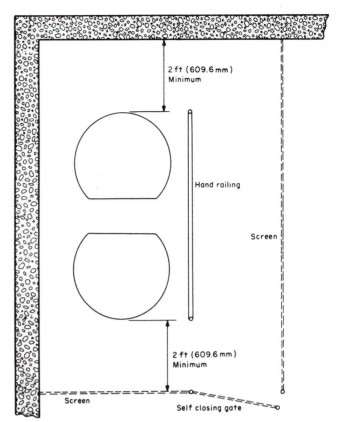

Figure 7-11. Screen enclosure for manlift floor openings (plan view). (Printed with permission from the American National Standards Institute.)

A control rope, not less than ⅜ in. (9.5 mm) of manila and/or cotton with a bronze wire center, should be provided within easy reach of both the up and down runs. When pulled in the direction of belt travel, the rope should cut off the power and apply the brake. An electro-mechanical device that will automatically shut off the motor power supply and apply an electric brake, if the rider fails to alight at the top landing, should be installed. Reset buttons, located at the top and bottom terminals will permit restarting the manlift after it has been shut off by the electrical safety devices.

A secondary safety control located on the top operating floor is required. It should be set to operate when the belt has traveled 6 in. (15 cm) beyond the point of operation of the primary safety switch, in case the latter fails. The device should stop the manlift before the loaded step reaches a point 24 in. (0.6 m) above the top landing.

A fixed metal ladder, accessible from both the up and down runs, should be provided for emergency exit where the vertical distance between landings exceeds 20 ft (6 m). It should meet the requirements of the governing agency for ladders, except that an enclosing cage should not be provided, since the ladder should be accessible from either side throughout its entire run.

Inspections

Every manlift should be tested and inspected at least every 30 days (Figure 7-14). Indicators of a defect are such things as unusual or excessive vibrations, continual misalignments, or

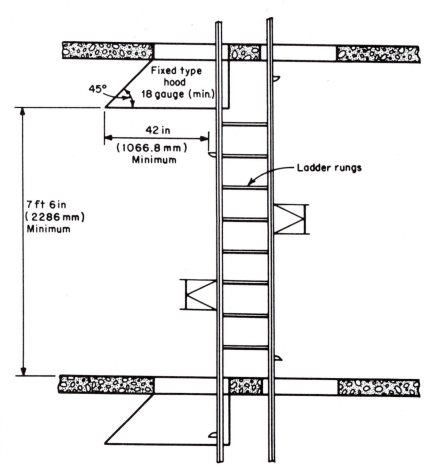

Figure 7-12. Drawing shows dimensions for a fixed flared-opening guard. (Printed with permission from the American National Standards Institute.)

"skips" when mounting steps (which indicates worn gears). On discovery of a defect, the manlift should be put out of operation *immediately* and not used until repaired. Each periodic inspection should cover, but not necessarily be limited to, the following items:

Steps	Lubrication
Step fastenings	Illumination
Rails	Warning signs and lights
Rail supports and fastenings	Signal equipment
Belt and belt tension	Drive pulley
Handholds and fastenings	Bottom (boot) pulley and clearance
Floor landings	Pulley supports
Guardrails	Motor
Limit switches	Driving mechanism
Electric switches	Gears
Belt splice joint	Brake
Step rollers	Key
Coupling	Keyway

The safety mechanism of a manlift should be inspected daily for free operation and for the accumulation of dirt and grease. A manlift should be completely dismantled once every year, and defective or excessively worn parts replaced.

A written record should be kept of findings at each inspection. Records of inspections should be available, in case a request to see such records is made by OSHA or state compliance officers.

General precautions

The maximum speed of a manlift belt should not exceed 80 fpm (0.4 m/s) and should be uniform on all manlifts throughout the plant. If the lift carries a great deal of traffic, a maximum speed of 60 fpm (0.3 m/s) is recommended.

Figure 7-13. In this installation the guard is counterbalanced and will yield slightly if hit. (Printed with permission from Humphrey Elevator Co.)

BELT MANLIFT INSPECTION REPORT
(Weekly & Monthly)

Location _____ Date _____

Manlift Make & Serial No. _____ Code _____

	ITEM	Weekly	Monthly		ITEM	Weekly	Monthly
1	OBSERVE MANLIFT IN OPERATION FOR POSSIBLE DEFECTS	( )	( )	19	TOP LANDING SAFETY SWITCHES	( )	( )
2	STEPS AND ROLLERS		( )	20	TOP BAR SAFETY	( )	( )
3	HAND HOLDS		( )	21	PHOTO EYE SAFETY (IF APPLICABLE)	( )	( )
4	BELT JOINT		( )	22	ON-OFF SWITCH & CONTROL ROPE (TEST UP & DOWN RUN)	( )	( )
5	BELT (LOOK FOR CUTS OR DAMAGE)	( )	( )	23	STEP CLEARANCE AT DRIVE BELT TRACKING AT DRIVE	( )	( )
6	BELT TENSION AND BELT TAKE-UP AT BOTTOM PULLEY	( )	( )	24	MOTOR		( )
7	STEP CLEARANCE AT BOOT	( )	( )	25	BRAKE		( )
8	BELT TRACK ON BOTTOM PULLEY	( )	( )	26	GEAR REDUCER & CHECK OIL LEVEL & CHANGE PER MANUAL		( )
9	BOTTOM BEARING LUBRICATION & SUPPORT		( )	27	COUPLINGS – COLLARS – KEYS		( )
10	GUIDE RAILS, PROPER ALIGNMENT FASTENINGS & SUPPORT		( )	28	HEAD SHAFT BEARINGS AND LUBRICATION		( )
11	FLOOR BRACES FOR GUIDE RAILS		( )	29	TOP PULLEY AND LAGGING	( )	( )
12	FLOOR HOODS AND OPENINGS		( )	30	OVERALL DRIVE ASS'Y AND SUPPORTS		( )
13	SAFETY SWITCHES ON MOVEABLE HOODS	( )	( )	31	SKIPPING WHEN MOUNTING STEP (CHECK DRIVE TRAIN)		( )
14	GUARD RAILS AND GATE OPERATION	( )	( )	32	VIBRATION OR MISALIGNMENT IN DRIVE		( )
15	FLOOR LANDINGS (CLEAR OF OBSTRUCTION)	( )	( )	33	GREASE BEARINGS PER MAINTENANCE MANUAL		( )
16	ILLUMINATION OF MANLIFTS AND FLOOR LANDINGS	( )	( )	34	GUIDE TRACK FREE OF FOREIGN MATERIAL AND LUBRICANTS	( )	
17	TOP & BOTTOM FLOOR WARNING SIGNS & LIGHTS	( )	( )	35			
18	BOTTOM SAFETY SWITCHES TREADLE &/ OR DROPOUTS	( )	( )	36			

IS MANLIFT BEING PROPERLY USED? _____ BY AUTHORIZED PERSONNEL? _____

ITEM NUMBERS ABOVE THAT WERE CORRECTED OR ARE IN NEED OF SUCH (GIVE COMPLETE DESCRIPTION ON BACK SIDE OF THIS FORM)

INSPECTED BY _____ DATE _____

Figure 7-14. Typical checklist for inspecting a manlift. (Printed with permission from the American National Standards Institute.)

Floors should be numbered with large figures in full view of both ascending and descending riders. A constantly illuminated sign, Top Floor—Get Off, in block letters at least 2 in. (5 cm) high, should be conspicuously placed not more than 2 ft (0.6 m) above the top landing, in full view of an ascending passenger. In addition, at least a 40-watt red warning light should be located immediately below the top landing to shine in an ascending rider's face.

Signs carrying instructions for use of the manlift should be displayed prominently at each landing. Only authorized employees should be permitted to ride manlifts, and signs so stating should be displayed at each landing.

The entire manlift should be illuminated at all times while it is in operation, with at least 1 foot-candle (11 lux) at all points and at least 5 foot-candles (54 lux) at landings.

Nothing which cannot be placed entirely inside a pocket, a sling, or a pouch should be carried by a rider on a manlift.

Employees, particularly new ones, should be carefully instructed in safe use of the manlift and should be instructed to report immediately any defects or irregularity in operation of the manlift or the safety devices. Corrective action should be initiated immediately by supervisors.

REFERENCES

American National Standards Institute, 1430 Broadway, New York, N.Y. 10018.
Elevators and Escalators (Inspectors' Manual), ANSI/ASME A17.2.
Elevators, Escalators, and Moving Walks, ANSI/ASME A17.1.
Belt Manlifts, ANSI/ASME A90.1.
American Society of Mechanical Engineers, 345 East 47th St., New York, N.Y. 10017.
ASME Boiler and Pressure Vessel Code, Section III, "Boilers of Locomotives," and Section VIII, "Unfired Pressure Vessels."
Association of American Railroads, 59 East Van Buren Street, Chicago, Ill. 60605.
Engineering Division (American Railway Engineering Association) *Manual of Recommended Practice*.
Federal Railroad Administration, U.S. Dept. of Transportation, Washington, D.C. 20590.
Laws, Rules and Instructions for Inspection and Testing in Steam Locomotives and Tenders and Their Appurtenances.
National Bureau of Standards, U.S. Department of Commerce, Washington, D.C. 20234.
National Electrical Safety Code, NBS Handbook H30. (Also adopted as American National Standard C2.)
National Fire Protection Association, 470 Atlantic Ave., Boston, Mass. 02210.
National Electrical Code, NFPA 70. (Also adopted as American National Standard C1.)
"National Fire Codes," Vol. 4, *Building Construction and Facilities*.
National Safety Council, 444 North Michigan Ave., Chicago, Ill. 60611.
Industrial Data Sheets
Escalators, 516.
Floor Mats and Runners, 595.
Manlifts, 401.

Principles of Guarding

FOUR MAJOR AREAS OF SAFETY must be considered with every machine:

- Maintenance work during which some part of the body may be in the path of a machine movement.
- Servicing, adjustment, maintenance, and repair, which must be done dynamically—perhaps with guards temporarily removed and other safety devices bypassed. For instance, maintenance personnel with adequate training and experience may need to search for a malfunction.
- Protection of the operator at the point of operation.
- Protection from moving machine parts, other than at the point of operation.

The first condition is perhaps the most hazardous, and will be discussed first.

GUARDING DURING MAINTENANCE

During maintenance, it is vital that the machine be put into a state in which the possibility of its making an unexpected and injury causing movement is reduced to a practical minimum. The procedure used for this purpose is called *lock-off, lockout, power lockout,* or *power lock-off.*

The difficulty with the locking-off concept is that its meaning varies greatly in each individual's mind. For instance, some people include *all* of the logical precautions that should be taken to shut down a machine for maintenance. Others take it to mean merely locking off the electrical disconnect switch.

Between these two extremes there are many interpretations. The result is that a machine locked off by one employee may not be in as safe a condition to work on as one that has been locked off by another. Yet the same term is used by both. (Note—Electrical lockout procedures are discussed in detail in Chapter 15, Electricity.)

An inadequately locked out machine might be capable of injuring or killing someone who is working in, on, or around it. It takes far more than merely locking out electrical energy to achieve maximum protection for machine operators and maintenance and service people. The locking-out procedure should include all energy sources:

- Electrical power
- Hydraulic fluids under pressure
- Compressed air
- Energy stored in springs
- Potential energy from suspended parts
- Any other sources that might cause unexpected mechanical movement.

All energy sources must be neutralized before any maintenance or setup work can be safely done.

Zero mechanical state (ZMS)

After all energy sources have been neutralized, the machine is in the "zero mechanical state" (ZMS). This state provides maximum protection against unexpected mechanical movement.

The ZMS concept includes not only locking out electrical energy, but also requires that all kinetic and potential energy be isolated, blocked, supported, retained, or controlled so the energy will *not* be unexpectedly released. The ANSI standard Z241.1-1981, sponsored by the American Foundrymen's Society, defines and explains ZMS (see Figure 8-1).

ANSI Z241.1 includes zero mechanical state (ZMS) and other safety concepts that are mandatory. Although this standard has

---

**Definition of Zero Mechanical State (ZMS) According to
American National Standard Z241.1-1981**

Standard Definition	Commentary/Explanation
2.69 Zero Mechanical State (ZMS). That mechanical state of a machine in which:	**E.2.69 Zero Mechanical State (ZMS)**—Over the years, changes in equipment requirements and design have incorporated the use of pressurized air, hydraulics, and electricity, as well as combinations of these media to perform certain functions on the equipment, with the result that the commonly used terminology of "Locked Out" or "Locked Off" does not describe as safe a condition as was originally intended. This committee proposes a new term, **Zero Mechanical State (ZMS),** which represents maximal protection against *unexpected mechanical movement of the machine* when setups are made or maintenance performed on the equipment by maintenance or authorized operator personnel. The ZMS concept does not apply only to foundry equipment, but has significant applications to all types of equipment used in industry. Special attention should be given to the definition and use of the ZMS concept in this standard.
1) Every power source that can produce a machine member movement has been locked off;	
2) Pressurized fluid (air, oil, or other) power lock-offs (shut-off valves), if used, will block pressure from the power source and will reduce pressure on the machine side port of that valve by venting to atmosphere or draining to tank;	This may be accomplished by more than one valve for each power source, provided that each valve can be locked off. This requirement is met by a three-way valve (or equivalent), properly connected. It will also prevent leakage, because of valve malfunction, from reaching the machine.
3) All accumulators and air surge tanks are reduced to atmospheric pressure or are treated as power sources to be locked off, as stated in paragraphs 1) and 2);	See commentary column opposite (5).
4) The mechanical potential energy of all portions of the machine is at its lowest practical value; so that opening of pipe(s), tubing, hose(s), or actuation of any valve(s) will not produce a movement that could cause injury;	Holding a machine member against gravity or a spring force by blocking, by suspension, or by brackets or pins designed specifically for that purpose is permissible to reach **Zero Mechanical State.**
5) Pressurized fluid (air, oil, or other) trapped in the machine lines, cylinders, or other components is not capable of producing a machine motion upon actuation of any valve(s); 6) The kinetic energy of the machine members is at its lowest practical value; 7) Loose or freely movable machine members are secured against accidental movement; 8) A workpiece or material supported, retained, or controlled by the machine shall be considered as part of the machine if the workpiece or material can move or can cause machine movement.	A machine in **Zero Mechanical State** may still have fluid (air, oil, or other) under pressure trapped in its piping. (In instructions for ZMS, proper consideration should be given to test to verify that the desired set of conditions has been reached. For instance, when the motor power disconnect has been locked out [off], this can be verified by pressing the **start** pushbutton to make sure the motor *does not* start. Part of the ZMS procedure should be test actuation of the **start** initiator after lock out.)

Figure 8-1. Definition of the zero mechanical state (ZMS). (Printed with permission from American Foundrymen's Society, Inc.)

been written in terms used in foundries, it has unique considerations that can be applied to reduce mechanical equipment accidents in many industries.

The significant terms with their definitions from the standard are stated, and then followed by applicable explanations from the body of the standard.

ZMS instructions are sometimes a required item of supply from the manufacturer of new machines. Machine users can write the ZMS instructions for existing machines, or they might purchase this service from the manufacturer of specific equipment.

Definitions

Initiator. A device that causes an action of controls or power actuator is defined as an *initiator.* Typical manual initiators are push-button, foot switches, manual starters, hand valves, and other valves with manual overrides. Typical nonmanual initiators are limit switches, pressure switches, temperature-actuated switches, flow switches, and cam-actuated valves.

Additional initiators. In an operation requiring more than one operator, separate hostage controls shall be provided for each operator.

Hostage control is a type of control in which the physical act of operating the initiator prevents operator exposure to the motion or response produced by the initiator. An example is to have an initiator located far enough away from the hazardous zone so the operator cannot return to the point of operation during the hazardous portion of the cycle.

Power off. The state in which power cannot flow to the machine is considered a power-off stage. Electrically—the disconnect switch is opened to turn power off. Pneumatically—an air valve is closed to turn power off. The use of a common term OFF is preferable to use of OPEN for electrical, and CLOSED for pneumatic lines. Where confusion might arise, both terms may be used—for example, switch locked *off* (open), valve locked *off* (closed).

Power locked off is the state in which the device that turns power off is locked in the off position with the padlock of every individual who is working on the machine.

Fluid power off. Means shall be provided for isolating fluid (air, oil, or other) energy sources from a machine, or group of machines, controlled as a system. These means shall have provisions for being locked in the isolating mode. Pressure build-up on the machine side port of the isolating means shall be eliminated, such as venting to atmosphere or draining to tank. This condition does not imply or require that the machine be in ZMS.

Primary protection. This is a protective means that cannot be deactivated by malfunction of a device or intentional bypassing, or the prevention of machine movement by direct control of the power source or by direct prevention through mechanical stops. Locking off the power to an electric drive motor using the main disconnect switch is an example of primary protection. Putting a machine in ZMS is an example of primary protection for a fluid-powered machine.

Secondary protection is a device or structure that either can be bypassed by foreseeable means or might malfunction and defeat the planned protection. All ordinary control devices are of this nature, because they do not directly prevent application of power, but do so only by controlling a device that does prevent power application. Limit switches are an important example of secondary protection. Where limit switches are used to prevent the operation of a motor—or through valves, operation of a cylinder—they might malfunction, might be frustrated by a ground or short circuit, or the component they control might be directly operated, thus defeating their purpose. The fact that operator-protective devices are secondary protection should be understood by the operator whose constant attitude should be cautionary. Protective devices should frequently be inspected. By the very nature of the device, a malfunction can go unnoticed until it is needed.

Responsibility

This section is written like a specification or instructions, thus the repeated use of the word "shall."

Manufacturer. It shall be the responsibility of the manufacturer to furnish operation and maintenance instructions with equipment covered by the standard. These shall include instruction for ZMS. Specific operating and maintenance instructions should be outlined in the operating and maintenance manuals. These instructions should be written to aid the operators and maintenance personnel in the safe operation and maintenance of the equipment. Adequate instructions may consist of written, illustrated, and audio recorded material. The procedure and sequence of lock-off provided by the manufacturer to attain ZMS shall be followed in its entirety.

Employer. It shall be the responsibility of the employer to specify maintenance procedures for the equipment covered by the standard. These procedures shall minimize hazards to operating and maintenance personnel. The employer should schedule frequent or periodic inspections that would reveal hazards caused by age, overloading, corrosion, fatigue, improper use, or improper installation. If the equipment cannot be immediately fixed, it should be removed from service.

It also shall be the responsibility of the employer to provide competent personnel for maintaining equipment covered by the standard. Competent maintenance personnel should be adequately trained and have the technical background to understand the information contained in maintenance manuals for the particular machines they are inspecting or maintaining.

The employer shall provide a means of monitoring the activities of employees engaged in troubleshooting, maintenance, or repair in isolated or hidden areas.

The employee shall use the monitoring means established by the employer to inform others of his or her presence when performing maintenance or setup work in hidden or isolated areas.

The employer should be particularly careful in the purchase of used equipment. In addition to possible structural and mechanical faults, the controls are not always in the original condition. Interlocks may have been removed and machine sequence may have been changed.

Employee.
- *Physical entry into machine or equipment.*—The employee shall put the machine or equipment in ZMS before placing any part of the body in the path of any movable machine or equipment member.
- *Troubleshooting with power on.*—When it is necessary to locate and define problems with the power on, the employee has authority to work on machines or equipment with guards removed, or to work within areas protected by barriers, if such action will not place any part of the body in the path of any movable machine or equipment element. A machine or piece of equipment may have to be stopped, locked out, or put in ZMS before removing a guard or barrier so the machine or equipment may subsequently be observed with power on.
- *Defeating protective devices.*—During necessary maintenance, the employee shall not remove, bypass, or alter any device that was provided to reduce hazardous conditions, other than temporarily. Before the equipment is returned to production, the employee shall replace all devices that may have been removed during maintenance.
- *Returning equipment to production.*—The employee shall verify that the machine or equipment is in adequate operating condition with all guards secured before releasing it for production use.
- *Use of special tools.*—The employee shall use those special tools, stop blocks, or slings provided by the employer for performing certain maintenance functions on equipment in the least hazardous manner. The desire to return equipment to productive operation should not prompt any personnel to use, encourage, or condone the use of hazardous methods or improper tools.

The ZMS instructions must be used in an intensive training program to teach employees that it can be very dangerous to enter a machine unless every requirement of ZMS is satisfied. Persons must not place any part of the body in the path of movable machine members during setup, adjustment, lubrication, maintenance, or installation. Operators must be trained to respect the possible danger from machine motions. The operator must be guarded by one or more of the devices or barriers listed under the Guarding sections of this chaper, and must be trained to detect possible defects or malfunctions of these guards. The operator must report malfunctions or defects at once.

WARNING	CAUTION
DO NOT CLIMB ON, ENTER, OR REACH INTO THIS MACHINE IF IT IS NOT IN ZERO MECHANICAL STATE (SEE WRITTEN INSTRUCTIONS)	THIS MACHINE MUST BE IN ZERO MECHANICAL STATE BEFORE ADJUSTMENT OR MAINTENANCE YELLOW TAGS INDICATE LINES WHICH MAY RETAIN PRESSURE AFTER POWER HAS BEEN LOCKED OFF.

Figure 8-2. An easy-to-read information plate (*left*) should be attached to all machines. This plate serves as a warning to visitors, service and maintenance personnel, supervisors, and all others passing near the machine. Another plate (*right*) is directed to service and maintenance personnel. A special set of tags is attached to pressure lines serving the machine.

Startup and shutdown procedures. It shall be the responsibility of the manufacturer to recommend a general startup procedure that minimizes hazards. The employer shall establish and follow a specific startup procedure, incorporating the manufacturer's recommendations, before any equipment covered by the standard is placed in regular operation.

It also is the responsibility of the manufacturer to recommend a shutdown procedure that minimizes hazards. The employer shall establish and follow a shutdown procedure before allowing any inspection, adjustments, or maintenance covered by the standard. It shall be the responsibility of the employee to follow the shutdown procedure established by the employer.

Written policy for ZMS

Maintenance work rarely follows a set pattern because jobs and companies vary. In a large plant, many crews perform similar maintenance and service tasks. Each crew works under a supervisor having individual ideas about safe practice.

It is essential that management standardizes maintenance procedures by developing a program including a written policy, specific procedures, and rules. The program should also cover supervisory follow-up policy covering positive locking out of switches and all other steps that are necessary in achieving ZMS for the safe performance of maintenance, setup, or other work on machinery and equipment. Specific requirements, such as those listed in this chapter, should be precisely and completely stated and incorporated into the standard operating procedures.

All too often the specific operation, maintenance, and ZMS instructions for a machine are left in a purchasing department or receiving department file and are not used to train and instruct the operators and maintenance personnel. These instructions must be made available to pertinent personnel.

The use of warning or advisory plates (Figure 8-2) is strongly recommended. They warn against hazards, and if the manufacturer's information has not been properly distrubted to personnel, they will initiate questions about the availability of instructions.

How to write a ZMS instruction. The manufacturer of new equipment or the employer with existing equipment will need the following to write ZMS instructions:

Pre-ZMS. First, there must be accurate and up-to-date electrical control schematic drawings, and hydraulic and pneumatic flow diagrams for the equipment. Second, a definition of the position each movable machine member must be in is

needed for each maintenance task. If a part is to be in an elevated attitude, the manner of supporting or restraining must be specified and the weight of each member that might require attention must be noted. Pressure retained by detented (closed) valves should be watched.

Third, the method and sequence of locking off each power source and reducing all pressure that could create or cause a machine movement is then detailed. Material in hoppers and workpieces in the machine must be considered along with restraining of any free-moving members.

Finally, after all phases of ZMS have been performed, each initiator and valve "manual override" must be tested to verify that all power sources have actually been disconnected and that all remaining pressure has been reduced to prevent unexpected motion of machine members. Particular attention must be given to detented valves and closed center valves.

Post-ZMS. After the maintenance task has been performed, remove all tools, restraining members, tooling, loose parts, ladders, and other maintenance equipment. Replace all guards. Test operator protection interlocks and devices. Be certain that all personnel are accounted for and out of the path of moving machine members. Restore power and test the machine before releasing the machine to production operation.

Other hazards. In addition to the ZMS and specific maintenance instructions, consider other hazards inherent in the equipment. Employees must be cautioned of the injury and pain that can result from misuse or negligent operation of equipment.

Secondary devices in a control circuit must be treated with caution. A limit switch may be used as in interlock to sequence a valve controlling cylinder motion. A limit switch is vulnerable. It can have an internal short; the wires might be pinched in the enclosure or in a conduit or the operating lever may be bent to prevent normal start up. In some cases the limit switch has been deliberately wedged in an actuated position or its terminals jumpered in the control cabinet.

Service personnel who must place themselves in a cylinder's path of motion should lock off primary power and not depend on the secondary control component.

The main source of each type of energy connected to a machine is considered the primary power, as explained earlier. If these power sources are locked off and the machine is vented or grounded, the chances of unexpected motion are reduced. However, potential or kinetic energy of mechanical parts must still be considered. Stored energy of the machine is still a possibility. Thus, locking off the primary power is desirable, but

GENERAL MACHINE SAFETY INSTRUCTIONS

General Instructions

1. Machine installation must be approved by employer before allowing employees to work on equipment.

2. Only trained, authorized, and supervised personnel, designated by the employer, may perform operational adjustment and maintenance functions on equipment.

3. Instructions and drawings provided by the equipment manufacturer must be used when performing adjustments and maintenance functions.

4. All tools and equipment (hoists, slings, chains, blocking, etc.) used in maintenance must be approved by the employer.

Adjustment and Troubleshooting (Equipment *not* in ZMS)

1. Work in this category includes only those tasks which by their nature must be accomplished while the equipment is operating or has *not* been placed in ZMS (Zero Mechanical State). *This condition requires special alertness and training of all employees and must be recognized as extremely hazardous.*

2. During debugging, adjustment, check-out and certain troubleshooting, it may be necessary to remove guards to observe machine functions while the equipment is *not* in ZMS. *This is extremely hazardous and requires special alertness.* Guards must be replaced as soon as the task which required their removal is complete.

3. In restoring equipment to its normal state following a malfunction or jam-up, it is frequently necessary to operate certain powered machine functions by use of manual control or manual valve override. *This must be recognized as an extremely hazardous procedure,* and all persons must know for sure that all persons are clear of the path of possible moving machine members.

Maintenance (Equipment in ZMS)

1. Work in this category includes:

    a. All tasks that require any part of the body of the maintenance employee to be placed in the path of possible moving machine members.

    b. All adjustment and troubleshooting tasks which can be accomplished with the equipment in ZMS (Zero Mechanical State).

    c. When locking off air or hydraulic power sources of the machine being worked on and the adjacent machines, this may cause other non-adjacent machines to become either partially or completely free of power sources. This does not, however, place that machine in ZMS.

    Repair or maintenance activity on any piece of equipment must be started only after the complete ZMS Instructions for the machine to be worked on have been satisfied.

2. Specific ZMS Instructions for the machine unit being worked on must be followed closely before placing any part of the body in the path of possible moving machine members.

3. Specific ZMS Instructions for the machine being worked on may also include instructions for adjacent machines to lock off power sources and reduce the mechanical potential energy of the adjacent components.

Figure 8-3. This ZMS policy statement can be used to preface company AMS instructions.

PROCEDURE FOR PLACING MACHINE 'A' IN ZERO MECHANICAL STATE

Sequence of Operations	Description of Operations	Key Safety Factors
1. Shut off main electrical power to machine	Turn disconnect switch located at column P-6 to the OFF (open) position	Deenergizes all electrical components on the machine
2. Lock-out disconnect switch in step no. 1	Insert lockout device through the switch and panel hasp holes and lock securely in place with your own personal padlock	Prevents the possibility of electrical shock and accidental startup of the machine to employees in the process of service or repair.

Figure 8-4. This ZMS procedure guide uses an easy-to-understand three-column format. (Printed with permission from John Deere Company.)

---

may not be adequate. The requirements of ZMS should be followed.

Examples of policies and instructions. Figure 8-3 is an example of a ZMS policy statement, which, if made more specific, could be used as a preface to a company's ZMS instructions.

Specific instructions. Once the proper procedure has been determined by analyzing the specific equipment, checking manufacturer's manuals, and reviewing recommended maintenance procedures, then the procedure must be completely and clearly written. Not only must the writing be clear, but the format should be an easy-to-follow sequence.

The procedure must be thoroughly tested at the work site to verify and evaluate its effectiveness.

Those who will use the procedure must be trained to make certain the procedure is clearly understood before implementation.

Once the procedure is put into practice, supervisors must follow up to make certain the employees are correctly using it.

Here is an example of a statement that could be common to a ZMS procedure: "Turn the main electric power disconnect switch located at Column P-6 to the OFF (open) position and lock it out." All terms in this statement should be familiar to employees using the procedure; the statement clearly states the location of the disconnect. However, the statement is incomplete because the method of locking out is missing. Some new employee might think that locking out means inserting a short piece of wire in the hasp holes and twisting it together; another might believe that simply pulling the fuse would be adequate.

Figure 8-4 illustrates a format used to detail the correct procedure. It is based on the familiar three-column format used for a job safety analysis (JSA).

- The first column is a sequential listing of all operations that must be performed by the employee.

- The second column, Description of Operations, explains how to do it.
- The third column explains why the procedure is important by identifying the potential hazards associated with (1) each operation, (2) the performance of that operation, or (3) the omission of that operation.

This ZMS procedure format is ideal for both classroom and on-the-job training of new employees, and retraining of experienced ones. The format also helps the supervisor audit employees to make sure that the procedure is correctly being followed. If visual aids would help make the procedure easier to understand, then by all means use them.

GUARDING OF HAZARDS

One of the major goals of OSHAct is to guard all machinery and equipment to eliminate hazards created by points of operation, ingoing nip points, rotating parts, and flying chips and sparks. These hazards have been responsible for countless injuries and fatalities.

In the past, the excuses for not guarding were almost as many as there were accidents. The most common reason, and one still heard today, was that the guard either interfered with or made it impossible to achieve production goals. Thus points of operation and power transmission were either never guarded, or the guards were removed and never replaced. In some cases, removal of guards was justified if they were poorly designed and created additional hazards to personnel. But in most instances guarding was not installed, or was removed and not replaced, simply because it was thought unnecessary or inconvenient to work around it during maintenance or adjustment.

Today, most of these reasons and the attitudes that went along with them, are not acceptable excuses to OSHA or anyone else. The regulations clearly state that points of operation and power transmission *shall* be guarded. Exceptions to this can be obtained only by applying for a variance from the regulations. Variances may be granted if it can be proven that the current technology is not capable of guarding the hazard without creating an additional hazard or without seriously degrading the production process. Even then, those applying for a variance must develop alternate safeguarding methods, which will be equivalent or better than the guard. Details are in Chapter 2 of the *Administration and Programs* volume, Governmental Regulation and Compliance.

It should be obvious that the words "shall be guarded" apply to most, if not all, machines and equipment that a company uses. How they apply may not be very clear. Some machines require specific methods of guarding and all machines are regulated by the general requirements.

Guarding terminology

As in many problems, a lack of agreement on definitions of terms causes much confusion and even disagreement. To avoid this, definitions of common terms follow. (Additional terms are defined in Chapter 11, Cold Forming of Metals.)

GUARDING. Any means of effectively preventing personnel from coming in contact with the moving parts of machinery or equipment which could cause physical harm to the personnel. When discussing power press guarding, the word "guard" is used exclusively for referring to barriers designed for safeguarding at the point of operation. Therefore, the word "enclosure" or "safeguard" should be used for a barrier or cover that protects other danger zones.

ENCLOSURES. Guarding by fixed physical barriers that are mounted on or around a machine to prevent access to the moving parts. They are most effective when designed as part of the machine, but they can be bolted or welded to the frame.

FENCING. Guarding by means of a locked fence or rail enclosure which restricts access to the machine except by authorized personnel. Enclosures must be a minimum of 42 in. (1.0 m) away from the dangerous part of the machine.

Both fencing and location are very limited as guarding techniques and are permitted only if restrictions can be met.

LOCATION. Guarding that is the result of the physical inaccessibility of a particular hazard under normal operating conditions or use.

POINT OF OPERATION. That area on a machine where material is positioned for processing or change by the machine, and where work is actually being performed on the material.

POWER TRANSMISSION. All mechanical components including gears, cams, shafts, pulleys, belts, clutches, brakes, and rods that transmit energy and motion from the source of power to the point of operation.

INGOING NIP POINTS OR BITES. A hazard area created by two or more mechanical components rotating in opposite directions in the same plane and in close conjunction or interaction.

SHEAR POINTS. A hazard area created by a reciprocal (sliding) movement of a mechanical component past a stationary point on a machine.

Built-in guards

Most machine manufacturers make power transmission guarding standard equipment on their machines. Most also will supply point-of-operation guarding if they are informed of the end use and other factors.

The enactment of the OSHAct has resulted in greater incentive for the employer to operate effectively guarded equipment. That incentive and the rise in the number of litigations alleging unsafe design account for the increased awareness about machine guarding.

Long before the OSHAct, however, conscientious employers operated, and reliable manufacturers produced, machines that met or exceeded current safety standards for guarding. Regardless of the reasons for guarding, it is generally true that manufacturer designed and installed guards can offer advantages to the user:

1. Guards designed and installed by the manufacturer are usually less expensive than those purchased and installed by the user.
2. Manufacturer built guards usually conform better to the design and operation of the machine.
3. Manufacturer built guards often serve more than one purpose.

The disadvantages would be that unanticipated circumstances of the process, or slight modifications in the process, after machine installation would render the guarding less effective.

Although new machines are increasingly equipped with manufacturer installed guarding, these machines are a small percentage of the machines currently in use. Many older machines are in violation of the OSHA regulation despite the fact that OSHA has existed for well over 15 years. It is safe to presume that many older machines are being operated with little or no guarding at hazardous points of operation and at the power transmission.

Checklists for guarding

Some conditions invite two consequences to owners of machines—accidents and penalties for OSHA violations. Most managements would like to avoid these consequences and they can by developing and following up an OSHA compliance checklist for all their machines and equipment.

Basically the checklist will individually list all the company's machines and equipment. For each unit, it will then list the points of operation and power transmission components. It should indicate if they are guarded or unguarded. If they are guarded, the kind of guarding should be noted—enclosure, fencing, or location, and whether the guard meets OSHA or other requirements. In some machines these facts may be apparent, but most machines are complex and require (1) a detailed analysis of the design, (2) the interaction of all components, and (3) the methods of servicing, including loading and unloading of materials as well as maintenance and repair. The total operation or process should be looked at as a system of interrelated parts and actions. A hazard point may appear well guarded when the machine is at rest, but when that machine is operating, portions of it may be exposed or new hazards created by one component interacting with or bypassing another component.

A poorly designed or maintained guard or safety device can interact with moving parts and produce a hazard point. Newer

Table 8-A. Point-of-Operation Protection

Type of Safeguarding Method	Action of Safeguard	Advantages	Limitations	Typical Machines on Which Used
ENCLOSURES OR BARRIER GUARDS				
Complete, simple fixed enclosure	Barrier or enclosure which admits the stock but which will not admit hands into danger zone because of feed opening size, remote location, or unusual shape.	Provides complete enclosure if kept in place. Both hands free. Generally permits increased production. Easy to install. Ideal for blanking on power presses. Can be used with automatic or semiautomatic feeds.	Limited to specific operations. May require special tools to remove jammed stock. May interfere with visibility.	Bread slicers Embossing presses Meat grinders Metal squaring shears Nip points of inrunning rubber, paper, textile rolls Paper corner cutters Power presses
Warning enclosures (usually adjustable to stock being fed)	Barrier or enclosure admits the operator's hand but warns him before danger zone is reached.	Makes "hard to guard" machines safer. Generally does not interfere with production. Easy to install. Admits varying sizes of stock.	Hands may enter danger zone—protection not complete at all times. Danger of operator not using guard. Often requires frequent adjustment and careful maintenance.	Band saws Circular saws Cloth cutters Dough brakes Ice crushers Jointers Leather strippers Rock crushers Wood shapers
Barrier with electric contact or mechanical stop activating mechanical or electric brake	Barrier quickly stops machine or prevents application of injurious pressure when any part of operator's body contacts it or approaches danger zone.	Makes "hard to guard" machines safer. Does not interfere with production.	Requires careful adjustment and maintenance. Possibility of minor injury before guard operates. Operator can make guard inoperative.	Calenders Dough brakes Flat roll ironers Paper box corner stayers Paper box enders Power presses Rubber mills
Enclosure with electrical or mechanical interlock	Enclosure or barrier shuts off or disengages power and prevents starting of machine when guard is open; prevents opening of the guard while machine is under power or coasting. (Interlocks should not prevent manual operation or "inching" by remote control.)	Does not interfere with production. Hands are free; operation of guard is automatic. Provides complete and positive enclosure.	Requires careful adjustment and maintenance. Operator may be able to make guard inoperative. Does not protect in event of mechanical repeat.	Dough brakes and mixers Foundry tumblers Laundry extractors, driers, and tumblers Power presses Tanning drums Textile pickers, cards
AUTOMATIC OR SEMIAUTOMATIC FEED				
Nonmanual or partly manual loading of feed mechanism, with point of operation enclosed	Stock fed by chutes, hoppers, conveyors, movable dies, dial feed, rolls, etc. Enclosure will not admit any part of body.	Generally increases production Operator cannot place hands in danger zone.	Excessive installation cost for short run. Requires skilled maintenance. Not adaptable to variations in stock.	Baking and candy machines Circular saws Power presses Textile pickers Wood planers Wood shapers
HAND REMOVAL/RESTRAINT DEVICES				
Hand restraints (hold-back)	A fixed bar and cord or strap with hand attachments which, when worn and adjusted, do not permit an operator to reach into the point of operation.	Operator cannot place hands in danger zone. Permits maximum hand feeding; can be used on higher-speed machines. No obstruction to feeding a variety of stock. Easy to install.	Requires frequent inspection, maintenance, and adjustment to each operator. Limits movement of operator. May obstruct space around operator. Does not permit blanking from hand-fed strip.	Embossing presses Power presses Power press brakes

Table 8-A (continued). Point-of-Operation Protection

Type of Safeguarding Method	Action of Safeguard	Advantages	Limitations	Typical Machines on Which Used
Hand pull-backs or pull-outs	A cable-operated attachment on slide, connected to the operator's hands to pull the hands back only if they remain in the danger zone; otherwise it does not interfere with normal operation.	Acts even in event of repeat. Permits maximum hand feeding; can be used on higher speed machines. No obstruction to feeding a variety of stock. Easy to install.	Requires unusually good maintenance and adjustment to each operator. Frequent inspection necessary. Limits movement of operator. May obstruct work space around operator. Does not permit blanking from hand-fed strip stock.	Embossing presses Power presses Power press brakes
		MISCELLANEOUS		
Limited slide travel	Slide travel limited to ¼ in. or less; fingers cannot enter between pressure points.	Provides positive protection. Requires no maintenance or adjustment.	Small opening limits size of stock.	Foot power (kick) presses Power presses
Presence-sensing device	Sensing field and brake quickly stop machine or prevent its starting if the hands are in the danger zone.	Does not interfere with normal feeding or production. No obstruction on machine or around operator.	Expensive to install. Does not protect against mechanical repeat. Limited to use on machines with means to quickly stop the machine during the operating cycle.	Embossing presses Power presses Press brakes
Type A and B gate devices	Encloses danger area before machine action starts. Stays closed until hazard ceases or stops machine if opened too soon.	Interlocked with operating cycle. Allows free access to load and unload machine. Fully encloses point of operation.	Usually limited to machines on which the part or material being processed is fully within the point-of-operation area.	Power presses Plastic injection-molding machines Compression-molding machines Die-casting machines
Special tools or handles on dies	Long-handled tongs, vacuum lifters, or hand die holders which avoid need for operator's putting hand in the danger zone.	Inexpensive and adaptable to different types of stock. Sometimes increases protection of other guards.	Operator must keep hands out of danger zone. Requires unusually good employee training and close supervision.	Dough brakes Leather die cutters Power presses Forging hammers
Special jigs or feeding devices	Hand-operated feeding devices of metal or wood which keep the operator's hands at a safe distance from the danger zone.	May speed production as well as safeguard machines. Generally economical for long jobs.	Machine itself not guarded; safe operation depends upon correct use of device. Requires good employee training, close supervision. Suitable for limited types of work.	Circular saws Dough brakes Jointers Meat grinders Paper cutters Power presses Drill presses
		TWO-HAND TRIP		
Electric	Simultaneous pressure of two hands on switch buttons in series actuates machine.			

Table 8-A (continued). Point-of-Operation Protection

Type of Safeguarding Method	Action of Safeguard	Advantages	Limitations	Typical Machines on Which Used
Mechanical	Simultaneous pressure of two hands on air control valves, mechanical levers, controls interlocked with foot control, or the removal of solid blocks or stops permits normal operation of machine.	Can be adapted to multiple operation. Operator's hands away from danger zone. No obstruction to hand feeding. Does not require adjustment. Can be equipped with continuous pressure remote controls to permit "inching." Generally easy to install.	Operator may try to reach into danger zone after tripping machine. Does not protect against mechanical repeat unless blocks or stops are used. Not generally suitable for blanking operations. Must be designed to prevent tying down of one button or control which would thereby permit unsafe one-hand operation.	Dough mixers Embossing presses Paper cutters Pressing machines Power presses Power press brakes
		TWO-HAND CONTROL		
Electric	Simultaneous pressure on two hand switches held down until dies close.	Can be adopted for multiple operators. Operators' hands away from hazard during die closing portion of stroke. No obstruction to hand feeding.	Hand buttons must be spaced far enough from point of operation to stop machine upon removal before hand can reach into the hazard. Control circuit must be designed to prevent tying down of one button. Buttons must be spaced far enough apart to prevent operation with one hand and another part of the body.	Dough mixers Embossing presses Paper cutters Pressing machines Power presses Press brakes

machines may appear well guarded, and may be, but sometimes their unitized, streamlined designs are more for appearance than for the safety of operation. A good checkout may take some time and effort to complete, but it is well worth the effort to pinpoint guarding problems and detect a course of action. Generally, high-hazard problems should be dealt with first, particularly where there is no guarding. As previously mentioned, OSHA does not specify whether guarding is to be accomplished by enclosure, fencing, or location of the hazardous part or operation. OSHA standards do require that the method chosen must be adequate to eliminate the hazard under normal operating conditions. Since fencing and location methods can be used only when a machine can be isolated, these methods have limited use. Most production machines will require some kind of enclosure of hazard points to comply with the OSHA regulations.

Types of guards

There are several types of enclosures that are suitable for guarding hazard points depending on how and where they are used.

The fixed guard is considered preferable to all other types and should always be used unless it has definitely been determined that it is not at all practicable. The fixed guard at all times prevents access to the dangerous parts of the machine.

Fixed guards may be adjustable to accommodate different sets of tools or kinds of work. Once adjusted, they should remain fixed, and there should be no movement nor detachment.

Some fixed guards are installed at a distance from the point of remote feeding arrangements. This makes it unnecessary for the operator to approach the danger point.

A table of the safe distance between a guard and the danger point along with the permissible size of openings in a fixed guard is given in the Point-of-Operation Protective Devices section of this chapter.

Interlocking guards. If a fixed guard cannot be used, the first alternative is an interlocking guard fitted onto the machine. Interlocking guards can be mechanical, electrical, pneumatic, or a combination of types.

The interlocking guard prevents operation of the control that sets the machine in motion until the guard is moved into position. This prevents the operator from reaching into the point of operation or the point of danger.

When the guard is open and dangerous parts are exposed, the starting mechanism is locked, and a locking pin or other safety device prevents the main shaft from turning or other mechanisms from operating. When the machine is in motion, the guard cannot be opened. It can be opened only when the machine has come to rest or has reached a fixed position in its travel.

An effective interlocking guard must satisfy three requirements; it must:
1. Guard the dangerous part before the machine can be operated
2. Stay closed until the dangerous part is at rest
3. Prevent operation of the machine if the interlocking device fails.

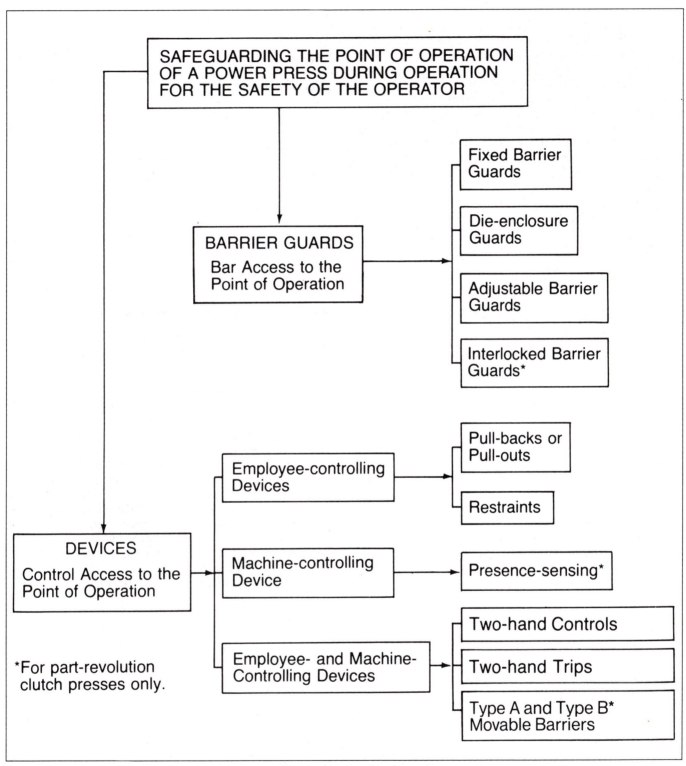

Figure 8-5. Automatic guards can protect the employee at the machine's point of operation.

Where neither a fixed guard nor an interlocking guard is practicable, mechanical interlocks may be used.

Automatic safeguards. An automatic guard can be used, subject to limitations outlined in Table 8-A, where neither a fixed guard nor an interlocking guard is practicable. Such a guard must prevent the operator from coming in contact with the dangerous part of the machine while it is in motion, or it must be able to stop the machine in case of danger (Figure 8-5 shows the relation between safeguarding devices and barrier guards).

An automatic guard functions independently of the operator, and its action is repeated as long as the machine is in motion. It is usually operated by the machine itself through a system of linkage or levers.

Whenever an automatic guard is used on a machine that is loaded and unloaded by hand, the operator should always use hand tools. See details in Chapter 11, Cold Forming of Metals.

All of the foregoing types of guards are suitable for protection at points of operation, but only the fixed guard can be used for guarding power transmission components (unless they are guarded by fencing or location as previously noted).

Safeguarding devices can be used in lieu of or in conjunction with barrier guards at the point of operation. Presence-sensing devices (either photoelectronic or radiofrequency), pullbacks, restraints, and two-hand controls can be used under specified conditions and constraints. (See Chapter 11, Cold Forming of Metals, for a fuller discussion of these devices.)

POINT-OF-OPERATION PROTECTIVE DEVICES

If all machines were alike, it would be simple to design a universal point-of-operation guard and install it during manufacture of the machine. But machines are not all alike, and to further complicate the problem, purchasers of the same model machine may use it in different ways and for different purposes, and these uses may change during the lifetime of the machine. This is why effective point-of-operation guards cannot always be installed by the manufacturer. In many cases, a guard can only be made and installed after the user has tested the machine through on-site use.

Whenever point-of-operation guards are needed, certain principles and conditions must be applied to the design and construction of the guard.

Guard openings

The following is a summary of these principles and other important data as found in *Safe Openings for Some Point-of-Operation Guards,* a publication based on research material developed by Liberty Mutual Insurance Company and the Alliance of American Insurers.

It has been common practice in the design of point-of-operation guards to consider that any opening not over ⅜ in. (10 mm) was relatively safe, as it would not permit entrance of

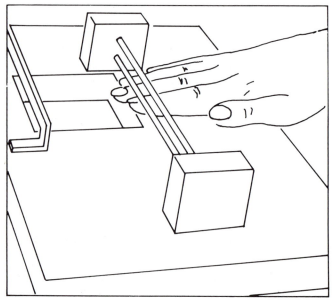

Figure 8-6. A ⅜-in. (10 mm) opening permits a small part of the fingers to slip past the guard. A smaller, ¼-in. (6 mm) opening would prohibit the fingers crossing into the danger zone.

any considerable part of a hand inside the guard (Figure 8-6). In many instances, however, a ⅜-in. opening is not a sufficient space for passing material to be processed through or under the guard. But as the width (or height) of the opening is increased to accommodate the material, the operator is able to reach farther inside the guard. Under such conditions, it is no longer possible to prevent entrance of some part of the hand within the guard. Now the problem is to stop movement of the hand *inside* the guard at a safe distance from the danger zone.

Figure 8-7 shows sketches of an inrunning roll hazard where a feed table is used. There are other types of point-of-operation hazards, however, where an opening larger than shown may be required in a guard design.

In Figure 8-8 the proper location of the guard distance X for

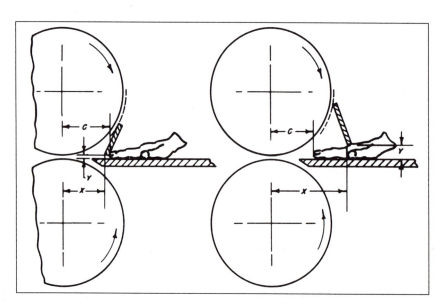

Figure 8-7. Left: ¼-in. opening on guard stops the hand. Right: Larger opening (Y) placed farther back can still stop the fingers from reaching the danger zone. Both illustrations show ends of fingers stopped approximately the same distance (C) from the danger zone. (Printed with permission from Alliance of American Insurers; Liberty Mutual Insurance Company.)

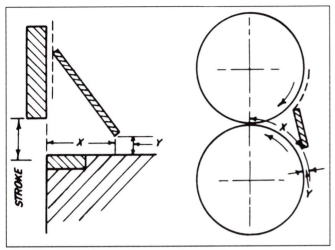

Figure 8-8. Left: Vertical shear hazard. Right: Inrunning roll hazard (no feed table used). (Printed with permission from Alliance of American Insurers; Liberty Mutual Insurance Company.)

the use of required opening Y must be determined. If the dimensions of the opening and its location (distance from the hazard) are properly selected, adequate safety for the operator can be established.

Some guard designers have made use of the formula:

$$\text{Maximum safe opening} = \frac{1}{4} \text{ in. } + \frac{1}{8} \text{ distance to guard from danger zone (in.)}$$

This formula is not intended for use where the distance from the guard to the danger zone exceeds 12 in. (30 cm).

Figure 8-9 illustrates a situation where it may be essential to have part of the hand and fingers extend through the guard to permit manipulation of the material inside the guard. A condition can exist where it is impossible to use hand tools or mechanical devices to carry on the operation.

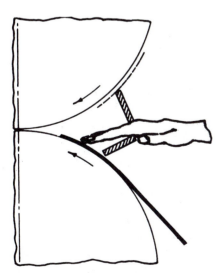

Figure 8-9. This guard protects the fingers while they position materials. (Printed with permission from Alliance of American Insurers; Liberty Mutual Insurance Company.)

Test results

For their design analysis, the Loss Prevention Department of Liberty Mutual Insurance Company constructed a number of test fixtures which approximated common guards. By testing different-sized openings and hands in these fixtures, information was developed to guide guard designers when specifying allowable openings and location of guards. These design data are shown in Figure 8-12, with a feed table, and Figure 8-13 without a feed table.

One set of guide data was established for general use. Because of variations in hands and inaccuracies to be expected in maintaining a ⅜-in. (10 mm) opening, it was decided that for the first 1½ in. (38 mm) away from the danger line, in Figure 8-10, no openings over ¼ in. (6 mm) could be considered safe.

Most men and women have finger tips which will not travel any considerable distance through a ⅜-in. opening. However, if a designer wishes to maintain a definitely safe zone beyond a ⅜-in. opening, the opening must be no more than ¼ in. wide within 1½ in. (38 mm) of the danger point, as shown in Figure 8-10.

Application of test findings

Vertical shear. The findings of the vertical shear portion of these tests, Figure 8-11, apply to:
1. All vertical openings in the guard proper, such as visibility slots, clearance slots for ejection devices and stop gages.
2. All horizontal openings in the guard proper as are necessary for feeding stock (front or sides) and for ejection of finished parts or scrap.
3. The installation or adjustment of a guard.

Inrunning rolls with feed table. In applying findings of tests to guards for use on inrunning rolls, it is necessary to take into account the characteristics of a nip point. In Figure 8-10, the danger line represents a hazardous contact; for vertical shear exposures, it is equivalent to the shear line.

On rolls, the hazard (nip or pinch zone) is not defined by a straight line. Therefore a ⅜-in. width of nip zone is considered the actual nip point through which the danger line DE is drawn (see Figure 8-11). The distance of the ⅜-in. width of nip zone from the contact point between the rolls is designated as dimension S. It is recommended that rolls held less than ⅜-in. apart be considered as rolls in contact.

Figure 8-12 shows an inrunning roll nip where a feed table is used. To design a properly placed barrier guard, it is suggested that the following procedure be used:
1. Draw a full-scale outline of the nip zone with the top surface of the feed table accurately shown. Indicate the clearance line on the top roll. If more than a ⅜-in. (10 mm) clearance is required, the top edge of the guard should be located in accordance with the safe opening layout shown in Figure 8-14 for layout on a roll surface.
2. Determine the distance S—the distance from the center line of the rolls to a point where a ⅜-in. space exists between the top of the feed stock and the surface of the upper roll.
3. At this distance, begin the layout of the safe opening dimensions, as shown in Figure 8-10, up to the opening necessary for the particular guard being designed. Outline the guard section (top edge on clearance line on upper roll and bottom edge at proper point on safe opening layout), and determine the necessary dimensions for installing the guard. (Width of guard can be determined in addition to locating distances.)

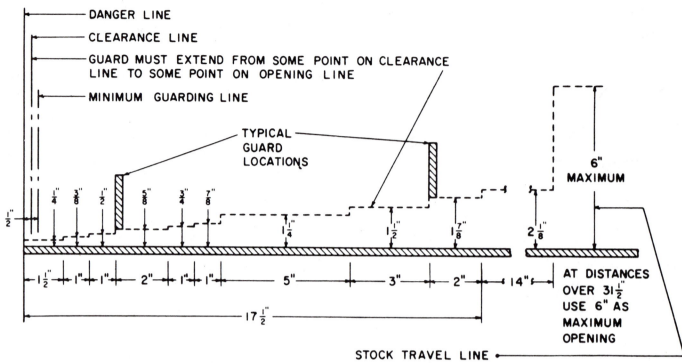

Figure 8-10. Every point-of-operation guard shall prevent entry of hands or fingers by reaching over, under, or around the guard into the point of operation. It should use fasteners not easily removable by the operator to minimize the possibility of misuse or removal of essential parts. It should offer maximum visibility of the point of operation consistent with other requirements. (Dimensions assume adult hands held flat as shown in Figure 8-7.) (Printed with permission from *Requirements for the Construction, Care, and Use of Mechanical Presses,* ANSI B11.1.)

4. Before the guard is put in operation, check carefully for hand travel under the guard, stability of mounting, and rigidity of construction.

Inrunning rolls with central feed (no feed table). Figure 8-14 shows an inrunning roll nip where no feed table is used and the stock processed runs into the nip at right angles to the center of the rolls. (If the run of the stock is slightly above or below the horizontal, the center of the guard opening should be shifted accordingly.)

To design a properly placed barrier guard fitted with an opening for passage of stock, the following procedure is suggested:
1. Draw a full-scale outline of the nip zone with the stock travel line accurately shown. Indicate the clearance line on both rolls. If more than a ⅜-in. (10 mm) clearance is required between the edges of the guard and the rolls, the edges of the guard should be located in accordance with the safe opening layout shown in Figure 8-14 for layout on a roll surface.
2. Determine distance S—the distance from the center line of the rolls to a point where a 3/16-in. (5 mm) vertical space exists on each side of the travel line of the stock (total nip width of ⅜ in.), as shown in Figure 8-13.
3. At this point, begin the layout of the dimensions shown centered on the travel line in Figure 8-10. Outline the guard sections, giving the required opening between sections. (One edge of each section will touch a clearance line; the other will touch the safe opening layout at the proper point to give the required opening.) Determine the necessary dimensions for properly locating the guard from this final layout. The width of the guard sections can also be determined.

4. Before the guard is put in operation, check carefully for hand travel under the guard, stability of mounting, and rigidity of construction.

Inrunning rolls—stock traveling over one roll before entering nip zone. Figure 8-14 shows an inrunning roll nip where the stock travels over a portion of one roll before entering the nip. The stock in such an arrangement is fed either under or over a barrier.

To design a properly placed barrier guard under such conditions, it is suggested that the following procedure be used:
1. Draw a full-scale outline of the nip zone with the travel line of the stock indicated on the roll. Indicate the clearance line on the top roll. If more than a ⅜-in. (10 mm) clearance is required, the top edge of the guard should be located in accordance with the safe opening layout.
2. Determine distance S—the distance of the center line of the roll to the point where there is a ⅜-in. space between the rolls.
3. At this distance begin the layout (on the roll with the stock travel) of the safe opening dimensions, as shown in Figure 8-10. (Layout can be made on roll surface with ½-in. [13 mm] divider steps.) Outline the guard section (one edge touching clearance line and the other touching the safe opening layout at the proper point to give the required opening). Determine the necessary dimensions for properly locating the guard from this final layout. The width of the guard can also be determined in addition to locating the dimensions.
4. Before the guard is put in operation, check carefully for hand travel under the guard, stability of mounting, and rigidity of construction.

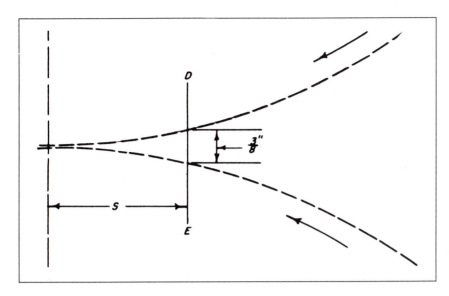

Figure 8-11. Application of test findings to guard design on inrunning rolls. See accompanying text for explanation of these figures.

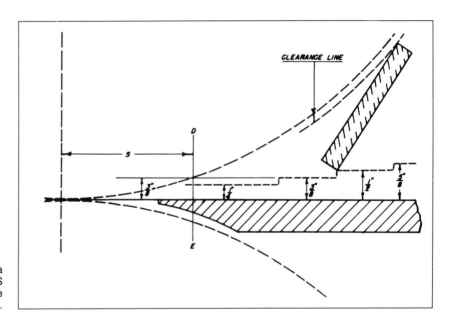

Figure 8-12. Inrunning roll nip where a feed table is used. DE is the stop line. S is the distance of ⅜-in. wide nip zone from contact point between rolls.

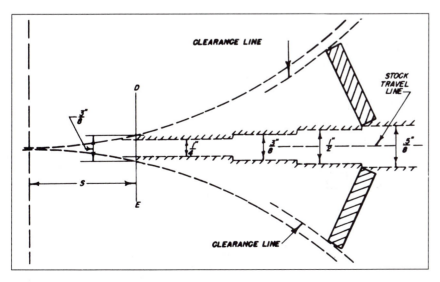

Figure 8-13. Inrunning rolls with central feed and no feed table.

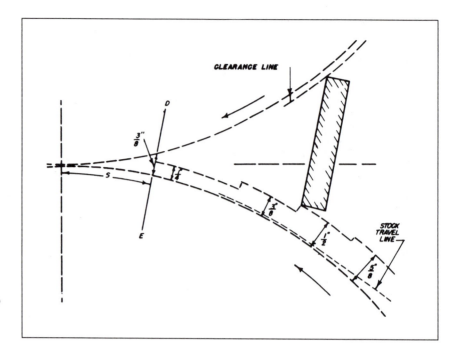

Figure 8-14. Inrunning rolls with stock traveling over one roll before entering nip zone. (Printed with permission from Alliance of American Insurers; Liberty Mutual Insurance Company.)

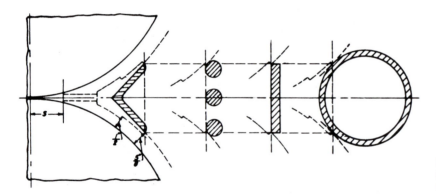

Figure 8-15. These designs for a roll nip guard can be used for feeding over or under a guard through a ⅝-in. opening.

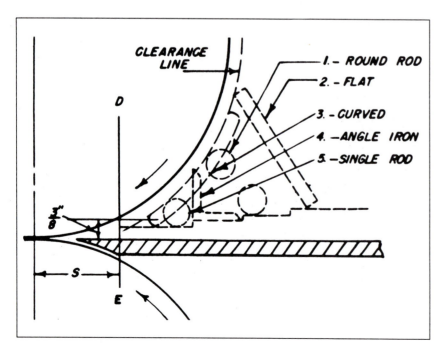

Figure 8-16. Within the outline are different designs for an effective guard on a roll nip with a feed table.

Table 8-B. Standard Materials and Dimensions for Machinery Guards

| | | SIZE OF FILLER MATERIALS | | |
Material 1 in. = 2.5 cm; 1 ft = 0.305 m	Clearance from Moving Part at All Points (inches)	Largest Mesh or Opening Allowable B (inches)	Minimum Gage (U.S. Standard) or Thickness	Min. Height of Guard from Floor or Platform Level (ft, in.)
Woven Wire	Under 2 2-4 4-15	3/8 1/2 2	No. 16⅜ in. No. 16½ No. 12-2	8-0* 8-0 8-0
Expanded Metal	Under 4 4-15	1/2 2	No. 18½ in. No. 13-2	8-0 8-0
Perforated Metal	Under 4 4-15	1/2 2	No. 20½ in. No. 14-2	8-0 8-0
Sheet Metal	Under 4 4-15	 . . .	No. 22 No. 22	8-0 8-0
Wood or Metal Strips Crossed	Under 4 4-15	3/8 2	¾ in. wood or No. 16 metal	8-0
Wood or Metal Strips Not Crossed	Under 4 4-15	½ the width One width		
Plywood, Plastic or Equivalent	Under 4 4-15	 . . .	¼ in. ¼ in.	8-0
Standing Railing	Min. 15 Max. 20	 . . .	 . . .	3-6

*Guards for rotating protuding objects should extend to a minimum height of 9 ft from the floor or platform.

Guard construction

To ensure that safe opening dimensions and guard effectiveness are maintained, it is important that guards be constructed to minimize the distortion or movement. All guard parts should be strong enough to withstand expected stress. Fastenings should be secure and designed to prevent the guard from shifting and unauthorized removal or movement. Any guard with openings larger than ¼ in. (6 mm) should be considered a precision construction and checked frequently for alignment and condition.

Depending upon the need for visibility and rigidity and the method of feeding for a particular guard, the designer can use the layouts shown in Figures 8-12, -13, and -14 to select the best suited guard sections.

Figures 8-15 and -16 show typical guard designs for a roll nip with and without a feed table. Working from a layout similar to this one, the designer can determine the location, size, and shape of the guard section.

Point-of-operation safeguarding is also discussed in Chapter 11, Cold Forming of Metals. Guarding woodworking and metal-working machinery is in Chapters 9 and 10.

GUARDING POWER TRANSMISSIONS

In general, the same principles used in designing point-of-operation guards apply to designing power transmission guards except for not having to consider openings for loading and unloading materials. The only openings allowed for power transmission guards are those for lubrication, adjustment, and inspection. Even these openings must have hinged, sliding, or bolted cover plates that cannot be removed (except for service or adjustment) and they must remain closed when not in use.

In general, power tansmission guards must cover all moving parts in such manner that no part of the body can come in contact with them. In many cases, a simple flat plate or box which covers the opening is all that is necessary, particularly if the parts are flush with or recessed within the frame of the machine. Where parts protrude beyond the frame, it may be necessary to cover the part; that is, build a guard which conforms to the dimensions and forms of the parts being guarded (Figure 8-17). In such cases, any openings to permit shafts or other components to pass into the machine must follow the requirements of the maximum size of permissible opening as related to the distance from the moving part. (See Table 8-B.)

Guarding materials

The preferable material for guards under most circumstances is metal. Framework of guards is usually made from structural shapes, pipe, strapping, bar, or rod stock. Filler material generally is expanded, perforated, or solid sheet metal or wire mesh (Table 8-B). Where visibility is required, use plastic, polycarbon on lexan, or safety glass.

Guards made of wood have limited application. Their lack of durability and strength, relatively high maintenance cost, and flammability are objectionable. Wood guards, particularly when they become oil-soaked, can be ignited by nearby welding operations, by overheated bearings, by rubbing belts and defective

Figure 8-17. These expanded metal covers safeguard protruding moving parts and power transmissions.

wiring, and by other sources of heat. Wood is also subject to splintering that can contaminate products or cause injury.

Where resistance to rust or possible damage to tools and machinery is an important factor, guards of aluminum, other soft metal, or plastic are sometimes used. Plastic guards are frequently used where inspection of the moving parts is necessary. Shatterproof glass is similarly used, particularly where visibility of guarded parts is a problem and where the flexibility of plastic is not required. Safety glass and plastic used where chips or other flying particles are likely to mar the surface may be protected by inexpensive and easily replaced cover glasses.

When a guard cannot be made to exclude lint, ample ventilation should be provided. Vents, too small to admit a hand, should be built into the bottom of larger guards to let lint or dust drop through. Larger guards also should have self-closing access doors for cleaning by brush, vacuum hose, or compressed air. Consideration should be given to latches interlocked with the power source to prevent operation of the machine while the door is open.

Whatever material is selected for a guard, it should also be substantial enough to withstand internal as well as external impacts of materials, parts under stress, or passing pedestrians or vehicles. Parts or materials in process can become lethal projectiles. Machines located close to heavy traffic aisles can be very vulnerable to damage, particularly if they overhang the aisle. If the machine cannot be relocated or the traffic rerouted, a regular enclosure guard may not be sufficient to protect both people and the machine from all possible hazards. Loaded lift trucks could snag on the overhang and damage the guard, machine, and even nearby personnel if the load topples. Such cases require a floor guard in addition to the enclosure guard to prevent vehicles from straying into the hazard area.

Planning for maintenance

A major consideration in designing enclosure guards is planning for routine maintenance. Failure to establish, enforce, and facilitate safe maintenance procedures is probably the major cause of failure to replace a guard, particularly if the maintenance has to be frequently done. Then, it becomes easier to permanently remove the guard. This condition now allows (and even promotes) an additional problem—maintenance of the machine while it is in operation. These conditions are highly hazardous for all personnel concerned, and can be solved in any one or combination of three ways.

- The first would be to apply engineering techniques that reduce or eliminate the job frequency. If, for example, the fittings for those parts needing service were relocated to the outside of the guard, this would make it unnecessary to remove the guard and it would permit maintenance during machine operation. Or, an oil or grease fitting might be lubricated by an extension through the guard. This procedure is highly recommended for operations that cannot be shut down for adjustment or maintenance.
- A second way, which is similar to the first, involves equipping the machine with automatic controls for lubrication, adjustment, or service. Sophisticated equipment like this may be costly, but for some machines the cost is offset by ongoing savings effected by better and more reliable adjustment and maintenance procedures.
- The third method requires interlocking all guards with the source of power so the machine cannot be operated unless all guards are in place. This method ensures guard replacement. Interlocking is effective by itself; and if used in conjunction with the other methods, makes the total guarding system almost foolproof.

Consideration should also be given to guard installation or mounting. They should, of course, be mounted securely, and the best ways are either tack welding or special bolting. Special bolting would require special tools for removal, and thus both methods eliminate unauthorized guard removal. They should be the only ways of mounting guards that are not interlocked.

GUARDS AND NOISE CONTROL

Earlier in this chapter it was mentioned that guards and safeguards can be designed to strengthen machines and serve other purposes. One of the best of these is noise control. A well-designed, well-made, and well-mounted guard can be very effective in reducing unwanted noise. In too many cases, the guard is an after-thought and it is cheaply made and installed. Incorrectly designed or inadequate guards can aggravate machine noise. This is one reason why manufacturer-built and installed guards are recommended over those made in house. However, unless properly designed for noise reduction, many manufacturer-built guards are no better than those made in house.

Noise primarily travels by conduction and vibration through air. It can be attenuated by barriers that effectively stop or restrict further activity either by absorption or by reflection and confinement of the sound waves. Since guards are usually positioned where machine noise originates, at either the point of operation or power transmission, they can be designed as a barrier against noise as well as a barrier against personal injury.

A guard can be designed to be either absorbent or reflective of sound waves. A common way to absorb sound is to cover the surrounding frame with a soft cellular material designed to soak up sound with dead-air spaces. Sometimes a thin layer of lead is sandwiched between two layers of soft material to further reduce sound transmission. However, if oil is used in the process, this could create a serious fire hazard in areas with heat buildup because some of these materials soak up oil. If the guard is made of heavier metal stock (16 gage [1.5 mm] or better), it provides a dense, vibrationless surface when it is correctly mounted. Proper mounting is essential for the success of either method, because both absorptive and reflective barriers require that the guard have an identical configuration with the surface on which it is attached. Some sort of gasket material should be used around the edges and any other place that metal-to-metal contact can occur. The entire guard should be secured with shakeproof fittings. In effect, this bonds the guard to the machine and contains the sound, thus reducing conducted noise.

SUMMARY

In summary, a complete guarding program is essential for any company. It prevents accidents to people and machines and facilitates production goals. It can be achieved only by studying and understanding the relationships between people and the machines they operate. Machines can produce only if they are used correctly—and they are used correctly only when all guards and safeguards are in place.

REFERENCES

Alliance of American Insurers, 1501 Woodfield Rd., Schaumburg, Ill. 60173. *Safe Openings for Some Point of Operation Guards*, Technical Guide 02-678.

American National Standards Institute, 1430 Broadway, New York, N.Y. 10018.

Bakery Equipment, Safety Requirements for, Z50.1-1983.
Baling Equipment, Safety Requirements for, Z245.5-1982.
Blown Film Take-Off and Auxiliary Equipment—Construction, Care, and Use, B151.4-1982.
Cleaning and Finishing of Castings, Safety Requirements for the, Z241.3-1981.
Coil Slitting Machines/Systems, Safety Requirements for Construction, Care, and Use of, B11.14-1983.
Cold Headers and Cold Formers, Safety Requirements for the Construction, Care, and Use of, B11.7-1985.
Construction, Care, and Use of Die Casting Machines, Safety Requirements for the, B152.1-1973.
Construction, Care, and Use of Horizontal Injection Molding Machines, Safety Requirements for the, B151.1-1984.
Construction, Care, and Use of Machinery Used in Envelope Manufacturing, Safety Requirements for the, B169.1-1975.
Construction, Care, and Use of Packaging and Packaging-Related Converting Machinery, Safety Requirements for, B155.1-1986.
Construction, Care, and Use of Riveting Setting Equipment, Safety Requirements for the, B154.1-1984.
Conveyors and Related Equipment, Safety Standards for, ANSI/ASME B20.1-1984.
Drilling, Milling, and Boring Machines, Safety Requirements for the Construction, Care, and Use of, B11.8-1983.
Extrusion Blow Molding Machines, Construction, Care, and Use of, B151.15-1985.
Film Casting Machines—Construction, Care, and Use, B151.2-1982.
Floor and Wall Openings, Railings, and Toeboards, Safety Requirements for, A12.1-1973.
Gear Cutting Machines, Safety Requirements for, the Construction, Care, and Use of, B11.11-1984.
Granulators, Pelletizers, and Dicers Used for the Size Reduction of Plastics—Construction, Care, and Use, B151.11-1982.
Grinding Machines, Safety Requirements for the Construction, Care, and Use of, B11.9-1975.
Horizontal Extrusion Presses, Safety Standard for the Construction, Care, and Use of, B11.17-1982.
Hydraulic Power Presses, Safety Requirements for Construction, Care, and Use of, B11.2-1982.
Industrial Robots and Industrial Robot Systems, Safety Standard for, ANSI/RIA R15.06-1986.
Iron Workers, Safety Requirements for the Construction, Care, and Use of, B11.5-1975(R1981).
Lathes, Safety Requirements for the Construction, Care, and Use of, B11.6-1984.
Machinery and Machine Systems for the Processing of Coiled Strip, Sheet, and Plate, Safety Requirements for the Construction, Care, and Use of, B11.18-1985.
Mechanical Power Presses, Safety Requirements for Construction, Care, and Use of, B11.1-1982.
Melting and Pouring in the Metal Casting Industries, Safety Requirements for, Z241.2-1981.
Metal Sawing Machines, Safety Requirements for Construction, Care, and Use of, B11.10-1983.
Mills and Calenders in the Rubber and Plastics Industries, Safety Requirements for, B28.1-1967.
Pipe, Tube, and Shape Bending Machines, Safety Standard for Construction, Care, and Use of, B11.15-1984.
Plastic Film and Sheet Winding Equipment—Construction, Care, and Use, B151.5-1982.

Pneumatic Conveying Systems for Handling Combustible Materials, ANSI/NFPA 650-1984.

Portable Pipe Threading Machines and Portable Power Drives, Safety Requirements for, B208.1-1982.

Power Press Brakes, Safety Requirements for the Construction, Care, and Use of, B11.3-1982.

Roll Forming and Roll Bending Machines, Safety Requirements for the Construction, Care, and Use of, B11.12-1983.

Safety Standard for Forging, B24.1-1985.

Sand Preparation, Molding, and Coremaking in the Sand Foundry Industry, Safety Requirements for, Z241.1-1981.

Screen Changers—Construction, Care, and Use, B151.3-1982.

Shears, Safety Requirements for the Construction, Care, and Use of, B11.4-1983.

Shops Fabricating Structural Steel and Steel Plate, Safety Requirements for, Z229.1-1982.

Single- and Multiple-Spindle Automatic Screw/Bar and Chucking Machines, Safety Requirements for the Construction, Care, and Use of, B11.13-1983.

Slit Tape and Monofilament Post-Extrusion Equipment—Construction, Care, and Use, B151.6-1982.

Wheels, Safety Requirements for the Use, Care, and Protection of Abrasive, B7.1-1978.

Woodworking Machinery, Safety Requirements for, 01.1-1975.

Woodworking Machinery, Safety Requirements for, 01.1a-1979.

Hall, Frank, B. "Historical Background of the Concept: Zero Mechanical State." *National Safety News,* 115:2 (February 1977).

Johnson, Earl A. "Zero Mechanical State Procedure Development." *1976 National Safety Congress Transactions,* vol. 12, Industrial Subject Sessions.

Mitchell, Robert D. "Development of the Concept: Zero Mechanical State." *National Safety News,* 115:5 (May 1977).

Mitchum, B. G. *Concepts and Techniques of Machine Safeguarding.* U.S. Dept. of Labor, OSHA Publication 3067. Washington, D.C.: U.S. Government Printing Office.

National Fire Protection Association, Batterymarch Pk., Quincy, Mass. 02269.
Electrical Standard for Industrial Machinery, NFPA 79.
National Electrical Code, NFPA 70.

National Safety Council, 444 N. Michigan Ave., Chicago, Ill. 60611.
Safeguarding Concepts Illustrated.
Power Press Safety Manual.

Royal Society for the Prevention of Accidents, London. *Industrial Accident Prevention Bulletin,* Vol. 11, No. 118.

U.S. Department of Health and Human Services, National Institute for Occupational Safety and Health, Publications Dissemination, DTS, Cincinnati, Ohio 45226.
Machine Guarding—Assessment of Need, HSM-99 73 71.
Occupational Safety and Health in Vocational Education, Pub. No. 79-125.

U.S. Department of Labor, Occupational Safety and Health Administration, Washington, D.C. 20210. *Principles and Techniques of Mechanical Guarding,* Bulletin 2057, 1972.

Woodworking Machinery

EACH PIECE OF WOODWORKING EQUIPMENT generates its characteristic accident problems. To reduce the possibility of serious injury, the worker should be provided with the right type of correctly guarded equipment. For maximum protection, adequate jigs or fixtures must be available. The hands should be kept as far as possible from the point of operation; this can be accomplished through guarding.

As in any type of occupation, the woodworker must be taught proper and safe procedures. The operator needs to be frequently observed to make certain the established procedures are being followed. The well-trained operator recognizes potential accident situations and knows what to do when they are seen. A change in noise, pitch, or any other operating characteristic of mechanical equipment will alert the trained worker to follow the approved procedures for reporting or correcting a potentially hazardous situation.

Because woodworking equipment is used in many industries, this chapter will be concerned with the equipment and not with specific types of operations in the wood industry.

GENERAL SAFETY PRINCIPLES

The workplace must be provided with equipment meeting the requirement of the existing standards and regulations (such as OSHA, ANSI, NFPA). Purchasers must specify on their purchase orders whatever optional or accessory components are necessary to meet the requirements of mechanical and electrical safeguarding.

The following general safety principles should be observed. (A summary of operating precautions is on page 234.)

- All machines should be constructed and maintained so that, while running at full or idle speed and with the largest cutting tool attached, they are free of excessive noise and harmful vibration.
- All machines, except portable or mobile ones, should be leveled and, where necessary, vibration dampened. They should be securely fastened to the floor or other suitable foundation to eliminate all movement or walking.
- Small units should be secured to benches or stands of adequate strength and design (see Table 9-A).
- Each machine should be designed so that tools larger than those for which the machine was designed cannot be mounted on it.
- All arbors and mandrels should be constructed so they have firm and secure bearings and are free from slip or play.
- All safety devices should be regularly checked for correct adjustment. Those involving electrical circuits should be actuated to make sure they operate properly. Operators should always stop and securely lock out machines before cleaning.
- Loose clothing, long hair, jewelry, and gloves should not be worn around rotating parts of machinery, and especially near nip points and the point of operation.
- Adjustments should not be made, if at all possible, while the machine is running. Chapter 8, Principles of Guarding, contains information on this subject.

Electrical

All of the metal framework on electrically driven machines should be grounded, including the motor, and should comply with the *National Electrical Code* (NFPA 70) and other applicable standards and codes.

Table 9-A. Typical Table Heights and Work Space

Machine	Table Heights (in.)	(cm)	Work Area
Band saws	46	115	On three sides—a radius equal to twice the band saw diameter (as measured from the point cut).
Circular saws	36 (Hand feed)	90	Clearance on the working
	32 (Power feed)	80	side should be 3 ft (90 cm) plus the length of the stock.
Jointers	33	85	3 ft plus the length of stock.
Lathes	41	100	Clearance of at least 30 in. (75 cm) from stand, with smaller distances on ends and backside allowable.
Radial saws	39	85	Ripping—saw table equal to twice the length of the stock. Crosscutting—saw table equal to length of the stock plus 3 ft.
Sanders	36	90	3 ft plus the length of stock.
Shapers	36	90	3 ft plus the length of stock.

Each machine shall be controlled in accordance with the provisions of the *Safety Requirements for Woodworking Machinery* (ANSI O1.1). There may be other provincial, federal, and local codes applicable to these machines that must be consulted.

- The machine shall have a cutoff device (stop switch) within reach of the normal operating position.
- Electrically driven equipment shall be controlled with a magnetic switch or other device that will prevent automatic restarting of the machine after a power failure, if in starting, the machine would create a hazard (see Figure 9-1).
- Power controls and operating controls shall be located within easy reach, and away from a hazardous area. They shall be positioned so the operator can remain at the regular work location while operating the machine.
- Each operating control shall be protected against unexpected or accidental activation (see Figure 9-1).
- Each machine operated by an electric motor shall be provided with a positive means for rendering the controls inoperative. If more than one person is involved in the maintenance or repair of the machine, each should install a separate padlock with a hasp. In addition to locking out the machine, it is a good idea to tag it out. If the machine does not have a power disconnect to lock it in the off position, unplug the cord and insert a small padlock through the holes in the plug (see Figure 9-2).
- Install an electronic motor brake on machines that have excessive coasting time (see Figure 9-3). This device can greatly reduce the exposure at the point of operation.

Guards

All belts, shafts, gears, and other moving parts must be fully enclosed or guarded, in a manner to present no hazard to the operator. (See *Safety Standard for Mechanical Power Transmission Apparatus,* ANSI/ASME B15.1-1984.) Because most woodworking operations involve cutting, it is necessary, although often difficult, to provide guards at the point of operation. (See *Safety Requirements for Woodworking Machinery,* ANSI O1.1-1975.) On most machines, the point-of-operation guard must be movable to accommodate the wood, balanced so as

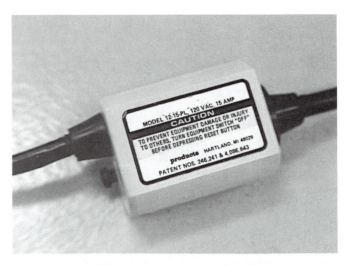

Figure 9-1. Top: When installed in the electrical cord of a machine, this device will prevent automatic restarting. Bottom: This magnetic switch has a ring guard around the start button to protect against accidental reactivation.

not to impede the operations, and yet strong enough to provide protection to the operator. Whenever possible, blades and cutting edges need to be completely covered at the point of operation. Not all such areas can be fully covered while the tool is in the work, for example, radial saws. (Also review the discussion in Chapter 8, Principles of Guarding.)

Work areas

There should be ample work space around the machine, as required by its type. (Suggested minimums are given in Table 9-A.)

The working surfaces of the machine should be at a height that will minimize fatigue. (See Table 9-A.) Adjustments should be made if the worker is taller or shorter than average. All accessory or feed tables should be at the same height as the working surface of the machine.

Floors should be well maintained to prevent splintering conditions and protruding nails. Floors should be kept even and free from holes and irregularities. The work area floor near the machines should have slip-resistant surfaces. Aisleways should be marked by paint, railings, or other approved markings.

Figure 9-2. Left: Typical machine lock out. Center: Lock out with hasp, separate padlocks, and a tag. Right: Lock out for machines without a power disconnect.

Figure 9-3. On machines with excessive coasting time, this electronic motor brake greatly reduces exposure at the point of operation.

Maintain good housekeeping to prevent dust and chip accumulations. A clean operation makes work easier and helps prevent fire and dust explosions.

The machines and the adjacent stock areas should be adequately illuminated. Generally 50 foot-candles will be needed for work, but fine work may require 100 or more foot-candles. General illumination of 80-100 foot-candles will pay dividends in both accident prevention and efficiency. There should be no shadows or reflected glare.

Material handling

The layout should encourage an even flow of materials and keep to a minimum back-tracking and crisscrossing. Operators should not have to stand in or near aisles.

The machines should be arranged so that the material handled by the operator and others requires a minimum of movement and change of heights. This applies to both incoming supply and outgoing stock.

Provision should be made for sawdust and scrap removal before they accumulate. Exhaust vacuum systems are desirable and effective.

The number of fires that originate and spread through ductwork justifies installation of automatic extinguishing systems in ducts, as well as in the collecting systems.

Working surfaces should be kept entirely free from scrap and waste (see Chapter 3, Manual Handling and Material Storage).

Inspection

Safety checks can be made by putting machines through trial runs before beginning an operation and after each new setup. This usually is the responsibility of the setup person.

The operator should inspect the machine at each new setup, and at the start of each shift. Machine inspection should follow the manufacturer's recommendations and workplace requirements and flow patterns. This would include inspecting the operating controls, safety control, power drives, and sharpness of cutting edges and other parts.

All cutting edges and tools must be kept sharp at all times. They must be properly adjusted and properly secured.

Health

Because some woodworking machines (especially saws) are noisy, sound level measurements should be taken by a qualified person. If the reading of the sound level dBA (slow response) exceeds 85, then action must be taken to protect the worker on the job. If the level is less than 85 dBA, legally (in the U.S.) no action will be required; however, it may be desirable to control employee exposure or to somehow reduce the noise level.

Some circular sawblades are specifically designed to reduce noise levels, or large, sound dampening washers can sometimes be used.

When a process creates fine dust, the amount should be sampled. The threshold limit values and maximum permissive levels have been established for many materials and these should be followed. Fine dust can be a health, fire, and explosion problem. Light duty respirators are available to reduce breathing certain kinds of nuisance dust. (See National Safety Council book *Fundamentals of Industrial Hygiene,* 3rd Edition.)

Personal protective equipment

All individuals in the work area should wear eye protection. Safety goggles complying with ANSI Z87.1 are excellent for operations that may generate flying objects. Face shields are not adequate if there are flying objects, but do help if there is dust. Safety glasses with side shields may also be effective.

Workers should not wear loose clothing, gloves, and jewelry (especially rings, bracelets, and chains), which can become entangled in moving machinery.

Hair nets or caps should be worn to keep long hair away from moving parts. Beards should be kept trimmed. Gloves or hand pads can be worn to protect hands from splinters and rough

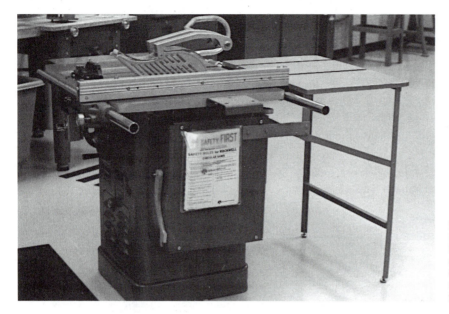

Figure 9-4. Features of this industrial model, tilting-arbor circular table saw include: (1) posted safety rules, (2) push stick, (3) tail-off table, (4) rip fence, (5) crosscut guide, (6) self-adjusting point-of-operation guard, and (7) enclosed power transmission.

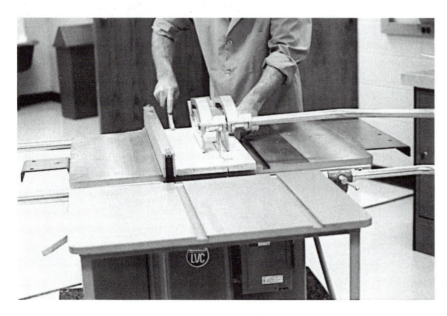

Figure 9-5. This operator is following safe operating procedures by: (1) standing to the side while ripping; (2) keeping sleeves rolled up; and (3) using the rip fence, blade guard, splitter and anti-kickback device, and tail-off table. When using a tail-off table, the operator has less tendency to reach over the blade to catch the stock before it falls to the floor.

lumber. However, gloves should not be worn near rotating parts of the machine.

Safety shoes should be worn when handling heavy material or when there is a danger of foot injury. Where there is danger of a kickback—especially in ripping operations—proper abdominal guard or anti-kickback aprons should be worn.

Standards and codes

There are a number of OSHA standards, such as §1910.213, that state required safety features for woodworking machines. Most of these standards are based on the ANSI standard O1.1 and O1.1a, *Safety Requirements for Woodworking Machinery*. Additionally, some states and other jurisdictions have codes that specify requirements. All of these sources should be consulted. The National Safety Council publishes data sheets on a number of woodworking machines. (See References.)

SAWS

Circular

Among the most frequently occurring types of accidents involving circular power saws are: (1) blade cuts or abrasions, and (2) kickbacks. These can be minimized by proper guarding and by establishing and enforcing safe work procedures.

Power saw operators often are injured when their hands slip off the stock while pushing it into the saw, or when holding their hands too close to the blade during the cutting operations.

Other personnel can be injured by coming into contact with the blade when removing scrap or finished pieces from the table.

Poor housekeeping practices and slippery floors are another source of accidents involving circular saws.

Circular saws are designed to permit a wide range of operations. A problem with saws, as with most multiple-use equipment, is the difficulty in designing one guard that offers maxi-

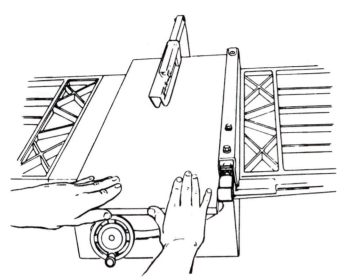

When the width of the rip is 6 in. or wider, use your right hand to feed the workpiece until it is clear of the table. Only use the left hand to guide the workpiece—do not feed the workpiece with the left hand.

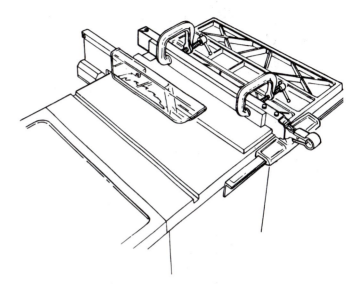

When the width of the rip is less than 2 in., the push stick cannot be used because the guard will interfere. Use the auxiliary fence-work support and push block. Use two C clamps to attach the auxiliary fence-work support to the rip fence.

When the width of rip is 2 in. to 6 in., use the push stick to feed the work.

Figure 9-6. Safe ripping procedure. (Continued on page 222.)

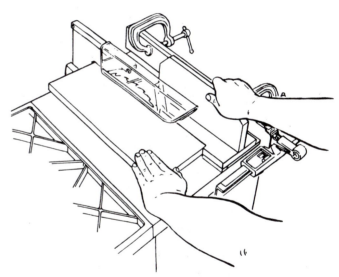

Feed the workpiece by hand along the auxiliary fence until the end is about 1 in. beyond the front edge of the table. Continue to feed using the push block. Hold the workpiece in position and install the push block by sliding it on top of the auxiliary fence-work support (this might raise the guard).

mum protection for all types of operations. Contact with the sawblade can be prevented through use of the proper type of hood guard, jigs, fixtures, combs, or other devices, along with safe operating procedures. Figures 9-4, -5, and -6 illustrate safety features and procedures.

Kickbacks. A kickback occurs during a ripping operation when a part or all of the workpiece is violently thrown back to the operator. Keep the face and body to one side of the saw-blade, out of line with a possible kickback. Kickbacks—and possible injury from them—usually can be avoided by:

1. Maintaining the rip fence parallel to the sawblade.
2. Keeping the sawblade sharp. Replace or sharpen anti-kickback pawls when points become dull.
3. Keeping sawblade guard, spreader, and anti-kickback pawls in place and operating properly. The spreader must be in alignment with the sawblade and the pawls must stop a kick-back once it has started. Check their action before ripping.
4. Not ripping work that is twisted or warped or does not have a straight edge to guide along the rip fence.
5. Not releasing work until it has been pushed all the way past the sawblade.

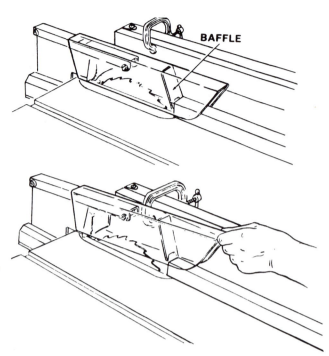

Figure 9-6. (Concluded) Narrow strips thicker than the auxiliary fence-work support may enter the guard and strike the baffle. Carefully raise the guard only enough to clear the workpiece. Use the push block to complete the cut.

Figure 9-7. Close up of a properly functioning splitter and anti-kickback device during a ripping operation.

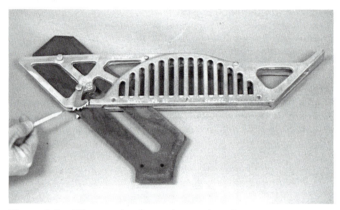

Figure 9-8. When sawing large pieces of stock, the support for the guard (see Figures 9-4 and -5) can be in the way and prevent sawing through the wood. The guard shown here permits sawing large pieces without interference. The splitter and anti-kickback dogs are built into the guard.

6. Using a push stick for ripping widths of 2 to 6 in., and an auxiliary fence and push block for ripping widths narrower than 2 in.

7. Not confining the cut-off piece when ripping or cross-cutting.

8. When ripping, apply the feed force to the section of the workpiece between the sawblade and the rip fence.

A safe ripping procedure is described in Figure 9-6.

Properly using the spreader and the anti-kickback devices, and following a safe procedure will reduce kickbacks. Carefully selecting dry, knot-free stock can also reduce kickbacks.

Besides using mechanical guards against kickback, several other precautions are advisable. It is particularly important to make certain the rip fence is parallel to the line of the saw so the stock will not bind on the blade and be thrown.

Guards (Figures 9-7 and -8) greatly reduce the likelihood of injury and are now considered standard equipment. If they are not furnished with a saw, it is necessary to supply them when the saw is installed. The protection gained by using guards makes them essential. It is, however, important that the guards be practical and correct for the job being done, or else operators may be tempted to remove them.

A circular table saw used for ripping should be provided with a spreader, which prevents wood with internal stresses from clamping down or binding at the outfeed edge of the sawblade. In this way a spreader helps prevent kickbacks. It also keeps chips and slivers away from the back of the saw where they might be caught by the saw teeth and thrown.

A spreader should be rigidly mounted, not more than ½ in. (13 mm) in back of the sawblade when the blade is fully elevated, and should be at least 2 in. (5 cm) wide at table level. It should conform to the radius of the saw as nearly as practicable and

be high enough above the table to penetrate the full thickness of the stock. The spreader should be attached so it will remain in true alignment with the sawblade, even when the table or arbor is tilted.

A circular table saw used for through-sawing should be guarded by a hood that completely covers the sawblade projecting above the table. Let the guard ride the stock being cut, adjusting to the thickness of the stock (Figure 9-9).

The hood shall be of adequate strength to resist any blows that could happen during reasonable operation, adjusting, and handling. It should be made of shatter-resistant material, and be no more flammable than wood. To be effective, the hood must remain in true alignment with the sawblade, even if the table or arbor is tilted.

The hood may be suspended from a post attached to the side of the machine, or supported on the spreader. The mounting must be secured and supported, so it will not wobble and strike against the sawblade. In strength and design, it must protect the operator against flying slivers or broken saw teeth. The mounting shall resist reasonable side thrust or force.

The part of the sawblade underneath the table should be guarded, so the operator cannot accidentally contact the blade.

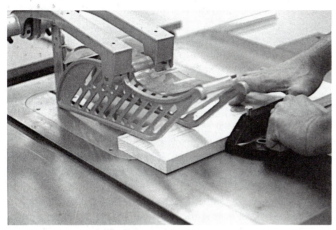

Figure 9-9. The overhead self-adjusting guard on this circular saw rides the stock as it is being cut and automatically adjusts to its thickness.

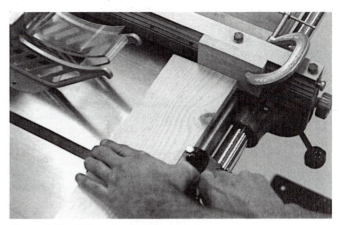

Figure 9-10. When crosscutting several pieces to the same length, it is important to use a small block of wood clamped to the rip fence to allow room for clearance when the piece is cut off. If this is not done, the piece being cut off will bind between the fence and the blade and be thrown back toward the operator.

The enclosure, which may be part of the exhaust hood, should be constructed with a hinged cover, so the sawblades can be easily changed.

Rabbeting and dadoing. It is impossible to use a spreader and often impracticable to use the standard hood guard when rabbeting and dadoing. These operations can be effectively guarded by a jig that slides in the grooves of the transverse guide. The work is locked in the jig, and the operator's hands are kept well away from the saws or cutting head.

Because of variation in rabbeting and dadoing jobs, special jigs may be needed (see Figure 9-10). The hazard in these operations justifies the effort, particularly when work is being done on small stock. If a shop does a lot of dadoing and rabbeting, one or more machines ought to be set aside for this work. This eliminates frequent removal of the standard guards from machines normally used for cutting and ripping.

Feather boards should be used to hold the work to the table and against the fence as it is fed past the dado head. They are suitable for short runs because they can be quickly set up and are inexpensive. A feather board should be made from straight-grained stock, preferably hardwood, and the parallel saw cuts (the comb) should be in the direction of the grain. The feather board should bear against the stock at an angle of 45 to 60 degrees.

Operating methods. Keep the hands out of the line of the cut when feeding a table saw. The guard is a protective device from the sides and from above, but not from the front.

When the operator is ripping with the rip fence close to the saw, he or she should use a push stick between the sawblade and the fence to keep fingers away from the saw. Push sticks or blocks of various sizes and shapes should be kept near the machine. The push sticks should be long enough (add 6 in. [15.2 cm] to the sawblade diameter) to keep hands well away from the blade.

Stock should be held against a gage, never sawed freehand. Freehand sawing endangers the hands and may cause work to get out of line and bind on the saw. When ripping stock with narrow clearance on the fence side, the operator can gain more clearance by clamping a filler board flat to the table between the fence and the sawblade and guiding the stock against it.

Use of a filler makes unnecessary the hazardous practice of removing the hood guard because of lack of clearance.

Because of kickbacks, the operator should stand out of the line of the stock he is ripping. A heavy leather or plastic apron or abdomen guard gives additional protection.

The best sawblade height above the workpiece depends on several considerations. They are as follows: high sawblade silhouette—meaning the blade is high as possible; low sawblade silhouette—meaning the sawblade just extends through the wood stock.

ADVANTAGES

High Saw Blade Silhouette	Low Saw Blade Silhouette
Reduced kickback potential	Less exposure of saw blade to operator
(1) Saw tooth cuts down nearly vertical to table	Smoother cut
(2) Saw blade closest to spreader	More (table) support for workpiece in front of saw blades
Less power needed	
Faster cutting	
Less saw blade wear	

Do not use a crosscut sawblade for ripping or a ripsaw for crosscutting. Using the wrong saw for the job makes the work harder and requires additional force when feeding the stock.

Using a general-purpose table saw for work that should be done on special machines is poor practice. For example, a table saw is often used for hand-feed ripping operations. If the work is done on a power-feed ripsaw instead, the danger of getting hands against the saw and having the work kick back is virtually eliminated.

Long stock is sometimes crosscut on a table saw. Unless it is adequately supported, this is a dangerous practice because stock extending beyond one or both ends of the table interferes with other operations and may be struck by persons or trucks. Also, it is difficult to guide long pieces, and the operator must exert considerable pressure while the hands are close to the saw. Such stock is more easily cut on a swing saw, pull saw, or radial saw.

Work that can be done on special or power-feed machines should not be done on hand-feed, general-purpose machines.

A circular table saw should be stopped when the operator leaves it. It is not sufficient to cut the switch and walk away, since amputations have been caused by saws still coasting with the power off. An electric brake attached to the motor arbor offers fast positive action.

Sawdust and slivers should be cleared away from the table with a brush or a stick, never with the hands.

Under no circumstances should the fence be adjusted while the saw is running. Parallel setting of the fence is particularly important.

To enable the operator to set the fence accurately, mark the top of the saw table with a permanent, distinctive line or other suitable device. The mark should be directly in front of and in line with the saw blade.

Circular sawblade selection and maintenance

The characteristics and conditions of sawblades are an important factor on the safety of the operators who use them. Saw manufacturers have published valuable information on the selection, use, and care of sawblades (see Figures 9-11 and 9-12).

During designing, building, and tensioning, the maker gives a saw enough rigidity and tensile strength to cut without harmful distortion. Altering its original design, operating it at other than the rated speed, or changing the balance or tension seriously affects the saw's efficiency and safety.

Conditions of sawblades that may cause unsafe, difficult, or unsatisfactory operation are:

- *Sawblade out of round.* If some teeth are longer than others, the long teeth do most of the work. An unequal strain is imposed on the saw, which may cause it to run out of line, to heat up, and to warp.
- *Sawblade not straight (out of plane).* Lumps or warps can be checked with a straightedge across the length of the diameter of the unmounted saw. However, because of blade tensioning, it may not be flat, except under power.
- *Improper hook or pitch of teeth.* Ripsaw and cutoff teeth differ in design for different kinds of wood and for different purposes. Combination saws may be used for both crosscutting and ripping. There are other blades for certain kinds of woods and wood material.
- *Improper or uneven set.* A sawblade has to cut a kerf thicker than the blade to give adequate clearance for the saw to pass through the wood. The teeth can be given set or swage by bending alternate teeth right and left or by spreading the point of every tooth so each is slightly wider than the sawblade.
- *Cracked blades.* As soon as a crack is detected, the blade should be removed from service. Sawblades should be inspected for cracks every time the teeth are filed or set. (Some cracks are so small that they may be invisible to the naked eye. A nondestructive testing method, such as Magnaflux, can be used.) If cracked sawblades are left in service, the crack frequently grows larger and eventually will cause partial fragmentation. Most cracks start in the gullets.
- *Dull blades.* Blades must be kept sharp if the saw is to work

at top efficiency and the operator is to exert minimum force when feeding the saw.
- *Gummed blades.* Blades also must be kept clean and free of pitch buildup to run at top efficiency and safety. A gummed blade can cause a kickback.

Whatever method is used, the sawblade must be retensioned after repairs have been made. This is a job for a sawsmith and, unless the company has the services of such a person, the blade should be repaired by the manufacturer.

Excessive heat and vibration cause saw blades to crack. To prevent cracking, these precautions should be followed:
1. The sawblade should be tensioned for the speed at which it will operate. Otherwise, the blade will wobble and vibrate and, as a result, will heat, expand, and crack.
2. The teeth must have sufficient clearance (set or hollow grinding) to prevent burning, with consequent heating and cracking.
3. The blade should be in perfect round and balance.
4. The blade must be kept sharp at all times. A dull blade will not cut; rather, it will pound or burn itself through the wood, so that vibration, heating, and then cracking result.

Proper operation. Only authorized persons should operate saws. A saw in good condition and running at correct speed should cut easily. The operator should not crowd the saw by forcing the stock faster than it can be easily cut. If the saw does not cut as fast as it should, or if it does not saw a clean, straight line, something might be wrong with the saw or the running speed. These conditions—potential sources of accidents—should be checked and corrected before proceeding with the job.

Most production hand- and power-feed saws run at 3600 rpm (actually about 3450 rpm). This will give a 12-in. (0.3 m) blade a rim speed of 10,839 sfm (55 m/s); a 16-in. (0.4 m) blade, 14,451 sfm (74 m/s) and an 18-in. (0.46 m) blade, 16,258 sfm (83 m/s). The manufacturer's instructions always should be followed.

When tightly clamped in position, only the outer edge of the collar should come in contact with the sawblade. If the inside of the collar has not been machined properly, it will force the rim of the saw out of line. When the sawblade comes in contact with the stock being cut, there will be a buckling effect on the saw. After the loose collar is securely fastened in place, it is well to test the saw with a straight-edge. This test is of considerable importance on circular saws as well as on edgers and trimmers.

A means should be provided to prevent the operator from placing a larger sawblade on the mandrel than is allowable for the speed of the mandrel.

Overhead swing saws and straight-line pull cutoff saws

Overhead swing saws and straight-line pull cutoff saws cause hand injuries because of several characteristics. Hands can be cut while the sawblade coasts or idles, when the operator attempts to remove a sawed section of board or a piece of scrap, and when he or she measures a board or places it in position for the cut.

The operator's hands can be struck by the saw if it bounces forward from a retracted position or moves forward if the return device fails. The operator may pull the saw against the hands or may suffer body cuts from a saw that swings beyond its safe limits.

CIRCULAR SAW BLADES FOR CUTTING WOOD

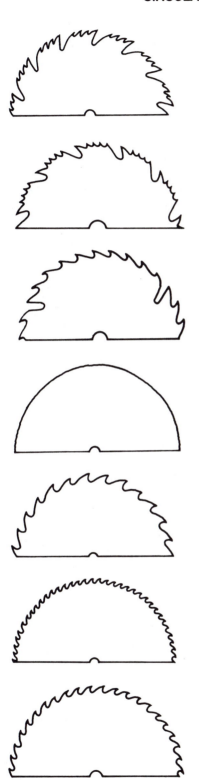

HOLLOW GROUND PLANER BLADES—The hollow ground planer blades are for precision cross cutting, mitering, and ripping on all woods, plywood, and laminates where the smoothest of cuts are desired.

MASTER COMBINATION BLADES—The master combination blades are for use on all woods, plywood, and wood base materials, such as fiberboard and chipboard. This type blade is better for cross cut and mitering than for ripping in solid woods. The teeth are set, and deep gullets are provided for cool and free sawing.

RIP BLADES—The rip blades are primarily intended for rip cuts in solid woods. The teeth are set and deep gullets are provided for cool and free cutting.

PLYWOOD BLADES—The plywood blade is a fine tooth cross cut type blade intended for cross cutting of all woods, plywood, veneers, and chipboard. It is especially recommended for cutting plywood where minimum of splintering is desired. The teeth are set and sharpened to give a smooth but free-cutting blade.

CHISEL TOOTH COMBINATION—The chisel tooth combination blade is an all-purpose blade for fast cutting of all wood where the best of finish is not required. Ideal for use in cutting of heavy rough timbers, in framing of buildings, etc. It cross cuts, rips, and miters equally well.

CABINET COMBINATION—The cabinet combination blade is for general cabinet and trim work in solid wood. It will cross cut, rip, and miter hard and soft wood to give good accurate cuts for moldings, trim, and cabinet work.

STANDARD COMBINATION—The standard combination blade is used for all hard and soft wood for cross cut, rip, or miter cut. It is especially recommended for use on power miter boxes and for accurate molding and framing work.

Figure 9-11. Choosing the correct sawblade for the job will increase operator safety. (Printed with permission from Sears, Roebuck and Company.)

METAL-CUTTING BLADES

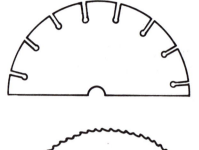

NONFERROUS METAL CUTTING BLADES—The nonferrous metal cutting blades are for cutting brass, aluminum, copper, zinc, lead, bronze, etc. Blades are taper-ground and need no set. Use wax or lubricant on the blades for the best results.

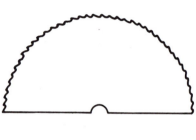

STEEL SLICER—The steel slicer blades are for cutting thin steel and sheet iron up to ³⁄₃₂ in. (2.4 mm) thickness. Not for use on nonferrous metals, wood, or plastic. This blade will give off sparks when cutting steel because it cuts by friction. Always keep sawdust chips free of machine to prevent fires.

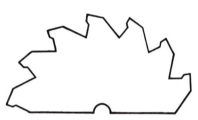

FLOORING BLADE—The flooring blade is a tungsten carbide-tipped blade especially designed for rough cutting where occasional nails, metal lathe, etc. will be cut. It is especially recommended for the professional carpenter or installer of air conditioning or heating ducts where it is necessary to cut through old walls and floors. Always wear safety goggles when cutting metal.

Figure 9-12. Metal-cutting sawblades.

Guards. Cutoff saws must be guarded with a hood guard. The hood shall extend at least 2 in. (5 cm) in front of the saw blade when the saw is in the back position. Some guards cover the lower half of the saw, when the saw is not cutting, and ride on the top of the stock as the saw cuts.

There should be a counterweight or other device to automatically return the swing saw to the back of the table without rebound when released. The counterweight must be secured by a device designed to hold twice its weight. The counterweight shall be guarded, if within 7 ft (2.1 m) of the floor.

There should be a limit chain or other device to prevent the saw from swinging beyond the back or front edge of the table. A device should likewise keep the saw from rebounding from its idling position. A latch with a ratchet release on the handle is positive, but in some instances a nonrecoil spring or bumper is adequate. A magnetic latch provides one method.

Stop and Start buttons should be located for quick and easy access.

Start buttons should be of the protected type so accidental contact will not cause the switch to start. A collar around the button extending ⅛-¼ in. above the top of the button is a recommended method. (See Figure 9-13 and additional examples in Chapter 13, Ergonomics, in the National Safety Council's book, *Fundamentals of Industrial Hygiene,* 3rd ed.)

Stop buttons should be easily contacted in an emergency. Mushroom type buttons are a recommended type.

The saw table can be provided with a wood bumper to pre-

Figure 9-13. Start button for a cutoff saw should be protected so accidental contact will not start the saw. (Printed with permission from *Machine Design Magazine,* June 23, 1977.)

vent bodily contact with the sawblade when it is extended the full length of the support arm.

Operating methods. If the saw is pulled by a handle, the handle should be attached either to the right or left of the saw rather than in line with it. The operator should stand to the handle side and pull the saw with the hand nearer it.

Thus, if the handle is on the right side of the saw, boards should be pulled from the right with the right hand, and the saw should be pulled with the left hand. This method (1) makes it unnecessary for the operator to bring the hand near the saw while it is cutting and (2) keeps the body out of line of the saw.

Saws may be ordered with either right or left handles. On a new installation, a saw should be ordered with the handle on the side from which the stock is to be pulled. If it is necessary to pull stock from the other side, the handle should be placed on that side so the operator can stand in the correct position.

To measure boards, the ends should be placed against a gage stop. When it is necessary to measure the board with a scale while it is on the table, move the board away from the sawblade.

At the completion of each cut, the operator should put the saw back to the idling position and make certain all bounce has stopped before putting his or her hand on the table.

No automatic or constant-stroking saws should be used unless the point of operation is guarded and there is no hazard to the operator.

Underslung cutoff saws

An underslung cutoff saw is usually operated by a treadle. Because its forward movement is fast, it should be completely enclosed in the noncutting position. For general work, it should also be covered by a movable hood guard that slides forward or drops to rest on the stock while the saw is cutting. A treadle guard ensures that use of the treadle is intentional.

Underslung cutoff saws are commonly used to cut knots out of such narrow pieces as flooring and molding. The stock is placed by hand, and the hands are customarily held close to the line of the cut on either side. The movable guard gives little protection because the saw action is so fast that the guard can ride over the top of the hands.

On either side of the line of travel, a barrier guard can be constructed with enough clearance between the guard and the table top to admit the stock, but not the operator's hands or fingers. With practice an operator can feed stock rapidly under this type of guard.

Radial saws

When crosscutting, radial saws cut downward and pull the wood away from the operator and against a fence. These saws, like straight-line pull cutoff saws, require many adjustments to permit their full use. These adjustments can create additional hazards for the user (see Figure 9-14).

The radial saw head can be tilted to cut a bevel, or the supporting beam and track can be swung at an angle to make a miter cut. Both adjustments may be used to cut a compound bevel or miter. Likewise, the head may be turned parallel to the length of the table and the saw used for ripping. In this case it is an overhead, stationary saw against which the stock is fed by hand.

Obviously, a saw with so many features should be operated only by a competent woodworker.

The principal sources of injury connected with radial saw operation are those common to other power-driven saw operations. They include cutting injuries to the arms and hands caused by the sawblade; by flying wood and chips; and by handling materials.

As with other power saws, prevention of injuries requires the equipment to be properly used. The operator should be trained, and be aware of possible hazards. He or she needs to know what to do when the machine is performing below standards.

The upper half of the saw, including the arbor end, always

Figure 9-14. Typical crosscut operation on a radial arm saw. The saw is equipped with an electronic brake motor that reduces the blade's coasting time after machine shutdown.

should be guarded. The lower half of the saw should have an articulating guard for 90-degree crosscut operations. The lower guard shall automatically adjust itself to the stock thickness and remain in contact with the stock being cut for the full working range. This prevents accidental contact with the sides of the sawblade (in an axial direction) when the cutting head is at rest (not in the cut) behind the fence and in the 90-degree crosscut mode. Under certain conditions, lower blade guards can cause additional hazards (refer to ANSI O1.1-1979).

Provide some means so the cutting head will not roll or move out on the arm away from the column because of gravity or vibration (see Figure 9-15).

For repetitive crosscut operations, provide an adjustable stop to limit forward travel of the cutting head to that necessary to complete the cut.

The saw table should be large enough to cover the sawblade in any position (miter, bevel, or rip). Therefore, the saw should never be operated with the sawblade in a position where it protrudes or extends beyond the table (see Figure 9-16).

Ripping. When ripping, the saw head is rotated 90 degrees to a position so the sawblade is parallel to the fence and is clamped in position. Then lower the sawblade until it will cut through the stock. Before ripping, position the nose of the guard, or dropdown guard, and the spreader and anti-kickback devices (see Figures 9-17 and -18). Feed the stock against the direction of rotation of the revolving blade from the nose of the

Figure 9-15. A spring return is installed on this saw so the cutting head will not roll or move out on the arm—away from the column—because of gravity or vibration.

Figure 9-16. When this sawblade is fully extended, it does not extend beyond the edge of the table. Note that the spreader with anti-kickback dogs is in the up position for crosscutting.

Figure 9-17. Before ripping, lower the guard on the infeed side until it almost touches the workpiece. Adjust the splitter-anti-kickback device on the outfeed side so the splitter rides within the saw kerf. Feed the workpiece against the blade rotation.

Figure 9-18. Warning sign on the outfeed side reads CAUTION—NEVER RIP FROM THIS END. Not to heed this warning would cause the workpiece to be pulled from the operator's grasp and sent flying across the room.

guard, the side at which the sawblade rotates upward toward the operator. For in-rip, feed material from right to left; for out-rip, feed from left to right (see Figure 9-19).

A spreader must be used in ripping. The spreader prevents the wood from immediately coming together after being cut. This action reduces the chances of the sawblade binding and causing a kickback. The spreader is mounted in direct line with the blade. When properly adjusted, it must prevent wrongway feed.

The anti-kickback device must be used in the ripping operation. This device is positioned so the fingers ride on the stock. The angle (height) of the fingers should be adjusted so if the stock is pulled out (by hand), it will jam under the fingers and the stock cannot be moved. The anti-kickback device is adjustable for different thicknesses of stock. The anti-kickback fingers should be regularly checked for sharpness and to make certain none of the fingers are bent and are in contact with the stock.

It can be disastrous if proper direction of rotation is not followed. The saw should always rotate upward toward the operator. Feeding from the wrong side (wrong-way feed) tends to grab the material away from the operator and throw it toward the

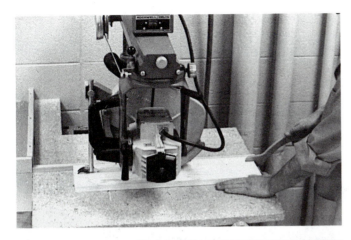

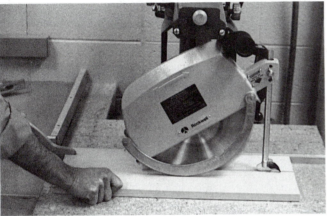

Figure 9-19. Top: In-ripping. Bottom: Out-ripping.

infeed end (nose) of the guard. Wrong-way feed is prevented if a spreader is properly installed (as required by ANSI O1.1) and positioned.

Two possibilities of severe injury arise from improper operation of this type. The sawblade's direction of rotation makes it easy for the operator's hands to be drawn into the revolving saw. There is the additional danger to the helper, and to other people on the opposite (infeed) side of the saw—flying stock can be thrown with enough force to drive the stock through a one-inch board. There have been serious and fatal injuries because of failure to observe this precaution.

Power feed rolls are available for ripping operations. These not only greatly reduce kickback, but also speed up production.

When feeding, work must be held firmly against the table and fence. The sawblade should be sharp and be parallel to the fence. Feed pressure must always be applied between the sawblade and the fence. A pushstick (length = sawblade diameter + 6 in. [15.2 cm]) must be used when ripping narrow or short stock.

The operator should never release feed pressure until the cut is completed and the workpiece has fully cleared the sawblade.

Special care must be exercised when ripping thin, lightweight, hard, or slippery surface materials, because of the reduced efficiency of anti-kickback devices. When ripping, the operator should wear an anti-kickback apron.

Crosscutting. Saws of this type are pulled across the cutting area by means of a handle located to one side of the

sawblade, rather than in line with the sawblade. Whenever possible, the operator should stand on the handle side, pull the cutting head with the head nearest the handle, and maneuver the lumber with the other hand. This position keeps the body out of line with the sawblade, and, at the same time, makes it unnecessary for an operator's hands to be near the sawblade's cutting area. The sawblade should never be pulled out beyond the point necessary to complete the cut, because the back of the blade could lift the workpiece and throw it back over the fence.

Short pieces should never be removed from the table until the saw has been returned to its position at the rear of the table. A stick or brush should always be used to remove scrap from the table—never the hands.

Under normal circumstances, measuring should be done by placing the boards to be cut against a stop gage; however, in instances where it is necessary to measure with a rule, turn the saw off until the measuring is completed.

At the conclusion of each cut, the operator should always return the cutting head to the full rearward position, behind the fence. The operator's hand should never be removed from the operating handle unless the cutting head is in this position.

Since the sawblade's direction of rotation and the feed's direction tend to cause the sawblade to feed itself through the work, the operator should develop the habit of holding the right arm straight from the shoulder to the wrist, to prevent grabbing and possible stalling of the sawblade while in the workpiece.

Power-feed ripsaws

Because long stock is often ripped on power-feed ripsaws, the clearance at each working end of the saw table should be at least 3 ft (0.9 m) longer than the length of the longest material handled. Feed rolls should be adjusted to the thickness of the stock being ripped. Insufficient pressure on the stock can contribute to kickbacks. (See Figure 9-20.)

Where multiple-cut power ripsaws are used on a production basis, it is advisable to have a dado head attachment installed alongside the last sawblade. This head disposes of the edging. The offbearer then does not have to handle any scrap pieces and can pay more attention to material coming from the saw.

A common accident occurs with a power-feed chain ripsaw using an overhead cutting saw with solid chain having a rabbeted center trough. Unless care is taken to maintain the rabbet in the trough, very thin slivers may drop from a ripped edge into the center trough of the chain—these can come flying out like bullets toward the operator. Regular inspection is necessary.

Band saws

Although injuries from band saws are less frequent and less severe than those from circular saws, they are not uncommon. The usual cause of band saw injuries is hands coming into contact with the sawblade. When hand feeding, the operator's hands must come close to the blade. Therefore, it is particularly important that the saw table be well lighted, but free from glare.

The band saw point of operation cannot be completely covered, but an adjustable guard, designed to prevent operator contact with the front and right side of the sawblade above the upper blade guides, should be set as close as possible to the workpiece (see Figure 9-21).

Figure 9-20. Anti-kickback fingers should spread at least the full width of the feed rolls between the operator and the sawblade on a power-feed saw. This saw is equipped with a double set of anti-kickback fingers. (Printed with permission from Western Electric Company.)

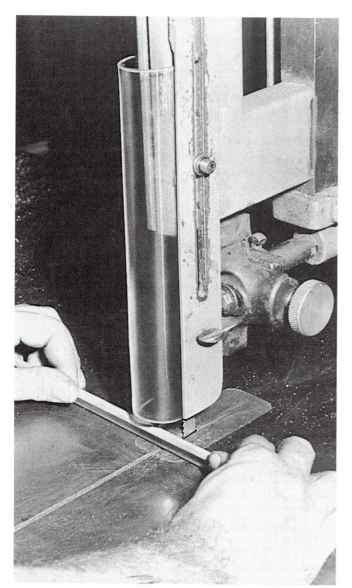

Figure 9-21. A clear plastic shield has been formed into a guard for this band saw.

The wheels and all nonworking parts of the blade should be encased and the outside of the enclosure should be of solid metal. The front and back of the enclosure should be solid or sturdy mesh material (Figure 9-22).

A band saw should have a tension control device to indicate proper blade tension. If it does not, the operator should test the blade for correct tension before beginning the operation. An automatic tension control device will help to prevent blade breakage. Another type of device also prevents the motor from starting if the tension on the blades is too much or too little.

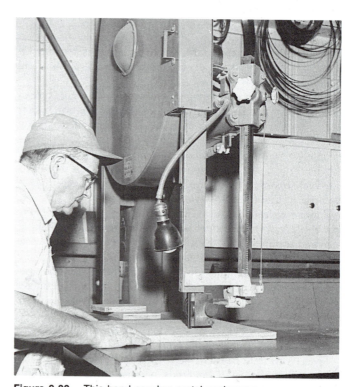

Figure 9-22. This band saw has metal enclosures over its wheels and blade. A light fixture with flexible connection provides the correct illumination. (Printed with permission from U.S. Naval Civil Engineering Laboratory, Port Hueneme, California.)

Band sawblades should periodically be examined for cracks and broken teeth.

Because a modern band saw, especially a large one, will run for a long time after the power is shut off, it should have a brake operating on one or both wheels to minimize the potential hazard of coasting when the machine is shut off and left unattended.

Serious injuries have occurred when operators took hold of the running blade, not realizing it was in motion. Another reason for brakes is to stop the wheels in case the blade should break. The safety device shown in Figure 9-15 stops the saw if the blade breaks.

On a band saw, a guard formed to the curvature of the feed rolls should cover the nip point. It should be installed so the edge is ⅜ in. (10 mm) from the plane that is formed by the inside face of the feed roll in contact with the stock.

When small pieces of stock are cut, a special jig or fixture should be used.

A technique for safely sawing a sharp radius is shown in Figure 9-23.

Jig saws

Jig saws are not normally considered hazardous, but occasionally they cause injuries too, especially to the fingers and hands.

Safe operating procedure requires the blade to be properly attached and secure, the threshold rest (slotted foot) to be on the stock, the guard to be in an effective position, and the operator to keep the hands a safe distance from the blade.

In addition, turn cuts should be made slowly, with no sharp or small radius turns if working with a wide blade, and clearance cuts should be planned to eliminate the need to back out of curves. The table should be cleaned with a long-handled brush after the blade has stopped.

WOODWORKING EQUIPMENT

Jointer-planers

Hand-feed jointers or surface planers are, second to circular saws, the most dangerous woodworking machines (see Figure 9-24). Most of the injuries are caused by the hands and fingers

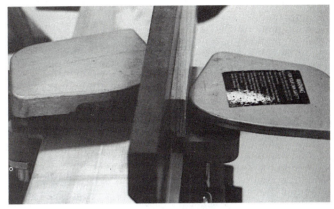

Figure 9-24. This jointer is equipped with two guards: one on the working side of the fence, and the second on the back side of the blade.

contacting the knives. Many of these accidents occur when short lengths of stock are being jointed.

A jointer should be equipped with a horizontal cutter head, the knife projection of which extends beyond the body of the head not more than ⅛ in. (3.2 mm). The clearance between the path of the knife projection and the rear table should not be more than ⅛ in. The clearance is measured radially from the path of the knife projection to the closest point on the table and with the rear table level with the path of the knife projection. The clearance between the path of the knife projection and the front table shall not be more than 3/16 inch (4.8 mm). This clearance is measured as described in the preceding sentence and with the tables coplanar. (See Figure 9-25.)

The openings between the table and the head should be just large enough to clear the knife, but not more than 2½ in. (63.5 mm) when the front and rear tables are set or aligned with each other for zero cut.

Cover the table opening on the working side of the fence with a guard that adjusts itself to the moving stock. For edge jointing, good protection can be achieved by installing a swinging guard or a guard that moves away from the fence along the axis of the other head. For surface planing, only the swinging guard

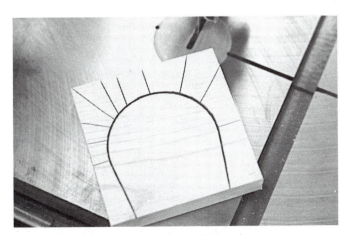

Figure 9-23. When sawing a sharp radius, make several release cuts up to the cutting line. This prevents sawblade binding and possibly breaking the blade or causing it to jump off the guide wheels.

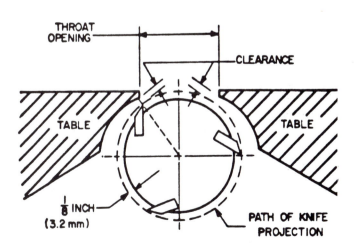

Figure 9-25. Table clearance for jointers. (Printed with permission from American National Standards Institute; Underwriters Laboratories Inc.)

Figure 9-26. Note the difference between a hold down (*left*) and a push block (*right*). The push block has a piece of wood acting as a positive stop against the end of the workpiece; the hold down is flat on the bottom. Both are used to keep the operator's thumbs and fingers away from the cutter head.

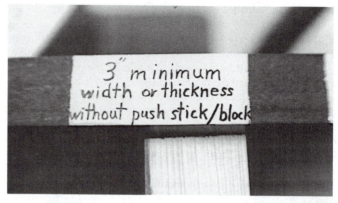

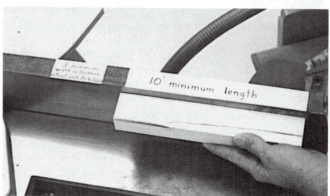

Figure 9-27. Painting a 3 in. (7.6 cm) strip (*top*) and a 10 in. (25 cm) strip (*bottom*) on top of the fence—and labeling them accordingly—serves as a quick reference for checking the width, thickness, or length of stock before jointing it.

should be used because it permits the use of holddown-push blocks (such as those pictured in Figure 9-26) to feed material smoothly over the other head. Holddown-push blocks should be used whenever the operator joints wood that is narrower than 3 in. (7.6 cm) (see Figure 9-27).

The unused end of the head (behind the fence) should be enclosed at all times. A sheet metal telescoping guard is acceptable for this purpose.

Jointer-planers are commonly used for planing off cupped or warped stock to make it flat. To do this job, power-feed attachments are available with resilient holddown devices that simulate the pressure of the hands. If stock is properly conditioned, there should be comparatively little work having to be trued in this manner.

For surfacing work on the jointer, the operator should have both hands on top of the stock (if thicker than 3 in.), never over the front or back edge where they can easily come in contact with the head.

Shapers

The principal danger in the use of the wood shaper is that hands and fingers might strike against the revolving knives. Severe accidents can also result when broken knives are thrown by the machine. When a shaper knife breaks or is thrown from the collar, the other knife is usually thrown too so that four or five pieces of heavy, sharp steel are thrown about the shop with sufficient speed to kill a person.

The danger from broken or thrown knives can be eliminated by using solid cutters that fit over the spindle. The initial cost is greater for cutters than for knives, but, on moderately long runs, cutters are less expensive. Carbide tip or solid carbide cutters are available. In all cases they are safer.

When knives are used, several precautions can be taken to keep them from breaking or flying:

1. Knives for shapers must meet rigid specifications for shaper steel.
2. Knives must be sharpened and installed only by a fully qualified person.
3. Knives and the grooves in the collars must fit perfectly and be free of dust.

4. The two knives must balance perfectly. They must be weighted against each other in a beam balance each time they are set.
5. A knife must not be used after it has become so short that the butt end does not extend beyond the middle point of the collar.
6. Deep cuts should be avoided. It is safer and more efficient to take two light cuts than one heavy cut.
7. During startup, the operator should apply the power in a series of short starts and stops to slowly bring the spindle up to operating speed. He or she should listen carefully for chatter and watch for other evidence that knives are out of balance.

Various types of safety collars are in use that help prevent shaper knives from flying. Nothing should be done to discourage the use of such collars, but they should not be considered substitutes for perfectly balanced and fitted knives of adequate length.

There should be some type of braking device to stop the spindle after the power is shut off. With double spindle shapers, there should be starting and stopping devices for both spindles, and the spindles should be started one at a time.

Only a long-handled brush should be used to remove chips and scraps from the work table.

A number of guards are available to protect the fingers. For straight-line shaping, the fence frame or housing should contain the guard. A fence should have as small as possible an opening for the knives and should extend as far as possible, at least

18 in. (46 cm), on either side of the spindle to assure good support at the start and finish of the cut. If the entire edge of the stock is to be shaped, the portion of the fence beyond the cutter should be adjusted so as to receive the thinner stock and provide a stable bearing. This cannot be accomplished with a flat, continuous fence, but it is best done with a split, adjustable fence.

For curve shaping, an overhead guard should be adjusted to just clear the stock.

The greatest number of accidents occurs when shaping narrow stock, which if held in the hand brings it close to the knives. Holddown-push blocks or jigs should be used. A cardinal rule is to always feed against the direction of rotation of the cutter—never back up the workpiece. A starting pin must be used when curve shaping.

Each shop should have a well-understood rule that stock narrower than a specified width must be held in a jig. Some shops put the limit at 6 in. (15 cm), others as high as 12 in. (30 cm).

The cutting head guard should be adjusted for a minimum head exposure. Provision should be made to contain wood dust and chips. Jigs and holddown clamps should be maintained to hold the stock securely.

Shaper work must be held against guidepins (curved shaping) or a fence (straight-line shaping). A feather board may be clamped to the fence (see Figure 9-28). The portion of the cutter or knives behind the fence should be covered.

When feeding the shaper, the operator must never forget that the direction of the cut must be made in the opposite direction of the rotation of the cutting head. Never back up the workpiece or a kickback can occur.

Power-feed (thickness) planers

Planer vibration can be reduced by anchoring the planer on a solid foundation and by insulating it from the foundation with cork, springs, or other vibration-absorbing material. The planer should have a three-point bearing and be bolted down without distortion, since distorting any woodworking machine will ultimately cause it to malfunction. Because of noise, it is desirable to isolate a planer in a separate room or build a soundproof enclosure for it. Helical cutterheads also will substantially reduce the noise levels. If neither soundproofing nor using helical cutterheads is practical, hearing protection should be worn by those in the immediate area.

Cutter heads should be completely enclosed in solid metal guards, which should be kept closed when the planer is running. There should be good local exhaust from the cutting heads.

Feed rolls, cutter heads, and cylinders must be stopped before the operator reaches into the bed plate to remove wood fragments, to make adjustments, or for any other reason. If planer parts are driven by belts running on the back side of the planer, belts and sheaves should be completely enclosed by sheet metal or heavy mesh guards, even though the planer is fenced at the back or is next to the wall.

Feed rolls should be guarded by a wide metal strip or bar that will allow boards to pass, but will keep the operator's fingers out of the rolls. Anti-kickback fingers should be installed and be operative on the infeed side across the entire throat (width) of the machine.

Danger of kickbacks cannot be entirely overcome by mechanical means. Therefore, the operator should always stand out of

Figure 9-28. Top: A pair of feather boards clamped to the table top hold the workpiece down on the table and in against the fence. The operator is using a push stick. Bottom: This ring guard is installed over the cutter at the point of operation.

the line of board travel. Other people should not work or walk directly behind the feeding end of the planer. A barrier or guardrail should be considered during operation of the machine. Operators must avoid feeding boards of different thicknesses at the same time because a thinner board is not held by the feed rolls and can be kicked back from the heads.

The operator should wear safety goggles. To keep workers from being struck by long, fast-moving boards, the space at the outrunning end should be fenced or marked off.

Sanders

Drum, disk, or belt sanding machines should be enclosed by an exhausting dust hood that encloses all portions of the machine except the portion designed for the work feed. Personnel who operate sanders should wear goggles and dust respirators during sanding operations and cleanup.

On a belt sanding machine, a guard should be placed at each inrunning nip point on both power transmission and feed roll parts. The unused run of the abrasive belt on the operator side of the machine should be guarded to prevent human contact.

All manually-fed sanders should have a work rest and be properly adjusted (1) to provide minimum clearance between the belt and the rest, and (2) to secure support for the work. Small pieces should be held in a jig or holding device.

Abrasive belts on sanders should be the same width as the pulley-drum. The drums should be adjusted to keep the abrasive belt taut enough to turn at the same speed as the pulley-drum.

SUMMARY OF SAFETY RULES FOR VARIOUS WOODWORKING TOOLS

Every operator should be trained in the safety rules covered in this chapter. As a summary, safety rules that demand close attention are listed below. Be sure operator checks the manufacturer's manual, understands the requirements, and follows the recommended procedures.

TABLE SAW

- Feed with body to side of stock
- Blade height
- Splitter and antikickback fingers for ripping
- Stock firm to fence
- Remove ripfence for crosscuts
- Blade guards

CIRCULAR SAW

- Blade guards
- Binding
- Blade—correct type
- Blade—tight on the arbor
- Firm support for work
- No obstructions
- Begin cut with motor at manufacturer's recommended speed for materials being cut

RADIAL ARM SAW

- Ripsawing—direction of (cut) feed and antikick fingers
- Blade guards
- Pull for cross cuts
- End plates on track—arm tight
- Clamp handles tight
- Material tight to fence
- Return cutter to rear of track

BAND SAW

- Feed with body to side of stock
- Guard height ⅛-in. clearance of material
- Tension and type of blade
- Release cuts before long curves
- Stop machine to remove scrap or pull out incomplete cut
- Flat stock

JOINTER/PLANER

- Depth of cut
- Length of stock
- Sharp cutters
- Don't pass hands over cutters
- Push stick for small stock
- Guard

WOOD SHAPER

- Clamping work piece
- Use correct guard
- Feed into knives—don't back off
- No feeding between fence and cutters
- Collar and starting pin work for irregular work—stock of sufficient weight
- Fence opening only enough to clear cutters
- Use stock as guard by shaping the underside of stock
- Spindle nut tight
- Shape only pieces 10 in. or longer

SANDER

- Keep hand from abrasive surface
- Ventilation
- Belt or disk condition
- Sand on downward side of disk

LATHE

- Stock without defects, glued joints dry
- (When using V-belt lathes, power should be off when changing speeds.)
- Toolrest close to stock
- Hold tools firmly in both hands
- Remove toolrest when sanding or polishing

Figure 9-29. Summary of rules for the safe operation of woodworking tools. (Printed with permission from Power Tool Institute, Inc.)

when material is brought into contact with the moving abrasive belt.

Inspect abrasive belts before using them. Those found to be torn, frayed, or having any excessive wear should be replaced.

Feed rolls should be guarded by a wide metal strip or bar that will allow boards to pass, but will keep the operator's fingers out.

Lathes

Rotating heads, whether running or not, should be covered as completely as possible by hoods or shields, which should be hinged so they can be thrown back when making adjustment.

Lathes used for turning long pieces of wood stock held only between the two centers should be provided with long curved guards extending over the tops of the lathes to prevent the work from being thrown out should they come loose. Safety goggles must be worn.

The operator should carefully inspect all parts for any defects so they can be corrected before operation. Workpieces containing checks, splits, cracks, or knots must not be used.

Extreme care is essential when selecting and training the lathe operator. The operator must give constant attention to stock being turned in order to discard any material likely to break. He or she must carefully place stock in the machine and feed the cutting tool slowly into the stock. Safety goggles are required, and also a faceshield if the operation is dusty. Long hair or loose clothing must not be permitted.

Figure 9-29 summarizes the safety rules for lathe use, and other woodworking tools.

Routers

Jigs with handles should be used to keep hands at least 6 in. (15 cm) away from the cutting bit. A ring guard should be mounted around the cutting bit to reduce contact with it.

ACKNOWLEDGMENT

In most cases, and unless otherwise noted, the photos used in this chapter were supplied by Wausau Insurance Companies.

REFERENCES

American National Standards Institute, 1430 Broadway, New York, NY 10018.
Safety Requirements for Woodworking Machinery, O1.1-1975.
Safety Standard for Mechanical Power Transmission Apparatus, ANSI/ASME B.15.1-1984.
Safety Standard for Stationary and Fixed Electric Tools, ANSI/UL 987.
National Fire Protection Association. Batterymarch Park, Quincy, Mass. 02269.
Fire Protection Handbook.
Prevention of Fires and Explosions in Wood Processing and Woodworking Facilities, NFPA 664.
National Safety Council, 444 North Michigan Avenue, Chicago, Ill. 60611.
Safeguarding Concepts Illustrated.
Industrial Data Sheets
Electric Hand (Circular Blade Type) Saws, 675.
Power Feed Planers (Wood), 225.
Radial Saws, 353.
Sawmill Edgers, 571.
Tilting-Table and Tilting-Arbor Saws, 605.
Wood Turning Lathes, 253.
Woodworking Band Saws, 235.
Power Tool Institute, Inc., 5105 Tollview Drive, Rolling Meadows, Ill. 60008.
U.S. Department of Health and Human Services, National Institute for Occupational Safety and Health, Cincinnati, Ohio 45226.
Health and Safety Guide for Manufacturers of Woodworking Machinery, DHEW (NIOSH) Publication 79-131.
Health and Safety Guide for Millwork Shops, DHEW (NIOSH) Publication 76-111.
Health and Safety Guide for Plywood and Veneer Mills, DHEW (NIOSH) Publication 77-086.
Health and Safety Guide for Prefabricated Wooden Building Manufacturers, DHEW (NIOSH) Publication 76-159.
Health and Safety Guide for Sawmills and Planing Mills, DHEW (NIOSH) Publication 78-102.
Health and Safety Guide for Wooden Furniture Manufacturers, DHEW (NIOSH) Publication 75-167.

Metalworking Machinery

INJURIES ON MACHINE TOOLS are most often caused by unsafe work practices or incorrect procedures—basically problems of training and supervision. The more hazardous equipment with exposed moving parts and cutting edges should only be operated by qualified and competent personnel. If equipment were properly maintained and operated, injuries from machine tools would be rare. Installation of effective guarding devices that do not hamper operation or lower production would further reduce the number of injuries. Of course, certain guards are required by OSHA, state or provincial, and local regulations.

More rarely injuries result when a machine fails mechanically or is operated after an unsafe condition develops.

Good housekeeping and good habits in maintaining a ship-shape work area can help establish good habits in machine operation, with a resulting reduction in accident rates.

Definition. Metalworking machinery tools include all power-driven machines that are not portable by hand, and are used to shape or form metal by cutting, impact, pressure, electrical techniques, chemical techniques, or a combination of these processes. Grinders, buffers, and similar machines are included in this definition.

The National Machine Tool Builders Association has classified some 200 types of machine tools into five basic groups—turning, boring, milling, planing, and grinding. Another classification is electrodischarge, electrochemical, laser, and machining. Some machines combine the functions of two or more groups.

Power presses, press brakes, and power squaring shears are also classified as metalworking tools. However, these are covered in Chapter 11, Cold Forming of Metals.

Portable power tools, normally hand-held during operation, are not considered to be machine tools. They are covered in Chapter 14, Hand and Portable Power Tools.

GENERAL SAFETY RULES

Great emphasis should be placed on safe operation. A policy to eliminate unsafe practices by operators should be established and should include these provisions:

1. Operation, adjustment, and repair of any machine tool must be restricted to experienced and trained personnel.
2. All personnel should be under close supervision during training.
3. Safe work procedures must be established, with shortcuts and chance-taking prohibited.
4. Supervisors must be responsible for the strict enforcement of this policy.
5. When purchasing new equipment, make sure that specifications conform to all applicable regulations concerning guarding, electrical safety, and other safeguards.
6. New equipment should be inspected and safety innovations made before allowing operator(s) to use the equipment.
7. Devote full-time attention to the work in progress. If the operator must leave the machine, it should be shut down unless the machine tool has been designed to operate in this mode. For example, have interlocked guarding all around and equip with automatic shutdowns, which work as soon as there is any deviation from normal operation.

Maintenance and repair personnel should comply with this seven-point policy.

A tool rack should be provided for the convenience of the operators and repair and maintenance personnel. All wrenches and tools needed for operation or adjustment should be included as standard equipment.

The National Electrical Code (ANSI/NFPA 70) and the *National Electrical Safety Code* (ANSI C2) should govern the installation of electrical circuits and switches.

Other metalworking standards can be found in NFPA 79 *Electrical Standard for Industrial Machinery.*

In addition to the manufacturer-installed electrical controls on machine tools, each machine must have a disconnect switch that can be locked in the OFF position to isolate the machine from the power source.

Maintenance or repair should not be permitted on any machine until the disconnect switch serving the equipment has been shut off, padlocked in the OFF position, and tagged. (See the discussion of Zero Mechanical State in Chapter 8.)

The following rules apply to the safe operation of any machine tool:

1. Machine tools should never be left running unattended, unless the machine has been designed to do so.
2. Operators should not wear jewelry or loose-fitting clothing, especially loose sleeves or shirt or jacket cuffs and neckties. Long hair, which could be caught by moving parts, should be covered. (See the next section, Personal protection.)
3. All operators should wear eye protection, as should others in the area, such as inspectors, stock handlers, and supervisors.
4. Throwing refuse or spitting in the machine tool coolant should not be allowed—such actions foul the coolant and might spread disease.
5. Manual adjusting and gaging (calipering) of work should not be permitted while the machine is running.
6. Operators should use brushes, vacuum equipment, or special tools for removing chips—they should not use their hands.
7. Operators should use the proper hand tools for each job.

One of the major causes of eye accidents on machine tools, especially on drilling equipment, is indiscriminate use of high-pressure compressed air to blow chips from machines or workers' clothing. Brushes provide a less dangerous method; as do hand-held air or electrically operated vacuum units.

In cases where neither a brush nor a vacuum system is practical, it may be necessary to use air, but the line pressure should be as low as possible. Many companies have found a nozzle pressure of 10 to 15 psig (70-100 kPa) is sufficient for most operations. (OSHA specifies that the pressure be less than 30 psig −207 kPa.)

The operation should be isolated so that employees nearby are not endangered. Baffles or chip guards should be placed around the machine to shield the operator. Only reliable and trained employees should be permitted to use compressed air, and they should be wearing cup-type goggles and other personal protective equipment. Nozzles meeting OSHA requirements are available.

Employees shall be prohibited from using high-pressure compressed air to blow dust or dirt from their clothing or out of their hair because damage to or injection of foreign materials into their ears and eyes can result.

Employees should understand the differences in machining ferrous and nonferrous metals. They should also know the health or fire hazards.

Personal protection

The machine tool operator's safety depends largely upon following established safe work procedures and wearing proper protective clothing and equipment.

Obviously, all machine tool operators should wear eye protection. In addition, there should be some effort made to confine or control the metal chips and coils removed from the stock being machined; otherwise, there is no way for an operator to judge or control their direction of flight.

Wearing closely fitting clothing is vitally important to the operator's safety. Many serious injuries and fatalities have resulted when neckties, loose shirt sleeves, or other clothing were caught in a belt and sheave, between gears, in a revolving shaft, or in the revolving work held in the chuck (Figure 10-1).

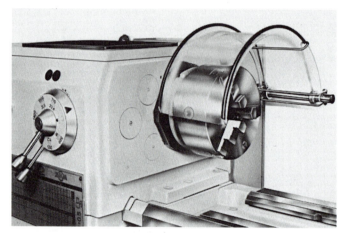

Figure 10-1. This lathe chuck shield has a semicircular construction and is made from high-impact-resistant, transparent acrylic. The shield is mounted to a chromium-plated extension tube fastened to the lathe headstock by mounting brackets. The shield can be lifted up and out of the way for quick and easy access to the piecepart.

Both male and female operators having long hair should wear caps, snoods, hairnets, or other protection that completely covers the hair.

There have been many instances of people being partially (or entirely) scalped when their hair became entangled in the moving parts of a machine. Operators also should not wear gloves, rings, necklaces or neckties, or jewelry.

Because most machine operations involve handling heavy stock or heavy machine parts, such as face plates and chucks, every operator should wear safety shoes.

Splash guards, shields, personal protective equipment, and other means may be practical to minimize worker exposure to the irritating cutting oils and mineral spirits used to clean metal parts. Barrier creams and personal hygienic measures will also minimize skin irritations.

TURNING MACHINES

Turning—shaping a rotating piece with a cutting tool, usually to give a circular cross section—is done on machines such as engine lathes, turret lathes, chuckers, semiautomatic lathes, and automatic screw machines.

Figure 10-2. A swinging, welded pipe fixture (*right*) supports the lathe chucks and face plates. This reduces operator strain when changes are made.

Engine lathes

To prevent accidents, an engine lathe must be correctly operated by qualified personnel. Injuries are likely to result from:

1. Contact with projections on work or stock, face plates, chucks, or lathe dogs, especially those with projecting setscrews
2. Flying metal chips
3. Hand braking the machine
4. Filing right-handed, using file with unprotected tang, or using the hand instead of a stick to hold an emery cloth against the work
5. Calipering or gaging the job while the machine is in operation
6. Attempting to remove chips when machine is in operation
7. Contact with rotating stock projecting from turret lathes or screw machines
8. Leaving chuck wrench in chuck; using a spring-loaded wrench can eliminate this hazard
9. Catching rings, loose clothing, or wiping rags on revolving parts.
10. Wearing gloves and catching the gloves on revolving parts.

Injuries may also occur on lathes because operators fail to keep the center holes of taper work clean and true and the lathe centers true and sharp. It is hazardous for operators to leave the machines running unattended, or to handle chips by hand instead of using a hooked rod.

Face plates and chucks without projections should be used whenever possible. Otherwise, a simple shield formed to the contour of the chuck or plate and hinged at the back should be installed to prevent contact with the revolving plate or chuck.

Safety-type lathe dogs are relatively inexpensive to install and should be substituted for those with projecting setscrews. Chip shields, particularly on high-speed operation, will control flying chips. Plastic or small-mesh screen chip guards have been successfully used because they allow visibility and confine the flying chips (Figure 10-1). Use of these shields does not eliminate the need for protective eye equipment. Shields may need

Figure 10-3. This automatic screw machine has a chip shield to protect against cutting oil splashes and flying chips. (Printed with permission from Delco-Remy, Division of General Motors.)

to be frequently replaced because of the chips' abrasive action.

Mechanical means, such as an overhead hoist or a special device (Figure 10-2), should be provided to lift heavy face plates, chucks, and stock on both lathes and screw machines.

Turret lathes and screw machines

Hazards associated with turret lathes and screw machines (Figure 10-3) are similar to those listed for other lathes. Additional hazards are caused by the operator not moving the turret back as far as possible when changing or gaging work, or using machine power to start the face plate or chuck onto the spindle.

Other hazards result when the operator fails to keep hands clear of the turret slide or permits a hand, arm, or elbow to strike the cutter while adjusting or setting up.

Splash shields, especially on automatic machines, should be installed and kept in good condition. Enclosure shields over the chuck confine hot metal chips and oil splashes, and also act as exhaust hoods for fume removal.

When steel and some other materials are turned on lathes, the chip produced is in a continuous spiral that frequently causes hand and arm injuries. Proper chip breakers will provide protection against this. Figure 10-4 illustrates two types of chip breakers.

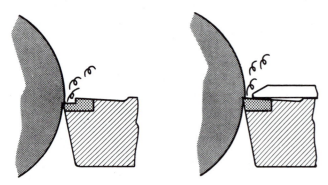

Figure 10-4. Chip breakers can be ground into the tool tip (*left*), or a piece of metal can be clamped or brazed onto the tool (*right*).

Screw machines can create noise exceeding OSHA standards. Consideration should be given to the noise level reduction through engineering methods, including commercially available sleeves for rotating bar stock. See details in the Council's publication *Industrial Noise and Hearing Conservation.*

Spinning lathes

A spinning lathe is a forming tool rather than a cutting tool and usually requires a specially skilled and qualified operator.

Unsafe practices that should be prohibited when operating a spinning lathe include:
1. Inserting blanks and removing the processed part without first stopping the machine
2. Failing to fully tighten the tailstock handle and risking that the blank will work loose or ruin the stock or tool
3. Allowing the swarf to build up into a long coil when trimming copper and certain grades of steel.

This last practice has caused hands to be severed and arms severely cut. One fatality occurred when a coil became snarled around the operator's neck. Operators should remove the tool, when necessary, to allow the swarf to break off.

The spinning lathe chuck is usually a form built up of hardwood, shaped exactly like the finished part. If the piece being worked must be "necked-down" so the chuck cannot be taken out after the piece is formed, the chuck is made in sections held together by locking rings or locking grooves.

Great hazard is possible if a chuck flies apart because the grooves have become worn or the rings break. In view of the speed of spinning lathes (500 to 2000 rpm), the danger of flying sections of the chuck is obvious. Prevention of injuries from this cause lies in frequent inspection and maintenance of chucks. Similarly, lathe tools should be frequently inspected for cracks in the tool or handle.

Vibration or worn parts can cause the tailstock to loosen during operation and the piece then will work loose or fly off. The solution is frequent inspection and maintenance.

BORING MACHINES

Boring consists of cutting a round hole using a drill, boring cutter, or reamer.

The drill press, in its various forms, is probably the best known machine tool because of its frequent use in home workshops. Drilling machines are equipped with rotating spindles, handles, and chucks that carry pointed or fluted cutting tools. Operations performed with drilling machines include countersinking, reaming, tapping, facing, spot facing, and routing.

Boring mills use a cutter, either single- or multi-edged, and mounted on a supporting spindle or shaft to true up or enlarge a hole that has already been rough formed by drilling, casting, or forging.

Drills

The most common hazards in drilling operations are:
1. Contacting the rotating spindle or the tool. (Do not touch the tool while using a quick-change clutch.)
2. Being struck by a broken drill
3. Using dull drills
4. Being struck by insecurely clamped work
5. Catching hair, clothing, or gloves in the revolving parts
6. Sweeping chips or trying to remove long, spiral chips by hand
7. Leaving key or drift in chuck
8. Being struck by flying metal chips
9. Failing to replace guard over speed change pulley or gears

To guard an operator from contact with a spindle, use a plastic shield, a simple wire mesh guard, or other barrier (Figure 10-5).

When necessary, the tool can be guarded by a telescoping guard that covers the end of the tool, leaving only enough of it exposed to allow easy placement into the piece being worked.

The telescoping drill shield shown in Figure 10-5 has a stationary, ribbed cage. The outer sleeve rides upward and downward on the inner cage. This shield provides high visibility.

A frequent cause of breakage is using a dull tool. A thin drill, smaller than 1/8-in. (3 mm) diameter will often break and cause injury. A larger drill may "fire up," freeze in the hole, and then break. Furthermore, a frozen tool may cause unclamped or insecurely clamped work to spin and injure the operator.

To avoid having a drill catch in thin material and spin it, the work can be clamped between two pieces of metal or wood

Figure 10-5. This transparent drill shield protects against flying chips and pieces of broken drills.

before drilling. When drilling thin ferrous stock, it is advisable to grind the drill point to an included angle of about 160 degrees, and thin the point of the drill by grinding the flutes. With nonferrous metals a negative rake will further reduce chances of the drill grabbing or digging in.

Drill press accidents are more likely to occur during odd jobs, because special jibs or vises for holding the work are not usually provided.

When deep holes are being drilled, the drill should be frequently removed and the chips cleaned out. If chips are allowed to pile up, the tool may jam, with results similar to those of freezing.

Counterweight chains should be maintained in good condition and a shield should be installed around the counterweight.

Radial drill accidents are frequently caused by incorrect control manipulation. The drill head and arm, as well as the workpiece, should be properly clamped prior to cutting metal.

Boring mills

Some common causes of injury in boring mill operations are:
1. Being struck by insecurely clamped work or by tools left on or near revolving table
2. Catching clothing or wiping rags in revolving parts
3. Falling against revolving work
4. Calipering or checking work while machine is in motion
5. Allowing turnings to build up on the table
6. Removing turnings by hand.

The same accident prevention measures are effective on both the table and floor types of horizontal boring mills. The operator should never attempt to make measurements near the tool, reach across the table, or adjust the machine or work while the machine is in motion.

Clamps and blocking should be frequently inspected to make certain the clamping is positive. Makeshift setups always should be avoided.

Before attempting to raise or lower a boring mill head, the operator should make sure that the clamps on the column have been loosened. Otherwise, the boring bar can be bent or the clamps or bolts broken, with possible damage to the machine and injury to the operator.

Before the boring bar is inserted into the spindle, the operator should make certain that the spindle hole and the bar are clean and free from nicks. No attempt should be made to drive the bar through the tail stock bearing with a hammer or other heavy tool.

A soft metal hammer should be used to drive the bar into the spindle. If a steel hammer or piece of steel must be used, the operator should hold a piece of soft copper or brass against the bar while driving it into the spindle.

The same procedures apply to the safe operation of a vertical boring mill. Each mill table, particularly those tables 100 in. (2.5 m) or less in diameter, should have the rim enclosed in a metal band guard to protect the operator from being struck by the revolving table or by projecting work. Such guards should be hinged so they can be easily opened during setting up and adjustment (Figure 10-6).

If the table is flush with the floor, a portable fence, usually of iron pipe sections, should be installed. Such fencing should conform to the state code or the specifications of the ANSI standard A12.1, *Safety Requirements for Floor and Wall Openings, Railings, and Toe Boards.*

The operator should never attempt to tighten the work or the tool, caliper or measure the work, feel the edges of the cutting tool, or oil the mill while it is in operation. He or she should never ride the table while it is in motion. There is an exception—on some large mills, like those used for boring turbine castings, the operator may have to ride the table in order to observe work progress. In such cases, he or she should always make certain that no portion of the body will come in contact with a stationary part of the mill.

Steps or stairs which provide access to the machine or to the work should have a pitch of not more than 50 degrees and should have non-slip treads. Stairs with four or more risers must have a handrail.

MILLING MACHINES

Milling—machining a piece of metal by bringing it into contact with a rotating multi-edged cutter—is done by horizontal and vertical milling machines, by gear hobbers, profiling machines, circular and band saws, and a number of other types of related machines (Figure 10-7).

Many milling machine accidents occur when operators unload or make adjustments. Other causes of injuries include:
1. Failure to draw the job back to a safe distance when loading or unloading
2. Using a jig or vise that prevents close adjustment of the guard
3. Placing the jig- or vise-locking arrangement in such a position that force must be exerted toward the cutter
4. Leaving the cutter exposed after the job has been withdrawn
5. Leaving hand tools on the worktable
6. Failing to securely clamp the work
7. Reaching around the cutter or hob to remove chips while the machine is in motion
8. Removing swarf (fines, turnings, or particles) by hand instead of with a brush
9. Adjusting the coolant flow while the cutter is turning
10. Calipering or measuring the work while the machine is operating
11. Using a rag to clean excess oil off the table while the cutter is turning

Figure 10-6. This guard for a vertical boring mill is made in two sections of sheet metal. The sections are hinged to the machine. Left: the guards are closed. Right: the guards are opened to allow setup or adjustment.

Figure 10-7. The adjustable shield on this vertical milling machine protects against both flying metal chips and splashing coolant.

12. Wearing gloves, rings, ties, or loose clothing
13. Using incorrectly dressed cutters
14. Incorrectly storing cutters
15. Attempting to remove a nut from machine arbor by applying power to the machine
16. Striking the cutter with hand or arm while setting up or adjusting stopped machine
17. Misjudging clearances between the arbor or other parts
18. Cleaning the machine while it is in motion
19. Catching fingers, gloves, or clothing in power clamps.

Basic milling machines

Regardless of the classification, direction of movement, or special attachments that make varied operations possible on a milling machine, the safeguarding requirements are basically the same. To guard the cutter, one of several methods can be employed (see Figure 10-8).

The hand-adjusting wheels, for quick or automatic traverse on some models, should be mounted on the shaft by either clutches or ratchet devices so the wheels do not revolve when the automatic feed is used. Or the wheels should be provided with removable handles with compression springs so the handles cannot remain in the wheels, unless held in place by the operator.

The horizontal milling machine should have a splash guard and pans for catching thrown cutting lubricant and lubricant running from the tools. The lubricant should be directed on the work so the distribution setup will not be drawn into the cut by the cutter rotation. When possible, all cuts should be made into the travel of the table, rather than away from the direction of travel.

Metal saws

A **circular saw** for cutting cold metal stock should have a hood guard at least as deep as the roots of the teeth, and the guard should automatically adjust itself to the thickness of the stock being cut.

A sliding stock guard should be used when tube or bar stock is cut. The portion of the saw under the table should be guarded with a complete enclosure that provides for disposal of scrap metal. A plastic or metal guard placed in front of and over the saw will provide protection against flying pieces of metal. No guard should be considered a substitute for eye protection.

Chapter 9, Woodworking Machinery, contains additional

Figure 10-8. Self-closing guard for milling machine cutter. Top: the cutter is completely enclosed when the table is withdrawn. Bottom: as the table moves forward, the guard automatically opens.

information about guarding circular, swing-type, and band saws.

In swing-type saws, the length of the stroke should be adjusted, so the blade will not pass the table at its most forward point. The control should be located so the saw can be operated with the left hand when fed from the left, or with the right hand if fed from the right. This will position the operator to the side away from the moving blade.

Band saws. The upper and lower wheels of metal-cutting band

saws should be completely enclosed with sheet metal or a heavy, small mesh screen mounted on angle iron frames. To make blade changing convenient and safe, provide access doors equipped with latches. The portion of the sawblade between the upper wheel and the saw table should be completely enclosed with a sliding fixture attached to the slide, except for the point at which the cut is made.

The length of blade exposed should not be more than the thickness of the stock plus ⅜ in. (1.0 cm). Flying particles of metal can be confined by a metal or transparent plastic guard installed in front of the saw. On a hand-fed operation, care should be taken at the end of a cut. Use a push block, not hands.

Gear cutters

During operation of gear cutters and hobbers (Figure 10-9), both the tool and the work move. As a result, the point of operation guard should be simple and easily adjusted.

On operations where the work (gear blank or rough-cut gear) is moved to the tool, a simple barrier guard, formed to cover the point of operation and sized to fit the work, is satisfactory. The guard can be mounted on a spindle that carries the work. This causes the guard to fit over the point of operation when the work is brought into position.

When the tool is brought to the work and when both the tool and work are adjustable, an encircling type of guard may be attached to the tool head. Such a guard can be an automatic dropgate device, which can be equipped with a release latch to open the guard enclosure and a spring release to return the guard to a position clearing the work. Each guard should have an automatic interlock, so the machine will not operate except when the guard is in place.

On some makes of machines, the lever that controls the direction of operation of the spindle is located so the operator's hand can be caught on the back gears driving the spindle. An auxiliary lever should be installed that can be operated at a point outside the danger zone created by these gears.

On large machines where the operator is not close to the regular control switch, a pendant switch, mounted on an arm or sweep, should be installed to operate a magnetic brake that can instantly stop the machine.

When the operator inserts an arbor into the spindle, he or she should be sure that both arbor and spindle holes are clean and free from nicks. The operator should draw the arbor firmly into place by a sleeve nut and securely tighten the nut. Before removing the arbor from the spindle, he or she should make certain the machine is at a standstill.

Electrical discharge machining

Electrical discharge machinery (EDM) is a process designed to perform a variety of machining operations. This process makes possible simple or complex through hole, boring, or cavity sinking in any electrically conductive work material, including carbide, high-alloy steels, and many types of hardened metals.

During the machining process, the workpiece is normally clamped to the table and an electrode is fastened to the vertical ram platen above the workpiece. The electrode is then brought near to the workpiece so an accurately controlled electrical discharge will take place between the electrode and the workpiece. This discharge removes metal from the workpiece at the point where the gap is smallest.

Figure 10-9. This gear hobbing machine has protective covers and chip and splash covers.

As metal is removed, the electrode tool is fed into the work-piece and held in the correct cutting relationship by electro-hydraulic servo control of the ram workhead. The servo control automatically maintains a gap distance between the electrode and workpiece.

A dielectric fluid is used during the operation and should completely cover the workpiece in the work tank. During electrical discharge machining, a flow of dielectric fluid through the machining gap should be employed whenever possible to increase machining efficiency. (See Figure 10-10.)

Only a qualified electrician designated to work on the machine tool circuits should maintain or hook up the electrical system. Before attempting any work, the electrician should read and completely understand the electrical schematics for the machine.

After the machine has been hooked up, the operator should test all aspects of the electrical system for proper functioning. Be sure the machine is properly grounded and check that all exposed electrical systems are covered properly. Place all selector switches in their OFF or neutral (disengaged) position; be sure that the machine pushbuttons, manual limit switches, or controls are set for a safe setup; and check that the doors of the main electrical cabinet are closed and that the main disconnect switch is in the OFF position before considering the hookup job completed.

Safety precautions for EDM machining. To keep the operator from accidentally brushing against the live electrode or platen when the machine is operating at high voltages, install a clear plastic safety shield on the work tank. The shield must be in place before power supply is operated. No attempt should be made to block out the wire around any electrical safety inter-

lock. Before removing any cover, or before working on this machine or power supply, turn off and lock out the main electrical disconnect device. Tag the devices and all START buttons with an OUT-OF-ORDER or DO NOT START tag.

Electrical hazards, interlocks, and safety switches. Always turn the electrical disconnect switches to the OFF position at the end of the working day.

Oil level must be maintained above the highest portion of the electrode workpiece working gap. Once the safe oil level above the part has been determined, the safety float switch should be adjusted to make certain that oil level is maintained.

Dielectric level should be maintained at a minimum of one inch (2.5 cm) per 100 amperes of average current for flat geometry work. After the part's safe oil level has been determined, the safety float switch should be adjusted to assure that oil level is maintained. The machine operator should be a qualified person trained in the operation of this machine as well as its safety requirements, and must be aware of the possibility of discharge gases igniting. The electrical power must be turned off so no additional gas or hot metal particles will be formed, and the flame will be extinguished. Smother an oil fire with a CO_2 (carbon dioxide) foam fire extinguisher only. It is necessary that a CO_2 extinguisher be kept in the vicinity of the EDM machine.

EDM machining is a heat-producing process. The dielectric oil removes the concentrated heat from the machining gap and distributes it in the available oil. An EDM machine should only be installed in a location with adequate ventilation. Air conditioners and electronic precipitators do not constitute adequate ventilation.

All discharge gases are flammable and should not be allowed

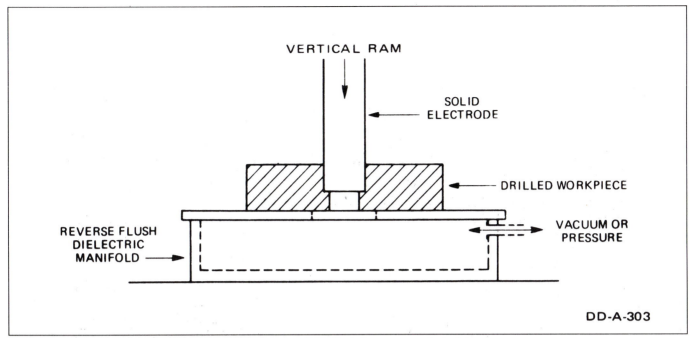

Figure 10-10. Electrical discharge machining (EDM) into manifolds (includes a reverse flush dielectric manifold), tanks, domes, or any other structures capable of trapping discharge gases. (Printed with permission from Cincinnati Milacron Company.)

to exist near a spark or flame. An EDM machine must be adequately ventilated. Discharge gases must always be allowed to escape without being trapped in a closed area. Any setup that can result in trapped gases is extremely dangerous and must be avoided. When the dielectric level in a storage tank is suddenly raised, by dumping a work tank for example, a large volume of discharge gas will escape into the outside air. Ignition of such displaced gases can cause a fire that could backlash into the enclosed area and cause an explosion. It is important that the operator and maintenance personnel read and completely understand all the precautions before operating, setting up, running, or performing maintenance on EDM metal removal. Failure to comply with instructions can result in serious or fatal injury.

PLANING MACHINES

A planer machines a metal surface—the cutting tool is held stationary while the work is moved back and forth underneath it. Shapers are generally classified as planing machines, but the process is reversed—the work is held stationary while the cutting tool is moved back and forth. Other machine tools within this classification are slotters, broaches, and keyseaters.

Planers and shapers

Planer accidents frequently result from unsafe practices caused by inadequate training and supervision:
1. Placing the hand or fingers between the tool and the work
2. Running the bare hand over sharp metal edges
3. Measuring the job while the machine is running
4. Failing to clamp the work or tool securely before starting the cut
5. Riding the job
6. Having insufficient workpiece clearance
7. Coming in contact with reversing dogs

8. Failing, when magnetic chucks are used, to make certain current is turned on before starting the machine
9. Unsafely adjusting tool holder on cross head.

The reversing dogs on planers and shapers should be covered. If the planer bed, when fully extended, or any stock on the bed being processed, travels within 18 in. (46 cm) of a wall or fixed objects, the space between the end of the travel and the obstruction should be closed by a barrier on either side of the planer (Figure 10-11).

The guard should also be constructed so it will not cause an accident during bed extension.

Shaper accidents have essentially the same causes as planer accidents. In addition, injuries frequently result from contact with projections on the work or with projecting bolts or brackets, especially when the table is being vertically adjusted. The shaper ram should be left projecting over the table, to alert the operator that the table is high enough.

Failure to properly locate the stops or dogs can injure shaper operators. The stops should be rigidly bolted to the table, especially on heavy jobs.

The shaper operator should make sure the tool is set so, if it shifts away from the cut, it will rise away from the cut and not dig into the work. The handle of the stroke-change screw should be removed before the shaper is started. Flying chips must be controlled to prevent injury to the operator and workers nearby.

Slotters

In slotter operations, the most serious accident that occurs is catching the fingers between the tool and the work, or between the ram and the table when the ram is at the end of the downstroke. Since the ram works at slow speed and the platen or machine table is small, the operator may instinctively reach across the table and under the ram to pick up a tool or other

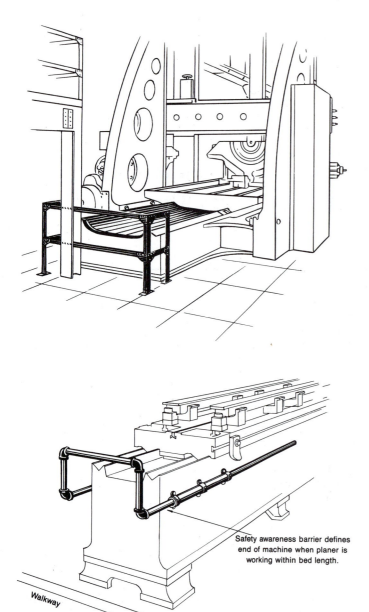

Figure 10-11. *Top:* A guardrail or similar barrier should close off any space 18 in. (46 cm) or less between a fixed object and parts of a fully extended planer or its stock. Any openings in the planer should be filled to eliminate shear hazards. *Bottom:* This self-adjusting planer table guard moves out with the table and is retained in position by friction sleeves.

object. The ram eccentric should be enclosed, preferably with a hinged guard made of sheet metal or cast iron.

Broaches

During normal operations, broaches, like heavy production machine tools, can be safeguarded through supplementary controls. The most widely used safeguard for electrically controlled broaches is a standard two-hand, constant-pressure control. An emergency stop button, preferably of the mushroom type,

should be installed adjacent to one of the two-hand controls. Another type of safeguard for broaches is a foot-operated emergency stop bar with a wide surface plate.

All pneumatically or hydraulically powered clamping equipment should be actuated by two-hand controls, located so the operator's hand cannot reach the pinch area before the clamps close. Controls should be shielded if they could be tripped by other parts of the body.

Tongs should be used for loading and unloading, if the hands are exposed in the clamping area.

The broach's rated capacity should be equal to or greater than the force required for the job. The centerlines of the work, the ram head puller, and the follow rest, if used, should all line up.

Fixtures should be checked to make sure the work is securely held. Trial runs at slow speeds are advisable to make certain the chips do not pack tightly between the teeth.

GRINDING MACHINES

Grinding—shaping material by bringing it into contact with a rotating abrasive wheel or disk—includes surface, internal, external cylindrical, and centerless operations. Polishing, buffing, honing, and wire brushing also are classed as grinding operations. Portable machines that use small, high-speed grinding wheels are discussed in Chapter 14, Hand and Portable Power Tools.

The text and illustrations in this section have been adapted with permission from American National Standard B7.1, *Safety Requirements for the Use, Care, and Protection of Abrasive Wheels.* Specifications for operation of grinding machines and construction of guards and safety devices are in this code.

Abrasive disks and wheels

An abrasive disk is made of bonded abrasive, with inserted nuts or washers, projecting studs, or tapped plate holes on one side of the disk, which is mounted on the machine face plate of a grinding machine. Only the exposed flat side of an abrasive disk is designed for grinding.

An abrasive wheel is made of bonded abrasive and is designed to be mounted, either directly or with adapters, on the spindle or arbor of a grinding machine. Only the periphery or circumference of many abrasive wheels are designed for grinding.

Hazards associated with grinding machines include:
1. Failure to use eye protection in addition to the eye shield mounted on the grinder
2. Incorrectly holding the work
3. Incorrect adjustment or lack of work rest
4. Using the wrong type, a poorly maintained, or imbalanced wheel or disk
5. Grinding on the side of a wheel not designed for side grinding
6. Taking too heavy a cut
7. Too quickly applying work to a cold wheel or disk
8. Grinding too high above the wheel center
9. Failure to use wheel washers (blotters)
10. Vibration and excessive speed that lead to bursting a wheel or disk
11. Using bearing boxes with insufficient bearing surface
12. Using spindle with incorrect diameter or with the threads cut so the nut loosens as the spindle revolves
13. Installing flanges of the wrong size, with unequal diameters, or unrelieved centers
14. Incorrect wheel dressing

15. Contacting unguarded moving parts
16. Using controls that are out of operator's normal reach
17. Using an abrasive sawblade instead of a grinder disk
18. Failure to run a wet wheel dry (without coolant) for a period of time before turning off the machine. The wet wheel can become unbalanced if coolant is allowed to accumulate on a portion of the wheel. This unbalanced condition can cause the wheel to disintegrate upon re-starting
19. Using an untested, broken, or cracked grinding wheel
20. Reaching across or near the rotating grinding wheel to load, unload, or adjust the machine during setup.

Inspection, handling, and storage

While they are being unpacked, abrasive disks and wheels should be inspected for shipment damage and then, given the "ring" test by a qualified person. This test can be used for both light and heavy disks or wheels that are dry and free of foreign material. To conduct the ring test, a light disk or wheel should be suspended from its hole on a small pin or the finger, and a heavy one should be placed vertically on a hard floor. Then the wheel or disk should be gently tapped with a light tool, such as a wooden screwdriver handle. A mallet may be used for heavy wheels or disks. The tap should be made at a point 45 degrees from the vertical centerline and about 1 or 2 in. (2.5-5 cm) from the periphery (Figure 10-12). A wheel or disk in good condition will give a clear, metallic ring when tapped. A clear ring indicates good condition. Wheels and disks of various grades and sizes give different pitches.

Daily inspection of grinding machines should include those points necessary to safe operation (Figure 10-13).

Abrasive disks and wheels require careful handling. They should not be dropped or bumped. Large disks and wheels should not be rolled on the floor. Disks and wheels too large or heavy to be hand-carried should be transported by truck or other conveyance that provides the correct support.

Store abrasive disks and wheels in a dry area not subject to extreme temperature changes, especially below-freezing temperatures. Wet wheels might break or crack if stored below 32 F

GRINDER CHECKLIST

TYPE _____ RPM _____
SIZE _____ PERIPHERAL SPEED _____

Item	OK
WHEEL GUARD: securely fastened	☐
properly aligned	☐
GLASS SHIELD: clean	☐
unscored	☐
in place	☐
WORK REST: within ⅛ in. (3.2 mm) of wheel .	☐
securely clamped	☐
FRAME: securely mounted	☐
no vibration	☐
WHEEL FACE: well lighted	☐
dressed evenly	☐
FLANGES: equal size	☐
correct diameter (½ wheel diam.)	☐
SPEED: correct for wheel mounted	☐
GUARD FOR POWER BELT OR DRIVE:	
in place	☐

DATE _____ DEPARTMENT _____
INSPECTED BY _____

Figure 10-13. A summary of checkpoints for safe grinder operation.

(0 C). Breakage can occur if a wheel or disk is taken from a cold room and work is applied to it before it has warmed up.

Abrasive disks and wheels should be stored in racks in a central storage area (Figure 10-14) under the control of a specially trained person. The storage area should be as close as possible to grinding operations to minimize handling and transportation.

The length of time abrasive disks and wheels may be stored and still be safe to use should be in accordance with manufacturers' recommendations. Disks and wheels taken out of long storage should be given the ring test. Follow this by a check for recommended speed, and a speed test on the machine on which they will be mounted. Check the speed of all grinding wheels against the spindle speed of the machine—some are designed only for low-speed use.

Wheel mountings

All abrasive wheels should be mounted between flanges; exceptions include: mounted wheels, threaded wheels (plugs and cones), plate-mounted wheels, and cylinder, cup, or segmental wheels mounted in chucks.

Flanges should have a diameter not less than one-third of the wheel diameter and preferably should be made in accordance with ANSI B7.1, Sect. 5. Flanges for the same wheel should be of the same diameter and thickness, accurately turned to correct dimensions, and in balance. The requirement for balance

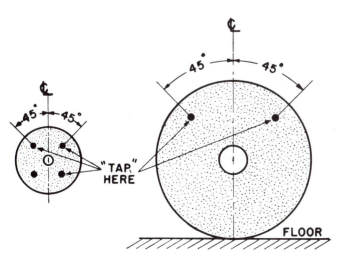

Figure 10-12. Tag points for the "ring" test.

Figure 10-14. Wooden racks for abrasive wheel storage. The heavier wheels are stored at the bottom and the upper tiers hold the smaller and lighter wheels. (Printed with permission from General Motors Corporation.)

does not apply to flanges made out of balance to counteract an unbalanced wheel.

The inner or driving flange should be keyed, screwed, shrunk, or pressed onto the spindle, and the bearing surface of the flange should run true with the spindle. The outer flange bore should have an easy sliding fit onto the spindle.

Flange inspections should be frequently scheduled. A flange found to be sprung, not bearing evenly on the wheel, or defective in any other way should be removed from the spindle at once and replaced with a flange in good condition.

Incorrectly mounting an abrasive wheel is the cause of much wheel breakage. Since rotational forces and grinding heat cause high stresses around the central hole of the wheel, it is important to follow safety regulations concerning size and design of mounting flanges and mounting techniques.

Before a wheel is mounted, it should be given the same inspection and ring test as was given when originally received and stored. The bushings, particularly on wheels that have been rebushed by the user, should be checked for shifting or looseness.

Compression washers should be used to compensate for unevenness of the wheel or flanges. Blotting paper (not more than 0.025 in. — 0.6 mm — thick) or rubber or leather compression washers (not more than 0.125 in. — 3.2 mm — thick) may be used for this purpose.

Allowance for the wheel mounting fit should be made in the wheel hole rather than in the arbor or wheel mount (Figure 10-15). The wheel should not be forced on the spindle, because forcing can loosen, or otherwise damage, the wheel bushing

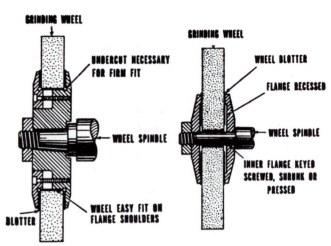

Figure 10-15. Correct methods of mounting abrasive wheels with small holes (*right*) and wheels with large holes (*left*).

or crack the wheel. A wheel that is too loose on the spindle will run off-center, causing stress and vibration. Spindle end nuts should hold the wheel firmly but not too tightly. Too much pressure can spring or distort the flange or even break the wheel.

If rebushing is necessary to make the wheel fit the spindle, the job should be done by the manufacturer or, if in the plant, by an experienced person with suitable equipment.

Immediately after mounting the wheel and before turning on the power, the operator should turn the wheel by hand for a few revolutions to see that it clears the hood guard and machine elements, such as work rests on work holding equipment.

When starting a grinding machine, stand to one side away from the grinding wheel. Allow at least one minute warm-up time before truing or grinding with the wheel. Always use coolant when truing the wheel or during normal grinding. Never allow coolant to flow on a stationary grinding wheel; coolant might collect on one portion of the wheel, causing an unbalanced condition. While the machine is running, *never* remove a guard fastener or guard. Do not touch any moving part of the machine or the rotating grinding wheel to determine its smoothness or condition. *Do not* attempt to physically operate a machine that is in its automatic mode.

Never alter or try to alter the machine, its wheel speed, or any of its safety equipment at any time. The guards are on the machine for the operator's safety; if they are removed, serious injuries to the operator or others can result.

Adjusting safety guards

The peripheral protecting member can be adjusted to the constantly decreasing diameter of the wheel by an adjustable tongue or similar device. Then the angular protection specified for bench and stand grinders will be maintained throughout the life of the wheel, and the maximum distance between the wheel periphery and the tongue or end of the peripheral band at the top of the opening will not exceed ¼ in. (6 mm) (Figure 10-16).

The guard should enclose the wheel as completely as the nature of the work will permit. It should be adjustable so that, as the diameter of the wheel constantly decreases, the protection will not be lessened. The maximum angular exposure varies with the type of grinding; specifications are shown in Figure 10-17.

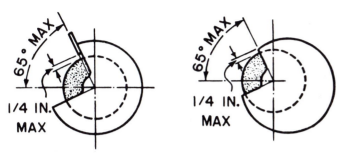

Figure 10-16. The correct wheel exposure can be maintained with an adjustable tongue (*left*) or a movable guard (*right*).

Safety guards should cover the exposed arbor ends.

On machines used for cutting, grooving, slotting, or coping stone or other materials, the safety guard or hood seldom offers adequate protection. On machines that permit a relative horizontal traverse between wheel and work greater than 10 in. (25 cm)

and those that use solid cutting wheels 10 in. (25 cm) or more in diameter, provide an auxiliary enclosure in addition to the guard.

This auxiliary enclosure can be a set of heavy screen panels, suspended from approximately 8 ft (2.5 m) above the floor to or below the work table. The panel screens should be ½-in. (13 mm) mesh or smaller, and the wire should be ⅛ in. (3.2 mm) in diameter or more. The framework of the panels should be made of 1- by 1¼-in. (2.5 by 3.1 cm) or heavier structural steel angles or channels.

Safe speeds

Abrasive wheels and disks should be operated at speeds not exceeding those recommended by the manufacturer. In particular, unmarked wheels of unsual shape, such as deep cups with thin walls or backs with long drums, should be operated according to the manufacturer's recommendations.

Before a wheel is mounted, the machine spindle should be checked for the correct size. Spindles, including those with adjustable speeds, should be changed only by authorized persons.

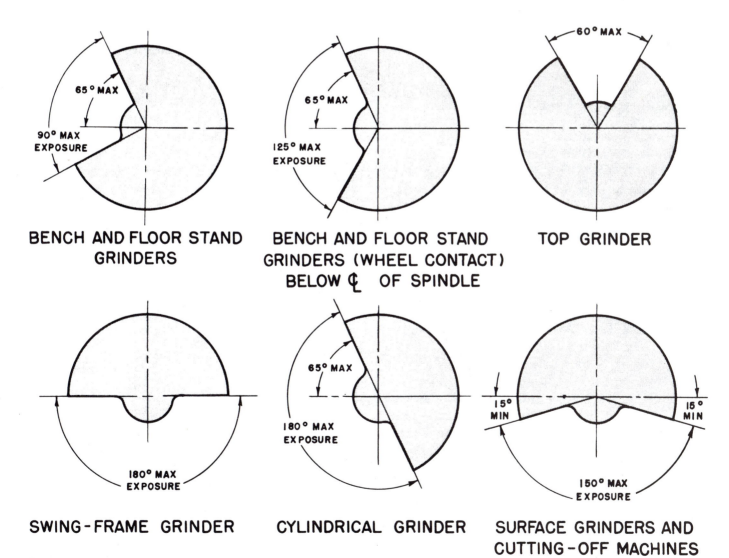

Figure 10-17. Maximum exposure angles for various grinding applications.

As the wheel wears down, the spindle speed (rpm) is sometimes increased to maintain the surface speed (sfpm). When the wheel is nearly worn down, the spindle is running at the highest rpm. When the worn wheel is replaced, spindle speed must be adjusted or the new wheel might break because the surface speed exceeds manufacturer's recommendations.

Grinding wheel and disk failures should be thoroughly investigated, preferably with the manufacturer's representative. This type of investigation, along with immediate corrective action, greatly reduces the possibility of recurrent failures.

Grinding equipment for high-speed operation should be specially designed, with particular reference to spindle strength, guards, and flanges for eliminating mounting stresses. The manufacturer's approval should be obtained for high-speed wheel and disk operation. Such things as side grinding pressure and the wheel shape must also be considered. Approval of special high-speed operation should apply only to the particular machine investigated. Also important to safe high-speed operation is equipment maintenance and protective devices.

Work rests

Many bench and floor stand grinder wheels have broken, causing serious injury, because work has become wedged between the work rest and the wheel. The work rest should be substantially constructed and securely clamped at not more than ⅛ in. (3.2 mm) from the wheel (Figure 10-18). The work rest position should be checked frequently. The work rest height must be on the horizontal center line of the machine spindle. The rest should never be adjusted while the wheel is in motion—the work rest might slip, strike the wheel, and break it, or the operator might catch a finger between the wheel and the rest.

To prevent work twisting and bending stress on the wheel, operators should use guides to hold the work in position when slot grinding or performing similar operations.

Abrasive wheel dressing

Abrasive wheels that are not true or not in balance (Figure 10-19) will produce poor work. They can damage the machine and injure the operator. Keeping the wheels in good condition eliminates these possibilities, decreases wheel wastage, and lengthens wheel life.

To recondition a rutted or excessively rough wheel, often it is necessary to dress it by removing a large area of the face. Wheel dressing tools should be equipped with hood guards over the tops of the cutters to protect the operator from flying wheel particles or pieces of broken cutters. The wheel dresser operator should use a rigid work rest set close to the wheel. The wheel dresser should be moved back and forth across the wheel face, while the heel or lug on the underside of the dresser head is firmly held against the edge and not on top of the work rest (Figure 10-20).

Wheels should be occasionally tested for balance and rebalanced if necessary. Wheels that are too worn or out of balance to be balanced by truing or dressing should be taken out of service.

Surface grinders and internal grinders

Operating requirements for surface grinders and internal grinders differ from those for other types of wheels. Insecurely clamped work and unenergized magnetic chucks are common sources of injury to surface grinder operators. Under these conditions, work can be thrown with considerable force.

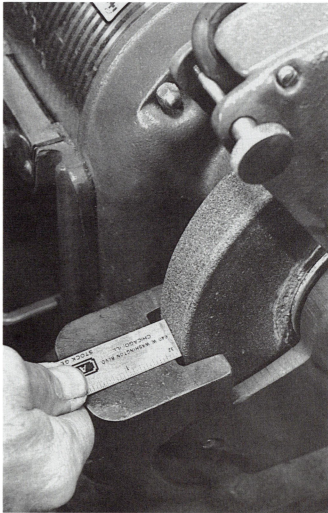

Figure 10-18. *Top:* With a properly adjusted work rest, the operator can keep the hands away from the wheel and still firmly hold the work in place. *Bottom:* There should be a safe space of no more than ⅛-in. (3.2 mm) between the tool rest and the wheel.

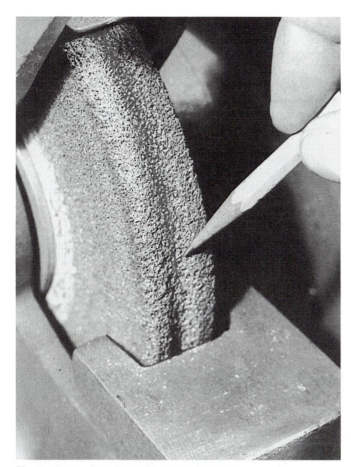

Figure 10-19. A badly rutted or out-of-balance grinding wheel should be taken out of service and dressed.

If the operator takes too deep a cut or too quickly traverses the table or wheel, the wheel can overheat at the rim and crack. Operators, therefore, should be trained and supervised to clamp work tightly. They must always properly adjust and turn a magnetic chuck on before applying the wheel, and have control of the work speed and depth.

Baffle plates on each end of a surface grinder are usually standard equipment, but should include some provision for exhausting the grinding dust (Figure 10-21).

Internal grinders often can be guarded with an automatic positioning hood that covers the grinding wheel when it is in the retracted or idling position.

Grindstones

In using grindstones, the manufacturer's suggested running speeds and operational procedures should be followed. Stones of unknown composition or manufacture should never be run at more than 2500 sfpm (12.5 m/s) and ordinarily not more than 2000 sfpm (10 m/s).

The size and weight of grindstones require a stand that is rigidly constructed, is heavy enough to hold the stone securely, and is mounted on a solid foundation to withstand vibration.

Since grindstones are run wet, all possible precautions should be taken to prevent slipping accidents near the stones. Rough concrete or other slip-resistant floor material is recommended.

Grindstones should be carefully inspected for cracks and other

---

BENCH AND STAND GRINDERS
DRESSING

1. Wear a face shield over your safety glasses for protection against heavy particles.
2. Use a dressing tool approved for the job. Never use a lathe cutting tool.
3. Inspect star dressers for loose shaft and worn disks.
4. Round off the wheel edges with a hand stone before and after dressing to prevent the edges from chipping.
5. Use the work rest to support and guide the tool. Use a tool holder if one is available.
6. Apply moderate pressure slowly and evenly.
7. Always apply diamond dressers at the center or slightly below the center, never above.

Figure 10-20. Wheel dressing operations should adhere to these safety measures.

Figure 10-21. Guarded horizontal surface grinder.

defects as soon as they are received from the manufacturer. Those not immediately used should be stored in a dry, uniformly heated room. During storage they should be placed so that they will not be damaged.

Many grindstone failures result from faulty handling and incorrect mounting. Grindstones should not be left partially submerged in water. This practice causes an unbalanced stone that can break when rotated. Wooden wedges should not be used on power-driven stones. Often, these wedges are too tightly driven or can become wet and swell. In either case, cracks start in the corners of a square center hole, radiate outward, and weaken the stone causing ruptures to occur when operated at normal speeds.

After the stone has been centered, the central space about the arbor should be filled with lead or cement. Double thicknesses of leather or rubber gaskets, rather than wood washers, should be used wherever possible. If wood washers are used between the flanges and the stone, the washers should be ½- to 1-in. (1.3 to 2.5 cm) thick and the flanges should be clamped in place by heavy nuts.

Work rests should comply with the same conditions as those for grinding wheels.

To remove dust and wet spray or mist when dressing or operating power-driven grindstones (either wet or dry), provide an adequate exhaust system.

Polishing and buffing wheels

Polishing wheels are either wood faced with leather or made of stitched together disks of canvas or similar material with a coat of emery or other abrasive glued to the periphery of the wheels.

Buffing wheels are made of disks of felt, linen, or canvas. The periphery is given a coat of rouge, tripoli, or other mild abrasive.

The softness of the wheel built up of linen, canvas, felt, or leather is determined by the size of the flanges used—the larger the flange, the harder the surface. When large flanges are used, it often is necessary to soften the working surface of the built-up wheels to conform to the contour of the object being polished. A safe procedure for softening is to place the wheel on the floor or other flat surface and pound the edges of the wheel with a hammer or mallet. The wheel should not be placed on the spindle with a file or other object held against it, because the object can catch in the wheel and be thrown with such force that the operator or nearby workers are injured.

Polishing and buffing wheels should be mounted on rigid and substantially constructed stands, heavy enough for the wheels used. Mounting procedures are the same as those for grinding wheels. Hood guards should be designed to prevent the operator's hands or clothing form catching on protruding nuts or the ends of spindles. If working conditions require a hood that does not give the needed protection, then use a spanner wrench to install smooth nuts over the spindle ends. A prick punch and hammer should never be substituted for a spanner wrench.

The polishing and buffing peripheral speed range is from 3,000 to 7,000 sfpm (15 to 35 m/s), with 4,000 sfpm (20 m/s) in general use for most purposes. If the motors that drive the polishing and buffing wheels are equipped with adjustable speed controls, the controls should be installed in a locked case and the speed changed only by an authorized person.

Exhaust hoods should be designed to catch particles thrown off by the wheels.

Gloves should not be worn by polishers and buffers because a glove can catch and drag the operator's hand against the wheel. Small pieces being polished or buffed can frequently be held in a simple jig or fixture. Some operators use a piece of an old linen or canvas wheel for holding small pieces. The operator should not attempt to hold a small piece against the wheel with the bare hands.

When rouge or tripoli is applied to a revolving wheel, the side of the cake should be held lightly against the wheel periphery. If a stick is used, the side of the stick should be applied on the off side so, if thrown, it will fly away from the wheel.

Wire brush wheels

Brushing or, more commonly, scratch wheel operation is used to remove burrs, scale, sand, and other materials.

Wheels are made of various kinds of protruding wires and with different thicknesses. Use flanges or nuts to hold scratch wheels rigidly in place.

The same machine setup and conditions that apply to polishing and buffing wheels apply to brush wheels. The speed recommended by the manufacturer should be followed. The hood on scratch wheels should enclose the wheel as completely as the nature of the work allows and should be adjustable, so protection will not lessen as the diameter of the wheel decreases. The hood should cover the exposed arbor ends, or a smooth nut should be installed on them. The work rest should be adjusted to about ⅛ in. (3 mm) from the brush wheel.

Personal protective equipment is especially important when operating scratch wheels because the wires tend to break off. It should be mandatory to wear aprons of leather, heavy canvas, or other heavy material and leather gloves, faceshields, and goggles.

REFERENCES

American National Standards Institute, 1430 Broadway, New York, N.Y. 10018.

Accuracy of Engine and Tool Room Lathes, B5.161952 (R1986).

Chucks and Chuck Jaws, B5.8-1972(R1979).

Electrical Standard for Industrial Machinery, NFPA 79-1985.

External Cylindrical Grinding Machines—Centerless, B5.37-1970.

External Cylindrical Grinding Machines—Plain, B5.33-1981.

External Cylindrical Grinding Machines—Universal, B5.42-1981.

Grinding Machines, Cutter and Tool, B5.53M-1982.

Grinding Machines, Surface, Reciprocating Table—Vertical Spindle, B5.32.1-1977.

Machine Mounting Specifications for Abrasive Discs and Plate Mounted Wheels, ASME B.35-1983.

Marking for Identifying Grinding Wheels and Other Bonded Adhesives, B74.13-1986.

Milling Machine Arbor Assemblies, B5.47-1972(R1984).

Milling Machines, B5.45-1972(R1984).

National Electrical Code, C1 NFPA 70, 1987.

Preferred SI Units for Machine Tools, B5.51M-1979.

Rotary Table Surface Grinding Machines, B5.44-1971(R1986).

Safety Requirements for the Construction, Care, and Use of Drilling, Milling, and Boring Machines, B11.8-1983.

Safety Requirements for the Construction, Care and Use of Gear Cutting Machines, B11.11-1984.

Safety Requirements for the Construction, Care and Use of Grinding Machines, B11.9-1975.

Safety Requirements for the Construction, Care, and Use of Iron Workers, B11.5-1985(R1981).

Safety Requirements for the Construction, Care, and Use of Lathes, B11.6-1984.

Safety Requirements for the Construction, Care, and Use of Metal Sewing Machines, B11.10-1983.

Safety Requirements for the Construction, Care, and Use of

Single- and Multiple-Spindle Automatic Screw/Bare and Chucking Machines, B11.13-1983.

Safety Requirements for Floor and Wall Openings, Railings, and Toeboards, A12.1-1973.

Safety Requirements for Use, Care, and Protection of Abrasive Wheels, B7.1-1978.

Specifications for Shapers and Sizes of Grinding Wheels, B74.2-1982.

Spindle Noses and Adjustable Adapters for Multiple Spindle Drilling Heads, B5.11-1964(R1982).

Spindle Noses and Tool Shanks for Horizontal Boring Machines, B5.40-1977(R1984).

Ventilation Control of Grinding, Polishing, and Buffing Operations, Z43.1-1966.

Grinding Wheel Institute, 14600 Detroit Ave., Cleveland, Ohio 44107. *Safety Recommendations for Grinding Wheel Operations.*

National Machine Tool Builders Association, 7901 West Park Dr., McLean, Va. 22102.

National Safety Council, 444 N. Michigan Ave., Chicago, Ill. 60611.

Safeguarding Concepts Illustrated.

Industrial Data Sheets

    Cadmium, 726.

    Coated Abrasives, 452.

    Cutting Oils, 719.

    Engine Lathes, 264.

    Gear-Hobbing Machines, 362.

    Horizontal Metal Boring Mills, 269.

    Hydraulic Fluids, 543.

    Lithium, 566.

    Magnesium, 426.

    Manganese, 306.

    Metal Planers, 383.

    Metal Saws (Cold Working), 584.

    Metal Shapers, 216.

    Metal-Working Drill Presses, 335.

    Metal-Working Milling Machines, 364.

    Portable Grinders, 583.

    Selenium and Its Compounds, 578.

    Vertical Metal Boring Mills, 347.

    Zinc, 267.

    Zirconium Powder, 729.

U.S. Department of Health and Human Services, National Institute for Occupational Safety and Health, Division of Technical Services, 4676 Columbia Pkwy., Cincinnati, Ohio 45226.

Control of Exposure to Metalworking Fluids, DHEW (NIOSH) Publication 78-165.

Health and Safety Guide for the Screw Machine Products Industry, DHEW (NIOSH) Publication 76-165.

Cold Forming of Metals

POWER PRESSES

Power presses are used with the attachment of many different dies for performing many types of metalworking operations required to produce a variety of products.

Since a power press machine (see definitions) is so versatile, safeguarding the point of operation is dependent upon (1) the die or tooling component, (2) the type of press selected to power the die, and (3) the selected method of inserting materials and removing parts, along with the method of scrap disposal.

▪ The first step in the production of parts from a press is the designing of the die or tool component. Consideration must be given to the material used, the configuration of the part, the method of feeding, the method of part removal, the disposal of scrap, and other pertinent factors, including possible noise reduction through proper die design.

▪ The next consideration is the selection of the press component. The design of the die component determines the physical characteristics of the press, such as tonnage required, die space dimensions, and speed and stroke of the press component.

▪ The method of feeding may be manual, semiautomatic, or automatic. Scrap disposal may be manual or automatic.

▪ The fourth necessary component is safeguarding. The method of safeguarding the point of operation depends upon the design of the die component, the selection of the feeding component and the type of press component. See the discussion on Guarding of hazards in Chapter 8.

Complete safety depends upon adequate safeguarding of the point of operation, the proper training of press operators, and enforcement of safe working practices by the employer's supervision. Setup personnel and maintenance personnel must be trained by the employer to assure their safety while working in or around a press.

Press operations fall into two basic categories, namely primary operations and secondary operations.

Primary operations (blanking)

A primary operation is one in which stock material is processed for producing suitably sized and shaped flat blanks. Generally these blanks require subsequent forming and shaping; that operation is referred to as a secondary operation.

Primary operations are the easiest to guard at the point of operation. The use of flat material permits guards to be constructed or adjusted to allow only sufficient opening for material to pass through the guard into the die. The trailing edge of strips can often be processed by pulling the stock through with the scrap skeleton or pushing it through with the leading edge of the next strip. If this is not possible, inward forming of a guard to meet the die at the point of stock entry can help to fully utilize the strip when advancing the strip manually. If this customizing of the guard cannot be accomplished, the balance of the material should be scrapped unless other appropriate point-of-operation protecton is used.

The hazard that remains occurs during die setup or repair when guards, necessary for the operator's safety, have been removed.

▪ On part-revolution clutch presses, this hazard can be controlled by operating the machine in the "Inch" or "Single Stroke" mode utilizing two-hand buttons for stroking of the press when the dies are being tested. Repair or modification work should never be done in the die when the machine is capable of being stroked. Protection can be provided by the use of interlocked safety blocks and assuring zero mechanical state (ZMS).

- On full-revolution clutch machines, the flywheel must be stopped. This is also recommended for partial-revolution clutch machines with the additional procedure of shutting off the control, as well as the motor, when work is to be done in the dies.

- If foot controls are being used for production, they should be disconnected and removed from the area while die work is being performed. Before turning on the main motor to start a full-revolution clutch press after the die work has been performed and at any other time as well, be certain that the slide adjustment is correct and that everything has been removed from the die, including one's hands. This precaution is necessary because if the foot pedal were inadvertently tripped while the machine was shut down, the press would stroke immediately when the motor is turned on.

Stock primary material

Coiled material. Stock material initially in coiled form is frequently fed directly into a press which, with one or more dies, produces a finished part. Coiled material is sometimes fed into a "cut-to-length" line which produces strips. It can also be fed into a blanking line where shaped and sized blanks are produced for use in secondary operations.

Strip material, from cut-to-length line, is processed as stock material for subsequent primary operation or a combination primary and secondary operation.

Scrap and drop-off material. Select scrap and drop-off material is considered by some to be a valuable source of primary material for production of some products. If such a choice is made, it must be recognized that the material extending from the die varies in length and position with each stroke resulting in a corresponding prepositioning of the operator's hands. All die hazards, pinch points, guide posts, etc., as well as the point of operation at which blanking is performed, are potential hazards that must be dealt with. A practical method to assure protection of the operator is to use a die that has attached to it its own complete guard that covers every hazard and allows only sufficient opening for material pass through, while still providing the ability for the operator to position the material close to the die for maximum yield. Such an operation requires positive ejection of parts and freedom of movement of the material within the die to make it absolutely unnecessary to remove guards except for die sharpening.

Material handling hazards. Material hazards exist in the handling of the sharp edge or burr on strips, coil stock, or scrap. A material handling hazard is frequently encountered during the unstrapping of stock. The sudden release of a strap on a bundle of strips or on a restrained coil can injure any part of the body (including the eyes). When handling stock material, use gloves, safety glasses, and arm gauntlets.

Machine feeding of material

Coil feeds. When coiled stock is used, it is placed into an uncoiler which will "pay out," on demand, the required length of stock. Some uncoilers are powered to maintain a free loop of material from which the powered rolls on roll feeds or gripper jaws on gripper feeds pull a measured amount of stock. Other uncoilers are simple reels from which material is pulled directly. Heavy coils, fast feeding, and "pay out" of long strips require powered uncoilers in order to keep the pulling effort to a minimum.

Strip feeds. Strip material is usually fed directly into the die manually; however, it can be machine fed. Although a roll feed can be used, generally gripper feeds are preferred. Gripper feeds can be automatically supplied with strips from an unstacker or each strip can be manually loaded by an operator.

Blank feeds. Blanks are usually manually inserted into the die but blank loading can be and is frequently done by use of programmed robots.

Feed-machine hazards. Pinch points, crush points on any type of feeder, and nip points at the in-running side of rolls or gripper jaws are feed-machine hazards that must be guarded. Openings for material should be only large enough for free movement of stock and too small to permit fingers to enter. To facilitate the guarding of in-running points of rolls, it is sometimes necessary to use bell-mouthed guides up to the rolls for threading in the end of the stock.

Hand-fed material. Material is sometimes hand fed into the die from a coil or from a stack of strips. In each case, material can be fed through a guard opening sufficiently small so that one's hands cannot enter into the point of operation. This type of operation is very easy to safeguard. Proper die guard design, with part ejection, will produce the ultimate in safety for the operator.

Part and scrap ejection. Removal of parts and scrap should be designed into the operation so that there is never any accumulation within the die that makes guards difficult to use or that would discourage their use. Efficient, reliable part and scrap ejection is an important consideration in a fully guarded die since access to the point of operation has been eliminated. Unless part and scrap removal is positively assured, the use of inadequate enclosures or the removal of guards will be encouraged. At the first evidence of any failure to eject, the operation should be shut down and the problem resolved. It is frequently necessary to provide inclined chutes and/or to incline the press if possible, so that part ejection and scrap removal can be assured.

The design of the die should permit material to be removed or retracted from the side in which it is entered as well as from the side at which it exits. Sometimes this requires beveled edges at both sides of component parts of the die so that edges in the material do not hang up on edges within the die which would prevent the withdrawal or backup of the strip or scrap.

In the event that a die is used to recover usable material from drop off and scrap, it is necessary that the material be free to be moved in any direction in the die without hangup. Consequently, all edges in the die need to be beveled. On some dies, it may be necessary to use supporting steel runners within the die.

Secondary operations

Many secondary power press operations are adaptable for feeding methods which require manual handling and, together with adequate safeguarding, do not allow the operator to reach into the danger area. One such method is through the use of gravity feed in which the part is placed on a chute and allowed to slide down into the lower die (Figure 11-1). Pins, gages, and stock guides should be provided to assure easy part-nesting in the proper position. Open back inclinable presses, because of their ability to be inclined, allow good application of this type of work because the part can be placed into the die by gravity and ejected out of the back of the press by gravity or air. Even if the opening in the chute is small enough to prevent a hand from entering into the die area, a full barrier guard can be used to prevent inadvertent reaching into the danger area. If the opening is large enough to allow the entry of a hand, there are safeguarding devices, such as Type "A" or "B" movable barriers, two-hand controls, presence-sensing devices, or pull-back devices, which can be used. (These are explained later.)

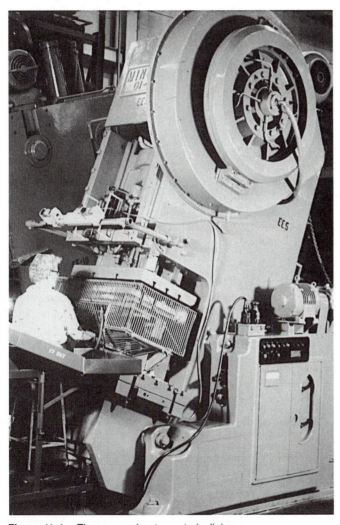

Figure 11–1. There are advantages to inclining a press. The fixed barrier guard is simple and economical to make. This operation allows either hand feed or automatic gravity feed direct from blanking operation (via the chute).

There are various adaptations of gravity-chute feeds which can be used to assure only one part being placed in the die at a time, such as through the use of a single-piece feeder. When oil or

other lubricant on the parts causes the parts to stick in the chute or slide, wire or metal rods can be installed in the chute to reduce friction and allow the parts to slide into the die without sticking.

A common type of semiautomatic feeding is the follow or push feed, which allows the operator to push the parts on a tray into the die by pushing one part behind the other or using a pusher stick to place the part into the die. This keeps hands well out of the danger area. This die is easily guarded with a die-enclosure guard. Even irregularly shaped parts can frequently be fed with a push feed to keep the hands out of the danger area.

The addition of a magazine on the push feed can considerably increase the production rate as well as minimize the manual handling of the blanks at the press operation. Blanks with various configurations are adaptable to magazine feeds. Magazine slide feeds can be adaptable to nearly automatic operation by actuating the slide through mechanical attachment to the press ram or crankshaft. The operator then needs only to keep the magazine filled with blanks to feed the press. Other ways of providing power movement of the slide feed include air or hydraulic cylinder actuation controlled by solenoid valves timed with the press cycle.

Finally, where parts must be placed and/or removed from the die manually, hand tools such as soft metal pliers, tongs, tweezers, and suction cups provide an effective aid for keeping the hands out of the danger area (Figure 11-2a and 2b). It is sometimes necessary to provide hand tool clearance holes or slots to assist the operator when inserting or removing the work pieces. Providing clearance will make the job of grasping the parts much safer and more efficient. Hand tools are not, by themselves, a means of safeguarding. For the operator's safety, a barrier guard or some type of safeguarding device must be used.

A power press safeguarding program should be based on an intelligent evaluation of the specific problems, and a definite company policy should be formulated to cover use of guards or devices for all operations, consideration of the safety factor in new operations, and enforcement of safe operating standards. See general principles in Chapter 8, Principles of Guarding.

Such a program includes adherence to ANSI Standard B11.1, *Safety Requirements for the Construction, Care and Use of Mechanical Power Presses,* and to applicable state, federal, and local codes.

Many press rooms are small, with only a foreman, a die setter, and a few press operators. In such a shop, a simple program can suffice, consisting of personal supervision, guarding some dies, and providing proper safeguarding devices and hand tools on others.

Large power press shops, which may be divided into departments under various supervisors and with several die setters, usually have split responsibilities; therefore, development of company-wide safeguarding standards and adherence to them are most necessary and practicable. The standards set up should apply to and be followed by all groups concerned with the problem: supervisors, operators, die setters, those who design and make dies, maintenance personnel, and electricians.

The number and productivity of the dies are major factors in deciding the number of point-of-operation guards to be installed. As a first step, high production (long run) dies can be guarded individually (Figure 11-3). Adjustable barrier safeguarding should be considered only for dies used infrequently and for short-run jobs (Figure 11-4).

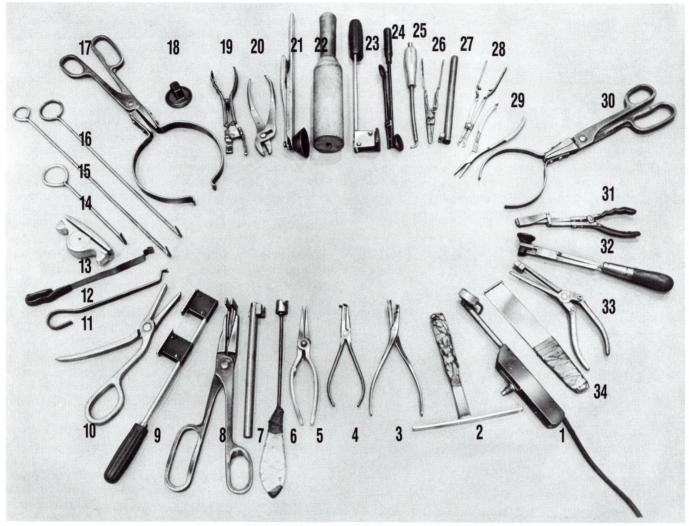

Figure 11-2a. Thirty-four "mechanical hands" show what can be accomplished through determination and ingenuity. These hand tools have been used for loading and unloading dies and work in 20 presses of one large plant and have contributed to making a "No Accident Month" last for more than 25 years. Tool description is below.

Tool Identification

Shown in the photograph above are 34 simple, safe hand tools—all of which can be made in a shop, and all of which can save the hands and fingers of power press operators.

The tools shown include pliers up to 12 in. (30 cm) long which permit a hand to be kept out of the danger zone. Other pliers are designed to grasp a vertical flange by means of bent jaws, to pick up thin pieces, or to hold material in work. Tools also include vacuum cups for handling sheet metal at slitting and shearing machines, permanent and electromagnets, pliers with magnets, steel hooks, and steel or brass pusher sticks.

Not only do tools save workers' fingers and hands but they also contribute to speed of operation. Studies show that it takes 1.4 seconds to load a press with a 12-in. pliers compared to 1.8 seconds by hand.

Key to numbers on the accompanying photograph:
1. 110-V electric magnet for picking up sheet metal.
2. Steel or brass pusher.
3. Pliers with extra-long handles.
4. Pliers with adapters for grasping vertical edges.
5. Pliers with long handles and long grip.
6. Alnico magnet on a stick.
7. Fiber stick with Alnico magnet.
8. Pliers with adapters for grasping vertical edges.
9. Fiber stick with Alnico magnet.
10. Pistol-grip pliers.
11. Push stick.
12. Push stick.
13. Sheet-edge gripper.
14-16. Hooks with 90-degree bend.
17. Pliers with adapter for grasping vertical edges.
18. Vacuum cup.
19-20. Pliers with high-pressure grip.
21. Releasable vacuum gripper.
22. Cylindrical holding tool.
23. Fiber stick with magnet.
24. Releasable vacuum gripper.
25. Hook with 90-degree bend.
26. Normally closed pliers.
27. Fiber stick with Alnico magnet.
28. Pliers with adapter for grasping vertical edges.
29. Long-nosed pliers.
30. Pliers with adapter for grasping vertical edges.
31. High-pressure pliers.
32. Vacuum cup with handle.
33. Adjustable-handled pliers with Alnico magnet.
34. Push stick.

Figure 11-2b Closeup views of one versatile hand-feeding tool in use. Tools may be purchased or custom made in company shop.

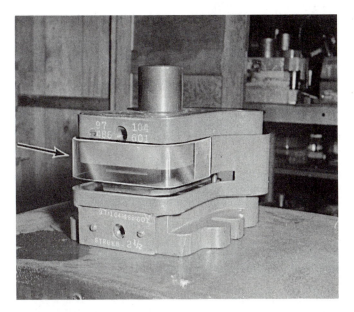

Figure 11-3. Die set with enclosure. Fully enclosed dies can be operated with complete safety to personnel. Die shown in top photo has plastic wrapped around three sides of die and is very effective. Bottom photo shows a die set with fixed barrier guard mounted directly on die. Note opening for feeding material.

If a shop has short runs and thus many die changes each day or uses dies which are the property of other manufacturers, safeguarding devices should be installed for secondary opera-

tions, with adjustable barrier guards for primary operations (see Definitions section later in this chapter).

Safe operation of power presses requires that attention be paid to facilities and factors other than guards and safeguarding devices, such as layouts of machines, machine and aisle space, light and visibility, containers for handling scrap and processed parts, and an effective preventive maintenance program.

Hydraulic and pneumatic presses

Adherence to ANSI Standard B11.2, *Safety Requirements for the Construction, Care, and Use of Hydraulic Power Presses,* should be included in a plant safeguarding program for hydraulic presses.

Equally as important as feeding or placing the part into the die is the ejection or removal of the part from the die. This should be accomplished without requiring the operator to reach into the danger area. Some means is frequently necessary to first strip the part from the upper die or punch or to lift the part from the lower die before actually removing the part. The part can then be ejected by gravity with an inclined press or by other means, such as air blowoff to force the part from the die. Liftout fingers in the upper die can remove cup-shaped parts from the lower die that tend to adhere or stick in the cavity. Spring strippers in the upper or lower die also will loosen the part in the die. Knockouts in the upper die are frequently used as a mechanical means of freeing the part to eliminate reaching in to pull the part from the upper die. On larger parts, air cylinder ejectors or lifters, as well as mechanical knockouts, are used to lift or eject the part from the die. (See the section on Feeding and Ejecting Mechanisms later in this chapter.)

Shuttle or shovel-type extractors can be used to reach between dies to catch the part as it is stripped or knocked from the upper die and then transfer it from the die into a stock container.

Unloaders must be timed with ram travel and can be used on automatic or semiautomatic operations.

Hydraulic and pneumatic-powered presses have point-of-operation safeguarding requirements similar to mechanical power presses.

Safeguarding considerations

Guards or point-of-operation safeguarding devices must be selected for all operations. They should be installed, adjusted, and maintained to prevent an operator from inadvertently reaching into the point of operation during the down stroke of the press. (Examples are given in Figure 11-3.)

An employer should provide the maximum in safety for all employees. A "no hands in the die" policy lessens the exposure of the operator and should be used whenever size and configu-

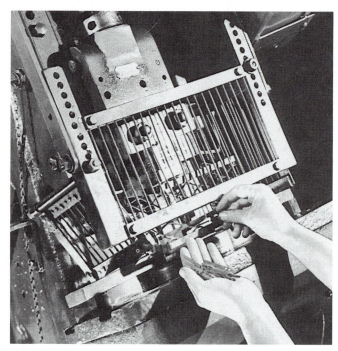

Figure 11-4. Adjustable barrier fits any size die used. This barrier is especially practical for short-run jobs.

ration of the part permits a die to be designed for "no hands in the die" operation.

While there are a number of ways to achieve maximum safety in power press operation, the best is to barrier guard the point of operation and use automatic or semiautomatic loading and unloading methods, or else design the feed opening to permit entry of material but not a hand. If neither of these two methods can be used, then hand-feeding tools or some other feeding method should be used to keep hands out of the die area.

Definitions

It is important to the proper understanding of this subject to know the definitions of various terms used in this chapter:

Antirepeat. The part of the clutch/brake control system designed to limit the press to a single stroke if the actuating means is held or stuck on "operate." Antirepeat requires release of all actuating mechanisms before another stroke can be initiated. Antirepeat is also called "single-stroke reset" or "reset circuit."

Bolster plate. The plate attached to the top of the bed of the press component; it has drilled holes or T-slots for attaching the lower die or die shoes.

Brake. The mechanism used on a mechanical power press component to stop the slide or slides, or hold the slide or slides, or both, either directly or through a gear train, when the clutch is disengaged.

Brake monitor. A sensor that has been designed, constructed, and arranged to monitor the effectiveness of the press braking system.

Clutch. The coupling mechanism used on a mechanical power press component to couple the flywheel to the crankshaft in order to produce slide motion, either directly or through a gear train.

Full-revolution clutch. A type of clutch that, when tripped, cannot be disengaged until the drive mechanism (usually a crankshaft) has completed a full revolution and the slide, a full stroke.

Part-revolution clutch. A type of clutch that can be disengaged at any point before the drive mechanism (usually a crankshaft) has completed a full revolution and the press slide, a full stroke.

Direct drive. The type of driving arrangement wherein no clutch is used; coupling and decoupling of the driving torque is accomplished by energizing and deenergizing a motor. Even though not employing a clutch, direct drives match the operational characteristics of part-revolution clutches because the driving power may be disengaged during the stroke of the slide.

Concurrent. Acting in conjunction, and used to describe a situation wherein two or more controls exist in an operated condition at the same time.

Continuous. Uninterrupted multiple strokes of the slide without intervening stops (or other clutch control action) at the end of individual strokes.

Counterbalance. The mechanism that is used to balance or support the weight of the connecting rods, slide, and slide attachments.

Device. A press control or attachment that (1) restrains the operator from reaching into the point of operation, (2) prevents normal press operation if the operator's hands are within the point of operation, or (3) automatically withdraws the operator's hands, if the operator's hands are within the point of operation as the dies close.

Presence-sensing device. A device designed, constructed, and arranged to create a sensing field or area and to deactivate the clutch control of the press component when an operator's hand or any other body part is detected within such field or area.

Type A movable barrier device. A self-powered movable barrier, which in normal operation is designed to: (1) close off access to the point of operation in response to operation of the press-tripping control; (2) prevent engagement of the clutch prior to closing of the barrier; (3) hold itself in the closed position; and (4) remain in the closed position until the slide has stopped at top of stroke.

Type B movable barrier device. A self-powered movable barrier, which in normal single stroke operation is designed to: (1) close off access to the point of operation in response to operation of the press-tripping control; (2) prevent engagement of the clutch prior to closing of the barrier; (3) hold itself in the closed position during at least the downward portion of the stroke while the slide is in motion, but be permitted to open during the downward portion of the stroke if the slide is stopped due to clutch control action; and (4) in normal single-stroke operations, it may be able to open during the upstroke of the slide.

Holdout or restraint device. A mechanism, including attachments for operator's hands, that when anchored and adjusted, prevents the operator's hands from entering the point of operation.

Pullback device. A mechanism attached to the operator's hands and connected to the upper die or slide that is intended (when properly adjusted) to withdraw the operator's hands as the dies close.

Two-hand control device. A two-hand trip that further requires concurrent pressure from both hands of the operator during a substantial part of the die closing portion of the stroke.

Die. The complete (or a portion of) tooling component used for cutting, forming, or assembling material within its point of operation. This tooling component is used with the press and other components to comprise a complete machine capable of producing a part.

Die builder. Any person who builds dies for power press machines.

Die set. A tool holder held in alignment by guide posts and bushings and consisting of a lower shoe, an upper shoe or punch holder, and guide posts and bushings.

Die setter. An individual who places or removes tooling components in or from a mechanical power press machine; and who, as a part of the duties, makes the necessary adjustments to cause the tooling and the safeguarding to function properly and safely.

Die setting. The process of placing or removing tooling components in or from a mechanical power press machine, and the process of adjusting the dies and other tooling and safeguarding means to cause them to function properly and safely.

Die shoe. A plate or block upon which a die holder is mounted. A die shoe functions primarily as a base for the complete die assembly and, when used, is bolted or clamped to the bolster plate or the face of slide.

Employer. Any person who contracts, hires, or is responsible for the personnel associated with power press operations.

Face of slide. The bottom surface of the slide to which the punch or upper die is generally attached.

Feeding. The process of placing or removing material within or from the point of operation.

Automatic feeding. Feeding wherein the material or part being processed is placed within or removed from the point of operation by a method or means not requiring action by an operator on each stroke.

Semiautomatic feeding. Feeding wherein the material or part being processed is placed within or removed from the point of operation by an auxiliary means controlled by operator on each stroke.

Manual feeding. Feeding wherein the material or part being processed is handled by the operator on each stroke of the press.

Foot control. The foot-operated control mechanism designed to be used with a clutch or clutch/brake control system.

Foot peddle (Treadle). The foot-operated lever designed to operate the mechanical linkage that trips a full-revolution clutch.

Guard. A barrier that prevents entry of the operator's hands or fingers into the point of operation.

Die-enclosure guard. An enclosure attached to the die shoe or stripper, or both, in a fixed position.

Fixed barrier guard. A die space barrier attached to the press component frame or bolster plate.

Interlocked barrier guard. A barrier attached to the press frame and interlocked so the press stroke cannot be started normally unless the guard itself, or its hinged or movable sections, enclose the point of operation.

Adjustable barrier guard. A barrier requiring adjustment for each job or die setup.

Guide post. The pin attached to the upper or lower die shoe, operating within a bushing on the opposing die shoe, to maintain the alignment of the upper and lower dies.

Hand feeding tool. Any hand held tool designed for placing within or removing from the point of operation material or parts to be processed.

Inch. An intermittent motion imparted to the slide (on machines using part-revolution clutches) by momentary operation of the inch operating means. Operation of the inch operating means engages the driving clutch so that a small portion of one stroke or indefinite stroking can occur, depending upon the length of time the inch operating means is held operated. Inch is a function used by the die setter for setup of dies and tooling, but is not intended for use during production operations by the operator.

Jog. An intermittent motion imparted to the slide by momentary operation of the drive motor, after the clutch is engaged with the flywheel at rest.

Knockout. A mechanism for releasing material from either the upper or the lower die.

Liftout. The mechanism also known as knockout.

Maintenance personnel. Those individuals who care for, inspect, and maintain mechanical power press machines.

Manufacturer. Any person who constructs, reconstructs, or modifies either the mechanical power press machine or any of its components.

Operator. Any individual performing production work on the mechanical power press and controlling the movement of the slide.

Operator's control station. The complement of controls used for stroking the press machine.

Owner. Any person who owns the power press machine.

Person. An individual, corporation, partnership, or other legal entity or form of business operation; the term does not include individuals who are employed as die setters, maintenance personnel, or operators as defined here.

Pinch point. Any point of the power press machine, except the point of operation, at which it is possible for a part of the body to be caught between the moving parts of a press component, or auxiliary equipment component(s), or between moving and stationary parts of a press component or auxiliary equipment component(s), or between the material and moving part or parts of the press component or auxiliary equipment component(s).

Point of operation. The area of the die (tooling component) where material is actually positioned and work is being performed during any process such as cutting, forming, or assembling.

Press or mechanical power press.

Press component, or mechanical power press component. The basic press (component) of the mechanical power press machine; that portion devoid of the tooling component, the safeguarding component(s), and auxiliary feeding components.

Press machine or mechanical power press machine. The combination of the press component, tooling component, safeguarding component(s) and feeding components; thus a complete machine existing in a state capable of processing the specific job requirement for which it is outfitted by the user, i.e., the production system.

Primary operation. Any preliminary press machine operation with respect to material to be subsequently processed.

Repeat. An unintended or unexpected successive stroke of the slide resulting from a malfunction.

Run. Single stroke or continuous stroking of the slide.

Safeguarding. This is an umbrella term that needs much explanation. The next section of this chapter explains it.

Safety block. A prop that, when inserted between the upper and lower dies or between the bolster plate and the face of the slide, prevents the slide from falling of its own deadweight.

Secondary operation. Press machine operations in which a preworked part is further processed.

Shall. Used to indicate a mandatory requirement.

Should. Used to indicate a sound safety practice but does not in every case indicate a mandatory requirement.

Single stroke. One complete stroke of the slide, usually initiated from a full-open (or up) position, followed by closing (or down), and then a return to the full-open position.

Single-stroke capability. An arrangement wherein the operating means (lever, pedal, switch, or buttons) when held depressed, normally does not result in more than a single stroke of the slide. Release and reapplication of the operating means is required to obtain a successive stroke. Single-stroke capability is provided by antirepeat or by a single-stroke mechanism.

Single-stroke mechanism. A mechanical arrangement within the clutch assembly, or part of the trip mechanism on the full-revolution clutch, designed to provide single-stroke capability.

Slide. The main reciprocating press component member. A slide may be called a ram, plunger, head, or platen.

Stop control. An operator control designed to immediately deactivate the clutch control and activate the brake to stop slide motion.

Top stop control. An operator control designed to delay its action after being operated so as to stop slide motion at a predetermined point in stroke, this predetermined point being independent of the instant at which button actuation was made.

Stripper. A mechanism or die part for removing the parts or material from the punch.

Stroking selector. The part of the clutch/brake control that determines the type of stroking when the operating means is actuated. The stroking selector generally includes positions for "off" (clutch control), "Inch," "Single Stroke," and "Continuous" (when Continuous is furnished).

Trip or tripping. Activation of the drive mechanism to run the press machine.

Turnover bar. A bar used in die setting to manually turn the crankshaft of the press machine.

Two-hand trip. An actuating means requiring the concurrent use of both hands of the operator to trip the press machine.

Unitized tooling. A type of die in which the upper and lower members are incorporated into a self-contained unit so arranged as to hold the die members in alignment.

POINT-OF-OPERATION SAFEGUARDING

Safeguarding the point of operation as described in the following paragraphs is limited to those means that will protect operating personnel (including helpers) after the dies have been installed, tested, and operated, and are ready for production.

Protection of the "die setter" who must have access to the point of operation while setting dies is not being discussed in this section. For safe die-setting procedures and die setter protection, see the later section on Power Press Setup and Die Removal.

Protection of passersby from these same hazards must also be considered by the employer. Administrative measures may be necessary as dictated by the hazard created by the location of the press and the traffic conditions near the press.

As the term "point of operation" implies, this discussion deals with hazards in the die where material is being processed and worked. Of necessity, all die hazards must be considered, extended to wherever relative motion exists in the die space that may crush, cut, punch, or sever, or otherwise injure personnel.

"Safeguarding the point of operation" is a term under which all safeguarding methods fall and all types of die hazards exist. Two basic categories exist under this term. One category is the guard; the second category is the device. (Refer to Figure 8-5, in Chapter 8, Principles of Guarding.)

A guard (or barrier guard) is best understood as a physical barrier that *prevents access* to a die hazard when it is in place and while it remains in place during a productive run of successive cycles. If a barrier allows access to a die hazard during the productive run, it is *not* a guard; it then serves only as an inadequate enclosure. An inadequate enclosure *always* requires a device in conjunction with its use to form an acceptable safeguarding system. Recommended permissible guard openings are shown in Figure 8-10, in Chapter 8, Principles of Guarding. (Some devices require the use of enclosures to form an acceptable system.)

A safeguarding means is best understood as a device that *controls access* to the point of operation. Devices can be divided into three types: (1) press-controlling devices, (2) operator-controlling devices, and (3) devices that control both the operator and the power press.

Full-revolution clutch power presses

It then becomes clear from the controlling devices listed that a power press whose stroking cannot be interrupted (controlled) during the closing or opening of the stroke cannot use a press-controlling device, although a properly installed two-hand trip can serve as both an actuating means and a safeguarding means. This is true of a press equipped with a full-revolution clutch which is simply "tripped" to initiate a stroke and must make a full revolution before it is forcibly disengaged through cam action. On such power presses, a selection must be made from the following list of safeguarding systems.

Guard.
1. Die enclosure guard
2. Fixed barrier guard
3. Adjustable barrier guard

Remember, if hands can reach through, around, over, or under a "guard" to allow access to the point of operation, it is an inadequate enclosure and not acceptable by itself.

Devices (operator-controlling type).
1. Restraints (properly adjusted)
2. Pullbacks (properly adjusted)
3. Type A movable barrier (with an enclosure to prevent access through areas not protected by the movable barrier)
4. Two-hand trip (located at a distance exceeding the applicable safety distance for the particular press)

Part-revolution clutch power presses

On a press whose stroke can be interrupted during the closing or opening of the stroke, a press-controlling device can be used as well as an operator-controlling device. On such power presses, a selection must be made from the following list of safeguarding systems:

Guard
1. Die enclosure guard
2. Fixed barrier guard
3. Adjustable barrier guard
4. Interlocked press barrier guard

Again remember that if hands can reach through, around, over, or under a "guard" to allow access to the point of operation, it is an inadequate enclosure and not acceptable by itself.

Devices (operator-controlling type, single-stroke operations only)
1. Restraints (properly adjusted)
2. Pullbacks (properly adjusted)

Devices (machine-controlling type)
1. Presence sensing

Devices (operator- and machine-controlling devices)
1. Two-hand control
2. Type A movable barrier
3. Type B movable barrier

Guards

Die-enclosure guards. A fixed-die enclosure guard provides the most positive protection for the operator since the die is completely enclosed and the guard is a permanent part of it. Die enclosures can be used on many types of press operations to prevent operators from placing their hands into the point of operation. (See Figures 11-5, 11-6, and 11-7.)

Die enclosures are attached to the die shoe or stripper in a fixed position and are designed so that hands cannot reach over, under, through, or around the guard into the point of operation.

It is important that this type of guard be constructed to permit easy feeding of the stock, ejection of the part, and scrap removal. It should also afford good visibility at all times.

A minimum clearance of 1 in. must be provided between the top edge of the guard and the slide or any projection on the slide. In addition, the guard should extend at least 1 in. above the bottom of the punch holder to prevent a shearing hazard caused by travel of the slide.

Enclosure guards can be built of various types of material. Preslotted material is often used. A metal frame can be built, and rod stock can be welded or otherwise fastened to it. Openings should run vertically to lessen eye fatigue.

Transparent polycarbonate also may be used. When properly maintained, it has the advantage of affording good visibility. However, this material scratches easily and is damaged by oil and grease. It should be at least ¼ in. thick and will give longer service if mounted in a metal frame (see Figure 11-7).

The expense of designing and installing permanent die enclosures is very small and yields positive safety results. In many plants in which this type of guarding is applicable, each die is so guarded before it leaves the tool room.

Fixed barrier guard. A fixed barrier guard, when used, shall be attached to the frame of the press or to the bolster plate. (See Figure 11-1.)

Interlocked press barrier guards. Barrier guards (see Figure 11-8) may be designed with a pivoting, sliding, or removable section to allow ready access to the die. The pivoting or sliding section must be interlocked with the press clutch control if operation of the machine is to be prevented when the section is open.

This type of interlocked fixed barrier guard can be used successfully on automatic presses where the point of operation must occasionally be exposed so that jams can be relieved. As a further safety measure, in the event that the interlock fails to function, hand tools or picks should be used to relieve jams, or the slide should be supported by use of safety blocks. The pivoting, sliding, or removable section is not permitted to be used for feeding on single stroke applications, since the simple interlocking is too easily bypassed.

Adjustable barrier guards. Where a die enclosure guard or fixed barrier guard would take some time to complete or its provision is considered functionally impractical, or both, an adjustable barrier device can be provided on each press. This type of guard can be used on many operations to prevent the operator's hands from entering the point-of-operation zone. (See Figure 11-4.)

An adjustable barrier guard must fit any size die used. This barrier is especially practical for short run jobs.

Adjustable barrier guards are available commercially or may be made in the plant. They are attached to the frame of the press and have front and side sections which can be adjusted for dies of almost any size. (See Figure 11-9.)

This type of guard is usually constructed of rod stock or perforated metal. Any pivoting or sliding sections used for occasional access should be interlocked with the press control for maximum safety.

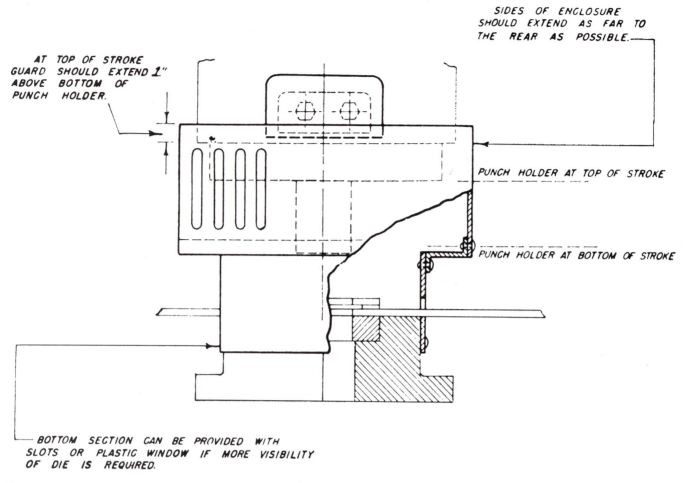

AT TOP OF STROKE
GUARD SHOULD EXTEND 1"
ABOVE BOTTOM OF
PUNCH HOLDER.

SIDES OF ENCLOSURE
SHOULD EXTEND AS FAR TO
THE REAR AS POSSIBLE.

PUNCH HOLDER AT TOP OF STROKE

PUNCH HOLDER AT BOTTOM OF STROKE

BOTTOM SECTION CAN BE PROVIDED WITH
SLOTS OR PLASTIC WINDOW IF MORE VISIBILITY
OF DIE IS REQUIRED.

Figure 11-5. Die enclosure guard for strip or coil stock showing the vertical clearances required. (Reprinted courtesy of Liberty Mutual Insurance Co.)

Figure 11-6. Die guard made of preslotted material can be easily fabricated in the shop.

Unless feeding or ejection is automatic, it may be necessary to leave an opening in the barrier for insertion of a tool to remove the piece from the die. Any opening should not be wide enough for a hand or finger to extend into the point of operation.

In the case where the whole guard is removed to change dies or for job adjustment, the guard is usually not interlocked with the press control. When adjusting the sections of this type of guard, die setters should be instructed to follow the dimensions for permissible openings given in Figure 8-10. The operator should never be allowed to make changes in the adjustments without the supervisor's approval.

Point-of-operation safeguarding devices

Type A movable barrier device. The Type A movable barrier device, when used, protects the operator by enclosing the point of operation before a press stroke can be initiated, and remains in this enclosed position until motion of the slide has ceased at the top of stroke. (See Figure 11-10.)

Type B movable barrier device. The type B movable barrier device, when used, protects the operator by enclosing the point of operation before a press stroke can be initiated, being so applied as to make impossible an act of reaching into the point

Figure 11-7. Slide feed allows loading of die outside the danger zone. Permanent plastic barrier guard permits full visibility of operation. Note separation of guard at top to permit die maintenance. Overlap of guard at separation eliminates shear hazard during travel of slide. (Reprinted courtesy of Allis-Chalmers Mfg. Co.)

of operation prior to die closing during the downward stroke. The Type B device is permitted to open on the up stroke of the slide, or during the downstroke, if the slide is stopped on the down stroke. Type B devices should not be used on full revolution machines.

Two-hand tripping device. A two-hand tripping device shall meet *all* of the following requirements:
1. Have the individual operator's hand controls protected against unintentional operation
2. Have the individual operator's hand controls arranged by design and construction, by separation, or both, to require the use of both hands to trip the press
3. Use a control arrangement requiring concurrent operation of the individual operator's hand controls
4. Incorporate a single stroke capacity
5. Be spaced far enough from the point of operation that the hands cannot reach into the point of operation before the dies close after hands are removed from controls. This distance is mandated by federal regulation and is prescribed to protect the operator from reaching into the point of operation before the down stroke is completed.

A two-hand trip device is used on a full revolution clutch machine.

Two-hand control device. A two-hand control device must meet the protected button requirements for a two-hand trip and it must *also* be arranged to require concurrent pressure from both hands during a substantial part of the die-closing portion of the stroke. (See Figures 11-11 and -30.) The devices must be

located to require operation by hands and no other means. A two-hand control device can only be used on a part-revolution clutch machine, so long as that machine is controllable (can be stopped), during the closing portion of the downstroke.

Federal regulations mandate a safety distance for mounting controls away from the point of operaton, and, if hands-in-dies feeding is used, brake monitoring and additional control reliability are required.

Where more than one operator is required on a press, two-hand controls must be provided for each operator.

Pullback devices. Pullbacks or pullouts have the advantage that, when properly used, adjusted, supervised, and maintained, they always remove the hands from the point-of-operation zone as the slide descends. This type of device is usually limited to secondary operations and jobs where the operator can remain at the feeding position. Pullback devices should be attached to the press slide or upper die shoe. Because of the variation of the size of the operators' hands and characteristics of various dies, *it is very important that the pullback device be adjusted to fit each operator and after each die change.*

Restraints. A restraint or holdout is used to prevent the operator from reaching into the point of operation at any time. This is achieved by providing wrist straps and firmly anchored restraint cords or cables.

Where more than one operator is required on a press, pullback and restraint devices should be provided for each operator. (See Figure 11-34.)

Presence-sensing device. A presence-sensing device is designed, constructed, and arranged to create a sensing field or area and to deactivate the clutch control of the press when an operator's hand or any other body part is within such field or area.

A presence-sensing device can only be used on a part-revolution clutch machine. Federal regulation mandates that the sensing field be located at a safety distance from the nearest point of operation hazard depending on the slide stopping time. Additionally, if hands-in-dies feeding is used, there must be brake monitoring and additional control system reliability.

These devices should always be supplemented by hand tools or feeding and ejection devices so that the operator need not place hands in the point-of-operation zone. These devices cannot protect against repeats.

AUXILIARY MECHANISMS

Feeding and extracting tools

A variety of special tools has been developed for feeding and extracting parts. These tools are made of soft metal, aluminum, or magnesium (some are magnetized), and include pushers, pickers, pliers, tweezers of various types, forks, and suction disks (Figure 11-2).

Tools provide protection only if they are always used by the operator. They should never be permitted to substitute for proper safeguarding.

One way to make sure tools are used properly is to provide storage convenient to the workplace. Pegboard can be used to mount various tools; a silhouette of each tool can be painted on the board to indicate where it belongs. This encourages keeping tools in their correct place.

Figure 11-8. Rear view of automatically fed straight side press. Interlocked barrier guards are manually operated.

Foot control and shielding for protection

Foot operation, by itself, is only an operating means and provides no protection for the operator unless it is located at a safe distance from the point of operation. When the operator is within reach of the point of operation, safeguarding of the point of operation by a separate means must be used. A foot-operated mechanism must be shielded and positioned to control its accessibility, and to afford protection under circumstances, such as die setting, when safeguarding is necessarily removed from the point of operation.

Full-revolution clutch presses. A mechanical foot pedal trips the clutch whenever it is depressed. It remains tripped until a stroke is made. If the press is not running, it will stroke immediately when started. Controlled accessibility to the pedal is therefore required to prevent unintended actuation.

An electric foot switch can be turned "off" to reduce the possibility of accidental actuation during off periods. However, during operating periods, it is more easily actuated than a pedal and, therefore, must be shielded in a manner to inhibit unintended foot entry from actuating the machine.

Part-revolution clutch presses. Foot switches, when used with clutches capable of being disengaged anywhere in the stroke, should be controlled through an antirepeat circuit to prevent a second stroke if the foot switch is held depressed for the full cycle and withholding time so that release of the foot switch on the down stroke will cause the ram to stop.

Covers, such as the one shown in Figure 11-12, should be large enough to allow room for the operator to place a foot in the operating position without undue fatigue and without striking or scraping the leg against sharp edges of the cover. A split length of ordinary garden hose around the opening of the pedal guard will protect against such injury.

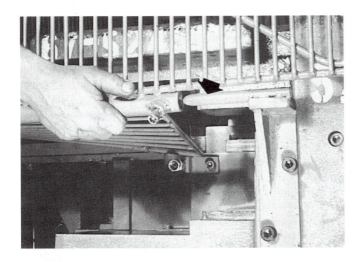

A foot pedal should travel about 3 in. (7.6 cm). The clutch rod should be connected to the pedal lever so that the distance between the clutch rod and the rear pivot on the pedal will be approximately one-third the length of the pedal. The travel of the connecting point will thus be about 1 in. (2.5 cm) when the pedal travel is 3 in.

The pedal lever should not have side play, and the tension of the pedal should be maintained as recommended by the manufacturer. Springs or counterweights should not be added to the pedal shaft or to the pedal shaft lever.

Measures to prevent accidental tripping of the press (in addition to placing a cover over the pedal) include:

1. A safety spring on the trip rod.
2. Aisles of ample width for trucking material to the presses.
3. Aisles adjacent to the press used only for necessary trucking of dies and stock.

Figure 11-9. Press guarding can be simplified by designing "universal" barrier guards. Two sections completely enclose the point of operation—the upper guard encloses any operation done on the press; the lower guard conforms to the upper guard, the feed device, and the lower die contour. A mounting pin is shown at lower right and also at black arrow in top photograph. Tack welding clips the bottom to the top and keeps employees from making their own adjustments. The guard shown here can be used for several different operations with similar feeding methods and die characteristics.

Figure 11-10. A movable barrier device protects the operator by enclosing the point of operation before a press stroke can be initiated.

4. Enough working space between adjacent machines in the same line.
5. Unauthorized persons kept away from the area.
6. Operators protected from distraction.
7. The operator, and all other persons who are not concerned, kept away from the press when it is being set up.
8. A closely supervised, mandatory rule prohibiting the operator from "riding" the pedal. The operator should remove the foot from the pedal immediately after each time the press is tripped.
9. Flywheel brakes. Flywheels—especially on large presses—continue to coast from some time after the power source has been turned off, and they have been known to cycle a press when actuating control (or controls) was tripped. A flywheel brake will stop the flywheel within 10 to 20 seconds.

Single-stroke attachments

A press with a full-revolution clutch should have a single-stroke attachment which disconnects the pedal or operating lever after each stroke.

A single-stroke spring device should depend on spring action only if it is a compression spring encased in a close-fitting tube or closely wound on a rod. The type wound on a rod is preferable because it permits easy detection of a broken spring. Be sure that the space between the spring coils is less than the diameter of the wire.

Of necessity, a single-stroke device is made inoperative when the press is used on continuous operation. In this case, the die should be completely enclosed, regardless of the method of feeding. When the press is set for other than continuous work, it is of the utmost importance that the single-stroke device be reconnected.

FEEDING AND EJECTING MECHANISMS

A significant percentage of power press injuries (such as puncture wounds, lacerations, strains, and amputation or crushing of the hands or feet) occur during the handling of work. Thus any mechanism which will eliminate handling of the work should reduce exposure to the hazards, particularly in operations which put the operator's hands in a danger zone.

Primary operations

Primary operations are normally conducted with random length strip stock or coiled stock. Strip stock is usually manually fed. Coiled stock can be fed either manually or by means of a roll feed or hitch feed.

Automatic roll feeds are often used on continuous operations of blanking from strip stock. Small gears on the roll feed should be enclosed and the run-in nip point of the feed rolls should also be guarded. These precautions are especially important where operators with long hair are employed, even though they wear hair nets or close-fitting caps.

Automatic push or pull fees are similar to roll feeds and are used mostly in blanking larger pieces.

When coiled stock is used with a reel and roll feed or hitch feed, a feed table is generally unnecessary. Where coiled stock is used with a stock reel but not with a roll feed, a feed table helps the operator backgage the stock and feed it to the die with minimum effort.

In manual feeding of strip stock, a feed table eliminates unnecessary motion and reduces operator fatigue. The feed table should be adjusted to the height at which the operator can work with minimum effort.

Oiling rolls or pressure guns can be used instead of a paintbrush system to lubricate strip or roll stock. Automatic or manual control pressure guns can be provided to lubricate the punch and the die.

Secondary operations

On secondary operations, selection of feeding and ejection methods which make it unnecessary for the operator to place hands in the point-of-operation zone is more difficult.

When automatic feeding methods are used, it is still necessary to safeguard the point of operation by limiting the slide stroke to ¼ in. (6 mm) or less, or to provide a guard or device, according to ANSI Standard B11.1, *Safety Requirements for the Construction, Care and Use of Mechanical Power Presses*.

Semiautomatic feeding mechanisms place the piece being processed under the slide by a mechanical device that requires the attention of the operator at each stroke of the press. Such

Figure 11-11.   Two-hand controls. Center button is STOP switch.

feeds have a distinct advantage in that the operator is not required to reach into the point-of-operation area to feed the press, and the feeding method permits complete enclosure of the die. This type of feeding may not be adaptable for certain blanking operations or for nesting of odd-shaped pieces.

Semiautomatic feeds

The six principal types of semiautomatic feeds are chute (both gravity and follow), slide or push, plunger, sliding dies, dial, and revolving dies. Each one will be discussed in turn.

Chute feed. Of the six semiautomatic methods, the chute feed is probably the most widely used. It is a horizontal or inclined chute into which each piece is placed by hand. The pieces then slide or are pushed one at a time into position in the lower die. The entire die may be enclosed, since it is unnecessary for the operator to place hands in the point-of-operation

area if automatic ejection is also provided (Figure 11-13).

It is customary to use a soft metal pick or rod or a wood stick to remove pieces that jam in the die. *A steel rod should never be used. If caught in the die, it could shatter, causing injury to the operator and damage to the die.*

Many hand-fed dies can be changed to chute feed by reversing the dies and inclining the press. A chute feed on an inclined press not only helps center the piece as it slides into the die, but also simplifies the problem of ejection.

Slide or push feed. A variation of the chute feed, the slide or push feed is combined with magazines and plungers. The pieces are stacked in the magazine. As each piece reaches the bottom, it is pushed into the die by means of a hand-operated plunger (Figure 11-14).

Plunger feed. A variation of the push feed, the plunger feed may be semiautomatic or manual in operation. The semiauto-

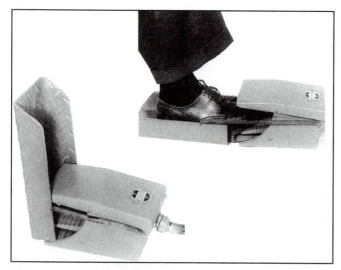

Figure 11–12. One method of controlling a foot switch to which a spring-loaded shield has been attached in such a way that it closes automatically when foot is removed.

matic plunger feed is a magazine or chute in which blanks or partly formed pieces are placed. The blanks or pieces are fed, one at a time, by a mechanical plunger or other device which pushes them under the slide.

Manually operated plunger feeds are used for individual pieces which, because of their irregular shape, will not stack in a magazine or will not slide easily down a gravity chute. Each piece is placed in the nest in the pusher and moved to the die by manual operation of the pusher. To get correct location of the part in the die, an interlock is sometimes necessary so that the press cannot be operated until the pusher has spotted the part accurately (Figure 11-15).

Sliding die feed. The die is pulled toward the operator for safe feeding and then pushed into position under the slide (ram) for the downward stroke. The die may be moved in and out by hand or by means of a foot lever. Regardless of how the die is actuated, it should be interlocked with the press to prevent tripping when the die is out of alignment with the slide (Figure 11-16), and stops should be provided to prevent the die from being inadvertently pulled out of the slides.

Dial feed. Two or more nests arranged in dial form revolve with each stroke of the press so that the operator can feed the machine safely. The part to be processed is placed in a nest on the dial which is positioned in front of the die. The dial is indexed with each upstroke of the press to deliver the next nested part into the die (Figure 11-17).

The best method of ejection is usually by pickup fingers or compressed air. However, in many installations the operator both nests the part on the outside and removes it when it is returned on the dial.

Two operators are sometimes used on a dial fed press, one to feed the press and the other to remove the processed parts.

Revolving die feed. Operating on the same principle as the dial feed, the revolving die feed may consist of two dies or multiple dies.

When a dial feed is used, the point-of-operation area should be enclosed. Motion of the dial when hands are free may create

a hazard. An idle station may be required beyond the load station to prevent injury at the station which enters the guard.

Ejector mechanisms

Properly designed and installed ejector mechanisms will eliminate many common hazards. Ejector mechanisms can automatically clear the press faster than human hands can and with greater safety.

In the development of ejection facilities, two problems must be solved: how to strip the piece from the punch or the die, and how to eject the piece from the die and the press to a container or conveyor. The problems can be solved in many ways. In some cases, a single mechanical means performs both stages. In other cases, a separate method is used for each stage.

Some of the more common methods which are used singly or in combination to strip pieces from the die are:
1. Positive stripper plates (Figure 11-18A).
2. Spring-pressure pads or pins (Figure 11-18B).
3. Latch-type mechanical lift dogs (Figure 11-18C).
4. Compressed air jets (Figure 11-18D).
5. Pneumatic or hydraulic lift pins or pickup fingers (Figure 11-19).

Single or multiple air jets can be used for effective removal of small pieces. Air ejection can be combined with other mechanical release means or with gravity removal. All jets should be anchored securely to direct the air stream effectively and to prevent jet tubes from shifting into the die working area.

Both air consumption and noise can be reduced considerably by incorporation of one or more of the following:
- Locate the jet discharge as close as possible to the piece to be removed. Optimum positioning is essential for maximum effectiveness and frequently can be accomplished through die design.
- Limit the duration of air discharge to the minimum peroid required to remove the piece.
- By means of flow valves or pressure regulators, reduce the discharge pressure to the minimum required.

   Reducing the discharge pressure will also result in greater operator safety from flying particles. However, this statement should not be taken to mean that the operator need not wear eye protection. There are always hazards from flying particles in press operations, and the addition of air ejection increases the hazards. Using compressed air also increases the overall noise level. Be sure all operators wear eye and ear protection at all times.
- Use an air-ejector nozzle with several orifices (Figure 11-20).

Pneumatically powered cylinders operating sweeps or kickout pins timed with the upstroke of the press are more effective than air jets for removal of large and heavy pieces. For safety, when jams must be cleared or tryout operations performed, the pneumatic equipment should have a valve to shut off the flow of air and bleed any residual air pressure between valve and cylinders.

Clamp and pan shuttle extractors are used on presses of all sizes. Pivoting or straight-line clamps grip the part and remove it from the die area to a pallet, bin, or conveyor for further processing. A pan shuttle (Figure 11-21) catches the piece as it is stripped from the upper die by knockout pin or other means and removes it from the die area.

While some extractors are actuated mechanically by connection with the press slide or shafting, independently powered

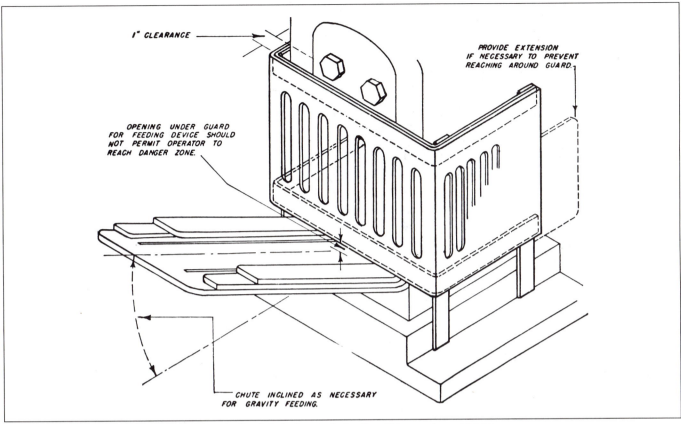

Figure 11-13. A die enclosure guard with an inclined chute for gravity feeding. This guard may also be used on an inclined press but with a straight chute. (Reprinted courtesy of Liberty Mutual Insurance Co.)

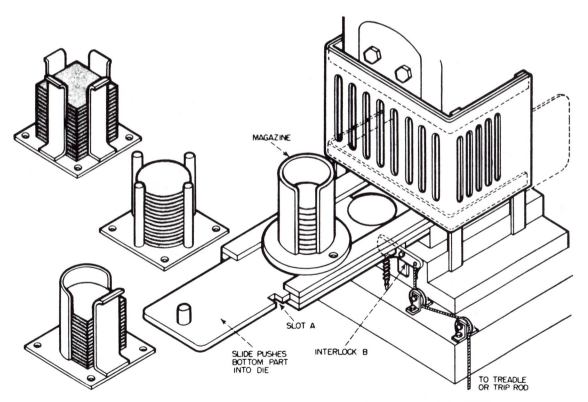

Figure 11-14. This magazine on a push feed enables the operator to catch every press stroke. Slot A in the pusher must be in alignment with interlock B before the press can be tripped. This feature assures proper positioning of the part in the die. (Reprinted courtesy of Liberty Mutual Insurance Co.)

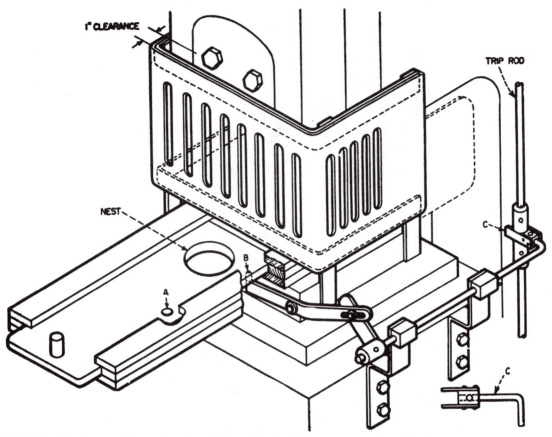

Figure 11-15. A manually operated plunger feed. Note that the press cannot be tripped until the part is pushed to the next location. At this point, hole A is directly over tapered pin B; it can rise and release yoke C so that the press can be tripped.

extractors are used more frequently because they are more versatile and can be used with several presses. Every independently powered extractor should be interlocked with the press control circuit so that the press cannot operate unless the extractor is in the "home" or "out" position.

Simple sheet steel chutes generally are provided in conjunction with air ejection to guide the parts into a container. Chutes or slides also can be used for a controlled movement of pieces to subsequent operations. Sharp bends and inside projections, which may cause parts to pile up, should be eliminated from enclosed chutes.

Elevators and power and gravity conveyors can be used to transport pieces of work to containers or to subsequent press operations.

All conveyors should be supported for ample side and end stability. A mobile conveyor, which can be moved from station to station, should have base supports of a width not less than one-half the height of the conveyor. Conveyor sections, especially those of the gravity type, should be adjustable so that angles of slope can be set as the primary or auxiliary factor to control movement of parts at safe speeds and with orderly spacing.

The drive mechanisms of power conveyors and elevators should be enclosed as completely as possible, and pinch points formed at pulleys or conveyor structures by moving belts should be eliminated or covered. Electrically driven conveyors should

be provided with equipment grounds and with "on-off" controls located conveniently for emergency use.

In every plant, there will probably be times when the safety procedures just outlined will not be followed and parts will be moved manually. Not only is manual removal less efficient but it also involves greater operator fatigue and exposure to injury than does mechanical removal. However, the basic safety principle of keeping the operator out of danger zones must be followed even if parts are removed manually. This can only be accomplished by proper guarding of the danger zones and the use of suitable hand tools by the operator.

HAND- OR FOOT-OPERATED PRESSES

All power presses have counterparts in presses which are hand or foot powered. There are still many shops using kick presses, foot shears, hand folders, and hand rollers. Hazards do exist even though the operator is the source of power. Such presses can be guarded.

Kick presses are used principally for piercing, notching, forming, and shearing of small parts, eyelet closing, and subassembly work.

Although simple to operate, the kick press presents serious hazards. Since even the most experienced operator is subject to occasional failure of hand-foot coordination, the short job that is often performed without the use of guards is especially

Figure 11–16. An adjustable barrier with a sliding die. A locating mechanism is provided on the slide bolster to locate and lock the die slide in alignment with the punch holder. An interlock should be provided to prevent tripping of the press until the die slide is in the proper location.

hazardous. All kick presses should have guards installed and all operators should be thoroughly trained.

Foot press operations can be effectively safeguarded by an interlocking tripping mechanism which requires the simultaneous use of both hands of the operator to release the slide head before the pedal can be used. At all other times, the slide head should come to rest in the top position.

Types of injury

The principal types of injury resulting from unsafe operation of kick presses include finger amputations and finger punctures from unguarded points of operation and fatigue or abdominal strain resulting from pressure required to perform the work or from improper posture. Strains from lifting materials and eye injuries caused by small flying particles are also possible.

General precautions

When a new kick press is installed or an old one is moved to a new location, it should be securely fastened to the floor or bench. Otherwise, as the press is operated, it will shift. The gib should be clean and lubricated and the gib screws tightened.

Good general lighting should be provided, and in some cases local lighting may be required. Good visibility not only will increase production and reduce fatigue, but will lessen the possibility of injuries. Visibility can be improved by painting the press in contrasting colors.

Guarding

When a kick press is used for piercing and similar operations, the point of operation should be guarded. A guard that fits

around the punch can be made from a small piece of perforated sheet metal, transparent plastic, or other material. The guard should allow enough room for the work to be inserted into the press, but not enough to permit the fingers to come within the danger zone.

On some assembling jobs, when the operator must place hands under the punch, the point of operation cannot be guarded and a two-handed safety device should be used.

The mechanism shown in Figures 11-22 and -23 is known as the ratchet type. It was designed to be a thoroughly flexible protective device for kick presses. It operates this way: A steel plate with ratchet teeth is securely fastened to the movable slide head. A pivot pin is anchored securely to each side of the head casting. Two-hand levers having shaped points to fit the ratchet teeth in the plate are so designed that from the pivot point the top portion of the levers, when the levers are not held back, bears against the ratchet plate. The slide head is thus positively locked from downward motion.

To operate the press, the operator must use both hands to disengage the pawls from the ratchet and thus permit downward travel of the slide head. Should the operator release either hand, the pawls will engage the ratchet and stop the punch instantly. During the loading and unloading periods, the ram is positively locked in the top position.

This type of device not only prevents the operator from absentmindedly placing fingers beneath the punch, but also provides a brace when the operator applies pressure on the treadle.

Another simple, yet effective, type of two-hand device can be easily installed at small expense. One-half inch holes are drilled through the sides of the frame and into the swinging arm of the ram. Then, when the ram is in top position, automatic locking pins that can be withdrawn by the operator are inserted in the holes. Levers with springs attached to these pins are located conveniently for the operator to lock the ram when it is in top position.

Eye protection

Eye protection should be worn by operators on kick press operations. When hard materials, such as brass and spring steel, are being shaved or notched, small particles thrown off with great force may cause serious injury. Chipping goggles or side-lens spectacle goggles are required on such operations.

Reducing fatigue and strain

Since most kick presses are operated from a sitting position, a seat or chair should be provided that will enable the operators to work with minimum fatigue. They should not have to make a long reach with the foot to perform the operation, nor should the seat be arranged so that they sit in a cramped position. The seat should have a comfortable back rest, and the chair legs should have a wide enough spread so that the chair will not tip over easily.

Operator fatigue will be greatly lessened by proper adjustment of the counterweight and stop. Proper adjustment reduces the travel distance of the pedal and results in a balanced operation.

If the pedal is worn smooth, the operator's foot can readily slip, giving a quick strain or painful ankly injury. Flanges on both ends of the pedal and rubber or abrasive surfacing on the pedal will minimize this hazard.

Figure 11-17. Well-guarded double-dial feed. Note plastic coating (arrow) on lower part of metal guard, which eliminates sharp edges and reduces glare. (Reprinted courtesy of Automatic Electric Co.)

There is a possibility that the operator may suffer abdominal strain from continued heavy operation of a kick press. Occasionally, heavy operations are performed that cause unusual fatigue. In such cases, the operator should be permitted to alternate kick press work with light work.

Only employees who are physically capable of such work should be allowed to do it. No employee who is known to have weak abdominal walls, such as from a recent operation, should be permitted to operate a kick press.

Lifting of heavy containers of parts may result in severe muscular strain, particularly in continued operations where fatigue may make lifting increasingly difficult. Conveyor belts which

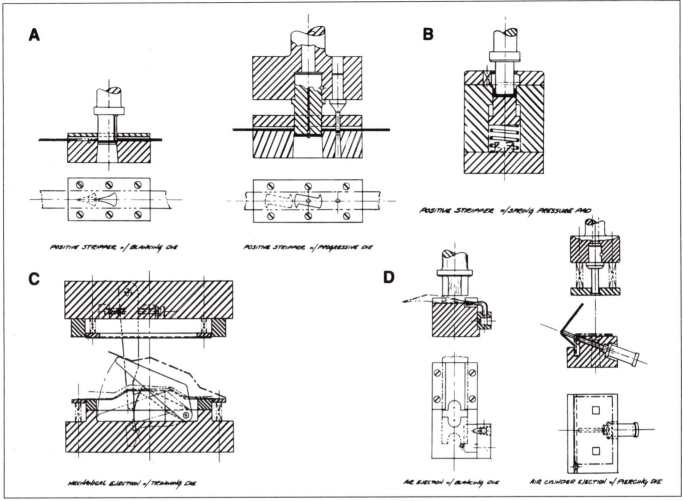

A

POSITIVE STRIPPER w/ BLANKING DIE

POSITIVE STRIPPER w/ PROGRESSIVE DIE

B

POSITIVE STRIPPER w/ SPRING PRESSURE PAD

C

MECHANICAL EJECTION w/ TRIMMING DIE

D

AIR EJECTION w/ BLANKING DIE

AIR CYLINDER EJECTION w/ PIERCING DIE

Figure 11-18. Mechanical methods for stripping finished pieces from power press dies.

Figure 11-19. Expanded metal guard protects the point of operation as well as the mechanical transfer mechanism. Note the electrical interlock on the die block.

carry the material along the bench to the machine will eliminate this hazard.

If the volume of work involved does not warrant the use of a conveyor, brackets fastened to the press at truck height will make it possible to slide the containers from the truck to the press and back again with little effort.

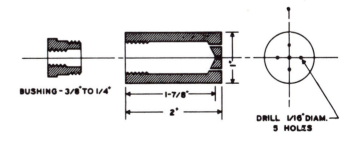

BUSHING - 3/8" TO 1/4"

1-7/8"

2"

DRILL 1/16" DIAM.
5 HOLES

USE 1-INCH ROUND STOCK - STEEL, ALUMINUM, BRASS
DRILL 37/64 INCH DIAM. HOLE TO ACCEPT 3/8"-18 N.P.T. TAP

Figure 11-20. A multi-orifice air ejector nozzle designed for noise reduction.

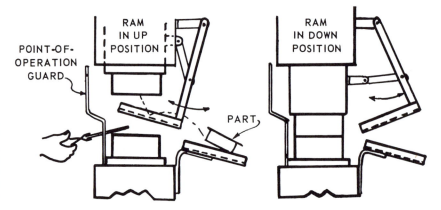

Figure 11-21. Pan shuttle mechanism moves under finished part on upside of press slide. Shuttle then catches part stripped from slide by knockout pins and deflects part into chute. When press slide moves down toward next blank, pan shuttle moves away from die area.

Maintenance

The kick press and its safeguard should be checked frequently, and all working parts lubricated and adjusted. Parts of the press or the guard showing wear or defects should be replaced immediately.

POWER PRESS ELECTRIC CONTROLS

Properly designed, applied, and installed electrical controls are an important element in press safety. This is particularly true when a two-hand control device or a two-hand trip is used for point-of-operation safeguarding.

Electrical controls may range from simply a power disconnect switch and a motor starter on a small full-revolution clutch press to an extensive system on a large part-revolution clutch machine. The latter typically consists of:

1. A main disconnect switch or circuit breaker for isolating the machine from electric power.
2. Starters for all motors including main drive, slide adjustment, lubrication, and any auxiliary drives.
3. One or more control transformers with isolated secondary windings to step down line voltage to a level of not more than 120 volts AC for control circuits and solenoid coils.
4. Overcurrent and ground fault protective equipment or circuitry.
5. Control relays or the electronic equivalent for energizing magnetic valves, interlocking, sequencing, and checking of safety features.
6. Operator controls, set-up controls, indicating lights, limit switches, pressure switches, and other sequence, checking, and protective components.
7. Magnetic air valves for clutch/brake, flywheel brake, and auxiliary equipment.

The following information is to guide users of mechanical power presses in the application of two-hand controls and two-hand trips as point-of-operaton safeguarding devices. The contents of the following paragraphs apply in all installations regardless of the means of point-of-operation safeguarding.

I. Installation

A. Controls of all machines should be built in accordance with *Electrical Standard for Industrial Machinery,* NFPA 79, and recognized industry standards. Installation of machines should conform to *National Electrical Code,* NFPA 70, and any applicable local codes. The purpose of these codes and standards is to safeguard personnel and equipment.

B. The machine control panel should contain the main disconnect switch, motor starters, clutch control relays, and control transformer. This panel should be on or immediately adjacent to the machine it serves, but out of the operator's way during normal operation. It should be properly identified and accessible for maintenance (see Figure 11-24).

The main disconnect switch handle should be on the exterior of the control enclosure and operable from floor level. The recommended maximum height for the centerline of the switch handle is 6½ feet above the floor. The switch must be provided with a means for positive lockout.

C. Cases or frames of all control components should be grounded to the press frame. Mounting bolts are satisfactory for grounding provided that all paint and dirt are removed from the joint surfaces before assembly. Movable control components not secured to the press frame should be grounded by means of a grounding conductor in the connecting cable or conduit.

D. The rotary limit switch is a vital element in the control system (see Figure 11-25). Mechanical or electrical failure of this switch may cause the press to malfunction. The switch-driving mechanism—including all couplings, gears, sprockets, and associated parts—should be securely assembled and be of a rugged design. A positive, keyed connection is preferred to set screws, as a single set screw has a tendency to vibrate loose.

The use of two separately driven switches or a drive failure detector is recommended. All cam switches should be of a type in which the switch contacts are forced open by the actuating cam.

E. Where a flywheel brake is used, it should be electrically interlocked with the main drive starter to prevent simultaneous operation.

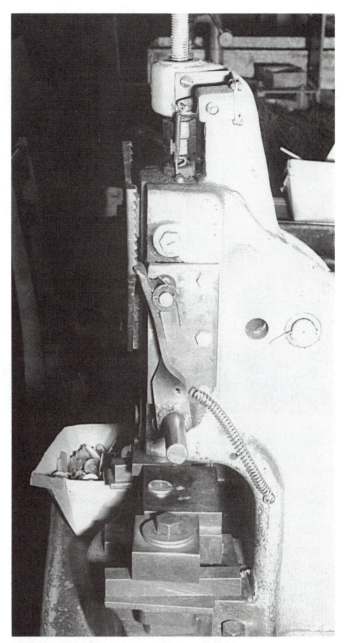

Figure 11-22. A ratchet mechanism for kick presses. The operator must use both hands to disengage the pawls from the ratchet. The ratchet teeth are cut into the plate attached to the slide head. Below the ratchet teeth, the plate tapers to the width of the slide and is secured to the slide by two hex-head bolts and washers. (Reprinted courtesy of General Electric Co., Pittsfield, Mass.)

Figure 11-23. Side view of ratchet mechanism for kick presses shows the simple ratchet principle of the guard. (Reprinted courtesy of General Electric Co., Pittsfield, Mass.)

should never be used for this purpose because they have pistons sliding in close-fitting sleeves and can easily stick due to dirt, corrosion, or improper lubricaton. (Also see Section V paragraph C.2, regarding self-checking double valves.)

II. Definitions

A. Two-hand trip. A clutch- or clutch/brake-actuating method requiring the momentary concurrent use of both hands of each operator to initiate a complete stroke of a mechanical press, generally of the full-revolution clutch type. Because only momentary pressure is necessary on hand buttons or other mechanisms, two-hand trip can be used as a point-of-operation

F. Safety blocks should be provided with interlock plugs which disconnect the clutch and motor control circuits when blocks are removed from their storage pockets.

G. All magnetic air valves used for actuating the clutch should be of the three-way, normally closed, poppet design. Spool valves

Figure 11–24. Presses must also be made safe for electricians, millwrights, pipe fitters, and others who must maintain them in a safe condition. Identification of components of this machine control panel are given in the chart on inside of door. Components are readily accessible for maintenance.

safeguarding device only for single stroke operations and only when the location of the buttons or other hand mechanisms is of sufficient distance from the point of operation to prevent moving either hand into a point-of-operation hazard prior to die closing. Two-hand trip is generally impractical on presses having speeds of less than 120 strokes per minute (SPM). (See Section IV for design and construction requirements.)

B. Two-hand control. A clutch/brake-actuating and stroke-controlling method requiring the concurrent use of both hands of each operator on "run" buttons or other mechanisms during a substantial portion of the die-closing stroke of a part-revolution clutch press. Two-hand control can be used as a point-of-operation safeguarding device for the single stroke mode of operation when properly designed, installed, and adjusted as outlined in Section V.

III. Safeguarding exclusions for two-hand systems

A. Continuous stroking. Two-hand control or two-hand trip does not qualify as a point-of-operation safeguarding device in the continuous mode of stroking. The use of a die-enclosure guard or a fixed barrier guard is recommended to keep operators' hands out of the point of operation. The operator hand controls can be used simply as a stroke-initiating means.

B. Use of foot control. If a foot control is connected in place of any hand-operated button or buttons, the operator's foot control cannot be used for point-of-operation protection because one or more hands are free to enter the die area at any time. Use of a die-enclosure guard, fixed barrier guard, or a type A movable barrier device is recommended for point-of-operation safeguarding on power presses. (See Section IV.)

IV. Construction features necessary in all two-hand trip systems

A. Run buttons or other hand-tripping mechanisms. The individual operator's hand controls for tripping the clutch must meet all of the following requirements.

1. Each must be protected against unintentional operation. Protective rings around run buttons or suitable barriers are recommended. (See Figure 11-11).

2. Each pair must be arranged by design and construction or separation, or both, to require the use of both hands. This means buttons must be far enough apart and so located that one hand and elbow of the same arm, a knee, or any other part of the body will not be used instead of both hands.

3. When two-hand trip is used on a multiple-operator press, each operator must have a separate set of two buttons or other hand mechanism. The bypassing of buttons not needed for particular operations must be done in complete sets of two, not by individual buttons, and must be by a means that is capable of supervision.

4. When two-hand trip is used as a device for safeguarding the point of operation, buttons must be fixed in position with a safety distance from the point of operation to each button that is great enough to prevent moving either hand from a button into a point-of-operation hazard prior to die closure. Minimum safety distances are shown in Table 11-A and are based on the following formula:

D_m = 63 inches per second $\times$ T_m where
D_m = Minimum safety distance in inches,
63 inches per second = Possible hand movement speed,
T_m = Maximum time in seconds the press takes for die closure after it has been tripped.

Figure 11-25. Rotary cam limit switch. All such switches should be of the type in which the contracts are forced open by the actuating cam.

For full-revolution clutch presses with only one engaging point, T_m is equal to the time necessary for 1½ revolutions of the crankshaft. For full-revolution clutch presses with more than one engaging point, T_m is calculated as follows:

$$T^m \; ½ \; + \; \left(\frac{1}{\begin{array}{c} \text{number of} \\ \text{engaging points} \\ \text{per revolution} \end{array}} \right) \times \left(\begin{array}{c} \text{time in seconds} \\ \text{necessary to complete} \\ \text{one revolution of} \\ \text{crankshaft} \end{array} \right)$$

B. System requirements. The system must include the following features:

1. Concurrent operation of all run buttons or other hand mechanisms must be necessary.
2. An antirepeat feature must be incorporated for single-stroke operation.
3. Electrical control circuit and valve coil voltages must not exceed a nominal 120 volts AC or 240 volts DC isolated from higher voltages.
4. Ground fault detection to detect an accidental ground, and circuitry to prevent false operation due to an accidental ground.
5. Design features to minimize failures that would cause an unintended stroke.
6. On a multiple-operator press, the control system must prevent actuation of the clutch if all operating stations are bypassed.

V. Construction features necessary in two-hand control systems

A. Run buttons or other hand-control mechanisms. The individual operator's hand controls for engaging the clutch/brake must meet all of the following requirements:

1. Each must be protected against unintentional operation. Protective rings around run buttons or suitable barriers are recommended (see Figure 11-11).
2. Each pair must be arranged by design and construction or separation, or both, to require the use of both hands. This means buttons must be far enough apart and so located that one hand and elbow of the same arm, a knee, or any other part of the body will not be used instead of both hands.
3. When two-hand control is used on a multiple-operator press, each operator must have a separate set of two buttons or other hand mechanisms. The bypassing of buttons not needed for particular operations must be done in complete sets of two, not by individual buttons, and must be by a means that is capable of supervision.
4. When two-hand control is used with a part revolution clutch press as a device for safeguarding the point of operation, buttons must be fixed in position with a safety distance from the point of operaton to each button great enough to prevent moving either hand from a button into a point-of-operaton hazard while the slide (ram) is in motion during the die-closing portion of a stroke.

Minimum safety distances shown in Table 11-B are based on the following formula:

D_s = 63 inches per second $\times$ T_s, where

D_s = Minimum safety distance in inches,

63 inches per second = Possible hand movement speed,

T_s = Longest stopping time of the slide (ram) in seconds usually measured at approximately the 90-degree position of crankshaft rotation.

Table 11-A. Minimum Safety Distances for Two-Hand Trip Between Hand Controls and Nearest Point-of-Operation Hazard

Press speed in strokes per minute	Time in seconds (Tm) for one revolution of crankshaft	Full revolution clutch engaging points per revolution					Part revolution clutch
		1	2	3	4	14	Infinite
30	2.0	189	126	105	95	72	63
45	1.33	126	85	70	63	48	42
60	1.0	95	63	53	48	36	32
75	0.8	76	51	42	38	29	26
90	0.67	63	42	35	32	24	21
105	0.57	54	36	30	27	21	18
120	0.5	48	32	27	24	18	16
135	0.44	42	28	24	21	16	14
150	0.4	38	26	21	19	15	13
165	0.36	35	23	19	18	13	12
180	0.33	32	21	18	16	12	11
210	0.29	27	18	15	14	11	9
240	0.25	24	16	14	12	9	8

With adjustable-speed drive, use slowest speed. (Based on OSHA regulations 1910.217 (c) (3) (viii) of December 3, 1974, for presses with full- or part-revolution clutches.)

The stopping time includes the operating time of the press control system, air exhaust from the clutch/brake, and braking time of the slide. Stop time measuring units are available or a stop watch can sometimes be used. If the minimum safety distance is not practical for production, a guard or another safeguarding device can be used.

B. Stroking control system requirements.
1. *Stop control*

A red color stop control must be provided with the clutch/brake control system. Momentary operation must immediately deactivate the clutch and apply the brake. This button must override all other controls and reactuation of the clutch must require use of the run buttons or other operating mechanisms. It is recommended that a stop button be available to each operator. At least one stop button must be connected and operative regardless of whether hand or foot control is being used.
2. *Stroking selector*

a. A means of selecting "Off," "Inch," "Single Stroke," or "Continuous" (when the continuous mode is provided) must be supplied with the clutch/brake control to select the type of operation. There can be other positions when required for special features, but hand-foot selection must be separate from the stroking selector.

b. Fixing of selection must be by a means capable of supervision. Selectors can be locked in position and key operated selectors are common methods used for this. Additional precautions are needed for continuous stroking, as described under Continuous Control, later in this section.

c. For standardization of the location of each mode of operation on rotary stroking selector switches, the following sequence is recommended as the handle, knob, or key is rotated clockwise: "Off"–"Inch"–"Single Stroke"–"Continuous" (when continuous is provided).
3. *Inch control*

a. The "Inch" mode of operation should never be used for pro-

duction. There is generally no antirepeat circuitry, automatic top stop, or drive motor interlock as in single stroke because such features could interfere with the type of slide movement often needed during setup or tryout. For example, it is sometimes necessary to "Inch" with the drive motor and flywheel coasting at reduced speed in order to get finer increments of slide motion.

b. To prevent exposure of personnel within the point of operation, the operating means should be two-hand controls requiring the use of both hands on control mechanisms to actuate the clutch, or a single hand button protected against accidental actuation and located so that the worker cannot reach into the point of operation while pressing the button.

c. A foot control should never be used for inching.
4. *Single stroke control*

In addition to the run button features listed above under A of Section V, Construction Features, the control system must provide run button holding time and interrupted stroke protection. Control reliability and brake monitoring are also required with hands-in-dies operations.

a. Holding time. If all run buttons are held down until the dies have closed sufficiently for safe release of the buttons, the slide must stop before entrance of a hand into the point of operation is possible. The control should provide for an adjustment of the rotary limit switch contacts to bypass the run buttons at a safe point in the stroke. This will vary according to speed, length of stroke, and the type of dies.

b. Interrupted stroke protection. This requires release of all run buttons before an interrupted stroke can be resumed (by again pressing all run buttons). Its purpose is to minimize the possibility of inadvertent restarting of the slide after a button has been released during the holding time. This feature is accomplished through proper design of the circuitry.

Every effort should be made to always use dies and feeding methods that do not require placing hands in the point of operation at any time. It is recognized, however, that there are

Table 11-B. Minimum Safety Distances for Two-Hand Control Between Hand Controls and Nearest Point-of-Operation Hazard

Stopping time of slide in seconds (T_s) at 90° point of stroke	Minimum safety distance (D_s) in inches
0.100	7
0.125	8
0.150	10
0.175	11
0.200	13
0.225	15
0.250	16
0.275	18
0.300	19
0.325	21
0.350	22
0.375	24
0.400	26
0.450	29
0.500	32
0.600	38
0.700	45
0.800	51
0.900	57
1.000	63

With adjustable-speed drive, use the stopping time for highest speed. (Based on OSHA regulations 1910.217 (c) (3) (vii) of December 3, 1974, for presses with part-revolution clutches only.)

instances in which it is not feasible to keep hands out of dies. In such cases, when two-hand control, presence-sensing, or a type B movable barrier is used for safeguarding part-revolution clutch presses, OSHA regulations also require the following features:

c. Control reliability. The control system must be constructed so that any control failure does not interfere with normal stopping action when required, but does prevent initiation of successive strokes. The failure must be easily detectable.

d. Brake system monitoring. A brake monitor must be incorporated into the control and prevent activation of a successive stroke if the stopping time or distance deteriorates to a point where the safety distance no longer complies with Table 11-B for two-hand control, or similar safety distances required for presence-sensing or type B movable barriers. The monitoring action must take place each cycle in single stroke operation and indicate any unsafe deterioration.

5. *Continuous control*

When a control system has provisions for both Single Stroke and Continuous functions, it must be designed for both of the following conditions:

a. Selection of the continuous mode may be supervised. A key-lock selector is usually used for this.

b. Initiation of continuous run must require a prior action or decision by the operator different from that used for Single Stroke, in addition to the selection of Continuous on the stroking selector, before actuation of the operating means will result in continuous stroking. A common method is the use of an "Arming" or "Continuous Setup" button which must be momentarily pressed within a few seconds prior to actuation of the run buttons or other operating means.

These steps, or equivalent actions, are necessary to prevent inadvertent starting of the slide on the continuous mode because there is no run button holding time and no automatic stop at the end of each stroke as in the single stroke mode of operation.

A die-enclosure guard or fixed barrier guard is recommended to keep operators' hands out of the point of operation, but other guards or safeguarding devices may be used.

6. *Multiple-operator stations*

On a multiple-operator machine, the control system must prevent actuation of the clutch/brake if all operator stations are bypassed. This is frequently called "dummy plug protection" because supervised dummy plugs or key-operated selectors are generally used to bypass stations not needed.

C. Stroking control system component failure protection

1. Stroking control systems (clutch/brake control circuits) must incorporate design features to minimize the possibility of failures that could cause a repeat (unintended) stroke. Components that are critical include:

a. Input elements, such as stroking selectors, run buttons, inch buttons, foot controls, and stop buttons.

b. Sensing elements related to stroking, such as rotary limit switches or equivalent apparatus, including their drive mechanisms; motion or position detectors; valve failure detectors; and any associated relays or electronic components.

c. Output elements to the slide operating mechanism, such as "valve" relays and clutch/brake valves.

d. Acceptable methods of controlling normal stopping of the slide include interruption of the clutch (and brake, if separate) valve current through the cycle-control rotary limit switch, or series contacts of two or more independent relays or static circuits. Adjustable stopping controls, using counters or timers, may be used with suitable limit switch backup.

2. The control of air clutch machines must be designed to prevent a significant increase in the normal stopping time due to a failure within the operating valve mechanism and to inhibit further operation if such failure does occur. A self-checking assembly of two valve elements in a common housing is usually the best way to meet this requirement (see Figure 11-26).

If a machine has a separate clutch and brake system requiring individual valves connected to a common manifold, both valves should be of the self-checking double type. Failure of a brake valve is just as dangerous as a faulty clutch valve, and sometimes is even more likely to cause a malfunction.

The protective valve arrangement is not needed on machines intended only for continuous operation with automatic feeding.

3. If a press operator is required to place a hand within the point of operation, see paragraphs B.4.c. and B.4.d., on Control reliability and Brake system monitoring, for additional requirements.

D. *Interlocks.*

The clutch/brake control must be automatically deactivated in event of loss of electrical power and, if the clutch is air operated, loss of proper air pressure. If the machine has air counterbalance cylinder(s), loss of proper counterbalance air pressure must also deactivate the control. Reactivation of the clutch must require restoration of normal electric and air supply, and use of the run buttons or other tripping means (see Figure 11-11).

The control must incorporate an automatic means to prevent initiation or continued activation of the single stroke continuous functions unless the drive motor is energized and in the

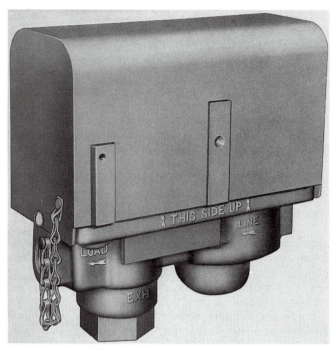

Figure 11-26. A multiple clutch/brake air valve. Its use reduces the possibility of press repeats due to malfunctioning of a single valve.

forward direction. This requirement can be met by connecting an auxiliary contact on the main drive forward contractor in the single stroke and continuous (when used) control circuit.

E. Control circuit voltage and ground protection.

1. The AC electrical control circuits and valve coils on all machines must be powered by not more than a nominal 120 volt supply obtained from a transformer with an isolated secondary winding. Voltages above 120 that may be necessary for particular mechanisms must be isolated from any control component handled by an operator. All DC control circuits must be powered by not more than a nominal 240 volt DC supply isolated from any higher voltage.

2. All clutch/brake control circuits must be protected against the possibility of an accidental ground in the control circuit causing false operation of the machine (see examples in Figure 11-27).

VI. Foot operation

If foot operation is provided as an alternate to two-hand operation, the following arrangements are necessary:

A. Point-of-operation protection.

Whenever a foot control is used, there must be a suitable means of providing point-of-operation protection. A die-enclosure guard, fixed barrier guard, or a type A movable barrier device is recommended for safeguarding presses.

B. Control requirements.

1. Foot switches or foot valves and any attached mechanism must be protected to prevent unintended operation caused by falling or moving objects, or by accidental stepping onto the foot control.

2. Selection of hand or foot operation must be by a means

capable of being supervised. A key-lock selector switch is the most common method of accomplishing this.

3. When two-hand operation has been selected, the foot control must be deactivated and vice versa. It should never be possible to use either run buttons or a foot control at the same time.

4. The antirepeat feature of the control system should be in effect during foot operation.

5. Foot controls should be deactivated in the "Inch" mode, and not used in place of hand controls.

POWER PRESS SETUP AND DIE REMOVAL

Power press dies must remain rigidly accurate in spite of the pressure and stress they transmit during metal stamping operations. As a result, they are usually strong, heavy, and difficult to handle. They vary in weight from a few pounds for small dies to several thousand pounds for large ones.

Handling, setting up, and removing these dies is hazardous unless proper equipment and methods are used. These operations should be entrusted only to experienced setup personnel whom the supervisor has instructed in detail about safe procedures.

Injuries likely to result when setting or removing power press dies are:

1. *Strains and hernias* from improper handling techniques.

2. *Foot injuries* from dies slipping off trucks, benches, bolster plates, or storage shelves.

3. *Crushing injuries* from placing part of one's body between the die and press.

4. *Hand injuries or amputations* from sudden descent of the ram through brake failure, premature tripping during tryout, failure to lock out the switch, or failure to assure that proper pressure is maintained on pneumatic clutches.

5. *Lacerations* from wrenches slipping off worn nuts or from use of incorrect tools.

6. *Eye injuries* from flying pieces of shattered parts, dies, or scrap.

Whenever the toolmakers or setup personnel suspect that the control mechanism is not in perfect order, they should take immediate steps to have the press shut down until repairs are made.

Transferring dies safely

Very light dies can be handled and carried manually. Dies weighing up to about 100 lb (45 kg) can generally be handled without lifting apparatus only if proper die trucks are provided (Figure 11-28). These trucks should have elevating tables which are adjustable to the heights of storage shelves and press bolster plates. When dies are being transported, they should be carried at the lowest elevation of the truck. Rollers or balls mounted in the top surface of the table will aid in sliding the die on or off the truck.

Heavy dies require more equipment for safe handling. Because they are often lifted and moved by hoists, these dies should have tapped holes and eye bolts (or hooks), drilled holes and pins, chain slots, cast lugs in lower shoes, or clamping lugs to facilitate hookup and transfer. When tapped holes are used for securing lifting hooks or eye bolts, they should be either ¾ or 1 in. (1.9 or 2.5 cm) in diameter. (The use of ⅞ in. (2.2 cm) holes is not recommended because ¾ in. bolts might be used in them and appear to fit, only to pull out during the lift.)

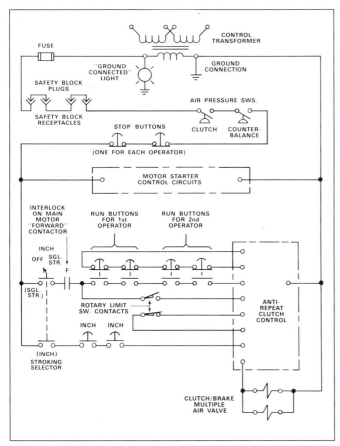

Figure 11-27a. Diagram of a typical single break, grounded press control connection.

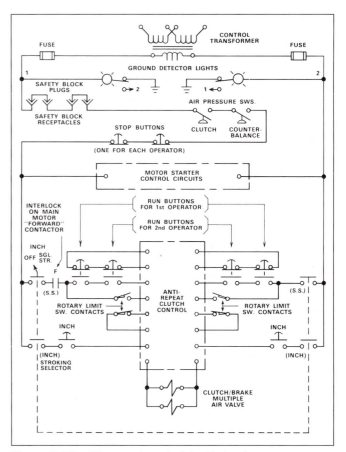

Figure 11-27b. Diagram of a typical double break, ungrounded press control connection.

The depth of the hole should be one and one-half times the diameter.

For moving dies that are about 1,000 lb (450 kg) or heavier, special die-handling power trucks are recommended. Of these trucks, special equipment for pushing or pulling the dies includes power winches, roller tables, and hydraulic or pneumatic clamps.

When transferring dies, place the truck close to the storage shelf or press; adjust the table to the same height as the storage shelf or bolster plate; and either chock the wheels, lock the brake, chain the truck to the press, or use some other method to prevent movement (Figure 11-30). Next, push the die off the truck onto the storage shelf or bolster plate. Or, to bring a die onto a truck, engage a hook in the die so it cannot slip and exert a steady pull—do not jerk or tug on the hook. Where lifting is needed to effect the transfer, use a hoist and never lift higher than is necessary for minimum clearance. At no time should an employee have hands, feet, or other body part underneath a suspended load. All signals should be given by the person in charge and by no other person.

When in transit, dies should be carried on trucks with tables in the lowest position. The dies should be secured to the tables and their halves should be fastened together. Large-diameter wheels on the trucks are desirable to minimize the effect of uneven floor surfaces.

Procedure for setting dies

Safe procedures for setting and removing dies vary slightly

dependent upon press size. Most of the difference occurs because the slide on the light presses can be moved manually by turning the flywheel or crankshaft, while on heavy presses it must be power operated. The following safe method for setting and removing power press dies may be followed for all presses, except where variations in the procedure are designated as being specifically for light presses or for heavy presses.

For light presses. Disconnect or shut off the power and lock out the switch. Bring the slide down to its lowest position by turning the flywheel or crankshaft. (If a safety bar is used as a lever to turn the crankshaft, it should have a spring and collar arrangement that will prevent its being accidentally left inserted in the crankshaft. See Figure 11-29.) Measure the clear height between the bottom of the slide and the bolster plate. This distance should be slightly greater than the height of the closed die. If not, raise the slide by means of the adjusting screw. When physical access is necessary, block up the slide with timber, metal blocks, or posts provided for this purpose. Blocks should be equipped with an electric receptacle plug to hold open the circuit when they are in the die. Thus, they will not be left in the press when power is applied.

For heavy presses. Jog or inch the press to bring the slide down to its lowest position and measure the clear height between it and the bolster plate. Again, if this distance is not slightly greater than the height of the die in the closed position, adjust the slide until such clearance is assured. Raise the slide, block it in position, shut off the power, and lock out the switch.

Figure 11-28. Die truck with adjustable-height and adjustable-angle die table can match angle of inclined presses. Truck is secured to press before moving die. Holes in upper die shoe indicate drilled-hole-and-pin method of lifting.

Dismantle or disconnect a point-of-operation safety device only if it is absolutely necessary. For example, enclosure guards fastened to the press, gate or barrier guards, and other types of guards will have to be removed, but not, in general, pull-back devices or presence-sensing devices. All parts of any disconnected safety devices should be stored so that they may be reinstalled in good condition as soon as the new dies have been installed.

Clean off the bolster plate, preferably with a brush. All bolt holes must be clear of all obstructions. (A pencil-shaped magnet is valuable for removing iron or steel particles.) Burrs can be removed with a file.

Check the die to make sure it contains no chips, tools, or parts, and that it is in good operating order. Then transfer the die from the truck to the press. (See Transferring dies safely, earlier in this section.)

Line up the die in the correct operating position and remove the posts or blocks from under the slide. For heavy presses, reconnect the power to operate the slide. For light presses, manually turn the flywheel to move the slide, using a safety bar if necessary (Figure 11-29). Then, lower the slide until it fits firmly against the top die. It is extremely important not to put too much pressure on dies. Tighten all bolts and clamps as necessary to secure the top half of the die. Bolting the die to the ram with bolts through holes in the upper die shoe is recommended (Figure 11-30). On heavy presses, if air-cushion pads are used, keep the ram close to the top of the die until the die is properly seated on the cushion pins.

Shim or block up the lower half of the die to the proper level; bolt and clamp it to the bolster plate. (Make sure that the bolts, bolt heads, and wrenches are of proper size and in good condition.) Bolting the die shoe to the bolster produces the most secure die setup. If clamps must be used (Figure 11-31), their outer ends should be blocked up slightly higher than the die surface on which their inner ends will rest. Clamps should be of minimum length. Clamp fastening bolts should be closer to the die than to the block end of the clamp. (See Figure 11-31 again.)

Check all bolts and clamps to see that they are tight and that dies are securely fastened in the press. Remove all tools and equipment from the dies, bolster plates, or other areas on the press. For forming or drawing operations, adjust the ram down

to almost its proper depth. For a pierce or trim die, raise the ram slightly so that the punch will not shear when the crank is again brought over and the die entered. Turn light presses several times by hand, checking to see that they are in satisfactory adjustment. (On heavy presses, do not check adjustment at this point.)

Raise the ram to its highest point and block it in this position. (On heavy presses, disconnect the power before proceeding further.) Wipe out the die and remove the safety blocks. Replace the safety device and check it for adjustment and operation. When used, pullout devices must be adjusted properly so that an accident could not occur to an operator who has longer hands or arms than the previous operator.

Reconnect the power and try out several actual operations, using the proper stock, on the press. Make any necessary adjustments only after shutting off the power and blocking up the ram. After completing the adjustments, turn on the power and again try several actual operations on the press to assure safe operation.

Removing dies

Safe die removal methods must be followed at all times. Although modifications may be necessary in special cases, the safe procedure is as follows:

1. Make sure that the working space is cleared of all stock, containers, tools, and other items.
2. Disconnect or shut off the power and lock out the switch. Turn the flywheel by hand or by means of the bar until the ram is at the bottom of the stroke. If the press cannot be turned over manually, jog it under power, then shut off the power and lock out the switch.
3. Dismantle or disconnect the point-of-operation safety devices as required. Store the parts of the dismantled safety device so that it can be reinstalled in good condition when the new die is in place.
4. Clean off the bolster plate, preferably with a brush.
5. If the die is to be operated with an air pad, shut off the air supply and open the release valve to permit the pins to go down. Also shut off the air supply to the automatic blowout system used in the die.
6. Remove bolts and clamps holding the die in the press.

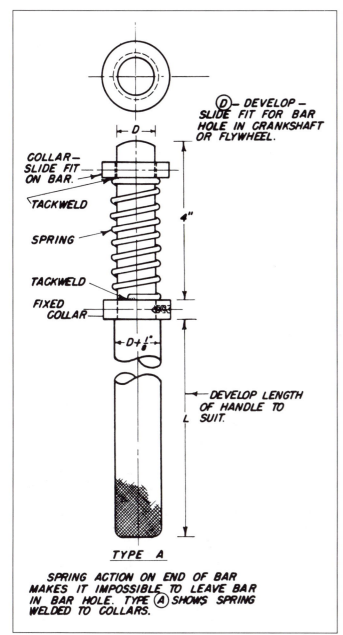

- D - DEVELOP -
SLIDE FIT FOR BAR
HOLE IN CRANKSHAFT
OR FLYWHEEL.

COLLAR-
SLIDE FIT
ON BAR.

TACKWELD

SPRING

TACKWELD

FIXED
COLLAR

DEVELOP LENGTH
OF HANDLE TO
SUIT.

TYPE A

SPRING ACTION ON END OF BAR
MAKES IT IMPOSSIBLE TO LEAVE BAR
IN BAR HOLE. TYPE Ⓐ SHOWS SPRING
WELDED TO COLLARS.

Figure 11-29. Spring-action safety bar for turning power press flywheels during setups. (Reprinted courtesy of Liberty Mutual Insurance Co.)

7. Make certain that the die is loose and that bolts, nuts, clamps, and other obstructions have been removed.

8. Raise the ram slowly—manually on light presses, and by jogging or inching under power on large presses—and make sure that the die does not hang in the slide.

9. Block the ram in its highest position and if power was used, shut off the power and lock out the switch.

10. Place the die truck close to the press, adjust the table to the same height as the lower bolster plate, and chock the wheels or set the brake of the truck to prevent movement. To pull the die onto the truck, use a device engaged so that it cannot slip.

11. Dies should be inspected, repaired, and protected before being stored for the next run, inspection should include

the pins and bushings. Hardened dies and punches should be stored in closed position with a piece of soft wood between the edges to protect them. Injuries have occurred when hardened dies were handled when open or partially open, because a sudden jolt can cause them to close and pinch a hand or finger.

INSPECTION AND MAINTENANCE

The best press safety program in the world cannot succeed, nor can maximum production be met, without good inspection and maintenance of both the press and its safeguards.

Proper inspection, adjustment, and repair of power presses and related equipment can only be done by competent, thoroughly trained employees. These employees must be fully familiar with the construction and operation of the equipment for which they are responsible. They also must be equipped with the proper tools and equipment to assure good workmanship with safety for all.

The work area should be cleared of all personnel not directly involved in the maintenance. Flashing warning lights or other barriers or barricades should be used to mark the temporary area of maintenance to all crane operators and other material and equipment handlers.

Zero mechanical state

When inspection or maintenance work is to be performed on a machine, it is recommended that the following zero mechanical state condition (ZMS) be followed. (Full details are given in Chapter 8, Principles of Guarding.)

Zero mechanical state (ZMS) is defined as that mechanical state of a machine in which:

- Every power source that can produce a machine member movement has been locked out in the OFF position. Over the years, changes in equipment requirements and design have incorporated the use of pressurized air, hydraulics, and electricity, as well as combinations of these media, to perform certain functions on the equipment, with the result that the commonly used terminology of "locked out" or "locked off" does not describe a safe condition. Zero mechanical state, on the other hand, represents maximal protection against unexpected mechanical movement of the machine when setups are made or maintenance performed on the equipment by maintenance or authorized operator personnel. The ZMS concept applies to all types of equipment used in industry.

- Pressurized fluid (air, oil, or other) power lockoffs (shut-off valves), if used, will block pressure from the power source and will reduce pressure on the machine side port of that valve, by venting to the atmosphere or draining to a tank. This may be accomplished by having more than one valve for each power source, provided that each valve can be locked off. This requirement is met by a three-way valve (or equivalent), properly connected. It will also prevent leakage, because of valve malfunction, from reaching the machine.

- All accumulators and air surge tanks are reduced to atmospheric pressure or are treated as power sources to be locked off, as stated in the first two items of this section.

- The mechanical potential energy of all portions of the machine is at its lowest practical value so that opening of pipe(s), tubing, hose(s), or actuation of any valve(s) will not produce a movement which could cause injury.

Figure 11-30. Mounting bolts pass directly through ram to fasten upper die securely. Lower die is clamped to bolster.

Holding a machine member against gravity or a spring force by blocking, by suspension, or by brackets or pins designed specifically for that purpose is permissible to reach ZMS.
- Pressurized fluid (air, oil, or other) trapped in machine lines, cylinders, or other components is not capable of producing a machine motion upon actuation of any valve(s).

  Caution: A machine in ZMS may have fluid (air, oil, or other) under pressure still trapped in its piping.
- The kinetic energy of the machine members is at its lowest practical value.
- Loose or freely movable machine members are secured against accidental movement.
- A workpiece or material supported, retained, or controlled by the machine shall be considered as part of the machine if the workpiece or material can move or can cause machine movement.

As part of the instructions for ZMS, proper consideration should be given to test to verify that the desired set of conditions has been reached. For instance, when the motor power disconnect has been locked out (off), this can be verified by pressing the START button to make sure the motor does not start. Part of the ZMS procedure should be test actuation of the START initiator after lockout.

To attain ZMS condition, the procedure and sequence of lockoff provided by the manufacturer shall be followed in its entirety.

Complex equipment may require that the movable members be in a particular physical position prior to actuation of power lockoff.

Manufacturer. It should be the responsibility of the manufacturer to furnish operating and maintenance instructions with equipment. This should include instruction for ZMS. Specific instructions should be outlined in the operating and maintenance manuals to aid personnel in the safe use and upkeep of the equipment. Adequate instructions may consist of written, illustrated, and audio recorded material. Where hazards to personnel associated with moving parts cannot be eliminated by design or protection, the manufacturer should warn against the hazard.

Physical entry into machine or equipment. The employee should put the machine or equipment in ZMS before placing any body part in the path of any machine or equipment member which may move.

Troubleshooting with power on

When it is necessary to locate and define problems with the power on, the employee can work on machines or equipment with guards removed or work within areas protected by barriers, *if such action will not place any body part in the path of any movable machine or equipment element.*

A machine or piece of equipment may have to be stopped, locked out, or put in ZMS before removing a guard or barrier so that the machine or equipment may subsequently be observed with power on.

Power presses, like all machinery, are subject to wear, breakage, and malfunction. Therefore, to prevent costly accidents and repairs and to promote maximum production, the entire machine and its related equipment must be inspected periodically and required adjustments and repairs made. The frequency of inspections should be determined by the type of press, its related equipment, and usage.

All power presses must be inspected periodically and the employer shall maintain records of these inspections and the maintenance performed. OSHA requires that each press must be inspected weekly to determine the condition of the clutch/brake mechanism, antirepeat feature, and single-stroke mechanism. Necessary maintenance or repair or both shall be performed and completed before the press is operated. The weekly inspection is not required on those presses meeting the requirements for "hands in the dies."

A checklist detailing the frequency of inspection and maintenance should be set up for each piece of equipment in order to give immediate knowledge of the condition of the equipment, and facilitate better production scheduling with less down time due to equipment failures. These checklists need not be complicated, but should at least provide an assigned inspection frequency for all of the following items.

Frame

The press frame should be visually inspected for cracks and broken parts. Check the fastenings of all brackets, guides, cylinders, covers, and other auxiliary parts.

The tierods and nuts should be checked for fractures or stretching. Since the tierod nuts on top of the press may fall if the tierod fractures, it is advisable to chain these nuts to the press frame. A metal strap under each bottom tierod nut will prevent the rod from dropping, should it fail.

If a machine is bolted to a foundation, hold-down fasteners should be checked for looseness and fractures.

Bearings

Crankshaft, pinion shaft, eccentric gear, and toggle link bearings should be checked for snug, nonrotating fit, and any loose caps or fastenings. Journal bearings, especially those supporting gearing, normally should be replaced when worn to a shaft looseness of 0.0025 inch per inch of shaft diameter for bearings up to 12 in. (25.5 cm) diameter. Larger bearings and some toggle-link bearings may require replacement when worn to a looseness of 0.0015 to 0.002 inch per inch of diameter. Badly scored bearings should be replaced. Check for proper lubrication.

The slide or ram guide and gib surfaces should be examined for dirty or clogged lubricant grooves. Check for proper running clearances. These vary with the type of work, but generally should total about 0.002 inch per foot of ram guide length, for any two opposing clearances such as front and back. This is minimum for cast iron and steel which requires more clearance than bronze or brass and steel due to a greater tendency to "pick up." Tighten all screws and lock nuts to hold setting.

Antifriction bearings should be checked for proper lubrication. Overlubrication can cause swelling of oil seals and overheating of the shaft. This can result in binding and shaft failure. Both this failure and the failure of flywheel bearings can cause unexpected descent of the ram. If lubrication of noisy flywheel bearings fails to silence bearings, replace bearings immediately.

Motor or power source and drive

Drive belts should be properly adjusted to prevent excess slippage or excess load on the bearings and shaft, possibly causing

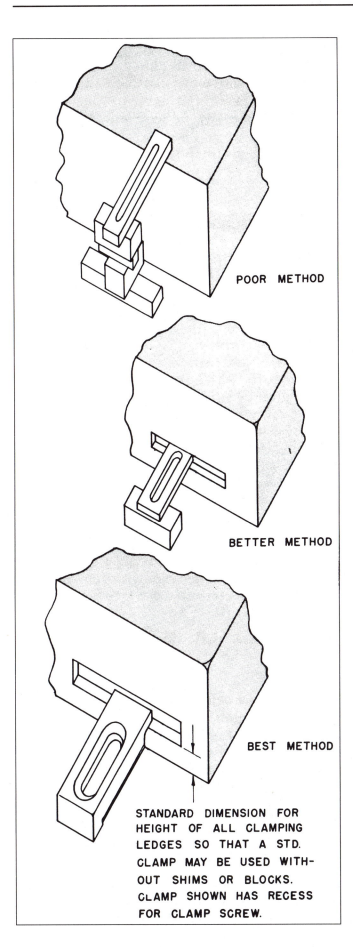

POOR METHOD

BETTER METHOD

BEST METHOD

STANDARD DIMENSION FOR HEIGHT OF ALL CLAMPING LEDGES SO THAT A STD. CLAMP MAY BE USED WITHOUT SHIMS OR BLOCKS. CLAMP SHOWN HAS RECESS FOR CLAMP SCREW.

Figure 11-31. Three methods of clamping dies are compared. It is always important to use correct-size clamps. The method illustrated at the top shows a practice that is seriously hazardous.

premature failure. Properly adjusted drive belts should slip slightly on initial start up of motor but not during press operation. Check all pins, slides, turnbuckles, jack screws, or other means of motor adjustment for secureness. Tighten all hold-down screws. Attach the motor to the press by a chain or wire rope for maximum safety. Inspect and lubricate motor shaft bearings.

Gears should be checked for worn, pitted, and broken teeth, and for proper lubrication. Check bores and shafting for bad keys and keyways.

Whenever a crankshaft or shaft carrying the flywheel or clutch and brake is removed from a press, it should be inspected for fatigue cracks. Some companies inspect all drive shafts once a year for cracks, bending, or deformation. Fatigue cracks can be detected by numerous methods such as ultrasonic, radiographic, magnetic, or dye-penetrant techniques.

Selection of turnover bar operation shall be by means capable of being supervised by the employer. A separate push button shall be employed to activate the clutch, and the clutch shall be activated only if the driver motor is deenergized. Turnover bars shall be spring loaded to prevent the possibility of leaving the bar in the bar hole.

Electrical controls

Check all operating buttons for proper operation. All buttons must be depressed to start the press cycle and released at the end of stroke. "Holding time" should be adequate for the operation and tooling involved.

Check for defective lamps in ground detector circuits where provided. Check ground connection on grounded controls or ground detector connection on ungrounded controls.

Check physical condition of wiring, relays, rotary-limit switch drives, pressure switches, valves, and other electrical and pneumatic devices. Follow manufacturer's recommendations for preventive maintenance.

Ram or slide

Check slide structure visually for cracks and check die mounting surface for evidence of overload or improper die-mounting damage. Inspect the slide adjustment lock provisions to be sure die setting can be maintained. On a machine with a motorized ram adjustment, the motor should be checked for loose mounting bolts, loose drive chain or gears, excessive grease in the motor, and worn or frayed flexible electrical lead-in wires for motor control. Check slide adjustment limit switches for proper operation.

Knockout bars should be pinned or chained to prevent their falling. If clearance between slide face and right-to-left stationary support bar is less than 1½ in. (3.8 cm) and the bar is within reach from the floor, finger guards should be installed along top of bar.

If the ram is counterbalanced by springs, they should be checked for breaks. If the ram is counterbalanced by air, a check should be made for air leaks, air line restrictions, correct operating pressure, loose piston rods, lubrication, and proper operation of pressure switches. All brackets should be tight. Note air pressure rise on downstroke of press. Any rise in excess of 20 percent probably indicates a surge tank filling up with condensate and lubricant. Periodic draining of all surge tanks is recommended.

Visually check for fatigue cracks in the ram-adjusting screw

and the connection. Check for secure fastening of the slide to the adjustment mechanism and the connection cap to the connection. Look for evidence of ram having been adjusted too high with subsequent interference with frame. All mechanical power presses are capable of producing an overload force several times the press tonnage at the top of the stroke as well as the bottom. Sudden failure of any of the parts attaching the ram to the crank may cause an equally sudden and dangerous dropping of the slide.

Clutch and brake

Dry friction type. All modern friction clutch units are air engaged and spring released. All modern brake units, except constant-drag types, are spring set and air released. Any other arrangement is not failsafe. When the clutch and brake are combined into a single unit operated by a single air cylinder, only one unit can be engaged at a time due to the mechanical interlock. When the clutch is separated from the brake and each has its own air cylinder, it is possible to engage both at the same time. This must be prevented by either restricting air flow into the clutch and out of the brake, or by limit switch and/or pressure switch timing of two air valves. Check to see that both clutch and brake are not engaged at the same time.

The unit should be checked for loose fasteners, broken parts, lubrication leaks, air leaks, faulty or loose wiring, excessive accumulation of particles on the friction lining, and broken springs. Springs that have changed in free length more than five percent should be replaced.

During inspection, the action of the clutch and brake, both at rest and in motion, should be checked. Travel on the friction disks will indicate the amount of wear, and the disks should be adjusted in accordance with the manufacturer's instructions. If not adjusted properly, they can cause a malfunction.

It is also important that the brake be properly adjusted. There should be little or no coasting of the press slide (ram) when the brake engages. The clutch and brake should operate smoothly and engage and disengage quickly.

If the press is equipped with a brake monitoring device, it should be checked for proper operation in accordance with the manufacturer's recommendations.

Sliding surfaces that keep parts in alignment should be checked for excessive wear that might allow the parts to cock or wedge.

Inspection of the clutch and brake units will readily disclose leaks in the air cylinder packings and in air glands. Air line filters, lubricators, and moisture traps will increase the life of these packings and are very necessary for safe operation. Traps and strainers should be checked and cleaned frequently. Plastic oil reservoirs on air line lubricators should not be cleaned with solvent because some plastic reservoirs are adversely affected by solvents. Lubricators must be refilled regularly with the proper type of oil recommended by the press manufacturer.

Air valves are usually the magnetically operated type. They should operate smoothly without sticking or leaking. Valves may stick because of dirt or scale in the air line. A leaky valve packing should be replaced. Valves should be inspected and cleaned according to the manufacturer's recommendations.

Electrical controls, although usually not part of the clutch and brake unit, affect the operation of the unit. Initiators (pushbuttons), limit switches, relays, and contactors should be inspected for excessive wear, broken springs, loose parts, loose or broken wires, peened magnet field surfaces, badly burned contacts, and dirt. Circuit-grounding connnections should be checked for continuity. Badly worn contacts should be replaced. Specifically, inspect the rotary cam limit-switch drive since failure of this mechanism may result in dangerous repeat strokes.

A convenient method for checking a clutch/brake system is to measure the stopping time of a press. An increase in stopping time for a given press indicates the need for clutch/brake system maintenance. Modern instrumentation is available for stop-time measurement.

Oil-wet type. This newer type of air-actuated friction clutch has many of the same type of maintenance characteristics as the dry friction clutch with the following exceptions: (1) friction surfaces last much longer; (2) units are usually physically smaller due to higher r.p.m. and may contain a set of planetary or other type gears; (3) the unit may require an oil cooler and pump. Lubricating oils should meet the manufacturer's requirements to prevent damage to the unit. These units may be serviced by qualified maintenance personnel.

Electric clutch. The eddy current electrical type has no friction surfaces to maintain but does have slip rings and a special electrical control to maintain the torque and slip characteristics. Proper maintenance consists of maintaining lubrication bearings and electrical apparatus according to manufacturer's recommendations.

Full-revolution clutch presses. A full-revolution clutch is one which, when tripped, cannot be disengaged until it has completed its cycle. It is known by many names, such as pin, jaw, dog, positive, key, or spline, depending on the type and the manufacturer. Usually associated with this type of clutch is a drag brake on the crankshaft.

Typically, this clutch couples the flywheel to the press crank by the release of a spring-loaded coupling-means such as a pin, rolling key, or jaw arrangement. These are normally disengaged by a cam action which extracts the engagement member disconnecting the flywheel from driving the crank. The crank is held in its disengaged position by a braking system.

The clutch should be examined for loose parts, worn pins, worn dogs, broken or weak springs, damaged lubrication seals, and excessive wear in the bearings. Worn or broken parts should be replaced and the clutch adjusted to throw out at the top or just before the top center stroke. This adjustment will affect the brake setting.

On some presses, the clutch is tripped by a foot or hand lever or levers with spring return. Other methods of tripping the clutch include all-electric, all-air, or a combination of air and electricity. All elements of the clutch trip should be examined. Sources of trouble include broken or weak springs; worn pins and bushings; loose fasteners; leaking air packings, connections, or valves; loose or broken wires; poor electrical contacts; and defective relays and limit switches. After defective parts have been replaced, the tripping mechanism should be readjusted and checked for smooth operation.

Drag or band brake. The continuous band brake is used on a large majority of full-revolution clutch presses. If the brake is not set or operating properly, the press may repeat and cause a serious accident. Therefore, worn, glazed or oil-soaked brake linings should be replaced. Rivets should not project above the linings. Anchor and drag force applying bolts, studs, springs,

and other parts should be removed, inspected, and replaced if found defective.

Pneumatic die cushions and springs

Pressure pads and die cushions should be examined for foreign or scrap material between the pressure pad and the bolster. A check should also be made for faulty air packings, air leaks, improper lubrication, and fasteners on the supporting rods or plates.

Rolling-bolster die carriages

Die carriages vary from small, two-position, manually moved die holders to large capacity, eight-wheeled, four-directional holders. The latter are self-powered for motion and wheel manipulation to change tracks.

The most common source of power is pneumatic; less common is electrical. Inspect for normal operation of all hydraulic and air valves, hydraulic and air cylinders, and air or electric motors. Inspect air line filters and oilers. Inspect and lubricate all gears and bearings. Inspect all springs and latches for wear and breakage.

Inspect locating keys in bed, retractable locating pins in carriage, locating pin holes in floor or bed, and tracks.

Lubrication

Proper press and air line lubrication is essential. Failures producing accidents and down-time are frequently directly traceable to either lack of lubrication or overlubrication. It is commonly understood that lack of lubrication cannot be tolerated in machine parts. Sometimes the results of overlubrication are less well known. Excess lubrication to flywheel and shaft bearings in the vicinity of dry friction clutch and brake linings is undesirable due to loss of capacity to do work and to stop in emergencies.

Overlubrication of air cylinders can lead to sluggish clutch and brake action. Improperly mounted clutch and brake surge tanks will accumulate oil and water. This results in increased clutch slippage and wear. For specific lubrication information, refer to the press manufacturer's service manual.

Guards and safeguarding devices

All gearing, belting, or other drive parts which can be inadvertently contacted should be covered.

Many safeguarding devices are synchronized with the action of the press. Most of these will get out of adjustment through wear and vibration and require periodic checking.

Wire ropes, leather straps, or steel springs used as parts of safeguarding devices will in time need replacing due to wear. In such cases, nothing except the proper replacement parts should be used. The manufacturer's recommendations for maintenance and adjustment should be followed.

All guards and covers on a power press must be in place and properly adjusted after completion of inspection and necessary repairs.

Check each press out for all modes of operation before releasing for production.

METAL SHEARS

Power squaring shears

The power squaring shear should be equipped with safeguarding that will either prevent the operator from placing hands into the point of operation, sound an alert when hands are approaching the point of operation, or prevent or stop the operation of the shear if any part of the body of the operator approaches the point of operation.

It is recommended that the safeguarding requirements in ANSI Standard B11.4, *Safety Requirements for the Construction, Care, and the Use of Shears,* latest revision, be followed.

The point of operation in ANSI B11.4 is defined to include the area between the upper and lower blades and the area between the hold-down (clamping mechanism) and the shear table.

The safeguarding should provide visibility into the point of operation (knife area) for positioning material for shearing to a scribed line. Removal of the safeguarding is normally necessary for changing the blades or adjusting them. The safeguarding should be reinstalled when this is completed. If a guard is used, sufficient clearance should be provided to allow for feeding the material. Normally double the metal thickness of the material being sheared is a recognized guideline. The design of the fixed guard should meet the dimensional requirements of ANSI B11.4–1983. See illustration 12 and Table 1 from this ANSI Standard. This table provides dimensional guidelines for a guard that will prevent the operator from placing hands into the point of operation.

When it becomes impractical to adhere to the guarding dimensions given in Table 1, ANSI B11.4 provides for another method of safeguarding for plate shears. This is an awareness barrier. The dimensional guidelines are in illustration 13 and Table 2 of ANSI B11.4. The design should assure that the movable sections are of sufficient weight that the operator will be aware if hands enter the safeguarding. The operator must be made aware that this safeguarding may not prevent him or her from forcing hands into the point of operation.

On shears with a throat, a guard should be provided. It may be removable to provide for slitting material longer than the shear, but must be replaced when the slitting work is completed.

A work chute or conveyor should be provided to discourage or eliminate the need for employees to be at the rear of the shear while it is being operated. No one should be positioned at the rear of the shear, within the area of the moving machine parts while it is being operated.

It is recommended that new shears be manufactured to comply with the construction and safeguarding requirements of ANSI B11.4 and machines already in the field be updated to meet these requirements (see Figure 11-37).

This article does not cover all the considerations necessary for the safe operation of squaring shears. The user of squaring shears should evaluate the equipment and operating methods in light of the recommendations in ANSI B11.4.

The operator should never place hands into the point of operation. The next piece to be sheared or a tool or pry bar should be used to move small pieces that are on the shear table and beyond the safeguarding.

Some suggested guidelines for the safe operation of squaring shears are:

- Be sure you know your shear—its capacity, controls, operating modes, safeguarding.
- Properly install adequate safeguarding.
- Keep blades sharp and clearance correct.
- Keep clamping mechanism/hold-downs operating satisfactorily.
- Keep work area clear, both front and rear.

- Keep shear table free of loose tools and materials.
- Keep hand tools and personal protective devices available, and use them. Safety glasses, gloves, safety shoes, and snug-fitting clothes should be worn.
- Keep your hands out of point of operation.
- Do not place your hands between the material and shear table.
- When leaving your shear, turn the power off and have the controls inoperative.
- Make certain all personnel are away from the shear table before operating.
- Keep alert—keep your mind on your job.
- Always turn power off when servicing, maintaining, or removing jammed material.

Check the shear at the start of each shift. Some suggested items are:

- Safeguarding at point of operation should be in proper adjustment. It must operate properly.
- Pinch point guarding properly installed.
- Operator station working properly.
- Operating modes functioning properly.
- Ram starting and stopping properly.
- Warning plates clean and easily read.
- Electrical wiring in good condition.
- Caution painting in good condition and clearly visible.
- Auxiliary equipment checked and working properly.
- Hand tools and personal protective equipment in good order and readily available.
- Safety manuals or operator manuals attached to machine.
- Scheduled normal maintenance work completed.

Alligator shears

Alligator shears perform a great variety of cutoff operations. Their principal use, however, is in salvage yards where they are used in forging operations for cutting rods and bar stock to length.

Because the alligator shear operates continuously, the operator must be trained to time movements with the opening and closing of the cutter. Because the machine is relatively simple and comparatively slow in its movment, management and operators often disregard the hazards created. Consequently, alligator shears are responsible for far more injuries than their inherent hazards or frequency of use warrants.

The principal hazards from operating an alligator shear are: finger or hand amputation at the shear point, finger injuries from being cut, hand or arm lacerations from handling material, eye injuries from flying bits of metal, foot and leg injuries from falling material, and injury to the operator or damage to the machine from failure to maintain the shear in a good condition.

If possible, a long bench should be built to the right or left of the shear, depending upon the type of machine, and the material slid along and through the cutter. Because the ragged edges are hazardous to handle, care should be used in piling the material onto the bench.

Most finger and hand injuries can be avoided by the installation of an adjustable guard. The wide variety of sizes and shapes of material to be cut makes it difficult at times to guard closely the point of operation. However, such a guard can often be used. When it is, it should be set close enough to prevent the fingers from entering the danger zone (Figure 11-32).

When stock size is such that the end held by the operator may fly up and strike him, hold-down guards should be used. Such a guard is illustrated in Figure 11-33. It can be adjusted to fit any type of shear.

Material to be cut should be kept within the capacity of the machine, and no attempt should be made to cut hardened steel since such action would probably result in damage to the machine and injury to the operator.

PRESS BRAKES

A power press brake is also sometimes called a bending brake or a brake press. Its design has evolved over the years from the hand or folder brake, resulting from the need for a power machine with capacity sufficient to bend thick sheets and heavy plate products. The primary fuction of the press brake is to cold bend angles, channels, and curved shapes in plates, strip, or sheet metal stock. Press brakes may also be used for punching, trimming, embossing, corrugating, and notching, when machined and arranged to do so, even though these operations are considered power press operations.

Power press brake beds are typically long and narrow and are located in front of, and often extend beyond the machine side frames. The frames are gapped (cut-out) to permit full-length utilization of the bed and ram. The piece part component typically extends in front of the press brake and moves during the bending operation. Both the bed and the ram are equipped with a die-clamping arrangement along the full length to accept a standardized die tongue. Press brake beds are often equipped with an adjustable die holder that provides for alignment and adjustment of the upper and lower dies.

Back gages or material-position gages and stops (front or rear) are used with power press brakes to gage the distance from the edge of the piece part component blank to the forming or bend line.

Follow ANSI Standard B11.3, *Safety Requirements for the Construction, Care and Use of Power Press Brakes*. Appropriate local, state or provincial, and federal regulatory agencies should be consulted along with this standard when working with press brakes.

Power press brakes have been classified into two basic categories. One of these is the general-purpose press brake, the other the special-purpose press brake. General-purpose press brakes, both mechanical and hydraulic, are operated by one individual with a single operating control station. Special-purpose press brakes include all other types having mechanical, hydraulic, and other drive arrangements.

Mechanical press brakes

A limited range of strokes is available in mechanical press brakes. Smaller units may have as little as 1½ in. (3.8 cm) stroke. Large press brakes can have as much as a 6 in. (15 cm) and even longer stroke. Ram position is adjustable to accommodate the closed height of dies. This is accomplished by changing the length of the connections from the drive to the ram through the use of die-height adjustment screws.

Hydraulic press brakes. On a hydraulic press brake, stroke length is variable and can be as long as 24 in. (61 cm) or more. Speed change from high speed advance to low speed press and upper and lower limits of ram travel are generally established by limit switches. Operating strokes per minute are usually less

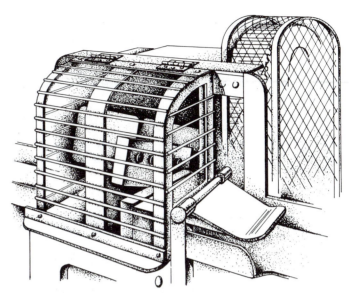

Figure 11-32. A well-designed guard for an alligator shear should permit easy maintenance and adjustment. Hinged section of the guard should be interlocked electrically to prevent shear operation if the hinged section is not in place. (Reprinted courtesy of Jones & Laughlin Steel Corp.)

than for a mechanical press brake of equal capacity; however, rated tonnage can be exerted through the full down stroke on a hydraulic press brake.

Planning the production system. A power press brake is but one part of a production system. All parts of that system must be brought together to perform any metal-working operation on a piece part component. This can also be called a human factors engineering approach to provide for the most efficient and safest method for performing a piece part bending operation. The power press brake is the power component of such

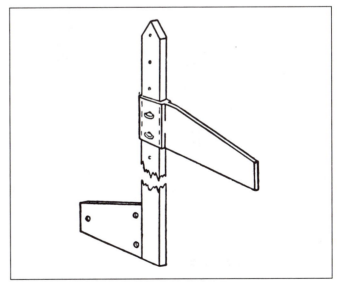

Figure 11-33. Adjustable hold-down bar used on an alligator shear when cutting heavy or long material.

a system. Depending on the tooling component selected by the user, press brakes can bend, form, notch, punch, pierce or perform other operations on the piece part component.

Depending on the piece part component and the product being produced, the feeding component of the production system must be established. Feeding can be either mechanical or manual. Included in this element of the system is removal of parts and scrap.

The final component necessary to complete a functioning production system is the safeguarding component. Vital to the selection of a suitable functioning safeguarding component is a thorough hazard analysis performed by the user and resulting from consideration of all the elements of the production system. These elements include the tooling component, feeding and removal of the piece part component, and any scrap created, and so on. Each new combination of production system elements requires that the user perform a new hazard analysis to select a suitable safeguarding means.

A safe combination of components for one production system may not be a safe combination of components for another piece part production system. It should be pointed out that it may be necessary to change more than one component to provide a safe piece part bending production system, once it is determined that a change must be made.

General-purpose press brakes. A general-purpose press brake is designed and built so as to be operated by one operator, who controls the speed and movement of the ram by skillful use of the operator control (usually a mechanical foot pedal). The ability and skill of the operator to control the speed of the ram permits slow bending of wide sheets using general-purpose dies (tools) without fast "whip-up" of the extended edge of the sheet. Precise ram speed control is also required to permit the operator to control the ram to a partially closed position for "line" gaging, that is, bending to a previously scribed line. On single-speed mechanical press brakes, this is accomplished by slipping the mechanically actuated partial-revolution friction clutch to bring the ram to a partially closed position. Variable speed and two-speed mechanical drive units and general purpose hydraulic press brake units permit the same type of control. Operating a press brake at reduced speed also facilitates handling of the piece part component by the operator and can minimize the exposure of the operator to sheet or piece part whip-up. A foot-operated machine should be operated only from a safe distance determined by the size and shape of the piece part unless point-of-operation safeguarding is provided.

Stroking control on a general-purpose mechanical press brake is by foot pedal. The foot pedal must be maintained above a "step-high" position to minimize accidentally stepping onto it. This activating force should be adjusted to require enough force to inhibit accidental operation of the foot pedal and to return the linkage to its normal off position. Stroking control on a hydraulic press brake may be either by foot control or two-hand operator station. A foot control should be positioned so that the operator is not able to reach into the point of operation unless safeguarding is provided.

Special-purpose press brakes

Special-purpose mechanical or hydraulic press brakes can be constructed with many operational features or stroking options to suit the intended use determined by the employer/user. The user must select a press brake component that has the features that are suitable for its safe intended use in each and every piece part operation.

Like the general-purpose press brake component, this special-purpose press brake can be used to bend, form, notch, punch, pierce, and so on, if it is machined and constructed to do so. Special-purpose press brake components, however, can be operated by one or a number of operators, each of whom shall be provided an operator control station of a type appropriate to the piece part production system in use. In this way, each operator and helper is able to concurrently exercise control of the press brake ram cycle by activating an operator control station.

A variety of stroking controls and drive options are available for special purpose press brakes, such as hydraulic electric, air electric, hydraulic mechanical, single speed, and two-speed. Hydraulic electric controls generally are two-speed having a high-speed ram advance, a slow-speed work-forming portion of the stroke, followed by a high-speed ram return. Limit switches are used to control the speed changeover points in the stroke. Hydraulic electric controlled press brakes have infinitely variable stroke lengths within their range.

Air electric clutch/brake controls are generally utilized on mechanical press brakes, both single and two-speed. The mechanical two-speed drive is similar to the hydraulic stroke control; however, the stroke length is constant. The changeover point from high speed to slow speed is also adjustable.

Operating modes. Special purpose press brakes have various employer/supervisor controlled modes of operation designated in the ANSI Standard B11.3 as "Off," "Inch," "Single Stroke," and "Continuous." "Off" shuts off the operator control station and stops the press brake. The "Inch" control is for use by the die setup person; it is not to be used in production. In this mode, the ram may be "inched" down and up but only with a two-palm-button operation, or a single control maintained, firmly secured and located a safe distance from the point of operation.

The "Single Stroke" mode is the standard production operating mode which can be initiated by the foot control or by a dual palm-button control. Determination of which of the controls is to be used is the responsibility of the user after considering the various components of the production system in use. In this mode, the press brake is under operator control in the descent portion of the stroke, and may automatically return to the top position where it must stop. The operator control station must be deactivated and then reactivated in order to initiate the next stroke.

"Continuous" operation control, used only with automatic feeds, is such that the press brake does not stop after each stroke but will operate continuously until the STOP button is activated. There are several methods of initiating operation in this mode, each of which is aimed at requiring a positive separate advance action on the part of the operator to positively minimize inadvertently placing the special-purpose press brake in this operating mode. These operating mode controls are always maintained under strict employer/supervisor control.

Tooling

Press brake tooling or dies are generally divided into two categories. These are general-purpose tooling and special-purpose tooling. General-purpose tools are the universal dies, widely available, used to perform bending and forming operations on a wide variety of piece parts and products. Special-purpose tools (dies) are those which are designed and built to perform work specialized on a specific piece part of product. Their sizes and forms are too numerous to mention. Many of them are designed so as to eliminate the necessity for an operator to hold the piece part component blank while one or more forming operations are performed on it during a ram stroke of the press brake component.

Safeguarding at the point of operation

ANSI B11.3, *Safety Requirements for the Use of Power Press Brakes,* should be consulted to help determine when to properly use the various safeguards described.

In the proper operation of a general-purpose press brake, protection of the operator's hands is accomplished by locating them along the extended edge of the piece-part component blank at a safe distance from the point of operation. ANSI B11.3–1982, in its explanation of section 6.1.4.3, Point-of-Operation Safe Distance, specifically states: "A dimension value has not been assigned to the minimum safe distance." The operator controls should only be in place or operable when it is the operator's intention to activate the machine. At other times, the foot pedal should be removed and/or the operating linkage locked to prevent actuation.

A general-purpose press brake is not recommended for performing operations such as narrow pieces, angles, channels, Z-Bars, etc. If a general-purpose press brake is used because a special-purpose press brake with two-hand controls is unavailable, safeguarding such as pullbacks or restraints can be used with hand tool feeding. Tool setups for these types of parts require arrangements that support the part before and after forming to prevent its falling behind the die so that an operator is not inclined to reach between the dies for a part that has fallen. Material gages that locate front and back edges are frequently required to make it unnecessary for an operator to have hands near the point of operation during forming. If forming is done to a scribed line, the use of supports and a back stop is recommended to assist the operator in maintaining control over the part and prevent hands from entering the die by following a part that could otherwise slip past the scribed line.

When the piece-part component is large and extends some distance in front of the die, the operator must hold the sheet at a "safe distance" (Figure 11-34) so that hands or fingers are not exposed to (1) injury due to impact by the moving piece part, or (2) pinch points or the point of operation while the bend is being made.

In the proper operation of a special-purpose press brake, protection of the operator is accomplished by either a two-hand operation control station or a safeguarding means while utilizing a foot control (see Figure 11-35). Depending upon the piece-part bending system, a presence-sensing gate or movable barrier device or other safeguarding means may be utilized (Figure 11-36).

A presence-sensing device may be useful for large or small parts for point-of-operation safeguarding provided piece-part movement or requirements for holding during forming do not interfere with its function.

Special-purpose mechanical press brakes are provided with air-electric clutch control mechanisms which provide a base for the adaptation of many safeguarding means. It is, therefore, easier for the employer/user to provide and use many of those

safeguarding means not available in a general-purpose press brake.

In this regard, the employer/user shall make a determination and is responsible for having the general-purpose press brake modified with air-electric constructional features and controls (in accordance with ANSI B11.3) to establish a special-purpose power press brake. These required modifications can be costly; however, they will extend use of the machine by permitting point-of-operation safeguarding devices, with piece-part bending systems. The employer will also gain by providing for operator protection, by saving any injury costs which may be attributable to an accident, and by higher productivity in most cases.

In addition to the foregoing described safeguarding devices, the following precautions should be observed:

- Periodic and thorough checks of the entire machine should be made to assure that all parts are in good working condition.
- Worn punches and dies should be retooled or discarded; use of worn tools throws excessive strain on the press brakes.
- Provide barrier guards or devices wherever possible and provide hand-feeding tools for inserting or removing stock (see Figure 11-35).
- Establish safe operating procedures and make sure that all operators follow the procedures. Do not allow any personnel to circumvent or bypass the built-in safety features. There can be no "occasional disregard" for safety.
- Locate machines with adequate floor space for each worker, and without interference from or hazard to other workers. Provide adequate illumination for all machines and auxiliary lighting where necessary.
- Work tables or receptacles that may be required to store material being processed should in no way interfere with freedom of the operator or access to the foot pedal or machine controls. These tables or receptacles should be stationary and constructed so that material on them cannot be knocked to the floor and possibly actuate the press brake control.
- Adequate clearance for the changing shape of the stock being processed should be assured before press brake action is initiated.
- Care should be taken to assure that the (back) material position gage fingers are of sufficient height to minimize the possibility of the material being gaged slipping over them during operating procedures.
- Select only personnel who have good vision, hearing, and physical condition as operators. Untrained personnel should never be permitted to operate press brakes. The operator should be instructed to "test operate" the press brake a few times at the beginning of each working period and after each setup to make certain that it is operating properly.

Responsibility for guarding and safeguarding

ANSI standard B11.3, Section 4.2.1, Hazards to Personnel Associated with Moving Parts (Other than Point-of-Operation Hazards), points out in the explanation of the standard some of the hazards which may exist. These hazards require installation of protective covers or other suitable means of protecting operators and others in the vicinity. They include:

1. Rotating components, such as flywheels, gears, sheaves, and shafts in close proximity to operating personnel (see Figure 11-37).
2. Run-in pinch points associated with meshing gears, belts, and chains.

Figure 11-34. Restraints are used as a means of safeguarding the point of operation of this press brake. The operator at right controls the manual clutch by means of a foot pedal while both operators support the workpiece. (Reprinted courtesy of Cincinnati Incorporated.)

3. Pinch points between the moving and stationary components of the power press brakes or auxiliary equipment.

The standard provides that the manufacturer shall eliminate a hazard by design where possible, or provide protection against it, and where a hazard cannot be eliminated by design or protection the manufacturer shall warn against the hazard.

Section 4.2.2 of the B11.3 standard, entitled "Hazards to Personnel Associated with the Point of Operation," states, "Point-of-Operation hazards and employer's responsibility regarding those hazards are covered under 6.1.1." Explanation of 4.2.2 points out that hazards at the point of operation are an entirely different problem from those that can be considered in the construction of the basic press brake component and its accessory equipment. Thus point-of-operation hazards must be considered separately.

Section 6.1.1, entitled "Hazards at the Point of Operation," stipulates:

1. The employer shall make an evaluation of each and every operation before any material is worked (formed, etc.) to determine either:

a. How a point-of-operation die guard can be used, or
b. How a point-of-operation device can be used to protect the operator(s) from injury at the point of operation. If a point of operation guard or device *can* be used, it *shall* be used.

2. When a point-of-operation die or device cannot be used, protection for the operator shall be provided by either:
a. Using hand tools to feed the part or
b. Maintaining a safe distance between the operator(s) and the point of operation, determined by the dimensions of the part being formed.

The nature of the work performed on power press brakes does not allow the use of point-of-operation die guards or devices in many instances. In these situations, the use of hand-feeding tools or maintaining a safe distance between the operator's body or positions thereof and the point of operation will suffice.

If the press brake is being operated to stamp, punch, or pierce material (similar to a power press), the point of operation should be safeguarded in accordance with (a) and (b) of 6.1.1 (1) of ANSI B11.3.

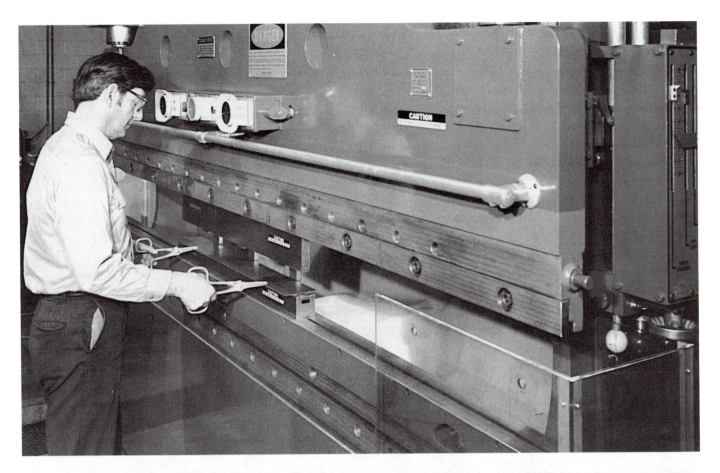

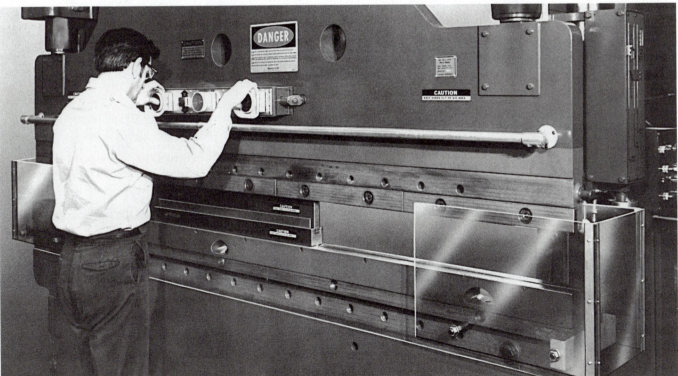

Figure 11-35. Top photograph shows hand tools being used for inserting and removing small piece parts. Lower photograph shows operator using a two-hand control station as a point-of-operation safeguarding device. Note the use of partial barriers that guard the unused portion of the point of operation of the brake. (Reprinted courtesy of Cincinnati Incorporated).

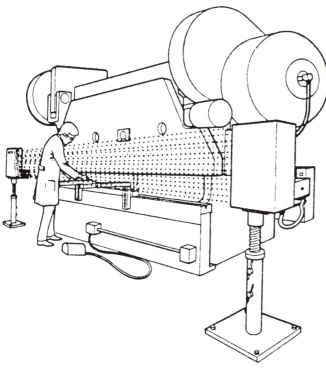

Definitions

Due to differences in press brake machines and power presses, the following definitions are interpreted and presented as applied specifically to press brakes.

Brake. The mechanism used to stop the motion of the power press brake ram; when engaged, it holds the power press brake ram in a stopped position.

Clutch. As assembly that connects the flywheel to the crankshaft either directly or through a gear train; when engaged, it imparts motion to the power press brake ram.
Part-revolution clutch. A type of clutch that can be disengaged at any point before the drive mechanism (usually a crankshaft) has completed a full revolution and the press brake ram a full stroke.

Combined stroking control systems. Two independent control systems on the same power press brake, only one of which is operable at a time.

Figure 11-36. Presence-sensing device uses light rays to safeguard the point of operation. (Drawing shows how it works.) Operator's hands must be outside the light beam before the foot switch can be activated. (Reprinted courtesy of Cincinnati Incorporated.)

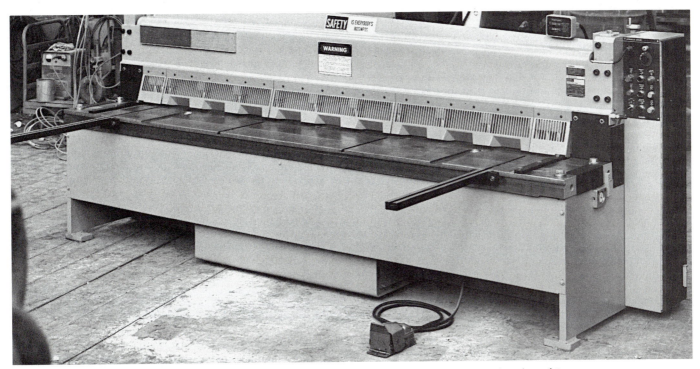

Figure 11-37. This hydraulic shear has fixed guarding for the point of operation, a throat or gap guard, and a safety cover on the foot pedal.

Concurrent. Acting in conjunction, this term is used to describe a situation wherein two or more controls exist in an operated condition at the same time. The use of the term concurrent is not intended to imply that the individual two-hand controls must be actuated simultaneously.

Connection. The part of the power press brake that transmits motion and force from the revolving crank or eccentric to the power press brake ram. (Also see Pitman.)

Continuous means uninterrupted multiple strokes of the ram without intervening stops (or other clutch control action) at the end of the individual stroke.

Counterbalance. The mechanism that is used to balance or support the weight of the connections, ram, ram attachments, and installed tooling components.

Cover guard. An enclosure that covers moving machine parts (excluding point of operation).

Cycle. A *full cycle* is the complete movement of the ram from its open position through its closed position and back to its previous open position.

A *partial cycle* is the movement of the ram from its open position to its closed position.

Device (Point of Operation). A press brake control or attachment that does one of the following:

1. Restrains the operator from reaching into the point of operation.
2. Inhibits normal press brake operation if the operator's hands are within the point of operation.
3. Automatically withdraws the operator's hands if they are within the point of operation as the dies close.
4. Maintains or restrains the operator or his or her hands during the closing portion of the ram strike or at a safe distance from the point of operation.

Hostage control device. A device designed, constructed, and arranged on a special-purpose mechanical and/or hydraulic power press brake to restrain and maintain the operator(s) at a control station which may be fixed by location at a safe distance from the point of operation or maintained by hand during the closing portion of the stroke.

Presence-sensing device. A device designed, constructed, and arranged to create a sensing field or area and, when used with a special-purpose mechanical or hydraulic power press brake, deactivates the clutch control of the power press brake when an operator's hand or any other body part is within such field or area.

Pullback device. When used on general- or special-purpose mechanical and/or hydraulic power press brakes, a pullback device is a mechanism attached to the operator's hands and connected to the moving portion of the die or ram of the power press brake that will, when properly adjusted, withdraw the operator's hands if they are inadvertently within the point of operation as the dies close.

Restraint device. When used in conjunction with a general- or special-purpose mechanical and/or hydraulic power press brake, a restraint device is a mechanism, including attachments for operator's hands, that when anchored and adjusted, inhibits the operator's hands from entering the point of operation.

Two-hand control device. A stroke control-actuating means on a special-purpose mechanical and/or hydraulic power press brake requiring the concurrent use of both of the operator's hands to start the ram movement, and requiring concurrent pressure from both of the operator's hands during a substantial part of the die-closing portion of the power press brake stroke.

Die(s). A term commonly used to describe the complete die, consisting of an upper and a lower component, also known as *tooling,* which are the components used in a power press brake for bending or forming the piece-part material.

General-purpose dies are the universal dies used to perform bending and forming operations on a variety of piece parts or products.

Special-purpose dies are designed to perform work not normally done on general purpose dies or for performing a common bending or forming operation in a manner that eliminates piece part whip up or the need for the power press brake operator to hand hold the piece-part component.

Die builder/employer. Any person who builds dies for power press brakes. This is the person or entity totally responsible for the initial design of the die. This term or definition is not to be applied to a person that merely fabricates a die from completely detailed drawings furnished by another company or entity.

Die holder. The heavy plate or rail to which the lower portion of the die is attached.

Die set. A tool holder held in alignment by guide posts and bushings and consisting of a lower shoe, an upper shoe, guide posts, and bushings.

Die setter. An individual who places or removes tooling components in or from a power press brake; and who, as a part of the job duties, makes the necessary adjustments to cause the tooling to function properly and safely in conjunction with the piece-part bending operation to be performed.

Die setting. The process of placing or removing tooling components in or from a power press brake and the process of adjusting the dies, material position stops (backgages), tooling, safeguarding means, if required for the piece part to be bent and for the operator to function properly and safely.

Die space. The adjustable distance between the bed and the ram. Into this space, the employer/user installs the tooling required for a specific piece part bending operation. Once the tooling has been installed and the proper adjustment for the material thickness made through the die space adjustment mechanism, it creates a point of operation, that is, the place where the actual piece-part bending will take place.

Eccentric. The offset portion of the crankshaft that governs the stroke or distance the ram moves on a mechanical power press brake.

Ejector. A mechanism for removing work or material from between the dies.

Employer. Any person who contracts, hires, or is responsible for, the personnel associated with power press brake operations.

Feeding. *Automatic feeding.* Feeding wherein the material or piece part being processed is placed within, or removed from, the point of operation by methods or means not requiring action by an operator on each stroke of the power press brake.

Manual feeding. Feeding wherein the material or piece part being processed is located and placed in the point of operation by the operator after the power press brake has completed each stroke. Manual feeding, for inserting or removing piece parts, shall only be used when the hands or other bodily members can be kept a safe distance from the point of operation.

Hand feeding. A type of manual feeding wherein the material is placed within, and processed parts removed from, the point of operation by use of a hand-feeding tool.

Push or slide feeding (hand operated). A pusher or slide can be used to feed a blank under the upper die and is withdrawn after the operation is performed. The pusher or slide may have a machined nest to fit the shape of the part. If the part neither drops through the die nor is ejected by other means, it can be withdrawn by the pusher or slide.

Semiautomatic feeding. Feeding wherein the material or piece part being processed is placed within, or removed from, the point of operation by an auxiliary means controlled by the operator on each stroke of the power press brake.

Foot control. The foot-operated control mechanism (other than mechanical foot pedal) designed to control the movement of the ram on mechanical, hydraulic, or special-purpose power press brakes.

Foot pedal (mechanical). The foot-operated lever designed to operate the mechanical linkage that requires a raising of the foot on the mechanical foot pedal and a significant amount of foot pressure and travel to actuate and engage the clutch and disengage the brake on a mechanical power press brake while the mechanical foot pedal is held depressed.

Foot treadle bar. A bar that is moved in a vertical direction when depressed by the foot of the operator at any point along its length. This bar is attached to two lever arms pivoted from the outside surface of the frame and is connected through linkage to the clutch and brake.

Gate or movable barrier device, constructed or arranged on a special-purpose mechanical and/or hydraulic power press brake, encloses the point of operation before the press brake ram can be actuated. This device interlocks in the press brake control system to inhibit or stop activation of the ram movement whenever the device is prevented from being in a full-close position.

Gibs. The parts which are used for guiding the ram.

Guard. A barrier that prevents entry of the operator's hands, or any other body part, into the point of operation.

Helper. Any person who assists a power press brake operator.

Holding distance (means of operating). The closing distance traveled by the ram during which time the operator(s) is compelled to hold the operating means depressed. Release of an operating means before this holding distance is traveled will stop the ram motion, whereas release of an operating means after this holding distance is reached will allow the ram motion to automatically continue as determined by the selected mode of operation.

Hostage operator control station. A type of operator control station which physically maintains and restrains the operator at a sufficient distance from the point of operation. A sufficient distance is one where the operator cannot reach near the point of operation during the closing portion of the power press brake cycle. The use of the term "near the point of operation" means no closer than the distance referred to as "safe distance."

Housing. The stationary portion of the power press brake structure on which the ram is guided and to which the bed, crown, and drive are attached.

Human factors engineering, as it applies to this specific chapter, is an analysis of the piece-part bending system to determine the most efficient and safest method of performing the operation. A piece-part bending system is an orderly arrangement of components that act to perform a specific task to a given piece part. The components of a piece-part bending system are as follows:

1. The specific piece part
2. The tooling designed or determined to perform the required bend or function
3. The power press brake to be utilized along with its operating control stations
4. The power press brake operator's function for loading, operating, and unloading the piece part
5. The safeguarding means itself which is the last but most important component necessary within this system.

Maintenance personnel. Individuals who care for, inspect, and maintain mechanical power press machines or press brakes.

Manufacturer. Any person who manufacturers, reconstructs, or modifies power press brakes for installation in the United States.

Material position gage. A stop against which the material is placed to properly locate it at the point of operation.

Not readily removable refers to using fastening procedures requiring effort and time to remove rather than quick-release fasteners such as wing nuts, etc.

Operating means. The mechanism depressed by the operator(s) that controls electric current or air flow (etc.), but excludes a mechanical foot pedal and other directly controlled mechanical linkage systems.

Operator. An individual who performs production work and who controls the movement of the ram.

Operator's control station. The control mechanism provided to each person for the purpose of initiating the starting and stopping of the power press brake ram stroking. The control station may be fixed to the power press brake or it may be portable.

Owner. Any person who owns the power press brake.

Pitman. That portion of the connection assembly that couples to the eccentric.

Point of operation. The area of the tooling or dies where material is actually positioned and work is being performed.

Ram (slide). The powered movable portion of the power press brake structure, with die attachment surface, which imparts the pressing load through dies, piece part, and against the stationary portion of the press brake bed.

Run. A single stroke or continuous stroking of a power press brake.

Safeguarding. "Safeguarding the point of operation" is an umbrella term, under which all safeguarding methods fall and all types of die hazards exist. Two basic categories exist under this umbrella term: the guard and the device.

A guard is best understood as a physical barrier that absolutely *prevents access* to a point-of-operation die hazard when it is in place and while it remains in place during a production run of successive cycles. If a barrier allows any access to a point-of-operation die hazard during a production run, it is not a guard; it then serves only as an inadequate enclosure. An inadequate enclosure always requires a device in conjunction with its use to form an acceptable safeguarding system. (Some devices require the use of enclosures to form an acceptable system.)

A device is best understood as a safeguarding means that *controls access* to the point of operation. These are (1) press brake controlling devices, (2) operator-controlling devices, and (3) devices that control the operator and the power press brake.

Safe distance is defined as a minimum distance from the press brake operator's hand (or hands) to the point of operation, which must be a minimum of 12 in. (30 cm).

Shut height. The distance from the bed to the ram when the ram is at the bottom of its stroke.

Shut height adjustment screws. The screws used to set the shut height of the power press brake.

Single stroke. One complete stroke of the ram, usually from a full-open position through a closed position back to a full-open position.

Stroke (up or down). The vertical movement of the power press brake ram during half of the cycle, from full-open to full-closed position or vice versa.

Stroking selector. That part of the control that determines the type of ram stroking when the operating means is actuated. Stroking selectors are normally furnished on hydraulic and special-purpose mechanical power press brakes.

REFERENCES

American National Standards Institute, 1430 Broadway, New York, N.Y. 10016.
Safety Requirements for the Construction, Care, and Use of Power Press Brakes, B11.3.
Safety Requirements for the Construction, Care, and Use of Mechanical Power Presses, B11.1.
Safety Requirements for the Construction, Care, and Use of Shears, B11.4.
National Safety Council, 444 N. Michigan Ave., Chicago, Ill. 60611.
Safeguarding Concepts Illustrated.
Industrial Data Sheets
  Alligator Shears, 213.
  Cutting Oils, Emulsions, and Drawing Compounds, 719.
  Electrical Controls for Mechanical Power Presses, 624.
  Handling Steel Plate for Fabrication, 565.
  Inspection and Maintenance of Mechanical Power Presses, 603.
  Kick (Foot) Presses, 363.

Metal Squaring Shears, 328.
Power Press Point-of-Operation Safeguarding: Concepts, 710.
Power Press Point-of-Operation Safeguarding: Movable Barrier Devices, 715.
Power Press Point-of-Operation Safeguarding: Presence-Sensing Devices, 711.
Power Press Point-of-Operation Safeguarding: Pullbacks and Restraint Devices, 713.
Power Press Point-of-Operation Safeguarding: Two-Hand Control and Two-Hand Tripping Devices, 714.
Press Brakes, 419.
Removing Piece Parts from Power Press Dies, 534.
Scrap Ballers, 611.
Setting Up and Removing Power Press Dies, 211.
Power Press Safety Manual.

United Auto Workers, 8000 E. Jefferson Ave., Detroit, Mich. 48214. (General.)

Wilson, Frank W. (ed.). *Handbook of Fixture Design.* New York, N.Y., McGraw-Hill Book Co., 1962.

12

Hot Working of Metals

THIS CHAPTER DISCUSSES HOW TO CONTROL materials handling hazards and environmental stresses (dust, fumes, gases, heat, and noise) that are present in foundries and permanent mold and die-casting plants. Also covered are safeguarding methods and safe operating practices for forging and hot metal stamping operations. A survey of the use of nondestructive testing methods supplements these discussions.

FOUNDRY HEALTH HAZARDS

Metals are formed into finished castings in foundries and permanent mold and die-casting plants. The overall foundry operation usually includes a pattern shop and sometimes a machine shop.

The principles and practices used for the recognition, evaluation, and control of health hazards are discussed in the Council's book *Fundamentals of Industrial Hygiene*. See this book for details on toxicity and flammability hazards, general and local exhaust ventilation, and specific problems such as silicosis, dermatoses, and radioactivity hazards. A person who is technically familiar with these subjects should survey foundry operations and make appropriate recommendations that should be followed when designing and installing control equipment. This technician should also be consulted for test procedures, analysis of new processes, and similar problems that arise in the course of the hygiene program. Other sources of professional help include the National Safety Council, the American Foundrymen's Society, National Institute of Occupational Safety and Health, American Industrial Hygiene Association, American Conference of Government Industrial Hygienists (ACGIH), insurance carriers, and state or provincial and local departments of industrial health.

Hazardous materials

The types of materials that present a health hazard and their mode of production in foundry operations are discussed below.

Dust is generated in many foundry processes and presents a two-fold problem: cleaning to remove deposits and control at the point of origin to prevent further dispersion and accumulation.

Vacuum cleaning is the most satisfactory method for removing accumulated dust in foundries, and the special equipment needed is well worth the investment. Once dust has been removed, further accumulation can be prevented through the use of local exhaust systems which remove it at the point of origin.

Solvents include many different substances, each of which must be evaluated on the basis of its chemical ingredients. Proper labeling, limitations on quantities in use, and other methods of control can help minimize the toxic and flammable hazards involved in solvent use.

Other materials.

Acrolein occurs in foundry operations as a result of the thermal decomposition of core oil.

Aluminum is not usually a toxic hazard in casting processes, but does present a fire and explosion hazard in dust-collecting systems.

Antimony is usually an unimportant contaminant in foundry operations.

Beryllium may produce a typical pulmonary disease, such as reported in one plant casting a 1 percent beryllium-copper alloy.

Carbon, as sea coal, is a common ingredient of molding sand used for facing. Carbon dust may cause anthracosis, which produces characteristic lung shadows in an X ray, but is a relatively harmless condition.

Carbon monoxide is generated during some cycles in the operation of a cupola, and after pouring into green sand molds.

Chromium is encountered in stainless steel casting as the element or the oxide. Exposures occur during melting, gate and head burning, and grinding.

Fluorides, sometimes in the form of cryolite (sodium aluminum fluoride), are used in the manufacture of ductile iron and magnesium castings.

Iron oxide fumes and dust are created during melting, burning, pouring, grinding, welding, and machining of ferrous castings. Exposure may be particularly high where manganese steel castings or oxygen-lancing of the furnace is involved. Local exhaust can be used (Figure 12-1).

Lead is a health hazard in nonferrous foundries. It forms the oxide in melting, pouring, and welding operations. Elemental lead dust is produced in cleaning and machining operations.

Magnesium dust or chips create serious fire and explosion hazards. Physiological effects are confined to a form of "metal fume fever" from the inhalation of finely divided magnesium. Magnesium oxide fumes will be generated when burned; the Threshold Limit Value (TLV) is 10 mg/m^3.

Manganese is usually associated with steel castings and bronze alloys in foundry work and presents no special control problem.

Phosphorus is used in the production of phosphor-copper. Acute cases of poisoning have not been reported and chronic cases are rare. The drying of phosphor-copper shot may produce phosphine gas.

Resins—phenol-formaldehyde and ureaformaldehyde—are used in shell molding and create several hazards. The phenol-formaldehyde type contains hexamethylenetetramine ("hex"), which is a skin irritant and highly explosive. This type of resin also decomposes on heating to give a mixture of phenol and formaldehyde vapors. Urea decomposes to give ammonia and carbon dioxide. In practice, however, vapors from resins are a nuisance, because the concentrations needed to produce toxic effects usually cannot be tolerated by workers.

Resin dust, especially "hex," is highly explosive when suspended in air and therefore requires wet dust collectors.

Alcohol, sometimes used for cold coating sand with resin, must also be controlled to keep its concentration well below the lower flammable limit.

Silica is usually encountered in the use of silica flour in molding sand or in core washes and sprays. This material is often 100 percent free-silica. Zircon, which is more dense and therefore settles more rapidly, is an effective substitute for silica flour in some applications. The Threshold Limit Value/Time-Weighted Average (TLV/TWA) is 0.1 mg/m^3, respirable dust based on 100 percent pure silica.

Sand handling and conditioning systems, shake-out operations, and sand-slinging constitute other sources of exposure to silica dust.

Silicones are used as mold-release agents in shell-molding. The hydrolyzing types are highly corrosive and hazardous on skin contact and inhalation. Care in handling can eliminate the dangers of skin or eye contact. However, the nonhydrolyzing types (methyl, mixed methyl, and phenylpolysiloxane) are the better choice because they can be just as effective as mold-release agents and are much less toxic.

Sulfur dioxide is the result of the oxidation of sulfur used in magnesium castings, and, in concentrations normally present in foundries, is an irritant nuisance.

Medical program

The health protection program for foundry workers should be based on the recommendations and guidance of an industrial hygienist and/or a physician. Such a program includes these integral functions:

1. Preplacement physical examinations including chest X rays and pulmonary function tests to place applicants in jobs most suitable to both their safety and health and that of other employees.
2. Periodic examinations including chest X rays and pulmonary function tests to keep track of the employees' health, detect incipient disease, and reclassify workers as necessary.
3. Adequate first aid facilities and employee training in first aid work.
4. Medical examinations if respirators must be worn.
5. Industrial hygiene monitoring where needed.

Personnel facilities

Coreroom workers whose hands and arms may be exposed to sand and core oil mixtures are candidates for dermatitis. Prolonged contact with oil, grease, acids, alkalis, and dirt also can produce dermatitis. Frequent washing with soap and water should be encouraged and adequate facilities should be installed.

Recommendations for toilets, washrooms, shower and locker rooms, and food service are detailed in the *Administration and Programs* volume, Chapter 18, Industrial Sanitation and Personnel Facilities. Sanitary food preparation and service is always important but especially so in nonferrous foundries where eating is prohibited in work areas where toxic materials are handled.

FOUNDRY WORK ENVIRONMENT

Housekeeping

For best housekeeping results it is advisable to hold each individual responsible for maintaining order in his or her own work area. A specific time should be set aside for housekeeping. Necessary housekeeping equipment should be available, and trash cans and special disposal bins should be kept handy and emptied regularly.

Each worker should:

1. Clean the machines and equipment used after each shift and keep them reasonably clean while working.
2. Put all trash in the proper trash bins.
3. Keep the floors and aisles in work areas free and unobstructed.
4. Properly stack and store materials.

The foundry floor should be arranged for efficient operating economy and to prevent accidents, especially those caused by spills and "run-outs" of molten metal.

Floors should be cleaned frequently and kept in good condition, firm, and level. Worn spots, holes, or other defects should

Figure 12-1. Fume-diverting baffles for barrel furnaces in a foundry. Hoods roll on a trolley so that correct positioning is made easy. (Printed with permission from American Brake Shoe Co.)

be reported and repaired immediately. Special types of flooring are necessary where fires or explosions may occur or other serious hazards may exist.

Floor loading should be controlled. Many buildings are being used for purposes for which they were not designed. Mechanized material movement introduces floor load problems because of the heavy dead weight of platform and lift trucks. The suspension of overhead cranes and hoists from wood ceiling joists severely taxes roof and floor members. Insurance engineers, local building inspectors, or private consulting engineers can help determine safe load limits.

Ventilation

Control of air contaminants is the primary purpose of ventilation in foundries. The degree to which air contaminants should be controlled by a ventilating system is determined by local and national standards and codes and need for employee comfort.

The need for ventilation control may be determined by one or more of the following:

1. Reference to applicable standards, codes, and recommendations.
2. Comparison with similar operations in a like environment.
3. Collection and analysis of representative air samples taken by qualified personnel in the breathing zone of workers.

Illumination

Good illumination is difficult to achieve in foundries because of the nature of the operations. Where craneways are used, lighting fixtures must be placed high and at considerable distances from the work areas. Nevertheless, good lighting should be provided for each work area. See Chapter 19, Safety Engineering Tables, for recommended levels of illumination. Foundries having difficulty in maintaining recommended light intensities can call on their local power companies or illuminating consultants for expert information.

Inspection and maintenance

Standard inspection and maintenance procedures should be followed in foundries. Maintenance personnel should be carefully selected and trained in safe practices, particularly in procedures for locking out controls and bringing equipment to its zero mechanical state (ZMS), as discussed in Chapter 8, Principles of Guarding. Also refer to *Safety Requirements for Sand Preparation, Molding, and Coremaking in the Sand Foundry Industry*, ANSI Z241.1.

Plant structures

Entrances and exits. In some cases, it is desirable to provide entrances to heated buildings with vestibules or enclosures constructed or located to prevent air drafts from reaching employees and sized to permit the passage of conveyances regularly used inside the plant. This provision does not apply to entrances used for railroad or industrial cars handled by locomotives or for traveling cranes, trucks, and automobiles.

All doors, particularly double-acting swinging doors, should have a window opening approximately 8 by 8 in. (20 by 20 cm), located at normal eye level, to permit a view beyond the door.

Stairways. All permanent and all portable stairways having four or more risers should be provided with substantial handrails, standard guardrails, and toeboards. See *Safety Requirements for Floor and Wall Openings, Railings, and Toeboards*, ANSI A12.1.

Floors and pits. The floor beneath and immediately surrounding foundry melting units should be pitched away from the melting units to provide drainage. The floor should be kept free from pools of water to prevent an explosion hazard. Where water is needed to reduce dusty operations, only enough should be used to hold down the dust.

Pits and other containers in which molten metal is handled or poured should be free from dampness because of the danger of explosion.

Pig molds and receiving stations for excess molten metal from ladles should be located clear of passageways and at least 1 ft (0.3 m) above floor level. Having pig holes in the floor near pouring areas is unsafe and should never be allowed.

Pits connected with ovens or furnaces and floor openings should be protected with either a cover or a standard guard rail when not in use.

Where tram or standard gage railroad tracks run into or through a foundry, the top of the rails should be flush with the foundry floor, which should be maintained at this level.

Galleries. Where molten metal is poured into molds, the galleries should be provided with solid, leakproof floors (concrete or sheet steel covered with sand) and with partitions of sheet steel. The partitions should be approximately 42 in. (1.0 m) high and should be installed on the open sides of such galleries.

Where floor space is cramped, it is sometimes desirable to construct galleries on which to store ladles, flasks, flask boards, and other equipment. These galleries should be equipped with standard handrails and toeboards and should have sturdy stairways instead of ladders.

Gangways. Every gangway should be kept in good condition, sufficiently firm to withstand the travel for which it is intended, uniformly smooth, without obstructions, and free from pools of water.

Concrete pavements around pouring floors should be kept coated with sand during pouring operations to reduce spalling of cement in case of a spill.

Aisles in which molten metal is handled should be kept in good condition, clear of obstructions, firm, uniformly smooth, and free from pools of water at all times.

Compressed air hose

The compressed air hose presents another foundry hazard. Improper use of the hose and "horseplay" have caused severe injuries to internal organs and eardrums. Compressed air shall be reduced to less than 30 psig (210 kPa).

Such unsafe practices as blowing and brushing sand from new castings without regard for the cloud of dust produced, blowing dust off patterns, and using compressed air to remove parting compounds and other light materials should be prohibited. Vacuum methods should be substituted for compressed air cleaning of molds. Workers should be carefully instructed in the safe use of air hoses if the latter method is employed. Use of nonsilica parting eliminates the possibility of silicosis from this source.

Fire protection

Foundries should organize brigades to make periodic fire inspections and to perform emergency fire fighting before the arrival of the local fire department. A fire brigade will also aid the safety program by keeping its members, as well as other employees in the foundry, safety-conscious.

MATERIAL HANDLING IN FOUNDRIES

Improper material handling results in a wide variety of injuries in foundries. Cuts and crushing injuries involving fingers, hands, toes, and feet, fractures of legs and arms, back strains, ruptures, and burns can occur during the manual handling of scrap metals, pig iron, and similar materials.

Some of the precautions that can be taken to prevent material handling accidents include:

1. Instructing workers in the safe methods of manual and mechanical handling of material.
2. Providing personal protective equipment—eye protection,

safety hats, faceshields, leather mitts or gloves, preferably studded with steel (unless hot metal is to be handled), hand pads, aprons, foot protection including metatarsal shoes, and other items such as flame-retardant clothing—to be worn as required.

3. Planning the sequence and method of handling materials to eliminate unnecessary handling.
4. Safeguarding mechanical devices and setting up inspection procedures to assure their proper maintenance.
5. Keeping good order at storage piles and bins, and piling materials properly.
6. Keeping ground and floor surfaces level so that workers handling materials will have good footing.
7. Installing side stakes or side boards on tramway or railroad cars to prevent materials from falling off.
8. Chocking railroad cars and flagging tracks as required. Using dock plates with box cars when loading or off loading.

Many foundries have replaced hand methods with material transporting machinery and thus reduced exposure to manual handling hazards. However, the machinery usually brings with it hazards of its own. For instance, lifting magnets are dangerous if swung over areas where people are working. A break in the magnet circuit would cause the load to be released without warning or pieces dangling from the magnet could be jarred loose. In no case should suspended loads be carried over workers' heads.

Handling sand, coal, and coke

Certain hazards in the handling of materials such as sand, coal, and coke can be avoided as follows:

1. Falls through hoppers while unloading bottom dump railroad cars can be prevented by requiring the use of safety belts and lifelines. Observers should always be on the scene and should be prepared to effect rescues in emergencies.
2. Using hopper car door safety ratchet wrenches can keep doors from swinging and striking workers.
3. Hand and foot injuries can be prevented by using safety car movers instead of ordinary pinch bars to spot cars by hand. (Where a locomotive is available, it should be used.)
4. By using warning targets, derails, and red lanterns at night, and by locking switches and using car chocks, workers can keep cars they are working on from being moved.
5. Prohibiting the undermining of piles and avoiding overhangs can reduce the danger of cave-ins of loose material.
6. Electric shock can be prevented by grounding portable belt conveyor loaders.

Some foundries have eliminated double handling by having the raw materials put directly from the cars, storage piles, or bins into unit charging trays or boxes, which are taken to the point of use and dumped mechanically. Trays or boxes to be carried overhead must be properly trimmed.

Ladles

Ladles for distributing molten metal or reservoir and mixing ladles, which are mounted on stationary supports or trucks or handled by overhead crane or monorail and which have a capacity of not more than 2,000 lb (900 kg), may be of the hand-shank type and should be provided with a manually operated safety lock (Figure 12-2). Shanks should be made from solid material, and shields should be installed on them. Suitable covers should be provided on portable ladles.

Figure 12-2. Tilting ladle equipped with manually operated antitilt level. (Printed with permission from the American Foundrymen's Society, Inc.)

Ladles which have a capacity over 2,000 lb should be of the gear-operated type. Such ladles and those which are mechanically or electrically operated should be equipped with an automatic safety lock or brake to prevent overturning or uncontrolled sway.

The rim or lip on hand or bull ladles should be built up above the top of the metal shell with fire clay no more than ½ in. (1.3 cm) if the refractory ladle lining is less than 1½ in. (3.8 cm) thick at the rim. In any case, the maximum height of the rim or lip should be not greater than 1 in. (3.5 cm).

Ladles should be thoroughly dried out before being used. Local exhaust should be provided to control vapors or fumes produced during ladle drying. Some foundries control both vapors and excessive noise produced by gas torches used for drying by performing all operations in a ladle-drying shed located outside the foundry.

Monorail ladles and trucks used to transport molten metal ladles should be equipped with warning devices (bells or sirens) to be used whenever molten metal is being transported.

Trunnions and the devices used to attach them to flasks, buckets, ladles, and other equipment should be constructed with a factor of safety of at least 10. The diameter of the head on the outside end of the trunnion shaft should be not less than one and one-half times the diameter of the trunnion shaft. The inside corners where the trunnion shaft joins the base and the head should be filleted to prevent the sling or hook from riding the trunnion base or head.

Inoculation or treatment of molten metal to desulfurize it or to change its composition or type, as in the making of an alloy or ductile iron, is done in the reservoir or in a pouring ladle. A hood can be made to cover this operation so that the workers are effectively shielded from possible spatters of metal caused by the violence of the reaction. The resulting fumes are drawn off and exhausted through a baghouse filtering system and then through a stack.

Hoists and cranes

Hoists and cranes that handle molten metal require a preventive maintenance program, conducted by personnel trained and thoroughly familiar with the equipment. (This is in addition to observations and inspections made by supervisors and operators.) The degree to which the program should be carried out depends on both the equipment being used and the tonnage moved. The program should be geared to making sure the operation is safe, not to a minimum compliance with existing regulations.

For example, a program for a 300-employee, ordinary gray iron foundry could require weekly visual inspection of crane and hoist structures, and an inspection of wire ropes and hooks before every shift.

Because of the severe stresses and demanding service required in some high-tonnage operations, inspection programs become more elaborate and should be done on a weekly basis by trained specialists. Some programs regularly schedule nondestructive testing—ultrasonic testing of crane hoist shafts and parts, and dye penetrant inspection for surface cracks on bales, dumping chains, clevises, and pins. (See the discussion under Nondestructive Testing later in this chapter.)

Conveyors

Conveyor systems are now being used in a number of foundries. Sand mixed in the mixing room is carried by belt conveyor to hoppers at molding stations, where each hopper is filled by means of a movable plow. Surplus sand is carried to the end of the belt and returned by bucket to the storage bin.

An endless conveyor is used to handle molds. Empty flasks come from the shake-out to the molders, who remove them and make their molds on molding machines, taking sand from the overhead hoppers as required. Spilled sand goes through a grating onto a belt conveyor which returns it to the mixing room. Thus all shoveling operations are eliminated.

The molds are then placed on a conveyor and passed into the pouring area. The pourers get their metal from the cupola or other furnaces and then step onto a moving platform geared to an endless single-rail conveyor moving at the same rate of speed, on which hand ladles can be supported. Pouring is done as the workers move along, and they are as safe as they would be if they were standing still.

An electric switch should be installed near the end of the conveyor, so that if workers ride that far, their feet will come into contact with the switch and the conveyor will stop.

The mold conveyors then pass into a cooling zone. Weights can be removed from the molds by a mechanical device and returned by another conveyor to the place where they originally are used.

The molds then move to the shake-out, where sand is dumped from flasks onto a vibrating grating. The sand falls to a belt conveyor, which returns it to the mixing room. Using a hook, a worker pulls the castings onto another conveyor, which takes them to the tumbling barrels. Empty flasks are brought back to the molding section by still another conveyor.

This is a complete system for mass production where each person has one job to do rather than several. In the installation of such a system, shear points, crush points, and moving parts must be effectively guarded.

Where conveyor systems run over passageways and working areas, the employees beneath should be protected by screens, grating, or guards strong enough to resist the impact of the heaviest piece handled by the conveyor.

Where chain conveyors operate at various levels other than in a fixed horizontal plane, a mechanism of safety dogs should be installed in accordance with applicable standards on both the upgrade and the downgrade. In case of chain failure, the safety dogs will hold the chain and prevent the load from piling up at the bottom of the incline.

Scrap breakers

Shears should be guarded to protect operators and passersby from flying particles. The working floor should be clear and level.

Use of a drop to break castings or scrap inside the foundry buildings during working hours should be prohibited unless such operations are performed within a permanent enclosure of planking or equivalent strong enough to withstand the most severe impacts from the demolition weight or flying scrap. The enclosure should be high enough to protect workers in the vicinity from flying fragments of metal.

If a rope is used, it should extend over pulleys to a point clear of the breaking area to assure that the operator will be at a safe distance and to prevent entanglement.

Storage

Foundry materials and equipment not in regular use should be stored in the space provided in a safe, orderly manner on level and substantial foundations. When workers remove materials from bins located at floor level or from storage piles, they should not undermine the piles and thereby cause cave-ins.

Hopper bins containing material which is fed out at the bottom, either by hand or by mechanical means, should be covered with a grating which will prevent workers from entering the bin. No one should be allowed to get on the rails of a bin nor to enter a hopper to break down bridged material.

A worker who must enter a bin should wear a safety harness with attached lifeline, tended by a worker similarly equipped.

Pattern storage buildings should have racks and shelves substantial enough to hold the loads placed upon them. The floors and stairways should be well designed and kept in good condition. The storage area should be well lighted. Pattern keepers should be provided with a sound ladder so that they can safely reach the patterns. The pattern keeper, who is likely to be alone in the pattern storage area, should report at regular intervals to the supervisor.

Flammable liquids should be stored in accordance with National Fire Protection Association's *Flammable and Combustible Liquids Code,* NFPA 30. See Chapter 16, Flammable and Combustible Liquids.

Slag disposal

Furnaces and pits should be designed to have removable receptacles into which slag and kish (separated graphite) may flow or be dumped. Unless it is disposed of in the molten state, enough of these receptacles should be provided so that slag can solidify before it is dumped.

To decrease the amount of slag that goes into the slag pits, slag or cinder pots should be used. The pots can be set aside and allowed to cool, to eliminate the danger of explosion when they are emptied.

The slag should be dumped where there is absolutely no water or dampness because water might cause an explosion if some of the slag is still molten. Before breaking up slag, it should be allowed to stand for several hours so that there will be no molten slag in the center.

CUPOLAS

Charging

The dangers in the charging of cupolas are principally confined to the handling of material. Barrows or buggies should never be unevenly loaded or overloaded. "Tip-up" barrows used for charging coke are sometimes so poorly balanced that they will not stay in the tipped-up position after being emptied. Instead, they fall back on the chargers' feet at the slightest touch. Lowering the center of gravity will prevent this hazard.

The use of mechanical devices for charging cupolas not only saves labor but also reduces the number of material handling accidents. Most foundry cupolas are now charged by either fully automatic charging machines equipped with crane and cone bottom buckets or by lift trucks equipped with tilting boxes.

The charging opening on some cupolas is covered by a door or chain curtain, which should be kept closed except during charging.

The space underneath the cupola charging elevators, machines, lift hoists, skip hoists, and cranes should be railed off or safeguarded to prevent material from dropping onto workers during charging operations.

Occasionally, during idle periods, workers may be resting under the charging platforms, close to the warm chambers and flues. This practice should be prohibited because of the danger of objects falling from the platform and the possibility of carbon monoxide escaping from the flues. Clothes lockers should not be located under these platforms.

Scrap cylinders, tanks, drums, and the like should be broken open before charging to prevent explosion in the cupola. Be sure that these containers are empty.

Charging floor

Charging floors should be kept free from loose materials, and storage racks should be provided for equipment not in use.

Steel floor plates should be substantial enough not to turn up and should be securely fastened in place. Steel plate flooring in the immediate vicinity of the furnace becomes extremely hot. Therefore, brick flooring laid on substantial steel framework is more satisfactory.

Standard railings 42 in. (1.0 m) high and 4-in. (10-cm) toeboards should be provided around all floor openings. See OSHA 1910.179(d)(3),(4)(ii).

Because railings on the charging floor are likely to be subjected to considerable abuse, it is usually best to construct them of angle iron, which is more easily repaired than pipe railings. At the tapping platforms, hinged gates or chains which may be hooked in place must often be provided.

Where cupolas are manually charged, a guardrail should be placed across the charging opening. Where cupolas are charged with wheelbarrows or cars, a curb of a height equal to the radius of the wheel of the barrow or car should be provided to prevent the barrow or car from pitching over and falling into it.

Carbon monoxide

Carbon monoxide (CO) is generated during some cycles in the operation of a cupola. It is an explosion hazard if it gets into the wind boxes and blast pipes when the blowers are shut down. To eliminate this hazard, adequate natural or mechanical ventilation should be supplied in back of the cupola, and two or more tuyeres should be opened after shutdown.

The large amount of blast air in the cupola generally carries the carbon monoxide out the stack. In some cases, it is burned in the stack before it can be discharged. Sometimes, however, the gas may escape. It gives no warning, so CO indicators that light and give a loud sound should be located about the cupola as well as training signs showing the proper procedures to follow should the CO indicator alarm sound.

Approved respiratory protection should be at hand, in good condition, and workers should be trained in its use. If concentration of CO is more than 200 ppm (0.2 percent), positive pressure, self-contained breathing apparatus or an air line respirator with emergency escape bottle should be provided. However, the eight hour TWA is 50 ppm and can peak at 400 ppm for 15 minutes.

Blast gates

Blast gates and explosion doors are used successfully to prevent damage from gas explosions. They are sometimes placed in front of the tuyeres, so that they can be opened to admit fresh air when the blowers are shut down. They should never be closed until the blast has entered the wind box and driven out all gas.

Blast gates should be provided in the air blast pipe that supplies air to the melting equipment. They should be closed when the air supply fails or when the melting equipment is shut down, in order to prevent the accumulation of combustible gases in the air supply system. In the cupola, the blast gate may be omitted if alternate tuyeres are opened to permit air circulation.

Blast gates should be so located in relation to the cupola wind box that the duct volume will be kept to a minimum. Motorized dampers may be installed at centrifugal blowers so that they will close automatically when the air supply fails.

Positive-pressure blowers should be equipped with safety valves having liberal discharge areas. If these are not provided, clogging of the cupola with slag or quick closing of the gate or damper may produce sufficient pressure in the blast pipe to burst it.

Every cupola should have at least one safety tuyere, with a small channel 1½ or 2 in. (4 or 5 cm) below the normal tuyere level. This channel has a fusible plate that will melt through should the slag and iron rise to an unsafe level.

Tapping out

Tapping out with safety requires skill and should be done only by experienced and dependable operators. In "botting-up" the hole, the bott should not be thrust directly into the stream of molten metal or it will cause spattering. To eliminate this, place the bott immediately over the stream of metal close to the hole and aim it down toward the hole at a sharp angle.

A supply of botts ready for use should be kept within convenient reach of the operator who does the tapping.

When the cupola is tapped, the back end of the tapping bar should be held below the level of the hole, to prevent puncturing the sand bed and causing run-out through the bottom.

A tilting spout placed with one end directly beneath the stationary cupola spout and mounted on trunnions on a stand increases safety and efficiency. It can be tilted back and forth by a foot lever. The rear end of the tilting spout is closed so that when that end is tilted down, it forms a reservoir to receive the molten metal from the cupola. When the supplementary spout is tilted forward, the metal runs from it into the waiting ladle. At the same time, more metal continues to run into the spout from the cupola. Thus, the stream of metal runs from the cupola continuously, and the tilting spout acts as a reservoir between ladle loads.

The slag spout of the cupola should be equipped with a shield or guard to protect workers from sprays of molten slag and to form a hood to collect slag wool. The slag wool is sometimes collected through a wet slagging system wherein slag is thrown off into a water-filled container or trough and flushed away.

Dropping cupola bottom

When the cupola is in operation, its bottom doors should be supported by one solid prop and two adjustable screw props (of the required structural strength) on a metal prop base set on a concrete or other fabricated footing of equivalent strength (Figure 12-3).

Temporary supports (timbers, blocking, etc.) should be placed under the cupola bottom doors to prevent their falling on employees while the metal props are being adjusted to proper height. Mechanical means for raising the bottom doors of the cupola should be provided (Figure 12-4).

Dropping the bottom doors of a cupola requires extraordinary care. One of the best methods is to use a block and tackle with a wire rope and chain leader attached to the props which support the doors. The props can then be pulled out by means of the block and tackle from a safe distance or from behind a suitable barrier.

Before the bottom is dropped, the area underneath the cupola should be carefully inspected to see that no water has seeped under the sand. One worker should make sure that no one is in the danger zone and that workers stay away during the operation. Employees should be warned by means of a whistle or other signal before the bottom is dropped.

If the cupola bottom doors fail to drop or if the remaining charge inside the cupola bridges over, employees should never be permitted to enter the danger zone to force the doors or relieve the bridging.

The bridging may be relieved by turning on the blast fan. The vibration produced usually corrects the condition. A mechanical vibrator attached to the bottom doors is also effective. Another method is to drop a demolition ball from the charging door. If these methods fail, the doors must be flame cut with a lance, but only after the cupola has cooled to a safe temperature.

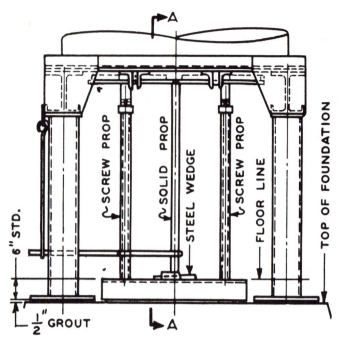

NOTE :— SCREW PROPS MUST BE REMOVED
BEFORE REMOVING SOLID PROP

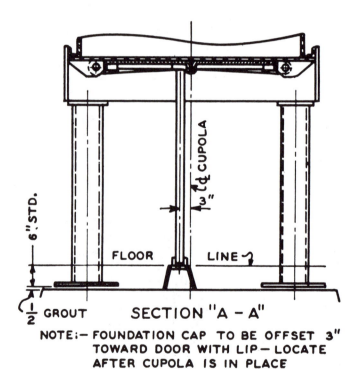

SECTION "A – A"

NOTE:— FOUNDATION CAP TO BE OFFSET 3"
TOWARD DOOR WITH LIP — LOCATE
AFTER CUPOLA IS IN PLACE

Figure 12-3. Proper method of supporting cupola bottom doors.

Special bottom door locking devices may be used if the cupola drop is to be caught in a container, car, or skid.

Repairing linings

Only careful and experienced workers should be allowed to make repairs inside cupolas. The precautions to be followed are:

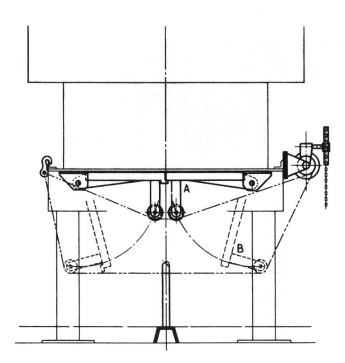

Figure 12-4. Suggested method of raising bottom doors of cupola by mechanical means.

1. Install a substantial guard over the cupola charging door to protect workers against falling objects. Such a guard should be constructed of heavy gage metal. It should be covered with a screen or equivalent. The guard should be securely supported by means of overhead slings or underpinning to resist falling objects (Figure 12-5). An alternate method is to place a solid steel plate on the bucket ring in the cupola at charging door level.
2. Require all workers who enter a cupola to wear safety hats and glasses.
3. Provide approved respiratory equipment for the workers relining the cupola, or place a blower fan in the bottom to keep the dust moving away from them. Dust should be exhausted out the stack.
4. Place warning signs or crossbars at the charging door to indicate that workers are working in a cupola.
5. Before relining of a cupola is begun, break all loose slag and bridges and allow them to drop to the bottom of the cupola.
6. Check the condition of the shell while the cupola is down for relining. A weak shell is likely to increase the risk of a gas explosion.
7. Leave ample clearance (at least ¾ in. or 2 cm) between the new brick lining and the shell to allow for expansion. This space should be filled with dry sand to serve as a cushion to protect the shell against severe stresses.
8. Before the cupola is started up, check to make sure that all personnel have made their exit from within the cupola and the area beneath it, that the lining is thoroughly dry, and that all tools and other equipment have been removed.
9. Request all cupola tenders and other individuals working apart from a group to report to their supervisor at regular intervals as a safety check practice.

Figure 12-5. A screen placed over the charging door prevents falling objects from dropping on a worker who is repairing cupola lining. (Printed with permission from Hamilton Foundry & Machinery Co.)

CRUCIBLES

The principal danger in handling refractory clay crucibles is that one may break when full of molten metal.

All new crucibles should be inspected for cracks, thin spots, and other flaws by a trained inspector. Those showing signs of dampness should be put aside to be returned to the manufacturer. Examination of the packages and car in which they were shipped may reveal whether or not they were exposed to moisture in transit.

Crucibles should be stored in a warm (about 250 F or 120 C), dry place and protected from moist air as much as possible. It is generally best to place them in an oven built on top of a core oven or at some other point where waste heat can be utilized. If all crucibles in stock cannot be kept in ovens, those stored elsewhere should be accurately dated and the oldest and best-seasoned crucibles should be selected first.

In the annealing process, the crucibles are brought up to red heat very slowly and uniformly, usually over a period of eight to ten hours. Crucibles should not be allowed to cool before being charged, because as they cool they may again absorb moisture. Moisture in the walls of crucibles that are heated quickly is converted into steam, which expands and causes cracks or ruptures and may also cause pinholes or "skelping." Do not use damp or high-sulfur coke or coal or fuel oil containing excessive moisture to heat crucibles.

Too high a percentage of sulfur in the fuel used in the drying or annealing process is likely to cause fine cracks, sometimes called "alligator cracks." Too little oil or too much air or steam used at the burners of oil furnaces tends to oxidize a portion of the graphite in the crucible wall, leaving the binding material somewhat porous.

Charging and handling

Proper care of crucibles is good economy as well as good safety. Since crucibles are costly, they should be made to last through as many heats as possible.

Crucibles should be charged carefully. Ingots should not be thrown in with such force that they may bend the bottom or walls of the crucible out of shape. Neither should the ingots be forced into the crucibles so as to become wedged or jammed.

New crucibles should be heated very slowly for the first few runs, especially the first. Because crucibles are soft at white heat and easily forced out of shape, they must be handled with great care.

In hoisting larger crucibles with air or electric hoists, it is preferable to have one person at each sling and one operator controlling the hoist.

Tongs of the proper size and shape for the particular crucible should be selected to prevent damage to it. Tongs should fit well around the bilge or belly of the crucible and should extend to within a few inches of the bottom.

Before tongs are applied, the sides of the crucible should be checked to see that no clinkers are adhering to them. Tong rings should never be driven down tight with a skimmer or other tool, because this practice is almost certain to squeeze the crucible out of shape and produce cracks and fissures.

At least two pairs of tongs should be provided for each size of crucible so that if one pair becomes bent, the other will be available.

The blacksmith should have a complete set of cast iron forms in the exact shapes of the crucibles used. Then the smith will have only to heat the tongs to red heat, clamp them onto the forms, and bring them into exact shape with a heavy hammer.

Avoid ramming the fuel bed around a crucible. If this becomes necessary, it should be done cautiously and only by experienced workers. Crucibles should be supported on foundations or pedestals of firebrick, graphite, or other infusible material.

The removal of heavy crucibles from furnaces calls not only for special skill but also for physical strength. If the workers are overstrained, serious accidents are likely to result. Where possible, therefore, a mechanical device should be employed for removing heavy crucibles—those exceeding 100 lb (45 kg) in combined weight of crucible, tongs, and metal.

Crucible furnaces

The operation of crucible furnaces is relatively hazard-free if suitable exhaust hoods are installed on all furnaces used for melting metals which give off harmful fumes.

Upright furnaces having crown plates more than 12 in. (30 cm) above the surrounding floor should be equipped with metal platforms having standard rails. Such platforms should extend along the front and sides of the furnace flush with the crown plates and clear of all obstructions. Crucibles containing molten metal should be lowered from these platforms mechanically.

Many crucible furnaces are oil fired. Unless the motors driving the oil pump and the air supply for the furnaces are connected to the same source of power, a considerable quantity of oil may flow onto the floor if the power goes off the air line.

One remedy is to put a gate valve in the oil supply line so that in case of an air failure, oil can be shut off from the entire battery of crucible furnaces in one operation. If the pilot and burner are electrically controlled, this valve can be so arranged that it will be similar to an interlock.

A mechanical shutoff can be made by the installation in the oil line of a gate lever-operated valve which is closed by the release of a weight. Because the weight is held up normally by an air cylinder, the oil supply is stopped at once when the air goes off. A small hole may be drilled in the cylinder so that air in it may be released promptly when the oil pressure goes off.

OVENS

The principal hazards in the construction and operation of core ovens and mold drying ovens are excess smoke, gases, and fumes given off into the foundry atmosphere. Other unsafe conditions are unprotected firing pits, unguarded vertical sliding doors or their counterweights (which may drop on workers), and flashbacks from fire boxes.

Firing pits into which people may fall should be guarded with substantial railings or with grating covers having trap doors to give access to the steps leading down into the pits.

Safe types of vertical sliding doors should be installed. Wire ropes and chains having a high factor of safety, substantial fastenings, large sheaves to prevent undue wear of ropes, and guarded counterweights are also necessary. All sliding doors should be thoroughly inspected at frequent intervals. Safety dogs may be used to hold the doors in raised position.

Many vertical core ovens are 60 to 70 ft (18 to 21 m) high and have a driving mechanism located at the top. Some foundries have installed steel stairways with standard treads wherever possible. Where metal ladders must be used, not less than 7 in. of clearance (18 cm) should be maintained from the center of each ladder rung to the side of the oven.

A caution sign should be placed on the wall near core ovens, giving necessary precautions and manufacturer's instructions to be observed when ovens are being lighted.

Gas-fired ovens

Gas-fired ovens should, whenever possible, be located in a room separated from the molding floors and also from the coremaking room by a partition. This measure will help prevent equipment failures caused by sand in the controls.

Blast tip pipe burners should be equipped with baffles to keep sand out of the tip and also to spread the flame. Tips should also be horizontal to protect them from sand.

Safety pilot valves will prevent the flow of unburned gas into the oven combustion chamber should the burner pilot light go out, or should a cock or burner be accidentally opened. A valve of this nature should be installed on every gas-burning furnace or oven.

A bleeder valve in the line between two control valves close to the burner is an additional safety device. The operator can then allow gas to escape safely into the atmosphere instead of to the fire box, should there be leakage past the main control valve when the burner is not in use.

Ventilation

Where fumes, gases, and smoke are emitted from drying ovens, hoods and ducts, exhaust fans, or other means for removal should be provided near the oven doors to keep exposure concentrations below toxic and irritant levels. Consideration must be given to the composition of any emissions discharged to the outside so that there is compliance with air pollution ordinances.

To prevent flashbacks, flues should be adequate in size and kept free from soot. Where oil burners are used, the type of equipment and the arrangement and control of drafts should ensure as perfect combustion as possible. In some installations, forced draft equipment may be needed.

Core ovens should be equipped with explosion vents. Lightweight panels may be installed on the top of the oven, or the oven may have hinged doors with explosion latches.

Natural draft ventilation is usually considered adequate for ovens under 500 cu ft (14,000 liters) in volume, but the doors must be arranged so that they have to be wide open when burners are being lighted.

Larger ovens, particularly those with vertically sliding doors and other heavy construction, should have forced draft ventilation. The ventilation system should be interlocked with the gas supply through a time relay arranged to allow for at least three complete changes of air in the oven before the burners are lighted.

Inspection

Before foundry core ovens are lighted, the ovens and burners should be given a thorough inspection. Only trained and qualified personnel should do this work. This equipment should already be covered by a preventive maintenance and inspection program.

The following checklists represent an example procedure, and should be helpful as a guide that includes many of the points of good inspection practice. First, the main valve controlling the fuel supply should be shut off (which automatically turns off the gas line to the burner pilots), and the following items should then be checked.

1. Do the safety cutoff valves for gas and oil close when the pilot burners fail? If not, check for stuck valves or unusually long flame electrodes.
2. Do the red signal lights on the flame-detecting device light up? If not, check first for burned-out signal lamps and then for defective relays on the flame-detecting device.
3. Does the oven-tender warning light on the back of the control panel light up? (This light is mounted so that it can readily be seen throughout the core room.)
4. Does the warning horn blow? Can it be shut off by manipulating the safety cutoff valve handle? (It should not go off.)

Other points to be checked with the main fuel valve and burner pilots turned off include:

1. Are spark plugs clean and ignition wires in good condition?
2. Are OPEN and SHUT markings on gas and oil safety shutoff valves legible?
3. Are there core oil resin deposits on safety shutoff valves or the pilot gas solenoid valve? (Extensive deposits may cause valves to stick.)

4. Are all cover-retaining screws in place on flame-detecting devices and flame electrode holders to prevent entry of dust or core oil resins?
5. Are all motor controller, relay box, and wiring junction box covers tightly in place and clearly marked as to what they control?
6. Is there excess fuel oil or an oil-soaked accumulation of dirt or rags on the burner deck?
7. Are fuel oil atomizing heads removed from the burners, cleaned if necessary, and checked for burned tips?
8. Is atomizing air tube removed and are swirl vanes and the main burner nozzle checked for burned parts or accumulations?
9. If the oven is being operated on oil, is the three-way cock between the gas control valve and the burners turned to permit outside air to be pulled into the normal gas inlet to the burners?
10. Are there any broken refractory burner blocks?
11. Are gas, oil, and air control valves, valve control motor, and valve control linkage checked for loose valve adjustment screws or broken valve cam springs?
12. Is the combustion chamber checked for defective or burned-out arches, side walls, combustion wall, or metal tie bars which may span the combustion chamber above the arches?
13. Are all electrical contacts in disconnect switches, motor controllers, and relays inspected? (After inspection, replace all enclosure covers securely.)
14. Are temperature controller, controller relays, and control resistance lamp (if one is used) checked? Make sure that the burner control motor returns to low burner position when the controller disconnect switch is in OFF position.
15. Are covers from all air flow switches removed and gummy or worn parts, damaged wires, or damaged mercury tube elements removed?
16. Has performance of flow switches been checked by starting the proper fans and observing the flow switch operation?
17. As fans are started, have they been checked for V-belt squeal or fan vibration?
18. Is it certain that fans can be started only in proper sequence as follows: main circulation fan, power exhaust fan for drying zone, combustion air blower, similar fans for additional burner units, if used, and cooling zone supply and exhaust fans?
19. Has the operation of the purge duct damper motor and linkage at the start and end of the purge cycle been checked?
20. The purge cycle time should be 10 minutes on all ovens (except horizontal drying oven burner units, which have 5-minute cycles). Have red signal lights of the flame-detecting device come on at the end of the purge cycle?
21. Do spark plugs and flame electrodes project approximately 2 in. (5 cm) beyond the burner-pilot casting? (The flame electrode should be centered in the pilot orifice and set to one side about 1/4 in. (6 mm) at its tip.)
22. Have all explosion doors been checked for proper operation using a light to check for evidence of damage to internal structures or ductwork? (Make sure that the explosion panel in the roof is intact.)

Next, the main fuel valve should be turned on (which opens the pilot gas valve), the ignition button pressed, and the following points checked:

1. Do the pilot lights ignite readily?
2. Are there fuel oil leaks in the piping around the burners?
3. Is there an odor of leaking gas around the burners, the gas regulators, or the gas valves? (The various gas valves should be tested for leaks periodically with a soap solution.)
4. Is there any indication of flame flashback to the outside from either the pilot burner or the main burners?

All fans should be inspected for these items:

1. Using a tachometer, check the rpm against the proper speed shown on the metal tag on the fan drive guard. If fan speed is too low, tighten the belts and recheck the speed.
2. Check all V-belt drives for proper belt tension. (A tag showing the proper V-belt size should be mounted on each belt guard.)
3. All belt guards should be in good condition and securely fastened in place.
4. Fans should be checked for unnecessary vibration.
5. Air leakage at fans on attached ductwork should be overcome.
6. Fan motors should be clean and securely bolted in place.
7. Damper motors, control valves, and control linkages must be securely in place.

FOUNDRY PRODUCTION EQUIPMENT

On production-line equipment, moving parts (such as belts, pulleys, gears, chains, and sprockets) and other common machine hazards (such as projecting setscrews) should be fully guarded in accordance with standard practice (see Chapter 8, Principles of Guarding).

Repairs should only be done on equipment that is locked in the OFF position. Electrical equipment should be grounded to eliminate shock hazards.

Some operations require mills, mixers, and cutters of such size that an employee can enter the machine for cleaning or repair. In these cases, a lockout procedure must be set up and enforced. See the section titled Lockouts in Chapter 15, Electricity, and Zero Mechanical State, discussed in Chapter 8, Principles of Guarding.

Sand mills

The principal danger of sand mills (mullers) exists when operators reach in for samples of sand or attempt to shovel out sand while the mill is running. In doing so, they may be caught and pulled into the mill. To protect against this hazard, one or more of the following measures may be used:

1. Provide screen enclosures for charging and discharging openings of mills.
2. Install self-discharging mills, or equip mills with discharge gates or scoops.
3. Provide sampling cones for taking samples of the sand during the mixing operation.
4. Prohibit the shoveling of sand out of mills while they are running.
5. Install an interlocking device so that the mill cannot be operated until the doors are closed.

Dough mixers

To prevent operators from reaching into a dough mixer while the blades are in motion, the top of the mixer should be covered with a substantial grating made of ⅜-in. (1.5-cm) round bars or the equivalent. Another method is to attach an interlocking device so arranged that the cover cannot be opened nor the bowl tilted until the blade drive mechanism has been shut off and the blades cannot be set in motion again until the cover is in place.

Where the mixer is driven by an individual motor, a small steel cable can be attached to the cover and be extended over a pulley to a counterweight. This cable is attached to a ring on the motor control switch handle so that when the cover of the mixer is lifted a predetermined distance, the switch is pulled open and cannot be closed again until the cover is back in place.

Sand cutters

Sand cutters throw sand and pieces of tramp metal with bullet-like force, sometimes causing serious puncture wounds. If a guard that would not seriously impair the efficiency of the operation cannot be devised, then suitable personal protective equipment should be worn by the operator.

Because it is often difficult to operate a power-driven cutter on a sand floor, parallel concrete strips can be installed to act as runways for the cutter wheels.

Sifters

Rotary sand sifters should be guarded by enclosures or by angle iron or pipe railings 42 in. (1 m) high placed from 15 to 20 in. (38 to 50 cm) from the sifter. Belt shifters and motor control switches should be placed within convenient reach. The control switches should be so designed that they cannot be actuated accidentally.

Portable sand sifters equipped with pneumatic vibrators are usually slower moving than those equipped with electric vibrators, and oscillation of heavy parts causes the entire machine to move around the floor in jerky fashion. Often, machine travel is limited only by the air hose. If the hose coupling breaks, the hose flails around and blows sand in every direction, presenting a hazard to workers' eyes. Anchoring the sifter with a rope a little shorter than the hose helps to prevent such accidents.

Molds and cores

Letting flasks down on feet, pinching fingers between flasks, dropping heavy core boxes on feet, cutting hands on nails and other sharp pieces of metal in the sand, and stepping on nails constitute the principal hazards in hand molding and core making. Hand and foot injuries can be minimized if workers use proper methods of handling flasks and core boxes and wear good foot protection with stout soles. Screening or magnetic separation to remove nails and other sharp metal from the sand is also essential to safety.

In general molding and core making, gagger rods (pieces of iron used in a mold to keep the sand or core in place) and core wires are cut, straightened, and bent using hammers and cutting sets. This operation presents danger from flying pieces of metal and dirt. Machines are available for performing this work, but many of their hazards are similar to those found in the use of hand tools.

Heavy cores in large molds must be carefully braced as the work progresses to keep the cores from toppling over. Working underneath molds suspended from cranes should be prohibited. Substantial tripod supports or wooden or steel horses will provide greater safety and efficiency.

Venting molds properly is essential to avoid explosions during pouring. However, when the sand in an undried mold is too wet, metal can boil and explosions may occur even though the molds are well vented.

In ramming a mold, the peen of the rammer should not be placed too close to the pattern. Otherwise a hard spot in the sand will be made and molten metal coming in contact with it will boil and tear the sand away to the depth of the hard spot. This will also occur if a gagger iron is rammed against a pattern and the sand between is pressed into a hard spot. When the molten metal reaches the wet gagger iron, an explosion usually results.

Molding machines

Three types of molding machines are used in modern foundries: straight, semiautomatic, and automatic. All molding machines should be equipped with two-hand controls for each operator assigned to the machine.

The carry-out person should stand clear of the squeeze at the back of the machine. Operators should never touch the frame while it is moving.

When patterns are to be changed, the frame should be blocked to prevent the table from falling and trapping pattern setup workers, should the dog fail or the stripping frame operate accidentally.

On automatic molding machines, shields or apron-type metal guards should be used to protect pinch points.

Where molding machines have sand delivered by elevator buckets, the side on which the buckets return should be enclosed or railed off with 42–in. (1–m) double railings placed 15 to 20 in. (30 to 50 cm) from the moving parts.

The jolt squeezer machine is used as a molding machine. The main hazard of this machine is that workers may keep their hands on the edge of the flask or place a hand between the head of the machine and the flask. Dual safety squeeze controls and a knee valve jolt control will eliminate this hazard.

Core-blowing machines

Straight, semiautomatic, and automatic core-blowing machines are used in foundries.

To prevent sand blows, parting lines should be maintained in good condition. Also, the parting line of the core box can be guarded by a dike seal (Figure 12-6).

Vents should be replaced when ruptured. Good feeding systems with vibrators should be installed to keep the magazine full. Core box and cover seals should be kept free of sand.

When a core machine is to be repaired, the air lines should be shut off, the controls should be locked in the OFF position (being sure that the machine is in Zero Mechanical State), and the air should be bled from the machine.

Where practical, two-hand operating controls should be provided to prevent the operator from placing a hand or fingers between the top of the core box and the ram. Where two operators are employed on a core-blowing machine, four-hand control buttons should be provided.

All core boxes should be equipped with handles so that employees can move their boxes without placing their hands on top of them.

If driers are located above the rollover area for each core, they should be placed high enough so that they will not become entangled during the rollover process.

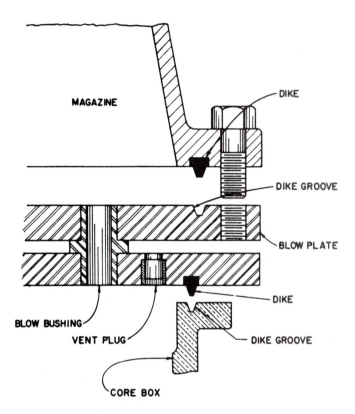

Figure 12-6. Section of core box shows rubber dike seal, which prevents sand blows and abrasion of the box. (Printed with permission from Dike-O-Seal Corp.)

On semiautomatic and automatic machines, core box push cylinders, counterweight cable pulleys, wheel guides, and table adjusting footpads should be guarded, and an automatic barrier guard should be installed between the operator and the machine. If the drier is lowered automaticallly from the rollover and then pushed and raised toward the operator, there can be a pinch point between the lowering table and the raise table. This hazard is eliminated by an automatic barrier guard.

Automatic and semiautomatic core machines should be equipped with double solenoid valves. The slide valve should be well maintained and lubricated to prevent recycling or other malfunction.

Materials used in cleaning core boxes may be toxic and therefore should be removed by a properly installed ventilation system.

Some core dips contain substances capable of producing dermatitis on sensitive persons. Rubber golves and plastic sleeves usually provide adequate protection. Employees engaged in core dip operations should be checked at frequent intervals for sensitivity to the core dip solution.

Flasks

Iron or steel flasks are preferable to wood flasks, which become worn, burned, or broken so that they do not fit together well and may let molten metal run out during pouring.

Flasks should be carefully inspected at frequent intervals by competent inspectors with authority to have the defective ones destroyed or sent to the repair shop.

It is unwise to leave defective flasks in the foundry building or in yard storage piles because they may be put back into service without first having been repaired.

Flask trunnions should have end flanges at least twice the diameter of the trunnions to minimize the danger of hooks slipping or jumping off. Trunnions preferably should be turned, or otherwise be smooth castings.

It is sometimes best to cast the trunnions separately and bolt or weld them in place. This procedure facilitates the machining operations and permits reuse of trunnions recovered from broken flasks. Trunnions cast separately should be of steel.

When trunnions are bolted or welded in place, the nuts should be inside the flask. If they project on the outside, slings are likely to catch on them and slip off with a jerk, which subjects both the sling and the trunnion to severe strain.

Large flasks should have loop handles of wrought iron. On steel flasks, handles should be cast in place at frequent intervals to facilitate chaining.

Trunnions and handles should be designed for the loads they are to carry and constructed with a safety factor of at least 10. The bolts with which they are fastened to the flasks should be of sufficiently heavy construction also.

The diameter of the button should be equal to the diameter of the groove plus one and one-half times the diameter of the sling used to handle the flask. Inside corners shall be well filleted. To prevent the sling from sliding off and riding the button, the radius of the corner between groove and button should be approximately equal to the radius of the sling used and the remainder of the inside edge of the button should be straight.

Sandblast rooms

Each foundry should have dust-tight sandblast rooms. The doors to these sandblast rooms should be kept closed, and castings should be dusted before they are removed from the rooms. Even small cracks in the walls or under doors will allow fine dust to escape and contaminate air in the foundry.

Tumbling barrels

Tumbling barrels need frequent care to keep them dust-tight. Barrels that cannot be maintained dust-tight should be enclosed in booths connected to an exhaust system. Modern barrels are equipped with exhaust ducts through the trunnions. A removable guard rail should be placed around the machine. While barrels are being loaded or unloaded, they should be locked in stationary position.

Shake-outs

The operation of shaking out castings presents the danger that hands and feet may be crushed or arms and legs broken. If steel hooks or rakes are used to pull castings from the screen, workers should be instructed to stand so that one foot is kept behind, so they can keep their balance in case the hook slips from the casting while they are pulling. Workers must wear steel-toed or metatarsel guard foot protection.

Because this operation is also often a source of dust, it should be hooded, and local exhaust should be provided to draw the dust to a collector. In fact, many foundries perform shake-out operations at night so that as few people as possible will be exposed to dust.

Shake-out machines should be designed so that the flasks cannot fall off the plunger. Foundry workers should not attempt to retrieve gagger irons while these machines are in operation.

The hazards of sand conveyors are found at shake-outs since the sand is collected under the shake-out on a conveyor belt which conveys it to storage for reclaiming and reuse. The area around the shake-out should be kept free of sand and scrap, and the conveyor belt opening should be guarded at sides.

CLEANING AND FINISHING FOUNDRY PRODUCTS

Abrasive, polishing, and buffering equipment for foundry use should be installed and operated as recommended in Chapter 10, Metalworking Machinery.

Operators should be required to wear full eye and face protection specified for the operations. Excessive dust generated in the use of dry abrasive wheels may be a health hazard and should be effectively removed at the point of origin by an exhaust system. If castings are precleaned in a barrel, mill, or abrasive chamber, dust from grinding can also be minimized. The space around the machines should be kept dry, clean, and as free as possible of castings and other obstructions.

Abrasive grinding wheels should be mounted and changed by qualified personnel. Use of correct washers and wheel-mounting procedures must be closely supervised.

Required wheel guarding must be kept intact. Eye, hand, and foot protection are necessary. Some companies speed-test new wheels before allowing them to be used on the job.

Magnesium grinding

The fundamental hazard in the grinding of magnesium lies in the possibility of fire or explosion. To eliminate this hazard, a proper dust-collection system must be used (Figure 12-7).

In a magnesium dust-collecting system, the dust should be precipitated by a heavy spray of water and immediately washed into a sludge pit in which it is collected under water. Sludge pits or pans must be well ventilated, because reaction of the collected dust with water evolves hydrogen. Sludge pits or pans should be cleaned frequently, and the sludge should be immediately mixed with fine sand or earth and then buried. Wet magnesium dust must not be allowed to stand and become partially dried.

The following safeguards should also be provided:
1. A means for immediate quenching of sparks from grinding wheels, disks, or belts.
2. Dust-proof motors to prevent accumulation of static charges.
3. Explosion doors on the collection system.
4. An automatic interlocking control on the collection system to assure its operation whenever grinding is started.

The dust-collecting system must not have filters or obstructions that will allow accumulation of dust. Pipes and ducts should be installed, using the shortest possible route, in order to eliminate bends or turns in which magnesium dust or fines could collect. Pipes and ducts connecting the grinder and the collecting device should be cleaned daily and should be disconnected while wheels are being dressed.

Good housekeeping is essential for safe handling of magnesium. Accumulations of magnesium dust on benches, floors, window ledges, overhead beams and pipes, and other equipment should be prevented. Vacuum cleaners should not be used to collect the dust. It should be swept up and placed in covered, plainly labeled iron containers. It should then be mixed with fine dry sand and buried. It should not be mixed with regular floor sweepings.

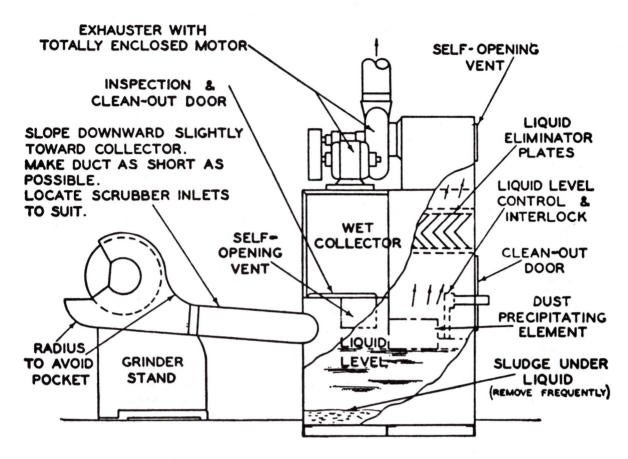

Figure 12-7. Magnesium dust-collection system converts dust into wet sludge for later removal. (Printed with permission from Dow Chemical Co.)

Because of the possibility of producing sparks, it is dangerous to use equipment involved in grinding magnesium to grind other metals. Equipment for magnesium grinding should be marked FOR MAGNESIUM ONLY.

Benches of wood grating are recommended for rough finishing operations.

An ample supply of powdered graphite in plainly labeled, covered metal containers should be kept close to each grinding unit. A scoop should be kept inside each container, and the lid should be kept loose for easy access.

Warning signs should be prominently displayed inside and outside the grinding rooms or areas. Post signs warning against smoking and against the use of water on fire and recommending the use of powdered graphite, limestone, or dolomite as an extinguishing agent.

To prevent fires and injuries, these safe work practices should be observed during the grinding of magnesium alloys.
1. The grinder and exhaust system should be started and run for a few minutes before grinding operations are begun.
2. Operators of grinding equipment should wear leather or smooth, fire-retardant clothing, not coarse-textured or fuzzy clothing. They should brush it frequently.
3. Goggles or a full-brim helmet with attached plastic faceshield should be worn. A skull cap may be worn under the helmet. Leather gloves with long gauntlets are recommended. A self-contained respiratory apparatus may also be used.
4. Machine tools must be sharp and properly ground for mag-

nesium alloys, or they may cause fire from friction.
5. Use only neutral mineral oils and greases for cooling and lubrication. Animal or vegetable oils, acid-containing mineral oils, or oil-water emulsions can be hazardous.

Chipping

Where castings are cleaned or chipped, tables, benches, and jigs or fixtures specifically designed and shaped to hold the particular casting should be provided. Screens or partitions should be installed to protect other employees from flying chips. These areas should be provided with hoods and exhaust systems to remove dust. Workers should wear eye and face protection.

Welding

Considerable welding is done when cleaning or reclaiming castings. The safe practices given in the next chapter, Welding and Cutting, should be followed.

One of the best noncombustible materials, sand, is plentiful in all foundries. To help prevent fire, it can be spread on the floor to a depth of 2 in. (5 cm) in areas where welding operations are conducted.

Powder washing, a method in which a stream of powdered iron oxide is introduced into the gas flame to intensify the heat produced, should be done according to the same safe practices as other carbon steel or cast iron welding and cutting. However, when this method is used to clean or cut sprues, gates, and risers from alloyed castings, exhaust ventilation is recommended.

Power presses

Power presses have wide application in finishing departments of foundries. Sufficient aisle space, good housekeeping, and effective lighting should be provided for safety in power press operation. Machines should be properly guarded and maintained in good working order. Operators should be carefully selected and should be trained in efficient and safe machine operation. Mechanical feed and ejection equipment should be used where practical. These topics are fully discussed in Chapter 11, Cold Forming of Metals.

NONDESTRUCTIVE TESTING

Visual observation, even with magnification, cannot locate all small, below-the-surface defects in cast and forged metals, or in weldments, such as found in pressure vessels, boilers, and nuclear components. Proper nondestructive testing will, however, reveal all such indications without damage to the parts being tested. Nondestructive testing methods will locate defects inherent in the metal (nonmetallic inclusions, shrinkage, porosity), defects that result from processing (high residual stresses, cracks and checks caused by handling, spruing, or grinding of casting and forgings), and in-service defects (sharp changes in section, corrosion, erosion).

The types of testing most commonly used for forged and cast metals are:
1. Magnetic particle
2. Penetrant
3. Ultrasonic
4. Triboelectric
5. Electromagnetic
6. Radiographic

These methods, as well as others applicable to nonmetallic substances, are fully discussed in the Council's Industrial Data Sheet *Ultrasonic Nondestructive Testing for Metals, 662.*

Recommendations for installation, inspection, and maintenance of the electrical equipment used in many of these testing procedures are given in Chapter 15, Electricity.

Magnetic particle methods

Magnetic particle inspection is the most widely used testing method for forgings. It utilizes magnetism to attract and hold very fine magnetic particles right on the part itself. If a defect is present, it interrupts the magnetic field and is clearly shown by the pattern made by the particles. The part is magnetized in suitable directions by DC line voltages transformed to low-voltage (4 to 18 volts), high-amperage AC, half-wave current, or three-phase full-wave current.

Inspection materials are finely divided ferromagnetic particles selected, ground, and controlled to provide mobility and sensitivity. Materials are available in several forms and colors. The type of indication to be located and the condition of the surface to be inspected determine which form of material and which method—dry, wet, or fluorescent—should be used. Color is selected to provide maximum contrast with the surface of the part.

All electrical circuits should be installed and grounded according to the *National Electrical Code,* NFPA 70.

Local exhaust is required to control the dust particles used for testing. If local exhaust is not feasible, operators should wear respiratory protective equipment. Eye protection should also be worn to guard against the irritating effects of the dust particles and arcing. Personal protective equipment should be worn to protect against possible skin irritation from dry powder and wet material used with this testing method.

Penetrant inspection

Penetrant inspection methods are useful for revealing cracks, pores, leaks, and similar indications which are open to the surface in a metal or other solid material. First, the part to be inspected is cleaned. A penetrant is then applied to the surface and within a few minutes is drawn into defects by capillary action. The penetrant is removed from the surface but will remain in the surface opening until removed by the developer.

Depending upon the sensitivity of the material, the penetrant is removed by water wash, a solvent cleaner, or an emulsifier followed by water wash.

Fluorescent penetrants may be used to reveal defects under ultraviolet ("black") light. Ultraviolet equipment should be effectively shielded or filter lenses of the correct shade should be worn. Indications may also be detected by a dye penetrant which contrasts with the surface color.

Most penetrants are organic compounds that may cause dermatitis. Skin contact should be avoided with good personal hygiene practices strictly followed. Smoking, smoking materials, food, or drinks should not be used or stored in the test area. Exposed skin should be washed before smoking, eating, or drinking.

Ultrasonic methods

Ultrasonic waves (above the audible range of 20,000 Hz) are created by an electronic generator which supplies high frequency voltage to a piezoelectric crystal.

Three basic ultrasonic methods have been developed: the reflection method, the through-transmission method, and the resonant-frequency method.

Reflection. That portion of the ultrasonic beam that strikes a flaw or discontinuity in the material is reflected; the rest of the beam goes on. The piezoelectric crystal transducer radiates these waves through a coupling medium into the material and also acts as a receiver to detect reflections, which are then picked up by an electronic amplifier and applied to a cathode ray oscilloscope on which the time intervals between the outgoing and the incoming waves can be measured.

Through-transmission. A beam or wave is directed through a piece of material. If a flaw or discontinuity is encountered, the energy is absorbed, and the beam or wave does not get through. Because fluids such as water, oil, and glycerine give better coupling than air, they are generally used as the coupling medium between the transmitter, material, and receiver. In some applications, however, air or other gases can be used.

Resonant-frequency. This method is used primarily to measure the thickness of material. The equipment consists of an electronic oscillator which supplies voltage of ultrasonic frequencies to a piezoelectric transducer. The transducer is pressed into contact with the part to be tested and includes longitudinal vibrations in the test piece under the area of contact.

One type of resonance instrument displays the thickness reading on a cathode ray tube as a pip on a calibrated scale. Equipment should be disconnected from the power supply and the condensers discharged whenever a cathode ray tube must be adjusted or removed.

Triboelectric method

The triboelectric method is used to detect minute quantities of current generated when two metallurgically or chemically unlike conductors are moved into frictional contact. If the conductors are alike, no current will be generated.

The equipment consists of a control unit and a portable sorting head connected by means of a cable. The sorting head contains all the main controls for actuating the test and is designed for one-handed operation. This method is designed to sort and identify metal parts of not more than four alloy types.

Electromagnetic methods

Two types of electromagnetic tests are currently being used in industry: magneto-inductive and eddy current. A third type, employing radar frequency, is also being used, but only to a limited extent.

- **Magneto-inductive.** This method uses variations in the permeability of magnetic materials to create variations in a pickup coil or probe.
- **Eddy-current.** The second and most common type of electromagnetic testing utilizes alternating current in a coil or probe to induce eddy current into the part being tested. Defects and variations in properties or geometry cause changes in the strength and distribution of the eddy current. Readout is presented on a cathode ray tube, on a meter, by audible or visible alarm, or by a combination of these methods.
- **Radar frequency.** This third method uses high-frequency radar waves to measure the electromagnetic properties of thin coatings and surface layers of material. To make such tests, a wave guide or cavity oscillator is coupled to the test object; high-frequency waves are then reflected from the object, providing an indication of the surface electrical resistance and of the thickness of the nonconducting coatings.

In some radar-frequency testing installations, operators have been burned internally when they passed between the object being tested and the testing device. Special regulations should be formulated and enforced, and barriers should be set up to prevent operators and other workers from entering such areas. Recommendations by the manufacturer of the equipment should be followed explicitly.

Radiography

Radiography uses X rays and gamma rays. X rays are unidirectional and their wavelengths can be varied (within certain limits) to suit the condition. Gamma radiography differs from X-ray radiography in that the gamma rays are multidirectional and their wavelengths, being characteristic of the source, cannot be regulated. Gamma rays for radiography usually are obtained from isotopes of cobalt-60 or iridium-192.

In some instances, gamma exposures are inferior to X-ray exposures in sensitivity and contrast. Gamma radiography, however, has several advantages. Because of the nature of isotopes, it frequently is possible to make a number of tests simultaneously, provided that specimens can be suitably located. Moreover, isotopes are independent of electrical power, their sources are portable, and the small size of the sources makes it possible to obtain radiographs in tight quarters.

Devices used to transform differences in intensity of the penetrating radiation into visible images are X-ray films, fluorescent screens, geiger proportional scintillation counters, and ionization gages.

All sources of ionizing radiation are potentially dangerous. X-ray and gamma ray sources may also produce hazardous secondary radiation. In addition, X-ray units involve both low- and high-potential electrical hazards.

FORGING HAMMERS

Open-frame hammers

Open-frame or Smith forging hammers are constructed so that the anvil assembly is separate from the foundation of the frame and operating mechanism of the hammer. They may be single or double frames.

Flat dies are generally used in Smith hammers, and the work done allows for more material to be machined off. Smith hammers are used where the quantity of forgings to be run is too small to warrant the expense of impression dies or where the forgings are too large or too irregular to be contained in the usual impression dies.

Gravity drop hammers

Drop forgings in closed impression dies are produced on gravity drop hammers (both board drop hammers and steam- or air-lift drop hammers) and steam hammers. Both types of gravity drop hammers shape the hot metal in closed impression dies. The impact of the hammer blows shapes the forging through one or more stages to the finished shape. On gravity drop hammers, the ram and the upper die are raised to the top of the hammer stroke, and the impact blow comes from the free fall of the ram and the die.

Forgings on gravity drop hammers may range in weight from less than an ounce (28 g) to 100 lb (45 kg) and may be made from any type of malleable metal, such as steel, brass, bronze, aluminum, or magnesium alloys. Gravity drop hammers are designed so that alignment between the dies can be maintained by the use of guides, die pins, die locks, or a combination of these devices.

The board drop hammer, the more common type, utilizes hardwood boards which are secured to the ram and held in it by wedges. The boards pass between rotating rolls which grip the boards and raise the ram and the upper die for the successive blows. At the top of the stroke, the rolls release and the ram is held in this position by clamps until released by depression of the treadle. The impact of a board drop hammer cannot be varied while the hammer is in operation.

A steam- or air-lift gravity drop hammer is controlled by a valve which admits steam or air under the piston into the cylinder in the head of the hammer to raise the ram and the die. At the top point, the air or steam is exhausted and has no effect upon the hammer blow, which is controlled entirely by the weight of the freely falling ram and die. Like the board drop hammer, the steam- or air-lift hammer is operated by treadles or pedals, levers, or air valves.

The height to which the ram is raised and consequently the impact can be varied during the operation of an air- or steam-lift hammer. The falling weight of the ram assembly and upper die of gravity drop hammers may range from 400 to 7,500 lb (80 to 3,400 kg).

Steam hammers and air hammers

Steam hammers are also classified as drop hammers. Most steam hammers are double-acting and use steam pressure (or air pressure) through the medium of a piston and cylinder to raise the ram and the die and to assist in striking the impact blow.

Because steam or air power is used in addition to the weight of the falling ram and die, the steam hammer will strike a heavier blow than will a gravity drop hammer using an equivalent falling weight.

The falling weight of the ram assembly and upper die of double-acting steam hammers ranges from 1,000 to about 50,000 lb (450 to 22,500 kg). They commonly produce forgings ranging from a few ounces to above 500 lb (225 kg), and sometimes up to 2,000 lb (900 kg).

Steam and air operated hammers are manufactured with many built-in safety features. Some of the most outstanding are:
1. Hammers are designed so that the distance between the floor and the die seat is approximately 36 in. (90 cm).
2. Safety latches prevent the ram from dropping when a job is being set up or when work is being done on the die. However, for complete protection, the steam or air should be shut off from the press and the ram should be blocked up.
3. A safety cylinder head protects against piston overtravel.
4. Many of the working parts are safely enclosed.
5. Operating levers are placed in a safe position.

Hammer hazards

For the most part, all types of forging hammers have identical hazards. The most frequent causes of injury are:
1. Being struck by flying drift and key fragments or by flash or slugs.
2. Using feeler gages to check guide wear or the matching of dies.
3. Using material handling equipment improperly, such as tong lifts.
4. Crushing fingers, hands, or arms between the dies.
5. Crushing fingers between tong reins.
6. Receiving kickbacks from tongs.
7. Using swabs or scale blowing pipes with short handles.
8. Being burned by hot scale.
9. Dropping stock on the feet.
10. Getting foreign objects in the eye (iron dust or scale).
11. Noise-induced hearing loss.

A hearing conservation program that includes proper hearing protection, annual audiometric examinations, as well as engineering controls, will greatly reduce or limit noise-induced hearing loss. See the Council's *Industrial Noise and Hearing Conservation.*

Injury may be incurred on a steam drop hammer when the ram pulls off a new piston rod. Sometimes the rod must be set in the ram several times before it holds. If the piston rod breaks, the ram will fall. This common hazard emphasizes the importance of the operator's using a safety prop to support the ram before reaching under it.

Operating a hammer with a worn cylinder sleeve is hazardous. When the sleeve is so worn that the swing of the ram cannot be controlled at the throttle control, the hammer should be shut down and repaired.

Operating a hammer with broken piston rings is also dangerous. A piece of broken piston ring passing through the steam ports and lodging in the throttle valve can cause the ram to drop out of control, often catching the operator's tongs or the transfer tool and causing serious injury.

Maintenance personnel, particularly, are exposed to crushing injuries when they remove and install parts on the top of the hammer and when they remove sow blocks, anvils, and columns. Means for locking out the power should be provided and used. Catwalks and guardrails should be installed on all hammers to provide safe footing (Figure 12-8).

Figure 12-8. Permanent catwalks installed along row of board drop hammers makes repair and servicing of hammers easy and safe.

Guarding

On gravity drop steam- or air-lift hammers, a hand lever is preferable to a treadle on cold restrike operations. Two-hand tripping controls should be provided where the material being forged is not held by the hands or by hand tools or where a safety stop or tripping lever cannot be installed.

On board drop hammers, a substantial guard should be provided around the boards above the rolls to prevent the boards from falling if they should break or come loose from the ram (Figure 12-9).

Other standard protective features for a board drop hammer include the ram stop and safety chain for tie bolt and nut.

A pneumatic key driver is superior to a manually operated driver and offers a far greater margin of safety (Figure 12-10). The driver should be made of steel of proper hardness so that it will not chip on impact.

A manually operated key driver should be sturdy and well balanced (Figure 12-10). The driving face should not be allowed

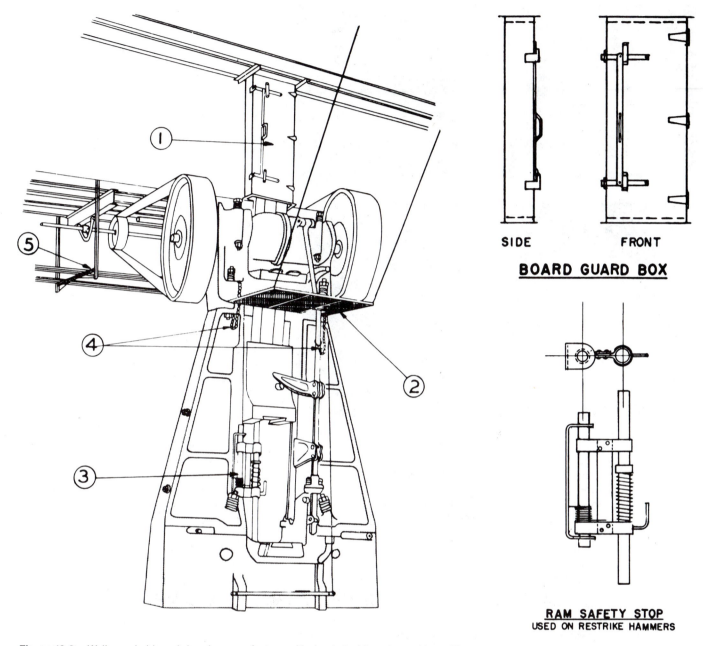

BOARD GUARD BOX

SIDE   FRONT

RAM SAFETY STOP
USED ON RESTRIKE HAMMERS

Figure 12-9. Well-guarded board drop hammer features: (1) sheet steel board guard box, (2) screen-platform made from No. 9 expanded metal, (3) steel ram safety stop that swivels on left column, (4) safety chain to restrain tiebolt and nut, and (5) catwalk and belt catcher. Details of board guard box and ram safety stop are shown in drawings at right. (Printed with permission from American Brake Shoe Co.)

to mushroom, but should be kept ground down—not burned off with a cutting torch.

A scale guard to confine pieces of flying scale should be standard equipment on the back of every hammer. The guard should allow ample clearance for the ram and easy access to the dies. It may be installed in one of the following ways:

1. Hinged on one side to an upright post so that the guard can be swung closed or open—out of positoin when access to the die area is required (Figure 12-11). This installation is considered the most efficient and is widely used throughout the industry.
2. Supported on a floor standard.
3. Suspended from the ceiling or anchored to a rail.

Treadles and pedals should have ample clearance and should be guarded to prevent accidental tripping by a falling object. Any portion of a treadle or pedal at the rear of the hammer should also be guarded so that scrap or other material cannot interfere with treadle action.

Several methods for interlocking the treadle have been developed. A simple mechanical interlock is effective. To lock the treadle when it is necessary to use a pry bar to remove stuck forgings from die cavities, the pry bar is kept in a sleeve-type holder with its weight resting on the actuator of an air valve. If the pry bar is removed, the air valve opens and the air cylinder drives

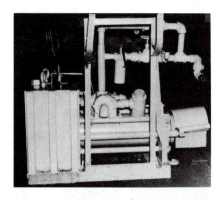

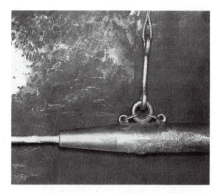

Figure 12-10. Mechanical key-driving rams, like the pneumatic model on left, are preferred to manual ones. A manually operated key driver is shown on right. (Right photo printed with permission from Tractor Works, International Harvester Co.)

Figure 12-11. This scale guard, closed in the left photo, is hinged to provide access to the drop hammer for housekeeping. The anchor post is set in a sleeve in the floor for easy removal. On the right, the hinged scale shield is open.

a wedge under the treadle arm to prevent its movement.

Flywheels or drive pulleys should be enclosed by a guard strong enough to prevent the pulley from falling to the floor if the shaft should break. In this installation, the strength and location of the guard bracket or the frame are more important safety factors than is the gage of the sheet metal used for the enclosure. Brackets should be bolted to the column of the hammer. In some instances, the guard enclosure is supported from the floor by an I-beam.

All cylinder bolts, gland bolts, guide bolts and liners, and the head assembly over the operator's working position should be restrained by wire ropes or chains (Figure 12-12).

Steam or air drop hammers should have a stop valve or quick opening and closing valve. Also, a safety head in the form of a steam or air cushion, if not already standard on the hammer, should be provided to prevent the piston from striking the top of its cylinder. The cylinder head and safety head bolts should be connected to an anchored wire rope.

If the hammer has no self-draining arrangement, a drain cock, preferably the quick acting type, should be installed in the lower part of the cylinder at the back of the hammer. This drain cock should be arranged so that it can be opened without danger to anyone, or it should be piped to discharge at a safe place. Steam lines should be well trapped.

If air or steam is used to remove scale, a quick shutoff valve should be provided so that the pressure can be regulated. The operator should adjust the scale guard to protect other employees from flying scale.

Props

Safety props equipped with handles at the middle should be provided and their use required when repairs or adjustments are to be made on the dies or when dies are to be changed. The props should be held in place while the power is released to permit the weight of the upper die and the ram to rest on the props. Operators should never place their hands on the top of a prop.

The props can either be chained to the hammer so that they cannot slip out of position or be hinged to the side of the hammer so that they are readily available and easily moved into and out of their blocking position. (Figure 12-9).

The material most commonly used for ram props is a hardwood timber not less than 4 by 4 in. (10 cm by 10 cm) in cross section with a ferrule on each end. Ram props may also be made of a section of structural shape carefully squared at the ends or a section of steel tubing not less than 2½ in. (6 cm) outside diameter with a wall thickness of ½ in. (1.3 cm). Structural shapes are usually made of magnesium or aluminum. Substitute materials must have the same load capacity and safety factor as the comparable hardwood prop.

Hand tools

Pliers, tongs, and other devices specially designed for the work to be handled should be used for feeding the material so that the operator's hands need not be placed under the hammer at any time. Tongs should be long enough that they can be held at the side of the body rather than in front. Tongs should fit the shape of the materials being held for forging.

Oil swabs and scale brushes or pipes should have handles long enough to make it unnecessary for the operator to place hands or arms underneath the die.

Figure 12-12. Safety ropes keep cylinder head, tie plate bolts, and gland bolts from falling if they break. Note the master rope which circles the base of the steam chest. The tie plate bolts and the gland bolts are secured to the master rope. The gland bolt safety rope should be tight enough to prevent a broken gland bolt from swinging down and striking the ram.

Die keys

Die keys should be made of a suitable grade of medium carbon alloy steel that has been properly heat treated so that it will not crack or splinter. Both ends of the key should be tapered for clearance in driving and removing the key. Mushroomed keys should never be used.

Die keys must be the correct length, so they do not project more than 2 in. (5 cm) in front and 4 in. (10 cm) at the back of the hammer. If necessary, shims should be used. If keys project farther, they become a hazard to the operator working in front and may break off while the hammer is operating and fall between the dies in back.

An adequate supply of die keys should be stocked so that drifts will be needed only when the end of a key becomes distorted and must be cut off before the key can be driven out. The drift must be blocked or held securely with a drift holder.

Safe operating practices

The supervisor who directs the activities of the people in the hammer crew should be made responsible for their following safe work practices.

All guards should be in place when the hammer is in operation and should be kept in good repair. Material and tools should be moved from the aisles and the operator's working space and stored in the proper place. The floor area around hammers should be free of scale, oil, water, and other material, to ensure safe footing.

Before starting work, the hammer crew should make its own inspection to see that the equipment and working conditions are in good order. A frequent check should be made for breakage at all critical points which are subject to severe strain and can therefore often fail. If an unsafe condition is found, the supervisor should be informed of it at once.

Drop hammers should never be operated when the dies are cold. Dies should always be preheated by hot steel placed between them.

No adjustments, repairs or service should be permitted until all energy sources (electric, air, steam, hydraulic and compression springs, etc.) have been reduced to zero mechanical state and locked out, the treadle blocked to prevent accidental tripping, and the ram propped.

When dies are being set on a board or steam- or air-lift gravity drop hammer, it is best to fit the dowel in the upper die and the ram with as few shims as possible. The bottom die, which should have enough shims so it can be lifted easily, should be moved first. (This procedure is the opposite of that followed in setting dies in a double-acting steam drop hammer.)

On steam hammers, a prop should be placed between the ram and the shank of the top die before the die is moved. When it is necessary to move the bottom die, the prop should be placed between the sow block and the ram on the side containing the dowels, so that the die can be moved.

On steam drop hammers, the spool bolt should not be adjusted until the main steam valve has been turned off and the treadle blocked, to prevent the ram from accidentally picking up while the operator is adjusting the bolt.

Laying liner stock between the dies to jar loose a stuck forging is dangerous. When a forging sticks, the safe method is to stop the hammer, remove the forging, relieve the die, and then continue the operation. On some operations where this method is not practicable, a safety liner made of soft steel may be used.

Whenever an operator leaves the hammer, even if only for a few moments, the upper die should be left resting on the lower die, to prevent accidental tripping.

Flywheel speeds must be carefully observed and should not as a rule be permitted to exceed the numbers of rpm's given on the specification sheets since this is the speed upon which proper operation of the press is based.

Personal protection

Operators of forging hammers and other employees in the vicinity of this equipment should wear suitable personal protective equipment—full eye protection, safety hats, foot protection, leather leggings and aprons, and hearing protection in accordance with plant policy.

Cotton fabric gloves should be worn and when wet, should be removed and allowed to dry. Because perspiration may cause steam burns, leather gloves should not be worn.

Setup and removal of dies

Dies are usually heavy and hazardous to handle without proper equipment. Uniform holes may be drilled in both sides of each heavy die block so that pins can be inserted to make lifting and moving easy. The diameter of the pin and hole will depend upon the weight of the die. Standard practice in many companies is to have the pin $\frac{1}{16}$ in. (1.6 mm) smaller in diameter than the hole. The depth of the hole and the diameter of the hole and the pin should be uniform in dies of a certain weight group to assure that there will be sufficient pressure to prevent the pin from falling out.

Transfer boards should not be used to transfer dies between the work bench and the machine. Transfer trucks, preferably the elevating type, are safer and more efficient. The top or table of the truck should be covered with sheet metal at least $\frac{1}{4}$ in. (6 mm) thick and should be securely fastened in place.

Power lift trucks or die trucks should be used for transporting and installing dies. Lift trucks should be well blocked or secured to the base of the hammer when dies are to be set or removed. Otherwise, the truck may slip away from the hammer, causing the die to slip and fall. In addition, these trucks should be kept in good working condition and the operator should check all the safety controls prior to starting the die set. The operator must be trained in the safe operation of this equipment.

When forge dies are set up or removed, the hammer operator should act as leader of the group and should see that all efforts are coordinated and all safety rules are observed so that the work will be done efficiently and safely.

Setting up dies. The immediate area around the hammer must be clean and clear of obstructions. Maintenance work should not be performed on the equipment during a die set-up.

The hammer crew should make a complete check of the equipment between setups: the hammer should be in good working order; the seats of both the sow block and the ram should be flat and clean; dowels and die keys should be inspected for galls and burrs; dies should be checked for burrs and other defects, such as cracks or sharp corners.

It is important that the overall height of the dies be greater than the shut height of the hammer.

Good lighting is essential for accurate setting of dies and will permit the operating crew a better view of potential hazards. Portable lights may be used, and should have heavy-duty cords, with bulbs protected by heavy screen guards.

If lift trucks are used, the floor should be level, in good condition, and free of obstructions. If cranes are used, the lift chains should be in good condition and the die pins should have a snug, but free fit.

Many methods are used in setting dies in hammers. The type and size of the dies and the type of hammer determine the method to be selected. The first action in setting up dies should be to prop the ram securely and to shut off and lock out the power, whether steam, electricity, or air.

Die dowels made of a grade of steel that will not splinter or crack should then be driven into the dowel holes in the die shank. Dimensions should be accurate to ensure a tight driven fit.

After the bottom die of a steam hammer has been set in place, the bottom key should be driven to help line up the die and partially tighten it. The top die should then be inverted and set in position so that the dies are face to face with the match lines lined up. This procedure should be reversed for a gravity drop hammer: the top die is set and keyed first. Sometimes both dies can be set at once.

The safety prop should be removed from between the ram and sow block at this time, and the ram allowed to descend slowly until it engages the top die.

If shims are used on the dowel in the top die, an extra hazard is created. Normalized spring steel is used to shim dowels, which must be set so that they will fall into place when the ram engages the die. The number of shims and their location (whether front or back) should be recorded so that succeeding shifts or different hammer crews can refer to the record of the setup for that specific set of dies.

If a die must be moved to match, a prop should be used after the ram is raised and before the operator reaches under the hammer to reset the shims. This prop must be strong enough to support the ram and long enough to extend from the top of the die to the ram.

If allowance is made for moving the dies, the allowance on a steam hammer should be made in the top die only and the bottom die should have a tight fit. On gravity drop hammers, the general safe practice is to have the top die tight, allowing for movement in the bottom die.

At this stage of the setup, the die keys should be driven with a hand sledge only. Common practice is to drive the keys up tight with a sledge or light ram, and then "bounce" the dies. The safe procedure for bouncing is as follows: bounce, shut off the power, ram the key, bounce, shut off the power, ram the key, etc. The impact helps align the dies, but creates an additional hazard if lock dies are being set.

Extra precautions and special equipment may be required for abnormally large or long dies. In setting such dies, the regular safety procedures for propping and handling may have to be changed. Any change should have approval of management.

After the die keys have been driven and before adjustments are made to the gibs or column wedges, heaters should be applied to the dies if they have not been preheated. On deep impression jobs, it is good practice to preheat dies in special low-temperature furnaces, in hot water baths, or with hot scrap steel before they are set up. Driving die keys too tight when the dies are cold may crack the shanks, sow blocks, or rams.

The hammer crew should use any waiting time to make a final check and get ready for production. The dies should be checked for proper alignment and for proper wing clearance. (Tight wings may cause breakage.) A check can be made for necessary tools, and they can be put in their proper work stations. Scale guards can be moved into position, and final adjustments can be made to the billet heating furnace.

After the dies have been heated to approximately 300 F (149 C), die keys should be driven tight again by means of either a pneumatic ram or a light suspended ram (Figure 12-10). If any further adjustment to the hammer is required, it can be done after a tryout forging has been made.

Removing dies. Before dies are removed, the immediate area around the hammer should be cleared of overhead trolleys, suspended tongs, portable conveyors, tool and billet stands, and other equipment.

Overhead trolleys should be tied down so that they will not creep back into the working area. The scale guard should be moved back, and accumulated scale that would interfere with safe footing should be removed. Forgings should be moved away from the unit immediately and sent to the next work station.

If another set of forging equipment has been delivered, it should be placed nearby but not directly in the area where the hammer crew will work. Service personnel, such as truckers, crane operators, and hookers, should be familiar with the proper procedure so that unnecessary handling can be eliminated.

The hammer's energy sources (electrical, air, steam, hydraulic) must be shut off and locked before the die keys are loosened. The top key is generally loosened first, usually with a mounted pneumatic ram (Figure 12-10). A light, well-balanced ram suspended from a cross beam or from an overhead crane or chain fall can also be used successfully.

Using a manually held drift pin or a knockout on a die key after it has been loosened and driven to a position even with the face of the ram or the sow block is a hazardous operation. A special type of adjustable knockout (Figure 12-13) that is held in position mechanically rather than manually should be used.

After the die keys have been driven out, the ram should be raised and propped at once. The prop must be in good condition and must be placed on a clean surface. On a gravity drop hammer, a jack should be used and a special prop may be required.

No attempt should be made to raise the hammer to propping level if the top die has a tendency to "hang." The die should be freed first within the shortest possible distance from the face of the bottom die. The ram on an air-drop hammer should be propped with special care. After the prop is positioned securely under the ram, the power should be shut off and locked out.

Special platform trucks with winches are employed for this operation. They are practical and safe because the dies are winched out or pulled out horizontally directly onto the table of the truck. Dumping the dies out of the hammer onto the floor should not be permitted.

After the dies have been removed from the hammer, the dowels must be extracted. It is important that the two workers who drive out the dowels have proper tools, usually a drift and a sledge. These tools should be in good condition and have sound handles. Because there is metal-to-metal contact, the workers must be extremely careful and properly protected.

After being removed, dies should be loaded on low steel pallets and taken from the vicinity as soon as possible. If dies need repair or modification, the hammer operator should notify his supervisor. The supervisor can then send this information to the die-servicing department so that the condition can be remedied before the next run.

Design of dies

Hammer dies are usually made of chrome, nickel, or molybdenum stellite—materials that have high resistance to heat, shock, and abrasion. Die blocks are supplied commercially in four different hardnesses. Selection of the proper die steel in the correct range of hardness is of utmost importance in controlling checking and breakage of dies.

Size, amount of striking surface, and height are all pertinent factors in the safe design of dies.

The correct amount of striking surface must be allowed in relation to the size of the die. Too little striking surface may cause breakage or an undersized forging when the dies pound down. Too much striking surface, especially if it is unbalanced, may cause a pull or misalignment.

Correct die height must be specified, especially for resinking forge dies. Normal work level should be maintained for the benefit of the hammer operator, and the height should never be below a specific minimum, to prevent the bottom of the cylinder from being knocked out. Maximum and minimum operating heights should be labeled on each unit for use by the design engineer.

The dies must be made so that they meet in precise alignment. Layout of the dies should provide for the major portion of the heavy forge work to be done in the center of the die under the center of the ram, where the maximum hammer force is transmitted. Each impression in the die must be backed up with enough die material to reduce the possibility of breakage, especially where multiple impressions or nesting methods are employed.

Preliminary or breakdown operations must be designed so that they accomplish the specific purpose and must be systematically arranged so that they do not create a hazard for the operator as the forging cycle is completed.

Radical bends or severe reductions in volume that might tend to jerk the tongs from the hands of the operator should be avoided. Such operations should be modified or completed in additional operations.

The thickness and width of flash, gutter, and sprue should be ample so that flash or tong holds are not sheared off. The size of the gates is important. They should be designed in relation to the size of the stock and the tongs used. Gates should have enough width, depth, and clearance to allow safe handling.

Some dies, especially for smaller hammers, are designed with cutoffs which shear the completed forging from the end of the bar. If possible, such cutoffs should be placed on one of the rear corners of the die for the operator's safety. If cutoffs are placed on the front of the die and are used by placing the stock across the knife portion at an angle, enough clearance must be provided between the die and the hammer frame or gib.

Hammer die maintenance is important because of the nature of forging work and the abnormal abuse to which the dies are subject. For example, fillet and corner radii in the impressions may be enlarged or sharpened by impact pressures, scale, abrasion, and wear. It is therefore necessary to inspect these radii and correct any alteration that may be hazardous.

Sometimes the face of a forging die may be welded either to correct some defect or to maintain specifications. A hard, brittle, or thin-skinned weld can become a flying hazard under impact. The rod and the preheating and postheating methods should be selected with extreme caution.

Provisions for storing dies, such as racks and rails, are essential to safety, good housekeeping, and efficiency.

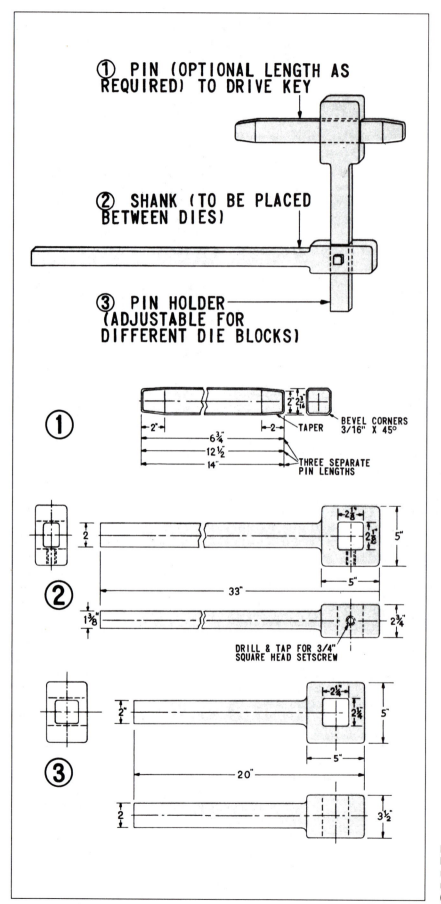

① PIN (OPTIONAL LENGTH AS REQUIRED) TO DRIVE KEY

② SHANK (TO BE PLACED BETWEEN DIES)

③ PIN HOLDER (ADJUSTABLE FOR DIFFERENT DIE BLOCKS)

①
BEVEL CORNERS 3/16" X 45°
TAPER
2"
6¾"
12½"
14"
THREE SEPARATE PIN LENGTHS

②
2
2⅛"
2⅛"
5"
5"
33"
1⅜"
2¾"
DRILL & TAP FOR 3/4" SQUARE HEAD SETSCREW

③
2
2¼"
2¼"
5"
5"
20"
2
3½"

Figure 12-13. A key knockout, which can be made easily, is held in position mechanically. The only machining required is to drill and tap the ¾-in. (2 cm) hole for the setscrew.

Dies preferably should be stored in an area separate from the forge shop and away from vibration.

Maintenance and inspection

A well-planned preventive maintenance program for forging hammers will help to reduce the number and severity of accidents by minimizing part breakage and wear. Regular inspections will disclose production units that are not operating up to standard so that repairs or adjustments can be made.

The results of a good maintenance program can be measured in reduced operating costs which include:

1. Cost of machine downtime, breakage, and lost production
2. Cost of replacement parts and labor
3. Cost of accidents due to faulty equipment

Maintenance checklists for hammers (Figures 12-14 and 15) can be the basis for formulating a definite, planned inspection program. A written checklist avoids the errors resulting from verbal reports that are often forgotten or misunderstood.

A work schedule for repairs may be set up on the basis of the data recorded on the checklists. The data may be transferred to the permanent records of the equipment and used for future planning in the maintenance program as well as for comparison of costs.

Steam hammers constitute a considerable portion of the forging equipment. Many hammers are not kept in as efficient condition as possible, usually because management fails to realize

BOARD HAMMER MAINTENANCE CHECK

Date_____ Hammer No._____ Location_____

ITEM CHECKED	CONDITION	TYPE OF REPAIR	EST. HRS. TO REPAIR
TIE BARS & SPRINGS			
FRAME STUDS & SPRINGS			
DIE KEYS & SHIMS			
SOW BLOCK KEY & SHIMS			
RAM CLEARANCE			
GUIDE BOLTS & ADJ.			
MOTOR MOUNTS			
MOTOR COUPLING			
ROLLSHAFT BEARINGS			
WIRING & CONTROLS			
FLYWHEEL & BEARINGS			
DRIVE GEARS			
ROLL ADJUSTMENT			
LUBRICATION			
STEAM LINES			
BOARDS & WEDGES			
AIR LINES			
AIR FOOT SWITCH			
AIR CYL. & LINKAGE			
TREADLE & LINKAGE			
BOARD CLAMPS & LINKAGE			
DOGS & STOPS			
KNOCKOUT ARM			
FRICTION RODS			
SAFETY RODS			
REMARKS:			

Figure 12-14. Maintenance checklist for board drop hammers.

STEAM HAMMER MAINTENANCE CHECK

Date_____ Hammer No._____ Location_____

ITEM CHECKED	CONDITION	TYPE OF REPAIR	EST. HRS. TO REPAIR
CYL. HEAD BOLTS			
MOTION VALVE STEM			
MOTION VALVE CRANK			
MOTION VALVE CONNECT.			
WIPER BAR & CRANK			
THROTTLE CRANK			
THROTTLE LINKAGE			
RAM & SOW BLOCK			
DIE KEYS & SHIMS			
SOW BLOCK KEYS & SHIMS			
GUIDE BOLTS			
GUIDE WEAR			
GUIDE ADJUSTING BOLTS			
GUIDE WEDGE POSITION			
HOUSING BOLTS & SPRINGS			
TREADLE			
TIE PLATE LINER			
PISTON ROD GLAND PLATE			
GLAND BOLTS			
TREADLE PLATFORM			
STEAM CUSHION LINE			
SCALE HOSE			
SAFETY LINER			
SAFETY PROP			
STEAM LINES			
STEAM SHUTOFF VALVES			
REACH ROD			
COLUMNS FLAT ON BASE			
COLUMN WEDGE & BOLTS			
SPOOL BOLT & PIN			
SAFETY CABLES			
CRACKS IN BASE			
REMARKS:			

Figure 12-15. Maintenance checklist for steam hammers.

that operating and upkeep costs are higher for units in poor condition. Waste of steam usually results from worn piston rings or sleeves, loose heads, blown head gaskets, and leaky glands.

Replacement of worn rings is good economy. Worn piston sleeves and sloppy linkage make the hammer hard to control and create a hazard. Loose cylinder heads also are dangerous.

Frequent periodic inspection of every forging hammer will help ensure proper condition of bolts, screws, keys, valves, and other parts which may be loosened by vibration. Similarly, thorough periodic inspection and adjustment should be made of all parts of the treadle or pedal, clutch, and other operating mechanisms.

Worn or loose treadle linkage, motion arm, crank arm, and treadle can cause the hammer to go out of control. These parts especially should be kept in good repair.

The clutch is a vital part of the forging press and must be kept in good condition if the press is to operate satisfactorily. A broken spring or part which shows wear should be replaced at once.

Lead casts

Lead casts are taken in practically every conceivable manner in the forging industry. Casts should be taken only in an isolated area, if possible, where there is no likelihood of interference from or injury to other workers. The die impressions should be dry since hot metal that contacts water produces flying particles of molten metal.

Lead pots should be ventilated properly. See the Council's *Fundamentals of Industrial Hygiene.*

FORGING UPSETTERS

The upsetter is a horizontal forging machine which forges hot bar stock, usually round, into a great many forms by a squeezing action instead of impact blows, as in the case of forging hammers. Although numerous hazards are involved in the operation of an upsetter, the most serious problems are encountered in changing the dies.

The entire machine should be enclosed as much as possible, except for the feed area. Heavy wire mesh or expanded or sheet metal reinforced with structural steel should be used. Doors may be cut into the enclosure for servicing flywheel, brake, and other moving parts. A guard should be installed over the operating pedal.

Safe operating conditions call for the area around the machine to be clean and clear of obstructions and litter. It is especially important that the top of the machine be clear of any objects, such as loose bolts, bars, nuts, or shims, which might fall into it or from it.

The operator should shut off the power, lock the main power switch, and after the flywheel has stopped completely, immobilize the flywheel before attempting to adjust dies, heading tools, stock gages, or backstops.

Design of dies

Dies and heading tools used in an upsetter or horizontal forging machine are not usually subjected to such severe abuse as are hammer dies. A good grade of chrome, nickel, or molybdenum steel of the correct hardness is recommended for gripper dies.

It is common practice in die shops when resinking upset dies to use inserts rather than to cut down the dies. The inserts most commonly used are made of a high chrome, nickel, or molybdenum steel with silicon added.

Tool steels with a vanadium and tungsten analysis are also used for inserts and headers, but they must be selected to satisfy operating conditions because of their limited use where coolants are employed.

Front or back stock gages or backstops should be designed and located to serve the specific purpose and should not be a handicap to the operator. Special care should be taken to eliminate all hazards, especially pinch points.

Jobs which are abnormally heavy or which would create an unbalanced condition when running should be designed to employ balancing equipment to facilitate handling and reduce operator fatigue. The grip sometimes provided on upset die impressions should be sufficient to hold the stock securely, and should be checked after every run. This precaution is important for the operator's safety, especially on jobs where the heading tool could push the stock bar out of the impression toward the operator.

Setup and removal of dies

At the end of a run and before further work is done, all skids of stock or forgings should be moved out of the area to allow as much room as possible when the dies are changed.

Die setters should not try to make a complete setup dimensionally by measuring headers, strokes, and dies, without inching the header slide forward. This practice should be prohibited particularly on a worn machine where special shimming is required for proper alignment.

The proper sequence and procedure are as follows:

1. Check to make sure that all tools of correct size and other equipment are at hand.
2. Check dies and headers for defects which could develop into hazards.
3. Shut off all power and lock the master switch off.
4. Turn off the water or header lubricant and set the air brake. If the upsetter does not have an air brake, wait until the flywheel has stopped completely.
5. Loosen all setscrews, lock nuts, and hold-down bolts on the die clamps. Do not remove hold-down bolts or die clamps until the die has been secured to the crane, hoist, or other device that will be used to remove the die. Even then, care should be taken to avoid putting oneself into a hazardous position concerning falling or swinging dies.
6. Remove the dies by means of a swivel or arm crane. (Eyebolt holes 2 in. (5 cm) deep should be drilled into all dies. Shouldered eyebolts or swivels should be stocked only in ¼-in. (6 mm) graduations to a minimum of ¾ in. (2 cm) in diameter.)
7. Remove all die packing and thoroughly clean the die set seat; inspect for burrs, especially along the die keys.
8. Remove headers and dummies and any shims which may have been used.
9. Measure the new dies to determine the amount of packing needed. At least an ⅛-in. (3-mm) liner should be used to help protect the die seats.
10. Set in the new stationary die. Check to make sure that it is seated properly and packed correctly.
11. Tighten the holddowns by hand before unhooking the die from the lifting device.
12. Turn on the power, open the safety, release the brake and the flywheel. Set the machine on "inch" and slowly inch the header slide forward to bring the tool holders into correct position for the headers.
13. Disconnect the power and lock out the master switch. Make certain that the flywheel has completely stopped.
14. Assemble the new headers, tool dummies, and tool holders. Be sure that the headers to the die cavity are correctly matched and on correct center. It is good practice to set up according to a die layout which shows all principal dimensions on the equipment.
15. Check for the need to use shims behind the headers. Shims should be the washer type, not the horseshoe type, which fit around the shank of the header and cannot fall out.
16. Make a complete check of the assembly before finally tightening the header and dummy setscrews.
17. Insert the moving die, make sure that it is properly located, and then tighten the holddown bolts, before die is unhooked from the lifting device.
18. Turn on the power. Set the machine on "inch" and slowly close the dies.
19. Check for match alignment and proper amount of packing. If too much packing is used, the safety pin should open the dies.

20. Allow for expansion of the dies when they become hot. It may be necessary to remove a shim from back of the die.
21. If everything checks out, shut off the power, lock out the master switch, and allow the flywheel to come to a complete stop.
22. Tighten hold-down bolts, lock nuts, and setscrews.
23. If needed, attach a front gage or backstop gage.
24. Turn on the power and try out one forging. Check dimensionally. If dies or headers need further adjustment, again turn off the power, lock out the master switch, and wait until the flywheel comes to a complete stop.

Inspection and maintenance

Because worn or defective upsetters can be dangerous to operate, these machines must be kept in excellent working order. A definite program of inspection and maintenance is essential.

The maintenance crew should check all working parts for wear and proper adjustment at least once a week. *The air clutch and brake should be inspected daily.*

The operator should also make daily inspection of air gages, air lines, water lines, water valves, belts, pulleys, and tools. The performance of the machine should be checked on a daily basis and any abnormal functioning reported immediately.

All equipment for handling dies, such as chains, cables, and eyes or swivels, should be inspected periodically as well as at each use.

Upsetters require daily lubrication. If possible, automatic lubrication should be installed.

Auxiliary equipment

All devices essential for safe operation, such as stock gages, should be designed for the particular job. There are three basic types of stock gages: the front gage (locates and swings away), the backstop gage (locates and helps hold the stock fixed), and the special tong gage or finger gage (locates and helps control the stock).

Tools should be in good condition. A complete set of wrenches to fit all sizes of nuts or bolts on the machine should be provided.

The jaws of tongs should conform to the shape of the stock being handled. Tongs should be made of tough, low-carbon steel so that they will not harden from repeated quenching in water.

Oil swabs and scale removers should have handles long enough to enable the operator to reach the full length of the dies without having to put an arm or hand between the dies.

Air, electrical, water, and oil lines should have distinctly marked shutoffs. Safety valves or switches should be located in a spot convenient for the operator.

FORGING PRESSES

The forging press, because of its basic design, is similar to the power presses discussed in Chapter 11, Cold Forming of Metals. However, since the work it performs is quite different from the conventional cold stamping operation, it has its own operating technique, die setup, and maintenance problems.

The increase in use of the forging press has been greater than that of any other type of forging equipment. Presses range from 500-ton capacity to 6,000-ton capacity (454 to 1,360 kg).

Compared to the cold stamping press, the forging press is designed with a faster-acting slide. The speed of the downward motion and the pick-up of the slide is one of the factors which determine the life of the forging dies. Another is control of the temperature of the dies. It is therefore important that the action of the press be fast enough to minimize the length of time that the dies are exposed to the forging temperature of the billet.

Basic precautions

This rapid action of the slide creates certain hazards which must be recognized and controlled by the operator. The size and shape of the forgings and the method of moving them into and out of the dies limit application of point-of-operation guarding. The most important single factor in the prevention of accidents, therefore, is the operator's control of the tools and methods used.

Because of the temperature of the forging dies and the forging temperature of the billet being forged, the operator must move the billet into the dies by means of tongs or special handling tools.

Tong handles should not be held in front of the body, and the fingers should not be placed between the handles. Die swabs must have handles long enough to keep hands and arms from between the dies.

Tongs, special handling tools, and die swabs are ordinarily the only tools used by an operator during the forging operation. It is important that they be kept in good condition at all times.

Proper clearance must be maintained at the front of the press. The operator must have enough hand room to allow for upward or downward motion of the tongs, which can be caused by improper spotting of the billets or tongs at the striking surfaces of the forging dies.

Steam lines, air lines, water headers, and scale and oil splash aprons must be properly located. If too near the working area of the operator's hands, they create pinch and shear points when tong handles are forced against them.

Unless properly confined, hot scale produced during the forging operation can cause serious burns and eye injuries. Air or steam curtains at the front of the die and directed onto the die facing will prevent scale from coming out the front of the press. Combination scale and smoke exhaust hoods should be installed at the back of the press to confine the scale and exhaust the smoke created by die lubricant.

Safety hats, hearing protection, eye protection, and foot protection should be worn by operators at all forging press operations.

All forging presses should be equipped with pedal guards and nonrepeat devices. A forging press should never be operated on continuous stroke. The operating controls should require depression of the pedal for every cycle of the slide. The press controls should also permit inching of the slide for die setting or other press adjustments.

Die setting

Because the setting of dies in forging presses is very different from the setting of dies in cold stamping presses, certain extra precautions not required on the latter type must be taken.

Forge press dies are set in die holders designed in sets, each set consisting of an upper and a lower section. The upper section of the holder is secured to the press slide by threaded bolts, and the lower section to the press bed by threaded bolts. The forging dies are set or nested in die pockets recessed in both the upper and lower sections of the holder.

The problem of setting the dies in the recessed pockets and of removing them can be solved by the use of special equipment consisting of an eyebolt attachment at the face of the die and a die truck with boom attachment. Pry bars and blocking should be carefully used since the pry bar may easily slip from the die and injure the operator's hands or fingers.

The dies are secured in the pockets of the holder by sectional flat clamps and a series of cap screws through the holder at all four sides of the die. These screws are used primarily to shift the die for matching and to prevent the die from floating after it has been properly aligned.

Suitable blocking should be installed to hold the top die in the pocket before the clamps which secure the top die in the holder are removed. The adjustment cap screw should not be relied upon to hold the die after the clamps have been removed.

The press should be equipped with safety props interlocked with the press motor circuit so that the motor cannot be started when the props are in use.

All die clamps, bolts, washers, blocking, wrenches, and handling equipment should be in good condition at all times.

The table wedges which are used for making vertical adjustment of the dies must be kept free of scale so that the wedge can be raised and lowered with minimum pressure. Trucks or driving rams should not be used for this operation. If wedges are kept free of scale, they can be raised or lowered by hand or by air motor wrench.

Maintenance

Maintenance and servicing of forging presses requires the same precautions used in maintenance and service work on power presses used in cold stamping. The operating energy source should be locked out so that the equipment cannot be energized while work is being performed. Safety props should be placed under the ram and surge tanks and pressure lines should be bled so that unanticipated actuation of the press parts will not occur. If adjustments have to be made while the press is energized, the work must be carried out under the direct supervision of the maintenance supervisor. Good communication is a must and each person must use caution to avoid hazardous positions.

Suitable permanent work platforms should be installed for making brake adjustments and doing repair work at the surge tank and booster cylinders. To prevent falls, maintenance crews should not use portable straight ladders nor stand on parts of the press, such as the press crown or the backshaft.

Major repair work on forging presses usually requires removal of bulky, heavy parts. Tearing down of the rolling type clutch, flywheel, slide, pitman, and crankshaft requires special heavy-duty rigging used by skilled workers trained in this type of work.

The brake should be adjusted properly and kept in good repair. Precautions must be taken to prevent the brake lining from becoming flooded with oil. A sheet steel disk about 2 in. (5 cm) larger than the brake wheel, placed on the eccentric shaft back of the brake wheel, will help reduce the amount of lubricant on the brake.

Pitman bolts should be securely tightened. It may be necessary to design and make a socket or box end wrench with a strong heavy handle for the specific press.

Flywheel hubs, clutch spline hubs, pinion gears, and brake wheels should not be permitted to become loose on the pinion shaft. If they do become loose and run that way very long, keys and keyways will be ruined and the keys will break out a por-

tion of the pinion shaft on one side of the keyway. The keys on each side of the pinion shaft should be inspected periodically and driven tight if they are loose. When a key will no longer stay tight, it should be replaced with a carefully fitted new key.

When the hub or gear on the shaft becomes too loose, it will be impossible to hold the keys tight and the shaft will have to be turned down. The inside of a steel hub should be welded and bored to a shrink fit on the new shaft size. The keyways should be remachined and new keys made. If the hub is cast iron, a tapered sleeve can be made or a new hub fitted to the shaft.

Experienced repair crews warn against welding the flywheel end of the pinion shaft. If normalizing of the welded section is not complete, a crack may start and allow the shaft to break and drop the rotating flywheel.

The friction slip on a press should be tightened so that it will not creep even on heavy jobs. If the friction slip is allowed to move a small amount with every forging, this movement will polish or glaze the friction surfaces and soon the friction slip will move more with each forging. This movement will cause the friction surface to score the hub and clamp surfaces.

Rotation of the flywheel on the friction hub will frequently wear or cut the inside diameter of the flywheel so that it runs off center. To correct this condition, the inside diameter of the flywheel can be bored and fitted with a bronze bushing.

Sometimes the eccentric shaft becomes cracked in the fillet on each side of the eccentric. This defect is usually noticed when the shaft is taken out so that the main bearings can be checked or the shaft can be machined. This crack should be examined carefully to determine its length, depth, and direction of travel. If the crack is in a critical area where failure would permit the rolling clutch to drop off, then the shaft must be replaced.

OTHER FORGING EQUIPMENT

Hot trimming dies can be made from hardened chrome, nickel, carbon, and vanadium steels or from medium carbon steels with cutting edges which have been hard faced with a rod similar to a Haynes Stellite No. 6 rod.

Stellite should be used on the trimmer only, and not on both the trimmer and the punch. A stellite punch should never work against a stellited punch die. One or the other, preferably the punch, should be softer.

Cold trim dies and punches are usually made from a high carbon, high chrome, molybdenum steel hardened and drawn to a 60 to 62 Rockwell C hardness. In both hot and cold trimmers it is important for the designer to equalize the trim so that stresses are equally distributed.

Proper working clearance must be maintained to facilitate correct location for unloading. If possible, cold trimmers should be designed to work on guide pins for proper alignment.

Padding, bending, or straightening equipment is sometimes required to complete a forging. This equipment can be made from a good grade of wear-resistant carbon, nickel steel. Hot pad dies are usually made with an opening between the die faces so that the hot forging acts as a cushion. Additional clearance should be provided in a hot pad or restrike die because the forging is constantly cooling and shrinking.

Cold coin dies are made from high carbon, high chrome steels hardened to 62 to 64 Rockwell C hardness. The forging is coined cold. For best result, either interlocking dies or dies with guide pins are used. Additional safety measures such as magazine type loaders are sometimes made an integral part of the die for the operator's convenience.

The faces of the coining dies should be ground as smooth as possible. No lubricant should be used because it may cause the forging to stick to one die face. The opening between the dies is sometimes limited mechanically so that no more than one forging can be loaded at a time.

Bulldozers. The greatest danger in the operation of a bull-dozer is the possibility of a worker's being caught between the dies. There are several ways to decrease this hazard. A guard may be attached to the side of the moving head and travel with it past the stationary head, or telescoping rods or rails may be used to serve the same purpose. The base plate can be notched out to leave room for the operator's leg. The clutch should be kept in good order so that the machine will not repeat. The power transmitting mechanisms should be guarded.

Cold heading and similar machines should have screen shields to protect workers from flying pieces. Relief springs should be guarded to prevent the bolts and nuts from being thrown out in case of breakage.

Bolt headers and riveting machines should have treadle guards to prevent accidental operation. The machines should be stopped and blocked before dies are changed or adjustments are made.

Hot saws should be provided with tanks of water placed below the saws and 8-in. (20-cm) sheet metal guards positioned to stop the flying sparks.

REFERENCES

Forging

American National Standards Institute, 1430 Broadway, New York, N.Y. 10018. *Safety Requirements for Forging Machinery,* B24.1.
National Safety Council, 444 North Michigan Ave., Chicago, Il. 60611.
Safeguarding Concepts Illustrated.
Industrial Noise and Hearing Conservation.
Industrial Data Sheets

Handling Materials in the Forging Industry, 551
Handling Steel Plate for Fabrication, 565.
Mechanical Forging Presses, 728.
Setup and Removal of Forging Hammer Dies, 716.
Steam Drop Hammers, 720.
Upsetters, 721.
United Auto Workers, 8000 E. Jefferson Ave., Detroit, Mich. 48214. (General)

Foundries

American Foundrymen's Society, Golf and Wolf Rds., Des Plaines, Ill. 60016.
Engineering Manual for Control of In-Plant Environment in Foundries.
Health Protection in Foundry Practice.
Safety in Metal Casting.
Recommended Practices for Grinding, Polishing, and Buffing Equipment.
American National Standards Institute, 1430 Broadway, New York, N.Y. 10018.
Safety Requirements for Floor and Wall Openings, Railings, and Toeboards, A12.1.
Safety Requirements for Melting and Pouring of Metals in the Metalcasting Industry, Z241.2.
Safety Requirements for Sand Preparation, Molding, and Coremaking in the Sand Foundry Industry, Z241.1.
Safety Requirements for the Cleaning and Finishing of Castings, Z241.3.
Ventilation Control of Grinding, Polishing, and Buffing Operations, Z43.1.
National Safety Council, 444 North Michigan Ave., Chicago, Ill. 60611.
Fundamentals of Industrial Hygiene.
Industrial Data Sheets
Coated Abrasives, 452.
Magnesium, 426.
Industrial Noise and Hearing Conservation.

Nondestructive testing

McMaster, Robert. *Nondestructive Testing Handbook.* New York, Ronald Press.
National Safety Council, 444 North Michigan Ave., Chicago, Ill. 60611.
Industrial Data Sheets
Beta Particle Sealed Sources, 461.
Ultrasonic Nondestructive Testing for Metals, 662.

13

Welding and Cutting

THE DETAILED SPECIFICATIONS FOR WELDING AND CUTTING operations are available in standards listed in the reference section of this chapter. In addition, city, state or provincial, and federal regulations, where applicable, should be consulted.

This chapter defines "welder" and "welder operator" as applicable to the individual only. The machine performing the welding operation is referred to as the "welding machine." Equipment supplying current for electric welding is referred to as either a "welding generator" or a "welding transformer." These definitions are in accordance with the American Welding Society's definitions.

HAZARDS

The hazards generally associated with welding are hot sparks, arc radiation, air contamination, electrical shock, chipping slag, and the handling of compressed gases. In addition there is also the potential for fire or explosion in the welding area which puts the operator at risk.

Air contamination

The most significant hazard in the welding process is the generation of fumes and gases. The amount and type of fumes and gases involved will depend on the welding process, the base material, the filler material, and the shielding gas, if any. The toxicity of the contaminants depends primarily upon their concentrations, and upon the physiological responses of the human body. Sampling by a qualified person may be necessary to fully identify the fumes and gases actually being given off in a specific operation. A listing of substances that may evolve while welding commonly welded metals and some of their effects upon the human body are described.

GASES

Effects of toxic gases

Exposure to various toxic gases generated during welding processes may produce one or more of the following effects:
1. Inflammation of the lungs (chemical pneumonitis).
2. Pulmonary edema (swelling and accumulation of fluids).
3. Emphysema (loss of elasticity of the lungs). A very small percentage of emphysema is caused by occupational exposure.
4. Chronic bronchitis.
5. Asphyxiation.

The major toxic gases associated with welding are classified as primary pulmonary and nonpulmonary.

Primary pulmonary gases

Ozone. Ozone is formed by electrical arcs and corona discharges in the air or by ultraviolet photochemical reactions. Welders who have had a severe acute exposure at an estimated 9 ppm of ozone plus exposure to other air pollutants develop pulmonary edema. Ozone may be a problem when gas-shielded metal arc welding or arc gouging is conducted in enclosed or confined areas with poor ventilation.

Oxides of nitrogen. Oxides of nitrogen are very irritating to the eyes and mucous membranes. Exposure to high concentrations may immediately produce coughing and chest pain. Death may occur within 24 hours of development of pulmonary edema.

Phosgene. Phosgene is produced when metals that have been cleaned with chlorinated hydrocarbons are heated to the temperatures used in welding. Inhalation of high concentrations of gas will produce pulmonary edema frequently preceded by a latent period of several hours' duration. Death may result from respiratory or cardiac arrest.

Phosphine. Phosphine or hydrogen phosphide is generated when steel that has been coated with a phosphate rustproofing is welded. High concentrations of the gas are irritating to the eyes, nose, and skin.

Nonpulmonary gases

Carbon monoxide. In some welding processes, carbon dioxide is reduced to carbon monoxide. In the case of carbon dioxide-shielded metal arc welding, carbon monoxide concentrations exceeding the recommended levels have been detected in the fumes near the arc; however, concentration decreases rapidly farther from the arc. With adequate ventilation, the carbon monoxide concentration in the welder's breathing zone can be maintained at harmless levels.

Carbon dioxide. Carbon dioxide is not usually considered a toxic gas. It is present in the atmosphere in a concentration of about 300 ppm and is present in somewhat higher concentrations in occupied structures. The air in the lungs contains about five-and-a-half percent carbon dioxide.

Particulate matter

The deposition of particulate matter into the lungs is called benign pneumoconiosis. Benign pneumoconioses associated with welding are anthracosis (carbon), stannosis (tin), siderosis (iron), and aluminosis (aluminum).

Pulmonary irritants and toxic inhalants

The majority of the metal components contained in welding fumes do not produce radiographic changes in the lungs and, depending on the definition of pneumoconiosis, can be classified in the separate category of pulmonary irritant or toxic inhalants. These materials include cadmium, chromium, lead, magnesium, manganese, mercury, molybdenum, nickel, titanium, vanadium, zinc, and the fluorides.

Aluminosis. The inhalation of aluminum, aluminum oxide, and aluminum hydrate does not injure the pulmonary system, according to a review of the subject by the Kettering Laboratory. This conclusion is supported by both industrial experience and animal experimentation. Most of the complications associated with the inhalation of aluminum are probably produced by some co-inhalant such as silica.

Anthracosis. The term anthracosis refers to a blackish pigmentation of the lungs caused by the deposition of carbon particles. Some investigators still feel that this pneumoconiosis is due to impurities in coal such as silica. In most instances, inhalation of pure carbon does not produce a significant lung tissue reaction.

Berylliosis. The only nonfibrotic type of harmful pneumoconiosis is associated with the inhalation of beryllium. Inhalation of the beryllium dust or fumes may result in an acute or chronic systemic disease. Chronic beryllium poisoning has been reported as resulting from exposure in plants producing beryllium phosphors, in beryllium-copper founding, in ceramics laboratories, in metallurgical shops, and in plants producing beryllium compounds from the ore.

This disease has been reported as occurring among individuals exposed to atmospheric pollution in the vicinity of plants processing beryllium and in persons dwelling in the same household as beryllium workers. Inhalation of the dust of beryl, the beryllium ore, has produced to date no known cases of acute or chronic beryllium poisoning.

Cadmium. Operations that involve the heating of cadmium-plated or cadmium-containing parts, such as welding, brazing and soldering, may produce a high concentration of cadmium oxide fumes. Inhalation of these fumes may cause respiratory irritation with a tender, sore, dry throat and a metallic taste, followed by cough, chest pain, and difficulty in breathing. The liver or kidney and the bone marrow may be injured by the presence of this metal.

A single exposure to cadmium fumes may cause a severe lung irritation that can be fatal. Most acute intoxications have been caused by the inhalation of cadmium fumes at concentrations that did not produce warning symptoms of irritation. Continued exposure to lower levels of cadmium has resulted in chronic poisoning characterized by irreversible lung injury and urinary excretion of specific low molecular weight protein that may be associated with kidney impairment.

Chromium. The oxidation of chromium alloys can produce chromium trioxide fumes that are often referred to as chromic acid. These fumes react with water vapor to form chromic and dichromic acid. Contact with these fumes can produce small, painless cutaneous ulcers as well as dermatitis from primary irritation or allergic hypersensitivity. Inhalation of these fumes can produce bronchospasm, edema, and hypersecretion, bronchitis, and hyper reaction of the trachea. Perforation of the nasal septa can also occur.

Copperosis. Inhalation of copper fumes has been reported to produce fibrosis in animal studies as well as symptoms of metal fume fever in welders. In chronic exposure, the liver, kidneys and spleen may be injured and anemia may develop, although chronic poisoning like that from lead poisoning is unknown. Excessive exposure can cause nasal congestion and ulceration with perforation of the nasal septum.

Fluoride. The use of electrodes with fluoride-containing coatings offer a definite hazard. Fluoride compounds—fumes and gases—can burn eyes and skin on contact. When excessive amounts are inhaled, excretion lags behind the daily intake, resulting in a buildup of fluorides in the bones. If storage of fluorides continues over a sufficiently long period, the bones may show an increased radiographic density.

Lead. Lead poisoning in industry is almost always a result of inhalation of lead fumes from lead-containing materials or materials protected by lead-base paints. Red blood cell stippling and a formation of a lead line in the gums are the chief findings. Lead poisoning is confirmed by a blood and urine analysis.

Magnesium. The oxide fumes from magnesium can produce metal fume fever, which may result in irritation of mucous membranes. Experimental work with animals has failed to show any detrimental response in the lungs.

Manganese. The fumes from manganese are highly toxic and can produce total disablement even after exposure as short as a few months. Such exposure is usually caused by inhalation of manganese dioxide dust. Chronic exposure will cause tiredness, uncontrollable laughter, and micrographia. Manganese exposure is noted for the mask-like face reminiscent of parkinsonism, and may produce subacute edema of the respiratory system.

Mercury. The welding of metals coated with protective materials containing mercury compounds produces mercury vapors. No pulmonary effects from mercury vapors have been noted.

Molybdenum. Little information is known concerning the exposure of humans to molybdenum or its compounds. Animal studies indicate a lower order of toxicity.

Nickel. Nickel and its compounds are carcinogenic and toxic. A significant increase in cancer of the lungs and sinuses has occurred among employees in nickel smelting and refining plants. Nickel fumes have been known to cause severe pneumonitis. Nickel carbonyl can produce cyanosis, delirium, and death between four and 11 days after exposure.

Siderosis. Siderosis is a benign pneumoconiosis resulting from the deposition of inert iron oxide dust in the lung. In general, there is neither fibrosis nor emphysema associated with this condition unless, as often occurs, there is a concomitant exposure to silica dust. Siderosis does not result in disability nor does it show any predisposition to pulmonary tuberculosis or lung cancer.

Stannosis. The inhalation of tin oxide dust over a long period of time will produce a benign pseudonodulation in the lungs known as stannosis. It is considered to be nonprogressive and nondisabling.

Titanium. High concentration of titanium dioxide dust may produce irritations in the respiratory tract. Slight fibrosis, without disabling injury, has been observed in the lungs from industrial exposure.

Vanadium. Vanadium is present in some welding filler wires. Vanadium pentoxide is moderately hazardous for both acute and chronic exposures.

Zinc. Zinc oxide fumes are formed during the welding, brazing, or cutting of galvanized metals. The inhalation of freshly formed fumes may produce a brief self-limiting illness known variously as zinc chills, metal fume fever, brass chills, and brass founders fever. This condition is characterized by chills, fever, nausea, vomiting, muscular pain, dryness of mouth and throat, headache, fatigue, and weakness. These signs and symptoms usually abate in 12 to 24 hours with complete recovery following. Immunity from this condition is rapidly acquired if exposure occurs daily, but is quickly lost during holidays or over weekends. Because of this behavior, metal fume fever is sometimes known as Monday morning sickness.

Cleaning compounds

Cleaning compounds because of their properties often require special ventilation precautions and following the manufacturer's instructions.

Chlorinated hydrocarbons

Degreasing operations may involve chlorinated hydrocarbons; these liquids or vapors shall be kept away from molten weld metal or the arc. Also keep them away from ultraviolet radiation from welding operations.

Asbestos

If welding or cutting involves asbestos, the regulations of the agency having authority shall be consulted in advance of the job.

EXPOSURE AND STANDARDS

Certain materials, sometimes contained in the consumables, base metals, coatings, or atmospheres of welding or cutting operations, have low or very low permissible exposure limits. (American Conference of Governmental Industrial Hygienists, Threshold Limit Value, 1.0 mg/m^3 or less). Among these materials are:

Antimony	Chromium	Mercury
Arsenic	Cobalt	Nickel
Barium	Copper	Selenium
Beryllium	Lead	Silver
Cadmium	Manganese	Vanadium

Refer to Material Safety Data Sheets provided by the manufacturer to identify any of the materials listed above that may be contained in the consumable.

Whenever these materials are encountered as designated constituents in welding, brazing, or cutting operations, ventilation precautions shall be taken to assure the level of contaminants in the atmosphere is below the limits allowed for human exposure. Unless atmospheric tests under the most adverse conditions have established that exposure is within acceptable concentrations, the following precautions shall be observed.

Ventilation

Natural ventilation is acceptable for welding, cutting, and related processes where the necessary precautions are taken to keep the welder's breathing zone away from the plume and where sampling of the atmosphere shows that concentrations of contaminants are below the levels given above.

Natural ventilation often meets the conditions where the necessary precautions are taken to keep the welder's breathing zone away from the plume and all of the following conditions are met:

1. Space of more than 10,000 ft^3 (284 m^3) per welder is provided.
2. Ceiling height is more than 16 ft (5 m).
3. Welding is not done in a confined space.
4. Welding space does not contain partitions, balconies, or other structural barriers that significantly obstruct cross ventilation. Welding space refers to a building or an enclosed room in a building, not a welding booth or screened area which is used to provide protection from welding radiation.
5. Materials covered above are not present as deliberate constituents.

The only way to assure that airborne contaminant levels are within the allowable limits, however, is to take air samples at the breathing zones of the personnel involved.

Mechanical ventilation includes local exhaust, local forced, and general area mechanical air movement. Local exhaust ventilation is preferred.

Local exhaust ventilation means fixed or movable exhaust hoods placed as near as practicable to the work and able to maintain a capture velocity sufficient to keep airborne contaminants below the limits.

Local forced ventilation means a local air moving system (such as a fan) placed so that it moves the air at right angles (90°) to the welder (across from the welder's face). It should produce an approximate velocity of 100 ft per min (30 meters per min), and be maintained for a distance of approximately 2 ft (600 mm) directly above the work area. Precautions must be taken to insure that contaminants are not dispersed to other work areas.

Examples of general mechanical ventilation are roof exhaust fans, wall exhaust fans, and similar large area air movers. General mechanical ventilation is not usually as satisfactory for health hazard control as local mechanical ventilation. It is often helpful, however, when used in addition to local ventilation.

General mechanical ventilation may be necessary to maintain the general background level of airborne contaminants below the levels referred to or to prevent the accumulation of explosive gas mixtures.

Where permissible, air cleaners that have high efficiencies in the collection of submicron particles may be used to recirculate a portion of air that would otherwise be exhausted. Some such filters do not remove gases, however. Therefore adequate monitoring must be done to assure concentrations of harmful gases remain below allowable limits.

Several of the factors known to affect the degree of risk associated with welding, cutting, and related operations have already been mentioned. In addition to these factors, those listed below must also be considered and evaluated to determine ventilation requirements:

1. Dimensions and layout of working areas.
2. Number of welding stations or welders or both.
3. Rates of welding.
4. Tendency of air currents to dissipate or concentrate fumes in certain areas of the working space.

The size and arrangement of the working facilities are major factors in ventilation requirements especially when the working area is confined or divided into sections that limit air circulation. The number of welding stations and welders is also an important factor to be considered in designing ventilation systems. Welders who are paid on piece rate or incentive basis may work at significantly higher rates than those who are paid a fixed daily rate. As a rule of thumb general ventilation should be provided at a minimum rate of 2,000 cubic feet per minute per welder except where local exhaust hoods are available.

The local exhaust hoods can be fixed or movable and should be provided with an air flow sufficient to maintain a velocity of 100 feet per minute in the breathing zone of the welder.

In addition to conventional ventilating devices, special equipment has been developed to remove welding fumes at their source. For example, fume extraction nozzles utilizing high velocity low volume exhaust are designed to fit over the end of the hand held welding torch. The sleevelike fixture is usually connected to a small exhaust fan by a flexible hose.

Fume avoidance

Welders and cutters must take precautions to avoid breathing the fume plume directly. This can be done by positioning of the work, the head, or by ventilation which directs the plume away from the face. Tests have shown that fume removal is more effective when the air flow is directed across the face of the welder, rather than from behind.

Light rays

Electric arcs and gas flames both produce ultraviolet and infrared rays which have a harmful effect on the eyes and skin upon continued or repeated ultraviolet exposure. The usual effect of ultraviolet is to "sunburn" the surface of the eye, which is painful and disabling but temporary in most instances. However, permanent eye injury may result from looking directly into a very powerful arc without eye protection, due to the effect of visible and near infrared radiation. Ultraviolet may also produce the same effects on the skin as a severe sunburn.

Production of ultraviolet radiation is high in gas-shielded arc welding. For example, a shield of argon gas around the arc doubles the intensity of the ultraviolet radiation, and, with the greater current densities required (particularly with a consumable electrode), the intensity may be five to thirty times as great as with nonshielded welding such as covered electrode or gas-shielded metal arc welding.

Infrared rays

Infrared has only the effect of heating the tissue with which it comes in contact. If the heat is not enough to cause an ordinary thermal burn, there is no harm.

Whenever possible, arc welding operations should be isolated so that other workers will not be exposed to either direct or reflected rays.

Arc welding stations for regular production work can be enclosed in booths if the size of the work permits. The inside of the booth should be coated with a paint that is nonreflective to ultraviolet radiation and provided with portable flameproof screens similarly painted or with flameproof curtains. Booths should be designed to permit circulation of air at the floor level and adequate exhaust ventilation.

Noise

In welding and cutting and the associated operation, noise levels may exceed the permissible limits. Personal protective devices may be needed. (See Chapter 17, Personal Protective Equipment, in the *Administration and Programs* volume of this Manual.)

GAS WELDING AND OXYGEN CUTTING

A gas-welding process unites metals by heating with the flame from the combustion of a fuel gas or gases; the process sometimes includes the use of pressure and a filler metal.

An oxygen-cutting process severs or removes metal by the chemical reaction of the base metal with oxygen at an elevated temperature maintained with heat from the combustion of fuel gases or from an arc. In the metal powder-cutting process, a finely divided material, such as iron powder, is added to the cutting oxygen stream. The powder bursts into flame in the oxygen stream and starts cutting without preheating the material to be cut. Metal powder cutting is used on stainless and other steels, on many nonferrous metals, and on concrete in construction and demolition jobs. Plasma arc cutting is now replacing metal powder oxy-fuel cutting.

Welding and cutting gases

Pure oxygen will not burn or explode. It supports combustion;

that is, it causes other substances to burn when they are raised to the kindling temperature. Combustible materials burn much more rapidly in oxygen than in air. Oxygen forms explosive mixtures in certain proportions with acetylene, hydrogen, and other combustible gases.

Acetylene (C_2H_2) consists of 92.3 percent by weight of carbon and 7.7 percent by weight of hydrogen in chemical combination. It contains stored-up energy which is released as heat when it burns, as in the welding flame. This heat is in addition to that which would be obtained by combustion of equivalent amounts of elemental carbon and hydrogen.

Acetylene burned with oxygen can produce a higher flame temperature (approximately 6,000 F or 3,300 C) than any other gas used commercially. Acetylene, like other combustible gases, ignites readily, and in certain proportions forms a flammable mixture with air or oxygen. The range of flammable limits of acetylene (2.5 to 81 percent acetylene in air) is greater than that of other commonly used gases, with consequently greater hazard.

Other fuel gases are used with oxygen in torches, primarily for oxygen cutting. For example, propane, propylene, and their mixtures are supplied in cylinders in liquid form, generally under various trade names. These gases are discussed in Chapter 16, Flammable and Combustible Liquids.

Compressed gas cylinders

Most of the gas used for welding and cutting is purchased in cylinders. These cylinders should be constructed and maintained in accordance with regulations of the Department of Transportation (DOT). The purchaser should make sure that all cylinders bear DOT, ICC (Interstate Commerce Commission), or CTC (Canadian Transport Commission) specification markings. The contents should be legibly marked on each cylinder in large letters.

Oxygen is supplied in steel cylinders; the usual size for welding contains 244 cu ft (6.9 m³) of oxygen under pressure of 2,200 psi (15.2 MPa) at 70 F (21 C). A cap should be provided to protect the outlet valve when the cylinder is not connected for use.

Hydrogen is furnished in cylinders under a pressure of about 2,000 psi (13.8 MPa) at 70 F (21 C). It may ignite in the presence of air or oxygen when in contact with a spark, open flame, or other source of ignition. Hydrogen-air mixtures in the range from 4.1 to 74.2 percent hydrogen are flammable.

Acetylene for welding and cutting is usually supplied in cylinders having a capacity up to about 300 cu ft (8.5 m³) of dissolved acetylene under pressure of 250 psi (1.7 MPa) at 70 F (21 C).

Acetylene cylinders should be completely filled with an approved porous material impregnated with acetone, the solvent for acetylene. The porous material should have no voids of appreciable size so that acetylene can be safely stored at the prescribed full cylinder pressure. Because acetylene is highly soluble in acetone at cylinder filling pressure, large quantities of acetylene can be stored in comparatively small cylinders at relatively low pressures.

Acetylene is either supplied in cylinders or generated as needed. It is a product of the reaction between water and calcium carbide, a gray crystalline substance made commercially by fusing lime and coke in an electric furnace. Calcium carbide itself is neither flammable nor explosive. It is stored and sold in airtight and watertight cans or drums. If the drums are damaged in handling and if water comes in contact with carbide, acetylene will be generated and there is then danger of ignition and explosion.

LP-gas. Handling, storage, and use of LP-gas are discussed in Chapter 16, Flammable and Combustible Liquids.

Handling cylinders

Serious accidents may result from the misuse, abuse, or mishandling of compressed gas cylinders. Workers assigned to the handling of cylinders under pressure should be properly trained and should work only under competent supervision. Observance of the following rules will help control hazards in the handling of compressed gas cylinders.

1. Accept only cylinders approved for use in interstate commerce for transportation of compressed gases.
2. Do not remove or change numbers or marks stamped on cylinders.
3. Because of their shape, smooth surface, and weight, cylinders are difficult to carry by hand. Cylinders may be rolled on their bottom edge but never dragged. Cylinders weighing more than 40 pounds (18.2 kg) (total) shall be transported on a hand or motorized truck, suitably secured to keep them from falling.
4. Protect cylinders from cuts or abrasions.
5. Do not lift compressed gas cylinders with an electromagnet. Where cylinders must be handled by a crane or derrick, as on construction jobs, carry them in a cradle or suitable platform and take extreme care that they are not dropped or bumped. Do not use slings.
6. Do not drop cylinders or let them strike each other violently.
7. Do not use cylinders for rollers, supports, or any purpose other than to contain gas.
8. Do not tamper with safety devices in valves or on cylinders.
9. When in doubt about the proper handling of a compressed gas cylinder or its contents, consult the supplier of the gas.
10. When empty cylinders are to be returned to the vendor, mark them EMPTY or MT with chalk. Close the valves and replace the valve protection caps, if the cylinder is designed to accept a cap.
11. Load cylinders to be transported to allow as little movement as possible. Secure them to prevent violent contact or upsetting.
12. Always consider cylinders as being full and handle them with corresponding care. Accidents have resulted when containers under partial pressure were thought to be empty.

The fusible safety plugs on acetylene cylinders melt at about the boiling point of water. If an outlet valve becomes clogged with ice or frozen, it should be thawed with warm (not boiling) water, applied only to the valve. A flame should never be used.

Storing cylinders

Cylinders should be stored in an upright position in a safe, dry, well-ventilated place prepared and reserved for the purpose. Flammable substances, such as oil and volatile liquids, should not be stored in the same area. Cylinders should not be stored near elevators, gangways, stair wells, or other places where they can be knocked down or damaged.

Oxygen cylinders should not be stored within 20 ft (6 m) of cylinders containing flammable gases or the location of other highly combustible materials. If closer than 20 ft, cylinders should be separated by a fire-resistive partition at least 5 ft (1.5 m) high; having a fire-resistance rating of at least 30 minutes.

Acetylene and liquefied fuel gas cylinders should be stored with the valve end up. If storage areas are within 100 ft (30.5 m) distance of each other and not protected by automatic sprinklers, the total capacity of acetylene cylinders stored and used inside the building should be limited to 2,500 cu ft (70 m³) of gas, exclusive of cylinders in use or connected for use. Quantities exceeding this total should be stored in a special room built in accordance with the specifications of NFPA 51, *Oxygen-Fuel Gas Systems for Welding, Cutting and Allied Processes,* in a separate building, or outdoors. Acetylene storage rooms and buildings must be well ventilated, and open flames must be prohibited. Storage rooms should have no other occupancy.

Cylinders should be stored on a level, fireproof floor. One common type of storage house consists of a shed room with side walls extending approximately halfway down from the roof and a dividing wall between one kind of gas and another.

To prevent rusting, cylinders stored in the open should be protected from contact with the ground and against extremes of weather—accumulations of ice and snow in winter and continuous direct rays of the sun in summer.

Cylinders are not designed for temperatures in excess of 130 F (54 C). Accordingly, they should not be stored near sources of heat, such as radiators or furnaces, or near highly flammable substances like gasoline.

Cylinder storage should be planned so that cylinders will be used in the order in which they are received from the supplier. Empty and full cylinders should be stored separately, with empty cylinders being plainly identified as such to avoid confusion. Group together empty cylinders which have held the same contents.

Storage rooms for cylinders containing flammable gases should be well ventilated to prevent the accumulation of explosive concentrations of gas. No source of ignition should be permitted. Smoking should be prohibited. Wiring should be in conduit. Electric lights should be in fixed position and enclosed in glass or other transparent material to prevent gas from contacting lighted sockets or lamps and should be equipped with guards to prevent breakage. Electric switches should be located outside the room.

Using cylinders

Safe procedures for the use of compressed gas cylinders include:
1. Use cylinders, particularly those containing liquefied gases and acetylene, in an upright position and secure them against accidentally being knocked over.
2. Unless the cylinder valve is protected by a recess in the head, keep the metal cap in place to protect the valve when the cylinder is not connected for use. A blow on an unprotected valve might cause gas under high pressure to escape.
3. Make sure the threads on a regulator or union correspond to those on the cylinder valve outlet. Do not force connections that do not fit.
4. Open cylinder valves slowly. A cylinder not provided with a handwheel valve should be opened with a spindle key or a special wrench or other tool provided or approved by the gas supplier.
5. Do not use a cylinder of compressed gas without a pressure-reducing regulator attached to the cylinder valve, except where cylinders are attached to a manifold, in which case the regulator will be attached to the manifold header.
6. Before making connection to a cylinder valve outlet, "crack" the valve for an instant to clear the opening of particles of dust or dirt. Always point the valve and opening away from the body and not toward anyone else. Never crack a fuel gas cylinder valve near other welding work or near sparks, open flames, or other possible sources of ignition.
7. Small fires at the cylinder should be extinguished, if possible, by closing the cylinder valve. In case of a larger fire or if extinguishment is not possible, evacuate and use a heavy stream of water to fight fire.
8. Use regulators and pressure gages only with gases for which they are designed and intended. Do not attempt to repair or alter cylinders, valves, or attachments. This work should be done only by the manufacturer.
9. Unless the cylinder valve has first been closed tightly, do not attempt to stop a leak between the cylinder and the regulator by tightening the union nut.
10. Fuel gas cylinders in which leaks occur should be taken out of use immediately and handled as follows:

    Close the valve, and take the cylinder outdoors well away from any source of ignition. Properly tag the cylinder, and notify the supplier. A regulator attached to the valve may be used temporarily to stop a leak through the valve seat.

    If the leak occurs at a fuse plug or other safety device, take the cylinder outdoors well away from any source of ignition, open the cylinder valve slightly, and permit the fuel gas to escape slowly. Tag the cylinder plainly. Post warnings against approaching with lighted cigarettes or other sources of ignition. Promptly notify the supplier, and follow instructions for returning the cylinder.
11. Do not permit sparks, molten metal, electric currents, excessive heat, or flames to come in contact with the cylinder or attachments.
12. Never use oil or grease as a lubricant on valves or attachments of oxygen cylinders. *Keep oxygen cylinders and fittings away from oil and grease, and do not handle such cylinders or apparatus with oily hands, gloves, or clothing.*
13. Never use oxygen as a substitute for compressed air in pneumatic tools, in oil preheating burners, to start internal combustion engines, or to dust clothing. Use it only for the purpose for which it is intended.
14. Never bring cylinders into tanks or unventilated rooms or other closed quarters.
15. Do not refill cylinders except with the consent of the owner and then only in accordance with DOT (or other applicable) regulations. Do not attempt to mix gases in a compressed gas cylinder or to use it for purposes other than those intended by the supplier.
16. Before a regulator is removed from a cylinder valve, close the cylinder valve and release the gas from the regulator.
17. Cylinder valves shall be closed when work is finished.

Manifolds

Cylinders are manifolded to centralize the gas supply and to provide gas continuously and at a rate in excess of that which may be obtained from a single cylinder. Manifolds must be of substantial construction and of a design and material suitable for the particular gas and service for which they are to be used.

Manifolds should be obtained from, and installed under the supervision of, a reliable manufacturer familiar with safe practices in construction and use of manifolds.

Portable manifolds connect a small number of cylinders (usually not over five) for direct supply to a consuming device. The cylinders may be connected by individual leads to a single, common coupler block or individual cylinders may be connected to a common line with coupler tees attached to the cylinder valves. A properly supported regulator serves the group of connected cylinders.

Stationary manifolds connect a larger number of cylinders for supply through piped distribution systems. This type of manifold consists of a substantially supported stationary pipe header to which the cylinders are connected by individual leads (Figure 13-1). One or more permanently mounted regulators serve to reduce and regulate the pressure of the gas flowing from the manifold.

Figure 13-1. Well-designed manifold system for acetylene cylinders. (Printed with permission from Linde Co., Division of Union Carbide Corp.)

Oxygen manifolds should be located away from highly flammable material and oil, grease, and the like. They should not be located in acetylene generator rooms, or in close proximity to cylinders of combustible gases. There should be a 5 ft (1.5 m) high, 30 min. fire-resistant partition between an oxygen manifold and combustible gas cylinders, unless the manifold and such cylinders are separated at least 50 ft (15.2 m). Regulations of NFPA 51, *Oxygen-Fuel Gas Systems for Welding, Cutting, and Allied Processes* should be followed.

Distribution piping

All piping should be color coded or clearly identified as to type of gas. Distribution piping carrying oxygen from a manifold or other centralized supply should be of steel, wrought iron, brass, or copper, as outlined in NFPA 51.

All pipe and fittings for oxygen service lines should be examined before use, and, if necessary, tapped with a hammer to free them from dirt and scale. They should, in every case, be washed out with a suitable non-flammable cleaner—hot water solutions of caustic soda and trisodium phosphate are effective.

Only steel or wrought iron piping should be used for acetylene distribution systems. Under no circumstances should acetylene gas be brought into contact with unalloyed copper except in a torch treated to prevent chemical reaction. Joints in steel or wrought iron pipe should be welded or made up with threaded or flanged fittings. Flanged connections in acetylene lines should be electrically bonded. Grey or white cast iron fittings should not be used.

Joints in brass or copper pipe may be welded, threaded, or flanged. A socket joint may be brazed with silver solder or similar high-melting point material. Threaded connections in oxygen piping should be tinned or made up with litharge and glycerine or other joint compound approved for oxygen service.

In fuel gas distribution systems, a back-flow check valve or hydraulic seal should be used to prevent back flow at every point where gas is withdrawn from the piping system to supply a torch or machine. Such devices should be listed (or approved) by an agency such as Factory Mutual or Underwriters Laboratories Inc.

Portable outlet headers

Portable outlet headers are assemblies of valves and connections used for service outlet purposes and are connected to a permanent service piping system by means of hose or other nonrigid conductors. Devices of this nature are commonly used at piers and dry docks in shipyards where the service piping cannot be located close enough to the work to provide a direct supply. Their use should be restricted to outdoor locations and to temporary service where conditions preclude a direct supply, and they should be used in accordance with regulations in NFPA 51, previously referenced.

Regulators

Pressure regulators must be used on both oxygen and fuel gas cylinders to maintain a uniform gas supply to the torches at the correct pressure. The oxygen regulator should be equipped with a safety relief valve or be so designed that, should the diaphragm rupture, broken parts will not fly. Workers should stand to one side and away from regulator gage faces when opening cylinder valves.

Only regulators listed by agencies such as Underwriters Laboratories Inc., or Factory Mutual should be used on cylinders of compressed gas. If unlisted regulators are used, they should be fully checked by a competent welding engineer. Each regulator (oxygen or fuel gas) should be equipped with both a high-pressure (contents) gage and a low-pressure (working) gage.

High pressure oxygen dial gages should have safety vent covers to protect the operator from flying parts in case of an internal explosion. Each oxygen dial gage should be marked OXYGEN—USE NO OIL OR GREASE.

Serious, even fatal, accidents have resulted when oxygen regulators have been attached to cylinders containing fuel gas, or vice versa. To guard against this hazard, it has been customary to make connections for oxygen regulators with right-hand threads and those for acetylene with left-hand threads, to mark the gas service on the regulator case, and to paint the two types of regulators different colors. Cylinder valve outlet threads have been standardized for most industrial and medical gases; see *Compressed Gas Cylinder Valve Outlet and Inlet Connections* (ANSI/CGA V-1-1977). Different combinations of right-hand and left-hand threads, internal and external threads, and different diameters to guard against wrong connections are now standard.

The regulator is a delicate apparatus and should be handled carefully. It should not be dropped or pounded on. Regulators should be repaired only by qualified persons or sent to the manufacturer for repairs.

Leaky or "creeping" regulators are a source of danger and should be withdrawn from service at once for repairs. If a regulator shows a continuous creep, indicated on the low pressure (delivery) gage by a steady buildup of pressure when the torch valves are closed, the cylinder valve should be closed and the regulator removed for repairs.

If the regulator pressure gages have been strained so that the hands do not register properly, the regulator must be replaced or repaired before it is used again.

When regulators are connected but are not in use, the pressure-adjusting device should be released. Cylinder valves should never be opened until the regulator is drained of gas and the pressure-adjusting device on the regulator is fully released.

These procedures should be followed in detail when regulators or reducing valves are being attached to a gas cylinder.

1. To blow out dust or dirt that otherwise might enter the regulator, "crack" the discharge valve on the cylinder by opening it slightly for an instant and then closing it. On a fuel gas cylinder, first see that no open flame or other source of ignition is near; otherwise, the gas might ignite at the valve.

2. Connect the regulator to the outlet valve on the cylinder. Be sure the regulator inlet threads match the cylinder valve outlet threads. Never connect an oxygen regulator to a cylinder containing fuel gas, or vice versa. Don't force connections that do not fit easily. Be sure that the connections between the regulators and cylinder valves are gastight.

3. Release the pressure-adjusting screw on the regulator to its limit—turn it counter-clockwise until it is loose. Engage the adjusting screw and open the downstream line to the air to drain the regulator of gas.

4. Open the cylinder valve slightly to let the hand on the high pressure gage move up *slowly*. On an oxygen cylinder gradually open the cylinder valve to its full limit, but on an acetylene cylinder make no more than 1½ turns of the valve spindle.

5. Attach oxygen hose to outlet of oxygen regulator and to oxygen inlet valve on torch. Attach acetylene hose to outlet of acetylene regulator and to acetylene inlet on torch.

6. Test oxygen connections for leaks. Be sure torch oxygen valve is closed; then turn oxygen regulator pressure-adjusting screw clockwise to give about normal working pressure. Using soapy water or approved leaktest solution, check connections for leaks as in Figure 13-2. At the same time, check regulator for creeping indicated by an increase in the reading on the low pressure (delivery) gage. If the regulator creeps, have it replaced or repaired before it is used.

7. Test acetylene connections for leaks. Be sure torch acetylene valve is closed and proceed in manner similar to No. 6 above—except that acetylene regulator pressure-adjusting screw should be set to produce a pressure of about 10 psig (69 kPa) (see Figure 13-2).

8. If torch is to be used immediately, proceed as in No. 9. If not, close cylinder valves, open torch valves to release pressure on regulator and gages, close torch valves, and release pressure-adjusting screws on regulators.

9. To adjust pressures of oxygen and fuel gas prior to using torch, proceed as follows: with all torch valves closed, slowly open oxygen cylinder valve, open torch oxygen valve, turn in pressure-adjusting screw on oxygen regulator to desired pressure, then close torch oxygen valve. Open acetylene cylinder valve (1½ turns maximum), and with torch

acetylene valve *closed,* turn in pressure-adjusting screw on acetylene regulator to desired pressure.

10. Purge each line individually. Open oxygen torch valve and release oxygen to the atmosphere for a few seconds before closing the valve; then open acetylene torch valve and release acetylene to the atmosphere for a few seconds and close the valve.

11. Open torch acetylene valve, light flame, and readjust regulator. Then close torch acetylene valve. (Acetylene pressure should first be adjusted with torch valve closed to prevent release of acetylene to air.)

12. Open torch valves and light torch according to procedure described in instructions provided with the equipment. The procedure for operating one torch is not necessarily best or even satisfactory for another.

Hose and hose connections

Oxygen and acetylene hoses should be different colors or otherwise identified and distinguished from each other. Red is the generally recognized color for fuel gas hose and green for oxygen hose (Figure 13-3). Black is used for inert gas and air hoses. The hose connections are usually marked STD-OXY for oxygen and STD-ACET for acetylene. The acetylene union nut has a groove cut around the center to indicate left-hand threads.

Connections for joining the hose to the hose nipple on the torches and regulators may be either the ferrule or clamp type. Gaskets should not be used on these connections. Special torch connectors with built-in shut-off valves are available.

Following are suggestions for the safe use of hose in welding and cutting operations:

1. Do not use unnecessarily long hose—it takes too long to purge. When long hose must be used, see that it does not become kinked or tangled and that it is protected from being run over by trucks or otherwise damaged. Where long hose must be used in areas exposed to vehicular or pedestrian traffic, suspend it high enough overhead to permit unobstructed passage.

2. Repair leaks at once. Besides being a waste, escaping fuel gas may become ignited and start a serious fire; it may also set fire to the welder's clothing. Escaping oxygen is equally hazardous. Repair hose leaks by cutting the hose and inserting a splice. Don't try to repair leaky hose by taping.

3. Examine hose periodically and frequently for leaks and worn places, and check hose connections. Test for leaks by immersing the hose under normal working pressure in water. (Refer to Figure 13-2, top left).

4. Protect hose from flying sparks, hot slag, other hot objects, and grease and oil. Store hose in a cool place.

5. A single hose having more than one gas passage shall not be used. When oxygen and acetylene hoses are taped together for convenience, and to prevent tangling, not more than 4 in. (10 cm) of each 12 in. (30.5 cm) of hose should be taped.

6. The use of hose with an external metallic covering is not recommended. In some machine processes and in certain types of operations, hose with an inner metallic reinforcement, which is exposed neither to the gas passage nor to the outside atmosphere, is acceptable.

7. Flashback devices (Figure 13-3) between torch and hose can prevent burnback into hoses and regulators. If a flashback occurs and burns the hose, discard the burned section.

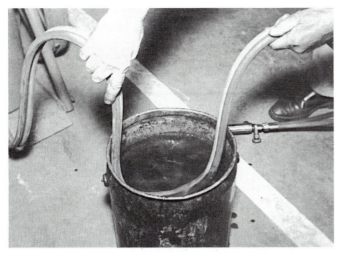

Figure 13-2a.

Figure 13-2b.

Figure 13-2c. Testing for leaks. With the pressure on and the torch valves closed, hold the hose (upper left) and the torch tip under water. Bubbles indicate leaks. Use soapsuds to test for leaks in the torch valves and hose-to-torch connections, as shown by arrows. Separately test the oxygen cylinder (upper right) and the acetylene cylinder and regulator connections (bottom photograph) for leaks at points marked by arrows. (Bottom photo printed with permission from J.I. Case).

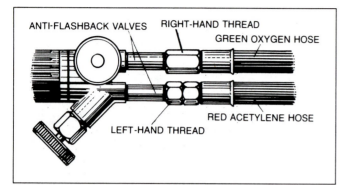

ANTI-FLASHBACK VALVES · RIGHT-HAND THREAD
GREEN OXYGEN HOSE

RED ACETYLENE HOSE
LEFT-HAND THREAD

Figure 13-3. When attaching hoses to a welding or cutting torch, use the red hose for acetylene and the green hose for oxygen; then test connections for leaks.

8. Hose which has been subject to flashback, or which shows evidence of severe wear or damage, shall be tested to twice the normal pressure to which it is subject, but in no case less than 300 psi. Defective hose, or hose in doubtful condition, shall not be used.

Torches

Torches are constructed of metal castings, forgings, and tubing. Usually, they are made of brass or bronze, but stainless steel may also be used. They should be of substantial design to withstand the rough handling they sometimes receive. It is best to use only those torches listed by an agency such as Underwriters Laboratories Inc. or Factory Mutual.

The gases enter the torch by separate inlets, go through valves to the mixing chamber, and then to the outlet orifice, located in the torch tip. Several interchangeable tips are provided with each torch and have orifices of various sizes according to the work to be done.

The cutting torch, unlike the welding torch, uses a separate jet of oxygen in addition to the jet or jets of mixed oxygen and fuel gas. The jets of mixed gases are for preheating the metal, and the pure oxygen jet is for cutting. The flow of oxygen to the cutting jet is controlled by a separate valve.

There are two types of torches in general use: the "injector" or low-pressure type, and the "pressure" or medium-pressure type. In the injector torch, the acetylene is drawn into the mixing chamber by the velocity of the oxygen. The acetylene may be supplied either from a low-pressure generator or a medium-pressure generator or from cylinders. In the medium-pressure torch, both gases enter under pressure; therefore, the acetylene is supplied from cylinders or from a medium pressure-generator.

In the operation of torches several precautions should be observed:

1. Select the proper welding head or mixer, tip or cutting nozzle (according to charts supplied by the manufacturer), and screw it firmly into the torch.

2. Before changing torches, shut off the gas at the pressure-reducing regulators and not by crimping the hose.

3. To discontinue welding or cutting for a few minutes, closing only the torch valves is permissible. If the welding or cutting is to be stopped for a longer period (during lunch or overnight) proceed as follows:

a) Close oxygen and acetylene cylinder valves.

b) Open torch valves to relieve all gas pressure from hose and regulator.

c) Close torch valves and release regulator pressure-adjusting screws.

4. Do not use matches to light torches. Use a friction lighter, stationary pilot flame, or other suitable source of ignition. When lighting, point the torch tip so no one will be burned when the gas ignites.

5. Never put down a torch until the gases have been completely shut off. Do not hang torches from a regulator or other equipment so that they come in contact with the sides of gas cylinders. If the flame has not been completely extinguished or if a leaking torch ignites, it may heat the cylinder or even burn a hole through it.

6. When extinguishing the flame, close the acetylene and oxygen valves in the order recommended by the torch manufacturer. If the oxygen valve is closed first, carbon soot will be deposited in the air. However, this ensures that the acetylene valve is closed tight when the flame is extinguished. If the acetylene valve is turned off first, no soot is formed, but there is no assurance that the fuel gas valve is closed and that it is not leaking.

Powder cutting

Powder-cutting processes for metal and concrete use similar equipment and gas supplies as do oxygen-cutting operations. The precautions previously discussed for safe handling and use of compressed gas equipment and cutting torches therefore apply. Manufacturers' recommendations for the operation and maintenance of the powder-dispensing apparatus—both pneumatic and vibratory—should be followed.

RESISTANCE WELDING

Because resistance welding equipment is normally permanently installed, the hazards are usually minimized if the equipment has been properly designed and safe operating practices have been established.

Certain hazards in the operation of this equipment—lack of point-of-operation guards, flying hot metallic particles, improper handling of materials, unauthorized adjustments and repairs—may cause eye injuries, burns, and electrical shock. Most of these hazards can be eliminated by safeguarding the equipment, by the wearing of protective clothing, and by strict control of operating practices.

Resistance welding is a metal-joining process in which the welding heat is generated at the joint by the resistance to the flow of electric current. The three fundamental parameters of resistance welding are current magnitude, current time, and tip pressure; each of these must be accurately controlled (Figure 13-4).

Power supply

Resistance welding usually employs 60-hertz alternating current which is fed to the primary of the water-cooled welding transformer. The primary can vary from 150 to 10,000 amp, at 240, 440, or 550 volts. The output, at the secondary of the transformer is a low voltage (max 30 volts) and high current (up to 200,000 amp) used for welding.

The welding current is sometimes furnished by the "stored energy" type of equipment, in which energy is built up and

Figure 13-4. An automatic resistance welding machine with a dial feed. The operator removes and places the work when the proper dial fixture comes to the front. (Printed with permission from General Electric Co.)

stored either in capacitors or in a combination transformer-reactor during the nonwelding period, and then is discharged to form the weld. This process involves low primary currents and high voltages, which must be guarded against.

To facilitate servicing the equipment, a safety-type disconnecting switch or a circuit breaker of the correct rating for opening supply circuits should be installed near the welding machine. Permanent injuries, and several fatalities have been caused by neglecting to use the line-disconnecting switch before

making adjustments inside of enclosures. The use of single-pole primary circuit breakers and electronic contactors which leave one line to the welder "hot" makes this precaution imperative.

Cables

Abuse of the cables for resistance welding is severe. The production requirements demand the utmost of the cable materials used, and even the best cables need frequent replacement. In use the cables are subjected to electrical pulsation, bending, and twisting, which leads to fatigue and eventual breakdown. This condition is minimized by the use of concentric cables.

The secondary voltage presents little shock hazard, since the maximum voltage is about 30 volts; but the operator can be hurt by a cable blowout such as is caused by steam pressure due to overheating from faulty water cooling circulation or from electrical failure.

A periodic check for weak spots in the cable covering is good practice. The use of concentric welding cables is now common because they do not have the undesirable features of the pulsating cables. Portable welding machines, including the cables, should have proper weight balance to permit operation without undue strain to the operator.

Machine installation

Installation of resistance welding equipment should conform to the NFPA *National Electrical Code,* Standard No. 70. Some items worthy of special attention are listed below.

1. Control circuits should operate on low voltage, not exceeding 24/36 volts maximum for portable spot welders.
2. Stored energy equipment (capacitor discharge or resistance welding) having control panels involving high voltage (more than 550 volts) should be completely enclosed. Doors should have locks and contacts wired in the control circuit to short circuit the capacitors when door or panel is opened. A manually operated switch will serve as an additional safety measure, assuring complete discharge of the capacitors.
3. Back doors of machines and panels should be kept locked or interlocked to prevent tampering.
4. A fused safety switch or circuit breaker should be located conveniently near the welding machine so that power supply circuits may be opened before servicing the machine and its controls.
5. The point-of-operation hazard should be eliminated by suitable guards. Enclosure guards, gate guards, two-hand controls, and similar standard guards as designed for punch press operations are applicable.
6. A flash welding machine should have a shield or hood to control flash and fumes, and a ventilating system to carry off the metallic dust and oil fumes.
7. Where flying sparks are not confined, the operators and nearby persons should be protected by shields of safety glass or other fire-resistant material or by the use of personal eye protection.
8. Foot switches, air or electrical, should be guarded to prevent accidental operation.

ARC WELDING AND CUTTING

For arc welding or cutting, two welding leads, the electrode lead and the work lead, are required from the source of current supply. Usually, one lead is connected to the work and the other to the electrode holder. The work lead (cable) is the most satisfactory means of providing the return (ground) circuit to the welding machine (Figure 13-5), but in some cases operating conditions may require the use of a grounded steel structure. The steel structure and connections should be capable of carrying the welding current.

Figure 13-5. One conductor of a three-conductor primary cable is permanently connected to the case of this AC welding unit and to the ground prong of a three-prong polarized plug, thus making grounding of the case both sure and easy.

Power supply

Either AC or DC may be used for arc welding or cutting of any kind. With small diameter electrodes used on thin sheets for manual arc welding, current values vary from 10 to 50 amp. For most manual welding, because the welder must withstand the heat, current values should not exceed 500 or 600 amp.

Automatic machine arc welding may use current values up to 200 amp or even higher on special applications.

If a gasoline-powered welding generator is used inside a building or in a confined area, the engine exhaust should lead to the outside atmosphere. Otherwise, carbon monoxide and other toxic gases may accumulate.

Voltages

The voltage across the welding arc varies from 15 to 40 volts,

depending on the type and size of electrode use. The welding circuit must supply somewhat higher voltage to strike the arc. This voltage is called the open circuit or "no load" voltage. After the arc is established, the open circuit voltage drops to a value about equal to the arc voltage plus the lead voltage drop. The open circuit voltages on DC welding machines should be less than 100 volts. Constant voltage power supplies (welder or converter) are now also being widely used.

For AC transformer welding machines, ANSI Z49.1, *Safety in Welding and Cutting,* prescribes a maximum open circuit voltage of 80 volts on manuals, 100 volts on automatics.

Heavy duty AC welding machines (ratings usually over 500 amp for automatic machine welding) are also built with open circuit voltages of 75 to 80 volts with a special tap to provide for 100 volts where necessary. The tap may be needed to obtain rated output from the machine if the line voltage is low or if the voltage of the secondary circuit drops. The tap should not be accessible to the welder for current adjustment but should be under the control of a responsible electrician or supervisor.

Open-circuit voltages should be as low as 50 volts on small AC welding machines used without expert supervision. Because these machines are often used with low arc voltage electrodes and with leads not more than 20 ft long (6 m), the probability of a large voltage drop in the welding circuit is small.

For other manual and automatic welding and cutting processes in which the work metal is connected electrically to one side of the circuit, open circuit voltage of 150 volts may be allowed if the following conditions are present:

1. All equipment and circuiting are fully insulated and the operator cannot make electrical contact other than through the arc itself, while the arc is maintained.
2. Disconnecting or voltage reducing devices operate within a time limit not exceeding one second after breaking the arc.

Where neither side of the circuit is electrically connected to the work, open circuit voltage of 300 volts is allowed if controls are present to prevent the operator from touching both sides of the circuit. One hand should be used to operate control devices. Also, the voltage should be disconnected automatically by a reliable switch instantly upon breaking the arc.

For AC or DC welding under electrically hazardous conditions, a reliable automatic control device for reducing no-load voltage is recommended.

Cables

Several lengths of welding cable may be used in one circuit. Substantially insulated connectors, of a capacity at least equivalent to that of the cable, should be used to splice or connect cables. Cable lugs used for ground and machine connections should be securely fastened to give good electric contact.

Welding cable is subjected to severe abuse if it is dragged over work under construction and across sharp corners, or run over by shop trucks. Special cable with high quality insulation should be used. The fact that welding circuit voltages are low may lead to laxity in keeping the welding cable in good repair. Operators and maintenance men should be instructed to see that defective cable is immediately replaced or repaired.

On large jobs, there is likely to be much loose cable lying around. Welders should keep this cable orderly and out of the way, preferably strung overhead to permit the passage of persons and vehicles. Welding cables should not lie in water or oil, in ditches or bottoms of tanks. Rooms in which arc welding is to be done regularly should be permanently wired with enough outlets so that extension cables will not have to be strewn about.

Electrodes and holders

Electrode holders for shielded-metal arc welding (SMAW) are used to connect the electrode to the welding cable supplying secondary current. Fully insulated holders (Figure 13-6) are preferred because there is less likelihood of shocking the welder or of accidentally striking an arc with such holders, particularly in close quarters.

Electrode holders will become hot during welding operations if holders designed for light work are used on heavy welding or if connections between the cable and the holder are loose.

If a holder of the correct size for the electrode cannot be used, an extra holder should be provided so that one can cool while the other is in use. *Dipping hot electrode holders in water is prohibited.*

On light or medium heavy work, where light, extremely flexible cables are used, holders may be attached directly to the work lead running to the machine. On heavier work, welders generally prefer a short length attached to the holder, which is more

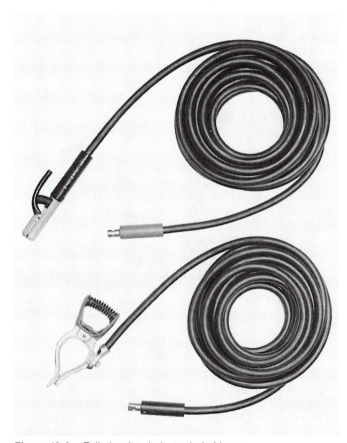

Figure 13-6. Fully insulated electrode holder on the electrode lead and the ground clamp on the work lead have insulated locking-type plugs for connection to receptacles on the welding machine. (Printed with permission from Westinghouse Electric Corp.)

flexible than the main work lead. Properly insulated cables of weight and flexibility which will not inconvenience the welder are available. Fully insulated connectors should be used to attach the short flexible cable to the main work lead.

Protection against electric shock

Although open-circuit voltages on standard arc-welding units are not high compared to those of other processes, they cannot be neglected as a potential hazard. Normally, the work setup is such that the work is grounded, and unless care is exercised, the welder or operator can easily become grounded.

The voltage between the electrode holder and the ground, during the "off" arc or "no-load" period, is the open circuit voltage.

The welder or welding operator shall be insulated from both the work and the metal electrode and holder. The bare metal part of an electrode or electrode holder should never be permitted to touch the operator's bare skin or wet clothing.

Consistent use of well insulated electrode holders and cables, dry clothing on the hands and body, and insulation from ground will be helpful in preventing contact.

Pacemaker wearers should check with the manufacturer or medical person regarding welding operations.

Some specific precautions for prevention of electric shock are:

1. In confined places, cover or arrange cables to prevent contact with falling sparks.
2. Never change electrodes with bare hands or wet gloves, or when standing on wet floors or grounded surfaces.
3. Ground the frames of welding units, portable or stationary, in accordance with the *National Electrical Code,* NFPA 70. A primary cable containing an extra conductor, one end of which is attached to the frame of the welding unit, can be used with a small welding unit. This ground connection can be carried back to the permanently grounded connection in the receptacle of the power supply by means of the proper polarized plug.
4. Arrange receptacles of power cables for portable welding units so that it is impossible to remove the plug without opening the power supply switch, or use plugs and receptacles which have been approved to break full load circuits of the unit.
5. If a cable (either work lead or electrode lead) becomes worn, exposing bare conductors, it may be repaired if the insulation repair on worklead cables is equivalent in insulation to the original cable covering.
6. Keep welding cables dry and free of grease and oil to prevent premature breakdown of the insulation.
7. Suspend cables on substantial overhead supports if the cables must be run some distance from the welding unit (Figure 13-7). Protect cables that must be laid on the floor or ground so that they will not interfere with safe passage or become damaged or entangled.
8. Take special care to keep welding cables away from power supply cables or high-tension wires.
9. Never coil or loop welding cable around the body.

Gas-shielded arc welding

Gas tungsten arc welding (GTAW). In gas-shielded tungsten arc welding, the electrode does not melt and is not used for filler

Figure 13-7. Cable and hose are suspended to keep them clear of work area on this gas-shielded metal arc welding operation.

metal. The electrode is tungsten, which is highly resistant to heat and nonconsumable in the welding process. Filler metal may be added by using a cold (nonelectrical) welding rod which is introduced into the arc or molten weld puddle.

Equipment. Either AC or DC welding units can be used for gas tungsten arc welding, depending on whether the weld is wide, deep, or narrow, whether the job is to be performed on a permanent fixture or a portable machine, and whether it is to be a manual or machine welding operation.

For gas metal arc, DC is used with a reverse polarity hookup, with current supplied by a generator or rectifier. The AC supply may be obtained through a transformer or high-frequency generator.

The manufacturer of the welding equipment should be consulted as to the specific job before the equipment is installed, especially when high-frequency AC is used in order to avoid interference with radio transmission.

Argon, helium, and gas mixtures are supplied by manufacturers in cylinders similar to oxygen cylinders. Since cylinder pressures range from 2,200 to 2,640 psig (15.2 MPa to 18.2 MPa), argon and helium cylinders should be stored and handled like other high-pressure gas cylinders.

Carbon dioxide, although not strictly an inert gas, is sometimes used as a shield gas when steel is welded by the gas metal arc welding process. It is usually supplied in partially liquid and partially gaseous form in cylinders at approximately 835 psig (5.8 MPa) pressure. These cylinders should therefore be handled like other high-pressure gas cylinders.

To supply gas to the welding torch, a regulator must be used to lower the pressure to 25 psig (172 kPa) or less, and a flowmeter should measure the volume of gas being used. If more than one torch is used from the same gas line, a flow-meter should be installed at each torch connection.

Air is used to cool the torch and electric current cables. Water is also used for cooling, generally where the welding current is more than 250 amp. The water supply line, even if city water is used, should be equipped with a strainer to keep out impurities which might get into and plug the water cooling passages.

In the gas-shielded tungsten arc torch (Figure 13-8), gas is conducted to the welding point through orifices in the torch around the electrode holder. Cooling water goes through passages through the torch handle and about the holder. In the smaller torches, ceramic cups are used. A torch for heavier work generally has a water-cooled gas cup.

OTHER WELDING AND CUTTING PROCESSES

There are several relatively new heat sources for welding and cutting, such as friction, ultrasonics, and lasers. Each of these special heat sources requires guarding and safe practices.

For example, the *laser* (Light Amplification by Stimulated Emission of Radiation) presents the hazard of eye damage from the optically amplified light beam, which because of its intensity, can do damage even at great distances. All employees should therefore be given preemployment and periodic follow-up eye examinations. Most companies require that employees work in pairs when using laser equipment.

To confine the laser beam, suitable shields should be developed and installed (Figure 13-9). Because a reflected laser beam is also hazardous, the work area should contain no glossy surfaces.

Power supplies for lasers and the electron beam are high-voltage equipment, which should be operated with the precautions developed for this type of equipment.

See discussion in the *Administration and Programs* volume, Chapter 17, Personal Protective Equipment, and in the Council's *Fundamentals of Industrial Hygiene.*

COMMON HAZARDS

Fire protection

Because portable welding and cutting equipment creates special fire hazards, it should be used in a permanent welding and cutting location which can be designed to provide maximum safety and fire protection. Otherwise, the welding and cutting site should be inspected to determine what fire protection equipment is necessary. See ANSI Z49.1 and NFPA 51B.

It is advisable, particularly in hazardous locations, to require "hot work" permits issued by the welding supervisor, a member of the plant fire department, or some other qualified person before welding or cutting operations are started. Specifications for hot work permits are outlined in Chapter 17, Fire Protection.

Floors and combustible materials

Where welding or cutting must be done near combustible materials, special precautions are necessary to prevent sparks or hot slag from reaching such material and starting fires. If the work itself cannot be moved, the exposed combustible material should, if possible, be moved a safe distance away. Otherwise, it should be covered with sheet metal. Spray booths and ducts should be cleaned to remove combustible deposits. Before welding or cutting is started, wood floors should be swept clean and, preferably, covered with metal or other noncombustible material where sparks or hot metal may fall. In some cases, it is

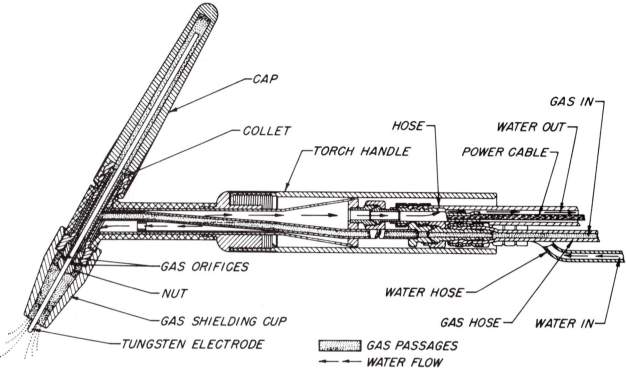

Figure 13-8. Gas-shielded tungsten arc welding torch, in which the electrode is nonconsumable. (Printed with permission from the Welding Handbook, American Welding Society.)

Figure 13-9. Self-standing safety shield provides close-quarter protection, rolls up for storage.

advisable to wet down the floor, though the wet floor increases the shock hazard to electric (arc and resistance) welders and necessitates special protection for them.

If gas welding or oxygen cutting is done inside a booth provided for arc welding, the gas cylinders should be placed in an upright and secured position away from sparks to prevent contact with the flame or heat.

Hot metal or slag should not be allowed to fall through cracks in the floor or other openings, nor into machine tool pits. Cracks or holes in walls, open doorways, and open or broken windows should be covered with sheet metal guards. Because hot slag may roll along the floor, it is important that no openings exist between the curtain and the floor. Similar protection should be installed for wall openings through which hot metal or slag may enter when welding or cutting operations are conducted on the outside of the building.

If it is necessary to weld or cut close to wood construction or near combustible material which cannot be removed or protected, a fire hose, water pump tank extinguisher, or fire pails should be conveniently located. Portable extinguishers for specific protection against Class B and C fires should also be provided (see Chapter 17, Fire Protection). Pails of limestone dust or sand may be useful. It is good practice to provide a fire extinguisher, either dry chemical, multipurpose chemical, or carbon dioxide, for each welder.

A fire watcher equipped with a suitable fire extinguisher should be stationed at or near welding or cutting operations conducted in hazardous locations to see that sparks do not lodge in floor cracks or pass through floor or wall openings. The fire watch should be continued for at least 30 minutes after the job is completed to make sure that smoldering fires have not been started.

Hazardous locations

Welding and cutting operations should not be permitted in or near rooms containing flammable or combustible vapors, liquids, or dusts, or on or inside closed tanks or other containers which have held such materials, until all fire and explosion hazards have been eliminated. All of the surrounding premises should be thoroughly ventilated, and frequent gas testing provided. Sufficient draft should be maintained to prevent accumulation of explosive concentrations. Local exhaust equipment should be provided for removal of hazardous gases, vapors, and fumes (present in the surroundings or generated by the welding or cutting operations) that ventilation fails to dispel.

Drums, tanks, and closed containers

Closed containers that have held flammable liquids or other combustibles should be thoroughly cleaned before welding or cutting. Sometimes containers which cannot be removed and handled properly for standard cleaning procedures are purged with an inert gas (Figure 13-10) or filled with water to within an inch or two of the place where the work is to be done, with a vent left open. Either of these two measures may also be employed as an added precaution after cleaning according to recommended methods. (See *Cleaning or Safeguarding Small Tanks and Containers*, NFPA 327. Also see Chapter 16, Flammable and Combustible Liquids, in this volume.)

The accepted method for preparing tanks and drums for welding is:

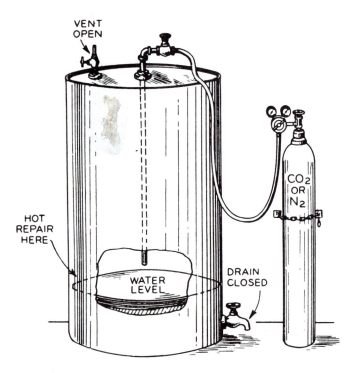

Figure 13-10. As an added precaution after cleaning, a container to be welded or cut may be filled with either carbon dioxide or nitrogen to dilute any combustible gas or vapor remaining—dilute it enough to render it nonhazardous. (Printed with permission from American Welding Society.)

1. Remove all sources of ignition (open flames, unguarded electric lights, etc.) from the vicinity of the drums to be cleaned.
2. Remove the bung with a special long-handled wrench.
3. Examine the inside for rags, waste, or other debris which might interfere with free draining. Use a portable electric hand lamp that is listed for hazardous locations, or an electric extension lamp protected by a guard of spark-resistant material.
4. Place the drums on a steam rack with the bung holes at the lowest possible point, and let the drums drain for 5 minutes.
5. Steam the drums for at least 10 minutes. Drums that have contained shellac, turpentine, or similar materials require longer steaming.
6. Remove the drums from the steaming rack, and fill them part way with caustic soda or soda ash solution. Rotate the drums for at least 5 minutes. Light hammering with a wood mallet will help to loosen scale.
7. Thoroughly flush the drums for at least 5 minutes with boiling water. A water spray nozzle placed 6 or 8 in. (15 to 20 cm) inside the drum can be used. Drums should be so placed that water can drain out the bung openings during this operation.
8. Wash down the outside of the drum with a hose stream of hot water.
9. Dry the drum thoroughly by circulating warm air throughout the inside.
10. Thoroughly inspect the interior of the drum, using a light that is listed for hazardous locations, and a small mirror. If it is not clean, repeat the cleaning process.
11. Test the container for the presence of flammable vapors with a combustible gas indicator. Test for toxic contaminants and for oxygen sufficiency or enrichment if personnel are to enter.
12. Make similar tests just before welding repair operations are performed. If the operations extend over an appreciable period of time, make repeated tests.

Precautions for employee protection during container cleaning operations include:

1. Wear head and eye protection, rubber gloves, boots, and aprons when handling steam, hot water, and caustic solutions. When handling dry caustic soda or soda ash, wear approved respiratory protective equipment, long sleeves, and gloves.
2. To handle hot drums, wear suitable hand pads or gloves. Steam irons or other hot surfaces which may be touched should be insulated or otherwise guarded.
3. Dispose of residue in a safe manner. In each instance, the method of disposal should be checked for hazards.
4. If a vessel must be entered, wear respiratory protective equipment approved for the exposure and a safety harness with attached lifeline tended by a helper who is similarly equipped and stationed outside the vessel. Rescue procedures should be tested for adequacy before beginning work.

Many containers that have held combustible or explosive material present special problems. Detailed information can be secured from the manufacturer of such materials. For gasometers or gas-holders for natural or manufactured gas, the American Gas Association should be consulted.

For cleaning and gas-freeing of tanks, bunkers, or compartments on board ship, refer to NFPA Standard No. 306, *Control of Gas Hazards on Vessels,* and to local fire marshal or Coast Guard marine inspection officer.

A word of caution. It may not be practical to purge and inert very large compartments due to volumes of gas needed and time required for such operations.

PERSONAL PROTECTION

Respiratory protection

If gases, dusts, and fumes cannot be kept below the TLV, welders should wear respiratory protective equipment certified for the exposure by NIOSH. Where oxygen is also deficient, self-contained breathing apparatus or hose masks with blowers are necessary.

Inert-gas shielded arc welding requires that precautions be taken to provide proper respiratory protection. Depending upon a number of factors, including the particular variety of gas-shielded arc welding to be done, the nature of the materials to be welded, and whether or not the work must be done in a confined space, there will be a need for positive ventilation, local exhaust removal, or approved respirator equipment, or a combination of these precautions (Figure 13-11).

Figure 13-11. Local exhaust for arc welding.

Medical control. A preplacement physical, including a chest X ray and lung examination, is recommended for all persons engaged in welding. Periodic reexaminations should be made as recommended by the company or plant physician.

Eye protection

Goggles, helmets, and shields that give maximum eye protection for each welding and cutting process should be worn by operators, welders, and their helpers. These items should conform to ANSI Z87.1, *Practice for Occupational and Educational Eye and Face Protection,* and Z89.1, *Protective Headwear for Industrial Workers.* Table 13-A is a guide for selecting the correct filter lens for various welding and cutting operations. Goggles or spectacles should have side shields. Guidance for lens

care is given in the *Administration and Programs* volume, Chapter 17, Personal Protective Equipment.

Protective clothing

Some of the items of protective clothing needed by welders (Figure 13-12) are:

1. Flame-resistant gauntlet gloves—leather or other suitable material. They may be insulated for heat.
2. Aprons made of leather or other flame-resistant material to withstand radiated heat and sparks.
3. For heavy work, fire-resistant leggings, high boots, or similar protection.
4. Safety shoes, wherever heavy objects are handled. Low-cut shoes with unprotected tops should not be used because of the spark hazard.
5. For overhead work, capes or shoulder covers of leather or other suitable material. Skull caps of leather or flame-resistant fabric may be worn under helmets to prevent head burns. Also, for overhead welding, ear protection (wool or rubber plugs or wire screen protectors) is sometimes desirable.
6. Safety hats or other head protection against sharp or heavy falling objects.

Operators and other persons working with inert-gas shielded arc welding machines should keep all parts of the body which could be exposed to the ultraviolet and infrared radiation covered to protect against skin burns and other types of injuries. Dark clothing, particularly a dark shirt, is preferable to light-colored clothing in order to reduce reflection to the operator's face underneath the helmet. Woolen clothing is preferable to cotton because it is more resistant to deterioration and is not readily ignited.

For gas-shielded arc welding, woolen clothing is also preferable to cotton. It is not readily ignited and protects the welder from changes in temperature. Cotton clothing, if used, should be chemically treated to reduce flammability. In either case, clothing should be thick enough to keep radiation from penetrating it.

Outer clothing should be reasonably free from oil and grease. Sleeves and collars should be kept buttoned. Aprons and overalls should have no pockets, in which sparks could be caught, on the front of them. For the same reason, trousers or overalls should not have turned-up cuffs.

Thermal insulated underwear is designed only to be worn under other clothing and should not be exposed to open flames, sparks, or other sources of ignition.

TRAINING IN SAFE PRACTICES

Welders and cutters should be well trained in the safe practices that apply to their work. The standards for training and qualification of welders set up by the American Welding Society are recommended. A training program should particularly emphasize that a welder or cutter can best provide for the safety of coworkers as well as the operator by observing safe practices which include:

1. For work at more than 5 ft (1.5 m) above the floor or ground, use a platform with railings, or with fall protection equipment.
2. Wear respiratory protection as needed and a safety harness with attached lifeline for work in confined spaces, such as

Table 13-A. Guide for Shade Numbers

Operation	Electrode size 1/32 in. (mm)	Arc current (A)	Minimum protective shade	Suggested* shade no. (comfort)
Shielded metal arc welding	Less than 3 (2.5)	Less than 60	7	—
	3-5 (2.5-4)	60-160	8	10
	5-8 (4-6.4)	160-250	10	12
	More than 8 (6.4)	250-550	11	14
Gas metal arc welding and flux cored arc welding		Less than 60	7	—
		60-160	10	11
		160-250	10	12
		250-500	10	14
Gas tungsten arc welding		Less than 50	8	10
		50-150	8	12
		150-500	10	14
Air carbon	(Light)	Less than 500	10	12
Arc cutting	(Heavy)	500-1000	11	14
Plasma arc welding		Less than 20	6	6 to 8
		20-100	8	10
		100-400	10	12
		400-800	11	14
Plasma arc cutting	(Light)**	Less than 300	8	9
	(Medium)**	300-400	9	12
	(Heavy)**	400-800	10	14
Torch brazing		—	—	3 or 4
Torch soldering		—	—	2
Carbon arc welding		—	—	14

	Plate thickness			
	in.	mm		
Gas welding				
Light	Under 1/8	Under 3.2		4 or 5
Medium	1/8 to 1/2	3.2 to 12.7		5 or 6
Heavy	Over 1/2	Over 12.7		6 or 8
Oxygen cutting				
Light	Under 1	Under 25		3 or 4
Medium	1 to 6	25 to 150		4 or 5
Heavy	Over 6	Over 150		5 or 6

* As a rule of thumb, start with a shade that is too dark to see the weld zone. Then go to a lighter shade which gives sufficient view of the weld zone without going below the minimum. In oxyfuel gas welding or cutting where the torch produces a high yellow light, it is desirable to use a filter lens that absorbs the yellow or sodium line in the visible light of the (spectrum) operation.

** These values apply where the actual arc is clearly seen. Experience has shown that lighter filters may be used when the arc is hidden by the workpiece. ANSI 249.1-1983

tanks and pressure vessels. The lifeline should be tended by a similarly equipped helper whose duty is to observe the welder or cutter and effect rescue in an emergency.

3. Take special precautions if welding or cutting in a confined space is stopped for some time. Disconnect the power on arc welding or cutting units and remove the electrode from the holder. Turn off the torch valves on gas welding or cutting units, shut off the gas supply at a point outside the confined area, and, if possible, remove the torch and hose from the area.

4. After welding or cutting is completed, mark hot metal or post a warning sign to keep workers away from heated surfaces.

5. Follow safe housekeeping principles. Don't throw electrode or rod stubs on the floor—discard them in the proper waste containers. Keep tools and other tripping hazards off the floor—put them in a safe storage area.

6. Use equipment in accordance to manufacturer's instructions and practices.

Operators and management shall recognize their joint responsibilities for safety. Management shall assure that welders and supervisors are trained.

Management shall establish and enforce procedures which

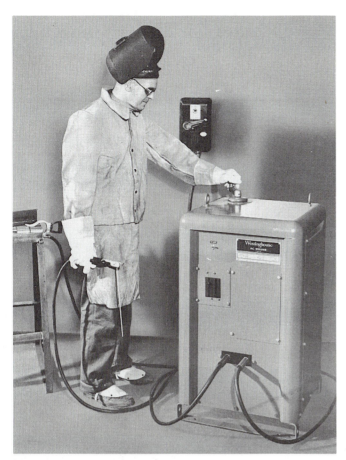

Figure 13-12. Proper personal protection for a welder—flash goggles worn under helmet, chrome leather jacket, apron, gauntlet gloves, and leggings. (Printed with permission from Westinghouse Electric Corp.)

are used at other than normal locations and/or in hazardous locations. Only approved welding equipment shall be used. Supervisors shall be responsible for the safe operation of welding equipment and operators.

REFERENCES

American Conference of Governmental Industrial Hygienists, Committee on Industrial Ventilation, P.O. Box 16153, Lansing, Mich. 48901. *Industrial Ventilation—A Manual of Recommended Practice.*

American Gas Association, 1515 Wilson Blvd., Arlington, Va. 22209.

American Insurance Association, 85 John St., New York, N.Y. 10038. *Lasers and Masers,* Special Hazards Bulletin Z-125.

American Petroleum Institute, 1220 L St. NW., Washington, D.C. 20005. *Cleaning Mobile Tanks Used for Transportation of Flammable Liquids,* Accident Prevention Manual, No. 13. 1958.

Cleaning Petroleum Storage Tanks, RP2015.

Gas and Electric Cutting and Welding, RP2009.

American Society of Mechanical Engineers, 345 East 47th St.,

New York, N.Y. 10017. *ASME Boiler and Pressure Vessel Code,* Section IX, "Qualification Standard for Welding and Brazing Procedures, Welders, Brazers, and Welding and Brazing Operators." 1965.

American National Standards Institute, 1430 Broadway, New York, N.Y. 10018.

Electric Arc Welding Power Sources, ANSI/NEMA EW-1. 1983.

Practice for Occupational and Educational Eye and Face Protection, Z87.1.

Practices for Respiratory Protection, Z88.2.

Safe Use of Lasers, Z136.1.

Safety in Welding and Cutting, Z49.1.

Protective Headwear for Industrial Workers, Z89.1.

Safety Standard for Mechanical Power-Transmission Apparatus, ANSI/ASME B15.1.

Safety Standard for Transformer-Type Arc-Welding Machines, ANSI/UL 551.

American Welding Society, P.O. Box 351040, Miami, Fla. 33135.

Fire Prevention in Arc Welding and Cutting Processes, 249.2.

Recommended Practices for Gas Tungsten Arc Welding, C.5.5–1980.

Recommended Practices for Plasma Arc Cutting, C.5.2.

Safety in Welding and Cutting, 249.1–1983. Z49.1-83.

The Welding Environment, 1973.

Welding Fumes, 1–1981, 2–1981, 3–1983.

Welding Handbook, 7th ed. Section 1. 1976.

Compressed Gas Association, Inc., 1235 Jefferson Davis Hwy., Arlington, Va. 22202.

Oxygen-Deficient Atmospheres, SB-2.

Regulator Connection Standard.

Safe Handling of Compressed Gases in Cylinders.

Specification for Rubber Welding Hose.

Linde Division, Union Carbide Corp., 39 Old Ridgebury Rd., Danbury, Conn. 06817. "Plasma-Arc Process" Bulletins.

National Fire Protection Association, Batterymarch Park, Quincy, Mass. 02269.

Bulk Oxygen Systems at Consumer Sites, NFPA 50.

Cleaning or Safeguarding Small Tanks and Containers, NFPA 327.

Control of Gas Hazards on Vessels, NFPA 306.

Cutting and Welding Processes, NFPA 51B.

Gaseous Hydrogen Systems at Consumer Sites, NFPA 50A.

Liquefied Hydrogen Systems at Consumer Sites, NFPA 50B.

National Electrical Code, NFPA 70.

Oxygen-Fuel Gas Systems for Welding, Cutting, and Allied Processes, NFPA 51.

National Safety Council, 444 North Michigan, Chicago, Ill. 60611.

Industrial Data Sheets

Acetylene, 494.

Cleaning Small Containers of Combustibles, 432.

Gaseous Oxygen, 472.

Soldering and Brazing, 445.

Solon, Leonard R. "Occupational Safety with Laser (Optical Maser) Beams." *Archives of Environmental Health,* 6:414-17 (March 1963).

U.S. Department of Health, Education, and Welfare, National Institute for Occupational Safety and Health, Division of Technical Services, Cincinnati, Ohio 45226.

Criteria for a Recommended Standard: Occupational Exposure to Ultraviolet Radiation, 73-11009.

Engineering Control of Welding Fumes, DHEW (NIOSH) Publication 75-115.

Safety and Health in Arc Welding and Gas Welding and Cutting, DHEW (NIOSH) Publication 78-138.

U.S. Navy, The Pentagon, Washington, D.C. 20350.
Bureau of Ships Manual, Chapter 92, "Welding and Allied Processes." NAVSHIPS 250-00-92.

Underwater Cutting and Welding Manual, NAVSHIPS 250-692-9.

14

Hand and Portable Power Tools

PREVENTING ACCIDENTS

Because of the widespread use and abuse of hand and powered tools and the severity of many tool injuries, it is important that the elimination of tool accidents be made a part of every safety program.

Each year hand tools are the source of about 6 percent of all compensable disabling injuries. Disabilities resulting from misuse of tools or using damaged tools include loss of eyes and vision; puncture wounds from flying chips; severed fingers, tendons, and arteries; broken bones; contusions; infections from puncture wounds; and injuries too numerous to mention.

Safe practices

Failure to observe one or more of the following five safe practices accounts for most hand and powered hand tool accidents.
1. Always wear safety goggles to protect eyes.

    In operations requiring the use of hand or portable power tools, particles may fly. Therefore, it is essential that safety goggles, or equivalent eye protection, be worn by the user and others in the immediate vicinity.
2. Select the right tool for the job.

    Examples of unsafe practices are: Striking hardened striking faces of hand tools together (such as using any hammer to strike another hammer or hatchet), using a claw hammer to strike a steel chisel, using a file or a screwdriver for a pry, a wrench for a hammer, and pliers instead of the proper wrench.
3. Keep tools in good condition.

    Unsafe tools include wrenches with cracked or worn jaws; screwdrivers with broken tips, or split or broken handles; hammers with chipped, mushroomed, or loose heads, broken or split handles; mushroomed heads on chisels (Figure 14-1); dull saws; and extension cords or electric tools with broken plugs, improper or removed grounding system, or split insulation.

Figure 14-1. Poorly maintained tools—split hammer handle, mushroomed chisel heads.

Figure 14-2. Tool boxes containing personal tools can be neatly and safely stored in a rack like this. Shelves are movable so employees can reach all sections to get their tool boxes out. (Printed with permission from Lackawanna Plant, Bethlehem Steel Co.)

4. Use tools correctly.

Some common causes of accidents are: Screwdrivers applied to objects held in the hand, knives pulled toward the body, failure to ground electrical equipment, and nail hammers striking hardened tools. (For proper uses of hand tools, see the section on this subject later in this chapter.)

5. Keep tools in a safe place.

Many accidents have been caused by tools falling from overhead and by knives, chisels, and other sharp tools carried in pockets or left in tools boxes with cutting edges exposed.

A safety program designed to control tool accidents should include the following activities:

a) Train employees to select the right tools for each job and see that they are available.

b) Establish regular tool inspection procedures (including inspection of employee-owned tools), and provide good repair facilities to make sure that tools will be maintained in safe condition.

c) Train and supervise employees in the correct use of tools for each job.

d) Establish a procedure for control of company tools, such as a check-out system at tool cribs.

e) Provide proper storage facilities in the tool room and on the job (Figures 14-2 and -3).

f) Enforce use of proper personal protective equipment.

g) Plan each job well in advance.

Each supervisor should also make a complete check of operations to determine the need for special tools that will do the work more safely than ordinary tools. Special tools should be kept readily available.

Centralized tool control

The principal advantage of centralized tool control from the standpoint of accident prevention is that it assures uniform inspection and maintenance of tools by a trained employee. Special features and equipment, available at a central location, permit uniformly good maintenance. The correct type of personal protective equipment, such as safety goggles, can and should be issued when a tool is distributed.

Centralized control facilities and the keeping of effective

Figure 14-3. Cabinet and shelves provide tool storage and availability at work site.

records on tool failure and other accident causes will help locate hazardous conditions. Central storage facilities will also assure more positive control than will scattered storage. Tools that are exposed to less damage and deterioration are not as likely to fail or create other hazards.

The tool room attendant can help promote safety by recommending or issuing the right type of tool, by encouraging employees to turn in damaged or worn tools, and by encouraging the safe use of tools.

A procedure should be set up so that the tool supply room attendant can send tools in need of repair to a department or service firm thoroughly familiar with methods of repair and reconditioning.

Some companies issue each employee a set of numbered checks which are exchanged for tools from the supply room. With this system the attendant knows the location of each tool and can recall it for inspection at regular intervals.

Companies performing work at scattered locations may find

that it is not always practical to maintain a tools supply room. In such cases, the supervisor should inspect all tools frequently and remove from service those found to be damaged. Many companies have each supervisor check all tools every week. A checklist for hand tools considered most hazardous can be helpful in systematizing inspection (Figure 14-24).

Some workers prefer to use their own tools even though tools are furnished by the company. In this case, supervisors should examine the tools frequently to prevent the use of those that are unsafe. If privately owned tools are found to be damaged or unsafe, supervisors should insist that they be replaced.

The federal Occupational Safety and Health rules and regulations (29 CFR Chapter XVII §1910.242) state that the employer is responsible for seeing that *safe* tools are used, including tools and equipment that are furnished by the employee.

Tool boxes

First of all, tool boxes are meant to hold tools, not to stand on, use as an anvil, a saw horse, or to store your lunch. Light weight tool boxes are made of plastic or steel, but strong, heavy duty tool boxes are made of steel.

Let's consider the portable type first. These may have up to five drawers, a lift-out tray, and possibly a cantilevered tray that automatically opens out when the cover is lifted. All seams should be welded and smooth with no protruding edges to catch clothing or hands. In addition to the handle on the top of the tool box cover, look for handles at each end for those boxes designed to hold an extra heavy load of tools. A good tool box will have a catch or a hasp at each end and should be able to be locked with either a padlock or its own built-in lock. Look for weather-proof construction that will allow rain to drain away without entering the tool box.

Tool chests are big brothers to tool boxes. They are usually heavier, stronger, and of course have a much greater capacity than tool boxes. The drawers (as much as ten and even more) on the better models can be secured with their own built-in locks. Some have a tote tray that can be removed for carrying only those tools needed for a particular job. Most tool chests are designed to be placed on top of tool cabinets.

Mobile tool cabinets—the kind on wheels—may have ten or more drawers and if they are designed to hold a chest, sometimes as many as twenty or more drawers. Look for a locking arrangement that will lock all drawers automatically and for construction that will allow drawers, no matter how heavily loaded, to roll out freely. Casters should be of ball-bearing construction with two wheels that can be locked by means of a brake to prevent rolling. A good tool cabinet should be adequately braced to prevent any possibility of swaying as the tool cabinet is rolled around.

Carrying tools

Employees should never carry tools in any way that might interfere with their freely using both hands on a ladder or while climbing on a structure. A strong bag, bucket, or similar container should be used to hoist tools from the ground to the job. Tools should be returned in the same manner, that is, not brought down by hand, carried in pockets, or dropped to the ground.

Mislaid and loose tools cause a substantial number of injuries. Tools are laid down on scaffolds, on overhead piping, on top of step ladders, and in other locations from which they can fall on persons below. Leaving tools overhead is especially hazardous where there is vibration or where people are moving about.

Chisels, screwdrivers, and pointed tools should never be carried edge or point up in a worker's pocket. They should be carried in a tool box, a cart, or in a carrying belt like that used by electricians and steelworkers (Figure 14-4), in a pocket tool pouch, or in the hand with points and cutting edges away from the body.

Tools should be handed from one employee to another, never thrown. Edged or pointed tools should be passed, preferably in their carrying case, with the handle toward the receiver.

Workers carrying tools on their shoulders should pay close attention to clearances when turning around and should handle the tools so that they will not strike others.

MAINTENANCE AND REPAIR

When metal tools break in normal use, there are usually detectable causes related to tool quality. Therefore, it pays to purchase tools of the best quality obtainable.

Inspection and control

The tool room attendant or tool inspector should be qualified by training and experience to pass judgment on the condition of tools for further use. No dull or damaged tool should be returned to stock. Enough tools of each kind should be on hand so that, when a damaged or worn tool is removed from service, it can be replaced immediately with a safe tool.

Efficient tool control requires periodic inspections of all tool operations. These inspections should cover housekeeping in the tool supply room, tool maintenance, service, number of tools in the inventory, handling routine, and condition of tools. Responsibility for such periodic inspections is usually placed with the department head and should not be delegated to others.

Hand tools receiving the heaviest wear, such as chisels, punches, wrenches, hammers, star drills, and blacksmith's tools, require frequent maintenance on a regular schedule.

Proper maintenance and repair of tools require adequate facilities and equipment—workbenches, vises, safety goggles, repair and sharpening tools, and good lighting. Employees specially trained in the care of tools should be in charge of these facilities; otherwise tools should be sent out for repairs to a qualified facility.

Redressing tools

The redressing and reshaping of tools having chipped, battered, or mushroomed striking or struck surfaces is not recommended. When a tool has reached this stage through normal use or abuse, it should be discarded.

There are three basic rules which apply to the redressing of dull cutting edges:
1. Rigidly support the tool being redressed.
2. Use a hand file or whetstone only, *never* a grinding wheel. File or stone away from the cutting edge.
3. Restore the original contour of the cutting edge.

Axes. Use a hand file for redressing. Start 2 or 3 in. (50 or 76 mm) back from the cutting edge and file to about ½ in. (13 mm) from the edge. Work for a fan shape, leaving reinforcement at corners for strength. File the remaining ½ in., blend-

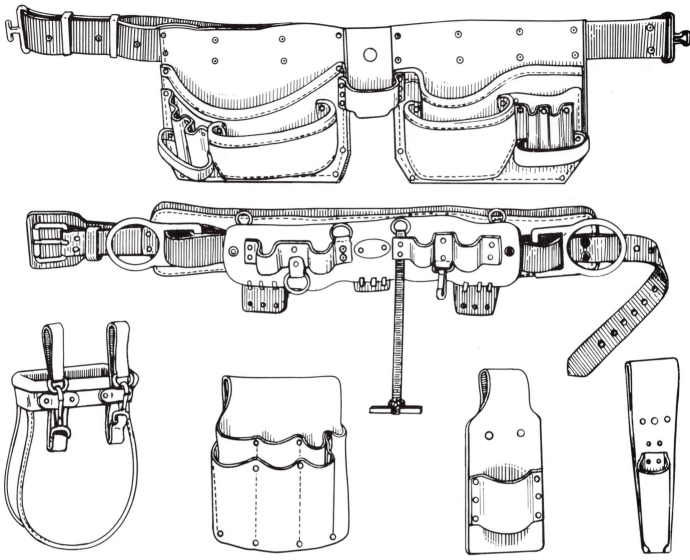

Figure 14-4. Miscellaneous tool holders and pouches. Equipment of this type frees hands while climbing and working on ladders, poles, and other elevated areas.

ing into previously filed area and preserving the original contour of the cutting edge. Remove all scratches with a whetstone or hone ½ in. back from the cutting edge. See cross section C illustration below for the "right" way to shape the edge in redressing axes. Other illustrations show "wrong" ways to redress axes.

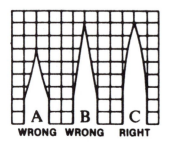

Hatchets. Hatchets with double bevels should be redressed as illustrated in A below. Hatchets with single bevels should be redressed as illustrated in B below. Use a hand file for redressing, removing scratches with a whetstone.

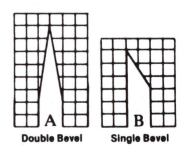

Cold chisels (flat). Cold chisels are hardened on the cutting edge. Redressing may be done with a hand file or whetstone, restoring to original shape or to an included angle of approximately 70 degrees (see illustration below).

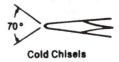

Other machinists' chisels. Other commonly used metal-working chisels are Round Nose, Diamond Point and Cape. Redressing instructions are the same as for flat cold chisels except that bevel angles are approximately as illustrated.

Hot chisels. These are handled tools used for cutting hot metal. Redresssing instructions are the same as for cold chisels.

Punches. The working end of pin and rivet punches and blacksmiths' punches should be redressed flat and square with the axis of the tool. The point of center punches should be redressed flat and square with the axis of the tool. The point of center punches should be redressed to an included angle of approximately 60 degrees; prick punches, to an included angle of approximately 30 degrees.

Bricklayer's tools. Bricklayers' tools should be redressed as follows: hammer blade, 40 degrees; brick chisel, 90 degrees; and brick set, 45 degrees. Bevel slightly to remove feather edge.

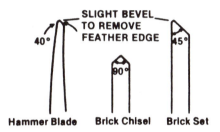

Woodchoppers' mauls and wood splitting wedges. The instructions for redressing axes apply also to these tools although they have heavier heads and thicker sections in the bit. Hand file the splitting edge to an included angle of approximately 70 degrees. See cold chisel illustration for 70 degree included angle.

Star drills. Hand file all cutting edges to an included angle of approximately 70 degrees. See cold chisel illustration for included angle of 70 degrees.

Prospecting picks. Redress with a hand file to restore original contour of the pick end.

Always follow the manufacturer's recommendations in the repair, shaping, and maintenance of tools.

Handles

Wooden handles of hand tools used for striking, such as hammers and sledges, should be of the best straight-grained material, preferably hickory, ash, or maple, neatly finished, and free from slivers. Alternate materials such as fiber glass or steel with a rubber sleeve may be used.

To make sure that they are properly attached, handles should be fitted to tools only by an experienced person. Poorly fitted handles make it difficult for the worker to control the tools and may be dangerous.

Loose wooden handles in sledges, axes, hammers, cold cutters, and similar tools create a hazard. No matter how tightly

Figure 14-5. The correct way to whet a wood chisel or plane iron on a stone to produce a sharp cutting edge. The bevel should be placed on the stone with the back edge slightly raised.

a handle may be wedged at the factory, both use and shrinkage can loosen it. In some cases, tapping the wedges will take up the shrinkage. In others, the head of the tool can be driven back on the handle, the wedges reset, and the protruding end of the handle cut off.

Eventually, any wooden handled tool will need handle replacement (Figure 14-6). A new equivalent handle is selected, and the old handle driven out. If the wood of the new handle does not bear against the head eye at all points, shave the handle until it fits snugly. Then replace the handle in the tool and sight along it to be sure that the head is properly centered. After the handle is firmly fitted into the head eye, wedge it according to the original pattern.

USE OF HAND TOOLS

The misuse of common hand tools is a source of injury to the industrial worker. In many instances, injury results because it is assumed that "anybody knows how" to use common hand tools. Observation and the record of injuries show that this is not the case.

A part of every job instruction program should, therefore, be detailed training in the proper use of hand tools. So important is this training that considerable attention is given, in the following pages, to those safe practices that characterize the competent and safe worker.

Workers should wear safety goggles and should set up a shield or screen to prevent injury to other people from flying particles. Exposed employees should wear safety goggles if a shield does not afford positive protection.

Screwdrivers

The screwdriver is probably the most commonly used and abused tool. The unsafe practice of using screwdrivers for punches, wedges, pinch bars, or pries *must* be discouraged. If used in such manner, they can cause injury and become unfit

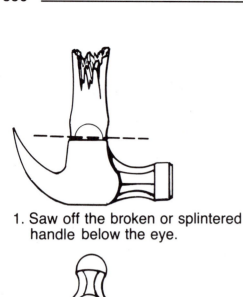

1. Saw off the broken or splintered handle below the eye.

2. Drive out the eye section of the old handle with a punch or chisel.

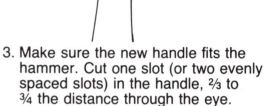

3. Make sure the new handle fits the hammer. Cut one slot (or two evenly spaced slots) in the handle, ⅔ to ¾ the distance through the eye.

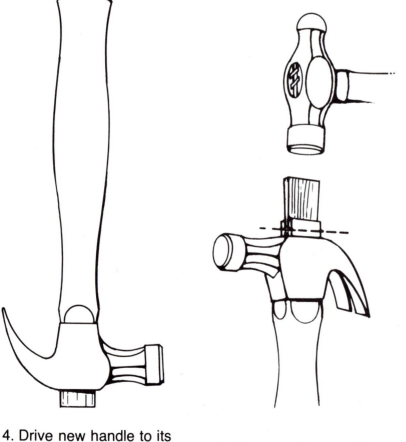

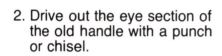

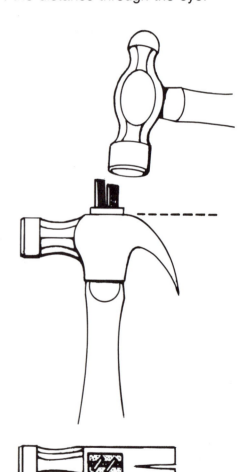

4. Drive new handle to its shoulder. It will project about ⅜ in. through the eye.

5. Drive wood wedge(s) into slot(s). Cut handle about ⅛ in. outside of head.

6. Form starting grooves for two metal wedges with a cold chisel. Drive the steel wedges as far as possible. With a hack saw, cut off excess portion of handle and wedges.

Figure 14-6.   How to replace the handle of a nail hammer.

for the work they are intended to do. Furthermore, a broken handle, bent blade, dull or twisted tip may cause a screwdriver to slip out of the slot and cause a hand injury.

A screwdriver tip should be selected to fit the screw (Figure 14-7). A sharp square-edged bit will not slip as easily as a dull, rounded one, and requires less pressure. By redressing the tip to original shape, it may be kept clean and sharp to permit a good grip on the head of the screw. Phillips screwdrivers and many other types are safer than the flat tip screwdriver because they have less tendency to slip.

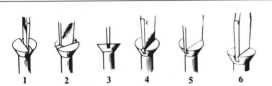

| 1 | 2 | 3 | 4 | 5 | 6 |

1. **This tip is too narrow for the screw slot; it will bend or break under pressure.**
2. **A rounded or worn tip. Such a tip will ride out of the slot as pressure is applied.**
3. **This tip is too thick. It will only serve to chew up the slot of the screw.**
4. **A chisel ground tip will also ride out of the screw slot. Best to discard it.**
5. **This tip fits, but it is too wide and will tear the wood as the screw is driven home.**
6. **The right tip. This tip is a snug fit in the slot and does not project beyond the screw head.**

Figure 14-7. Selecting the correct screwdriver to fit the screw.

When putting in a screw, the work should be held in a vise or laid on a flat surface. This practice will lessen the chance of injury to the hands if the screwdriver should slip from the work.

When it is necessary to work around electrical current-bearing equipment, use an insulated screwdriver. However, the handle, insulated with dielectric material, is intended only as a secondary protection. Insulated blades are also intended only as a protective measure against shorting out components. Be sure electrical current is off before beginning work.

Hammers

Hammers are made in various types and sizes, with varying degrees of hardness and different configurations for specific purposes. They should be selected for their intended use and used only for those purposes. Proper use of practically all types involves certain basic rules:
1. Safety goggles should always be worn to protect eyes.
2. A hammer blow should always be struck squarely with the hammer striking face parallel with the surface being struck. Always avoid glancing blows and over and under strikes.
3. When striking another tool (chisel, punch, wedge, etc.), the striking face of the hammer should have a diameter approximately ⅜ in. (9 mm) larger than the struck face of the tool.
4. Always use a hammer of suitable size and weight for the job. Don't use a tack hammer to drive a spike, nor a sledge to drive a tack.
5. Never use a hammer to strike another hammer.
6. Never use a hammer with a loose or damaged handle.
7. Discard any hammer if it shows dents, cracks, chips, mushrooming, or excessive wear. Redressing is not recommended.

Common nail hammers are designed for driving unhardened

common and finishing nails, and nail sets, using the center of the hammer face. Their shape, depth of face, and balance make them unsuitable for striking against metal, especially heavier objects, such as cold chisels.

Nail hammers are made in two patterns; curved claw and straight or ripping claw. The face is slightly crowned with the edges beveled, although certain heavy-duty patterns may have checkered faces designed to reduce glancing blows and flying nails. Handles may be wood, tubular or solid steel, or glass fiber. Tubular steel, solid steel, and glass fiber are generally furnished with rubber grips which are occasionally used also on wood handles.

When a nail is to be drawn from a piece of wood, a block of wood may be used under the head to increase the leverage, reduce the strain on the handle and prevent marring of the wood.

Ball peen hammers of the proper size are designed for striking chisels and punches, and for riveting, shaping, and straightening unhardened metal.

Sledge hammers of many types and weights are designed for general sledging operations in striking wood, metal, concrete, or stone. The manufacturer can make specific recommendations as needed.

Hand drilling hammers are designed for use with chisels, punches, star drills, and hardened nails.

Bricklayer's hammers are special-purpose tools. The striking face is flat with beveled edges. The blade has a sharp, hardened cutting edge. Handles may be wood, solid steel, or glass fiber and may be furnished with rubber grips.

Bricklayer's hammers are designed for setting and cutting (splitting) bricks, masonry tile, and concrete blocks, and for chipping mortar from bricks.

Never use a bricklayer's hammer to strike metal or to drive struck tools (including brick sets and chisels).

Riveting and setting hammers. Riveting hammers are designed for driving and spreading unhardened rivets on sheet metal work. The setting hammer is designed for forming sharp corners, closing and peening seams and lock edges, and glazing points. Consult manufacturers for specific needs in other types of hammers, such as scaling, chipping, soft face, nonferrous, magnetic, engineer's, blacksmith's, and spalling hammers and woodchopper's mauls.

Punches

Hand punches are made in various patterns from square, round, hexagonal, or octagonal steel stock. Punches are designed to mark metal and other materials softer than the point end, drive and remove pins and rivets, and align holes in different sections of materials.

When a hole is punched in sheet metal, some backing is necessary. A lead cake or the end grain of a block of wood may be used in order to minimize stretching the metal too much around the hole. A steel plate should not be used for backing because it does not allow the end of the punch to pass through the sheet metal and thus can ruin the punch. After the hole is punched, the metal may need to be reshaped around the hole; the sheet must then be turned over and reflattened on a flat iron surface with a mallet.

Never use a punch with a mushroomed struck face or with

a dull, chipped, or deformed point. Discard any punch if it is bent, cracked, or chipped. Redress cutting edge point to original contour as required.

Metal-cutting tools

Chisels. Cold chisels have a cutting edge at one end for cutting, shaping, and removing metal softer than the cutting edge itself (such as cast iron, wrought iron, steel, bronze, and copper) and a struck face on the opposite end.

A chisel can be used to cut metal in hard-to-reach places; it will cut any metal that is softer than its own cutting edge, which is hardened and kept sharp, at a 70-degree angle (for softer materials, a 60-degree angle is satisfactory because less pressure is required).

Four principal kinds of chisels are used in bench metal work (Figure 14-8):

1. A flat cold chisel is most commonly used for cutting, shearing, and chipping. The width of the cutting edge determines the size. For ordinary work, a ¾ in. (19 mm) chisel and a one-pound hammer are used.
2. A diamond point chisel cuts V-grooves and sharp interior angles.
3. A cape chisel is used for cutting keyways, slots, or square corners.
4. A round nose chisel cuts rounded or semicircular grooves and corners that have fillets; it can draw back a drill that wandered from its intended center.

The flat chisel should be ground with a slightly convex cutting edge; this reduces the tendency for its corners to dig into the surfaces being chiseled and concentrates the force directly to the material being cut.

Factors determining the selection of a cold chisel are the materials to be cut, the size and shape of the tool, and the depth of the cut to be made. The chisel should be heavy enough so that it will not buckle or spring when struck.

A chisel only large enough for the job should be selected so that the blade is used rather than only the point or corner.

As discussed earlier, the proper hammer for the job should be used. The striking face of the appropriate type and size hammer shall have a diameter approximately ⅜ in. (9.5 mm) larger than the struck face of the chisel.

Some workers prefer to hold the chisel lightly in the hollow of the hand with the palm up, supporting the chisel by the thumb and first and second fingers. They claim that if the hammer glances from the chisel, it will strike the soft palm rather than the knuckles. Other workers think that a grip with the loose fist, keeping the fingers relaxed (Figure 14-9), holds the chisel steadier and minimizes the chances of glancing blows. Moreover, in some positions, this is the only grip that is natural or even possible. For regular use, a sponge rubber pad, forced down over the chisel, provides a protective cushion for the hand (Figure 14-10).

Bull chisels held by one employee and struck by another require the use of tongs or a chisel holder to guide the chisel so that the holder will not be exposed to injury.

When shearing and chipping with a cold chisel, the worker should hold the tool at an angle that permits one bevel of the

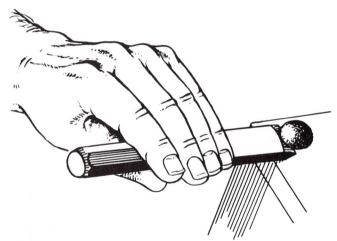

Figure 14-9. One way to hold a cold chisel.

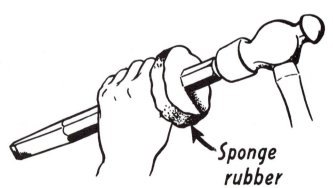

Figure 14-10. Inexpensive and effective hand protection is provided by a sponge rubber shield forced onto hammer-struck tools. Punch and chisel holders are commercially available.

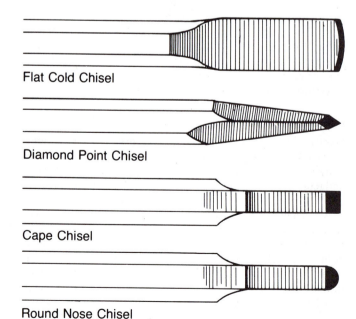

Flat Cold Chisel

Diamond Point Chisel

Cape Chisel

Round Nose Chisel

Figure 14-8. Four kinds of commonly used metal-working chisels.

cutting edge to be flat against the shearing plane. Protective holders are also commercially available.

Discard any chisel if it is bent, cracked, or chipped. Redress cutting edge or struck end to original contour as required.

When grinding a chisel, do not apply too much pressure to the head because the heat generated can draw the temper. Immerse the chisel in cold water periodically when grinding.

Stamping and marking tools of special alloy and design are available. If possible, marking tool holders should be used so that the worker does not have to hold fingers close to the face of the tool being struck.

Tap and die work requires certain precautions. The work should be firmly mounted in the vise. A tap wrench of the proper size should be secured. The hands should be kept away from broken tap ends. If a broken tap is removed by using a tap extractor or a punch and hammer, the worker should wear safety goggles. When a long thread is being cut with a hand die, the hands and arms should be kept clear of the sharp threads coming through the die.

Hack saws should be adjusted and tightened in the frame to prevent buckling and breaking, but should not be tight enough to break off the pins that support the blade. Install blade with teeth pointing forward.

Blades with 14 teeth to the inch should be used for cutting soft metal; 18 teeth for tool steel, iron pipe, hard metal, and general shop use; 24 teeth for drill rods, sheet metal, copper and brass, and tubing; and 32 teeth for thin sheet metal (less than 18 gage or 1.2 mm) and tubing. When thin metals are cut, make sure that at least two teeth are in contact with the surface being cut.

Pressure should be applied on the forward stroke only. Lift the saw slightly and pull back in the cut lightly to protect the teeth. Cutting speed of 40 to 60 strokes a minute is recommended. If the blade is twisted or too much pressure is applied, the blade may break and cause injury to the hands or arms of the user. Do not continue an old cut after changing to a new blade; it may bend and break because the set of the teeth on the new blade is thicker than that of a used blade.

Files. Selection of the right kind of file for the job can prevent injuries, lengthen the life of the file, and increase production. Files should not be used without a secure handle. Because the extremely hard and brittle steel of the file chips easily, the file should never be cleaned by being struck against a vise or other metal object—a file-cleaning card should be used.

For the same reason, a file should not be hammered or used as a pry. Such abuse frequently results in the file's chipping or breaking, causing injury to the user. A file should not be made into a center punch, chisel, or any other type of tool because the hardened steel may fracture in use.

Clamp the work to be filed in a vise at about waist height. The correct way to hold a file is to grasp the handle firmly in one hand and use the thumb and forefinger of the other to guide the point. This technique will give good control and ensure better and safer work. To file, push the file forward while bearing down on it. Release the pressure and bring the file back to its original position; if pressure is not released, the teeth will wear excessively.

A file should never be used without a smooth, crack-free handle; otherwise, if the file should slip or be struck by a revolving part of a machine, the tang may puncture the palm of the hand, the wrist, or other part of the body. Under some conditions, a clamp-on, raised offset handle may be useful to give extra clearance for the hands.

When work to be filed is placed in a lathe, the job should be done left-handedly, with the file and hands clear of the chuck jaws or the dog. A fine mill file or long-angle lathe file should be used. Take long, even strokes across the rotating work.

Hand snips should be heavy enough to cut the material so easily that the worker needs only one hand on the snips and can use the other to keep the edges of the cut material pulled aside. The material should be well supported before the last cut is made so that the cut edges do not press against the hands.

Jaws of snips should be kept tight and well lubricated. Select a hand snip that cuts easily and is not tiresome to use. Snips are divided into two groups—those for straight cuts and those for circular cuts. Snips for thicker sheets and harder materials have longer handles, alloy steel blades, and sometimes, special arrangements of levers to make cutting easier. Do not hammer on the handles or jaws of the snips.

Workers should wear safety goggles because small particles often fly with considerable force. Leather or heavy canvas work gloves will help prevent hand cuts or scratches due to handling sharp edges of the sheet metal. When cutting long sheet metal pieces, push down the sharp ends next to the hand holding the snips. File off any jagged edges or slivers after cutting. Hand snips are not designed to cut wire.

Cutters used on wire, reinforcing rods, or bolts should have ample capacity for the stock; otherwise, the jaws may be sprung or spread. Also, a chip may fly from the cutting edge and injure the user.

Cutters are designed to cut at right angles only. They should not be "rocked" to facilitate the cut because they are not designed to take the resulting strain. This practice can also cause the knives to chip.

Cutters require frequent lubrication. To keep cutting edges from becoming nicked or chipped, cutters should not be used as nail pullers or pry bars.

Cutter jaws should have the hardness specified by the manufacturer for the particular kind of material to be cut. By adjustment of the bumper stop behind the jaws, cutting edges should be set to have a clearance of 0.003 in. (0.76 mm) when closed.

Wood-cutting tools

Safety goggles should be worn when using wood cutting tools. Wood-cutting tools should be used so that, if a slip should occur, the direction of force will be away from the body. For efficient and safe work, wood cuting tools should be kept sharp and ground to the proper angle. A dull tool does a poor job and may stick or bind. A sudden release may throw the user off balance or cause a hand to strike an obstruction. Dressing of wood-cutting tools is covered earlier in this chapter under Maintenance and Repair.

Wood chisels. Inexperienced employees should be instructed in the proper method of holding and using chisels. Wood handles should be free of splinters. If the wooden handle of a chisel is designed to be struck by a wood or plastic mallet, the handle should be protected by a metal or leather band to prevent its splitting. Heavy-duty or framing chisels are made with solid or molded handles and can be struck with a steel hammer. Finish or paring cuts are made with hand pressure alone. Be sure the edge is sharp and that both hands are back of the cutting edge at all times. The work should be clamped or otherwise secured so it cannot move while cutting is being done.

The work to be cut should be free of nails to avoid damage to the blade of the chisel. Should metal be struck by the cutting edge, a chip from the chisel might fly.

The wood chisel should not be used as a pry or a wedge. The steel in a chisel is hard so that the cutting edge will hold, and may break if the chisel is used as a pry.

When not in use, chisels should be kept in a rack, on a workbench, or in a slotted section of the tool box so that the sharp edges will be out of the way and will not contact metal surfaces.

Saws should be selected for the work they must do. For cutting across the grain of the wood, use a crosscut saw; for cutting with the grain, use a ripping saw. The difference between them is their teeth angle and shape. For fast crosscut work on green wood, use a coarse saw (4 to 5 points per inch); for smooth, accurate cutting of dry wood, use a fine saw (8 to 10 points per inch). For ripping, use a coarse saw for thick stock and a fine saw for thin. The number of points per inch is stamped on the blade.

Saws should be kept sharp and the teeth kept well set to prevent binding. When not in use, saws should be wiped off with an oily rag and kept in racks or hung by the handle to prevent the teeth from being dulled.

Avoid sawing nails, or sheetrock with a saw—use a keyhole saw with a metal-cutting blade to cut the nail; use an old saw for cutting sheetrock.

Do not drop a saw because the handle can break or become loose or teeth can be nicked.

Axes and hatchets. The double-bit axe is usually used to fell, trim, or prune trees and to split and cut wood. It is also used for notching and shaping logs and timbers. The single bit axe, in addition to the above uses, can be used to drive wood stakes with the face.

The cutting edges of axes are designed for cutting wood and equally soft materials. They should never be struck against metal, stone, or concrete.

Never use an axe as a wedge or maul. Never strike with the sides, and never use an axe with a loose or damaged handle. Steel wedges are available for splitting wood, like splitting logs for the fireplace. Use a sledge hammer or maul for driving the wedges.

To use an axe safely, workers must be taught to lift it properly, to swing correctly, and to place the stroke accurately. The proper grip for a right-handed person is to have the left hand about 3 in. (7.6 cm) from the end of the handle, and the right hand about three-fourths of the way up. A left-handed person should reverse the position of the hands.

A narrow-bladed axe should be used for hard wood, and a wide axe for soft wood. A sharp, well-honed axe yields better chopping speed and is much safer to use because it bites into the wood (Figure 14-11). A dull axe will often glance off the wood being cut and may strike the user in the foot or leg.

A chopper should make sure there is a clear circle in which to swing the axe before starting to chop. Also, all vines, brush, and shrubbery within the range should be removed, especially overhead vines that may catch or deflect the axe. Safety shoes,

Figure 14-11. Good honing saves labor and makes an axe safer to use. The axe should be honed after each sharpening and each use. Correct honing motion is shown here. A double bit axe can be ground to different cutting bevels for various types of work.

safety goggles, and pants of durable material should be worn when using an axe.

Axe blades should be protected with a sheath or metal guard wherever possible. When the blade cannot be guarded, it is safer to carry the axe at one's side. The blade on a single-edged axe should be pointed down.

Adzes are hazardous tools in the hands of inexperienced employees and should be handled only by those skilled in their use. Workers should straddle the work or, as is necessary at times, stand at the side of the material. They should see that the adze is sharp and that the handle is sound and fitted securely in the head. Safety shoes, shin guards, and safety goggles should be worn. When not in use, the adze should be set aside in a safe place with its cutting edge covered or secured to protect passers-by or, temporarily, left stuck in the timber.

Hatchets are used for many purposes and frequently cause injury. For example, a worker attempting to split a small piece of wood while holding it in the hands may be cut on the hands or fingers.

To properly start a hatchet cut, it is a good practice to strike the wood lightly with the hatchet, then force the blade through by striking the wood against a solid block of wood.

Miscellaneous cutting tools

Planes, scrapers, bits, and drawknives should be used only by trained employees. These tools should be kept sharp and in good condition. When not in use, they should be placed in a rack on the bench, or in a tool box in such a way that will protect the user and prevent damage to the cutting edge.

Knives are more frequently the source of disabling injuries than any other hand tool. In the meatpacking industry, hand knives caused more than 15 percent of all disabling injuries, more than is caused by any other agency.

The principal hazard in the use of knives is that the hands may slip from the handle onto the blade or that the knife may strike the body or the free hand. A handle guard or a finger ring (and swivel) on the handle can reduce these hazards. Adequate guarding is important. Knives must be kept sharp. Knives with worn handles should be replaced. Retractable blade knives should be used wherever possible.

The cutting stroke should be away from the body. If that is not possible, then the hands and body should be in the clear, a heavy leather apron or other protective clothing should be worn, and, where possible, a rack or holder should be used for the material to be cut. Jerky motions should be avoided to help maintain balance.

Be sure employees are trained and supervised. Employees who must carry knives with them on the job should keep them in sheaths or holders. Never carry a sheathed knife on the front part of a belt—always carry it over the right or left hip, toward the back. This will prevent severing a leg artery or vein in the case of a fall.

Knives should never be left lying on benches or in other places where they may cause hand injuries. When not in use, they should be kept in racks with the edges guarded. Safe placing and storing of knives and other sharp hand tools is important to knife safety.

Supervisors should make certain that employees who handle knives have ample room in which to work so that they are not in danger of being bumped by trucks, the product, overhead equipment, or other employees. For instance, a left-handed worker should not stand close to a right-handed person; the left-handed person might be placed at the end of the bench or otherwise given more room.

Supervisors should be particularly careful about the hazard of employees leaving knives hidden under the product, under scrap paper or wiping rags, or among tools in work boxes or drawers. Knives should be kept separate from other tools to protect the employee as well as the cutting edge of the knife.

Work tables should be smooth and free of slivers. Floors and working platforms should have slip-resistant surfaces and should be kept unobstructed and clean. If sanitary requirements permit mats or wooden duck boards, they should be in good repair, so workers do not trip or stumble. Conditions which cause slippery floors should be controlled as much as possible by good housekeeping and frequent cleaning.

Careful job and accident analysis may suggest some slight change in the operating procedure which will make knives safer to use. For instance, on some jobs special rigs, racks, or holders may be provided so it is not necessary for the operator to stand too close to the piece being cut.

The practice of wiping a dirty or oily knife on the apron or clothing should be discouraged. The blade should be wiped with a towel or cloth with the sharp edge turned away from the wiping hand. Sharp knives should be washed separately from other utensils and in such a way that they will not be hidden under soapy water.

Horseplay should be prohibited around knife operations. Throwing, "fencing," trying to cut objects into smaller and smaller pieces, and similar practices are not only dangerous but reflect poor training and inadequate supervision.

Supervisors should make sure that nothing is cut that requires excessive pressure on the knife, such as frozen meat. Food should be thawed before it is cut or else it should be sawed. Knives are not a substitute for can openers, screwdrivers, or ice picks.

To cut corrugated paper, a hooked linoleum knife permits good control of pressure on the cutting edge and eliminates the danger of the blade suddenly collapsing as in the case of a pocket knife. Be sure hooked knives are carried in a pouch or heavy leather (or plastic) holder (Figure 14-4). The sharp tip must not stick out.

Ring knives—small, hooked knives attached to a finger— are used where string or twine must be cut frequently. Supervisors should make sure that the cutting edge is kept outside the hand, not pointed inside. A wall-mounted cutter or blunt-nose scissors would be safer.

Carton cutters are safer than hooked or pocket knives for opening cartons. They not only protect the user, but eliminate deep cuts that could damage carton contents. Frequently, damage to contents of soft plastic bottles may not be detected immediately; subsequent leakage may cause chemical burns, damage other products, or start a fire.

A brad awl should be started with the edges across the grain to keep the wood from splitting. Then it should be held at right angles to the surface to prevent its slipping.

Material handling tools

Crowbars. Whenever a crowbar is needed, the proper size and kind of bar for the job should be used.

The crowbar should have a point or toe of such shape that it will grip the object to be moved, and a heel to act as a pivot or fulcrum. In some cases, a block of wood under the heel will prevent the crowbar from slipping and injuring the hand.

Crowbars not in use should be secured if stood on end so that they will not fall; or if laid on the ground, placed where they will not create a stumbling hazard.

Hooks. Hand hooks used for handling material should be kept sharp so that they will not slip when applied to a box or other object.

The handle should be strong and securely attached and shaped to fit the hand. The handle and the point of long hooks should be bent on the same plane so that the bar will lie flat when not in use and not constitute a tripping hazard (Figure 14-12). The hook point should be shielded when not in use.

Shovels. Shovel edges should be kept trimmed and handles should be checked for splinters. Workers should wear heavy shoes with sturdy soles, preferably safety shoes. They should have their feet well separated to get good balance and spring in the knees. The leg muscles should take much of the load.

To reduce the chance of injury, the ball of the foot—not the arch—should be used to press the shovel into clay or other stiff

WRONG   RIGHT

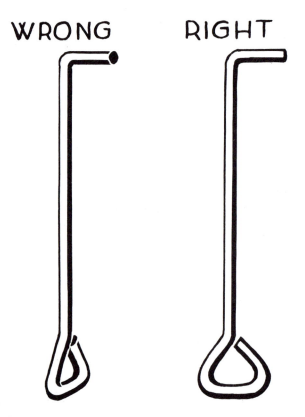

Figure 14-12. Hand hooks can be made safer if the hook is bent so that it will lie flat when set down on a flat surface.

material. If the instep is used and the foot slips off the shovel, the sharp corner of the shovel may cut through the worker's shoe and into the foot.

Dipping the shovel into a pail of water occasionally will help to keep it free from sticky material, making it easier to use and less likely to cause strain. Greasing or waxing the shovel blade will also prevent some kinds of material from sticking. Teflon-coated shovels prevent sticking of certain materials.

When not in use, hang up shovels, stand them against a wall, or keep them in racks or boxes.

Rakes. A rake should not be left with the prongs turned upward where they may be stepped on, causing foot injury or causing the handle to fly up. Rakes should be racked when not in use.

Torsion tools

Safe use of all wrenches requires the user always to be alert and prepared for the possibility that the wrench may slip off the fastener, the fastener may suddenly turn free, the wrench may break, or the fastener may break. Therefore, the user should always be braced in such a way that should the wrench become free for any reason, the user will not lose balance and be injured by falling into moving machinery or falling off a platform. The user should always inspect a wrench for flaws. Because a previous overloading or misuse of the tool may have weakened it to the point that it will not even carry a normal load, the user should be safely braced when pulling hard on a wrench.

Wrenches should not be ground to change their size or reduce their dimension to fit into close quarters. Instead, a wrench of the correct size and fit should be used. It is unsafe practice to try to make the wrong wrench fit by using shims.

The great variety of wrenches used for turning nuts, bolts, and fittings makes it important that workers know the purpose and limitations of each type and size.

Open-end wrenches (Figure 14-13a) have strong jaws and are satisfactory for medium-duty turning. They are susceptible to slipping if they do not fit properly or are used incorrectly.

Combination wrenches (Figure 14-13c) have a box end and an open end. They are very handy for speeding the turning with the open end and using the box end for initial loosening or final tightening.

Box and socket wrenches are indicated where a heavy pull is necessary and safety is a consideration. Box wrenches (Figures 14-13b and 14-14) and socket wrenches (Figure 14-15) completely encircle the nut, bolt, or fitting and grip it at all corners, as opposed to the two corners gripped by an open-end wrench. They will not slip off laterally, and they eliminate the dangers of sprung jaws.

Socket and box wrenches normally come in three styles of openings—single hex, double hex, or double square (Figure 14-14). On square-headed bolts, nuts, and fittings, either a sin-

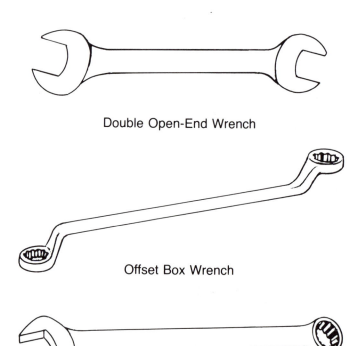

Double Open-End Wrench

Offset Box Wrench

Combination Wrench

Figure 14-13a–c. Three types of wrenches. The offset provides hand clearance and allows the worker to reach into recesses and over obstructions. The combination wrench has both jaws the same size. The greater gripping strength of the box is used to remove the nut quickly.

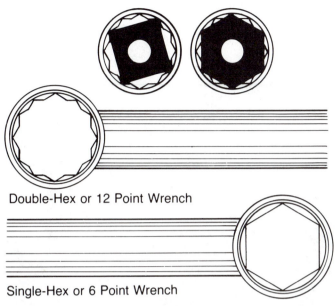

Double-Hex or 12 Point Wrench

Single-Hex or 6 Point Wrench

Figure 14-14. The ring of a box wrench has 6 or 12 points. The 12-point wrench need only be turned 30 degrees before engaging a new set of flats and is therefore handy for use in confined spaces. Never use a 12-point box wrench on a square nut or bolt because it may slip.

gle or double square design should be used. Single or double hexagonal wrenches are designed for hexagonal-shaped fittings, bolts, and nuts. Ratchet handles and universal joint fittings for socket wrenches allow them to be used where space is limited (Figure 14-15).

Never overload the capacity of a wrench by using a pipe extension on the handle or striking the handle of a wrench with a hammer. Hammer abuse weakens the metal of a wrench and can cause the tool to break. Special heavy-duty wrenches are available which can be used with handles as long as three feet. For extra stubborn bolts and nuts, heavy-duty, sledge-type box wrenches are available. These are of a heavy design, properly tempered, and have a striking surface for the hammer or sledge (Figure 14-16). Where possible, penetrating oil should be used to first loosen tight nuts.

There is a correct wrench for every nut and bolt. Oversize openings will not grip the corners securely and shims should not be used to compensate for an oversize opening. The use of the wrong size wrench can round the corners of the bolt, or cause slippage, as well as make it difficult to then apply the proper size. Be sure to use the proper tool and not a makeshift approach (Figure 14-17a and b).

Sockets should be kept clean of dirt and grime inside the socket. Dirt will prevent the socket from seating fully and the concentration of the pull force at the end of the socket opening (even with a moderate pull) can easily damage the socket or nut.

A common cause of socket and box wrench breakage is "cocking." Cocking is a situation in which the tool does not fit securely on the bolt or nut but fits at an angle. This concentrates the entire strain on a smaller area, making the tool or nut vulnerable to fracture.

Torque wrenches measure the amount of twisting force that is applied to a nut or bolt by means of a dial or calibrated arm.

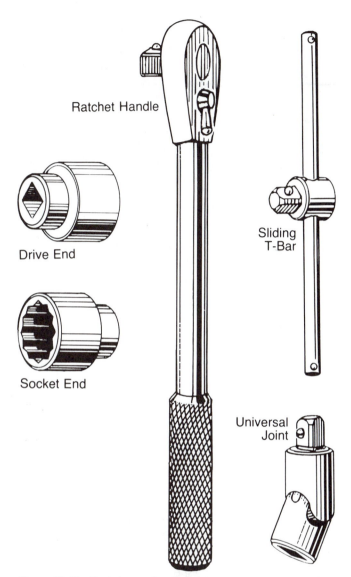

Ratchet Handle

Drive End

Socket End

Sliding T-Bar

Universal Joint

Figure 14-15. Socket wrenches (left) have a square hole in the drive end into which various handles can be fitted. The ratchet handle increases turning speed because the socket does not have to be removed from the nut between turns. The universal joint fitting provides flexibility where space is limited.

Torque is a turning force (see Figure 14-18), usually measured in foot-pounds or newton-meters, that is the force used times the length of the lever used to apply it. For example, a 15-lb force applied to a 2-ft lever give 30 ftlb torque. The metric unit is the newton meter (1 kgf · m = 9.6 N · m; 1 lbf · ft = 1.356 N · m).

Torque wrenches are used where the torque has been specified for the job or where it is important that all fasteners be fully and uniformly tight. If a torque has been specified for a particular application, it is unsafe not to measure that torque with a torque wrench. Torque wrenches must be used carefully and recalibrated frequently in accordance with the manufacturer's instructions.

Figure 14-16. Using a striking-face wrench designed to be hammered is a safe approach where large nuts must be set up tight or frozen nuts loosened.

Figure 14-17a. This makeshift is devised from the nearest tools at hand. One slip could cause injury.

Adjustable wrenches are generally recommended for light-duty jobs or when the proper size fixed-opening wrench is not available. They may slip because of the difficulty in setting the correct size and the tendency for the jaws to "work" as the wrench is being used. They possess one advantage in that they are easily adjusted to fit metric system nuts and bolts.

Unless the space in which the job is being done makes the method impractical, an adjustable wrench should be placed on the nut with the open jaws facing the user. With the open jaw facing the user, the pulling force applied to the handle tends to force the movable jaw onto the nut. For that reason, and for reasons of safety, wrenches should be pulled, not pushed (Figure 14-19). According to the manufacturers, the movable jaws on adjustable wrenches are weaker than the fixed jaws.

Pipe wrenches. Workers, especially those on overhead jobs, have been seriously injured when pipe wrenches slipped on pipes or fittings, causing the worker to lose balance and fall. Pipe wrenches, both straight and chain tong, should have sharp jaws and be kept clean to prevent their slipping.

The adjusting nut of the wrench should be inspected frequently. If it is cracked, the wrench should be taken out of service. A cracked nut may break under the strain, causing complete failure of the wrench and possible injury to the user.

Using a wrench of the wrong length is also a source of accidents. A wrench handle too small for the job does not give proper grip or leverage. An oversized wrench handle may strip the threads or break the work suddenly, causing a slip or fall.

A piece of pipe slipped over the handle to give added leverage also can strain a pipe wrench or the work to the breaking point. Using a makeshift as an extension to secure greater leverage may easily cause the wrench head to break. The handle of every wrench is designed to be long enough for the maximum allowable safe pressure. Handle extensions, also known as cheater bars, should not be used to gain extra turning power, unless the wrench is so designed.

A pipe wrench should never be used on nuts or bolts, the

Figure 14-17b. The correct tool at the right time makes an otherwise hard job safe and easy.

corners of which will break the teeth of the wrench, thereby making it unsafe to use later on pipe and fittings. A pipe wrench used on nuts and bolts ruins their heads. It should not be used on valves or small brass, copper, or other soft fittings which may be crushed or bent out of shape. A wrench should not be struck with a hammer nor used as a hammer unless it is a specialized type specifically designed for such use.

Pipe tongs should be placed on the pipe only after the pipe

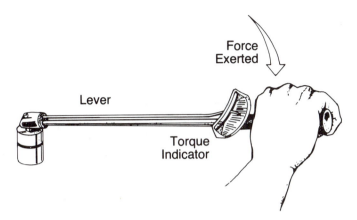

Figure 14-18. The torque, or twisting force, exerted on a nut or bolt is directly proportional to the length of the wrench handle and the pulling force exerted on it.

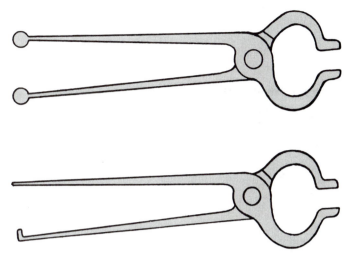

Figure 14-20. Ends of tongs can be upended, or projections can be welded on them to give clearance for user's hands.

Figure 14-19. Correct use of a wrench. The wrench is in good condition and securely gripped. Hand of worker is braced and clear in the event that the nut should suddenly turn. To protect worker's hand, wrench is pulled, not pushed.

has been lined up and is ready to be made up. A 3- or 4-in. (7.6 or 10 cm) block of wood should be placed near the end of the travel of the tong handle and parallel to the pipe to prevent injury to the hands or feet in the event the tongs slip.

Workers should neither stand nor jump on the tongs nor place extensions on the handles to obtain more leverage. They should use larger tongs, if necessary, to do the job.

Tongs usually are brought, but some companies make their own to perform specific jobs. Often they are designed in such a way that the hands are pinched when the tongs are closed. To prevent pinching, the end of one handle should be up-ended toward the other handle, to act as a stop (Figure 14-20).

Pliers are often considered a general-purpose tool and are often misused for purposes for which they were not designed. Pliers are meant for gripping and cutting operations; they are not recommended as a substitute for wrenches because their jaws are flexible and frequently slip when used for this work. Pliers also tend to round the corners of bolt heads and nuts and leave jaw marks on the surface, making it difficult to use a wrench at some future time.

Side-cutting pliers sometimes cause injuries when short ends of wire are cut. A guard over the cutting edge (Figure 14-21) and the use of safety goggles will prevent flying short ends from causing injuries.

Be certain that pliers used for electrical work are insulated. In addition, employees should wear electrician's gloves, if company policy requires them to. *Warning:* The cushion grips on handles are primarily for comfort. Unless specified as insulated handles, they are *not* intended to give any degree of protection against electric shock and should *not* be used on live electric circuits.

Special cutters for heavy wire, reinforcing wire, and bolts are safer than makeshift tools. It is important that the cutting edges apply force at right angles to the wire or other work being cut (Figure 14-22). The cutter should not be used near live electrical circuits and should be used only for the rated capacity specified by the manufacturer. Eye protection should be worn.

Special cutters include those for cutting banding wire and strap. Claw hammers and pry bars should not be used to snap metal banding material. Only cutters designed for the work provide safe and effective results.

Nail band crimpers make it possible to keep the top band on kegs and wood barrels after nails or staples have been removed. Use of these tools eliminates injury caused by reach-

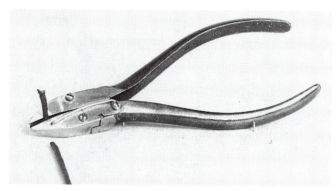

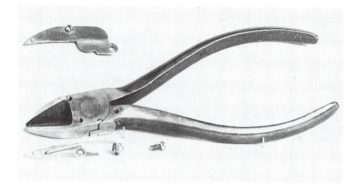

Figure 14-21. At left, the guard on a pair of pliers holds a wire end after it has been cut, thus preventing it from flying. At right, a disassembled view of the same pliers. (Reprinted with permission from Capco Co., Ottumwa, Iowa.)

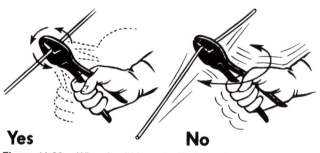

Yes   **No**

Figure 14-22. Wire should be cut with the cutter knives swung up and down, at right angles to the wire (as shown at left). Do not bend wire back and forth (as shown at right). Be sure to keep cut ends under control. Stand back when cutting wires so that if an end snaps back, it cannot reach you.

ing into kegs or barrels that have projecting nails or staple points.

Pullers are the only quick, safe, and easy way to pull a gear, wheel, pulley, or bearing from a shaft. Prybars and chisels should not be used because they concentrate the force at one point and tend to cock the part on the shaft. Select the correct-sized puller. The jaw capacity should be such that the jaws press tightly against the part being pulled. Use a puller with as large a pressure screw as possible.

Spark-resistant tools

So-called spark-resistant tools made of nonferrous materials (such as beryllium copper alloy) are sometimes advised for use where flammable gases, highly volatile liquids, and other explosive substances are stored or used. There is some question about the hazard of friction sparks igniting gasoline vapors and petroleum products. Nonferrous tools reduce the hazard from sparking but do not eliminate it. They need inspection before each use to be certain that they have not picked up foreign particles that could produce friction sparks, thereby obviating the value of these special tools.

Soldering irons

Soldering irons are the source of burns and of illness resulting

from inhalation of fumes. Insulated, noncombustible holders will practically eliminate the fire hazard and the danger of burns from accidental contact. Ordinary metal covering on wood tables is not sufficient because the metal conducts heat and may ignite the wood below.

Holders should be designed so that employees cannot accidently touch the hot irons if they reach for them without looking. The best holder completely encloses the heated surface and is inclined so that the weight of the iron prevents it from falling out. Also see National Safety Council Data Sheet 445, *Hand Soldering and Brazing.*

Fumes, from soldering can be toxic and/or irritating. Soldering fumes should be removed through local exhaust ventilation, especially with a continuous production operation. Lead oxides and chlorides are released when soldering with lead-tin solder and zinc chloride flux. Lead oxides and formaldehyde are released when soldering with rosin core solder. There are different types of solder and the hazards from each should be known before beginning work. Air sampling must be performed to determine if hazardous amounts of contaminants are present.

Lead solder particles should not be allowed to accumulate on the floor and on work tables. If the operation is such that the solder or flux may spatter, employees should wear faceshields or do the work under a transparent shield.

PORTABLE POWER TOOLS

Portable power tools are divided into five primary groups according to the power source: electric, pneumatic, gasoline, hydraulic, and powder actuated. Several types of tools, such as saws, drills, and grinders, are common to the first three groups; hydraulic tools are used mainly for compression work; powder-actuated tools are used exclusively for penetration work, cutting, and compression.

A portable power tool presents similar hazards as a stationary machine of the same kind, in addition to the risks of handling. Typical injuries caused by portable power tools are burns, cuts, and strains. Sources of injury include electric shock, particles in the eyes, fires, falls, explosion of vapors or gases, and falling tools.

The source of power should always be disconnected before accessories on a portable tool are changed, and guards should be replaced or put in correct adjustment before the tool is used again.

A tool should not be left in an overhead place where there is a chance that the cord or hose, if pulled, will cause the tool

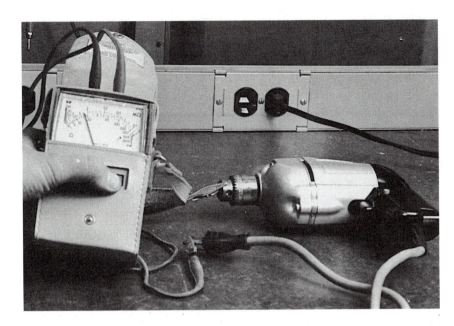

Figure 14-23. Insulation resistance tester has test leads attached to grounding pin of attachment plug and chuck of portable tool. Meter indicates condition of insulation of portable tool and permits identification of impending failures. (Reprinted with permission from Daniel Woodhead Company.)

to fall. The cord or hose and the tool may be suspended by a tool balancer which keeps them out of the operator's way. Cords and hoses on the floor create a stumbling or tripping hazard. They should be suspended over aisles or work areas, or if laid across the floor, protected by wooden strips or special raceways. Suspend them in such a way that they will not be struck by other objects or by material being handled or moved. An unexpected pull might cause the tool to jam or be dislodged from its holder and cause injury or damage to the tool.

Do not hang cords or hoses over nails, bolts, or sharp edges. They should also be kept away from oil, hot surfaces, and chemicals.

Because of the extreme mobility of power-driven tools, they can easily come in contact with the operator's body. At the same time, it is difficult to guard such equipment completely. There is the possibility of breakage because the tool may be dropped or roughly handled. Furthermore, the source of power (electrical, mechanical, air, hydraulic, or powder cartridge) is brought close to the operator, thus creating additional potential hazards.

When using powder load equipment for driving anchors into concrete, or when using air-driven hammers or jacks, it is recommended that proper hearing protection be used. All companies and manufacturers of portable power tools attach to each tool a set of operating rules or safe practices. Their use will supplement the thorough training which each power tool operator should receive.

Power-driven tools should be stored in secured places and not left in areas where they may be struck by passers-by or otherwise activated.

Selection of tools

Replacement of a hand tool by a power tool designed for the same purpose may mean merely a substitution of electrical or mechanical hazards for relatively less serious manual hazards. Therefore, the safety professional should anticipate the new problems and avoid as many as possible through insistence upon safe design and proper training.

The tool supplier should be given complete information about the job on which a tool is to be used so that the most appro-

priate tool can be recommended. Factors to be considered include clearance in the working quarters, type of job to be done, and the nature and thickness of materials.

Portable power tools designed to be used intermittently or on light work are generally designated as "home owners grade." Those intended for continous operation and production service or for heavy work are usually indentified as "industrial duty."

For safe operation of portable power tools, workers should be trained in the proper selection and limitations of the tools used, and they should be taught never to tackle a job with an undersized tool. A tool that is too light may not only fail, but may cause undue fatigue to the operator.

Inspection and repair

Periodic inspections are essential to the maintenance of power tools. In addition to uncovering operating defects and preventing potentially costly breakdowns, the inspection of tools and their timely repair may prevent hazardous conditions from developing. A schedule for systematic inspection, at least annually, and maintenance records for each tool should be set up. Defective tools should be tagged and withdrawn from service until repaired.

Electric tools should be checked periodically. (See Figure 14-23.) Some companies provide for a visual or external inspection at the tool room each time a tool is returned and for a thorough "knock down" inspection at specified intervals.

A colored tag can be used to identify when the equipment was last inspected. The important thing is to record the condition of the tool and to correct any unsafe conditions.

Employees should be instructed and trained to inspect tools and to recognize and report (and, if authorized, to correct) defects. The extent of this inspection and of the responsibility for correcting defects should be clearly outlined so that there is neither unnecessary duplication of effort nor misunderstanding about the responsibility for maintenance. A convenient reminder of points to check is a card similar to that shown in Figure 14-24. Employees must be warned not to do makeshift repairs and to do no repair work unless authorized.

Power tools should be cleaned with a recommended nonflam-

PORTABLE ELECTRIC TOOLS

Inspection Checklist

GENERAL
Low voltage or battery powered equipment used
in tanks and wet areas? ☐
Tools well maintained? ☐
Motors in good condition? ☐
Approved tools used in explosive atmospheres? ☐
Tools left where they cannot fall? ☐

CORDS
Insulation and plugs unbroken? ☐
Cords protected against trucks and oil? ☐
Cords not in aisles? ☐

GROUNDING
Ground wire fastener in safe condition? ☐
3-wire plug extension cord (if a 3-wire tool)? ☐
Ground wire used? ☐
Defects or minor shocks reported? ☐
Ground fault circuit interrupter used? ☐

GUARDING
Guards used on grinders and saws? ☐
Movable guards operate freely? ☐
Eye or face protection worn? ☐

Figure 14-24. An inspection checklist card can be used for portable electric tools. Such a card encourages workers to inspect equipment before and after use. More specific cards or tags simplify the prompt recording of defects and result in better maintenance records.

mable and nontoxic solvent. Air drying should be used in place of blowing with compressed air.

Electric tools

There are a limited number of rechargable battery-powered tools available. For safety from electric shock, they are the best possible tool. Added mobility, without need for an extension cord, is another safety feature.

Electric shock is the chief hazard from electrically powered tools. Types of injuries are electric flash burns, minor shock that may cause falls, and shock resulting in death.

Serious electric shock is not entirely dependent on the voltage of the power input. The ratio of the voltage to the resistance determines the amount of current that will flow and the resultant degree of hazard. The current is regulated by the resistance to the ground of the body of the operator and by the conditions under which he or she is working. It is possible for a tool to operate with a defect or short in the wiring. The use of a ground wire protects the operator and is mandatory for all but double insulated electric power tools.

Insulating platforms, rubber mats, and rubber gloves provide an additional factor of safety when tools are used in wet locations, such as in tanks or boilers and on wet floors.

Low voltage of 6, 12, 24, or 32 volts through portable transformers will reduce the shock hazard in wet locations. Standing orders should be issued to supervisors and employees to use any available safety low-voltage equipment when working in these locations.

Electric tools used in wet areas or in metal tanks expose the operator to conditions favorable to the flow of current through the body, particularly if a person is wet with perspiration. Most electric shocks from tools have been caused by the failure of insulation between the current-carrying parts and the metal frames of the tools.

Double insulated tools. Protection from electric shock, while using portable power tools, has been described as depending upon third-wire protective grounding. "Double insulated" tools, however, are available which generally provide more reliable shock protection without third-wire grounding. Paragraph 250-45 of the *National Electrical Code* permits "double insulation" for portable tools and appliances. Tools in this category are permanently marked by the words "double insulation" or "double insulated." Units designated to this category which have been tested and listed by Underwriters Laboratories Inc. will also use the UL symbol. Many U.S. manufacturers are also using the symbol

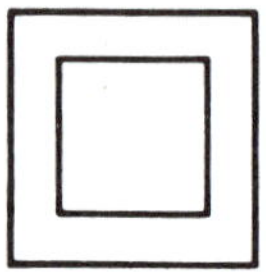

to denote "double insulation." This symbol has been widely used in most European countries.

Conventional electric tools have a single layer of functional insulation and are metal encased. For small-capacity tools, "double insulation" can be provided by encasing the entire tool in a nonconductive material which is also shatterproof. The switch is also nonconductive, so that no metal part comes in contact with the operator.

Large capacity electric tools require a more rigid design in order to provide for greater stress requirements where more power and high-torque gearing are involved. Double insulated tools with metal housings have an internal layer of protecting insulation completely isolating the electrical components from the outer metal housing (Figure 14-25). This is in addition to the functional insulation found in conventional tools. This means that in addition to the functional insulation, a reinforced or protecting insulation is also incorporated into the tool. This extra or reinforced insulation is physically separated from the functional insulation and is arranged so that deteriorating influences such as temperature, contaminants, and wear will not affect both insulations at the same time. Unless subject to immersion or extensive moisture which might nullify the double insultation, a double insulated or all-insulated tool does not require separate ground connections; the third wire or ground wire is not needed. (See Figure 14-26).

Failure of insulation is harder to detect than worn or broken external wiring, and points up the need for frequent inspection, testing with an insulation resistance tester, and thorough maintenance (see Figure 14-23). Care in handling the tool and frequent cleaning will help prevent the wear and tear that cause defects.

Grounding of portable electric tools and the proper use of ground fault circuit interrupters provide the most convenient way of safeguarding the operator. If there is any defect or short circuit inside the tool, the current is connected from the metal frame through a ground wire and does not pass through the operator's body; or, where a ground fault circuit interrupter is used, the current is shut off before a serious shock can occur.

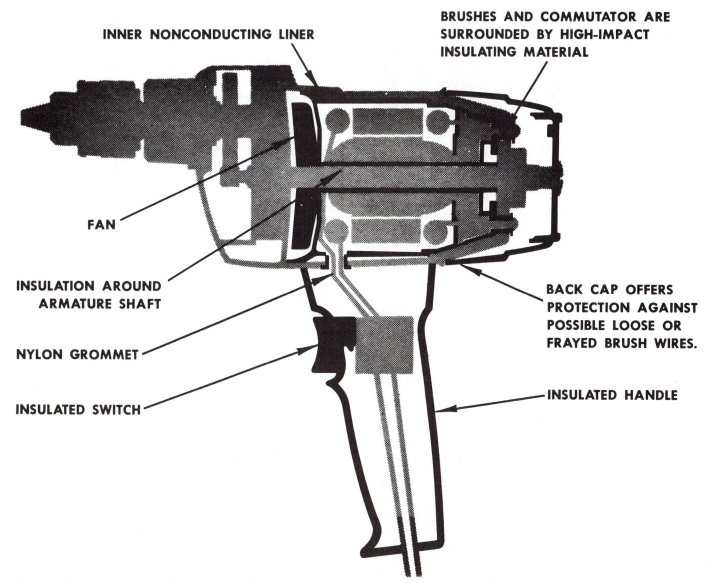

INNER NONCONDUCTING LINER

BRUSHES AND COMMUTATOR ARE SURROUNDED BY HIGH-IMPACT INSULATING MATERIAL

FAN

INSULATION AROUND ARMATURE SHAFT

NYLON GROMMET

INSULATED SWITCH

BACK CAP OFFERS PROTECTION AGAINST POSSIBLE LOOSE OR FRAYED BRUSH WIRES.

INSULATED HANDLE

Figure 14-25. On a double-insulated, shock-proof electric tool, an internal layer of protective insulation completely isolates the electrical components from the outer metal housing. (Reprinted with permission from Millers Falls Co.)

All electric power tools should be effectively grounded except the double insulated and cordless types. Correctly grounded tools are as safe as double insulated or low voltage tools, especially when used with a proper ground fault circuit interrupter. The continuity of the ground must be checked so that there will not be a false sense of security.

The noncurrent-carrying metal parts of portable and/or cord- and plug-connected equipment required to be grounded may be grounded either 1) by means of the metal enclosure of the conductors feeding such equipment, provided an approved grounding attachment plug is used, or 2) by means of a grounding conductor run with the power supply conductors in a cable assembly or flexible cord that is properly terminated in an approved grounding attachment plug having a fixed grounding contacting member (Figure 14-27). The grounding conductor may be uninsulated; if individually covered, however, it must be finished a continuous green color or a continuous green color with one or more yellow stripes.

There is an exception to 1) and 2) in that the grounding contacting member of grounding attachment plugs on the power-supply cord of portable, hand-held, hand-guided, or hand-supported tools or appliances may be of the movable, self-restoring type.

(3) By special permission, nonportable cord- and plug-connected equipment can be grounded by a separate flexible wire or strap, insulated or bare, that has been protected (as well as possible) against physical damage.

For more information on grounding refer to Chapter 15, Electricity.

Electric cords should be inspected periodically and kept in good condition. Heavy-duty plugs that clamp to the cord should be used to prevent strain on the current-carrrying parts if the cord is accidentally pulled. Terminal screws or connections on plugs and connectors should be covered with proper insulation. Employees should be trained not to jerk cords and to protect

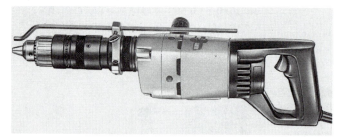

Figure 14-26. Double insulated hammer drill. A twist of the nose converts it into a regular drill. (Reprinted with permission from Millers Falls Co.)

them from sharp objects, heat, and oil or solvents that might damage or soften the insulation. To ensure the continuity of grounding, extension cords used with tools and equipment that require grounding should also be of the three-wire, grounded-connection type.

Electric drills cause injuries in several ways: a part of the drill may be pushed into the hand, the leg, or other parts of the body, the drill may be dropped when the operator is not actually drilling, and the eyes may be hit either by material being drilled or by parts of a broken drill. The use of proper eye protection will reduce the possibility of eye injury. Although no guards are available for drill bits, some protection is afforded if drill bits are carefully chosen for the work to be done, such as being no longer than necessary to do the work.

Where the operator must guide the drill with a hand, the drill should be equipped with a sleeve that fits over the drill bit. The sleeve protects the operator's hand and also serves as a limit stop if the drill should plunge through the material.

Oversized bits should not be ground down to fit small electric drills; instead, an adapter should be used that will fit the large bit and provide extra power through a speed reduction gear. Small pieces of work should be clamped or anchored to prevent whipping.

Electric saws must be equipped with guards above and below the faceplate—the lower guard must automatically retract to cover the exposed saw teeth. Employees must be trained to use the guard as intended. The guard should be checked frequently to be sure that it operates freely and encloses the teeth completely when it is not cutting. (See Figures 14-28 and -29.)

In the United States, OSHA standards require that frames and exposed metal parts of portable saws and other portable electric woodworking tools operating at over 90 volts should be grounded, unless double insulated or battery powered.

Circular saws should not be jammed or crowded into the work. The saw should be started and stopped outside the work. At the beginning and end of the stroke, or when the teeth are exposed, the operator must use extra care to keep the body out of the line-of-cut. The saws must be equipped with "dead-man" controls or a trigger switch that shuts off the power when pressure is released. Such a saw cannot run when not in use.

Grinding wheels, buffers, and wire brushes have special uses and require extra care when used. However, grinding wheels present unusual hazards to the unwary. Consequently, the storage, mounting, and use of grinding wheels requires thorough training and extensive knowledge. Persons using or maintaining grinding wheels should observe safety recommendations conforming to ANSI Standards B7.1, *Safety Requirements for the Use, Care, and Protection of Abrasive Wheels,* and B74.2, *Specifications for Shapes and Sizes of Grinding Wheels, and Shapes, Sizes, and Identification of Mounted Wheels.* Improper mounting, storage, or use of abrasive wheels could turn an important useful tool into a lethal device. More information can be obtained in Chapter 10, Metalworking Machinery.

Sanders of the belt and disk type cause serious skin burns when the rapidly moving abrasive touches the body. It is impossible to guard sanders completely; therefore, employees require thorough training in their use. The motion of the sander should be away from the body, and all clothing should be kept clear of the moving parts. Dust-type safety goggles or plastic face-

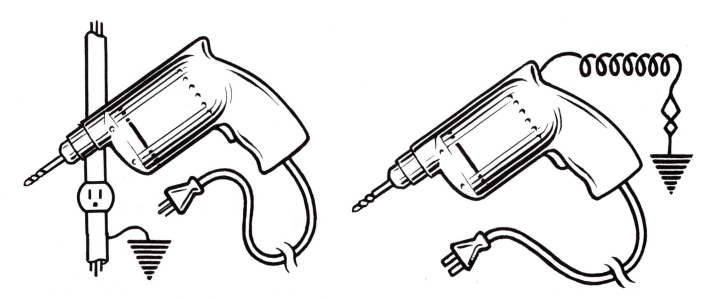

Figure 14-27. Electric drill at left is automatically grounded by means of built-in ground wire connected to rounded receptacle by means of three-prong plug. Drill at right is unacceptably grounded.

Figure 14-28. Portable band saw has guard that encloses the blade except for the actual work area. Hands are kept above the blade and are occupied.

Figure 14-29. Lower movable guard of this portable circular saw always returns to the guarded position. Operator should keep fingers away from the trigger when saw is not being used. (Reprinted with permission from Black & Decker Manufacturing Co.)

shields should be worn, and, if harmful dusts are created, a respirator approved for the exposure should be worn.

Sanders require especially careful cleaning because of the dusty nature of the work. If a sander is used steadily, it should be dismantled periodically, as well as thoroughly cleaned every day by being blown out with low-pressure air. To reiterate, compressed air used for cleaning must be 30 psig (200 kPa) or less; the operator should wear safety goggles or work with a transparent shield between his or her body and the air blast.

The fire and explosion hazard of wood sanding is considerable, and precautions should be taken to keep the dust to a minimum through adequate ventilation or, if the machine is so designed, with a dust collector or vacuum bag. Because of the extreme combustibility of wood dust, or woodwork finishes, such dust should not be disposed of in incinerators where it would burn with almost explosive force.

If much wood sanding must be done, electrical equipment designed for this exposure should be used to minimize the explosion hazard. Fire extinguishers approved for electrical or Class C fires should be available, and employees should be told what to do in case of fire.

See Chapters 9 and 10 for more details.

Air-powered tools

Air hose. An air hose presents the same tripping or stumbling hazard as do cords on electric tools. A number of manufac-turers offer self-storing recoiling air hoses that work well when suspended above work stations. Persons or material accidentally hitting the hose may unbalance the operator or cause the tool to fall from an overhead place. An air hose on the floor should be protected against trucks and pedestrians by two planks laid on either side of it or by a runway built over it. It is preferable, however, to suspend hoses over aisles and work areas.

Workers should be warned against disconnecting the air hose from the tool and using it for cleaning machines or removing dust from clothing. U.S. federal safety and health regulations mandate that air pressure in excess of 30 psig (200 kPa) must not be used for cleaning machines, and low pressure must be used only with effective chip guarding or personal protective equipment such as safety goggles. Brushing or using vacuum equipment is recommended for removing dust from clothing.

Accidents sometimes occur when the air hose becomes disconnected and whips about. A short chain attached to the hose and to the tool housing will keep the hose from whipping about if the coupling should break. In some cases, couplings should also have such chains between the sections of hose.

Air should be shut off before attempting to disconnect the air hose from the air line. Any air pressure inside the line should also be released before disconnecting.

A safety check valve installed in the air line at the manifold will shut off the air supply automatically if a fracture occurs anywhere in the line. However, it must be selected for compatibility with the air flow rate.

If kinking or excessive wear of the hose are problems, it can be protected by a wrapping of strip metal or wire. One objection to armored hose is that is may become dented and thus restrict the flow of air. This applies only to heavy-duty hose used in construction work.

Air power grinders require the same type of guarding as electric grinders. Maintenance of the speed regulator or governor on these machines is of particular importance in order to avoid over-speeding the wheel.

Regular inspection by qualified personnel at each wheel change is recommended.

Pneumatic impact tools, such as riveting guns and jackhammers, are essentially the same in that the tool proper is fitted into the gun and receives its impact from a rapidly moving reciprocating piston driven by compressed air at about 90 psig (600 kPa) pressure.

Noise levels from pneumatic impact tools should be determined to see if hearing protective devices for workers are necessary to comply with the time limits and sound level requirement of the federal standards.

Two safety devices are required. The first is an automatically closing valve actuated by a trigger located inside the handle where it is reasonably safe from accidental operation. The machine can operate only when the trigger is depressed. The second is a retaining device that holds the tool in place so that it cannot be fired accidentally from the barrel (Figure 14-30).

Figure 14-30. Chipping hammer safety retainer prevents the discharge of the tool. (Reprinted with permission from Chicago Pneumatic Tool Co.)

A good safety rule to impress on all operators of small air hammers is—do not squeeze the trigger until the tool is on the work (Figure 14-31).

Because in the use of all pneumatic impact tools there is a hazard from flying chips, the correct eye protection for the hazard must be worn. If other employees must be in the vicinity, they should be similarly protected. Where possible, screens should be set up to shield persons nearby where chippers, riveting guns, or air drills are being used.

Because excessive noise is a definite possibility, consideration should be given to isolating the noisy operation, substituting quieter methods, or providing hearing protection to all exposed workers.

Two chippers should work away from each other, that is, back to back, to prevent face cuts from flying chips. Workers should not point a pneumatic hammer at anyone, nor should they stand in front of operators handling pneumatic hammers.

Handling of heavy jackhammers causes fatigue and may even cause strains. Jackhammer handles should be provided with heavy rubber grips to reduce vibration and fatigue, and operators should wear metatarsal-type safety shoes to reduce the possibility of injury should the hammer fall.

Figure 14-31. Air hammer tool must be against work surface before starting operation. (Reprinted with permission from Bosch Power Tools.)

Many accidents are caused by breaking of the steel drill because the operator loses his balance and falls. Also if the steel is too hard, a particle of metal may break off and strike him. The manufacturer's instructions for sharpening and tempering the steel should be followed.

Special power tools

Flexible shaft tools require the same type of personal protective equipment as do direct power tools of the same type. Abrasive wheels should be installed and operated in conformance with ANSI standard B7.1 and federal regulations. The flexible shaft must be protected against denting and kinking, which may damage the inner core and shaft.

It is important that the power be shut off whenever the tool is not in use. When the motor is being started, the tool end should be held with a firm grip to prevent injury from sudden whipping. The abrasive wheel or buffer of the tool is difficult to guard and, because it is more exposed than the wheel or buffer on a stationary grinder, extra care should be exercised to avoid damage. Wheels should be placed on the machine or put on a rack, not on the floor.

Hydraulic power tools are used in some industries, notably the electric utility industry where employees work aloft from a hydraulically powered aerial lift device. The power is obtained from the source used to operate devices such as hydraulic chain saws and compression devices. Some compression devices have an incorporated small hydraulic press that is pumped by the operator. A hazard in the use of such equipment is small leaks in the hydraulic hose or around fittings. There have been instances where employees have put a hand over a pinhole leak and had oil forced into their finger by the high pressure. Care must also be taken to always use hose built for the pressure involved because a rupture can cause serious consequences.

Gasoline power tools are widely used in logging, construction, and other heavy industry. The best known and most commonly used tool is the chain saw. Persons using chain saws must be trained in the operation of the equipment according to the manufacturer's specific instructions.

Hazards. Operators of chain saws are exposed to hazards similar to those encountered by workers using hand saws. In addition, there are some special hazards, such as:

- Falling while carrying a saw or when sawing
- Sprains and strains from carrying a heavy saw and working with it
- Being cut by contact with the chain while it is in motion
- Being cut by the chain not in motion, either on or off the saw
- Injuries from starting the gasoline engine
- Inhaling exhaust gases
- Electric shock from electrically operated saws
- Being struck by wood from overhead due to vibration of tree
- Sawdust in the eyes, especially when holding the tail stock or "stinger" end of a saw above the head
- Burns from contact with hot muffler or cylinder head
- Injuries due to saws binding and "kicking back" at an operator
- Injuries from falling trees and snags or rolling logs resulting from inability to hear because of gasoline engine noise. When electric saws are used by line crews, only qualified linemen should tap into powerlines. (The saw frame should be grounded.)

Proper selection of saw operators is as important as for other skilled occupations. The inexperienced employees should work, during their training period, with an experienced faller under the constant supervision of the bull bucker.

Prevention of fires. Gasoline should be taken out to the job in a substantial capped container painted red and labeled GASO-LINE. The container should have a suitable spout for pouring gasoline into the tank, or a funnel should be provided for this purpose. *Under no circumstances should the gasoline tank be replenished while the engine is running. Any gasoline spilled on the tank or engine should be carefully wiped off before starting the engine.*

Precautions to prevent fires are:

- Gasoline containers with a spout or funnel should be used
- Tanks should be filled only in an area of bare ground
- Motors should not be filled where a tank was previously filled
- Saws should be kept clean of gasoline, oil, and sawdust
- Mufflers should be in good condition
- Spark plugs and wire connections should be kept tight
- Fire extinguishers should be near power saws at all times
- Flammable materials should be kept clear and away from point of saw cut.

Personal protective equipement. Workers operating chain saws and carrying them through the woods must be surefooted. Good sharp calked boots should be worn. In some sections hobnailed shoes are preferable, and in the winter rubber soled shoes. Among the most common accidents are falls, so the wearing of proper footwear to minimize this hazard should be mandatory.

Safety-toe footwear is a good investment for members of a cutting crew.

Protective helmets should always be worn by fallers and buckers and others working in the woods. It is important that safety hats be worn. Many people have been killed by falling wood, and many lives have been saved by using safety hats. *Their use should be mandatory.*

Employees should wear eye protection where flying particles present a hazard. Some companies require eye protection for all power saw operators.

A ballistic nylon patch covering part of the leg has reduced accidents to that part of the body. The patches are held in pockets or otherwise attached to the work pants. The patches increase the time for an operator to shut off, in the event a saw chain should come against the leg.

First aid kits should be provided at the jobsite. One individual in each work crew should be trained in first aid.

In areas where poisonous snakes are known to exist, a snake bite kit should be provided.

The employer should be aware of the noise level of a chain saw and associated equipment. It may be necessary to provide noise attenuating devices (earmuffs or plugs).

Powder-actuated tools are used to make forced-entry fastenings in various construction materials (Figure 14-32). The systems are simple to use; however, there are precautions and safeguards that must be observed. Only qualified trained operators

Figure 14-32. Powder-actuated tool drives studs into concrete slab. Operator should wear eye, ear, and head protection. Note holster (*left*) for carrying the tool. (Reprinted with permission from Hilti Inc.)

should be allowed to use this equipment, and then under close supervision. To become a qualified operator, a person must be thoroughly trained under supervision of a manufacturer's authorized instructor. Upon completion of the background training, the person is required to demonstrate competence through use of the system in varied applications and to pass a written examination. Upon successful completion of the examination, the instructor will issue a Qualified Operator's Card (Figure 14-33), records of which are kept by certain regulatory agencies.

Also, the possessor of a Qualifed Operator's Card must be familiar with any state regulations applicable to the use, maintenance, and storage of the systems. More information regarding the use of these systems can be obtained by writing to the

```
┌─────────────────────────────────────────────────────────────┐
│  QUALIFIED OPERATOR OF POWDER-ACTUATED TOOLS                 │
│                                                               │
│  Make(s)_____    Model(s)_____          │
│                                                               │
│  This certifies that _____         │
│                            (NAME OF OPERATOR)                 │
│                                                               │
│  Card No._____  Soc. Sec. No._____         │
│                                                               │
│  Has received the prescribed training in the operation        │
│     of powder actuated tools manufactured by                  │
│  _____        │
│                  (NAME OF MANUFACTURER)                       │
│                                                               │
│  Trained and issued by _____         │
│                   (SIGNATURE OF AUTHORIZED INSTRUCTOR)        │
│                                                               │
│  I have received instruction in the safe operation and        │
│  maintenance of powder actuated fastening tools of the        │
│  makes and models specified and agree to conform to all       │
│  rules and regulations governing their use. Failure to        │
│  comply shall be cause for immediate revocation of this       │
│  card.                                                        │
│                                                               │
│  _____     _____                   │
│      (SIGNATURE)                 (DATE)                       │
└─────────────────────────────────────────────────────────────┘
```

Figure 14-33. Wallet card for qualified powder-actuated tool operators is available from tool manufacturers. List of instructors should be maintained by each manufacturer. Based on American National Standard A10.3 1984.

Powder-Actuated Tools Manufacturer's Institute (see References), and from the National Safety Council's *Powder-Actuated Hand Tools,* Industrial Data Sheet 236.

Power-drive sockets are widely used tools in garages, repair shops, field work, and manufacturing industries during disassembly and assembly operations. Electric and pneumatic power are commonly used. Pneumatic power is favored in heavy-duty operations. Electric power is used most often with single-drive socket operations. On single-socket operations, electric extension cords and air hoses cause tripping hazards when lying on the floor. On repetitive operations such as assembly lines, a common practice is to suspend the units above the point of operation or with a spring-loaded overhead on trolleys, or similar balancing device to reduce operator fatigue. Devices suspended on balancing units could cause injury by striking the upper body during a pendulum motion swing.

Stationary units are designed to have the material moved to a predetermined location underneath the unit. Contact with a limit switch starts an automatic cycle. This type of unit can be enclosed with an electrically interlocked barrier guard to stop the conveyor from moving forward. Also a relief valve should be attached to the guard to automatically bleed the air cylinders. The main air supply valve to the unit should be shut off first. Also, all permanent, quick-disconnect fittings and hoses should be inspected periodically to prevent sudden failure which could inflict a serious injury to a worker by whipping action.

Power-drive sockets are subject to extremely heavy use during repetitive operations. The sockets should be of high quality, high strength alloy steel specially treated to be able to withstand the shock and pounding during bolt running and nut setting.

Safety pins are usually designed to shear at definite preset pressure levels. Nails, bolts, or other makeshift items should not be used as substitutes. Pins should be secured by retainers.

Substitutions of poor quality shear pins, inadequate design or improper use of sockets could cause sudden failure and result in flying missiles.

Pneumatic-powered tools frequently generate high impact noise levels. If noise cannot be controlled to meet the permissible OSHA requirements, then hearing protection must be worn. Also, one hundred percent eye protection should be enforced near the work station.

PERSONAL PROTECTIVE EQUIPMENT

Gloves, ties, loose clothing, and jewelry should not be worn by workers using revolving tools such as drills, saws, and grinders. The weight of most power tools makes it advisable for users to wear safety shoes to reduce chances of injury should the tools fall or be dropped.

Clothing should be free of oil, solvents, or frayed edges to minimize the fire hazard from sparks.

When power tools are used in overhead places, the operator should wear a fall protection harness to minimize the danger of falling should the tool break suddenly or shock the operator. It is a good practice to attach a safety line to the tool to keep it from falling on persons below if it is dropped.

Approved dust-type respirators should be worn on buffing, grinding, and sanding jobs that produce harmful dusts.

For powder-actuated tools or jack hammers, hearing protective equipment should be used if more positive noise controls are not possible.

Eye protection

In all operations where striking and struck tools are used, or where the cutting action of a tool causes particles to fly, eye protection (conforming to ANSI Z87.1) is needed by the user of the tool and by others who may be exposed to flying particles. The hazard can be minimized through the use of nonferrous, "soft" striking tools and through shielding the job by metal, wood, or canvas. However, safety goggles are still required.

Safety goggles or faceshields should be worn when woodworking or cutting tools, such as chisels, brace and bits, planes, scrapers, and saws, are used with the chance of particles falling or flying into the eyes.

Safety goggles or faceshields should be worn for work on grinders, buffing wheels, and scratch brushes because the unusual positions in which the wheel operates may cause particles to be thrown off in all directions. For this reason, protective equipment is even more important than it is for work on stationary grinders.

Eye protection should also not be overlooked on such jobs as cutting wire and cable, striking wrenches, using hand drills, chipping concrete, removing nails from lumber, shoveling material or working on the leeward side of a job, using wrenches and hammers overhead, and on other jobs where particles of materials or debris may fall.

REFERENCES

American National Standards Institute, 1430 Broadway, New York, N.Y. 10018.
Practice for Occupational and Educational Eye and Face Protection, Z87.1
Portable Air Tools, B186.1
Safety Requirements for the Use, Care, and Protection of Abrasive Wheels, B7.1.
Safety Requirements for Ball Pein Hammers, ANSI/HTI B173.2.
Safety Requirements for Heavy Striking Tools, ANSI/HTI B173.3.
Safety Requirements for Nail Hammers, B173.1.
Boyd, T. Gardner. *Metal Working.* Homewood, Ill., The Goodheart-Willcox Co., Inc., 1978.
Budzik, Richard S. *Precision Sheet Metal Shop Theory.* Indianapolis, Ind., Howard W. Sams & Co., Inc., 1969.
Compressed Air and Gas Institute, 1230 Keith Building, Cleveland, Ohio 44115.
Editors of Consumer Guide. *The Tool Catalog.* Skokie, Ill., Publications International, Ltd., 1978.

Grinding Wheel Institute, 712 Lakewood Center, N., 14600 Detroit Ave., Cleveland, Ohio 44107.
Hand Tools Institute, 25 N. Broadway, Tarrytown, N.Y. 10591. *Hand Tool Safety—Guide to Selection and Use.*
Jackson, Albert, and David Day. *Tools and How to Use Them—An Illustrated Encyclopedia.* New York, Alfred A. Knopf, Inc., 1978.
Kratfel, Edward R. *Introduction to Modern Sheet Metal,* Reston, Vir., Reston Publishing Co., Inc., 1976.
McDonnell, Leo P., and Kaumeheiwa, Alson. *The Use of Hand Woodworking Tools,* New York, N.Y., Litton Educational Publishing, Inc., 1978.
National Safety Council, 444 N. Michigan Ave., Chicago, Ill. 60611.
Industrial Data Sheets
Air Powered Hand Tools, 392
Brush Cutting Tools, 427.
Electric Hand Saws, Circular Blade Type, 675.
Equipment Grounding, 684.
Hand Soldering and Brazing, 445.
Portable Power Chain Saws, 320.
Powder-Actuated Hand Tools, 236.
Powder-Actuated Tool Manufacturers' Institute, 100 S. Third St., St. Charles, Mo. 63301. *Basic Training Manual.*
TPC Training Systems, 1301 S. Grove Ave., Barrington, Ill., 60010. *Maintenance Fundamentals: Hand Tools and Portable Power Tools,* 1980.
Underwriters Laboratories Inc., 333 Pfingsten Rd., Northbrook, Ill. 60062. *Electric Tools,* UL 45.
U.S. Bureau of Naval Personnel. *Tools and Their Uses,* Rate Training Manual, NAVPERS 10085-B. Dover, Del., 1973.
U.S. Department of Commerce, Office of Technical Service, Washington, D.C. 20234. *Sparking Characteristics and Safety Hazards of Metallic Materials,* Technical Report No. NGF-T-1-57, PB 131131.
Walker, John R. *Exploring Metalworking: Basic Fundamentals.* South Holland, Ill., The Goodheart-Willcox Co., Inc., 1976.
Willson, Frank W., ed. *Tool Engineers Handbook,* 2nd ed. Detroit, Mich., American Society of Tool and Manufacturing Engineers, 1959.
Zinngrabe, Claude J., and Fred W. Schumacher. *Sheet Metal Hand Processes.* Albany, N.Y., Delmar Publishers, a Div. of Litton Educational Publishing, Inc., 1974.

15

Electricity

PROPERLY USED, ELECTRICITY is our most versatile form of energy. Failure to ensure that safe design considerations, work practices, procedures, servicing, and maintenance operations are established often results in bodily harm (including fatalities), property damage, or both.

In most cases, electrical and electronic equipment can be designed for both maximum safety and efficiency. Potentially hazardous conditions such as inadvertent contact with hazardous voltages, do, however, often exist in performing engineering analysis (commonly referred to as "debugging"), installation, servicing, testing, and maintenance of electrical and electronic equipment. Minimizing exposure to the majority of these potential hazards is neither difficult nor expensive if the safeguards and procedures are introduced in the design cycle. If hazard reduction is ignored, however, serious accidents may result.

This chapter provides an overview of basic safety considerations to minimize employee exposure to hazardous low voltages. Power distribution systems above 600 volts are not addressed.

DEFINITIONS

There are several terms which should be well understood before continuing this chapter. Three of these basic terms—current, voltage, and resistance—are defined using the analogy that electricity flowing through a circuit can be likened to the flow of water through a pipe. If this analogy is kept in mind, these terms are not troublesome.

Current may be thought of as the total volume of water flowing past a certain point in a given length of time. Electric current is measured in amperes. The measurement used, in relation to electric shock, therefore, is the milliampere (0.001 ampere).

Voltage may be thought of as the pressure in a pipeline; it is measured in volts. Low voltage is a rather ambiguous term depending upon whether it is being used by a safety professional, a plant electrician, or a lineman. For the purposes of this chapter, low voltage is 600 volts or less. Potentially hazardous voltage is between 24 volts and 600 volts. Potentially lethal voltage is 50 volts and above.

Resistance is any condition which retards current flow; it is measured in ohms.

ELECTRICAL INJURIES

Current flow is the factor that causes injury in electric shock; that is, the severity of electric shock is determined by the amount of current flow though the victim (Table 15-A).

Because current flow depends on voltage and resistance, these factors are important. In addition to current flow, other factors affecting the amount of damage are the parts of the body involved, the duration of current flow through the victim, and the frequency (if alternating current).

Experimental and field data from authoritative sources (see References for articles written by Charles F. Dalziel) indicate that, in general, an alternating current of 100 milliamperes (100 mA) at commercial frequency of 60 cycles per second (60 Hz) may be fatal if it passes through the vital organs. Similarly, it is estimated that a current of 16 mA is the average current at which an individual can still release an object held by the hand.

Table 15-A. Effects of Electric Current on Humans

Effect	Current in Milliamperes					
	Direct		Alternating			
			60 Hz		10,000 Hz	
	Men	Women	Men	Women	Men	Women
Slight sensation on hand	1	0.6	0.4	0.3	7	5
Perception threshold	5.2	3.5	1.1	0.7	12	8
Shock—not painful, muscular control not lost	9	6	1.8	1.2	17	11
Shock—painful, muscular control not lost	69	41	9	6	55	37
Shock—painful, let-go threshold	76	51	16	10.5	75	50
Shock—painful and severe, muscular contractions, breathing difficult	90	60	23	15	94	63
Shock—possible ventricular fibrillation effect from 3-second shocks	500	500	100	100		
Short shocks lasting *t* seconds			$165/\sqrt{t}$	$165/\sqrt{t}$		
High voltage surges	50*	50*	13.6*	13.6*		

*Energy in watt-seconds or jouls

Note: Data is based on limited experimental tests, and is not intended to indicate absolute values.

Such current flow can easily be received on contact with low-voltage sources of the ordinary lighting or power circuit.

A person's resistance to current flow is mainly to be found in the skin surface. Callous or dry skin has a fairly high resistance, but a sharp decrease in resistance takes place when the skin is moist (Table 15-B). Once the skin resistance is broken down, the current flows readily through the blood and body tissues.

Table 15-B. Human Resistance to Electrical Current

Body Area	Resistance (ohms)
Dry skin	100,000 to 600,000
Wet skin	1,000
Internal body—hand to foot	400 to 600
Ear to ear	(about) 100

Note: Data is based on limited experimental tests, and is not intended to indicate absolute values.

Whatever protection is offered by skin resistance decreases rapidly with increase in voltage. A high voltage alternating current of 60 Hz causes violent muscular contraction, often so severe that the victim is thrown clear of the circuit. Although low voltage also results in muscular contraction, the effect is not so violent. Low voltage is dangerous, however, because it often prevents the victim from breaking the contact with the circuit.

Death or injury by electric shock may result from the following effects of current on the body:

1. Contraction of the chest muscles, which may interfere with breathing to such an extent that death will result from asphyxiation when the contact is prolonged.
2. Temporary paralysis of the nerve center, which may result in failure of respiration, a condition which often continues until long after the victim is freed from the circuit.
3. Interference with normal rhythm of the heart, causing ventricular fibrillation. In this condition the fibers of the heart muscles, instead of contracting in a coordinated manner (which causes the heart to act as a pump), contract separately and at different times. Blood circulation ceases and (unless proper resuscitation efforts are made) death ensues. The heart cannot spontaneously recover from this condition. It has been estimated that 50 milliamperes is sufficient to cause ventricular fibrillation.
4. Suspension of heart action by muscular contraction (on contact with heavy current). In this case the heart may resume its normal rhythm when the victim is freed from the circuit.
5. Hemorrhages and destruction of tissues, nerves, and muscles from heat due to heavy current along the path of the electric circuit through the body.

In general, the longer the current flows through the body, the more serious may be the result. Considerable current is likely to flow from high-voltage sources, and in general only very short exposure can be tolerated if the victim is to be revived.

Injuries from electric shock are less severe when the current does not pass through or near nerve centers and vital organs. In the majority of electrical accidents in industry the current flows from hands to feet. Since such a path involves both the heart and the lungs, results are usually serious.

Another type of injury is burns from electric flashes. Such burns are usually deep and slow to heal and may involve large areas of the body. Even persons at a reasonable distance from the arc may receive eye burns.

Where high voltages are involved, flashes of explosive violence may result. This intense arcing is caused by short circuits between bus bars or cables carrying heavy current, failure of knife switches, opening knife switches while they are carrying a heavy load, or pulling fuses in energized circuits.

Other injuries include falls from one level to another. A worker receiving a shock from defective or malfunctioning equipment, because of the resulting muscular contraction, may lose balance and fall.

C. F. Dalziel's statistics (see References) indicate that only a small percentage of those who recover from electric shock show permanent disability. In many cases, the victim may be saved by prompt application of cardiopulmonary resuscitation (CPR), since a common result in electrical accidents is failure of the nervous system which controls breathing.

It is, therefore, essential that persons engaged in electrical work involving hazardous energy levels be instructed in CPR. Training is provided by the Heart Association, the Red Cross, and St. Johns Ambulance in Canada.

Resuscitative techniques should be immediately applied to a victim of electric shock and should be continued until he or she revives, or until death is diagnosed by a physician. The possibilities of successful revival of a victim in terms of elapsed time before the start of resuscitation are shown in Figure 15-1.

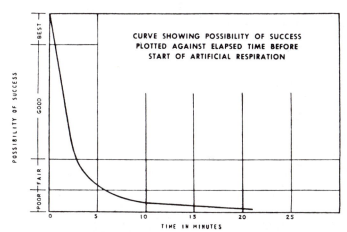

Figure 15-1. The possibility of successful revival decreases with time. (Reprinted with permission from Edison Electric Institute Resuscitation Manual.)

ELECTRICAL EQUIPMENT

Selection

Most items of electrical equipment are designed and built for specific types of service. They will operate with maximum efficiency and safety only when used for the purposes and under the conditions for which they are intended.

In the selection of equipment, it is advisable to follow the recommendations of the various codes and standards which have been established. In addition to the *National Electrical Code*, NFPA 70, the state and local codes should be checked for industrial zoning requirements. Adherence to most of the provisions of the *National Electrical Code* (NEC) is not only required by the federal Occupational Safety and Health Administration but also by insurance companies.

Engineering consultant services, recommendations of manufacturers, and publications from the following agencies will answer practically all questions concerning electrical equipment:

American National Standards Institute
1430 Broadway
New York, N.Y. 10018

Canadian Standards Association
178 Rexdale Blvd.
Rexdale, Ontario M9W 1R3
Canada

Factory Mutual System
1151 Boston-Providence Turnpike
Norwood, Mass. 02062

Illuminating Engineering Society of North America
345 East 47th St.
New York, N.Y. 10017

National Fire Protection Association
Batterymarch Park
Quincy, Mass. 02269

Underwriters Laboratories Inc.
333 Pfingsten Rd.
Northbrook, Ill. 60062

When copies of codes or standards are ordered from the publishers, give full information on general types of equipment under consideration, the application, and operating conditions. For special problems, include photographs or blueprints to clarify the physical conditions at the installations.

Installation

Transformers, dead-front control boards, switches, motor starters, and other electrical equipment should be installed so that the possibility of accidental contact with energized conductors is reduced to a minimum.

When an interlock is used as a safety device, it should be fail-safe; that is, it must be a device that will automatically compensate for failure of its mechanism and will not jeopardize the safety of personnel depending upon it. Interlocks selected should meet the following criteria:

1. Fail-safe features. Failure of the interlock mechanism, loss of power, short circuit, or malfunction of equipment will cause the circuit to be interrupted.
2. A visible disconnect in the primary power circuit.
3. A locking arrangement which makes attempts to circumvent the interlock impractical.

Where space and operating requirements permit, electrical equipment should be placed in the less congested areas of the plant or, where practical, in special rooms to which only authorized persons have access. When electrical equipment is located in the production areas of the plant, enclosures should be built around those parts of the equipment having exposed conductors and warning signs should be posted. Floor curbing or heavy steel barriers may be necessary if there is even a remote danger of industrial trucks striking critical electrical equipment.

Barriers prevent accidental contact with electrical equipment (Figure 15-2). Frames may be made of wood, rolled metal shapes, angle iron, or pipe. Wood strips, sheet metal, perforated metal, expanded metal, wire mesh, plastic, or shatterproof transparent material may be used for filler. Dry wood and many plastics have the advantage of being nonconductors. But unless specially treated, wood is combustible and may be objectionable on that account. Metal frames or guards should be grounded.

In addition, warning signs should be displayed near exposed current-carrying parts and in especially hazardous areas, such as high-voltage installations. These signs should be large enough to be read easily and should be visible from all approaches to the danger zone. They should meet ANSI *Specifications for Accident Prevention Signs, Z35.1.*

Figure 15-2. An expanded metal screen guards the V-belt drive, belts, and sheaves of this injection pumping unit. The 5-cylinder pump is powered by a 200-hp explosion proof electric motor. The screen guard allows easy maintenance. (Reprinted with permission from Service Pipe Line Co., Tulsa, Okla.)

In many respects, standard machine guarding practices can be applied to electrical equipment. There are, however, certain hazards peculiar to electricity which must be given consideration when the overall plant guarding program is planned.

Wiring should be installed in accordance with the NEC unless more restrictive local requirements apply. The type of wiring that should be used in an industrial plant depends upon the type of building construction, the size and distribution of electrical load, exposure to dampness or corrosive vapors, location of equipment, and various other factors. For most plant conditions, grounded metal conduit is satisfactory.

The varous types of insulation used on electrical conductors and their application provisions are given in the NEC.

Among wiring methods which may be used under certain circumstances are armored cable, nonmetallic sheathed cable, flexible metal conduit, raceways, open wiring and insulators, and concealed knob and tube wiring. When new installations are made or existing circuits changed, applicable requirements of national and local wiring codes should be observed.

Wires having insulation designed for the type of service and location should be used, but the insulation alone should not be considered sufficient protection against shock, especially in high-voltage circuits.

Frequent inspection of equipment and competent supervision of maintenance crews are extremely important, and the subsequent reports will aid the safety program considerably. Often in their routine work throughout the plant, maintenance personnel can spot hazards before they cause injuries.

Motors should be mounted so that they do not intefere with the normal movement of personnel or materials. Motors not enclosed should be in areas free from dust, moisture, and flammable or corrosive vapors, and in some instances, they can be isolated from personnel by being mounted on overhead supports, installed below the floor level, or placed in special motor rooms.

If current-carrying parts must be exposed, they should be made inaccessible by elevating them at least 8 ft (2.4 m) above the work area, or enclosures, barriers, or guards should be provided to prevent contact.

Switches

Among the several types of switches are pushbutton switches, snap switches, knife switches, and enclosed externally operable air break switches. Many switches are designed for a specific function, such as the enclosed type for controlling individual motors and machine tools and for lighting and power circuits. However, all switches, regardless of their function, must have approved voltage and ampere ratings compatible with their intended use.

An open knife switch is hazardous and should not be used because of the exposure of live parts and because of the arc formed when the switch is opened. Knife switches should be enclosed in grounded metal cabinets having control levers that operate outside the cabinets. A further safeguard is the safety switch whose cover must be closed before the switch can be used.

A knife switch should be mounted so that the blades are dead when the switch is open and it should be installed so that gravity will not close it. Knife switches installed in power switching circuits should be electrically interlocked so that they cannot be opened when the circuit is energized, unless switches are of the load-break type.

Where it is necessary to use disconnect switches for high voltage, heavy current feeder circuits to test floors, or other service installation, they should be located out of normal reach and should be operated only by insulated switch sticks. Where it is impractical to locate the switch out of normal reach, it must be protected against accidental contact by completely enclosing it or placing it behind a suitable fence or barricade. If placed behind a barrier, it can be opened or closed remotely by an insulated switch stick.

Pushbutton or snap switches are recommended because the live parts are enclosed. Flush switches should be installed in boxes, and surface switches, used in open wiring and molding work, should be mounted on porcelain, composition, or other insulating subbases. These switches must have indicators to show open and closed position.

Pendant switches are used primarily where switches have not been installed on walls and where it would be difficult to put them there, or where the switches on walls or in cabinets control two or more circuits and it is desired to subdivide them still further so that each circuit can be controlled individually.

Pendant switches, and in particular pendant pushbutton control stations, should be provided with exterior strain relief at the point of suspension from the ceiling, and at the point of entry into the enclosure; in other words, there should be no pull or harmful friction on electrical wiring or connections.

A convenient toggle switch with a button automatically illuminated as soon as the circuit is open will aid in the location of the switch in darkened areas. Such a switch, however, should not be relied upon for any illumination.

Protective devices

The safe current-carrying capacity of conductors is determined by their size, the material of which they are made, the type of insulated covering, and the manner in which they are installed. If they are forced to carry more than the maximum safe load or if heat dissipation is limited, excessive heating results. Overcurrent devices, such as fuses and circuit breakers, open the circuit automatically in the event of excessive current flow from accidental ground, short circuit, or overload.

Overcurrent devices must be installed in every circuit, and they should be of a size and type that will interrupt the current flow when it exceeds the capacity of the conductor. The selection of the properly rated equipment not only depends on the current-carrying capacity of the conductor, but also on the rating of the power-supply transformer or generator and its potential short-circuit-producing capacities. Protection of this kind, both for personnel and for equipment, is one of the important features of an electrical installation. Where higher interrupting capacity is required, special high-capacity fuses or circuit breakers are needed.

Among the many types of fuses are link, plug, and cartridge. Each should be used only in the type of circuit for which it is designed. Use of fuses of the wrong type or the wrong size may cause injury to personnel and damage to equipment. Overfusing is a frequent cause of overheated wiring or equipment—and resultant fires.

Before fuses are replaced, the circuit should be locked-out and tested to be certain it is deenergized. Testing can save lives. An investigation should be made for the cause of the short circuit or overload. Blown fuses should be replaced by others of the same type and size; fuses should never be inserted in a live circuit.

A link fuse, as its name indicates, is a strip of fusible metal between two terminals of a fuse block. If not enclosed, it may scatter metal when it blows. Replaceable link fuses should be replaced only under the direction of qualified maintenance personnel.

Plug fuses are used on circuits which do not exceed 30 amperes at not more than 150 volts to ground. In plug fuses the fusible metal is completely enclosed. The type that cannot be bridged inside the holder is recommended.

The cartridge fuse, which is widely used in industrial installations, has a fusible metal strip enclosed in a tube. Cartridge fuses that indicate when the fuse is blown, and renewable types in which the fusible element may be replaced, are available.

It is recommended that a switch be placed in any circuit that can be opened to deenergize the fuses to be handled. As an additional precaution, insulated fuse pullers should be used.

Circuit breakers have long been used in high-voltage circuits with large current capacities. In recent years their use has become more common in many other kinds of circuits. They are available in a variety of types and sizes. They may be instant in their operation or equipped with timing devices, and may be manually operable or power operable.

Circuit breakers fall into two general categories—thermal and magnetic.

- The thermal circuit breaker operates solely on the basis of temperature rise, and consequent variations in the temperature of the room in which the circuit breaker is installed will affect the point at which it interrupts the circuit.
- The magnetic circuit breaker operates only on the basis of the amount of current passing through the circuit; this type has considerable advantage where a wide fluctuation in ambient temperature would ordinarily require overrating the circuit breaker or where undue tripping frequencies are experienced.

Circuit breakers should be checked regularly by experienced maintenance personnel to make sure thay are in good operating condition at all times.

Ground fault circuit interrupters

A ground fault circuit interrupter (GFCI) is a fast-acting electric circuit-interrupting device that is sensitive to very low levels

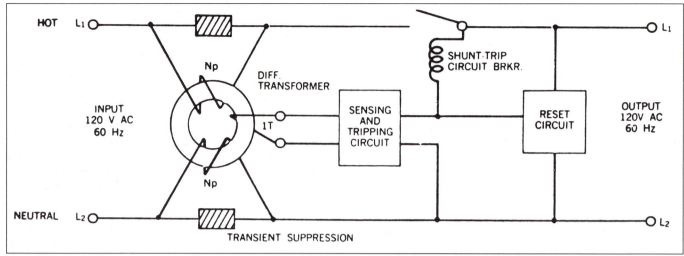

Figure 15-3. Differential ground fault interrupter. Current-carrying conductors pass through circular iron core of doughnut-shaped differential transformer. If the transformer senses even a portion of the current flowing to ground, it causes the circuit breaker to operate and open the circuit.

of current flow to ground. The interrupter is designed to sense leakage currents of a magnitude that could cause serious injury. The unit operates *only* on line-to-ground fault currents, such as insulation leakage currents or currents likely to flow during accidental contact with a "hot" wire of a 120 volt circuit and ground. *It does not protect in the event of a line-to-line contact.*

A receptacle differential GFCI is also available which functions the same as the circuit breaker type, but protects only the associated integral receptacle, plus a limited number of additional receptacles connected downstream.

- The differential ground fault interrupter (Figure 15-3) has current-carrying conductors passing through the circular iron core of a doughnut-shaped differential transformer. As long as all the electricity passes through the transformer, the differential transformer is not affected and does not trigger the sensing circuit. If a portion of the current flows to ground and through the fault-detector line, flow of electricity through the sensing windings of the differential transformer causes the sensing circuit to open the circuit breaker. These devices are designed to interrupt a circuit for currents of as little as 5 mA flowing to ground.

- Another unit design is the isolation ground fault circuit interrupter (Figure 15-4). This unit combines the safety of an isolation system with the response of an electronic sensing circuit. In this setup, an isolating transformer provides an inductive coupling between load and line; both the hot and neutral wires connect to the isolating transformer. There is no continuous wire between.

In this type of interrupter, the ground fault must pass through the electronic sensing circuit, which has sufficient resistance to limit current flow to as little as 0.2 mA—well below the level of human perception.

Although neither of these types protects an individual in the event of line-to-line contact, both will operate effectively in line-to-ground situations. They should significantly reduce electric shock accidents that presently account for about 1,100 fatalities each year.

Control equipment

Switchboards with lockout capabilities for both alternating- and direct-current distribution circuits should be arranged so that the controls are easily accessible to the operator. Likewise, instruments should be readable and equipment adjustable from the working area. Boards should be placed so that the operator will not be endangered by live or moving parts of machinery. The space behind the switchboard should not be used for storage and should be kept clear of rubbish. The control board should be placed in a special room or be made inaccessible to other than authorized personnel by screen enclosures. Doors to the enclosure should be kept locked.

Good illumination should be provided for the front and rear of switchboards and maintained ready for use at all times. An emergency source of illumination should also be provided.

Switch and fuse cabinets should have close-fitting doors which should be kept locked. To warn employees that electrical parts are exposed and to remind them to close all doors of equipment, the inside of the doors may be painted orange.

The switchboard framework and metal parts of guards should be grounded in accordance with code regulations.

Connections, wiring, and equipment of switchboards and panel boards should be arranged in an orderly manner. Switches, fuses, and circuit breakers should be plainly marked and arranged to afford ready identification of circuits or equipment supplied through them. It is good practice to keep a diagram of switchboard connections and devices posted near the equipment.

Protection against accidental shock from live electrical parts on switchboards, fuse panels, and control equipment can be minimized by insulating the floor area within range of the live parts. Where equipment provides exposure to 600 volts or less, special insulating mats or dry wood floors with no metal parts are used. The insulating mats should be made of a material having proper moisture resistance and nonconductive properties, and should be able to withstand mechanical abuses which may be encountered in such service. (The J6 specifications, covering rubber protective equipment for electrical workers, are available in one booklet from ANSI.)

Circuits initiated by pushbuttons should be low voltage (600 volts or less). However, pushbuttons may be used to control a high-voltage circuit if step-down transformers or relays are

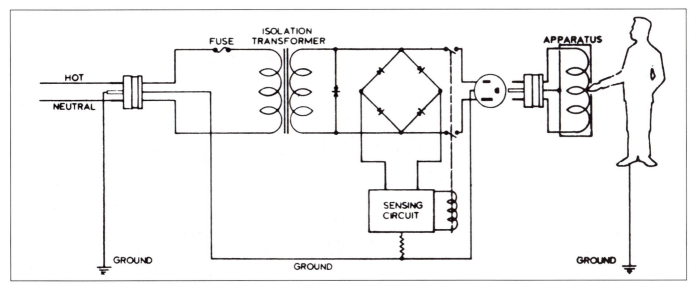

Figure 15-4. Isolation ground fault interrupter. An isolation transformer provides an inductive coupling between load and line; both hot and neutral wires connect to the isolating transformer. There is no continuous wire between load and line.

provided to prevent the voltage in the control part of the circuit from exceeding 250 volts.

Motors

For maximum safety each motor should be of the type and size required for the load and for the conditions under which it must operate. Excessive overloading shall be avoided. Only motors suitable for use in hazardous locations shall be used in areas containing flammable vapors or dusts.

The following are common motor problems in industry: dust, stray oil, moisture, misalignment, vibration, overload, and friction.

- In plant operations, dust is constantly settling on motors, housings, windings, slip rings, and commutators, and trying to work its way to the bearings. The problem of dust is at a minimum in the case of totally enclosed, fan-cooled motors.

  Dust is best removed from motors before it has a chance to unite with water or oil to form a gummy mess. Motors, and their housings, slip rings, and commutators, must therefore be wiped off on regular inspections. Occasionally, dust must be blown out of the wire-wound section with not greater than 30 psig of pressure (207 kPa) from a compressed-air unit. When this operation is performed, there must be good ventilation to prevent accumulation of dust. In some cases, it is advisable to use vacuum cleaning equipment or a hand bellows. The person doing the dusting should wear eye protection.

- Oil harms commutators by deteriorating the mica insulating segments between the bars. It causes excessive sparking and deteriorates the insulation on the windings, with immediate danger of a burnout or breakdown. Oil also tends to hold dust, lint, and other material that increases the fire hazard.

  When oil and dust have been allowed to build up, the motor should be taken out of service and cleaned in accordance with the manufacturer's directions.

- Not all electrical insulation acts as a perfect barrier to moisture. Some types become porous with age and absorb moisture. The electrical resistance may then drop to a point where the leakage burns through the insulation to ground and causes a short circuit with risk of fire or shock hazard.

  Careless handling of liquids in the manufacturing process, wetting of motors during cleanup operations in plants where floors must be frequently washed with water, and overflowing of water softener tanks, filters, dye cups, or other liquid reservoirs can cause failure and shock hazards.

  Prevention can be accomplished by personnel training and relocation of equipment. If the moisture hazard cannot be wholly avoided or is inherent in the work for which the motor is used, manufacturers can provide drip-proof, splash-proof, or totally enclosed designs.

  After a motor has been subjected to moisture and the resistance is found to be at a dangerous point, the motor should be dried, preferably by an organization or person specializing in this type of work.

- Belts and chain drives should be checked for tension, and gearing should be checked for binding. Misalignment of the motor shaft can cause it to spring or break, the bearings to burn out, and the motor itself to fail as a result of an overload.

- Misalignment is one of the significant causes of motor vibration. Other causes include careless servicing that puts motors out of balance (loose motor mounting bolts and worn bearings which allow the shaft to oscillate).

- A motor can be overloaded by (1) excessive friction within the motor itself, (2) using the motor for the wrong kind of job, (3) an obstruction in the driving or driven machine, or (4) pushing the machine to perform beyond the capacity of the motor.

  If one of these conditions causes the current in a motor to exceed its nameplate current rating, heating may increase as much as the square of the current increase. Insulation may be burned, soldered connections melted, or bearings burned out.

- To safeguard against such overloads, motors have various forms of overload protection. In most cases, a thermal element is connected in the power circuit to the motor. This is accomplished with either (1) integral over-temperature protective devices that will open the motor circuit in the event of overheating or (2) current sensitive protective devices that generate heat from an excessive current so as to operate an overload relay, opening the circuit to the motor. The current sensitive thermal element should be of the proper capacity as listed in the *National Electrical Code.* No unit of more than the recommended value should be used.

Motor output and production methods should be revised to control overload conditions that trip or blow these protective devices.

- To reduce friction and to prevent excessive wear, overheating of bearings, and possible fires from faulty lubrication, the manufacturer's lubrication charts and instructions on types and grades of lubricants, the frequency of lubrication, and other practices should be followed.

Grounding

In order to understand grounding properly, it should first be noted that both equipment grounding and the grounding of the electrical system itself are included under this term. Equipment grounding is the bonding of all conductive materials which enclose electrical conductors. These enclosures are not normally current-carrying; the bonding prevents a difference in voltage between these materials and ground. The electrical system itself is grounded in order to prevent the occurrence of excessive voltages from such sources as lightning, line surges, or accidental contact with higher voltage lines. Both the electric system and metallic enclosures are grounded in order to cause overcurrent devices to operate in the event of a ground fault occurring from insulation failure.

System grounding. Alternating current systems operating at 50 volts or more are required to be grounded under a variety of voltage conditions. Grounding is accomplished by bonding the identified conductor (identification takes the form of white or neutral grey-colored insulation) to a grounding electrode by means of an unbroken wire called a grounding electrode conductor.

In a domestic occupancy, the utility will ground the identified conductor at the transformer by means of a plate attached to the bottom of the supporting pole. This identified or grounded circuit conductor is then once again grounded when it is brought to the service equipment within the home. This grounding takes the form of a grounding electrode conductor run to a metal underground water pipe, a metal building framework, a bare copper conductor encircling the building, or a concrete-encased steel reenforcing bar system. If none of these forms of electrodes is available, buried plates, pipes, rods or other metal underground structures are to be used.

Industrial installations are grounded in a similar manner except that the transformer may be very large and mounted on a concrete pad; in that case, the grounding electrode will be a carefully placed metal grid within the ground or within the concrete pad itself. Like the home, the grounded circuit conductor brought to the service equipment within the plant or commercial building must be bonded to a grounding electrode that is specified in the NEC.

Much has been said about the resistance of the grounding electrode and many erroneous ideas are prevalent concerning its use. Metal water piping systems, building structures, other buried tanks and piping systems, as well as concrete-encased reinforcing power systems, all provide an adequate resistance to ground providing careful attention is paid to the workmanship of the original installation.

It is only when resorting to a "made" electrode such as rod, pipe, or plate that higher resistance values are encountered. In many parts of the country, low resistance values for made electrodes are impossible to obtain, so the NEC simply requires in this case that one additional electrode be used. Because the same term is applied both to system grounding and equipment grounding, it is often thought that the grounding electrode must have a very low resistance in order to dissipate ground faults. This is not true. Most of the current from a ground fault finds its way back to the source transformer from the service equipment via the neutral conductor, not the earth.

Although it is not intended to carry large fault currents, a grounding electrode provides a point of equalization so that large voltage differences both inside and outside the building are minimized. As stated previously, some of the primary causes of such voltage differences are lightning surges or lightning strikes on the building itself.

Once the grounded circuit conductor is taken beyond the service equipment, it must not be grounded at any point. The major exception to this rule is the case of additional transformers being used within the establishment in order to step down voltages at various locations. In such instances, the secondary conductors constitute a new or "separately derived system" and the identified circuit conductor is to be grounded to a grounding electrode consisting of the nearest available effectively grounded water piping system or building structural metal.

One of the most frequent abuses of this rule prohibiting the subsequent grounding of the grounded circuit conductor is at the terminals of an ordinary parallel U-blade receptacle. Many well meaning but uninformed maintenance personnel will connect the white grounded circuit conductor to both the silver terminals and the green equipment grounding terminals on the receptacle, feeling that they have thereby achieved redundant grounding. In addition to violating the provisions of the NEC, they have set up a circumstance whereby, if the neutral is interrupted anywhere between the receptacle and its point of attachment to ground, any equipment enclosures which have been grounded by means of this receptacle now will be energized at full line voltage.

It should be understood that not all systems are required to be grounded. Providing trained electrical personnel are present, some manufacturing processes are dependent on ungrounded systems or high-impedance grounded systems to provide a higher degree of safety by not interrupting strategic equipment when a first fault to ground occurs. The presence of fault-indicating equipment together with the necessary personnel who can make repairs quickly will ensure against costly equipment down time and hazardous conditions arising from it.

Equipment grounding. When the insulation fails on conductors within undergrounded metal enclosures, the enclosures are raised to line voltage and constitute a serious hazard for personnel. However, if the metal enclosure is attached to the main bonding jumper and the service equipment with an equipment grounding conductor, this voltage difference will not occur.

Moreover, if the fault itself has a low resistance, and the equipment grounding conductor has been properly installed and well maintained, then a large amount of current can flow and the overcurrent device protecting the circuit will deenergize the circuit.

Fixed equipment which is to be grounded includes the exposed noncurrent-carrying metal parts that are likely to become energized that are (1) within 8 ft (2.4 m) vertically or 5 ft (1.5 m) horizontally of ground, (2) located in a damp or wet location and not isolated, (3) in electrical contact with metal, (4) in a hazardous location, (5) supplied by a metal clad, metal sheathed, or metal raceway wiring method, or (6) operated with any terminal in excess of 150 volts to ground. Additionally, exposed noncurrent-carrying metal parts regardless of voltage, of certain motor frames, controller cases for motors, the electric equipment in garages, theaters, and motion picture studios, accessible electric signs and associated equipment, and switchboard frames and structures must be grounded.

Also the frames and tracks of electrically operated cranes, the metal frames of nonelectrically driven elevator cars which have electrical conductors, hand operated metal shifting ropes or cables of electric elevators, metal enclosures around equipment carrying voltages in excess of 750 volts between conductors, and mobile homes and recreational vehicles must all be grounded.

In respect to cord-connected equipment, most items that are operated in hazardous locations or at more than 150 volts to ground must be grounded. Additionally, such items as refrigerators; freezers; air conditioners; clothes washing and drying equipment; dishwashers and sump pumps; portable, handheld, motor-operated tools and appliances such as drills, hedge clippers, lawnmowers, wet scrubbers, sanders, and saws; and all devices used in damp or wet locations or by persons standing on the ground or on metal floors or working inside of metal tanks or boilers must be grounded.

Tools, appliances, or portable handlamps proteced by a system of double insulation or a low-voltage isolating transformer are not required to be grounded.

Unlike the grounded circuit conductor, the equipment grounding conductor may be grounded continuously along its length and may be a bare conductor, the metal raceway surrounding the circuit conductors, or an insulated conductor. If the conductor is insulated, it must have a continuous green cover or have a green cover with a yellow stripe in it. The equipment-grounding conductor is always attached to the green hex-head screw on receptacles, plugs, and cord connectors. Where an approved metal conduit system is used as an equipment-grounding conductor, the receptacle must be bonded to the box by means of a separate jumper, or the receptacle must be listed by Underwriters Laboratories as being constructed in such a manner as to provide self-grounding (Figure 15-5).

Grounding of portable equipment is accomplished by means of a separate green insulated equipment-grounding conductor within the portable cord. Use of a separate external-grounding conductor is much less reliable, and the NEC permits this form of equipment grounding only where protected against physical damage and where part of the equipment.

Some attachment plugs are manufactured today of high-impact transparent plastic material so as to provide ready inspection of the terminations of the plug without having to disassemble it.

Connections to the equipment-grounding conductor and the grounded circuit conductor should be made in such a manner that removal of receptacles, switches, lighting fixtures, and other devices will not require the interruption of either of these two conductors. The proper use of jumpers and suitable pressure connectors will provide the necessary flexibility where main-

Figure 15-5. Standard receptacle and plug recommended by the National Electrical Manufacturers Association. The receptacle is designed to receive a plug having three blades—a U-shaped or round grounding blade with two standard parallel polarized blades. This type of fixture also permits use of the nonpolarized standard plug.

tenance is required, while still maintaining the integrity of both conductors. Solder is not permitted in the equipment-grounding circuit nor are switches, fuses, or any other interrupting device. The equipment-grounding conductor should always be grounded in the same enclosure with the conductors of the circuit which it protects.

A suitable attachment plug listed for equipment-grounding circuits is available in noninterchangeable configurations for all common voltage, current, and phase combinations. General-duty devices may be used where they are not subjected to rough service or moisture, but the more rugged and watertight equipment should be selected where this kind of exposure can be expected. Careful selection will ensure against loss of grounding continuity due to damaged components or corrosion.

Careful attention should be paid to the size of the equipment-grounding conductor, since it is this conductor that will be required to carry fault current in the event that a ground fault develops. Unless sufficient current can get to the overcurrent device and cause it to operate, the circuit will not be protected. Moreover, the heating effects that occur at points of high resistance in a poor equipment-grounding system may cause fire and explosion and result in loss of life and property, not to mention loss of valuable production time.

The NEC provides the proper sizing for equipment-grounding conductors used on various sizes of circuits. Conductor size should be increased where long runs are encountered and voltage drop exceeds an overall value of 5 percent.

One of the most overlooked aspects of good equipment grounding is that of workmanship. The code provides that "Connection of conductors to terminal parts shall ensure a thoroughly good connection without damaging the conductors. . . ." Where this provision is not observed, a point of low thermal capacity will occur. Even though the resistance of the equipment-grounding circuit is measured and determined to be quite low, if the system does not have adequate thermal capacity, high fault currents will cause these points to overheat and possibly result in arcing and fires as well as inoperative protective equipment.

Maintenance of grounds. Only electrically knowledgeable individuals should install or repair electrical equipment. They should make certain that the green insulated equipment grounding conductor is attached to the hexagonal green binding screw, and that the white grounded circuit conductor is attached only to the silver-colored binding screw.

Good maintenance will make sure of an electrically continuous equipment-grounding path from the metal enclosure of the portable equipment through the line cord, plug, receptacle, and grounding system that terminates at the bonding jumper at the service equipment or in the enclosure of the separately derived system.

Portable testing devices, such as a three-light neon receptacle tester, provide the most convenient means for checking polarity and other circuit connections. Other metered instruments are available to measure the actual impedance of the grounding circuit.

A receptacle tension tester may be used to inspect receptacles for deteriorating contacts. The tester employs little pointers which indicate the amount of tension that each receptacle contact will produce on the minimum size Underwriters Laboratories listed attachment plug. By using this form of inspection, the maintenance department can replace receptacles before they produce an ineffective equipment-grounding contact or a fire results in the power contacts.

When a program of testing is planned, the use of one or more of these instruments is recommended, since the entire test program can be handled by one person. A log of the results obtained at each test point will help predict deteriorating trends.

The condition of portable tools can easily be checked with the use of an insulation resistance tester which provides for the application of 500 v DC between the motor windings and the metal enclosure. Some insulation resistance meters provide an additional function which permits checking into continuity of the equipment grounding conductor as well. For the industrial user who has a planned tool maintenance program the use of a portable appliance tester will provide automatic cycling of the portable tool through a series of timed tests. Such test apparatus should be used in the tool room to check portable tools before they are issued or upon their return. A record of each tool will help predict insulation failure.

Three-wire adapters. Maintaining a good ground and making sure that it is used properly can be a difficult task in abusive environments.

In the hands of the untrained operator, the adapter is rarely connected properly. (The NEC now requires that all adapters have a wide neutral blade, and that the equipment ground be made by means of a rigid tab under the cover screw on the bottom of the adapter rather than by means of a flexible pigtail.) If the grounding pin on the attachment plug is clipped off, the operator now holds a potentially lethal device.

Double-insulated tools. As an alternate to equipment grounding, the NEC recognizes the use of double-insulated appliances. Such appliances are constructed with two separate systems of insulation so that the probabilities of insulation failure are reduced to the lowest practical minimum. Double-insulated tools must be handled carefully. Dropping a tool can destroy the effectiveness of the insulation. Double-insulated appliances are of particular value in the domestic situation; many industrial users have, likewise, found them to be an effective means of reducing exposure to the hazards of electric current. It should be noted that in plants which have instituted a strong safety program where an employee looks for the grounding pin on all attachment plugs, the double-insulated device now constitutes an exception to the rule. The ensuing confusion may result in the loss of some degree of safety. (For details, see Chapter 14, Hand and Portable Power Tools.) For protection against electric shock and to eliminate the need to ground the equipment, use self-contained battery powered tools.

Extension cords

Before using an extension cord, it should be found listed by the Underwriters Laboratories. The extension cords should be inspected regularly. Kinking or excessive bending of the cord should be avoided to prevent the wire strands from breaking because the broken strands may pierce the insulated covering and become a shock or short-circuit hazard. (Figure 15-6).

Ordinary twisted, household lamp cords should never be used as extension cords or for lamps in boilers, in tanks, or on damp or metallic floors, or where they will be exposed to mechanical wear.

Cord for use with portable tools and equipment is made in several grades, each of which is designed for a specific type of service. Jacketed cord should be used with portable electric tools and with extension lamps in boilers, tanks, or other grounded enclosures. Special types of rubber or plastic covering should be considered when the cord is to be used in areas where it may come in contact with oils or solvents.

Cord for heating devices, such as electric irons and water heaters, is made with an insulated covering which contains flame-retardant or thermosetting compound, such as neoprene.

Figure 15-6. Automatic cord take-up reel eases movement during inspection and repair work and helps prevent kinking and excessive bending of the cord. This particular unit has a transformer (mounted at top) to step down to "safety" low voltage. (Reprinted with permission from Daniel Woodhead Co.)

They are designed to resist high temperatures and, in the case of neoprene, dampness.

The various types of flexible cords and cables and their approved use are given in the NEC.

Because the metal frames of portable electrical equipment should be grounded, cord with a green-covered ground conductor should be used with a polarized plug and receptacle.

Flexible cords should be so connected to devices and fittings that tension will not be transmitted to joints or terminal screws. This is accomplished by special fittings, a knot in the cord, or winding with tape. All plugs that are attached to cords must have the terminal screw connections covered by suitable insulation.

Handles of portable handlamps must be made of nonconductive material, and there should be no metallic connection between the lamp guard and the socket shell. For use near exposed live parts, such as the rear of switchboards, the guard itself should be of nonconductive material.

Since extension lamps are sometimes used under conditions in which 120-volt shock may prove fatal, it is essential that safe cords and lamp holders be provided and that they be maintained in excellent condition. In some plants the shock hazard of portable lamps is eliminated by using small portable transformers which reduce the lamp voltage to 12 volts. Special lamps for these units are rated at 75-watt capacity and give sufficient working light.

Test equipment

Most electrical equipment is designed for safe operation under limited overload conditions for varying periods of time. Designers, installers, and operators should be thoroughly familiar with the limitations of their equipment and should be trained to observe and report abnormal conditions.

Continued overload may introduce additional operating hazards by causing fire, short circuits, circuit failures, or machine failures.

Through the use of various types of electrical and electronic testing equipment, many of these conditions can be detected before they get out of control and cause damage. Qualified technicians or specially trained maintenance personnel should make these tests.

The following types of equipment are standard and should be considered as essential items of test equipment: split-core ammeter, voltmeter, ammeter, megohmmeter, ground fault indicators and locators, wattmeter, industrial analyzer, recording instrument, and specialized testing instruments.

Specialized testing instruments, such as volt-ohm-milliammeter, oscilloscopes, and cable testers may be used. Instruments such as these are generally fitted for detailed engineering work or for use where ordinary recording and indicating instruments are not accurate enough.

Specialized processes

Among the many types of electrical equipment used in industry are the electric furnace, auxiliary heating devices, high-frequency heating equipment, electric welding equipment, X-ray lasers, ultraviolet, and infrared installations. Each of these devices may introduce special operating hazards, but protection from their electrical hazards may be secured through the same procedures recommended for use with the more common types of electrical equipment.

Because high-frequency heating installations range in power capacity from a few hundred watts to several hundred kilowatts, safety considerations are of prime importance.

The resistance of the body to the flow of high-frequency current is not dependent upon the skin. At frequencies of 200 kHz to several hundred megahertz, currents flow in a very thin shell on the surface of the conductor. This tendency of high-frequency current to flow on the surface is known as "skin effect" and increases as the frequency increases. Should the skin of a human being be punctured, the currents still flow on the surface and do not penetrate to the vital organs of the body.

A person coming into contact with high-frequency electrical energy will in general be burned because of the natural tendency to pull away, thereby setting up an arc. High frequency burns can also result from radio frequency antennae or waveguide exposure. In many cases, the burn will occur and nothing will be felt until after the exposure. These burns are painful and usually take longer to heal than burns from the more common thermal-heat sources.

The following methods may be used to prevent accidental contact and high-frequency burns to operators:

1. Interlocks, or other devices which remove power from the equipment whenever an access door is open, are connected into the control circuit to protect the operator from contacting any electrical or high frequency energy.
2. Material to be heated by induction is carried by revolving hopper feeds or conveyor to the heating coil. The heater coil is enclosed with a shield and cannot be reached by the operator.
3. Locate high-frequency generators some distance from the work position. The high-frequency energy is then conveyed by a waveguide or transmission line.
4. Insulate conductive coils or equipment surfaces with compatible insulation materials suitable for the application. Insulation protects the user in case of accidental contact with the coils or equipment surfaces.

CHOOSING EQUIPMENT FOR HAZARDOUS LOCATIONS

Standard electrical apparatus considered safe for ordinary application is obviously unfit for installation in locations where flammable gases, vapors, dusts, and other easily ignitable materials are present. Sparks and electric arcs originating in such fittings have been the igniting medium in costly fires and explosions.

Before electrical equipment and its associated wiring is selected for a hazardous location, it is necessary to determine the exact nature of the flammable materials present. For instance, an electrical fitting or device that is found by test to be safe for installation in an atmosphere of combustible dust may be unsafe for operation in an atmosphere containing flammable vapors or gases.

The *National Electrical Code* should be checked for these types of hazardous materials, which are described in detail in Articles 500 through 503. (A summary of the hazardous location classifications described in this code is given in the following section.)

A study should be made of the machine or devices to be used and of the processes involving liquids, gases, or solid substances, and their ratings should be checked. When hazards have been determined and the layout of the building inspected, it can be decided whether one small section of the plant should be classified as hazardous or whether the hazardous conditions extend to all parts.

The results of this study should be presented to a manufacturer of explosion-proof apparatus so that electrical equipment and wiring can be selected for safe installation.

Leading manufacturers of electrical equipment often maintain staffs of engineers to guide buyers in the purchase of explosion-proof fittings. Because of the highly technical nature of this field, competent engineers should always be consulted for these needs.

Hazardous (classified) locations

Articles 500 through 503 of the 1987 *National Electrical Code* (NEC) assign general requirements for electrical installations in hazardous (classified) locations; Articles 511 through 517 prescribe definitive requirements for specific types of occupancies such as commercial garages, aircraft hangars, and bulk-storage plants where hazardous locations prevail.

Determining if a hazardous situation exists in an industrial location is seldom difficult. It is common practice in industries in which hazardous locations normally exist to have a formal procedure for classifying areas and for reviewing equipment and processes to determine the degree of potential hazard. And once a hazardous location has been identified and classified—and its limits defined—NEC requirements for wiring methods and equipment compliance are relatively clear. Code requirements for electrical installations in hazardous locations are spelled out rather well.

A major problem is *defining the limits* of the hazardous location. How far above, below, and outward from the source does the hazardous location extend? For circumstances not covered in Articles 511 through 517, a general rule applies for determining these limits: The limits of the hazardous locations are those mutually agreed upon by the owner, the owner's insurance carrier, and the authority enforcing the code.

Definitions. Hazardous (classified) locations are areas in which explosive or flammable gases or vapors, combustible dust, or ignitable fibers are present or likely to become present. Such materials can ignite as a result of electrical causes only if two conditions co-exist:

1. The proportion of the flammable substance to oxygen (air) must permit ignition, and the mixture must be present in a sufficient quantity to provide an ignitable atmosphere in the vicinity of electrical equipment.
2. An electric arc, a flame escaping from an ignited substance in an enclosure, heat from an electric heater, or other source, must be present at a temperature equal to or greater than the ignition point of the flammable mixture.

Hazardous locations are classified as Class I, Class II, or Class III, depending on the physical properties of the combustible substance that might be present. These classes are subdivided into Divisions 1 and 2, depending on the degree of likelihood that an ignitable atmosphere might be present. And combustible substances are arranged into seven groups (A through G), depending on the nature of their behavior upon contact with an ignition source—highly explosive, moderately incendiary, etc. (These are explained later.) Equipment installed in hazardous locations must be approved for the applicable class, division, and group.

- *Class I* locations are those in which flammable gases or vapors are present or likely to become present.

- *Class II* applies to combustible dusts.
- *Class III* locations are those in which easily ignitable flyings such as textile fibers are present but not likely to be in suspensions in the air in sufficient concentrations to produce an easily ignitable atmosphere.
- In a *Division 1 location,* an ignitable atmosphere is to be expected to prevail at any time in the course of normal operations. Such an area represents a "worst case" condition, because one of the two requirements for ignition is likely to be present at any time, awaiting only a spark to ignite the flammable substance.
- In a *Division 2 location,* no ignitable atmosphere exists under normal operating conditions, but equipment malfunction, operator error, or other *abnormal* circumstances might create a hazardous environment.

Requirements for Division 2 electrical installations are less stringent than for Division 1 locations, because two possible, but improbable, circumstances must coincide for ignition. And, any accidental formation of an ignitable atmosphere in a Division 2 location can usually be quickly stemmed and the ignitable atmosphere dispersed, so any exposure to fire and explosion is usually of short duration.

Class I Locations are areas in which flammable gases or vapors are present or are likely to become present in the air in quantities sufficient to produce explosive or ignitable mixtures. In general, most hazardous locations in industrial plants fall in the Class I category.

Many common flammable substances, such as acetylene and naphtha, have been used in industrial plants for many years, and in recent years, many uses have been found for less common ignitable gases and liquids. Hydrogen, for example, has many uses and must be given special consideration. Because of its wide explosive-mixture range, high flame-propagation velocity, low minimum ignition energy level, and low vapor density, a hazardous atmosphere can develop far above the hydrogen source.

On the other hand, heavier-than-air vapors evolving from liquefied petroleum gases (LPG), such as propane and butane, also create special problems. LPG released as a liquid is highly volatile and has a low handling temperature; it readily picks up heat and creates large volumes of vapor. When released at or near ground level, the heavy vapors will travel along the ground for long distances, affecting the extent of the hazardous location in the horizontal plane. Some flammable liquids with flash points below 100 F (38 C) may produce large volumes of vapor that may spread much farther than might normally be expected.

Lighter-than-air gases will usually dissipate rapidly because of their relatively low densities. Unless released in confined, poorly ventilated spaces, low-density gases seldom produce hazardous mixtures in zones close to grade where most electrical equipment is located.

In the case of hydrocarbons—most of which are heavier than air—the problem is not to establish the existence of a Class I location, but to define the limits of the Division 1 and Division 2 areas. Anywhere that hydrocarbons are handled, used, or stored, there is a high degree of probability that flammable liquids, gases, and vapors will be released in sufficient quantities to constitute a hazard. Vapor can disperse in all directions as governed by the vapor density and air movement in the area.

A very mild breeze can extend the limits of the hazardous locations quite far in the direction of air movement, but the combustible mixture will not be dispersed significantly.

A Class I, Division 1 location is an area in which one or more of these conditions exist:

- Hazardous concentrations of flammable gases or vapors exist under *normal* operating conditions.
- Hazardous concentrations of flammable gases or vapors may exist frequently because of leakage, repair, or maintenance operations.
- Breakdown or faulty operation of equipment might release hazardous concentrations of flammable gases or vapors and might also cause simultaneous failure of electrical equipment.

See the accompanying Table 15-C for some examples of Class I, Division 1 locations.

Table 15-C. Some Examples of Class I, Division 1 Locations

Locations where volatile flammable liquids or liquefied flammable gases are transferred from one container to another.

Interiors of paint spray booths and areas adjacent to paint spray booths and other spraying operations where volatile flammable solvents are used.

Locations containing open tanks or vats of volatile flammable liquids.

Drying rooms or compartments for the evaporation of flammable solvents.

Cleaning and dyeing areas where hazardous liquids are used.

Gas generator rooms and portions of gas manufacturing plants where flammable gas may escape.

Inadequately ventilated pump and compressor rooms for flammable gas or volatile flammable liquids.

All other locations where hazardous concentrations of flammable vapors or gases are likely to form in the course of normal operations.

A Class I, Division 2 location is an area in which one or more of these conditions prevail:

- Volatile flammable liquids or flammable gases are handled, processed, or used, but are confined in closed containers or systems from which they can escape only in case of *accidental* rupture of the container or breakdown of the system, or in the event of abnormal equipment operation.
- Hazardous concentrations of gases or vapors are normally prevented from forming by positive mechanical ventilation, and ignitable concentrations can form only if the ventilation system fails.
- The area is adjacent to a Class I, Division 1 area from which hazardous concentrations might spread because of inadequate or unreliable ventilation supplied from a source of uncontaminated air.

Because maximum explosion pressures and safe operating temperatures vary widely for hazardous-location electrical installations, the electrical equipment must be approved for the specific flammable material responsible for the hazardous location designation. Hazardous atmospheres are categorized into seven groups—A through G. Substances that can contribute to Class I atmosphere fall into Groups A through D. A list of these substances can be found in NFPA 497M, *Classificaiton of Gases, Vapors and Dusts for Electrical Equipment in Hazardous (Classified) Locations.* Among the more common substances listed are:

Group A—acetylene

Group B—hydrogen or equivalent vapors and gases, such as manufactured gas

Group C—ethyl-ether vapors, ethylene, cyclopropane, and similar substances

Group D—gasoline, naphtha, benzine, hexane, butane, propane, alcohol, acetone, lacquer-solvent vapors, natural gas, and similar substances.

(Groups E, F, and G are described under Class II hazardous locations, next.)

Electrical equipment installed in hazardous locations must be approved not only for the class and division, but also for the group associated with the substance responsible for the location classification.

Class II hazardous locations are those locations where combustible dusts are present or likely to become present. A potential dust-explosion hazard exists wherever combustible dusts accumulate, are handled, or are processed. Many dusts fall into the "combustible" category. A list of these dusts is also contained in NFPA 497M.

Like all classified locations, Class II locations also have two divisions, as described earlier.

A Class II, Division 1 location is one that meets one or more of these criteria:

- Combustible dust is in suspension in air during the course of *normal* operations in quantities sufficient to produce explosive or ignitable mixtures.
- Mechanical failure or abnormal operation of machinery or equipment might create explosive or ignitable dust mixtures and might also provide a source of ignition through simultaneous failure of electrical equipment, operation of protective devices, etc.
- Combustible electrically conductive dusts may be present.

Following are some examples of Class II, Division 1 hazardous locations:

- Working areas of grain handling and storage plants
- Areas near dust-producing machinery and equipment in grain-processing plants, starch plants, sugar-pulverizing plants, flour mills, malting plants, hay-grinding plants
- Areas in which metal dusts and powders are produced, processed, handled, packed, or stored
- Coal bunkers, coal-pulverizing plants, and areas in which coke, carbon black, and charcoal are processed, handled, or used.

In a Class II, Division 2 location, ignitable concentrations of combustible dusts are not expected to be found in the course of *normal* operation; but, *abnormal* conditions may allow combustible dusts to accumulate in sufficient quantities and dust/air concentrations to permit ignition. Abnormal conditions could include failure of dust-control equipment, process equipment, or containers, chutes, or other handling equipment. The following are usually considered to be Class II, Division 2 locations:

- Areas containing only closed bins or hoppers, and enclosed spouts and conveyors
- Areas containing machines and equipment from which appreciable quantities of dust would escape only under abnormal conditions
- Warehouses and shipping rooms in which dust-producing materials are handled or stored only in bags or containers
- Areas adjacent to Class II, Division 1 locations.

Because dust that is carbonized or excessively dry is susceptible to spontaneous ignition, electrical equipment installed in Class II locations should be capable of operating at full load without developing surface temperatures high enough to cause excessive dehydration or carbonizations of dust deposits that might form.

Electrical equipment installed in Class II locations must be approved for class and division and the applicable group. For purposes of equipment approval, dusts are classified into three groups based on their conductivity:

Group E—metal dusts
Group F—carbon black, charcoal, coal, coke dust
Group G—flour, starch, grain dust.

Class III locations are those in which easily ignitable fibers or flyings are present, but are not likely to be in suspension in air in quantities sufficient to produce an ignitable atmosphere. Single fibers of organic materials such as linen, cotton tufts, and fluffy fabrics, however, are quite vulnerable to a localized heat source such as an electric spark. In pure oxygen, single fibers of cotton can be ignited by a 0.02 joule spark.

Textiles such as those used in clothing can be ignited and burned with repetitive or sustained high-energy electric sparks. Cotton and wool fabrics can be ignited in pure oxygen with a spark of 2.3 joules; in normal air, a spark of 193 joules or more is required for ignition. Silk and polyester fibers are more difficult to ignite than cotton or wool.

Fibers contaminated with oily substances can be ignited with much weaker sparks than clean fibers. Typically, only 0.0001 of the energy required to ignite a clean fabric is required for an oily sample of the same fabric. In general, the burning characteristics of fibers will be affected by the specific gravity of the substance, size and shape of the sample, air circulation in the area, oxygen concentration, and relative humidity. The burning characteristics of some common fibers whose presence can cause an area to be designated as Class III are given in Table 15-D.

Table 15-D. Burning Characteristics of Some Common Fibers

Substance	Specific Gravity	Approximate Ignition Temperature, C	Burning Characteristics
Acetate	1.32	525	Melts ahead of flame
Acrilan	1.17	560	Burns readily
Arnel	1.3	525	Melts ahead of flame
Cotton	1.54 to 1.56	400	Burns rapidly
Nylon 6	1.14	530	Melts and burns
Wool	1.3	600	Melts ahead of flame

Adapted from *Wellington Sears Handbook of Industrial Textiles*, New York, N.Y., Wellington Sears Co., Inc., 1963.

A *Class III, Division 1* location is one in which easily ignitable fibers or materials that produce combustible flyings are manufactured, handled, or used. This classification usually includes:
- Plants that produce combustible fibers
- Portions of rayon, cotton, or other textile mills
- Flax-processing plants
- Clothing manufacturing plants
- Woodworking plants

A *Class III, Division 2* location is one in which easily ignitable fibers are stored or handled, but are not manufactured or processed. An example is a textile warehouse.

There are no group designations associated with Class III locations, and electrical equipment installed in Class III locations need only be approved for the applicable class and division. The maximum equipment surface temperature under normal conditions shall not exceed 329 F (165 C) for equipment not subject to overloading, and 248 F (120 C) for equipment such as transformers and motors that are subject to overloading.

Establishing the limits. The decision to classify an area as hazardous for purposes of NEC compliance is based on the possibility that flammable liquids, vapors or gases, combustible dusts, and easily ignitable fibers or flyings may be present. After the decision has been made to classify an area as hazardous, the next step is to determine the degree of hazard. Should the area be classified as Division 1 or Division 2? Also, the limits of the hazardous location must be defined—how far above, below, and outward from the source of hazard does the hazardous location extend?

Table 15-E provides guidelines for classifying hazardous locations. Use it with Figure 15-7, explained later in this section.

The safest electrical installation in a hazardous location, of course, is one that is not there at all. Electrical equipment, as much as is practical, should be situated outside the area defined as hazardous. It is, however, seldom possible or practical to locate all electrical equipment outside the hazardous area. The plant engineer is responsible for ensuring that all electrical equipment in the hazardous area, and its associated wiring, conforms to the NEC and does not significantly increase the probability of explosion.

Two ways to reduce the probability of explosion from electrical sources are (1) remove or isolate the potential ignition source from the flammable material, and (2) control the atmosphere at the ignition source. For an explosion to occur, two conditions must coexist:

1. Combustible material present in a sufficient amount and the proper concentration to provide an ignitable atmosphere
2. An ignition source of sufficient incendivity to ignite combustible materials that are present.

If either of these two conditions is eliminated, the explosion hazard is reduced to zero.

The most difficult part of planning an electrical installation to conform to NEC rquirements for hazardous locations is to define the limits of the hazardous area. How far should the hazardous location be considered to extend to ensure that safety is served, without taking unnecessary and expensive precautions? No hard-and-fast rules can be applied. Experience on comparable projects and an understanding of specific conditions at the job site provide a far better basis for defining limits than any theoretical study of flammable vapors, gases, dusts, or fibers.

The environmental aspects of an installation—prevailing winds, site topography, proximity to other structures and equipment, and climatic factors—can significantly affect the extent of a hazardous location. Among the factors that should be evaluated in establishing the limits are:
- Size, shape, and construction features of the building
- Existence and location of doors and windows, and their manner of use

Table 15-E. Guidelines for Classifying Hazardous Areas

DETERMINING THE NEED FOR CLASSIFICATION
A need for classification is indicated by an affirmative answer to any of the following questions.

Class I	Class II	Class III
■ Are flammable liquids, vapors, or gases likely to be present? ■ Are liquids having flash points at or above 100 F likely to be handled, processed, or stored at temperatures above their flash points?	■ Are combustible dusts likely to be present? ■ Are combustible dusts likely to ignite as a result of storage, handling, or other causes?	■ Are easily ignitable fibers or flyings present, but not likely to be in suspension in the air in sufficient quantities to produce an ignitable mixture in the atmosphere?

ASSIGNMENT OF CLASSIFICATION
Classification is determined as indicated by an affirmative answer to any question.

Class I, Division 1	Class II, Division 1	Class III, Division 1
■ Is a flammable mixture likely to be present under normal operating conditions? ■ Is a flammable mixture likely to be present frequently because of repair, maintenance, or leaks? ■ Would a failure of process, storage, handling, or other equipment be likely to cause an electrical failure coinciding with the release of flammable gas or liquid? ■ Is the flammable liquid, vapor, or gas piping system in an inadequately ventilated location, and does the piping system contain valves, meters, or screwed or flanged fittings that are likely to leak? ■ Is the zone below the surrounding elevation or grade such that flammable liquids or vapors may accumulate?	■ Is combustible dust likely to exist in suspension in air, under normal operating conditions, in sufficient quantities to produce explosive or ignitable mixtures? ■ Is combustible dust likely to exist in suspension in the air, because of maintenance or repair operations, in sufficient quantities to cause explosive or ignitable mixtures? ■ Would failure of equipment be likely to cause an electrical system failure coinciding with the release of combustible dust in the air? ■ Is combustible dust of an electrically conductive nature likely to be present?	■ Are easily ignitable fibers or materials producing combustible flyings handled, manufactured, or used?

Class I, Division 2	Class II, Division 2	Class III, Division 2
■ Is the flammable liquid, vapor, or gas piping system in an inadequately ventilated location, but not likely to leak? ■ Is the flammable liquid, vapor, or gas handled in an adequately ventilated location, and can the flammable substance escape only in the course of some abnormality such as failure of a gasket or packing? ■ Is the location adjacent to a Division 1 location, or can the flammable substance be conducted to the location through trenches, pipes, or ducts? ■ If positive mechanical ventilation is used, could failure or improper operation of ventilating equipment permit mixtures to build up to flammable concentrations?	■ Is the combustible dust likely to exist in suspension in air only under abnormal conditions, but can accumulations of dust be ignited by heat developed by electrical equipment, or by arcs, sparks, or burning materials expelled from electrical equipment? ■ Are dangerous concentrations of ignitable dusts normally prevented by reliable dust-control equipment such as fans or filters? ■ Is the location adjacent to a Division 1 location, and not separated by a fire wall? ■ Are dust-producing materials stored or handled only in bags or containers and only stored—not used—in the area?	■ Are easily ignitable fibers or flyings only handled and stored, and not processed? ■ Is the location adjacent to a Class III, Division 1 location?

DEFINING THE LIMITS OF THE CLASSIFIED LOCATION

The limits of the classified location—outward, upward, and downward from the source—must be determined by applying sound engineering judgment, experience gained on similar projects, and information from handbooks and other sources. Figure 15-7 provides an example of recommended clearances for the most common industrial-type classified locations—those involving heavier-than-air liquids, vapors, and gases.

Table is based on recommendations of the National Fire Protection Association and American Petroleum Institute.

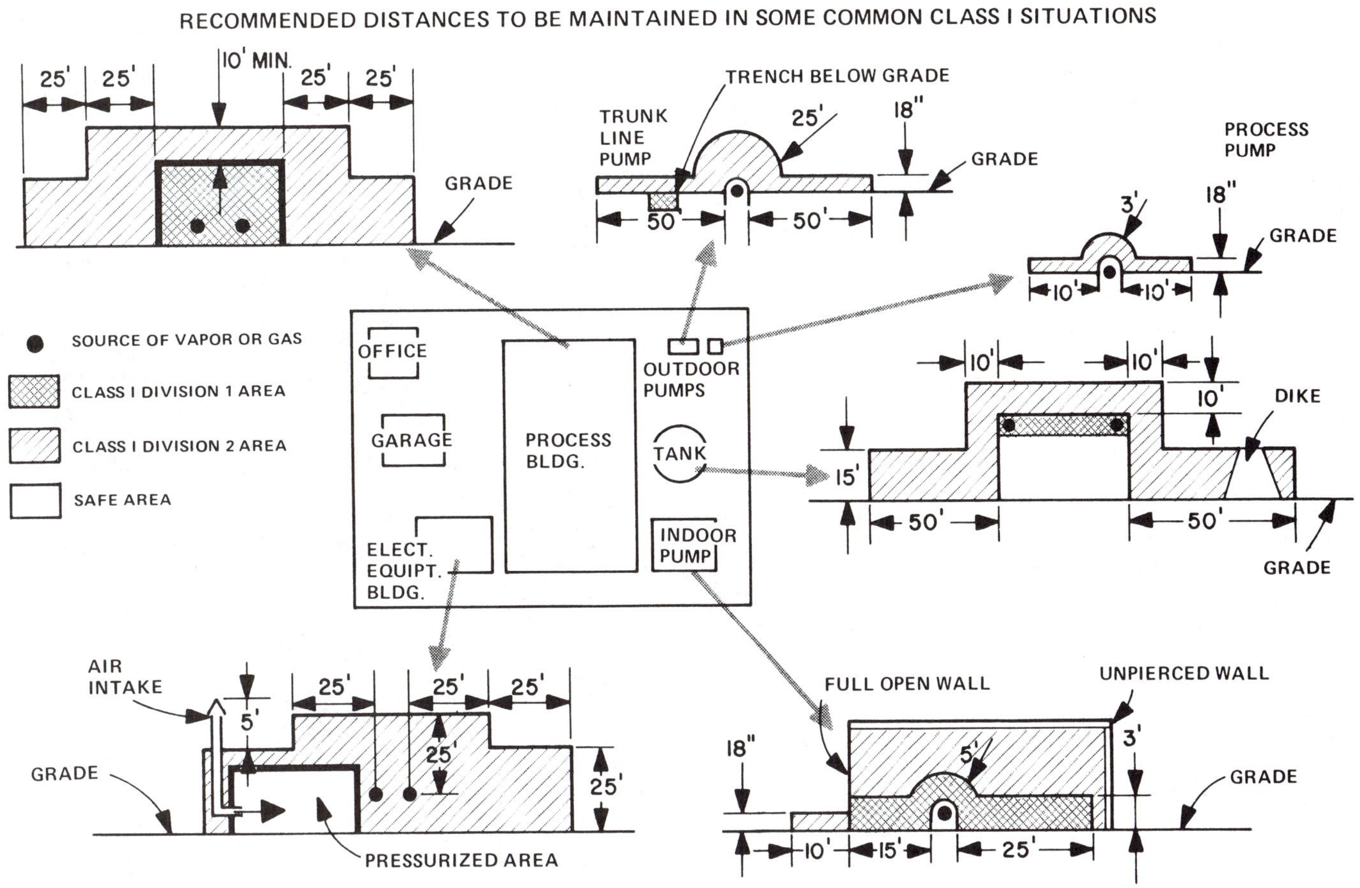

RECOMMENDED DISTANCES TO BE MAINTAINED IN SOME COMMON CLASS I SITUATIONS

Figure 15-7. Recommended distances to be maintained in some common Class I situations, developed from standards established by NFPA and API. (Reprinted with permission from *Plant Engineering Magazine*.)

- Absence or presence of walls, enclosures, and other barriers
- Existence and locations of ventilation and exhaust systems
- Existence and locations of drainage ditches, separators, and impounding basins
- Quantity of hazardous material likely to be released
- Locations of possible leakage
- Physical properties of the hazardous material—density, volatility, chemical stability, etc.
- Frequency and type of maintenance and repair work performed on the systems containing the hazardous substance and on other equipment in the area.

When the limits of hazardous locations are being established, the area surrounding each source of hazardous material must be considered as a location and its individual limits determined. For example, the following rules should be considered for a Class I location:

- In the absence of walls, enclosures, or other barriers, and in the absence of air currents or other disturbing forces, a gas or vapor will be distributed in a predictable fashion.
- For heavier-than-air gases and vapors released at or near grade level, potentially hazardous concentrations are most likely to be found below grade; heavier-than-air gases distribute themselves downward and outward. As the height above grade increases, the hazard decreases.
- For lighter-than-air gases, little or no potential exists for hazard at or below grade; lighter-than-air gases distribute themselves upward and outward. As the height above grade

increases, the potential hazard increases. For purposes of classification, gases with a density greater than 75 percent of the density of air at standard conditions should be treated as if they were both lighter and heavier than air, and the limits of the hazardous location defined accordingly.

Most hazardous locations in industrial plants fall into Class I and involve heavier-than-air gases and vapors. Figure 15-7 developed from recommendations established by NFPA and by American Petroleum Institute, provides recommendations for establishing limits for some typical situations involving such gases and vapors.

In Figure 15-7, a process pump handling flammable liquids at moderate pressures is at grade elevation. Because the source of hazard is in open air, there are no pockets below grade where flammable vapors can accumulate, and liquid can escape only in the event of equipment failure—such as a leaking gasket—the area surrounding the pump is classified Division 2.

A different situation prevails, though, for the trunk-line pump. Although also outdoors, the trunk-line pump operates at relatively high pressure, increasing the probability of equipment failure and the volume of liquid that might be released. A below-grade trench is in the vicinity of the trunk-line pump. For these reasons, the limits of the Division 2 location, surrounding the pump are extended, and the trench itself is classified as Division 1.

The outdoor tank in Figure 15-7 is installed with its base at grade level, and the tank is fitted with a floating roof. The space within the shell of the tank above the roof is classified as a Division 1 location. The area surrounding the tank is classified as Division 2; it extends 10 ft (3 m) above the tank. In the horizontal direction, the Division 2 area extends 50 ft (15.2 m) from the tank or to the dike, whichever is greater.

The indoor pump is in a building with one fully open wall but no through ventilation. The building also contains valves, meters, and other equipment and fittings that are likely to leak. The area adjacent to the pump and near grade is classified as Division 1. The rest of the building and the outdoor close-to-grade area adjacent to the open wall is classified as Division 2.

The process building has numerous sources of flammable materials and is not especially well ventilated. The entire interior is classified as Division 1, and the area surrounding the building is classified as Division 2. The Division 2 area extends horizontally from the building for 50 ft (15 m) in every direction, and at least 10 ft (3 m) above the roof or 25 ft (7.6 m) above the source, whichever is greater.

The electrical equipment room is separate from but immediately adjacent to the process building, and lies entirely within the Division 2 location that surrounds the process building for 50 ft (15 m) in all directions. No fire wall separates the electrical-equipment building from the process building.

The interior of the electrical-equipment building, however, is classified as a safe area, because its atmosphere is controlled to prevent ignition sources from contacting an ignitable mixture. A positive pressurization system is used to keep the building purged of any flammable mixtures. Note that the air intake for the pressurization system must be at least 5 ft (1.5 m) above the classified location. This minimum distance must under no circumstances be violated, nor may this space be invaded by any flammable mixture.

Each source of flammable material in Figure 15-7 was considered to be the focal point of a separate hazardous location. The positioning of sources of hazardous materials could cause an overlapping of Division 1 and Division 2 locations. In such cases, the more stringent classification will prevail; the area of overlap should be classified Division 1.

Many, many separate sources of information are available on the subject of classifying hazardous locations; this discussion has presented only some of the more basic and more important considerations.

Explosion-proof apparatus

Explosion-proof apparatus is defined in the *National Electrical Code:*

> Apparatus enclosed in a case that is capable of withstanding an explosion of a specified gas or vapor which may occur within it and of preventing the ignition of a specified gas or vapor surrounding the enclosure by sparks, flashes, or explosion of the gas or vapor within, and which operates at such an external temperature that a surrounding flammable atmosphere will not be ignited thereby.

Only fittings that have undergone exhaustive tests and meet the requirements of the Underwriters Laboratories Inc. for use in hazardous locations should be used (Figure 15-8). They must be of durable material, provide thorough protection, and be finished to be totally resistant to atmospheric conditions.

Explosion-proof fittings should be installed not only on new work, but also on old wiring systems where alterations are being

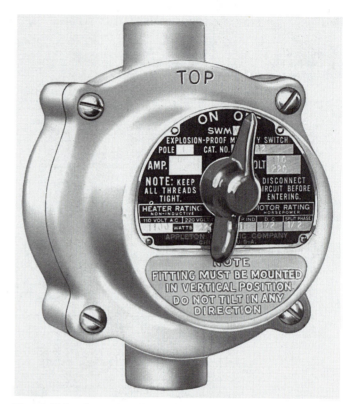

Figure 15-8. An explosion-proof and dust-tight mercury switch, listed by Underwriters Laboratories, which makes and breaks the circuit when the switch tilts the tube inside the mechanism. (Reprinted with permission from The Appleton Electric Co.)

made or new equipment is being installed. Observing code requirements will minimize dangers that might result from using ordinary fittings in hazardous locations.

These fittings are made of durable cast material, with roomy interiors for wiring and splices. They are capable of withstanding the high internal pressure that results from an explosion without bursting and without becoming loose. Furthermore, they are tested on the basis of a high safety factor as prescribed by the Underwriters Laboratories Inc.

It is impossible to prevent highly flammable gases from entering the interior of either an explosion-resisting or an ordinary wiring system. The gases will eventually enter the entire line through the joints and through the breathing of the conduit system caused by temperature changes. Furthermore, gaseous vapors will fill every crevice whenever covers are removed.

For these reasons, it is impossible to provide an entirely vapor-proof switch unit, regulate temperatures, or keep the air free from flammable gases inside the fittings.

To isolate that section of the industrial plant classified as a hazardous location, it is necessary to have positive confinement of the arc, heat, and explosion within the internal limits of the explosion-proof fittings. These fittings are constructed to imprison the dangerous arcing, intense heat, and subsequent explosion so that the gas-laden air outside does not become ignited.

Often control equipment can be located outside a room containing hazardous materials. In this case, conventional wiring equipment can be used, thus reducing cost and hazards in the installation.

INSPECTION

Whenever possible, equipment must be deenergized by the plant electrician or other authorized person. Live circuits and equipment left in the operable mode are always a hazard. It should always be assumed that a circuit is live until it is proved dead. To make certain, switches and circuit breakers carrying current to or from switchboards, buses, controls, and starting equipment should be *checked to see that they are open.* Tests should be made in the inspector's presence by the plant electrician, supervisor, or operator to determine that the parts to be worked on are *dead.* As an additional safeguard, the inspector should have the breakers and switches locked open, grounded, and tagged so that they cannot be energized until the tests have been completed. (A lockout procedure is detailed later in this chapter.)

Upon completion of the work, the worker should remove any grounds and lockout tags that were used. However, before the equipment is returned to service, clearance should be given by the chief electrician or supervisor or operator (whoever is in charge).

If the feed-in circuit to a switchboard, bus, or other equipment must be kept live on the incoming side of a breaker or switches, the electrician should provide ample clearance, barriers, or other protection between this section and the part being worked on. Work in cramped quarters should be avoided, if it is at all possible.

Live buses, conductors, and switches should be covered with insulated blankets, specially formed insulating shields, or isolated with barriers.

Electrically operated or remote-control circuit breakers or contactors should not be depended upon for protection. A ground or other disturbance on the control system may permit the circuit breaker to close. If no disconnects are installed ahead of or behind the circuit breaker, it will be necessary to block, rackout, or otherwise lock the breaker so that it cannot be closed.

If the electrician must leave the circuit being worked on before the test is complete, lockout procedures should be used. Upon returning, the electrician should check all markings, breakers, and switches to be sure that they have not been altered before continuing the tests. (Lockout procedures are described later in this chapter.)

Rotating and intermittent-start equipment

Before work is started on rotating machines or automatic and intermittent-start equipment that is presumed to be out of service and stopped, an inspection of all the electrical control and starting devices should be made with the chief electrician or the supervisor.

For example, when inspections or repairs are to be made on motors, generators, blowers, compressors, converters, any part of which is remotely controlled or which may automatically start, circuit breakers or switches should be locked out, and the fuses should be pulled.

Machinery connected to blowers, water wheels, or pumps, without check valves, may start turning even when the current to the motor has been disconnected. For this reason, the rotor or armature should be blocked before inspections are made.

Lockout procedures should be followed on generators driven by prime movers. The throttle, starting valve, or other means of controlling the energy to the driving part of the unit should be locked and tagged.

While the equipment is in operation, motor brushes should not be removed or adjusted, contacts or other parts of the electrical equipment should not be worked on, and no attempt should be made to clean, sandpaper, or polish commutators, or slip or collector rings.

When inspecting electrical equipment, employees should not wear loose clothing because it may become entangled in couplings, coils, or other moving parts. They should remove wrist watches, rings, metal pens and pencils, and they should not use metal flashlights.

High-voltage equipment

In general, high-voltage equipment is more carefully guarded than low-voltage equipment because of the greater inherent hazard. Work on this type of equipment should be done by well-qualified personnel. In newer installations of 2,300 volts or more, attempts have been made to insulate or armor apparatus so that casual contact with the current-carrying parts is not possible. In any case, only authorized, trained personnel should work on such equipment.

Persons working on high-voltage equipment should know that rubber gloves are not a good substitute for safety devices or proper procedures, but should be worn only as a supplementary measure. Before each use they should be checked for punctures, tears, or abrasions. For an on-the-job test, the cuffs should be rolled up and air forced into the fingers and palms of the gloves. If there is a leakage, the gloves should not be used. (Gloves are discussed in the *Administration and Programs* volume, Chapter 17, Personal Protective Equipment, and in the Council's Industrial Data Sheet No. 598, *Flexible Insulated Protective Equipment for Electrical Workers.* See References.)

Leather protectors should be worn over rubber gloves to protect the rubber from mechanical damage and from oil and grease.

An electrical glove testing service should be provided, with regular testing intervals determined by the amount of use of the gloves, the type of work, and the voltages they are subject to. If there is doubt about the insulating quality of the gloves, they should be discarded.

Private testing laboratories can be relied upon for accurate testing. In some cases they will keep individual company records on the physical condition of gloves and the time schedule for retesting. Where no laboratory is available, small concerns can usually get tests made at the local public utility.

MAINTENANCE

To be safe and to give the best service, electric equipment must be well maintained. Motors, circuit breakers, moving parts of switches, and similar current-carrying devices wear out, break down, and need adjustment. Repairs on electrical circuits and electrical apparatus should be made only by experienced electricians (Figure 15-9).

Only high-grade electrical equipment listed as standard by Underwriters Laboratories Inc., or other qualified authority, should be used in maintenance work. Inferior, unapproved equipment may become hazardous because of defective material or design or poor workmanship.

If practical, conductors should be deenergized when maintenance work on them is necessary. Before they begin work, maintenance personnel should make certain that the line is dead by testing it with the voltage testing devices provided. *It is not safe to test a circuit with the fingers or with makeshift devices.*

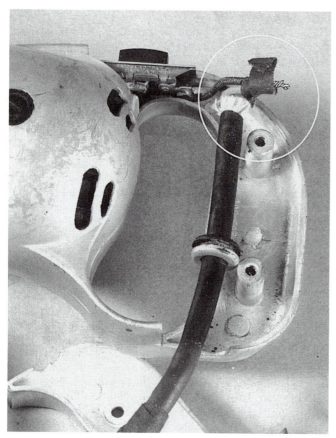

Figure 15-9. In the circle is an exposed electrical connection inside the handle of a portable electrical tool, creating a serious shock hazard. This sort of defect should be remedied only by an experienced electrician.

Figure 15-10. Typical multiple lockout designed for six individual locks.

Lockouts

Unexpected operation of electrical or electronic equipment that can be started by automatic or manual control may cause injuries due to electrical shock or mechanical energies. For example, the unexpected starting of motors may injure persons working on them or the unexpected energization of equipment can produce an electric shock.

For that reason, when electronic or electrical equipment must be repaired or modified, the circuit should be opened at the circuit contact, and the switch should be padlocked in the OFF position, tagged with a description of the work being done, the name of the person, and the department involved. Warning signs or tagging alone does not provide the positive protection of locking out equipment (Figure 15-10).

Because of the grave risk to life, lockout procedures should be drilled into all maintenance personnel, and the supervisor should see that the procedure is implemented with the necessary keys, locks, and arrangements.

Only locks made by a reputable lock company should be used. No two key configurations should be the same, and they should be checked to see that each key fits only one lock.

For identification, locks may be painted various colors to indicate types of craft or to differentiate shifts. Each lock should be stamped with the employee's name or clock number, or a metal tag should be attached.

Only one key should exist for each lock. Master keying is not recommended. In emergency situations, locks may be cut or torched off. If any duplicate keys exist, they must be kept under positive control at all times. A manager must use the duplicate key. Formal positive controls are required.

To make lockout systems operable, equipment should have built-in locking devices designed for the insertion of padlocks. In plants where older equipment is in use, or where explosion-proof or dust-tight equipment is installed, it may be necessary

Maintenance personnel should be instructed in the use of electrical testing equipment and meters: how to tap into the circuit, how to locate testing points from schematic diagrams, how to use insulated meter leads, clips, and probes, and how to fuse test circuits. Pliers, screw drivers, testing lights, and other tools used in electrical repair work should be insulated.

When maintenance or repair work must be done on energized conductors, it is advisable to have two or more employees work together. The supervisor should give detailed procedures to be followed, and see that maintenance crews are supplied with and use the right protective equipment.

The kind of protective equipment needed is determined by the type of circuit, the nature of the job, and the conditions under which the work must be done. Rubber gloves, sleeves, mats, line hose, insulating platforms, safety headgear, safety glasses, safety belts, fuse tongs, and insulated switch sticks are among the more common items of equipment used.

Good safety practices lie not only in using the proper protective equipment, but taking care of that equipment as well, such as inspecting it before use and checking it at frequent intervals.

Maintenance personnel must constantly be on guard when working around electrical equipment and circuits, in order to prevent accidental grounding and possible severe injury. Grounding can easily occur if an energized loose wire contacts a water pipe, conduit, metal fixture, another wire, or anything metallic that is connected in some way to the earth.

for the technicians or maintenance department to construct attachments to which locks can be applied. Electronic equipment may need plug control lockouts. Where more than one person or group will work on a piece of equipment or device, a multiple locking device should be employed so that each person must apply a personal lock to the equipment before it will operate.

No matter what method of locking electric switches OFF is used, effective control can be maintained only by constant supervision and training in the safe routine. Following is a lockout procedure which is generally acceptable.

1. Alert the operator and other users of the systems to be shut off.
2. Plan the shutdown to ensure that the system will be off.
3. Place your own padlock on the control switch, lever, or valve, even though someone has locked the control before you. *You will not be protected unless you put your own padlock on it.* Signs and blocks are no substitute for locking out the source of electrical power.
4. Test the lockout to be sure the system is really off.
5. Do work with peace of mind.
6. When through working, remove your padlock, your sign, and blocking. Never permit someone else to remove it for you, and be sure you are not exposing another person to danger. Verify that the equipment is clear and post a watch if necessary.
7. Reenergize the system.

Removing fuses

When it is necessary to remove a fuse, the operating switch should be opened to disconnect the electrical load. The fuse should be extracted with an insulated fuse puller. If the fuse is not protected with a switch, the supply end of the fuse should be pulled out first. When the fuse is replaced, the supply end should be put in first.

It is important that a fuse be replaced by one of the same type and size and that a copper wire or other conductor never be substituted.

For work on lines carrying more than 600 volts, the sources of energy can be opened by tripping the circuit breaker and then opening the disconnects. The conductors should be tagged, and, if possible, a substantial grounded conductor should be clamped to each leg of the circuit before the work is started. The grounding conductors should be removed just before the circuit is reenergized. As an added precaution, tests should be made with a voltmeter or other standard approved tester to make sure the circuit is clear.

In emergencies where the areas around fuse boxes are wet, wood platforms, insulated stools, or rubber boots should be used by maintenance workers.

Wiring

Electrical installations should be made in accordance with the *National Electrical Code* as a minimum standard. Each installation should be thoroughly inspected by a knowledgeable inspector. The use of temporary wiring should be discouraged even though it may be reasonably safe when first installed. As work progresses and repairs or installations are made, temporary wiring may become unsafe because it is not properly protected from mechanical injury and may be modified.

Technicians and maintenance personnel should make it a practice to inspect plugs and connections on portable electric or electronic equipment to see that the third wire frame ground is properly connected, that the grounding prong has not been removed, and that the cord is connected to the proper terminals. Where there is any doubt about the safe operation of a tool or instrument, it should be removed from service, tagged, and returned for inspection and repairs by qualified people.

When additional equipment is being installed or operated under temporary conditions, no taps should be made into an existing circuit unless an individual switch is installed in the branch line, with provision for locking the switch OFF and tagging it. This precaution prevents service interruptions in the main circuit when branch circuit power must be shut off. It also provides a means of preventing the accidental starting of connected equipment.

EMPLOYEE TRAINING

Deviations from the safe use and installation practices for electrical and electronic equipment often result in unnecessary hazards that injure or kill people. Consequently, the facility safety program must include thorough training of all employees who work with electrical and electronic equipment or operate electrical and electronic systems. In addition to instructions on the hazards of electricity, people should be trained in cardiopulmonary resuscitation, the use of warning signs, guards, and other protective devices, and in safe operational procedures. It is essential that each person be trained to handle emergency situations and be instructed to never work alone with potentially hazardous electrical energies.

The training program should be based on the facility's electrical system and use of electronic equipment. It should be applied to the specific operations and changes that may occur within the systems.

Supervisors should be given instructions to acquaint them with existing or probable electrical hazards, and they should be required to maintain close supervision over all operations which involve the use of electrical or electronic equipment. Supervisors must encourage employees to report immediately any electrical defect observed and respond by having it repaired or replaced at once.

Training is covered in the *Administration and Programs* volume, Chapter 9, Safety Training. Electrical safety training is a specialized process and must be approached positively and consistently. Design, fabrication, and installation people must be familiar with the equipment and its required operation.

REFERENCES

American National Standards Institute, 1430 Broadway, New York, N.Y. 10018.
"Dimensions of Attachment Plugs and Receptacles," C73 Series. (Also NEMA WD.1—1965.)
"Insulated Wire," C8 series.
Manual and Automatic Station Control, Supervisory and Associated Telemetering Equipment, C37.2.
Relays and Relay Systems Associated with Electric Power Apparatus, C37.
Specifications for Accident Prevention Signs, Z35.1.
Specifications for Rubber Insulating Tape, C59.6 (Also ASTM D119).

"Specifications for Rubber Protective Equipment for Electrical Workers," J6 Series.

Specifications for Weather-Resistant Wire and Cable (URC Type), C8.18.

Safety Requirements for Lockout/Tagout of Energy Sources, Z244.1.

Dalziel, Charles F. "Effect of Electric Current on Man." *Electrical Engineering,* February 1941.

"Scientific Facts Concerning Electrical Hazards." *National Safety News,* October 1947.

"Electricity—Good and Faithful Servant." *National Safety News,* September 1961.

National Fire Protection Association, Batterymarch Park, Quincy, Mass. 02269.

Classification of Gases, Vapors and Dusts for Electrical Equipment in Hazardous (Classified) Locations, NFPA 497M.

Electrical Safety Requirements for Employee Workplaces, NFPA 70E.

National Electrical Code, NFPA 70.

National Safety Council, 444 North Michigan Ave., Chicago, Ill. 60611.

Electrical Inspection Illustrated.

Industrial Data Sheets

Electric Plug and Receptacle Configurations, Applications of, 579.

Electrical Switching Practices, 544.

Electromagnets Used with Crane Hoists, 359.

Electrostatic Paint Spraying and Detearing, 468.

Flexible Insulated Protective Equipment for Electrical Workers, 598.

Grounding Electric Shovels, Cranes and other Mobile Equipment, 287.

Industrial Electric Substations, 559.

Lead-Acid Storage Batteries, 635.

Portable Reamer-Drills, 497.

Static Electricity, 547.

Temporary Electric Wiring for Construction Sites, 515.

Underwriters Laboratories Inc., 333 Pfingsten Rd., Northbrook, Ill. 60062.

"Electrical Appliance and Utilization Equipment Lists."

"Electrical Construction Materials List."

"Hazardous Location Equipment List."

U.S. Department of Commerce, National Bureau of Standards, Washington, D.C. 20234.

Safety Rules for the Installation and Maintenance of Electric Utilization Equipment. Handbook H33.

National Electrical Code, ANSI C2.

16

Flammable and Combustible Liquids

THE PRECAUTIONS COVERED IN THIS CHAPTER are directed to industrial operators who receive, store, handle, and use flammable and combustible liquids, and not to the manufacturers of flammable liquids. Because specific characteristics of flammable liquids vary as do their required handling precautions, this chapter does not attempt to cover the subject in all of its many details. It does present general information regarding flammable and combustible liquids and is intended to provide a broad view of the subject. It is advisable to consult federal and state or provincial laws, OSHA standards, fire underwriters, the National Fire Protection Association, product trade associations, state (provincial) and municipal authorities that have jurisdiction, and specific handbooks for detailed information. (See References at the end of this chapter.)

DEFINITIONS

As defined by OSHA, DOT, and the National Fire Protection Association Standard, *Flammable and Combustible Liquids Code,* NFPA 30, a flammable liquid is any liquid having a closed cup flash point below 100 F (37.8 C) and having a vapor pressure not exceeding 40 psia (1276 kPa) at 100 F. (The abbreviation "psia" stands for absolute pressure measured in pounds per square inch. Absolute pressure includes the influence of atmospheric pressure—about 14.7 psi [101 kPa]—on the measurement. If this effect is subtracted, then the measurement is called "gage pressure.") Combustible liquids are those with flash points at or above 100 F, but below 200 F (93.3 C), closed cup, and are divided into three classes (see NFPA No. 30). Although they do not ignite as easily as flammable liquids, they can be ignited under certain circumstances and so must be handled with caution. Flammable and combustible liquids are subdivided into classes as shown below (taken from NFPA 30 and 321, *Basic Classification of Flammable and Combustible Liquids*).

- CLASS I shall include those having flash points below 100 F (37.8 C) and may be subdivided as follows:
- CLASS IA shall include those having flash points below 73 F (22.8 C) and having a boiling point below 100 F.
- CLASS IB shall include those having flash points below 73 F and having a boiling point at or above 100 F.
- CLASS IC shall include those having flash points at or above 73 F and below 100 F.
- CLASS II shall include those having flash points at or above 100 F (37.8 C) and below 140 F (60 C).
- CLASS III shall include those having flash points at or above 140 F and may be subdivided as follows:
  CLASS IIIA shall include those having flash points at or above 140 F and below 200 F (93.4 C).
  CLASS IIIB shall include those having flash points at or above 200 F.

The more common flammable and combustible liquids are crude oils, coal tars, various hydrocarbons, alcohols, and their by-products. They are chemical combinations of hydrogen and carbon. The combination may also contain oxygen, nitrogen, sulfur, or other elements.

Manufactured liquids and fluid commodities that contain flammable liquids, such as paints, floor polish, cleaning solutions, driers and varnishes, should be considered flammable liquids and classed according to the flashpoint of the mixture. Handling and use precautions will differ according to their flash points, volatility, toxicity, and the percentage of flammable liquid in the mixture.

Flammable and combustible liquids vaporize and form flammable mixtures with air when in open containers, when leaks or spills occur, or when heated. The degree of danger is determined largely by (1) the flash point of the liquid, (2) the concentration of vapors in the air (whether the vapor–air mixture is in the flammable range or not), and (3) the possibility of a source of ignition at or above a temperature or energy level sufficient to cause the mixture to burst into flame.

When handling and using flammable liquids, exposure of large liquid surfaces to air should be prevented. It is not the liquids themselves that burn or explode, but rather the vapor–air mixture formed when they evaporate or are heated enough to cause the emission of flammable vapors. Therefore, handling and storing these liquids in closed containers and avoiding exposure of low flash point liquids during use are of fundamental importance. (Make sure all containers are correctly labeled.)

Appendix C, Chemical Hazards, of the Council's *Fundamentals of Industrial Hygiene,* 3rd ed., gives details about many liquids.

Terms used in this chapter will have the following definitions:

- **Auto-ignition temperature** is the minimum temperature at which a flammable gas– or vapor–air mixture will ignite from its own heat source or a contacted heated surface without necessity of spark or flame.

  Vapors and gases will spontaneously ignite at a lower temperature in oxygen than in air, and their auto-ignition temperature can be lowered by the presence of catalytic substances.

- **Flash point** means the minimum temperature at which a liquid gives off vapor in sufficient concentration to form an ignitable mixture with air near the surface of the liquid, within a vessel specified by appropriate test procedure and apparatus. For liquids having a viscosity less than 45 SUS (Saybolt universal seconds) and flash points below 200 F (93.4 C), the Tag Closed Tester is used (ASTM D-56-82 gives the procedure). For liquids of viscosity equal to or greater than 45 SUS and with flash points of 200 F or higher, the Pensky-Martins Closed Tester is used (ASTM D-93-85 gives the procedure). For specific liquids, such as aviation turbine fuels, paint, enamel, or varnish, alternate methods are available (see NFPA 321).

  Although other properties influence the relative hazards of flammable liquids, the flash point is the most significant factor. The relative hazard increases as the flash point lowers. The significance of this property becomes more apparent when liquids of different flash points are compared.

  At ordinary temperatures (under approximately 100 F) kerosene and No. 1 fuel oil, which have flash points above 100 F (43.3 C), do not give off dangerous quantities of vapor. On the other hand, gasoline gives off vapor at a rate sufficient to form a flammable mixture with air at temperatures of about −50 F (−46 C). (Gasoline is blended to have this characteristic so car engines will start in very cold weather.)

  Any combustible liquid, when heated to a temperature at or above its flash point, will produce ignitable vapors. Heavy fuel oil, when heated to several hundred degrees F, for example, may produce flammable vapors just as readily as gasoline. However, these vapors are less volatile and will condense. The characteristics of combustible liquids are also changed when they are atomized. When such liquids are heated above their flash points or atomized, they should be regarded as flammable liquids.

- **Flammable limits.** Gases and vapors of flammable liquids have a minimum concentration of vapor or gas in air below which propagation of flame does not occur on contact with a source of ignition. This is known as the lower flammable limit (LFL). There is also a maximum proportion of vapor or gas in air above which propagation of flame does not occur. This is known as the upper flammable limit (UFL).

  For example, a gasoline vapor-air mixture with less than approximately 1.0 percent of gasoline vapor is too lean, and propagation of flame will not occur on contact with a source of ignition. Similarly, if there is more than approximately 8 percent of gasoline vapor, the mixture will be too rich. Other gases, such as hydrogen, acetylene, and ethylene, have a much wider range of flammable limits.

  Flammable limits are determined at pressures of one atmosphere. Thus the range will increase with an increase of temperature or pressure, the UFL being more influenced than the LFL.

- **Flammable range** is the difference between the lower and upper flammable limits, expressed in terms of percentage of vapor or gas in air by volume. It is sometimes referred to as the "explosive range."

  For example, the limits of the flammable range of gasoline vapors are generally taken as 1.4 to 7.6 percent which is relatively narrow. Thus, a mixture of 1.4 percent gasoline vapor (and 98.6 percent air) is flammable, as are all the intermediate mixtures up to and including 7.6 percent gasoline vapor (and 92.4 percent air). The range is therefore the difference between the limits, or 6.2 percent.

- **Propagation of flame** is the spread of flame through the entire volume of the flammable vapor-air mixture from a single source of ignition. A vapor-air mixture below the lower flammable limit may burn at the point of ignition without propagating (spreading away) from the ignition source.

- **Oxygen limits.** Usually, to have an explosion, sufficient oxygen must be present along with a vapor concentration in the flammable range. Below 12 to 14 percent oxygen, flammable vapors will not burn; the actual limit depends upon the inert gas used to decrease the oxygen level. On the other hand, increasing pressures and temperatures above normal will reduce the oxygen limit.

- **Rate of diffusion** indicates the tendency of one gas or vapor to disperse into or mix with another gas or vapor, including air. This rate depends on the density of the vapor or gas as compared with that of air, which is given a value of 1. Whether a vapor or gas is lighter or heavier than air determines, to a large extent, the design of the ventilation system. If the vapor or gas is heavier than air, the intake duct should be slightly above floor level. Conversely, if the vapor or gas is lighter than air, the intake duct should be located just below ceiling level. Air supply duct openings should be located so they will be most effective in removal of the vapors.

- **Vapor pressure** is the pressure exerted by a volatile liquid, under any of the equilibrium conditions that may exist between the vapors and the liquid, as determined by the *Standard Method of Test for Vapor Pressure of Petroleum Products (Reid Method),* ASTM D 323-72.

■ **Volatility** is the tendency or ability of a liquid to vaporize. Such liquids as alcohol and gasoline, because of their well-known tendency to evaporate rapidly, are called volatile liquids.

■ **Oxygen deficiency** designates an atmosphere having less than the percentage of oxygen found in normal air. Normally, air contains about 21 percent oxygen.

When the oxygen concentration in air is reduced to approximately 16 percent, many individuals become dizzy, experience a buzzing in the ears, and have a rapid heart beat.

In addition to testing for the presence of toxic substances, the oxygen content of the atmosphere of a tank or similarly confined space should be determined before entry is permitted. Testing should be done with instruments approved for the purpose by NIOSH.

OSHA requires that no one enter a tank or enclosed space testing at less than 19.5 percent oxygen, unless wearing approved respiratory protective equipment, such as a fresh-air hose mask or self-contained or self-generating breathing apparatus.

Self-contained compressed air breathing apparatus, certified by OSHA-approved agencies, have proved satisfactory in oxygen deficient atmospheres. They are especially useful where it is difficult to run a hose line.

■ **Combustible liquids.** Combustible liquids are divided into three subclasses (III, IIIA, and IIIB). A combustible liquid has a flash point at or above 140 F (60 C). See earlier explanation under the heading Definitions.

The volatility of liquids is increased when they are heated. Actually, combustible liquids should be treated in accordance with the requirements for flammable liquids when they are heated, even though the same liquids when not heated are outside the range.

All tests should be made in accordance with methods either specified by OSHA or adopted by the American Society for Testing and Materials (ASTM) and approved by the American National Standards Institute (ANSI).

The Department of Transportation, in its *Hazardous Materials Regulations,* 49 CFR, Parts 170–179 (republished by the Bureau of Explosives in Graziano's Tariff No. 27), defined a flammable liquid as any liquid which gives off flammable vapors at or below a temperature of 80 F (26.7 C). This is of importance since the DOT Flammable Liquid Label is one means by which containers of flammable liquids can be identified for shipping, receiving, and transportation. However, this is not entirely foolproof since there are exceptions.

Further, other regulatory entities have adopted other definitions of what constitutes a "hazardous liquid." For example, the New York City Codes and the Federal Hazardous Substances Act have different definitions of flammability, both using the open cup test method. Such definitions are very important where an industry must comply with legal requirements. In cases of dispute, the testing method that is cited by the particular code must be used.

GENERAL SAFETY MEASURES

Preventing dangerous mixtures

Accidental mixture of flammable and combustible liquids should be prevented. A small amount of acetone accidentally put into a kerosene tank may lower the flash point of the contents, because of the relatively high volatility of acetone, and render the kerosene too dangerous to use. Gasoline mixed with fuel oil may change the flash point sufficiently to make the fuel oil hazardous in home-heating or similar use. In each case, the lower flash point liquid can act as a fuse to ignite the higher flash point material just as though it were a flammable mixture.

Fill, discharge openings, and control valves on equipment containing flammable and combustible liquids should be identified by color or tag, or both. In some plants, the pipelines are painted or banded with distinctive colors and show direction of flow. Each tank should be marked with the name of the product. Lines from tanks of different types and classes of products should be separated and, preferably, separate pumps for the different types and classes of products should be provided. (The NFPA system for identifying hazardous substances is described in Chapter 17, Fire Protection.)

A portable, FM- or UL-approved container should be used for handling flammable liquids in quantities up to five gallons. The containers should be clearly identified by lettering or color code.

Smoking

Smoking and carrying "strike anywhere" matches, lighters, and other spark-producing devices should not be permitted in a building or area where flammable liquids are stored, handled, or used. The size of the restricted area will depend on the type of products handled, the design of the building, local codes, and local conditions.

Suitable No Smoking signs should be posted conspicuously in buildings and areas where smoking is prohibited.

Static electricity

Static electricity is generated by the contact and separation of dissimilar material. For example, static electricity is generated when a fluid flows through a pipe or from an orifice into a tank. Examples of several methods of generating static electricity are shown in Figure 16-1. The principal hazards of static electricity are fire and explosion caused by spark discharges containing enough energy to ignite flammable or explosive vapors, gases, or dust particles. Also, a worker could be injured because of an involuntary reaction caused by static spark shock.

There is great danger from a static spark where a flammable vapor may be present in the air, such as the outlet of a flammable liquid fill pipe, at a delivery hose nozzle, near an open flammable liquid container, and around a tank truck fill opening or barrel bunghole. When the potential for static charge is present, a spark between two bodies can occur because there is not a good electrical conductive path between them. Hence, grounding or bonding of flammable liquid containers is necessary to prevent static electricity from causing a spark. The connections must be precisely made after consulting NFPA 77, *Recommended Practice on Static Electricity,* for details. A summary of that standard follows:

Bonding and grounding

The terms "bonding" and "grounding" are sometimes used interchangeably because of poor understanding of the terms. Bonding is done to eliminate a difference in static charge potential between objects. The purpose of grounding is to eliminate a difference in static charge potential between an object and ground (Figure 16-2). Bonding and grounding are effective only when the bonded objects are conductive.

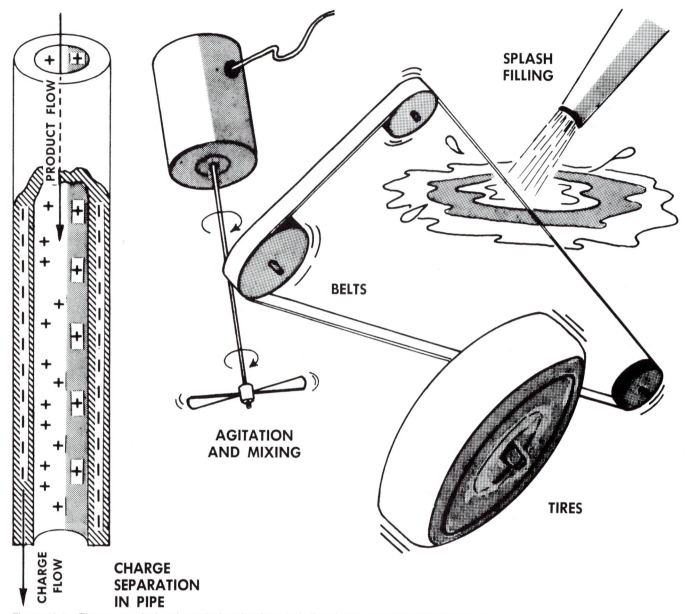

PRODUCT FLOW

CHARGE FLOW

CHARGE
SEPARATION
IN PIPE

AGITATION
AND MIXING

BELTS

SPLASH
FILLING

TIRES

Figure 16-1. These are typical static-producing situations, including charge separation occurring in pipes.

When two objects are bonded, the charges flow freely between the bodies and there is no difference in their charge. Therefore, the likelihood of sparking between them is eliminated.

Although bonding will eliminate a difference in potential between the objects that are bonded, it will not eliminate a difference in potential between these objects and the earth unless one of the objects possesses an adequate conductive path to earth. Therefore, bonding will not eliminate the static charge, but will equalize the potential between the objects bonded so that a spark will not occur between them.

An adequate ground will always continuously discharge a charged conductive body and is recommended as a safety measure when any doubt exists concerning a situation or when a governing authority requires it.

To avoid a spark from discharge of static electricity during flammable liquid filling operations, a wire bond should be provided between the storage container and the container being filled, unless a metallic path between the two containers is otherwise present.

When loading tank cars or tank trucks with flammable liquids through open domes, it is mandatory to use a down-spout long enough to reach the tank bottom to avoid flash during filling. Because of the possibility of stray currents, and in order to prevent a difference in static charge potential between the fill pipe and the tank car, loading lines should be bonded. Preferably the bonding should be connected to the rails to avoid operator errors.

For detailed information and exceptions to this generality, consult NFPA 77, *Static Electricity,* and API RP2003–82, *Protection Against Ignitions Arising out of Static, Lightning, and Stray Currents.*

Aboveground tanks used for storage of flammable liquids need not be grounded unless they are on concrete or on nonconductive supports (see NFPA 77). Ground wires should be uninsulated so they may be easily inspected for damage (Figure 16-3).

CHARGED AND UNCHARGED BODIES INSULATED FROM GROUND

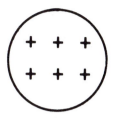

Charged body
insulated from
ground

Uncharged body
insulated from
ground

Charge (Q) = 6 microcoulombs
Capacitance (C) to ground = 0.01 microfarad
Voltage (V) to ground and
uncharged body = 600 volts

Charge (Q) = 0
Capacitance (C) = 0.01 microfarad
Voltage to ground (V) = 0

BOTH INSULATED BODIES SHARE THE SAME CHARGE

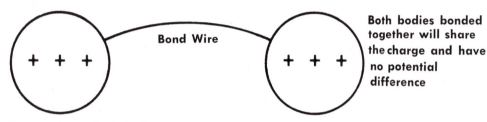

Bond Wire

Both bodies bonded
together will share
the charge and have
no potential
difference

Charge (Q) on both bodies = 6 microcoulombs
Capacitance (C) to ground for both bodies = 0.02 microfarad
Voltage (V) to ground = 300 volts

BOTH BODIES ARE GROUNDED AND HAVE NO CHARGE

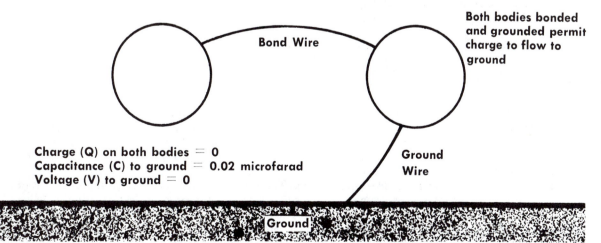

Bond Wire

Both bodies bonded
and grounded permit
charge to flow to
ground

Ground
Wire

Charge (Q) on both bodies = 0
Capacitance (C) to ground = 0.02 microfarad
Voltage (V) to ground = 0

Figure 16-2. Bonding eliminates the difference in static charge potential between objects. Grounding eliminates the difference in static charge potential between objects and the ground. Both bonding and grounding apply only to conductive bodies and, when properly applied, can be depended on to remove the charge.

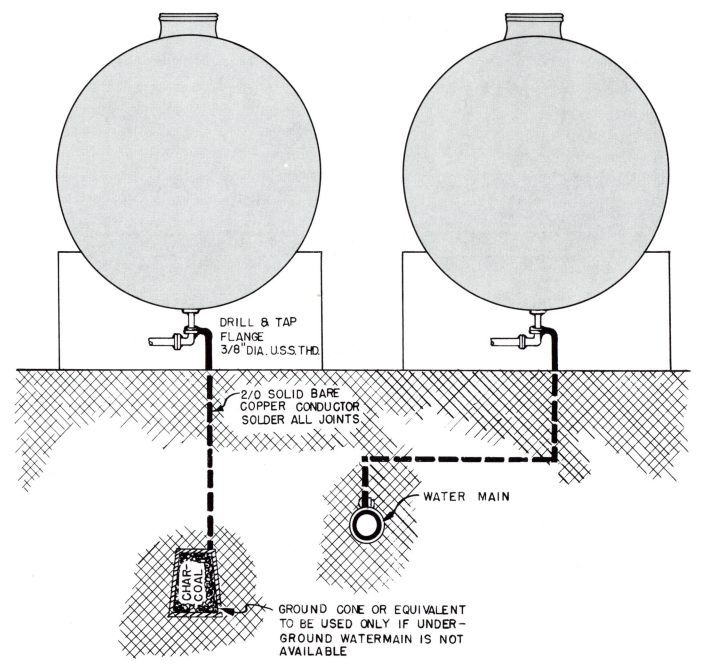

DRILL & TAP
FLANGE
3/8" DIA. U.S.S. THD.

2/0 SOLID BARE
COPPER CONDUCTOR
SOLDER ALL JOINTS

WATER MAIN

CHAR-COAL

GROUND CONE OR EQUIVALENT
TO BE USED ONLY IF UNDER-
GROUND WATERMAIN IS NOT
AVAILABLE

Figure 16-3. Because some aboveground flammable liquid storage tanks are not inherently grounded, such tanks should be grounded to a water main or to a ground cone or its equivalent.

Petroleum liquids are capable of building up electrical charges when they (1) flow through piping, (2) are agitated in a tank or a container, or (3) are subjected to vigorous mechanical movement such as spraying or splashing. Proper bonding or grounding of the transfer sytem usually drains off this static charge to ground as fast as it is generated. However, rapid flow rates in transfer lines can cause very high electrical potentials on the surface of liquids, regardless of vessel grounding. Also, some petroleum liquids are poor conductors of electricity, particularly the pure, refined products; and even through the transfer sytem is properly grounded, a static charge may build up on the surface of the liquid in the receiving container. The charge accumulates because static cannot flow through the liquid to

grounded metal as fast as it is being generated. If this accumulated charge builds up high enough, a static spark with sufficient energy to ignite a flammable air-vapor mixture can occur when the liquid level approaches a grounded probe or when a probe is lowered into a tank such as during sampling or gaging. This is particularly dangerous when the liquid is of intermediate volatility, for instance, toulene or ethyl alcohol.

This high static charge is usualy controlled by reducing the flow rates, avoiding violent splashing with side-flow fill lines, and using relaxation time.

Motor frames, starting or control boxes, conduits, and switches should be grounded in accordance with the requirements for installation of electrical and power equipment

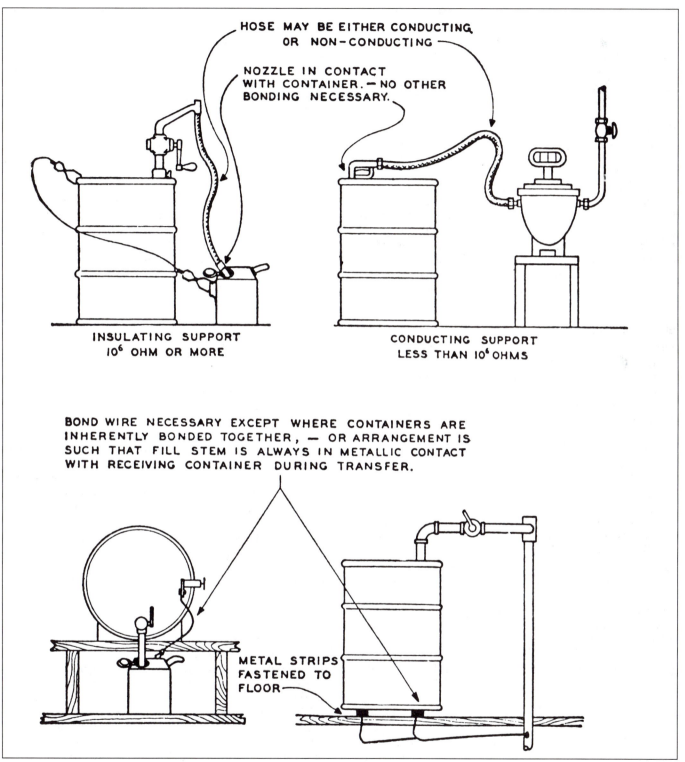

HOSE MAY BE EITHER CONDUCTING OR NON-CONDUCTING

NOZZLE IN CONTACT WITH CONTAINER. — NO OTHER BONDING NECESSARY.

INSULATING SUPPORT 10^6 OHM OR MORE

CONDUCTING SUPPORT LESS THAN 10^6 OHMS

BOND WIRE NECESSARY EXCEPT WHERE CONTAINERS ARE INHERENTLY BONDED TOGETHER, — OR ARRANGEMENT IS SUCH THAT FILL STEM IS ALWAYS IN METALLIC CONTACT WITH RECEIVING CONTAINER DURING TRANSFER.

METAL STRIPS FASTENED TO FLOOR

Figure 16-4. When a container is bonded during the filling process, any static electricity generated will be safely discharged. (Reprinted courtesy of National Fire Protection Association.)

outlined in the *National Electrical Code.* However, this is for personal protection rather than for control of static charges.

When flammable liquids are transferred from one container to another, a means of bonding should be provided between the two conductive containers prior to pouring, as shown in Figure 16-4. A flammable liquid tank truck or tank car must be bonded and grounded to the loading rack.

In areas where flammable liquids are stored or used, hose nozzles on steam lines used for cleaning should be bonded to the surface of the vessel or object being cleaned. Also, there should be no insulated conductive objects on which the steam could impinge and induce a static charge accumulation.

Flat moving belts also are sources of static electricity unless they are made of a conductive material or are coated with a conductive belt-dressing compound designed to prevent build-up of static charges. However, V-belts generally are not a cause of static charge.

Nonconductive materials, such as fabric, rubber, or plastic sheeting, passing through or over rolls also will create charges of static electricity. Static from these materials, as well as from the belts, can be discharged with grounded metal combs or tinsel collectors. Radioactive substances and static neutralizers using electrical discharges are also employed for this purpose.

Bonding and grounding systems should be regularly checked for electrical continuity. Preferably before each fill, the exposed part of the bonding and grounding system should be inspected for parts that have deteriorated because of corrosion or have otherwise been damaged. Many companies specify that bonds and grounds be constructed of bare-braided flexible wire because it facilitates inspection and prevents broken wires from being concealed, as discussed earlier.

Electrical equipment

Electricity becomes a source of ignition where flammable vapors exist, if the proper type of electrical equipment for these atmospheres either has not been installed or has not been maintained. The NFPA 30 and the *National Electrical Code,* as well as local codes, should be consulted when flammable liquids are used.

Spark-resistant tools

Although the hazard of ignition of flammable vapors or gases by sparks may at times be overemphasized, it must be recognized.

A summary of reports of experimental evidence and practical experience in the petroleum industry shows that no significant increase in fire safety is gained by the use of spark-resistant hand tools in the presence of gasoline and similar hydrocarbon vapors.

However, some materials such as carbon disulfide, acetylene, and ethyl ether have very low ignition energy points. For these and similar materials, special tools designed to minimize the danger of sparks in hazardous locations should be used as a conservative safety measure.

Leather-faced, plastic, and wood tools are free from the friction-spark hazard, although there is the possibility of metallic particles becoming embedded in them.

HEALTH HAZARDS

Flammable and combustible liquids create health hazards from skin contact. Irritation results from the solvent action of many flammable liquids on the natural skin oils and tissue. Intoxication and other health effects result from breathing vapors of flammable liquids. Atmospheres that are below flammable limits may well be too toxic to breathe.

Vapors from flammable and combustible liquids are heavier than air and will flow into pits, tank openings, confined areas, and low places where they will contaminate the normal air and, thus, cause a toxic as well as explosive atmosphere. Oxygen deficiency occurs in closed containers, such as a tank which has been closed for a long time and in which rusting has consumed the oxygen. *All* containers should be aired and tested for toxic and flammable atmosphere as well as the oxygen level before any entry. See *Threshold Limit Values,* published annually by the American Conference of Governmental Industrial Hygienists, and OSHA regulations (Title 29 CFR 1910.1000) for guidance; see listings in References. For background information, see *Fundamentals of Industrial Hygiene,* 3rd ed.

COMBUSTIBLE GAS INDICATORS

Unless tests prove otherwise, flammable and toxic mixtures should be assumed to be present in all tanks that have contained or have been exposed at any time to flammable and combustible liquids. Tests for flammable vapor-air mixtures in tanks and other vessels can be made by making a chemical analysis of samples or by using a NIOSH-certified or UL- or FM-approved combustible gas indicator.

One type of combustible gas indicator is an instrument operating on the principle that when a mixture of flammable vapor air is passed over a heated electric filament, the resistance of the filament will be increased in direct proportion to the amount of combustible vapor present. This is read on the scale usually as the percentage of the lower flammable limit of a calibration substance such as hexane.

A combustible gas indicator should be used only by experienced persons, and the operator should follow the manufacturer's instructions for balancing the unit. For an accurate determination, the reading should be corrected by using a correlation graph that adjusts the reading according to the tested atmosphere. Manufacturers supply correlation graphs for many combustible gases. It should be noted that combustible gas indicators calibrated for the LFL may not give any reading while sampling a vapor that is present in a very high concentration. The reason is that insufficient oxygen may be present to cause the vapor to burn in the instrument and thus give no reading. Negative readings will also be obtained when testing vessels containing high concentrations of inert gas.

The operator should check the operation and calibration of the instrument in accordance with the manufacturer's recommended procedures. The manufacturer generally will supply the appropriate test gas to be used for the calibration check. These test gases also can be used for daily operational checks if the instrument does not have internal circuitry to perform this function. The instrument should be thoroughly examined and tested at least once a year. This can be done by the manufacturer, by a recognized standards laboratory, by an outside consultant, or by the company.

The sampling hose should be kept where flammable liquid, steam, or water will not be drawn into the instrument. Any of these substances would put it out of service. Generally, combustible gas indicators should not be used to test vapors from heated combustible liquids. Such vapors condense in the sampling lines or combustion chamber and give results that are spurious.

A tank, tested before being cleaned, can be found to be vapor-free. However, if it contains sludge or scale, flammable and combustible liquid and vapor can be released as soon as the sludge or scale is disturbed. A safe rule to follow is that no tank should be considered free from the possibility of dangerous vapors as long as it contains sludge or scale.

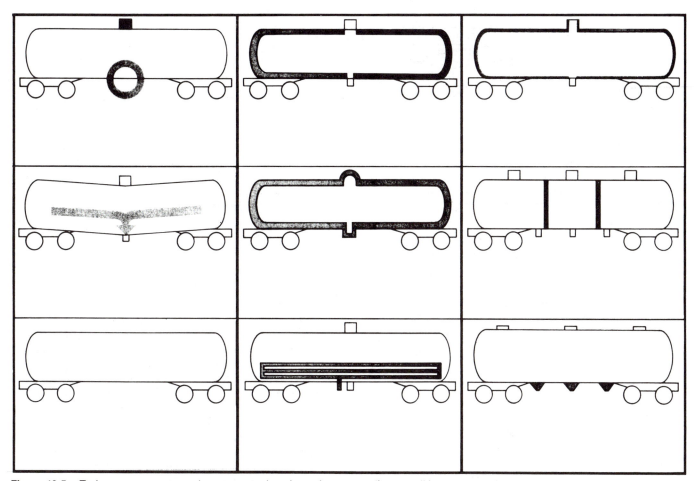

Figure 16-5. Tank cars are constructed to accommodate the various types of commodities transported. A common type of flammable liquid car is a straight shell with bottom outlet (not shown). (Reprinted courtesy of Union Tank Car Co.)

If workers leave a tank for a period of time, such as overnight, gas tests should be made the next morning before they are permitted to continue work. It is essential to use a written permit form when it is necessary for workers to enter vessels, especially those that have contained flammable or combustible liquids or those requiring "hot work" repairs.

Testing devices are available for carbon monoxide, benzene, hydrogen sulfide, and tetraethyl lead hazards. This last instrument is available from suppliers of tetraethyl lead.

LOADING AND UNLOADING TANK CARS

Only trained employees should load or unload tank cars containing flammable and combustible liquids. These workers should understand the possible dangers of fire, explosion, asphyxiation, and toxic effects from breathing flammable vapors. As an added precaution, unloading and pumping operations should be stopped during electrical storms.

Invoices and shipping papers on incoming tank cars should be closely checked to make certain they match the actual tank car numbers. In doubtful cases, the contents should be tested to prevent accidental mix-up of flammable and combustible liquids in plant tanks. See Figure 16-5 for typical tank car construction details.

Spotting cars

A car mover should be used when it is necessary to spot a car by hand. The worker should stand with the handle of the bar to one side of the track rail and face the car with his feet well apart so he cannot lose his balance.

After the car is spotted, with the dome outlet opposite the loading or unloading line, the brakes should be set and the car wheels blocked. The "blue flag" stop sign (at least 12 by 15 in. [30.5 by 38 cm], with words such as STOP—TANK CAR CONNECTED or STOP—MEN AT WORK) should be set before loading lines are connected or work is done on the car. The sign should be located about 25 ft (7.6 m) ahead of the car toward the main line, so train crews cannot come onto the track without seeing it. (Full details are given in Chapter 7, Elevators and Plant Railways.)

The siding should be bonded, grounded, and insulated from the main rail line for protection from stray electrical currents (Figure 16-6). Usually the spur track installation would be made or supervised by the serving railroad. For applicable specifications, refer to the Association of American Railroads (AAR) publication, *Signal Manual of Recommended Practice* (see References).

The relative location of electric power lines to the tank car unloading position is an important matter discussed in the AAR

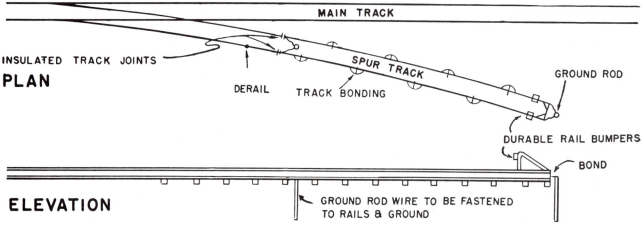

Figure 16-6. This siding is designed for unloading tank cars. It has rail joint bonding, insulated track joints, derail, and track grounding.

Bureau of Explosives Circular No. 17-E, *Recommended Practice for the Prevention of Electric Sparks That May Cause Fires in Tanks or Tank Cars Containing Flammable Liquids or Flammable Compressed Gases, Due to Proximity of Wire Lines.* Pertinent provisions contained in AAR 17-E include the following:

(A) GENERAL

1. Where any electric power line is within 20 ft (6 m) of the tank opening, the use of a metallic gaging rod is prohibited.

2. When the contents are being gaged or transferred, tank cars, wherever possible, shall not be located under or near any electric power lines.

3. Where tanks or tank cars (the contents of which are being gaged or transferred) are necessarily located under or near power lines having a span length of 150 ft (45 m) or less and operating at a voltage not exceeding 550 volts between conductors, the following rules shall be observed:

   a. Where power lines pass overhead, there shall be a minimum vertical clearance of 8 ft (2.4 m) at 60 F (16 C) between the wires and the tank.

   b. Where power lines pass nearby and do not have the minimum vertical clearance specified in paragraph 3-a, there shall be a minimum horizontal clearance of 8 ft between the wire lines and the tank.

   c. Openings in tanks shall be at least 6 ft (1.8 m) distant, measured horizontally from any overhead power lines.

4. Where tanks or tank cars (the contents of which are being gaged or transferred) are located under or near power lines having a span length in excess of 150 ft or operating at a voltage in excess of 550 volts between conductors, it is recommended that special studies be made by qualified persons and such additional clearance provided as is necessary to give adequate protection.

Inspection

The condition of the cars should be examined. Any missing or damaged parts or leaks should be reported to the railroad representative, so sources of ignition and locomotives can be kept away from the car.

Buckets or tubs should be placed under the leaks to prevent loss, and the accumulated drippings should be transferred to storage frequently so no large quantity of flammable liquid is exposed in an open container or allowed to spread over the ground. Consideration must also be given to complying with EPA rules governing hazardous substance spills.

The car should not be loaded or unloaded unless there are *no* exposed lights, fires, or other sources of ignition in the area.

Approved electrical equipment, installed in accordance with the *National Electrical Code* and subject to regular inspection, should be provided where flammable liquids are handled. A rack with a gangway or bridge to the car helps workers move safely when loading or unloading a car. The gangway or bridge should have guardrails.

Relieving pressure

The tank car should be relieved of interior pressure before the manhole cover or the outlet valve cap is removed. To relieve pressure, either the safety valves can be raised or the air valve can be opened a small amount at a time or the tank can be cooled with water. The vacuum relief valve can be depressed, if vacuum is the problem. Venting and unloading should be deferred if a dangerous amount of vapor collects outside the car.

If interior pressure is excessive, the car should be sprayed with water to cool it or should be allowed to stand overnight and be unloaded early the next morning.

Removing covers

A screw-type dome cover should be loosened by a bar placed between the lug and the knob on the top of the cover. Two complete turns should be taken so that the ½-in. (1.3 cm) vent holes in the threaded portion of the dome cover are exposed. If escaping vapors are heard, tighten the cover and release the pressure by raising the safety valve. Keep clear of the vapors by standing to windward.

Figure 16-7. These placards are used on vehicles and rail cars transporting flammable or combustible liquids. The Empty placard can be either a separate placard, printed on the reverse side of a placard, or a composite made by covering the top triangle with a black triangle having the word *empty* printed in white letters. Flammable and Combustible placards are red with white letters. They can be purchased through various suppliers. (Reprinted from *Title 49—Transportation,* Code of Federal Regulations, §172.525, .542, and .544.)

A cover should be removed with the employee facing the dome. With feet well braced, use short, vigorous pushes on the bar. The cover (if not provided with a chain) and loose tools should be removed to the walk platform or other safe place so they will not fall.

On the bolted type of dome cover, all nuts should be unscrewed one turn and the cover lifted to break any adhesion between the cover and the dome ring. If there is a sound of escaping vapors, the dome should be tightened and the venting operation repeated.

All dirt, cinders, and debris should be removed from around the interior type of dome cover before the yoke is unscrewed.

When the car must be unloaded through the bottom outlet valve, dome covers should be adjusted to allow for venting. The screw type of dome cover should be tightened just enough to expose the vent holes in the threaded portion of the cover.

A small, thin wood block should be placed under the bolted type of cover, and the interior type of dome cover should be tightened up in the yoke to within ½-in. (1.3 cm) of the closed position. A metal cover or wet burlap or canvas should be placed over the tank manhole to prevent entrance of sparks or other source of ignition.

Unloading and loading connections

The preferred method of unloading tank cars is through the dome, but they can be unloaded safely through either the top or bottom. Some states prohibit bottom unloading. Where a car is to be bottom unloaded, the tank's outlet valve must be closed before the outlet chamber cap or plug is removed. A pail or tub should be placed under the outlet chamber and the outlet cap unscrewed.

The condition of the outlet valve should be checked and, if there is no serious leak, the unloading can proceed. If the plug does not loosen easily with a 48-in. (1.2 m) wrench, the bottom outlet cap or plug can be tapped to loosen it. If the valve leaks so badly that a connection cannot be made without spilling the product, the car should be unloaded from the top.

In cold weather, if the chamber or valve becomes frozen or blocked with frozen liquid, the outlet chamber should be carefully examined for cracks before attempting the unloading con-

nection. If no crack is found, the connection for unloading can be made, but the outlet chamber should be wrapped with burlap or other rags, and hot water or steam should be applied.

The condition of the connection from the car to the storage tank should be carefully checked before the car valve is opened. The storage tank should be gaged and watched to prevent overflow, and the hose and unloading line should be frequently checked. The tank car should be examined, before loading lines are disconnected, to make certain it is completely empty.

A worker should be present throughout the entire loading or unloading operation and at all times while a car is connected. If it is necessary to discontinue operations before they are completed, the outlet valve should be closed and the dome cover and outlet chamber cap replaced. Tank cars should not be allowed to stand with loading or unloading connections attached after operations are completed.

If flammable liquids are spilled, spill areas should be covered with fresh dry sand or dirt, oil absorbents, or otherwise cleaned up. Never flush spills into public sewers, drainage systems, or natural waterways.

When heater pipes are used for unloading, the steam should be applied slowly until it begins to exhaust at the outlet pipe. Steam pressure should not exceed that needed to bring the contents to the desired temperature and should never be high enough to cause the contents to become overheated.

Care also must be exercised to avoid overheating of combustible liquids since this can result in evolution of flammable vapors. Also, overheating products containing certain additives can release dangerous quantities of flammable vapors.

Unloading of chemical tank cars was covered in Chapter 3. For specific details of unloading or loading tank cars, check with your car manufacturer or supplier.

Placards

The Department of Transportation provides very specific guidelines for the use of placards. Consult a current copy of the regulations (Title 49, *Code of Federal Regulations,* Part 172, "Hazardous Materials," Subpart F—Placarding) for appropriate requirements. (See Figure 16-7.) Placards, rags, waste, and blocks should not be disposed of in the tank or car body.

Fires

Tank car fires can be serious, especially if spilled product is burning under a tank car. Shut off all loading lines to the rack and apply water spray to cool the tank car, structural steel, and piping. If conditions permit, flush spilled fuel from under the tank body to an open area or into closed drainage system. Avoid getting too much water into the tank car and causing an overflow.

If a fire occurs at the dome of a tank car, and the dome is unobstructed by a fill pipe, close the dome cover to extinguish the fire. If a fill pipe is in the dome opening, do not remove the fill pipe because this could splash the burning liquid and spread the fire. Use a dry chemical or CO_2 extinguisher to put out the fire. In some cases, the flames can be "blown out" with a straight stream of water directed across the opening, taking care not to overflow the tank or splash the contents and spread the fire.

To protect nearby facilities, it is advisable to provide drainage to limit the spread of a spill and direct it to a safe area. The following Association of American Railroads publications are considered the authoritative sources on this subject:

"Recommended Good Practice for Handling Collisions and Derailments Involving Hazardous Materials in Transporation," BE Pamphlet No. 1.

"Recommended Good Practice for Handling Fires and Spills Involving Explosives and Other Dangerous Articles in Transportation," BE Pamphlet No. 2.

LOADING AND UNLOADING TANK TRUCKS

Tank trucks, tank trailers, and tank semi-trailers used for the transportation of flammable liquids should be constructed and operated according to Title 49, *Code of Federal Regulations,* Parts 177 and 397. Communications regulations (markings) are given in Part 172. Note: Revision and promulgation of these regulations are within the jurisdiction of the Department of Transportation.

Carriers handling hazardous materials are subject to federal regulations under the Hazardous Materials Act of 1974. Those not subject to the jurisdiction of the Department of Transportation may consult NFPA 385, *Tank Vehicles for Flammable and Combustible Liquids,* as an appropriate reference.

Inspection

Trucks should be kept in good repair and should be inspected daily with special emphasis placed on the condition of lights, brakes, horns, rear view mirror, bonding, tires, steering and motor.

Tanks and safety valves must be inspected and tested according to Title 49, *Code of Federal Regulations, Part 177.*

Smoking

Smoking by truck drivers or their helpers is not permitted while they are driving or attending to their unit. Drivers should keep smokers and other sources of ignition away from loading or unloading operations. Each tank vehicle should be provided with one or more portable fire extinguishers each having at least a 10 B, C rating.

Spotting trucks

When a tank truck is being loaded or unloaded the brakes should be set, the engine stopped, the lights turned off, and the bonding connection made before the dome cover is removed for inspection or gaging.

Where trucks must be started and moved before loading, all domes must be closed and latched, and the bonding connections should be removed before the engine is started.

Unloading and loading connections

Trucks with motor-driven pumps should be shut off before loading lines are connected or disconnected. The driver must remain at the tank truck controls. If he must leave, the flammable product discharge must be stopped and the hose removed. When an unloading line must be run across a sidewalk, suitable warning signs should be provided.

Drivers should make sure the correct product is being handled and that the tank has previously contained the same liquid. Personnel at the delivery site should be contacted before the discharge. To assure removal of trapped liquids, especially when changing from a low-flash liquid to a higher one, the tank should be completely drained and flushed.

They should determine that the tank is properly vented and that there is sufficient room to accommodate the quantity to be loaded or unloaded.

Leaks

Spills or overflows must be avoided. When they do occur, loading should be stopped, valves shut off, and the overflow cleaned up before loading is resumed. Allow adequate time for all flammable vapors to dissipate before the truck engine is started.

The driver should make every effort to park a damaged and leaking truck in such a way that it will not endanger traffic or property. The public should be warned to keep away, and the police and fire departments notified. The truck should be parked off the highway, if possible, in a vacant lot or, at least, away from buildings and away from areas in which there is a concentration of people. The truck should not be left unattended.

Carriers subject to the safety regulations of the Department of Transportation must report all broken, leaking, and damaged containers on Form F 5800.1 under Sections 171.15 and 171.16, Title 49, *Code of Federal Regulations, Part 171.*

The liquid should be trapped in containers or in a depression or pit if possible. In case of a large spill, especially in urban areas, rather than endanger lives, the best practice is to use portable hand pumps to discharge the product into drums or another tank truck. EPA spill regulations would apply.

Fires

In the event of a tank truck fire during loading or unloading, the fuel supply to or from the truck should be shut off if possible. If loading is being done, the spout should not be removed from the tank truck but, if possible, either the dome cover should be closed or the opening covered with a wet sack or blanket.

The fire should be attacked with carbon dioxide, foam, or dry chemical extinguishers or water fog. If a water supply is available, burning liquid may be flushed away to a safe place, but not into a public sewer, drain, or public waterway. Care must be exercised to prevent introducing water into the tank and displacing fuel, because this could cause more spillage.

Where burning vapors are escaping from leaks or vents, it may be better to let them burn until the source of escaping liquid or vapor can be controlled.

In all cases of fire, the fire brigade or the local fire department should be quickly informed. Fire fighting procedures should be prepared and, in the actual emergency, followed.

STORAGE

Class I and Class II liquids should not be kept or stored in a building used for public assembly, such as a school, church, or theater, except in approved containers (Figure 16-8). The approved containers should be kept in either a storage cabinet or a storage room that does not open to the public portion of the building. Quantities stored in such locations should be limited in accordance with NFPA 30, *Flammable and Combustible Liquids Code*. They should not be stored where they will limit the use of exits, stairways, or other areas used for safe egress. Neither should they be stored close to stoves or heated pipes, nor exposed to the rays of the sun or other sources of heat.

Storage of flammable and combustible liquids in open containers should not be permitted. Approved containers for flammable liquids should be closed after each use and when empty.

Bulk Class I liquids should be stored in an underground (buried) tank or outside a building. No outlet from the tank should be inside a building unless it terminates in a special room (Figures 16-9 and -10).

NFPA 30 states specifications limiting the quantity of each class of flammable liquids that can be stored in various loca-

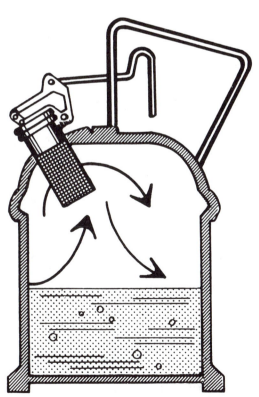

Figure 16-8. This spring-loaded cover is designed to open at 5 psi and relieve the internal vapor pressure. Only negligible losses are caused through evaporation of liquids stored in safety cans at ordinary temperature ranges.

tions on plant premises, along with data describing the required conditions and procedures relating to such storage.

Vehicles used on plant property to transport flammable and combustible liquids in sealed containers should be designed to minimize damage to the containers. (Figure 16-11)

When employees are filling tanks and other containers, they should allow sufficient vapor space (outage) above the liquid level so the liquid can safely expand when temperatures change.

For example, gasoline expands at the rate of about 1 percent for each 14-deg F (8-deg C) rise in temperature. The recommended outage space for gasoline is 2 percent of the tank or compartment capacity. High-levels should be permanently marked on the containers.

Tank construction

Tanks should be constructed and installed as recommended by the latest revision of NFPA code No. 30.

Vents

Storage tanks must be provided with vents that are of a type and size recommended in NFPA 30. Vent pipes of underground tanks, storing Class I flammable liquids, should terminate outside buildings, and be higher than the fill pipe opening and not less than 12 ft (3.7 m) above the adjacent ground level. They should discharge vertically and be located so flammable vapors cannot enter building openings or be trapped under eaves or other obstructions. Vent pipes from underground tanks, storing Class II or Class III liquids, should terminate outside buildings and be higher than the fill pipe opening. Vent outlets should be above normal snow level.

Some authorities have questioned the effectiveness of brass mesh or copper screens as flame arresters in vent terminals. Underwriters Laboratories state that "under some favorable conditions, screens of fine mesh can be effective in arresting flame, but their use is not dependable. Under service conditions, they are subject to clogging and freezing, and when attempts are made to clean them there is danger of mechanical injury. The displacement of one or more wires renders the screen useless as a flame arrester" (*Oil Tank Vents—Hazards of Screens*, Serial No. UL-31).

A well-located vent can provide more certain protection than screens. Vents should be safely located, so that escaping vapors will be properly dispersed and clear of ordinary sources of ignition. If the vent terminals are in locations where clogging by mud wasps is possible, a loosely attached screen of relatively coarse mesh can be provided. (See NFPA 30 for recommended vent sizes.)

Dikes

There are some locations where a flow of flammable or combustible liquid from a tank car might have serious consequences because of topography or neighboring property. In these cases, a curb, dike, or wall can be provided around a tank or group of tanks. Such a structure should be designed to contain any flammable liquids that could be released by overfilling tanks or by leaking tanks, pipelines, and valves. NFPA 30 gives construction details of drainage, dikes, and walls for aboveground storage tanks.

The need for dikes will be determined by local conditions. For example, if a tank is close to a building and the ground slope could cause liquid to flow toward the building, it is better

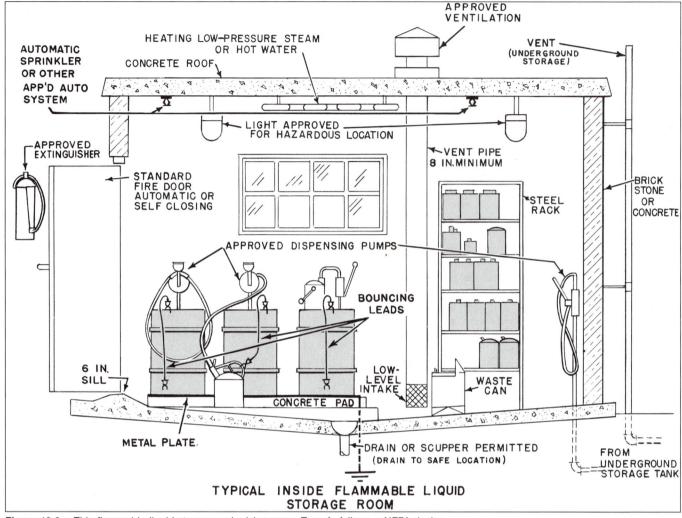

Figure 16-9. This flammable liquid storage and mixing room, Type A, follows a NFPA design.

to construct diversion walls to direct the escaping liquid to a safe location.

A gas-sealed storm drain often is installed for each dike enclosure. In larger plants, the drain often leads to an oil-water separator located far away from the main buildings. Here the flammable and combustible liquid can be separated from the water by flotation. Drains should be equipped with control valves, which should be kept closed unless water is being drained from the area. These drains should be designed to prevent flammable liquids entering natural water supplies, public sewers, or public drains.

Pump houses

Flammable liquid pumps should be located outside of buildings and diked areas wherever feasible.

Buildings housing equipment for transfer of flammable liquids should be of fire-resistive construction with ample ventilation, especially along the floor where vapors might be present. Pump houses should preferably not contain sunken pits, except small drain openings, because the danger of vapor concentrations in such areas is too great. The pump house should have a minimum of two exits, be kept clear of obstructions, and have doors swinging outward.

Where flammable liquids are handled, electrical equipment should be partitioned and sealed off from the rest of the pump house, or should be of a type approved for use in flammable atmospheres. The pump room should be kept well ventilated. It is good operating and safety practice to have a well-marked master cutoff switch outside the building. All electrical equipment inside the building must conform to the *National Electrical Code* and local codes for this use.

Valves and packing glands on pumps should be maintained in good operating condition to prevent leaks. The use of excess flow valves can also prevent large-scale spills and are sometimes required by local codes. Draining escaped liquids into a closed-pipe return system is recommended rather than simply catching materials that leak in drip pans. Leaks should be repaired as soon as they are noted.

If a centrifugal pump with priming bleeder is used, the outlet of the bleeder should be kept plugged when not in use or piped to a closed system for collection. Bearings should be well lubricated to prevent overheating due to friction.

Good housekeeping is of major importance, and approved containers should be provided for safe disposal of debris, rags, and other waste. These should be emptied daily. Tools other than those required to operate should not be stored in the pump house.

Figure 16-10. A well-designed flammable liquid storage room with both high and low level ventilation and automatic fire extinguishment. The ventilators are designed for 12 air changes per hour.

The pump house should not contain lockers or be used as a change house, and loitering by employees should not be allowed. All sources of ignition must be kept away from the pump house location.

Fire extinguishers should be located at convenient, easily identifiable points. It is advisable to have self-closing fire doors and the recommended type of automatic extinguishment.

Gaging

In installations requiring manual gaging, if a walkway is not provided, storage tanks should have a way of measuring the contents so employees need not walk across the tank roof. Remote-gage measuring equipment eliminates the need for tank roof walking or walkway installation.

NFPA 30 states that manual gaging openings independent of fill pipes should be equipped with vapor-tight caps or covers, when they are located in a building or buried under a basement. Each such opening should be protected against liquid overflow and possible vapor release by means of a spring-loaded check valve or other approved device. Manual gaging of tanks containing Class I liquids should be avoided. Substitutes for manual gaging include, but are not limited to, heavy-duty flat gage glasses, magnetic, hydraulic or hydrostatic reading devices, or sealed float gages.

The float-operated gage or the type operated by the liquid's pressure is often used on aboveground storage tanks. Some companies use a sounding weight and tape or, if the tank is small,

Figure 16-11. This truck conveniently transports safety cans and other flammable liquid containers to points of use throughout the plant. (Reprinted with permission from Western Electric Company.)

a wooden sounding rod through a gage hatch on the top of the tank. Personnel should stay off the tank roof while any pumping operation is in progress.

Gage glasses protected by metal cases are suitable when properly maintained. These are found on some flammable liquid tanks held under pressure.

Safe means of access should be provided for persons gaging tanks.

Tanks in flooded regions

In areas where there is danger from flood waters, tanks should be installed and anchored according to NFPA 30, Appendix B, "Protection of Tanks Containing Flammable or Combustible Liquids in Locations That May Be Flooded."

Underground tanks

When an underground tank is subject to heavy traffic over it, the tank should be protected with (1) at least 3 ft (0.9 m) of earth cover, (2) 18 in. (0.46 m) of tamped earth plus 6 in. (15 cm) of reinforced concrete, (3) 18 in. of tamped earth plus 8 in. (20 cm) of asphalt concrete.

If the underground tank is not subject to traffic, a protective cover will be provided by (1) covering with a minimum of 2 ft (0.6 m) of earth or (2) 1 ft (0.3 m) of earth plus 4 in. (10 cm) of reinforced concrete. The concrete cover should extend 1 ft beyond the outline of the tank.

An underground tank should have a firm foundation and be surrounded with at least 6 in. of noncorrosive, inert materials, such as well-tamped clean sand, earth, or gravel. The tank

Table 16-A. Location of Outside Aboveground Storage Tanks from Adjoining Property or Public Way For Operating Pressures No Greater Than 2.5 psig (17 kPa)

Type of tank	Protection	Minimum distance in feet from property line that may be built on, including the opposite side of a public way and not less than 5 ft (1.5 m)	Minimum distance in feet from nearest side of any public way or from nearest important building and shall be not less than 5 ft (1.5 m)
Floating roof	Protection for exposures	½ times diameter of tank	⅙ times diameter of tank
	None	Diameter of tank but need not exceed 175 ft (54 m)	⅙ times diameter of tank
Vertical with weak roof to shell seam	Approved foam or inerting system on the tank	½ times diamter of tank	⅙ times diameter of tank
	Protection for exposures	Diameter of tank	⅓ times diameter of tank
	None	2 times diameter of tank but need not exceed 350 ft (110 m)	⅓ times diameter of tank
Horizontal and vertical, with emergency relief venting to limit pressures to 2.5 psig	Approved inerting system on the tank or approved foam system on vertical tanks	½ times Table 16-B	½ times Table 16-B
	Protection for exposures	Table 16-B	Table 16-B
	None	2 times Table 16-B	Table 16-B

For Operating Pressures Greater Than 2.5 psig (17 kPa)

Type of tank	Protection		
Any type	Protection for exposures	1½ times Table 16-B but shall not be less than 25 ft (7.5 m)	1½ times Table 16-B but shall not be less than 25 ft (7.5 m)
	None	3 times Table 16-B but shall not be less than 50 ft (15 m)	1½ times Table 16-B but shall not be less than 25 ft (7.5 m)
Floating roof	Protection for exposures	½ times diameter of tank	⅙ times diameter of tank
	None	Diameter of tank	⅙ times diameter of tank
Fixed roof	Approved foam or inerting system	Diameter of tank	⅓ times diameter of tank
	Protection for exposures	2 times diameter of tank	⅔ times diameter of tank
	None	4 times diameter of tank but need not exceed 350 ft	⅔ times diameter of tank

From *National Fire Protection Association No. 30,* Flammable and Combustible Liquids Code.

should be anchored if there is any chance of its "floating" on rising ground water. Storage tanks inside buildings should have the fill and vent pipes located outside.

NFPA 30 specifies that "underground tanks or tanks under buildings should be so located with respect to existing building foundations and supports that the loads carried by the latter cannot be transmitted to the tank. The distance from any part of tank, storing Class I liquids to the nearest wall of any basement, pit, or cellar should be not less than 1 ft (0.3 m), and from any property line that may be built upon, not less than 3 ft (0.9 m). The distance from any part of a tank, storing Class II or Class III liquids, to the nearest wall of any basement, pit, cellar, or property line should be not less than 1 ft."

The Code also specifies that corrosion protection for an underground tank and its piping should be provided by one or more of the following methods: (1) use of protective coatings or wrapping, (2) cathodic protection, or (3) corrosion-resistant construction materials. Selection of the type of protection should be based on the area's corrosion history and a qualified engineer's judgment. Cinders or other acid-forming fills should not be used around the tank. Local, or other, authorities having jurisdiction should be notified in writing,

so approval will be obtained before the tank installation is covered.

Flammable liquids should be withdrawn by pump. The pump and piping system should be arranged so the liquid will flow back to the tank when the system is not in operation.

Aboveground tanks

NFPA 30 rigorously sets minimum distances from property lines, public ways, and nearest important buildings for aboveground tanks containing flammable or combustible liquids (Table 16-A). Consult NFPA 30 for distances for aboveground tanks containing unstable liquids. Vapor pressure or psig (pounds per square inch gage) is defined in NFPA 30 and explained at the beginning of this chapter.

Some recommended minimum distances are given in Table 16-B. Where end failure of a horizontal pressure tank or vessel can expose property, the tank shall be placed with the longitudinal axis parallel to the nearest important structure exposure.

Although not a frequent occurrence, two tank properties of diverse ownership might share a common boundary. In this case, if the owners and local authorities agree, normal minimum shall-to-shell spacing can be used, such as that within a tank farm.

Table 16-B. Reference Minimum Distances for Aboveground Outside Storage Tanks

Capacity Tank (gallons) (1 gal = 3.78 liters)	Minimum Distance in Feet from Property Line Which Is or Can be Built Upon, Including the Opposite Side of a Public Way	Minimum Distance in Feet from Nearest Side of Any Public Way or from Nearest Important Building on the Same Property
275 or less	5	5
276 to 750	10	5
751 to 12,000	15	5
12,001 to 30,000	20	5
30,001 to 50,000	30	10
50,001 to 100,000	50	15
100,001 to 500,000	80	25
500,001 to 1,000,000	100	35
1,000,001 to 2,000,000	135	45
2,000,001 to 3,000,000	165	55
3,000,001 or more	175	60

From *National Fire Protection Association No. 30.*

Table 16-C. Minimum Tank Spacing (Shell-to-Shell)

	Floating Roof Tanks	Fixed Roof Tanks	
		Class I or II Liquids	Class IIIA Liquids
All tanks not over 150 ft (45 m) diameter	⅙ sum of adjacent tank diameters but not less than 3 ft (0.9 m)	⅙ sum of adjacent tank diameters but not less than 3 ft (0.9 m)	⅙ sum of adjacent tank diameter but not less than 3 ft (0.9 m)
Tanks larger than 150 ft (45 m) diameter			
Remote impounding area	⅙ sum of adjacent tank diameters	¼ sum of adjacent tank diameters	⅙ sum of adjacent tank diameters
Impounding (diking) around tanks	¼ sum of adjacent tank diameters	⅓ sum of adjacent tank diameters	¼ sum of adjacent tank diameters

From *National Fire Protection Association No. 30.*

Tanks storing Class I, II, or IIIA flammable or combustible liquids must be separated in accordance with Table 16-C. There is, however, an exception for crude petroleum tanks situated at production facilities in isolated locations. When these tanks have individual capacities not exceeding 126,000 gal (3,000 barrels or 475,000 liters) they can be as close as 3 ft (0.9 m) apart.

Tanks used only for storing Class IIIB liquids can be spaced no less than 3 ft apart unless within a diked area or drainage path for a tank storing a Class I or II liquid, in which case the provisions of Table 16-C apply. For unstable liquids, the distance between such tanks shall be not less than one-half the sum of their diameters. When tanks are in a diked area containing Class I or Class II liquids, or in the drainage path of Class I or Class II liquids, and are compacted in three or more rows or in an irregular pattern, greater spacing or other means may be required. The authority having jurisdiction might require changes that would make tanks in the interior of the pattern more accessible for firefighting purposes.

There are a number of special cases for petroleum tanks in refineries and in producing areas, for unstable flammable or combustible liquids, for LP-gas, and for tanks arranged in a compact or irregular pattern. The NFPA Code should be consulted when special situations are involved.

It is recommended that truck loading racks, dispensing Class I liquids, should be at least 25 ft (7.6 m) and those dispensing Class II or Class III liquids should be at least 15 ft (4.6 m) from tanks, warehouses, other plant buildings, and the nearest property line.

Tanks should be located to avoid danger from high water levels. Tanks located on a stream without tide should, where possible, be downstream from burnable property.

Piping materials for aboveground tanks shall be steel as recommended in NFPA 30. Piping can be built of materials other than steel when used underground or protected against fire exposure and located so leakage would not result in a hazard. Piping other than steel may be required by the properties of the liquid handled. Cast-iron pipe is not recommended because it is brittle and can fracture under stress. When in doubt, the supplier or producer of the flammable or combustible liquid (or other competent authority) should be consulted about the suitability of the construction material being considered.

Piping from a storage tank should have a readily accessible shutoff valve at the tank. Provision should be available to drain or pump the contents of a tank into another tank, or to collect the liquid within dikes or retaining walls if the tank should leak or be overfilled.

A small tank is often provided with an emergency self-closing valve, located inside the tank or at the point of entrance. This valve would automatically stop the flow of liquid in case of fire. (See Factory Mutual System, Data Sheet 7-32, "Flammable Liquid Pumping and Piping Systems.") This valve should be closed except during loading and unloading operations. The rope or wire and the fusible link attached to an emergency valve, which is left open, should be in good condition and should be tested regularly for easy operation.

Besides the normal vents taking care of vacuum and pressure during pumping operations, an aboveground storage tank must have some form of emergency relief venting to prevent buildup of excessive internal pressure in case of fire surrounding the tank. In addition to a relief device, further protection should be provided by either a weak seam in the top, or at the joint between the top and the shell of the tank, or by some other recommended form. See NFPA 30 for guidance on sizing for normal and emergency vents.

Except in oil refineries and large marine terminals, tanks for storing Class I liquids should be labeled FLAMMABLE—KEEP FIRE AWAY with letters at least 2 in. (5 cm) high. NO SMOKING signs should be conspicuously posted. Similar signs should be posted for Class II and, if necessary, Class III liquids.

Good housekeeping should be maintained around storage tanks. Debris should not be permitted to accumulate, nor should the space adjoining the tank be used for storage of combustible materials. The grass around the tank (or under it, if it is off the ground) should be eliminated or, at least, kept cut.

Tanks used for storing flammable or combustible liquids should be made gas-tight to cut down evaporation losses and for fire prevention. Venting devices are arranged to be normally closed, but breathe automatically when liquid is pumped in or out of the tanks and as temperature fluctuates. (See API Standard

Figure 16-12a. Students practice the technique of shielding themselves from fire with water fog. (Reprinted with permission from Ansul Fire Protection.)

RP 2000–82, "Venting Atmospheric and Low-Pressure Storage Tanks.") Floating roofs also are effective in reducing evaporation losses and allowing expansion.

Large aboveground storage tanks should be equipped with stairways and platforms, preferably of steel. Tanks that stand more than 1 ft (0.3 m) aboveground should have foundations of noncombustible materials, except that wood cushions can be used. Supports for aboveground tanks should be of concrete, masonry, or steel protected by concrete or other approved fireproofing to prevent collapse in case of fire.

Overflowing of tanks presents a severe fire hazard because vapors might drift to a source of ignition. Operators responsible for filling tanks should maintain a constant watch on the filling rate and the level of liquid, so that operations can be stopped or liquid diverted to another tank when the required level is reached. Operators must also remain at the open valve while water is being drained from a tank.

It is common practice to paint flammable liquid tanks exposed to the sun with aluminum, pastel, or white paint in order to reflect the heat and help keep down the internal vapor pressure. In some states, use of such reflective paint is a requirement. In some installations, where highly volatile liquids are stored, an external water spray cooling system is used. Tanks can also be insulated.

Tank fires and their control

It would take only a few seconds for a fire inside a fixed-roof tank to either extinguish itself or blow open the tank roof. A tank fire most commonly occurs at one of the roof openings. It will occur only if vapor is being expelled from the tank because of either filling or heating, and it can usually be extinguished without difficulty if the cause of the vapor expulsion is removed by cooling or by shutting off the filling operation. Seal fires frequently occur on floating-roof tanks. These can

usually be extinguished by portable equipment and hand foam lines from the top of the tank or roof.

If a fire occurs near a tank, water should be applied immediately to the tank to cool it and reduce vaporization, not only to save stock but also to make ignition at the vents less likely. Vent fires extinguish themselves as soon as the outbreathing stops. So many variables are involved in tank fires that they should be fought only by personnel trained in their extinguishment. Figures 16-12 a through c show various ways of fighting fires.

Inside storage and mixing rooms

Flammable or combustible liquids in sealed containers present a potential, rather than an active hazard—the possibility of fire from without. Inside storage rooms are undesirable. If they must be used, they should be isolated as much as possible, located at or above grade, not immediately above a cellar or basement, and preferably along an exterior wall.

NFPA 30 states "Inside [storage] rooms shall be constructed to meet the selected fire resistance rating as specified in 4-4.1.4. Such construction shall comply with the test specifications given in NFPA 251, *Standard Methods of Fire Tests of Building Construction and Materials.* Except for drains, floors shall be liquidtight and the room shall be liquidtight where the walls join the floor. Where an automatic fire protection system is provided, as indicated in 4-4.1.4, the system shall be designed and installed in accordance with the appropriate NFPA standard for the type of system selected.

"Openings in interior walls to adjacent rooms or buildings shall be provided with:

"(a) Normally closed, listed 1½ hr (B) fire doors for interior walls with fire resistance rating of 2 hr or less. Where interior walls are required to have greater than 2 hr fire resistance rating, the listed fire doors shall be compatible with the wall rating. Doors may be arranged to stay open during material

Figure 16-12b. Demonstration of the use of dry chemical to attack an elevated fire which simulates a leaking fuel tank or pipe. (Reprinted with permission from Ansul Fire Protection.)

handling operations if doors are designed to close automatically in a fire emergency by provision of listed closure devices. Fire doors shall be installed in accordance with NFPA 80, *Standard for Fire Doors and Windows*.

"Noncombustible liquidtight raised sills or ramps at least 4 in. (10 cm) in height or otherwise designed to prevent the flow of liquids to the adjoining areas. A permissible alternative to the sill or ramp is an open-grated trench, which drains to a safe location, across the width of the opening inside of room.

"Wood at least 1 in. (2.5 cm) nominal thickness may be used for shelving, racks, dunnage, scuffboards, floor overlay and similar installations.

"Storage in inside rooms shall comply with the [specifications shown in Table 16-D].

"Electrical wiring and equipment located in inside rooms used for Class I liquids shall be suitable for Class I, Division 2 classified locations; for Class II and Class III liquids, shall be suitable for genral use. NFPA 70, *National Electrical Code*©, provides information on the design and installation of electrical equipment.

Table 16-D. Inside Storage Rooms

Automatic Fire Protection* Provided	Fire Resistance	Maximum Floor Area	Total Allowable Quantities Gal/sq ft of floor area
Yes	2 hours	500 sq ft	10
No	2 hours	500 sq ft	4
Yes	1 hour	150 sq ft	5
No	1 hour	150 sq ft	2

*Fire protection system shall be sprinkler, water spray, carbon dioxide, dry chemical, Halon, or other system approved by the authority having jurisdiction.

(10.7 sq ft = 1 m². 1 gal/sq ft = 40.5 liters/m²).

"Every inside room shall be provided with either a gravity or a continuous mechanical exhaust ventilation system. Mechanical ventilation shall be used if Class I liquids are dispensed within the room.

"Exhaust air shall be taken from a point near a wall on one

Figure 16-12c. Demonstration of how to simultaneously extinguish and secure a large gasoline spill fire using a dry chemical and aqueous film forming foam (AFFF). This technique is especially valuable in crash/rescue operations where the rapid extinguishing properties of dry chemical can be used to reach trapped victims while the AFFF application ensures that the fire will not flash back over the extinguished area. (Reprinted with permission from Ansul Fire Protection.)

side of the room and within 12 in. (30 cm) of the floor with one or more make-up inlets located on the opposite side of the room within 12 in. (30 cm) from the floor. The location of both the exhaust and inlet air openings shall be arranged to provide, as far as practicable, air movements across all portions of the floor to prevent accumulation of flammable vapors. Exhaust from the room shall be directly to the exterior of the building without recirculation.

"Exception: Recirculation is permitted where it is monitored continuously using a fail-safe system that is designed to automatically sound an alarm, stop recirculation, and provide full exhaust to the outside in the event that vapor-air mixtures in concentration over one-fourth of the lower flammable limit are detected.

"If ducts are used, they shall not be used for any other purpose and shall comply with NFPA 91, *Standard for the Installation of Blower and Exhaust Systems for Dust, Stock and Vapor Removal or Conveying.* If make-up air to a mechanical system is taken from within the building, the opening shall be equipped with a fire door or damper, as required in NFPA 91, *Standard for the Installation of Blower and Exhaust Systems for Dust,*

Stock, and Vapor Removal or Conveying. For gravity systems, the make-up air shall be supplied from outside the building.

"Mechanical ventilation systems shall provide at least one cubic foot per minute of exhaust per square foot of floor area (1 m³ per 3 m²), but not less than 150 CFM (4 m³). The mechanical ventilation system for dispensing areas shall be equipped with an airflow switch or other equally reliable method which is interlocked to sound an audible alarm upon failure of the ventilation system.

"In every inside room, an aisle at least 3 ft (0.90 m) wide shall be maintained so that no container is more than 12 ft (3.6 m) from the aisle. Containers over 30 gal (113.5 L) capacity storing Class I or Class II liquids shall not be stored more than one container high.

"Where dispensing is being done in inside rooms, operations shall comply with the provisions of Chapter 5 [of NFPA 30.]

Storage cabinets

"Not more than 120 gal (454 L) of Class I, Class II and Class IIIA liquids may be stored in a storage cabinet. Of this total,

Figure 16-13. A technician installs a multipurpose dry chemical piped system to protect a flammable liquid storage area. (Reprinted with permission from Ansul Fire Protection.)

not more than 60 gal (227 L) may be of Class I and Class II liquids and not more than three (3) such cabinets may be located in a single fire area, except that, in an industrial occupancy, additional cabinets may be located in the same fire area if the additional cabinet, or group of not more than three (3) cabinets, is separated from other cabinets or group of cabinets by at least 100 ft (30 m).

"Storage cabinets shall be designed and constructed to limit the internal temperature at the center, 1 in. (2.5 cm) from the top to not more than 325 F (162.8 C) when subjected to a 10-minute fire test with burners simulating a room fire exposure using the standard time-temperature curve as given in ASTM E152-81a. All joints and seams shall remain tight and the door shall remain securely closed during the fire test. Cabinets shall be labeled in conspicuous lettering, "FLAMMABLE—KEEP FIRE AWAY." The cabinet is not required to be vented.

"Metal cabinets constructed in the following manner are acceptable. The bottom, top, door and sides of cabinet shall be at least No. 18 gage sheet steel and double walled with 1½ in. (3.8 cm) air space. Joints shall be riveted, welded or made tight by some equally effective means. The door shall be provided with a three-point latch arrangement and the door sill shall be raised at least 2 in. (5 cm) above the bottom of the cabinet to retain spilled liquid within the cabinet.

"Wooden cabinets constructed in the following manner are acceptable. The bottom, sides and top shall be constructed of exterior grade plywood at least 1 in. (2.5 cm) in thickness, which shall not break down or delaminate under fire conditions. All joints shall be rabbetted and shall be fastened in two directions with wood screws. When more than one door is used, there shall be a rabbetted overlap of not less than 1 in. (2.5 cm). Doors shall be equipped with a means of latching and hinges shall be constructed and mounted in such a manner as to not lose their holding capacity when subjected to fire exposure. A raised sill or pan capable of containing a 2-in. (5-cm) depth of liquid shall be provided at the bottom of the cabinet to retain spilled liquid within the cabinet." (From NFPA 30.)

It is most important for each establishment to check with local fire authorities on the type of storage and handling it is using or proposes to use (see Figure 16-13).

Outside storage houses

If space permits, it is advisable to construct the flammable and combustible liquid storage room as a separate building set aside from the main plant. Construction can be similar to that described for inside storage rooms. The type of product stored and the proximity to other buildings and structures will determine the best design.

CLEANING TANKS

General precautions

Tanks and vessels that have contained flammable and combustible liquids should be cleaned, as required for repairs, entry, or change of product, only under the supervision of a competent employee who is familiar with fire and accident prevention recommendations and with first aid measures.

The following section covers fire, explosion, asphyxiation, and toxic material poisoning-prevention recommendations from the American Petroleum Institute Standards: RP 2015, *Cleaning Petroleum Storage Tanks;* API 2015 A, *A Guide for Controlling the Load Hazard Associated with Tank Entry and Cleaning;* and NFPA standard 327, *Cleaning or Safeguarding Small Tanks and Containers.*

An industrial plant without proper cleaning equipment and personnel trained in tank cleaning operations should consult the head of the service department of its flammable liquid supplier. Tank cleaning contractors with the necessary equipment and experienced crews can be found in many cities.

Before tank cleaning operations are started, the supervisor and crew should have the proper equipment. This can include supplied-air hose masks with blowers, self-contained breathing apparatus, suitable clothing, safety belts, safety lines, and tools. The masks and equipment should be the proper type, clean, and in good condition, and all workers must be properly instructed in their use.

Prior to beginning the repair or cleaning operations the tank must be purged of all flammable vapor through ventilation or other effective means. The inside of the tank must be checked with a combustible gas indicator and an oxygen-level meter before and frequently thereafter while work is in progress. If tank entry is required, the atmosphere must also be tested with instruments capable of readings below the TLV of the prior contents. An atmosphere that tests well below the LEL can be extremely toxic.

Remove all sources of ignition from the surrounding area. Obtain tank entry and hot work permits (see Chapter 17). Blank all piping, lock out all electrical equipment, and use only lighting approved for the specific atmospheric conditions in the tank. (Attaining zero mechanical state is discussed in Chapter 8, Principles of Guarding.)

Consider wind and weather conditions. Do not start work if wind might carry vapors into an area where they could create a hazard, or if an electrical storm is threatening or in progress. Prohibit smoking and carrying matches and lighters. All rules should be strictly enforced.

Sandblasting equipment must be bonded to the tank to prevent static sparks. Use of power chipping tools and rivet busters must be suspended when flammable vapors from tanks may be present in the area. Nozzles on steam lines used to free tanks of vapor must be bonded to the tank shells to prevent static accumulation. Steaming large storage tanks for gas freeing should be avoided. Motor trucks, gas engines, open flames, and portable electric equipment must be kept a safe distance from the tank being cleaned.

Protective equipment

It is considered good practice to wear impermeable rubber boots, thoroughly cleaned and in good condition, for tank cleaning. It is also advisable to wear impermeable gloves, but, if the work does not permit their use, a protective cream recommended for the exposure should be used. In an atmosphere that is dangerously irritant or corrosive to the skin, complete impermeable clothing must be worn.

Supplied-air hose masks with blowers and safety harnesses with lifelines, or self-contained or self-generating breathing apparatus, must be worn by workers entering tanks unless the person in charge has determined that the vapor concentration is below the threshold limit value and that no oxygen deficiency exists.

If a test of the tank shows it to be deficient in oxygen, whether or not it is otherwise immediately dangerous to life or health, the tank should be ventilated with fresh air and be checked for sufficient oxygen before entering without approved breathing apparatus. This can be either a compressed-air type of self-contained breathing apparatus, a self-generating type, or an air-supplied type. The fit of the facepiece should be carefully checked before the employee enters the tank. If the environment contains a substance that is dangerously irritant, corrosive to the skin, or seriously toxic via skin absorption (tetraethyllead, for example), the breathing apparatus should be supplemented by impermeable full-body protection.

Some medical authorities advise that individuals with broken eardrums not work in vapor-laden atmospheres, because no mask will protect these body openings.

The supplied fresh air hose mask with blower is generally the best type of respiratory equipment when working for extended periods in a tank.

Proper procedures

A tank should be free of hazardous gas before any work is performed inside it (see ANSI Z117.1).

A worker should not be allowed to enter a gassy or oxygen-deficient tank unless absolutely necessary, such as in an emergency. When this is necessary, the worker should be attended on the outside by another worker who is similarly outfitted with approved respiratory protection and with appropriate protective attire so the latter can rescue the worker in the tank if necessary. In some states this is a legal requirement.

Employees engaged in tank cleaning shoud be instructed in the method of giving cardiopulmonary resuscitation (CPR) and should be periodically retrained. A tank cleaner who is overcome by vapor or gas should be removed to fresh air immediately and, if necessary, given CPR until breathing resumes. A physician should be summoned, and the rescued worker kept quiet and warm until breathing and circulation are normal.

The sense of smell cannot be relied upon for an accurate estimate of the amount of flammable or toxic vapor in a tank. It does, however, give warning that vapors of some kind are present. Symptoms of intoxication, dizziness, nausea, and headache indicate that a dangerous concentration of flammable or toxic vapors is present. Exposed workers should immediately leave the contaminated area and not return until vapors have been cleared.

Workers should always have a clear path of escape from a tank and should be aware that they may have to use it in a hurry. A ladder always should be used when a tank must be entered from above, and it should be left secured in place until the last worker is out of the tank. A body harness and a lifeline are recommended to assist with rescue work.

Burning, welding, cutting, and spark-producing operations should not be permitted in a tank until: (1) the area to be heated has been thoroughly cleaned, (2) tests have determined that the tank atmosphere is vapor free, and (3) there are no flammable or combustible liquids or solids in the work area which could be ignited by cutting or welding operations. Where any vapor is present, further ventilation will be required to remove it from the tank. Heavy scale, if present, should be scraped and the scale probed for flammable vapors.

Hot work repairs (welding and cutting) should not be made whenever other means can be more safely used. Even after a tank has been freed of vapor, combustible mixtures may again be formed through admission of flammable vapors or liquids from other sources, such as an unblanked line or connection; a break in the bottom of the tank; sludge, sediment, or side-wall scale; or wood structures soaked with the liquid. The interior pontoons in floating roof tanks can trap quantities of vapor-releasing liquid. Also, the seals of some floating roof tanks and the covers of some internal floating roof tanks can be flammable or combustible. For this reason, special safeguards might be needed during hot work. Therefore, periodic tests with a combustible gas indicator should be made during the work.

Burning or cutting can release lead fumes from paint on either the inner or outer surface of the tank or from some other source. An approved mask for protection against lead may be necessary, or it may be advisable to exhaust the fumes. Wood supports and other combustible materials inside the tank should be protected or removed before hot work is begun.

Some companies use low-voltage transformers to reduce the hazard of electric shock, especially when employees are required to work in wet areas.

All portable electrical equipment should be grounded or ground fault circuit interrupters provided.

Cold work in a tank can present a dust problem, and approved dust respirators and goggles should be worn if tests definitely show that wearing a hose mask is not necessary.

Gas tests should be made frequently if the presence of a hazardous gas is suspected. The concentration of gas or vapor in a tank or vessel can increase as work progresses, especially if the inside to the tank is scraped or heated during the operation.

Forced ventilation can be required in many cases where repair work is done inside a tank.

Where fumes develop from welding or other repair work, mechanical air movers should be used or the workers should wear air-supplied breathing apparatus. On a large tank, a door sheet or sheets can be removed for ventilation after the tank has been made gas free. The opening also can be used to expedite removal of sediment and to increase illumination.

If a tank has been closed for some time, there can be a deficiency of oxygen caused by rusting (oxidation) of the metal of the tank. In that case, no one should enter the tank without an air supplied mask or, for a relatively brief stay, a self-contained or self-generating breathing apparatus unless the tank has been tested and found to be safe for entry.

Cleaning storage tanks

Here in step-by-step form is a summary of a typical procedure for cleaning tanks that have contained flammable or combustible liquids.

1. Remove all sources of ignition (matches, open flames, smoking, gas engines, welding, exposed electrical wiring and equipment) from the vicinity of the tank.
2. Empty the tank by pumping, draining, and floating tank contents with water. The water should be introduced through fixed tank connections.
3. Disconnect and blank all product, steam smothering, foam, and similar lines. Do not rely on valves.
4. Open all manholes and allow the tank to air thoroughly. Ventilate or steam the tank for the number of hours required by its size. If steam is used, cool and ventilate afterward, taking care that the ventilation is adequate to prevent a vacuum developing when the steam is turned off.
5. Have available the required personal protective equipment: fresh air hose masks, approved flashlights, safety belts and lifelines, flexible insulated boots and gloves.
6. If light is needed, use only flashlights or electric lanterns listed for combustible atmospheres by Underwriters Laboratories Inc.
7. Test for vapor content with a combustible gas indicator. No entry should be permitted if the vapor concentration is 20 percent (or more) of the LEL. For entering the tank, wear a fresh air hose mask, air supply tanks, and safety belt and lifeline, if the tank atmosphere contains more than the maximum acceptable concentration of vapor. Have ample help available outside for the number of workers inside.
8. If a storage tank has contained leaded gasoline subsequent to its last thorough cleaning, follow instructions of suppliers of tetraethyllead. If inert gas has been used for gas freeing, check for oxygen deficiency.
9. Bond steam lines and water wash nozzles to the tank.
10. Wash sludge, sediment, and scale from the tank. Let it drain or remove it with a pump. Thoroughly flush out and overflow the tank with water if necessary.
11. Make a gas-hazard test and, if the tank is found free of toxic or flammable vapors, check conditions inside the tank before issuing an okay for the work. Otherwise, further clean and ventilate as required.
12. Continue ventilation for the duration of the work in the tank, and make periodic tests for the presence of hazardous gases or flammable vapors as the work progresses.

Ventilation by air instead of steaming is more widely used where deposits of iron sulfide are not present and where a source of power supply for mechanical air removers is available. Special precautions are required for the servicing of tanks containing pyrophoric iron sulfide. (See the American Petroleum Institute's accident prevention manual, *Cleaning Petroleum Storage Tanks,* listed in the References.)

Cleaning small tanks and containers

General precautions. Work on containers that have held flammable or combustible liquids or gases shall be supervised by a trained supervisor capable of maintaining a high degree of safety during operations. If the container has held such unstable compounds as nitrocellulose, pyroxyline solutions, nitrates, chlorates, perchlorates, or peroxides, special precautions should be taken because the container may contain enough oxygen to support combustion. Contact the manufacturer or supplier for specific information regarding cleaning procedures and other precautions.

Small tanks and drums should be cleaned and steamed in an open-faced building with ample ventilation. It is even preferable to clean them in an outside area free from ignition sources. Pipes and nozzles should be electrically bonded to containers being steamed.

Covers, plugs, and valves should first be removed and the tank or drum permitted to drain into a container. The inside should be examined for rags, waste, or other debris that might interfere with draining or could retain flammable vapors. Only lights approved for use in the *National Electrical Code* (Class I, Division I, Group D hazardous locations) should be used for this inspection; see explanation in Chapter 15, Electricity. Mirrors can sometimes be used to reflect daylight into the tank.

Steaming. Steaming, hot chemical wash, water filling, and use of inert gas are among the common methods for cleaning and vapor-freeing small tanks and drums. If the inside of the container is clean and steam is available, the easiest method is to steam it. (See American Welding Society, *Safe Practices for Welding and Cutting Containers That Have Held Combustibles.* Also see NFPA 327, *Cleaning or Safeguarding Small Tanks and Containers.*)

The tank or drum should be allowed to drain by being placed on a steam rack or over a steam connection with the outlet holes at the lowest point. The tank or drum should rest against a steam pipe or should be bonded. An ample supply of live steam should then be applied for a period of not less than ten minutes.

The inside of the drum should be washed with hot water and, after being cooled, tested with a combustible gas indicator. If the drum tests vapor free, hot work repairs can then proceed. If not, it should be cleaned and again steamed. It is sometimes advisable to mark the drum to signify that it is vapor free.

If steaming will not clean the tank or drum, a cleansing compound of sodium silicate or trisodium phosphate (washing powder) dissolved in hot water and kept at a temperature of 170 to 190 F (77 to 88 C) should be used. Hot water should be added to overflow the container until no appreciable amount of volatile liquid, scum, or sludge appears.

Exceedingly dirty containers can require preliminary treatment with caustic soda solution that is agitated enough to make sure the interior surfaces are thoroughly cleaned. They can then be drained, washed, and steamed. Do not use this treatment on aluminum or zinc-coated drums that could generate copious quantities of hydrogen when in contact with a caustic soda solution.

If steam is not available to heat the water, a cold water solution with an increased amount of cleansing compound can be used. The solution should be agitated to ensure thorough cleaning.

To guard against burns, especially when using steam, hot water, and caustic soda, workers should be protected with suitable clothing, such as boots, gloves, faceshields, and rubber aprons.

Because of its low ignition temperature, drums having contained carbon disulfide should not be steamed. They should be made vapor-free with a cleansing compound and then gas tested.

Small tanks can be made safe by means of an inert gas, but this method is generally not considered as safe as steaming.

Portable inert gas generators are available for special jobs, but they should only be used by employees who have the train-ing and equipment to produce the proper atmosphere and safeguard against fire and explosion.

Carbon dioxide and nitrogen are sometimes used to make small tanks and drums that have contained flammable liquids safe for hot work repairs. However, use of carbon dioxide can lead to problems with static electricity. If entry is required, first check for oxygen deficiency.

When an inert gas like carbon dioxide or nitrogen is used, the tank or vessel is washed as free as possible of flammable liquids, and flushed thoroughly until the vessel overflows. When repairing, as much water as the work will permit should be left in the tank, and carbon dioxide introduced to produce a concentration of not less than 50 percent by volume. If the tank has contained hydrogen or carbon monoxide, an 80 percent concentration is required.

Nitrogen concentrations should be 60 percent or higher, depending on the previous contents of the container.

Abandonment of tanks

Tanks to be *permanently* abandoned should be thoroughly washed and made safe from flammable vapors, then dismantled, and removed from the premises. Underground tanks being abandoned but not removed should have all flammable liquid removed and be filled with a nonshrinking inert solid material. They should have all inlets and outlets capped.

Tanks, taken out of service for less than 90 days, should have fill lines, gage openings, and pump suction lines capped and should be secured against tampering. The vent lines should be left open.

DISPOSAL OF FLAMMABLE LIQUIDS

Unused, clean flammable liquids should be returned to the vendor, salvaged for resale, or used in some other way. When drummed and properly stored, most flammable liquids are stable and can safely be used for a period of several years.

Mixtures of clean flammable liquids sometimes need to be distilled before they are usable. Unless the user has the correct distillation facilities, it is best to have a recovery contractor do the separations.

Used or dirty flammable liquids also can be handled by a recovery contractor, who after cleanup can market the material.

If recycling or recovery is not feasible, the preferred method of disposal is burning in an EPA-approved incinerator.

Only as a last resort should flammable liquids be given to a disposal contractor who will discard the materials in a safe landfill. This is very expensive and can possibly contribute to future contamination problems at or near the landfill. All requirements of the Resource Conservation and Recovery Act must be observed.

COMMON USES OF FLAMMABLE AND COMBUSTIBLE LIQUIDS

Industrial use of flammable and combustible liquids involves all the general precautions outlined in this chapter. These include static, toxicity, storage, housekeeping, approved electrical installation, segregation and isolation of operations, ventilation, enclosure of operations, grounding sources of ignition, and fire and explosion precautions.

All rooms or portions of plants (and the equipment in them), in which flammable and combustible liquids or vapors are used or generated, should be constructed, installed, and operated as recommended by NFPA in its *Flammable & Combustible Liquids Code,* and as required by local ordinances.

The following paragraphs discuss several of the more common processes and items of equipment that use flammable and combustible liquids and the precautions applying to each.

Dip tanks

Dip tanks containing flammable liquids, subject to ignition at ordinary temperatures and giving off flammable vapors, present a severe fire and explosion hazard. (See NFPA 34, *Dip Tanks Containing Flammable or Combustible Liquids,* and OSHA regulations 29 CFR Chapter XVII, Section 1910.108).

Dipping operations should be conducted above grade in a detached one-story building of noncombustible construction or in a cutoff one-story section. The room should be as large as possible, adequately ventilated, away from sources of ignition, and conspicuously marked as a flammable liquid area.

Handling flammable and combustible liquids in open containers can be hazardous and should be avoided. The openings should be as small as possible and a cover should be provided. The cover should be either hinged-and-gravity closing or should slide on tracks and be held open by a fusible link or other heat-actuated device (Figure 16-14).

NFPA 34 states, "Mechanical ventilation shall be provided, and the ventilating system arranged to move air from all directions toward the vapor area origin and thence to a safe outside location. The ventilating system shall be so arranged that the failure of any ventilating fan shall automatically stop any dipping conveyor system."

Tanks with capacities over 500 gal (1890 liters) should have bottom drains unless the viscosity of the liquids they contain makes this requirement impractical.

Overflow pipes should be used to carry off any overflow liquids to a safe place, preferably outside the building. Tanks should also be protected by automatic fire extinguishing systems (Figure 16-15).

Japanning and drying ovens

Ovens used for evaporating varnish, japan enamel, and other flammable and combustible liquids can present serious fire and explosion hazards and should have ample provision for ventilation and explosion venting.

Drying ovens are of two types: the box oven, which is closed while in operation (it is commonly used in small-scale operations); and the continuous conveyor oven, which is open at both ends and is normally used for quantity production. NFPA 86A, *Ovens and Furnaces, Design, Location, & Equipment,* should be consulted for details of construction and operation, as should any jurisdictional authorities.

Ovens should be provided with the proper type of fire extinguishing equipment, and provision should be made so the fans and conveyor in the continuous oven will automatically stop in case of fire.

Oil burners

Oil burners should be of a type approved by a recognized testing laboratory. To prevent faulty ignition or accumulation

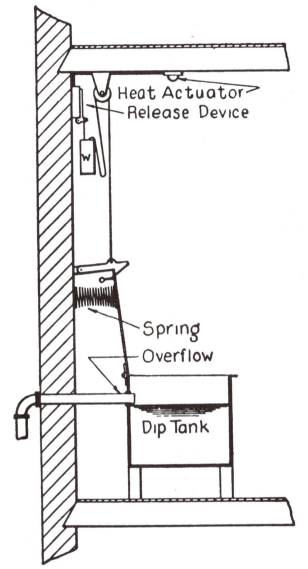

Figure 16-14. The cover for a medium-sized dip tank can be a hinged-and-gravity closing or sliding-track cover held open by a fusible link or other heat-activated device. (Reprinted with permission from National Fire Protection Association.)

of soot, with its attendant fire hazard, the correct type of fuel oil, as recommended by the approval agency, should be used.

Fuel oil should not have a flash point lower than 100 F (37.8 C). It should be a hydrocarbon oil, free from acid, grit, and foreign matter likely to clog or damage the burners or valves. Some plants use acid sludge for fuel, which requires special burning equipment and procedures.

Fuel for domestic burners can be stored in basement tanks, as provided by NFPA 31, *Installation of Oil Burning Equipment.* The supply tank preferably should be located outside the building and should be underground. The top of the tank should be below the level of all piping to which it is connected. This prevents discharge of oil through a broken pipe or connection by siphoning.

A gravity feed to burners should not be used unless special

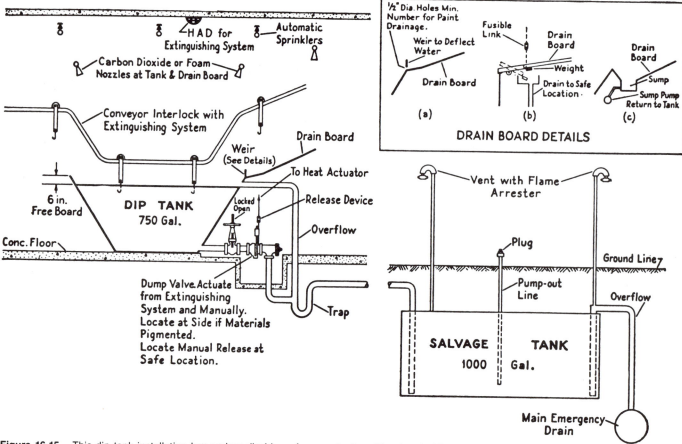

Figure 16-15. This dip tank installation has carbon dioxide or foam protection. (Reprinted with permission from National Fire Protection Association.)

safeguards are provided against abnormal discharge of oil at the burner. The primary hazard of oil burners is the possibility of discharge of unburned oil into a hot fire box where it can vaporize and form an explosive mixture. Approved automatic safeguards should be provided to control the hazard.

Cleaning metal parts

Stoddard solvent (over 100 F flash point) is safe for use in cleaning grease and oil from metal parts where ordinary ventilation is provided and the area is free of sources of ignition. Alkaline compounds, available under several trade names, are safe from a fire and toxic standpoint. Oil or grease should not be permitted to accumulate in the cleaning compounds. (See Factory Mutual Data Sheet 7-79, *Metal Cleaning.*)

The flammability of gasoline and the toxicity of carbon tetrachloride have been sufficient reasons for the general ban on these products for cleaning purposes. Even with high flash and low toxicity materials, ventilation should be provided to remove vapors.

Mixing carbon tetrachloride and a flammable solvent to reduce the fire hazard is not recommended, because this can cause both toxic and fire hazards.

Internal combustion engines

Good housekeeping should be practiced to prevent the accumulation of rubbish, oil or fuel, and rags around internal combustion engines. Proper receptacles should be provided for the refuse disposal. (See NFPA 37, *Stationary Combustion Engines and Gas Turbines.*)

Before a gas tank is filled, the engine should be shut down and hot exhaust pipes permitted to cool. Filling preferably should take place during daylight. Approved safety cans or a hand pump with a bonded filling hose should be used, and the main fuel supply kept in approved containers outside the building.

The fuel tanks of engines that operate continuously and cannot be shut down for filling should be located outside the engine room where vapors will not be exposed to hot engines or exhaust.

Lift trucks and other mobile equipment should be refueled outside buildings.

Engines should be kept clean, and insulation on electrical wiring kept in good repair.

Spray booths

Paint spraying operations should be done in detached buildings or cut off from other operations where possible. Where spraying is done in production areas, approved spray rooms or booths with adequate ventilation must be used. The enclosed area should not be made so small that explosive mixtures of vapor and air can easily be formed. Heating units and piping that might become coated with flammable materials should be eliminated or protected against such accumulations. (See NFPA 33, *Spray Finishing Using Flammable and Combustible Materials.*)

Fires in spray booths and spray booth operations most frequently result from spontaneous ignition of spray deposits. These fires can be prevented by a regular schedule of cleaning. The frequency is determined by the accumulation rate.

Water wash booths have proved to be safer from fire than the dry type of booth because they trap the excess spray before it can enter the exhaust ducts.

Electrical equipment in spraying areas should meet the requirements of the *National Electrical Code* for such locations.

Spraying operations lend themselves to automatic fire control. Automatic sprinklers or carbon dioxide systems most effectively confine fires. Protection should be provided for the exhaust duct as well as the spray booth. Discharge heads of such equipment must be protected from overspray. Protection of automatic sprinkler heads is usually achieved by covering the heads with ordinary paper bags.

Electrostatic spraying, usually automatic, introduces a possible source of ignition in the arcing of parts to the electrodes. To overcome this hazard, parts being sprayed can be held in tight-fitting fixtures instead of being permitted to hang or swing and, thus, come close enough to induce a spark.

Liquefied petroleum gases

Liquefied petroleum gases include any material that is predominantly composed of any of the following hydrocarbons, or mixtures of them: propane, propylene, butane (normal butane or isobutane), and butylenes. The gases liquefy under moderate pressure, but convert into a gaseous state upon relief of the pressure. LP-gas vapor presents a hazard comparable to that of any flammable natural or manufactured gas, except that since it is heavier than air, the matter of adequate ventilation requires some attention.

Liquefied petroleum gases are used as fuel gases, as raw materials in chemical processes, for example, in the making of hydrogen, and to form special atmospheres in heat-treating furnaces. It is important that employees understand the properties of these gases and that they be thoroughly trained in safe practices for handling, distribution, and operation. Detailed programs should be developed to handle any emergencies that might arise.

Systems should be designed and installed by experienced, reliable concerns thoroughly familiar with the hazards, with state and local codes, and with fire organization and insurance company recommendations. (See National Fire Protection Association, Factory Mutual Engineering and Research Organization, and National LP-Gas Association in References.)

REFERENCES

Alliance of American Insurers, 1501 Woodfield Rd., Suite 400 W., Schaumburg, Ill. 60173.
  Safe Handling of LP-Gas When Used as a Motor Fuel.
  Handbook of Organic Industrial Solvents.
American Conference of Governmental Industrial Hygienists, Bldg. D-7, Glenway Ave., Cincinnati, Ohio 45211.
  Threshold Limit Values (published annually).
American Petroleum Institute, 1220 L St., NW, Washington, D.C. 20005.
  Cleaning Petroleum Storage Tanks. Standard RP 2015.
  Protection Against Ignitions Arising Out of Static, Lightning, and Stray Currents. Standard RP 2003–82.
  Welded Steel Tanks for Oil Storage. Standard SID 650.
American Society for Testing and Materials, 1916 Race St., Philadelphia, Pa. 19103.
  Annual Book of ASTM Standards:
    Parts 23, 24 and 25, "Petroleum Products and Lubricants."
    Parts 27 and 28, "Paint."
    Part 30, "Soaps, Antifreezes, Polishes, Halogenated Organic Solvents, Activated Carbon, Industrial Chemicals."
  Test Methods for Flash Point by Pensky-Martens Closed Tester. Standard D93-85.
  Test Method for Flash Point by Tag Closed Tester. Standard D56-82.
  Test for Vapor Pressure of Petroleum Products (Reid Method). Standard D323-72.
American National Standards Institute, 1430 Broadway, New York, N.Y. 10018.
  Blower and Exhaust Systems, ANSI/NFPA 9.1.
  Fundamentals Governing the Design and Operation of Local Exhaust Systems, Z9.2.
  Safety Color Code for Marking Physical Hazards. Z53.1.
  Safety Requirements for Working in Tanks and Other Confined Spaces, Z117.1–1977.
American Welding Society, P.O. Box 351040, 550 LeJeune Rd., NW, Miami, Fla. 33135.
  Safe Practices for Welding and Cutting Containers That Have Held Combustibles.
Association of American Railroads, Bureau for the Safe Transportation of Explosives and Other Dangerous Articles, 50 F St., NW, Washington, D.C. 20001.
  Recommended Good Practice for Handling Collisions and Derailments Involving Hazardous Materials in Transportation, BE Pamphlet No. 1.
  Recommended Good Practice for Handling Fires or Spills Involving Explosives and Other Dangerous Articles in Transportation, BE Pamphlet No. 2.
  Recommended Practice for the Prevention of Electric Sparks That May Cause Fires in Tanks or Tank Cars Containing Flammable Liquids or Flammable Compressed Gases, Due to Proximity of Wire Lines. Bureau of Explosives Circular No. 17-E.
  Signal Manual of Recommended Practice.
Chemical Manufacturers Association, 2501 M St., NW, Washington, D.C. 20037.
  "Loss Prevention Data Sheets."
Factory Mutual Engineering and Research Organization, 1151 Boston-Providence Turnpike, Norwood, Mass. 02062.
  Factory Mutual System Approval Guide.
  "Factory Mutual Loss Control Data Books" (listing available).
Fawcett, H. H., and W. S. Wood, eds. *Safety and Accident Prevention in Chemical Operations,* 2nd ed. New York, N.Y.: Wiley-Interscience Publishers, 1982.
Industrial Risk Insurers, 85 Woodland St., Hartford, Conn. 06102.
  Recommended Good Practices (Supplements to NFPA Standards).
National Fire Protection Association, Batterymarch Park, Quincy, Mass. 02269.
  Basic Classification of Flammable and Combustible Liquids, 321.
  Cleaning or Safeguarding Small Tanks and Containers, 327.
  Control of Gas Hazards on Vessels, 306.

Dipping and Coating Processes Using Flammable or Combustible Liquids, 34

Drycleaning Plants, 32.

Fire Hazard Properties of Flammable Liquids, Gases, and Volatile Solids, 325M.

Fire Protection Handbook.

Flammable and Combustible Liquids Code, 30.

Hazardous Chemicals Data, 49.

Identification of the Fire Hazards of Materials, 704.

Installation of Oil Burning Equipment, 31.

National Electrical Code, 70.

Ovens and Furnaces, 86.

Protection of Tanks Containing Flammable or Combustible Liquids in Locations That May Be Flooded, No. 30, Appendix B.

Static Electricity, 77.

Spray Application Using Flammable and Combustible Materials, 33.

Storage and Handling of Liquefied Petroleum Gases, 58.

Storage and Handling of Liquefied Petroleum Gases at Utility Gas Plants, 59.

Tank Vehicles for Flammable and Combustible Liquids, 385.

National LP-Gas Association, 1301 W. 22nd St., Oak Brook, Ill. 60521.

National Safety Council, 444 North Michigan Ave., Chicago, Ill. 60611.

"Chemical Safety Slide Rule."

Industrial Data Sheets

Acetone, 398.

Carbon Bisulfide (Carbon Disulfide), 341.

Cleaning With Hot Water and Steam, 238.

Flammable and Combustible Liquids in Small Containers, 532.

Liquefied Petroleum Gases for Industrial Trucks, 479.

Liquid Degreasing of Small Metal Parts, 537.

Methanol, 407.

Static Electricity, 547.

Styrene Monomer, 627.

Vapor Degreasers, 718.

Underwriters Laboratories Inc., 333 Pfingsten Rd., Northbrook, Ill. 60062.

Classification of Hazards of Liquids. Research Bulletin No. 29.

Fire Protection Equipment List.

Flammable Liquids, Static Electricity Hazards. Serial No. UL-435.

Gas and Oil Equipment List.

Hazardous Location Equipment List.

The Lower Limit of Flammability and Autogenous Ignition Temperature of Certain Common Solvent Vapors Encountered in Ovens. Research Bulletin No. 43.

Oil Tank Vents — Hazards of Screens. Serial No. UL-31.

U.S. Bureau of Mines, Department of the Interior, 2401 E St., NW., Washington, D.C. 20241.

Flammability Characteristics of Combustible Gases and Vapors. Bulletin 627.

Limits of Flammability of Gases and Vapors. Bulletin 503.

Mine Gases and Methods for Their Detection. Circular No. 33.

U.S. Department of Commerce, Washington, D.C. 20230.

Sparking Characteristics and Safety Hazards of Metallic Materials. Technical Report No. NGF-T-1-57, PB 131131, Office of Technical Services.

Static Electricity in Nature and Industry. Bulletin No. 368.

U.S. Department of Labor, 200 Constitution Ave., NW., Washington, D.C. 20210.

OSHA Regulations 29 CFR, Subpart H, "Hazardous Materials," Sections 1910.106, 1910.107, and 1910.108.

U.S. Department of Transportation, 400 Seventh St., SW., Washington, D.C. 20590.

Hazardous Materials Regulations, 49 CFR, Parts 170 through 179.

Motor Carrier Safety Regulations, 49 CFR, Parts 393 and 397.

17

Fire Protection

THIS CHAPTER DISCUSSES MEASURES relating to safeguarding human life and preservation of property in the prevention, detection and extinguishment of fires. Identification of hazardous materials, vital to safe fire extinguishment, concludes this chapter.

Fire prevention is a term to indicate measures specifically directed toward avoiding the inception of fires.

Fire protection is usually understood to include fire prevention procedures, and in addition it includes fire detection and extinguishment. Fire prevention, detection, and extinguishment aim to protect employees and property, and to assure the continuity of operations. The information in this chapter is intended to develop an efficient fire protection program.

An effective fire loss control program must have as its objective the prevention of loss of life and personal injury, and the loss of property. Its primary purpose should be to prevent the inception of fire. If, nonetheless, a fire does start, it is vital:

1. To immediately detect the fire
2. To confine the fire
3. To extinguish the fire.

In general, this chapter will cover a fire protection program from the design of a structure through the development of a protection system and will include on-going management activities. It will discuss all parts of the fire protection chain—prevention, detection, attack, and extinguishment.

The important first step is a statement of fire safety objectives. Objectives might be stated in terms of safety to people and allowable downtime of the facility. Fire systems must exceed all codes, and be especially protective of high-priority areas—areas that are vital to the continuity of operations. Although fire codes aim to protect employees and property, they present minimum requirements. Actually, a greater measure of fire safety should be incorporated into building and process design. Architects and engineers must realize that fire conditions are a legitimate element of their design responsibilities. They must understand (1) the special loadings that fire puts on building elements and (2) the countermeasures that can be incorporated into their designs. The earlier that fire safety objectives are identified and design decisions are made, the more effective the fire protection system can be.

Here are some general facts about fire protection that must be kept in mind:

- No facility is absolutely fireproof. Nearly everything can burn—given ignition, adequate support, and sufficient oxygen.
- Heat (energy) is transmitted by convection, conduction, and radiation.
- Fire (and flame) will spread in a building both vertically and horizontally.
- The spread of the heat, smoke, and toxic gases is possibly the greatest single danger to life and takes place in much the same manner as the spread of fire. Smoke and toxic gases are responsible for 75 percent of fire deaths in buildings.
- On-site early detection of a fire is essential.
- The use to which a building is put (its occupancy) will influence the degree of fire hazard. The more hazardous a material, the more likely the chance for explosion.
- Contents of a facility are usually more important in fire origin (and continuance) than are the physical structures.
- Very often, there are only a few minutes between the beginning of combustion and the development of a destructive fire.
- What happens (or doesn't happen) in the first few minutes

FIRE-PREVENTION INSPECTION

INSTRUCTIONS TO INSPECTOR:	FILL OUT FORM WHILE MAKING INSPECTION. SEND COMPLETED FORM TO YOUR SUPERVISOR FOR NECESSARY ACTION. REPORT SHOULD BE HELD FOR REVIEW BY THE NEXT FACTORY MUTUAL ENGINEER.

PLANT	LOCATION	DATE

VALVE INSPECTIONS
INSPECT LOCKED VALVES AT LEAST MONTHLY.   INSPECT ALL UNLOCKED VALVES AT LEAST WEEKLY.

ALL INSIDE AND OUTSIDE VALVES CONTROLLING SPRINKLERS OR FIRE PROTECTION WATER SUPPLIES ARE LISTED BELOW. CHECK CONDITION OF VALVE AS FOUND. PHYSICALLY "TRY" GATE VALVES INCLUDING NONINDICATING AND INDICATOR POST GATE VALVES. DO NOT REPORT A VALVE OPEN UNLESS YOU PERSONALLY HAVE TRIED IT. FM APPROVED PIVA'S (POST-INDICATOR-VALVE ASSEMBLIES), IBV'S (INDICATING BUTTERFLY VALVES) AND STANDARD OUTSIDE SCREW & YOKE VALVES DO NOT HAVE TO BE TRIED BUT SHOULD BE VISUALLY CHECKED AT CLOSE RANGE.

NO.	VALVE LOCATION	AREA CONTROLLED	OPEN	SHUT	LOCKED	SEALED
1						
2						
3						
4						
5						
6						
7						
8						
9						
10						
11						
12						
13						
14						
15						
16						
17						
18						
19						
20						

THE FACTORY MUTUAL VALVE SHUT TAG SYSTEM IS USED TO GUARD AGAINST DELAYED REOPENING OF VALVES. FACTORY MUTUAL RED TAGS SHOULD BE USED EVERY TIME A SPRINKLER CONTROL VALVE IS CLOSED. WHEN THE VALVE IS REOPENED THE 2 INCH DRAIN SHOULD BE FLOWED WIDE OPEN TO BE SURE THERE IS NO OBSTRUCTION IN THE PIPING. THE VALVE SHOULD THEN BE RELOCKED.

WERE ANY VALVES OPERATED SINCE THE LAST INSPECTION	Yes ☐	No ☐
WERE FACTORY MUTUAL RED TAGS USED	Yes ☐	No ☐
WAS THE VALVE REOPENED FULLY AND A FULL FLOW 2 IN. DRAIN TEST MADE BEFORE THE VALVE WAS RELOCKED OR RESEALED	Yes ☐	No ☐

COMMENTS:

Figure 17-1. Typical inspection form for general industrial use. (Reprinted with permission from Factory Mutual System.)

INSPECT THESE ITEMS AT LEAST WEEKLY

S P R I N K L E R S	Auto-Matic Sprinklers	ANY HEADS DISCONNECTED OR NEEDED Yes ☐ No ☐				OBSTRUCTED BY HIGH PILING Yes ☐ No ☐		
		HEAT ADEQUATE TO PREVENT FREEZING (NOTE BROKEN WINDOWS, ETC.) Yes ☐ No ☐			Water Pressure	LB. AT YARD LEVEL		
	COMMENTS							

DRY PIPE VALVES	VALVE ROOM PROPERLY HEATED	No. 1 Yes ☐ No ☐	No. 2 Yes ☐ No ☐	No. 3 Yes ☐ No ☐	No. 4 Yes ☐ No ☐	No. 5 Yes ☐ No ☐	No. 6 Yes ☐ No ☐	No. 7 Yes ☐ No ☐	No. 8 Yes ☐ No ☐
	AIR PRESSURE	No. 1 Lbs	No. 2 Lbs.	No. 3 Lbs.	No. 4 Lbs.	No. 5 Lbs.	No. 6 Lbs.	No. 7 Lbs.	No. 8 Lbs.

W A T E R S U P P L I E S	FIRE PUMP	TURNED OVER Yes ☐ No ☐	GOOD CONDITION Yes ☐ No ☐
		AUTO. CONTROL TESTED Yes ☐ No ☐	FUEL TANK FULL Yes ☐ No ☐
		PUMP ROOM PROPERLY HEATED AND VENTILATED Yes ☐ No ☐	PRIMING TANK FULL Yes ☐ No ☐
	TANK OR RESERVOIR	FULL Yes ☐ No ☐	HEATING SYSTEM IN USE Yes ☐ No ☐
		TEMPERATURE AT COLD WATER RETURN (SHOULD BE 42°F MINIMUM)	CIRCULATION GOOD Yes ☐ No ☐

MFL WALL FIRE DOORS	CONDITION	OBSTRUCTED Yes ☐ No ☐	BLOCKED OPEN Yes ☐ No ☐

OTHER ITEMS

INSPECT THESE ITEMS AT LEAST MONTHLY

M A N U A L P R O T	EXTIN-GUISHERS	CHARGED Yes ☐ No ☐	ANY MISSING	ACCESSIBLE Yes ☐ No ☐	ATTENTION NEEDED (Give Location)
	INSIDE HOSE	IN GOOD CONDITION Yes ☐ No ☐		ACCESSIBLE	Yes ☐ No ☐
	YARD HYDRANTS & HOSE	CONDITION NO. 1 NO. 2	NO. 3 NO. 4	NO. 5 NO. 6	
		HYDRANTS DRAINED Yes ☐ No ☐	REMARKS:		

O C C U P A N C Y	GENERAL ORDER & NEATNESS	GOOD Yes ☐ No ☐	COMBUSTIBLE WASTE REMOVED ON SCHEDULE (PROMPTLY) Yes ☐ No ☐
			COMBUSTIBLE DUST, LINT OR OIL DEPOSITS ON CEILINGS, BEAMS OR MACHINES Yes ☐ No ☐
	ELECT. EQUIP.	DEFECTS NOTED Yes ☐ No ☐	DESCRIBE AREAS NEEDING ATTENTION INCLUDING YARD:
		SAFETY CANS USED Yes ☐ No ☐	
	FLAM. LIQUIDS	EXCESSIVE IN MFG AREAS Yes ☐ No ☐	DRAINAGE OBSTRUCTED Yes ☐ No ☐ VENT FANS ON Yes ☐ No ☐
	SMOKING REGULA-TIONS	LOCATIONS WHERE VIOLATIONS NOTED	
	CUTTING & WELDING	PERMITS ISSUED FOR ALL C&W OPERATIONS Yes ☐ No ☐	LISTED PRECAUTIONS TAKEN Yes ☐ No ☐
	STORAGE	WELL ARRANGED Yes ☐ No ☐	AISLES CLEAR Yes ☐ No ☐
		ADEQUATE SPACE BELOW SPRINKLERS Yes ☐ No ☐	CLEAR OF LAMPS, HEATERS Yes ☐ No ☐

DOORS AT CUT-OFF WALLS	CONDITION	OBSTRUCTED Yes ☐ No ☐	BLOCKED OPEN Yes ☐ No ☐

Sprinkler Alarms	TESTED Yes ☐ No ☐	OPERATION SATISFACTORY (IF "NO" - COMMENT BELOW) Yes ☐ No ☐

OTHER ITEMS

INSPECTED BY:		DATE
REVIEWED BY:	TITLE	DATE

SUBJECT: HOT-WORK PERMIT PROCEDURES
TO: ALL CONCERNED

Steps will be taken immediately to put into practice a permit tag system which should provide better protection against fire from welding and other hot work in hazardous areas. This program, as outlined below, is intended to be a practical one. It must be realized by everyone that for it to be effective, the wholehearted cooperation of all concerned must be secured.

1. After an inspection of the entire plant has been made by the fire chief and after a discussion with the various department heads, areas throughout the plant will be designated as hazardous for *any type of welding, burning, spark, open flame, or hot work.* Such areas will be prominently marked, and before hot work is done within any such area, permit tags must be secured in order to help ensure that the area will be as free as possible from fire hazards and that proper precautions will have been taken.

2. Tags have been prepared on which pertinent information must be filled out by the parties concerned. Each employee who may do hot work will be given a supply of tags to keep with equipment. When sent to perform work in a hazardous area, the employee and the immediate foreman will check the area together to determine if necessary precautions to prevent fire have been taken.

3. It will be the maintenance foreman's responsibility to notify the foreman in the department in which the work is to be done, and together they will sign the permit for the employee to do welding or other hot work.

4. The fire chief will then check the area for firesafe working conditions, see that standby fire extinguishers are present, and assign a fire watcher when necessary. When satisfied with the precautionary measures, the fire chief will sign the permit tag and return it to the employee who is to do hot work.

5. The signed permit card will be kept by the employee doing the work until the job is completed, at which time the employee will personally check the area for fire.

6. The area will subsequently be checked by the maintenance foreman who will see that any extinguisher that may have been used is designated for recharging.

7. The completed tag will be turned in at the end of the day to the maintenance department which will collect and forward all tags to the fire chief for filing and record purposes.

Figure 17-2. Example of policy statement that can be issued by a company inaugurating a hot-work permit program in order to instruct employees.

of a fire pretty well determines if it can be controlled or not.
- A fire is usually (1) controlled by built-in equipment and/or (2) put out by firefighters.
- Every firesafety protection device will be a compromise. A fire protection system will always represent some trade-off involving cost, reliability, and safety. Some risk will be assumed and there will be some loss. The optimum level of fire protection is that which minimizes the sum of cost and expected fire losses. (See the section describing Fire Risks.)
- The cost of fire protection should have a corresponding effect in reducing the amount of loss or risk involved.
- Planning for fire action is essential. Emergency planning involves basic concepts of action and reaction.
- Reliability of the system involving life safety should surpass 99.9 percent.
- An automatic sprinkler system is the best tool to reduce loss of life from fire.
- An absolute total life-safety system is not achievable.
- People and their actions are key elements. More than half of all fire losses are the result of human element problems—such as inadequate training, insufficient motivation, or improper action.

- Construction alone is not adequate protection insofar as life safety is concerned.

Fire protection engineering is a highly developed, specialized field in which special engineering disciplines are focused. The solution of many fire protection problems requires the special combination of training and perspective of an experienced fire protection engineer.

The safety professional who is faced with special fire problems should seek specific advice from a fire protection engineering consultant—names of experts can be obtained from the National Fire Protection Association (NFPA) and the Society of Fire Protection Engineers.

If no such specialist is available in the organization, the person responsible for overall industrial safety should be familiar with the sources of this information, if not also with the technology. Authoritative fire protection literature is available from such organizations as the National Fire Protection Association (NFPA), the American Insurance Association, and Factory Mutual Engineering Corporation. See the chapter on "Sources of Help" in the *Administration and Programs* volume, and the

NFPA *Fire Protection Handbook*. Also see the References at the end of this chapter.

To achieve the most efficient fire protection system requires the involvement of the architect, interior designer, urban planner, building contractor, electrical and structural engineers, fire-detection-system manufacturers, building safety engineer, and the local fire marshal.

General information can be obtained from trade associations, the fire insurance carrier, the local fire inspection bureau, the fire department, and governmental agencies having jurisdiction.

FIRE PREVENTION ACTIVITIES

This section dicusses the general fire protection and prevention guidelines that should be part of all ongoing plant loss prevention programs.

Complete steps to be taken for the prevention of fires are extensive. Be sure to refer to the codes and standards of the NFPA, as well as local building codes, in order to maintain the utmost in firesafety at new and existing installations.

As a first step, all companies and plants should hold design review meetings for all new construction and changes in process design. Every establishment should institute procedures and regulations that assure sufficient fire extinguishing equipment is on hand and that personnel are organized and trained to use the equipment correctly.

Regularly scheduled programs covering thorough inspection of all fire protection equipment are necessary. Inspections should be scheduled at regular intervals and a written record should be maintained.

Every plant should establish definite procedures to assure prompt and effective action by all employees at the time of a fire.

Fire prevention includes those activities specially directed toward preventing fire inception. Generally, these activities include inspection, fire drills, training, certain management procedures, and communications.

The management is the final authority in the operation and firesafety of the establishment or plant. Management determines the effectiveness of the firesafety prevention program.

Inspections

A system of periodic fire inspections should be set up for every operation. Some facilities, operations, and processes require daily inspection, while others can be inspected weekly, monthly, or at another interval as may be determined. Even if buildings are well designed and provided with protective devices and construction elements intended to render firesafety features, a periodic, detailed inspection program is needed to reasonably assure that they will maintain their intended value.

In addition to inspections made by insurance companies and by fire protection bureaus of fire departments, every industrial plant should include periodic self-inspection in its firesafety program. The function of these firesafety inspections is to check for proper placement and operation of fire-protection equipment and seek correction of common fire causes, such as poor housekeeping, improper storage of flammable materials, smoking violations, and excessive accumulations of dust or flammable material. Those who make the inspections need to be trained in inspection procedures and techniques.

In some establishments or plants, the responsibility for locating and reporting fire hazards is entrusted to the safety committee or, in larger plants, to one of its special subcommittees or to a person trained to manage fire risk.

The inspector, fire chief, or other individual in charge of fire prevention and protection should establish inspection schedules, determine the routing of reports, and have a complete list of all the items that should be inspected at regular intervals. Fire equipment inspection might cover at least the following items:

1. Control valves on piping that supplies water for fire protection
2. Hydrants
3. Fire pumps
4. Hose houses and associated equipment
5. Sprinkler system water supplies, including tanks
6. Sprinker heads
7. Special types of protection (carbon dioxide, foam, or other automatic systems)
8. Portable fire extinguishers
9. Fire doors, aisleways, exits
10. Special hazards and operation processes
11. Detectors
12. Control room or panel checkout
13. Alarm and communication system and routines within plant
14. Communication to fire department (or other mutual aids).

In addition to fire equipment, electrical equipment, machinery, and processing equipment, housekeeping conditions and other fire causes should be checked at regular intervals. Checklists can be used effectively. A typical inspection form for general industrial use is shown in Figure 17-1. It can be adapted for use in other types of properties. (The *Administration and Programs* volume covers inspections for housekeeping and hazards to safety.)

Special inspections should be conducted during and following any alterations in the plant or in a process. A complete seasonal check, done early enough for prompt replacement or repair, should be made of equipment which will be or has been exposed to freezing temperatures.

If the plant does not have its own fire protection expert, it is common practice to invite the local fire chief or marshal to inspect the plant facilities, fire equipment available, the fire hazardous materials used, their locations, and water supply available. It would be a good idea to invite the local fire chief to check how the municipal department can help in your specialized situation.

For greatest benefit, a written record is essential. To facilitate the inspection, the form should fit the conditions of the individual plant and be complete enough to make sure no part of the system is overlooked. It will also assist the inspector in making comments and bringing recommendations to the attention of those responsible. (An excellent guide to conducting inspections is the NFPA *Inspection Manual;* see References.)

Hot-work permits

In an effort to establish some control over operations using flames or producing sparks, many industrial firms have instituted hot-work permit programs, which require that authorization be secured before equipment capable of igniting combustible materials is used outside areas normally specified.

The first step in inaugurating a hot-work permit program is development of a policy statement by management, such as that given in Figure 17-2. The type and extent of the program will depend upon the size of the plant or facility, the complexity of the operations, and the degree of hazard present at the worksite and in surrounding areas.

Salient features of the program are to:

1. Inspect area where work is to be done
2. Establish fire watches, if hazards warrant
3. Provide fire extinguishing equipment, usually manned by a standby employee
4. Communicate with and coordinate all departments concerned
5. Isolate combustibles from sources of ignition
6. Limit unauthorized use of flame- or spark-producing equipment

A hot-work permit form or tag is generally used to administer the program. Although standard forms are available through insurance companies, many plants have developed special forms that specifically relate to their individual operation.

Flame-retardant treatments

The burning characteristics of a number of materials can be altered. A flame-retardant treatment that impregnates or coats a material will affect the ignition and/or spread-of-fire characteristics. Sometimes the treatment will also change other characteristics of the material. Unfortunately, the treatments are temporary.

Many substances, such as wood, many wood products, textiles, some plastics, and paper, can be successfuly made flame retardant. Even a Christmas tree can be sprayed (coated) to change its burning characteristics.

Manufacturers of fire-retardant treatments and products can supply a fire rating which can be used in comparison to the non-treated product. Only fire retardants listed by Underwriters Laboratories Inc. should be used.

Flame-retardant fabrics. Fabrics cannot be made noncombustible, nor even resistant to charring or to decomposition, but chemical treatment will reduce their flammability. (See NFPA 701, *Standard Methods of Fire Tests for Flame-Resistant Textiles and Films.*) Some treatments merely inhibit the rapid spread of flame, whereas other prevent flames and depress dangerous afterglow.

Where the only anticipated exposure is to small sparks, small flames, or temperatures up to 400 to 500 F (200 to 260 C), flame-retardant canvas is often preferred to asbestos or chrome-tanned leather, particularly where flexibility, durability, strength, and resistance to abrasion are required. Asbestos or chrome leather is required for more serious exposures, such as heavy welding, firefighting, or foundary work.

Several commercially treated fabrics are listed by Underwriters Laboratories Inc. Samples of treated materials should be tested periodically after each application to assure satisfactory performance.

Fire and emergency drills

Fire emergency planning is not an exact science. A practical emergency plan develops from an evaluation of the particular situation at hand. Often an emergency manual can be prepared to outline procedures and drills, and detail the responsibilities of each individual involved. The prevention of personal injury and loss of life should always be of prime importance.

Training employees to leave their workplace promptly on proper signal, and to evacuate a building speedily but without confusion, is largely accomplished through drills. Exit drills must be carefully planned and carried out periodically in a serious manner under rigid discipline. Up-to-date instruction sheets, including evacuation routes, should be posted and distributed to all employees, in conformance with the requirements of NFPA 101, *Life Safety Code.* Regardless of what method is used to alert employees, they should thoroughly understand what the signal means and how to respond in the safest possible manner.

Fire drills conducted at frequent intervals also demonstrate management's concern and sincere interest in all fire prevention activities. The drills should serve as a reminder to employees and supervisors that all fire prevention practices are important.

Emergency exit drills also serve as a valuable check on the adequacy and condition of the exits and alarm system. Any deficiency must be immediately and permanently corrected. Careful plans must be made to eliminate the possibility of panic in the event of an emergency regardless of cause, and to guarantee the smooth functioning of the emergency plan.

Fire brigades

Management cannot always depend wholly on automatic fire protection equipment, public fire departments, and/or mutual aid agreements to prevent fire losses. Because fires can get out of control before a municipal fire department could arrive, it is essential that a well-trained fire brigade be available to fight a fire as soon as it is discovered.

A brigade chief should be assigned. The qualifications should include having training and experience in fire protection and prevention, having ability to organize, and being a confident and capable fire prevention and firefighting instructor. Because fire protection demands competent and experienced leadership 24 hours a day, an assistant brigade chief or captain (one for each shift) is required. The assistant brigade chief or captain assignment demands the same qualifications indicated for the brigade chief.

A centrally organized plant fire brigade should be required for each plant. The fire brigade should be under the direction of the plant brigade chief or the brigade captain. Brigade members should be regular plant employees from all departments, thoroughly trained in using the plant firefighting equipment. Electricians, engineers, mechanics, safety and fire inspectors all have special skills which make them valuable members of the plant emergency team.

Regularly scheduled training of all members of the departmental unit and the emergency brigade should be conducted by the brigade chief and assistant brigade chief. Training sessions of a least one-hour duration per week are recommended. Continuous programs instructing every employee on proper procedures to avoid fire, both on and off the job, should be established.

The municipal fire department must be consulted when plans for a fire brigade are being considered. Go over with them what the in-plant hazards are, what outside help is available, and how the public fire department can help the brigade in other ways, such as training.

Protecting other buildings

When a fire breaks out in a building, adjacent buildings should be protected by (1) closing every window facing the burning building (2) stationing personnel with fire extinguishers or fire hoses at each window nearest the fire, and (3) stationing firefighters on the roof of the exposed building with hose lines to keep the roof wetted down, and with extinguishers to put out any burning embers.

Communications

Good communications are necessary once a fire has been detected, and, especially, in a disaster situation—first as a means of alerting occupants to the emergency, and second as a way to mobilize fire protection forces, be they a plant brigade, municipal fire department, or both. A coded fire alarm system, with alarm boxes and bells, horns, or other sounding devices suitably situated, is usually needed, except in very small plants, where a steam whistle or similar device might be adequate. In any case, the alarm system is no better than the level of training given employees in how to respond when the alarm is sounded. (See the discussion later in this chapter under Fire Detection.)

THE CHEMISTRY OF FIRE

Fire, the process of combustion, is extraordinarily complex. Combustion is an exothermic process (a release of heat energy), a self-catalyzed reaction involving either a condensed-phase fuel, a gas-phase fuel, or both. The process is usually associated with

Figure 17-3. "The fire pyramid." Oxygen, heat, fuel, and chain reactions are necessary components of a fire. Speed up the process and an explosion results.

rapid oxidation of a fuel by oxygen in the air. If the combustion process is confined so pressure can increase, it can result in an explosion. A similar process that takes place over long periods of time and at a lower temperature is called *oxidation*—rusting of metal is an example. A fire, then, is a combustion process intense enough to emit heat and light.

As shown pictorially in Figure 17-3, the process involves four basic needs for a fire to occur—fuel, oxygen, heat, and a chain reaction. Each of these four join in a symbiotic relationship.

In addition, a fire can be classified into two general forms or modes.
1. Flames. Direct burning of a gaseous (vaporized) fuel; this includes deflagrations. The rate of burning is usually high and a high temperature is produced. There are two types:
   a. Premixed flames. This condition exists in a gas burner or stove and is relatively controlled.
   b. Diffusion flames. This refers to gases burning on mixed vapors and air; control is difficult.
2. Surfaces burning. Fire occurring on the surfaces of a solid fuel is often called a "glow" or "deep-embered seated" fire; it takes place at the same temperature as open flames.

The surface fire is represented by the fire triangle—heat, fuel, and air, but *no* chain reaction. The flame fire is more accurately shown by the tetrahedron, which includes a chemical reaction. The two modes are not mutually exclusive and they may occur together or alone.

Control of fires

Knowing how and why a fire burns suggests ways to control and extinguish it. The surface fire has three components that can be controlled, while the gaseous flame has four components: heat can be taken away by cooling, oxygen can be taken away by excluding the air, fuel can be removed to an area where there is insufficient heat for ignition, and the chemical reaction (of the flame fire) can be interrupted by inhibiting the rapid oxidation of the fuel and the concomitant production of free radicals, the lifeblood of the flame reaction. They are described in the next column in the paragraph Interrupting the chain reaction.

Cooling. In order to extinguish a fire by cooling, it is necessary to remove heat at a greater rate than the total heat is being evolved by the fire; to do this the cooling agent must reach the burning fuel directly. The cooling action may also stop the release of combustible vapors and gases. The most common and practical agent is water applied in the form of a solid stream or spray, or incorporated in foam. In practice, the fire is literally drowned into submission by water.

Removing fuel. Often, taking the fuel away from a fire is not only difficult but dangerous; fortunately, there are exceptions. (1) Flammable liquid storage tanks may be arranged so their contents can be pumped to an isolated empty tank in case of fire. (2) When flammable gases catch fire as they are flowing from a pipe, the fire will go out if the fuel supply can be shut off. (3) In any mixture of fuel gases or vapors in air, adding an excess of air has the effect of diluting the fuel concentration below the minimum combustible concentration point.

Limiting oxygen. Extinguishment by separation of oxygen from fire can be accomplished through smothering the burning area with a noncombustible material, such as covering with a wet blanket (make sure the blanket is not made of highly combustible fibers), throwing dirt or sand on the fire, smothering it with inert gas, or covering it with a chemical or mechanical foam.

To be effective, the smothering blanket must be maintained long enough for all smoldering ignition to be extinguished. Further, smothering is ineffective on substances containing their own oxygen supply, such as ammonium nitrate or nitrocellulose.

Covering a fuel, however, can stop a fire. Many foams and some solids serve as an emulsion film or cover on the burning fuel, thus extinguishing the fire. If the contents of a wastebasket catch on fire, one can drop an empty wastebasket on top of it to smother the fire.

Smothering is ineffective on deep-seated materials like wood and rags and large rolls or skids of paper.

Extinguishment by diluting oxygen below the concentration necessary to support combustion is accomplished by discharging carbon dioxide or other inert agents into the fire. The fire will remain out (1) if the percentage of oxygen is reduced below the level of combustion for a sufficient period of time to allow the combustible materials to cool below their ignition temperature and (2) if no ignition sources are present. An inert gas can be used to purge operations involving flammable vapors and dusts in confined space where a source of ignition may exist. However, the flow of inert gas and/or the actual concentration of oxygen should be constantly monitored to prove the oxygen concentration remains low enough to prevent combustion.

Interrupting the chain reaction. In analyzing the anatomy of a fire, the original fuel molecules appear to combine with oxygen in a series of successive intermediate stages, called *branched-chain reactions,* in arriving at the final end products of combustion. It is these intermediate stages that are responsible for the evolution of flames.

As molecules fragmentize in these branched-chain reactions, unstable intermediate products called *free radicals* are formed. The concentration of free radicals is the determining factor of flame speed. The life of the free hydroxl radical ($-OH$) is very short, being in the order of 0.001 second, but long enough to be of vital importance in the combustion of fuel gases. The almost simultaneous formation and consumption of free radicals appears to be the lifeblood of the flame reaction.

It is the free radicals in these branched-chain reactions which are removed from their normal function as a chain carrier by dry chemical and halogenated hydrocarbon extinguishing agents. The effects that various dry chemical agents (sodium bicarbonate-base, potassium bicarbonate-base, and ammonium phosphate-base, and others) have on capturing free radicals depend upon their individual molecular structure. Potassium bicarbonate dry chemical is the most effective because of the large size of the potassium ion. In the case of halogenated agents, it is believed that they decompose when discharged into the fire and form free radicals of halogens (chlorine, bromine, or fluorine) that unite with the free radicals evolved in the branched-chain reaction. Again, the large halogen molecule proves an effective trap.

Some extinguishing agents do not perform according to only one of these four mechanisms. For example, both plain water fog (as compared with straight water streams) and carbon dioxide have the ability at flame temperatures to react with relatively slow-burning free carbon, producing carbon monoxide, with a resulting decrease in black-smoke production. Because the reactions are endothermic, they adversely affect the thermal balance of the fire as well as lowering the oxygen concentration. The examples use a chemical attack, but it is definitely not the chain-reaction-breaking phenomena just described.

To match the pace of newer and more potent fire extinguishing agents, more sophisticated tactics and techniques will be called for. A fire can be attacked from at least four different standpoints, but the use of any one of them does not necessarily result in the most rapid extinguishing time. A fire can be attacked with more than one agent to produce a synergistic effect. (This is done now in the joint use of "light water" and Purple K in aircraft-crash firefighting.)

Lest the reader feel that fire extinguishment sounds simple, rest assured that it is still very far from being an exact science. A fire is usually composed of more than one source of fuel, which will be fought with several different extinguishments, that, by working together, complement each other. Keep this in mind as you read through this chapter.

The four stages of fire

Most fires develop in four distinct stages—incipient, smoldering, flame, and heat. Detectors are available for each stage (see Fire Detection later in this chapter).

Incipient stage. No visible smoke, flame, or significant heat is developed, but a significant amount of combustion particles are generated over a period of time. These particles, created by chemical decomposition, have weight and mass, but are too small to be visible to the human eye. They behave according to gas laws and quickly rise to the ceiling. This stage can develop quickly or slowly over a period of minutes, hours, or even days. As discussed later in the section on Fire Detection, ionization detectors respond to these particles.

Smoldering stage. As fire development continues, the quantity of combustion particles increases to the point where they become visible—this is called "smoke." There is still no flame or significant heat developed. Photoelectric detectors "see" visible smoke.

Flame stage. As the fire develops further, the point of ignitioin occurs and flames start. The level of visible smoke decreases and the heat level increases. Infrared energy is given off by radiation to distant locations; this can be picked up by infrared detectors.

Heat stage. At this point, large amounts of heat, flame, smoke, and toxic gases are produced. The transition from the flame to the heat stage usually develops very quickly, as does the heat stage itself. Thermal detectors respond to heat energy.

CLASSIFICATION OF FIRES

Four general classifications of fires have been adopted by the National Fire Protection Association based on the types of combustibles and the extinguishing media necessary to combat each. (See NFPA 10, *Portable Fire Extinguishers.*)

Class A fires

Class A fires are those that occur in ordinary materials such as wood, paper, excelsior, rags, and rubbish. The quenching and cooling effects of water or of solutions containing large percentages of water are of first importance in extinguishing these fires. Special dry chemical agents (multipurpose dry chemicals) provide rapid knockdown of the flames and the formation of a coating that tends to retard further combustion. Where total extinguishment is mandatory, a follow-up with water is recommended.

Class B fires

Class B fires are those that occur in the vapor-air mixture over the surface of flammable liquids such as gasoline, oil, grease, paints, and thinners. The limiting of air (oxygen) or the combustion-inhibiting effect is of primary importance to stop incipient fires of this class. Solid streams of water are likely to spread the fire, but under certain circumstances water fog nozzles may prove effective in the control, but not the extinguishment. Generally, regular dry chemical, multipurpose dry chemical, carbon dioxide, foam, or halogenated agents are used.

Class C fires

Fires that occur in or near energized electrical equipment where nonconducting extinguishing agents must be used are called Class C fires. Dry chemical, carbon dioxide, or halogenated extinguishing agents are suitable.

Foam or a stream of water should not be used because both are good conductors of electricity and can expose the operators to a severe shock hazard. Water from a very fine spray can sometimes be used on fires in electrical equipment, as in transformers, since a spray is a poorer electrical conductor than a solid stream of water.

Class D fires

Fires that occur in combustible metals such as magnesium, titanium, zirconium, lithium, potassium, and sodium are classified under Class D. Specialized techniques, extinguishing agents, and extinguishing equipment have been developed to control and extinguish fires of this type. Normal extinguishing agents generally should not be used on metal fires due to the danger, in most cases, of increasing the intensity of the fire because of a chemical reaction between some extinguishing agents and the burning metal.

Other fires

Fires that involve certain combustible metals or reactive chemicals require, in some cases, special extinguishing agents or techniques. See NFPA 49 and 325M, *Hazardous Chemicals Data* and *Fire-Hazard Properties of Flammable Liquids, Gases, and Volatile Solids.*

FIRE RISKS

Fire protection measures can be effective only if they are based on a proper analysis and evaluation of the fire risk. A complete evaluation is important, since there is a wide variety of methods and equipment to provide protection. The optimum level of fire protection is that which minimizes both the costs and the expected fire losses. As safety to life is also involved, a value must likewise be placed on saving life. Ideally, the cost of fire protection should have a concomitant effect upon the reduction in the amount of loss exposure or risk involved. In time, production methods, hazards, and fire protection requirements may change;

these changed variables should be assessed periodically to assure continued optimum risk-to-protection relationship.

The fire risk evaluation process can be covered in three major groups:
1. Recognition of hazards and potential hazards
2. Evaluation of hazards and expected losses
3. Evaluation of the proposed countermeasures.

Fire hazards

The information assembled during the fire risk survey (which goes beyond the scope of the fire prevention inspection outlined earlier in this chapter) will serve as an excellent training resource for operating and emergency personnel and as a reference when modifications are made or for future design of similar facilities. Therefore, it is important to develop a format and list of references for organizing a systematic fire hazard survey.

Fire hazard analysis

The fire hazard analysis considers a number of points, presented here in outline form:
1. *Site*
   Location
   Accessibility for firefighting equipment (if not self-contained)
   Exposure—possibility
   Water supply available—amount
   Traffic—distance to other buildings, their type, and other characteristics
2. *Building construction*
   Framework
   Material fire-resistance rating
   Roof materials
   Interior walls—windows
   Exterior wall material
   Floor
   Interior finish
   Shape
   Heat, ventilating—air conditioning system
   Uses
   Concealed spaces
   Exterior doors and exits
   Elevators, stairways, chutes
   Fuel sources for heating or process operations
   Storage areas
   Raw stock
   Finish goods
   Electrical systems
3. *Building contents*
   Material
   Liquids—flammability, amounts, kind
   Solid material—amounts and kind
   Location of materials
   Basic processes
4. *Management factors*
   Design for ease in housekeeping
   Employee smoking policy
   Overall cleanliness
5. *People factor* (number, characteristics)
   Location in structure
   Exits—number, markings, accessibilitiy
   Activity of people—work, play, etc.
   Location of patrons, visitors, and employees

6. *Fire protection system* (if for an existing facility)
   Detection system
   Alarm system
   Communication
   Compartmentation
   Extinguishing means
   People problems
7. *After the fire*
   Follow-up/clean-up
   Emergency plan to keep the company, plant, or operation running

Evaluation of fire hazards

Evaluation of fire hazards should be based on the hazards and risk survey. Every time there is a change in the processes or operation or in the use of a facility, some evaluation should be made of the fire protection system, the degree varying with the degree of change.

The dollar value of potential fire losses should be determined. But other factors that must be considered may involve the loss of a key facility or operation, in which case the loss to the company may be greater than the actual dollar loss.

There are means of converting general fire hazards to some predetermined relative hazard scale, which then can be further matched based on some assumed levels of fire hazards or risks. However, there are many fire risk situations that do not clearly fit predetermined assessments.

An evaluation process can take a number of different directions or follow a number of procedures. The suggestions that are listed next are intended only to help determine whether a new fire-protection system or some modification of an existing one is more feasible. Some general principles or things to search for include:

- Fires usually start in contents or in operation rather than in the structure. Therefore ask:
  1. What materials are flammable?
  2. What materials, in some process or operation, are most likely to burn? ignite? explode?
  3. What is in the facility that can be a source of ignition? Are there some open sparks or flames present? Are there high temperatures involved in any operations?
  4. Where are flammable and combustible materials located? Are flammable materials brought or stored together? Are there indirect connections? If one should burn, could others easily ignite?
  5. Are there materials that might ignite due to convection or radiation?
  6. What toxic gases might evolve in a fire?
  7. How much time might it take for a fire to spread to other areas? To adjacent facilities?
- Smoke and toxic gases (and sometimes heat) are largely responsible for fire deaths. Determine:
  1. What toxic gases might evolve from the burning or smothering of contents?
  2. How many people are likely to be closely involved, or in adjacent facilities, or nearby in some other facility?
- Fire load relates to the total heat potential. In some analyses, the amount of fire load is also determined. Fire loads express the weight of combustible material per square foot ($0.09m^2$) of fire area. Paper and wood have a caloric value of 7,000-8,000 Btu per lb (16-20 MJ/kg). A typical office is likely to

have about 5 lb per square foot ($2.5 kg/m^2$). A flammable liquid has a heat-producing potential of about 14-15,000 Btu per lb (about 35 MJ/kg). Fire load does not account for the rate of heat liberation, nor for the distribution within the structure, and it is only a guide. The fire load can be used to develop some elements of the fire protection system.

Exterior and interior finish materials of the building structure can have a fire load factor. (For further information, see the next section and the NFPA *Fire Protection Handbook.*) Many other factors including ventilation system, building shape, degree or openness, and amount of compartmentation affect the fire load.

A review of causes of fire in a particular industry or operation can be of value in the analysis of hazard.

With some knowledge of the fire hazards in the facility and some ideas of what might happen in case of a fire, it is time to start planning the fire-protection system. Planning starts before a building or plant is built or remodeled; it should influence site planning, design, and construction, which are the subjects next discussed in this chapter.

FIRESAFE BUILDING DESIGN AND SITE PLANNING

This section is taken from *Principles of Fire Protection,* by Percy Bugbee, copyright 1978 by National Fire Protection Association, Boston, Mass. It is used with permission.

The object of firesafety is to protect life foremost and property secondly from the ravages of fire in a building. Building design and construction must take into account a wide range of firesafety features. Not only must the interiors and contents of buildings be protected from the dangers of fire, but the building site itself must have adequate water supplies and easy accessibility by the fire department. Architects, builders, and owners may assume that state codes provide adequate measures; however, these codes stipulate only minimal measures for firesafety. Planning and construction based upon such codes may limit firesafe design seriously.

Objectives of firesafety design

Before a building designer can make effective decisions relating to firesafety design, the specific function of the building and the general and unique conditions that are to be incorporated into it must be clearly identified. Decisions regarding the firesafety design and construction of the buidling have the same objectives as do all fire protection measures, namely (1) life safety, (2) continuity of operations, (3) property protection.

The art of probing to identify objectives is an important design function. The degree of risk that will be tolerated by the owner and the occupants is a difficult design decision; consequently, it should be identified in a clear, concise manner so that the designer can properly realize the design objectives.

Life safety. Design considerations for life safety must address two major questions: (1) Who will use the building? (2) What will the people using the building do most of the time?

The occupied building provides a great potential for fire because of the presence of large numbers of people, any one of whom could perform a careless or malicious act resulting in fire. Appliances and mechanical or electrical equipment are a potential hazard through misuse, failure, faulty construction, or substandard installation. Accumulations of combustibles, either waiting for disposal or in storage, frequently provide a ready means by which otherwise controllable fires could spread.

The identification of specific functional patterns, constraints, and handicaps is vital in designing specific fire protection features that recognize occupant conditions and activities.

Continuity of operations. Continuity of operations, the third major area of building design decision-making, must take into consideration those specific functions conducted in a building that are vital to continued operation of the business and that cannot be transferred to another location. In this regard, the owner must identify for the designer the amount of "downtime," or the amount of time an operation can be suspended without completely suspending total operations. The degree of protection required in firesafe building design varies with the number and scope of vital operations that are nontransferable.

Property protection. One of the most important questions to be asked about the design of buildings with regard to protection of property is: Is there any specific high-value content that will need special design protection? The requirements with regard to protection of property within a building are often fairly easy to identify. Materials of high value that are particularly susceptible to fire and/or water damage can usually be identified in advance of building design. For example, vital records that cannot be replaced easily or quickly can be identified in advance as needing special fire protection design considerations.

Fire hazards in buildings

When the designer and the owner either consciously or unconsciously overlook or ignore the possibility of fire in the building to be built, the building and its occupants are endangered. The broad approach to the firesafe design of a building requires a clear understanding of the building's function, the number and kinds of people who will be using it, and the kinds of things they will be doing. In addition, appropriate construction and protection features must be provided for the protection of the contents and, particularly for mercantile and industrial buildings, to assure the continuity of operations if a fire should occur. Too many fires disastrous to people and to property have occurred, and will continue to occur, because no one has given proper consideration to the threat of potential fire.

Smoke and gas. Studies of fire deaths in buildings indicate that about 75 percent of these deaths are due to the smoke and toxic gases that evolve as products of the fire. About 25 percent of the deaths result from heat or contact with direct flame. The carbon monoxide developed in many fires, particularly unventilated and smoldering fires, is probably the most common cause of death. Carbon monoxide can be neither seen nor smelled. Exposure to this gas, even in small quantities, can cause impaired mental behavior. When inhaled in large quantities, the smoke given off in most fires can lead to pneumonia and other lung troubles. Smoke also obscures visibility and thus can lead to panic situations when occupants cannot see and use escape routes.

Heat and flames. As has already been stated, heat and flames account for 25 percent of fire deaths. Although heat and flame injuries are much fewer than those caused by smoke and toxic

gases, the pain and disfigurement caused by burns can also result in serious, long-term complications.

Building elements and contents. If the building on fire has combustible furniture, flames and toxic gases may spread so rapidly that occupants may not be able to escape. Poor construction practices, such as failure to protect shafts and other vertical openings, make the vertical spread of fire more rapid and the work of firefighters more difficult.

Although the collapse of structural elements has not resulted in many deaths or injuries to building occupants, it is a particular hazard to firefighters. A number of deaths and serious injuries to firefighters occur each year because of structural failure. While some of these failures result from inherent weaknesses, many are the result of renovations to existing buildings that materially, though not obviously, affect the structural integrity of the support elements. A building should not contain surprises of this type for firefighters.

Elements of building firesafety

The firesafety of a building will depend first on what is done to prevent a fire from starting in the building, and second on what is done through design, construction, and good management to minimize the spread of fire if it happens. Good housekeeping is one of the major factors in both fire prevention and control. Keeping the fuel load down not only lessens the amount of material that can be ignited, but it also provides less material that can be consumed if a fire breaks out.

Once a fire has started, its spread will depend on the design of the building, the materials used in construction, building contents, methods of ventilation, detection and alarm facilities, and fire suppression systems, if any. Table 17-A describes the building design and construction features that influence safety. These elements are within the decision-making authority of various members of the design team, based on the assumption that their firesafety objectives are clearly defined by management, the owners, or other responsible parties, both public and private. The design and construction elements are organized in a manner that can give a quick overview of the major aspects that must be considered for firesafety. They show features that include both passive and active design and construction considerations. (A passive design element is one which requires no action to function, such as a fire wall. An active design or construction element is one which requires an action in order to function, such as a fire door which must be closed.)

The persons responsible for fire prevention are not the same ones responsible for the building design. Table 17-B describes the elements that comprise firesafety from a prevention consideration. Decisions concerning these elements are predominantly under the control of the building owner or occupant, or both. Table 17-B includes the elements of emergency preparedness in case of fire that are the responsibility of the owner and/or occupant.

Firesafety planning for buildings

Two major categories of decisions should be made early in the design process of a building in order to provide effective firesafe design. Early considerations should be given to both the interior building functions (discussed in this section) and exterior site planning (discussed in the next section). Building fire defenses, both active and passive, should be designed in such

Table 17-A. Elements of Building Firesafety

Building Design and Construction Features Influencing Firesafety

1. Fire Propagation
   a. Fuel load and distribution
   b. Finish materials and their location
   c. Construction details influencing fire and products of combustion movement
   d. Architectural design features (vertical/horizontal openings allowing fire spread)
2. Smoke and Fire Gas Movement
   a. Generation
   b. Movement
      —Natural air movement
      —Mechanical air movement
   c. Control
      —Barriers
      —Ventilation
      —Heating, ventilating, air conditioning
      —Pressurization
   d. Occupant protection
      —Egress
      —Temporary refuge spaces
      —Life support systems
3. Detection, Alarm, and Communication
   a. Activation
   b. Signal
   c. Communication systems
      —To and from occupants
      —To and from fire department
      —Type (automatic or manual)
      —Signal (audio or visual)
4. People Movement
   a. Occupant
      —Horizontal

      —Vertical
      —Control
      —Life support
   b. Firefighters
      —Horizontal
      —Vertical
      —Control
5. Suppression Systems
   a. Automatic
   b. Manual (self-help; standpipes)
   c. Special
5. Firefighting Operation
   a. Access
   b. Rescue operations
   c. Venting
   d. Extinguishment
      —Equipment
      —Spatial design features
   e. Protection from structural collapse
7. Structural Integrity
   a. Building structural system (fire endurance)
   b. Compartmentation
   c. Stability
8. Site Design
   a. Exposure protection (to facility and by facility to public)
   b. Firefighting operations
   c. Personnel safety
   d. Miscellaneous (water supply, traffic, access, etc.)

Fire Emergency Considerations

1. Life Safety
   a. Toxic gases
   b. Smoke
   c. Surface flame spread

2. Continuity of Operations
   a. Structural integrity

   b. Limiting of value or separation (by passive system) of like operations

3. Structural
   a. Fire propagation
   b. Structural stability

From NFPA *Fire Protection Handbook,* 14th ed.

Table 17-B. Fire Prevention and Emergency Preparedness

1. Ignitors
   a. Equipment and devices
   b. Human accident
   c. Vandalism and arson

2. Ignitable Materials
   a. Fuel load
   b. Fuel distribution
   c. Housekeeping

3. Emergency Preparedness
   a. Awareness and understanding
   b. Plans for action
      —Evacuation or temporary refuge
      —Handling extinguishers
   c. Equipment
   d. Maintenance—operating manuals available

From NFPA *Fire Protection Handbook,* 14th ed.

a way that the building itself assists in the manual suppression of fire.

Interior layout, circulation patterns, finish material, and building services are all important firesafety considerations in building design. Building design also has a significant influence on the efficiency of fire department operations. As a result, all fire suppression activities should be considered during the design phases.

Firefighting accessibility to building's interior. One of the more important considerations in building design is access to the fire area. This includes access to the building itself as well as access to the building interior.

In larger and more complex buildings, serious fires over the years have brought improvements in building design to facilitate fire department operations. The larger the building, the more important access for firefighting becomes. In some buildings, firefighters cannot function effectively—the spaces in which adequate firefighting access and operations are restricted because of architectural, engineering, or functional requirements should nonetheless be provided with effective protection. A complete automatic sprinkler system with a fire department connection is probably the best solution to this problem. Other methods which may be used in appropriate design situations include access panels in interior walls and floors, fixed nozzles in floors with fire department connections, and roof vents and access openings.

Ventilation. Ventilation is of vital importance in removing smoke, gases, and heat so that firefighters can reach the seat of a blaze. It is difficult, if not impossible, to ventilate a building unless appropriate skylights, roof hatches, emergency escape exits, and similar devices are provided when the building is constructed.

Ventilation of building spaces performs the following important functions:
1. Protection of life by removing or diverting toxic gases and smoke from locations where building occupants must find temporary refuge.
2. Improvement of the environment in the vicinity of the fire by removal of smoke and heat. This enables firefighters to advance close to the fire to extinguish it with a minimum of time, water, and damage.
3. Control of the spread or direction of fire by setting up air currents that cause the fire to move in a desired direction. In this way occupants or valuable property can be more readily protected.
4. Provision of a release for unburned, combustible gases before they acquire a flammable mixture, thus avoiding a backdraft or smoke explosion.

Connections for sprinklers and standpipes. Connections for sprinklers and standpipes must be carfully located and clearly marked. The larger and taller the building becomes, the greater the volume and pressure of water that will be needed to fight a potential fire. Water damage can be very costly unless adequate measures such as floor drains and scuppers have been incorporated into the building design.

Confinement of a fire in a high-rise building can only be accomplished by careful design and planning for the whole building. As buildings increase in size and complexity, more dependence on fire detection and suppression systems is necessary. Such systems are described in detail in the NFPA references at the end of this chapter.

Firesafety planning for sites

Proper building design for fire protection includes a number of factors outside the building itself. The site on which the building is located will influence the design, especially traffic and transportation conditions, fire department accessibility, water supply, and the exposure this facility has on the public. Inadequate water mains and poor spacing of hydrants have contributed to the loss of many buildings.

Traffic and transportation. Fire department response time is a vital factor in building design considerations. Traffic access routes, traffic congestion at certain times of the day, traffic congestion from highway entrances and exits, and limited access highways have significant effects on fire department response distances and response time, and must be taken into account by building designers in selecting appropriate fire defenses.

Fire department access to the site. Building designers must ask the question: Is the building easily accessible to fire apparatus? Ideal accessibility occurs where a building can be approached from all sides by fire department apparatus. However, such ideal accessibility is not always possible. Congested areas, topography, or buildings and structures located appreciable distances away from the street make difficult or prevent effective use of fire apparatus. When apparatus cannot come close enough to the building to be used effectively, equipment such as aerial ladders, elevating platforms, and water tower apparatus can be rendered useless.

The matter of access to buildings has become far more complicated in recent years. The building designer must consider this important aspect during the planning stages. Inadequate attention to site details can place the building in an unnecessarily vulnerable position. If its fire defenses are compromised by preventing adequate fire department access, the building itself must make up the difference in more complete internal protection.

Water supply to the site. Another important question that a building designer must ask is: Are the water mains adequate and are the hydrants properly located? The more congested the area where the building is to be located, the more important it is to plan in advance what the fire department may face in its attack if a fire occurs on the property. An adequate water supply delivered with the necessary pressure is required to control a fire properly and adequately. The number, location, and spacing of hydrants and the size of the water mains are vital considerations when the building designer plans fire defenses for a building.

Exposure protection

Still another consideration in the design of the building is the possibility of damage from a fire in an adjoining building. The building may be exposed to heat radiated horizontally by flames from the windows of the burning neighboring building. If the exposed building is taller than the burning building, flames coming from the roof of the burning building can attack and damage the exposed building.

The damage from an exposing fire can be severe. It is dependent upon the amount of heat produced and the time of exposure, the fuel load in the exposing building, and the construction and protection of the walls and roof of the exposed building. Other factors are the distance of separation, wind direction, and accessibility of firefighters.

Fire severity is a description of the total energy of a fire, and involves both the temperatures developed within the exposing fire and the duration of the burning. NFPA 80A, *Protection from Exposure Fires,* describes estimated minimum separation distance under light, moderate, or severe exposures. The severity of the exposure is calculated on the width and the estimated fire loadings of the buildings involved. Building designers should be aware that effective separation distances between the exposing buildings can be reduced by blank walls, closing wall operatings, use of automatic deluge water curtains, and use of wired glass instead of ordinary glass.

Interior finish

The way a building fire develops and spreads and the amount of damage that ensues are largely influenced by the characteristics of the interior finish in a building. The types of interior finish used in buildings are numerous, varied, and serve many functions. Primarily they are used for aesthetic and/or acoustical purposes. However, insulation and/or protection against wear and abrasion are also considered major functions by building designers. The following statements from the National Commission on Fire Prevention and Control's report titled "America Burning" point out the need for greater concern and attention to the potential fire hazards of interior finishes.

"The modern urban environment imparts to people a false sense of security about fire. Crime may stalk the city streets, but certainly not fire, in most people's view. In part, this sense of security rests on the fact there have been no major conflagrations in American cities in more than half a century. In part, the newness of so many buildings conveys the feeling that they are invulnerable to attack by fire. Those who think only of a building's basic structure (not its contents) are satisfied, mistakenly, that the materials, such as concrete, steel, glass, aluminum, are indestructible by fire. Further, Americans tend to take for granted that those who design their products, in this case buildings, always do so with adequate attention to their safety. That assumption, too, is incorrect."

Types of interior finish. Interior finish is usually defined as those materials that make up the exposed interior surface of wall, ceiling, and floor constructions. The common interior finish materials are wood, plywood, plaster, wallboards, acoustical tile, insulating and decorative finishes, plastics, and various wall coverings.

While some building codes do not include floor coverings under their definitions of interior finishes, the present trend is to include them. Rugs and carpets are not subject to test and regulation under the Flammable Fabrics Act administered by the Department of Commerce. Wool carpets present no particular hazard, but some of the fluffy rugs and carpets made out of synthetics are a factor in fire spread.

Many codes exclude trim and incidental finish from the code requirements for wall and ceiling finish. Interior finishes, however, are not necessarily limited to the walls, ceilings, and floors of rooms, corridors, stairwells, and similar building spaces. Some authorities include the linings or coverings of ducts, utility chases and shafts, or plenum spaces as interior finish as well as batt and blanket insulation, if the back faces a stud space through which fire might spread.

Plastics. Lower cost and aethestic considerations make the use of plastic building materials desirable. However, all plastics are combustible and presently there is no known treatment that is able to make plastics noncombustible.

Cellular plastics sprayed on walls for insulation have become popular. Fire retardants can be incorporated in many of these plastics so they can meet building code requirements. However, some plastics containing polyurethane or polystyrene have been involved in serious, rapidly spreading fires.

Wood. The physical size of wood and its moisture content are important factors that determine whether this material will provide reasonable structural integrity. Wood is the most common material used in the construction of dwellings. If a wood-frame house is subjected to a serious fire, either from burning combustibles inside the house or from an exposure fire, it will not withstand much heat and will have little structural integrity.

Heavy timber construction can resist fire very well. The timbers will char, and the resulting coating of charcoal provides an insulation for the unburned wood. Heavy timber maintains its integrity during a fire for a relatively long time, thus providing an opportunity for extinguishment. Much of the original strength of the members is retained and reconstruction is sometimes possible.

Because untreated wallboards and paneling are highly combustible, fire-retardant treatments are required by most codes. Without such fire-retardant treatments, combustible wallboards not only enable a fire to spread so fast that people may become trapped, but also contribute fuel to the fire and create hazardous concentrations of smoke and toxic gases.

Steel. The most common building material for larger buildings is structural steel. While steel is noncombustible and contributes no fuel to a fire, it loses its strength when subjected to the high temperatures that are easily reached in a fire. The normal critical temperature of steel is 1,000 F (590 C). At this temperature, the yield stress of steel is about 60 percent of its value at room temperature. Buildings built of unprotected steel will collapse relatively quickly when exposed to a contents fire or an exposure fire. The lighter the steel members, the quicker will be the failure.

Another property of steel that influences its behavior in fires is expansion when the steel is heated. Walls can collapse from the movement caused by expansion of steel trusses.

Because unprotected structural steel loses its strength at high temperatures, it must be protected from exposure to the heat produced by building fires. This protection, often referred to as "fireproofing," insulates the steel from the heat. The more common methods of insulating steel are encasement of the member, application of a surface treatment, or installation of a suspended ceiling as part of a floor-ceiling assembly capable of providing fire resistance. In recent years, additional methods, such as sheet steel membrane shields around members and box columns filled with liquid, have been introduced.

In recent years, intumescent paints and coatings have been used to increase the fire endurance of structural steel. These coatings intumesce, or swell, when heated, thus forming an insulation around the steel.

Structural steel members can also be protected by sheet steel membrane shields. The sheet steel holds in place inexpensive insulation materials, thus providing a greater fire endurance. In addition, polished sheet steel has been used in recent tests to protect spandrel girders (the horizontal supports beneath the windows on many modern high-rise buildings). The shield reflects radiated heat and protects the load-carrying spandrel.

Concrete. The resistance of reinforced concrete to fire attack will depend on the type of aggregate used to make the concrete, the moisture content, and the anticipated fire loading. In general, lightweight concrete performs better at elevated temperatures than normal-weight concrete.

Usually, reinforced concrete buildings resist fire very well; however, the heat of a fire will cause spalling (chipping and peeling away), some loss of strength of the concrete, and other deleterious effects.

Prestressed concrete is stronger than reinforced concrete and provides better fire resistance. However, prestressed concrete has a greater tendency to spall with the result that the prestressing steel may become exposed. The type of steel used for prestressing is more sensitive to elevated temperatures than the type of steel that is usually used in reinforced concrete construction.

In addition, the steel used for this type of reinforced concrete construction does not regain its strength upon cooling.

Glass. Glass is a commonly used building material. Modern high-rise buildings, particularly, contain large amounts of glass. Glass is used in three primary ways in building construction: (1) for glazing, (2) for glass fiber insulation, and (3) for glass fiber-reinforced plastic building products.

Glass used for windows and doors has little resistance to fire. Wire-reinforced glass provides a slightly higher resistance to fire, but no glazing should be relied upon to remain intact in a fire.

Glass fiber insulation is widely used in modern building construction. Glass fiber is popular because it is fire resistant and is an excellent insulator. However, glass fiber is often coated with a resin binder that is combustible and that can spread flames.

Glass fiber-reinforced plastic building products such as translucent window panels are becoming more common. The glass fiber acts as reinforcement for a thermosetting resin. Usually this resin, which is combustible, comprises about 50 percent or more of the material. Thus, while the glass fiber itself is noncombustible, the product is highly combustible.

Gypsum. Gypsum, as reflected in products such as plaster and plasterboard, has excellent fire-resistive qualities. Gypsum is widely used because it has a high proportion of chemically combined water, which makes it an excellent, expensive, fire-resistive building material that is far superior to highly combustible fiberboards.

Masonry. Masonry (such as brick, tile, and sometimes concrete) provides good resistance to heat, and usually retains its integrity. Because of the prevalence of brick construction in European dwellings, as compared to American wood-frame construction, the dwelling fire record in Europe is much more favorable than the dwelling fire record in America, according to the NFPA.

For specific details about how to use these materials for firesafe construction, see *The BOCA Basic Building Code* (References).

CONSTRUCTION METHODS

Fire-resistive construction

It is important to remember that so-called "fire-proof" or "fire-resistive" materials used in buildings provide a definite advantage, but they should not be confused with "firesafe." A building constructed with fire-resistive materials can withstand a burnout of its contents without subsequent structural collapse. Firesafe, on the other hand, indicates that if a fire starts it can be confined and extinguished without jeopardizing life and property elsewhere in the structure.

Fire-resistive construction actually describes a broad range of structural systems capable of withstanding fires of specified intensity and duration without failure. The materials are relatively noncombustible and are given a numerical rating as to their fire resistance (discussed below).

Common high fire-resistive components include masonry load-bearing walls, reinforced concrete or protected steel columns, and poured on precast concrete floors and roofs.

Although fire-resistive structures do not, in themselves, contribute fuel to a fire, combustible trim, ceilings, and other interior finish and furnishings (discussed in the previous section) may produce an intense fire; because they pose a serious threat to life

safety, they must be considered in providing fire protection. The use of approved building materials will lower the flame-spread ratings and limit those that can contribute to available fuel, especially in buildings without automatic sprinkler systems.

Nearly every building material has a fire-resistive rating. The rating is a relative term or number that indicates the extent to which it resists the effect of a fire. Ratings are usually available for most building components, such as column, floors, walls, doors, windows, and ceilings.

Heavy timber construction is characterized by masonry walls, heavy timber columns and beams, and heavy plank floors. Although not completely immune to fire, the great bulk of the wooden members slows the rate of combustion. Moreover the char which forms on wooden surfaces serves as an insulator for the wood within.

Noncombustible/limited-combustible construction includes all types of structures in which the structure itself (exclusive of trim, interior finish, and contents) is noncombustible but not fire resistant. Exposed steel beams and columns, and masonry, metal, or asbestos panel walls, are the most common forms.

Because of the tendency of steel to warp, buckle, and collapse under moderate fire exposure, noncombustible construction is relatively vulnerable to fire damage. It is, therefore, most suitable for low-hazard occupancies or ordinary hazard occupancies.

Ordinary construction consists of masonry exterior-bearing walls, or bearing portions of exterior walls, that are of noncombustible construction. Interior framing, floors, and roofs are made of wood or other combustible material whose "bulk" is less than that required for heavy timber construction.

If floor and roof construction and their supports have a one-hour fire-resistance rating, and all openings through floors (including stairways) are enclosed with partitions having a one-hour fire-resistance rating, then it is known as "protected ordinary construction." Its occupancy should be limited to light or moderate hazards.

Even when sheathed, ordinary construction (unlike fire-resistive or noncombustible construction) still has combustible materials in concealed wall and ceiling spaces. Fire frequently originates in these concealed spaces or enters into them through openings and then spreads rapidly throughout the entire room and building.

To prevent the free passage of flame through concealed spaces or openings in event of fire, (1) trim all combustible framing away from sources of heat, (2) provide effective fire barriers against the spread of fire between all subdivisions and all stories of the building, (3) provide adequate fire separation against exterior exposure, and (4) firestop all vertical and horizontal draft openings to form effective barriers to stop or slow the spread of fire.

Wood frame construction consists primarily of wood exterior walls, partitions, floors, and roofs. Exterior walls may be stuccoed or sheathed with brick veneer or metal, cement-asbestos, or asphalt siding.

Although generally inferior to other types of construction from a firesafety standpoint, it can be made reasonably safe for light-hazard, low-density occupancies. The safety-to-life factor can be greatly increased by suitable protection against the horizontal and vertical spread of fire, the provision of safe exits, and the elimination of combustible interior finishes. Sufficient fire detectors should be installed to alert all occupants. Automatic sprinkler protection can greatly improve the firesafety outlook in wood frame buildings.

Vertical and horizontal cut-offs

Regardless of the type of building construction, stair enclosures are necessary to provide a safe exit path for occupants, and serve to retard the upward spread of fire. Under certain conditions, such as where large areas or high values are involved, buildings may be divided horizontally by fire walls. Fire walls must be designed to rigid specifications in order to withstand the effects of severe fire and building collapse on one side without failure. All openings must be protected with approved closures at the same or greater fire exposure rating as the fire walls to prevent the passage of heat.

Fire confinement

Traditionally, building compartmentation provided functional units or offered occupants some degree of privacy. From the point of view of firesafety, however, compartmentation is regarded as the means for breaking up total building volume into small cells where, with an efficient protection system, fires will remain localized and be suppressed. To prevent fire from spreading from one compartment to another, various building codes require compartments be made structurally sound enough to withstand full fire exposure without major damage, and that boundaries be capable of acting as nonconducting heat barriers.

Fire doors are the most widely used and accepted means of protection of both vertical and horizontal openings. Fire doors are rated by testing laboratories as installed in the building. The fire doors usually have a rating of ¾ to 3 hours and may be constructed of metal or metal-clad treated wood materials, and be hinged, rolling (sliding), or curtain type. Single or double doors may be specified. When new construction is planned, it is important to select the proper types of fire doors, as they can be expected to perform properly only in the uses for which they were designed.

The fire door can be expected to perform properly only if it is installed with an approved frame, latching device, hardware, and closing device. The effectiveness of the entire assembly as a fire barrier may be destroyed if any component is omitted or one of substituted quality is used. To assure proper protection of openings, fire doors should be installed in accordance with NFPA 80, *Fire Doors and Windows*.

Fire doors are of value only if they will close or be closed at the time of a fire. Blocking or wedging fire doors open defeats their purpose; this practice should be prohibited. To ensure compliance with this policy, fire doors should always be checked during plant or building loss prevention inspections to assure that door openings and the surrounding areas are clear of anything that might interfere with the free operation of doors.

Like all emergency equipment, fire doors require periodic inspection and maintenance to assure that they will give protection. Hinges, catches, latches, closers, and stay rolls are especially subject to wear; they should be inspected frequently.

Doors with automatic closing devices should be operated regularly to assure proper operation. Several problems are frequently encountered in this critical test:

1. Chains or wire ropes may have stretched.
2. Hardware may be inoperative.
3. Guides and bearings may need lubrication.
4. Binders may be bent, obstructing the doorway.
5. Stay rolls may have accumulations of paint.
6. Fusible links may be painted.
7. The hoods over rolling steel doors may be bent, interfering with door operation.
8. Where swinging doors are used in pairs, coordinators may need adjustment.

The immediate repair of any defect which would interfere with proper operation of fire doors is an important responsibility of good loss prevention.

Smoke control

Smoke movement within a structure is determined by many factors, including building height, external wind force and direction, ceiling heights, venting, and suspended ceilings. Smoke control is the capability to confine smoke, heat, and noxious gases to a limited area or to exhaust them, thus preventing their spread to other areas in order to minimize fatalities. This usually is accomplished by dilution, exhaust, or confinement. Most control systems involve a combination of these methods. It is common to integrate these smoke control systems with the building heat and ventilating system.

One method involves a physical barrier, such as a door, wall, or damper that blocks smoke movement. Smoke control doors may be operated manually or by some automatic detection device coupled to a door closure. An alternative is to use a pressure differential between the smoke-filled area and the protected area.

Smoke and control venting. Smoke and hot gases generated by an uncontrolled fire, if confined within a building, can seriously impair firefighting operations, cause sickness and death, and can spread the fire under the roof for considerable distances from the point of origin. Venting is a planned and systematic removal of smoke, heat, and fire gases from the building. It is most effective to approach smoke control during the design stage of the building by specifying vents, curtain boards, and windows.

One method of smoke control is the use of smoke and heat venting systems consisting of curtain boards to protect heat-banking areas under a roof and automatic or manual roof vents for the release of smoke and heat through the roof.

Vents are most applicable to larger areas in one-story buildings lacking sprinklers and adequate subdivision. They are also useful in windowless and underground buildings. They are not a substitute for automatic sprinkler protection. Vents and draft curtains are effective for small special housing.

So many variables affect the burning of combustible material that no exact mathematical formula can be used for computing the amount of heat venting required. However, vent sizes and ratios have been developed from limited experiments in test buildings without sprinklers and theories about actual fire experiences. Also, there are a variety of prefabricated vents available on the market. They can be designed to open automatically at predetermined temperatures and/or smoke concentrations.

Exits

Of the many factors involved in securing safety to life from fire, the building exit facilities rank as the most important. Although

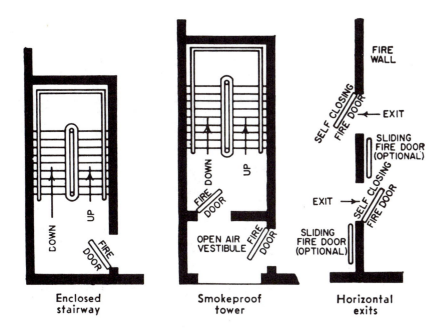

Figure 17-4. Plan views of types of exits. Stair enclosure prevents fire on any floor trapping persons above. Smokeproof tower is better because opening to air at each floor largely prevents chance of smoke on stairway. Horizontal exit provides a quick refuge and lessens the need of a hasty flight down stairs. Horizontal sliding fire doors provided for safeguarding property values are arranged to close automatically in case of fire. Swinging doors are self-closing. Two wall openings are needed for exit in two directions. (Reprinted with permission from National Fire Protection Association.

there is general acceptance of the importance of exit facilities, they remain inadequate at many facilities.

Management, architects, and others entrusted with the safety of their employees must consider many problems when planning emergency evacuation of buildings. In many cases panic causes more loss of life than fire. While fire is the most common cause of panic, such things as boiler, air receiver or other explosions, fume releases, or structural collapse may also threaten safe, orderly evacuation.

Exit design should be considered in the building's total fire-safety system criteria. The population and degree of hazard will be the major factors.

It is recommended that every building or structure, and every section or area in it, shall have at least two separate means of egress, so arranged that the possibility of any one fire blocking all of them is minimized (Figure 17-4).

The design of a means of egress involves more than a study of numbers, flow rate, and population densities. Safer exits from a building require provision of a safe path of escape from fire, arranged for ready use in case of emergency and sufficient to permit all occupants to reach a place of safety before they are endangered by the fire or smoke and toxic combustion products.

No construction plans or physical alterations should be made which may prevent safe evacuation under any circumstances that might arise.

NFPA 101, *Life Safety Code,* provides a reasonable and comprehensive guide to exit requirements. Where local, state (provincial), or federal codes containing more rigid recommendations are in existence, they will, of course, take precedence.

Evacuation

The following general provisions must be considered in planning for building evacuation.

1. Design of exits and other safeguards for life safety must not depend solely on any single safeguard, but additional safeguards shall be provided in case any single measure is ineffective due to some human or mechanical failure.

2. Exit doors must withstand fire and smoke during the length of time for which they are designed to be in use. Vertical exitways and other vertical openings must be enclosed or protected to afford reasonable safety to occupants while using exits.

3. Provide alternate exits and pathways to them in case one exit is blocked by fire.

4. Provide alarm systems to alert occupants in case of fire or emergency.

5. Exits and paths of travel to reach them shall be provided with adequate illumination.

6. Exits shall be marked by a readily visible sign. Access to exits shall be marked by readily visible signs in all cases where the exit or way to reach it is not readily visible.

7. Safeguard equipment and areas of any unusual hazard that might spread fire and smoke, endangering the safety of persons on the way out.

8. Practice an orderly exit drill procedure.

9. Control psychological factors that can lead to panic.

10. Select an interior finish and contents that prevent a fire from spreading fast and trapping occupants.

11. Maintain adequate aisles for exit access.

12. Provide adequate space at the point of building egress. Not more than 50 percent of the occupants should be discharged to a street.

13. Capacity of a door (minimum 28 in. [70 cm] wide) is suggested to allow 60 persons per minute, and a stairway (downward) at 45 persons per minute per unit of width (of 22 in. [56 cm] width). This can be used to calculate evacuation time.

FACTORS CONTRIBUTING TO INDUSTRIAL FIRES

To eliminate causes of fire, we must first know the many ways in which it can start, especially the most common ones. These are summarized in Table 17-C.

Table 17-C. Ignition Sources of Fires

To eliminate the causes of fire, it is important to know how and where fires start. The following summary of known causes is based on an analysis of more than 25,000 fires reported to the Factory Mutual Engineering Corporation from 1968 through 1977. The causes are arranged in order of their frequency throughout industry, but this is not necessarily a measure of their relative importance at any particular plant or property.

Electrical—22 percent. The leading cause of industrial fires. Most start in wiring and motors. Most prevented by proper maintenance. Special attention needed for equipment at hazardous processes and in storage areas.

Incendiarism—10 percent. Fires maliciously set by intruders, juveniles, disgruntled employees, and arsonists. Prevented by watch and guard service; install fences and other security measures.

Smoking—9 percent. A potential cause of fire almost everywhere. A matter of control and education. Smoking strictly prohibited in dangerous areas, such as those involving flammable liquids, combustible dusts or fibers, and combustible storage. Permitted in clearly designated safe areas.

Hot Surfaces—9 percent. Heat from boilers, furnaces, hot ducts and flues, electric lamps, irons, and hot-process-metal igniting flammable liquids and ordinary combustibles. Prevented by safe design and good maintenance of flammable-liquid piping and by ample clearances, insulation, and air circulation between hot surfaces and combustibles.

Friction—7 percent. Hot bearings, misaligned or broken machine parts, choking or jamming of material, and poor adjustment of power drives and conveyors. Prevented by a regular schedule of inspections, maintenance, and lubrication.

Overheated Materials—7 percent. Abnormal process temperatures, especially those involving heated flammable liquids and materials in dryers. Prevented by careful supervision and competent operators, supplemented by well-maintained temperature controls.

Cutting and Welding—7 percent. Sparks, arcs, and hot metal from cutting and welding operations. Prevented by the use of the permit system and other recognized precautions.

Burner Flames—6 percent. Improper use of portable torches, boilers, dryers, ovens, furnaces, portable heating units, and gas- or oil-burner flames. Prevented by proper design, operation, and maintenance; by ade-quate ventilation and combustion safeguards; and by keeping open flames away from combustible material.

Spontaneous Ignition—5 percent. In oily waste and rubbish, deposits in dryers, ducts and flues, materials susceptible to heating, and industrial wastes. Prevented by good housekeeping and proper process operation. Remove waste daily, frequently clean ducts, flues, and isolated storages subject to spontaneous heating.

Exposure—4 percent. Fires communicating from nearby properties. Blank walls furnish the most effective barriers. Protect wall openings with open sprinklers or wired glass, depending on severity of exposure.

Combustion Sparks—3 percent. Sparks and embers, released from incinerators, foundry cupolas, furnaces, fireboxes, various process equipment, and industrial trucks. Use well-designed equipment and well enclosed combustion chambers with spark arresters as needed.

Miscellaneous—3 percent. Unusual causes and relatively unimportant causes not included in the other classifications.

Mechanical Sparks—2 percent. Sparks from foreign metal in machines, particularly at cotton mills, and grinding and crushing operations. Prevented by keeping stock clean; remove foreign material by magnetic or other separators.

Molten Substances—2 percent. Fires caused by metal escaping from ruptured furnaces or spilled during handling; also by glass and tempering salts. Prevented by proper operation and maintenance of equipment.

Static Sparks—2 percent. Ignition of flammable vapors, dusts, and fibers by discharge of accumulation of static electricity on equipment, materials, or the human body. Prevented by grouding, bonding, ionization, and humidification.

Chemical Action—1 percent. Chemical processes getting out of control, chemicals reacting with other materials, and decomposition of unstable chemicals. Prevented by proper operation, instrumentation, and controls; and by careful handling and storage, particularly avoiding conditions of heat and shock.

Lightning—1 percent. Direct lightning strokes, sparks from one object to another induced by nearby lightning stroke, and induced surges in circuits and electrical equipment. Prevented by lightning rods, arresters, surge capacitors, and grounding.

Reprinted with permission from Factory Mutual Engineering Corp.

Electrical equipment

Electrical equipment should be installed and maintained in accordance with the *National Electrical Code*, NFPA 70. Overheating of electrical equipment and arcs resulting from short circuits in improperly installed or maintained electrical equipment are two of the leading causes of fire in buildings.

Only approved (UL-listed or FM-approved) equipment should be used where flammable gases or vapors may be present. Electrical equipment necessary in locations that are hazardous because of the presence of flammable liquids, gases, or dusts should be designed for the particular hazardous atmosphere.

Hazardous locations are divided into three classes depending on the material involved. Further division is by degree or severity of the hazard. Complete definitions of the several classes and divisions of hazardous locations and types of equipment for each are contained in the *National Electrical Code*, and are discussed in Chapter 15, Electricity.

Temporary or makeshift wiring, particularly if defective or overloaded, is a very common cause of electrical fires; it should not be used unless absolutely necessary and should be removed as soon as possible. Overloaded or partially grounded wiring may also heat up enough to ignite combustibles without blowing fuses or tripping circuit breakers.

Portable electrical tools and extension cords should conform to the *National Eectrical Code* and should be inspected at frequent intervals and repaired promptly. Waterproof cords and sockets should be used in damp places, and explosion-proof fixtures and lamps should be used in the presence of highly flammable gases and vapors.

Figure 17-5. Here is a closeup of the ceiling of a flammable liquids storage and dispensing room. Note explosion-proof light fixture (*upper left*), two explosion panels, and pipes for a fire-protection system.

All electrical equipment, particularly portable electrical tools, should be grounded or double insulated for the protection of persons using it. Switches, lamps, cords, fixtures, and other electrical equipment should be listed by a recognized testing and certifying agency, and should be used only in applications for which the approval or listing was granted.

Lamp bulbs should be protected by heavy lamp guards or by adequately sealed transparent enclosures, kept away from sharp objects, and secured to prevent falling. Bare bulbs should never be used when exposed to flammable dusts or vapors. Lamp bulbs must be considered as potential hazards in such areas; they must be safeguarded accordingly. (See Figure 17-5.)

Employees should be instructed in the use of electrical equipment and should be prohibited from tampering, blockng circuit breakers, using wrong fuses, bypassing fuses, and installing equipment without authorization.

Electrical installations and all electrical equipment should be periodically inspected and tested to assure continued satisfactory performance and to detect deficiencies.

Smoking

Carelessly discarded cigarettes, pipe embers, and cigars are a major source of fire. Smoking should be prohibited especially in woodworking shops, textile mills, flour mills, grain elevators, and places where flammable liquids or combustible products are manufactured, stored, or used.

Although it might be desirable to eliminate smoking completely in a plant, such a rule is difficult to enforce. It is better to allow smoking at specified times and in a safe place where supervision can be maintained than to have it done surreptitiously in out-of-the-way places.

"No smoking" areas should be marked with conspicuous signs and their exclusive use rigidly enforced—everyone, including supervisors and visitors, must adhere to the regulation. It may be necessary to use more than signs to draw attention to the No Smoking areas. Lines drawn on the floor or illuminated barriers placed around areas or processes have been effective. Small stickers, such as the National Safety Council's self-adhesive stickers, and POP posters can be placed on containers, storage cabinets, or doors to areas where there is danger of fire or explosion.

In many high-hazard occupancies, smoking is permitted in special firesafe rooms or during periods when there is no exposure. In such cases, instructions or warning signs to that effect should be posted. Where the exposure is severe, employees should even be prohibited from carrying matches, lighters, and smoking material of any kind into the danger areas.

In any case, the use of "safety" matches should be encouraged, and smoking should be allowed only in firesafe areas. Where carrying matches into the plant is prohibited, special lighters should be provided in smoking rooms.

Even in plants where there is little fire hazard, employees should be encouraged to discard matches and smoking materials in a safe container rather than on the floor. This practice will discourage employees from throwing matches and cigarettes into places which may not be free of hazards.

Friction

Excessive heat generated by friction causes a very high percentage of industrial fires. A program of preventive maintenance on plant machinery can avert fires resulting from inadequate lubrication, misaligned bearings, and broken or bent equipment.

Fires frequently result from overheated power transmission bearings and shafting in buildings where dust and lint accumulate, such as grain elevators, cereal, textile, and woodworking mills, and plastic and metalworking plants. Frequent inspection should be made to see that bearings are kept well oiled and do not run hot. Accumulation of flammable dust or lint on them should be held to a minimum.

Drip pans should be provided beneath bearings and should be cleaned frequently to prevent oil from dripping to the floor or on combustible material below. Oil holes of bearings should be kept covered to prevent dust and gritty substances from entering the bearings to cause overheating.

Frictional heat sufficient to cause ignition can result from the jamming of process material during production. Another common problem, frequently overlooked, is the tension adjustment on belt-driven machinery. If the belt is either too tight or too loose, excessive friction can cause serious overheating.

Foreign objects or tramp metal

Every precaution should be taken to keep foreign objects which might strike sparks from entering machines or processes where there are flammable dusts, gases, or vapors or combustible material, such as cotton lint or metal powder. Screens or magnetic separators are commonly used for this purpose, as in textile mills, grain elevators, and other operations which have explosive mixtures of dusts.

Open flames

Although open flames are probably the most obvious source of ignition for ordinary combustibles, and one would think they could be most easily avoided, they still account for a large percentage of industrial fires. Heating equipment, torches, and welding and cutting operations are principal offenders.

Air heaters (gas- and oil-fired). Air heaters are commonly used on construction work and often cause fires because of:
1. Overheating of the air heater with resultant radiation igniting nearby combustible materials, such as concrete form work, tarpaulins, wood structures, paper, straw, and rubbish.
2. Failure to insulate air heaters from floors or other combustible bases.
3. Failure to provide a substantial spark shield and to use fuels which do not produce high flame or sparks. (See NFPA 211, *Chimneys, Fireplaces, Vents, and Solid Fuel Burning Appliances.*)
4. Failure to provide a secure base or to anchor properly.

Air heaters in unventilated rooms or enclosures should be vented to the outside by means of an overhead hood and flue so as to remove the toxic products of combustion and any unburned gas.

Torches. If gasoline, kerosene, liquefied petroleum gas, acetylene, or alcohol torches are used, they should be placed so that the flames are at least 18 in. (46 cm) from wood surfaces. They should not be used around flammable liquids, paper, excelsior, or similar material.

Portable furnaces, blow torches, and the like should have overhead clearance of at least 4 ft (1.2 m). Combustible material overhead should be removed or protected by noncombustible insulating board or sheet metal and preferably by a natural-draft hood and flue of noncombustible material.

Welding and cutting. When possible, welding or cutting should be done in special firesafe areas or rooms with concrete or metalplate floors. Flame impingement on concrete may cause it to spall; consequently work should be kept off the floor or else the floor should be protected by a metal shield. (See NFPA 51B, *Cutting and Welding Processes.*)

In cases where welding and cutting operations are performed outside the special firesafe areas, hot-work permit programs (described earlier in this chapter) should be used to promote maximum firesafe working conditions.

If welding must be done over wood floors, they should be swept clean and wetted down, preferably covered with flame-resistant blankets, metal, or other noncombustible covering. Hot metal and slag should be kept from falling through floor openings and igniting combustible material below.

Sheet metal, flame-retardant tarpaulins, or flame-resistant curtains should be used around welding operations to prevent sparks from reaching combustibles nearby.

Welding or cutting should not be permitted in or near rooms containing any flammable liquid, vapor, or dust unless special precautions are observed and the hazards are well understood. No welding or cutting should be done on a surface until combustible deposits have been removed. Neither operation should be performed in or near closed tanks or other containers that have held flammable liquids, until the containers have been thoroughly cleansed, filled with water or purged with an inert gas, and combustible gas indicator tests show that no trace of a flammable gas or vapor is present. (See NFPA 327, *Cleaning or Safeguarding Small Tanks and Containers,* and National Safety Council Industrial Data Sheet No. 432, *Cleaning Small Containers of Combustibles.*)

The tests should be repeated periodically to determine if any trace of such a gas or vapor is released during the welding or cutting operation. If further tests show any trace, the work should be stopped until all flammable gas or vapor is dispelled.

Welding is sometimes necessary on equipment containing flammable material; however, this requires highly specialized procedures and should be avoided to the extent practical.

Fire extinguishing equipment suitable to the type of exposure should be within easy reach of welding and cutting operators. If the hazard is justified, and welding must be done outside of a shop or area designated for welding, watchers should be stationed to prevent sparks or molten slag from starting fires or to extinguish fires. The watchers should remain at the work location for at least thirty minutes after the welding or cutting is completed because some fires escape detection at the time of their inception.

Shields should be installed around spot welders to prevent sparks from reaching combustible materials nearby or from injuring employees.

Spontaneous ignition

Spontaneous ignition results from a chemical reaction in which there is a slow generation of heat from oxidation of organic compounds that, under certain conditions, is accelerated until the ignition temperature of the fuel is reached. This condition is reached only where there is sufficient air for oxidation but not enough ventilation to carry away the heat as fast as it is generated.

It is a condition usually found only in quantities of bulk material packed loosely enough for a large amount of surface to be exposed to oxidation, yet without adequate air circulation to dissipate heat. Exposure to high temperatures increases the tendency toward spontaneous ignition.

The presence of moisture also can advance spontaneous heating unless the material is wet beyond a certain point. Materials like unslaked lime promote spontaneous ignition, particularly when wet. Such chemicals should be stored in a cool, dry place away from combustible material.

It is generally agreed that at ordinary temperatures some combustible substances oxidize slowly and under certain conditions can reach their ignition point. These include vegetable and animal oil and fats, coal, charcoal, and some finely divided metals. Rags or wastes saturated with linseed oil or paint often cause fires too.

The best preventives against spontaneous ignition are either total exclusion of air or good ventilation. With small quanti-

Figure 17-6. This sheet metal waste can is constructed to prevent opening the cover more than 60 degrees from the horizontal, thus ensuring automatic closing.

ties of material, the former method is practical. With large quantities of material, such as storage piles of bituminous coal, both methods have been used with success.

Temperatures of 140 F (60 C) are considered dangerous in coal piles. If temperatures rapidly approach or exceed that figure, it is generally advisable to move the pile or rearrange it to allow better circulation of air.

Certain agricultural products are susceptible to spontaneous ignition. Sawdust, hay, grain, and other plant products such as jute, hemp, and sisal fibers may ignite spontaneously, especially if exposed to external heat or to alternate wetting and drying. Here again, the best preventives are circulation of air, removal of external sources of heat, and storage of material in smaller quantities.

Fires in iron, nickel, aluminum, magnesium, and other finely divided metals are sometimes attributed to spontaneous ignition believed to result from the oxidation of cutting or lubricating oils or possibly from chemical impurities.

Iron sulfide, commonly referred to as iron pyrite or pyrophoric iron, occurs as a result of lubrication of sulfur compounds in petroleum coming in contact with iron in pipes and vessels. As long as air is excluded or it is kept moist, iron sulfide does not constitute a hazard. However, when equipment containing iron

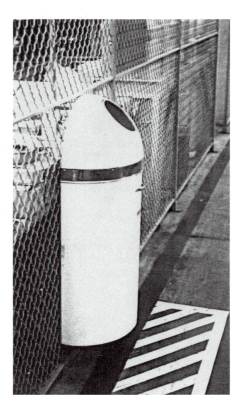

Figure 17-7. *Left:* Aerosol can is placed in puncturing device. *Center:* Lever comes down and can is punctured. *Right:* Waste receptacle for more public area.

sulfide is opened to the atmosphere and dried out, it will burn. Where iron sulfide may be present, provision should be made to keep the inner surface wet when opening equipment. The disposal of accumulated deposits of iron sulfide should be handled carefully and promptly so it will not start burning in an area where fire would create a hazard.

Housekeeping

Collection and storage of combustibles. Many industrial fires are the direct result of accumulations of oil-soaked and paint-saturated clothing, rags, waste, excelsior, and combustible refuse. Such material should be deposited in noncombustible receptacles, having self-closing covers that are provided for this purpose and removed daily from the work areas (Figures 17-6 and -7).

Exhaust systems of effective design will remove gases, vapors, dusts, and other airborne contaminants, many of which may be fire hazards. Exhaust systems and machine enclosures will help prevent accumulation of combustible materials on floors or machine parts. Such materials are most hazardous when airborne rather than when they have settled out.

Clean waste, although not as dangerous as oil-soaked waste, is readily combustible and should be kept in metal cans or bins with self-closing covers. Excelsior, cotton, kapok, jute, and other highly combustible fibrous material should be stored in covered noncombustible containers and, if large quantities are kept on hand, in fire-resistant rooms equipped with fire doors and automatic sprinklers. Portable extinguishers, hose lines, or other extinguishing equipment for Class A fires should be available for use at such storage places.

A schedule for safe collection of all combustible waste and rubbish should be a part of the fire prevention program. The safety professional should be certain that janitorial personnel

or others involved in collection of waste paper in offices and service areas have firesafe collection containers. It is important to check collection practices to be sure that ash trays, which may contain smoldering material, are not emptied into combustible bags or cartons or into containers of combustibles.

Accumulations of all types of dust should be cleaned at regular intervals from overhead pipes, beams, and machines, particularly from bearings and other heated surfaces. It must be understood that all organic as well as many inorganic materials, if ground finely enough, will burn and propagate flame. Roofs should be kept free from sawdust, shavings, and other combustible refuse. Such cleaning preferably should be done by vacuum removal, because blowing down with air may disperse dusts into dangerous clouds.

No such material should be stored or allowed to accumulate in air, elevator, or stair shafts, tunnels, in out-of-the-way corners, near electric motors or machinery, against steam pipes, or within 10 ft (3 m) of any stove, furnace, or boiler.

Rubbish disposal. Federal, state (provincial), and local laws forbid certain methods of waste disposal such as open burning, evaporation, flushing to sewers, etc. In addition, fires are often caused by burning rubbish in yards near combustible buildings, sheds, lumber piles, fences, and grass, or other combustible materials. If it must be burned, the best and safest way is with a well-designed incinerator that meets the requirements of the environmental pollution control laws.

Flushing and dumping waste materials into sewers may also be prohibited. In many cases, preplanning would include trap tanks which can be pumped out and the waste material disposed of properly. Chemically altering a waste material may be considered before disposal.

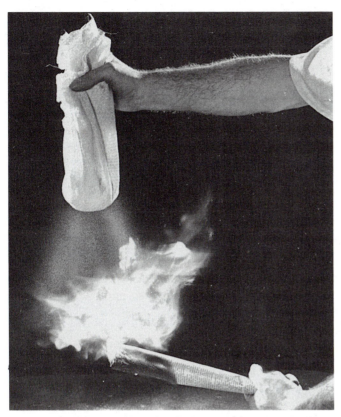

Figure 17-8. Demonstration of explosive properties of flour dust. Sifting flour through a cloth into the flame produces fiery bursts that illustrate what happens in grain elevators and other dust-laden atmospheres when dust clouds are ignited.

Table 17-D. Some of the More Common Potentially Explosive Dusts

Type	Example
Carbon	Coal, peat, charcoal, coke, lampblack
Fertilizers	Bone meal, fish meal, blood flour
Food products and by-products	Starches, sugars, flour, cocoa, powdered milk, grain dust
Metal powders	Aluminum, magnesium, zinc, iron
Resins, waxes, and soaps	Shellac, rosin, gum sodium resinate, soap powder, waxes
Spices, drugs, and insecticides	Cinnamon, pepper, gentian, pyrethrum, tea fluff
Wood, paper, tanning materials	Wood flour, wood dust, cellulose, cork, bark dust, wood extract
Miscellaneous	Hard rubber, sulfur, tobacco, many plastics

Source: U.S. Department of Agriculture.

In any event, the requirements of federal, state (provincial), and local laws must be ascertained when planning a process or a plant. Investigation should include plans for emergency conditions, such as spills, release of vapors, and fire.

Locker rooms. Lockers in which oil-soaked clothing, waste, or newspapers are kept are always a serious fire hazard. Every precaution should be taken to prevent such combustible accumulations. Lockers should be made of metal and have solid, fire-resistant sides and backs, but doors should have some open-

work for ventilation. Lockers should be large enough so that air can circulate freely around clothing hung in them. Employees should not be permitted to leave clothing saturated with oils or paints in lockers.

Where automatic sprinklers are used, locker tops may be covered with screening or be made of perforated metal so that water can reach burning contents if a locker fire occurs. Heavy paper pasted over the tops will serve to keep out dust. Lockers that have sloping tops and stand flat on the floor will prevent the accumulation of rubbish both above and below.

Explosive atmospheres

Dusts. A dust explosion hazard exists wherever material that will burn or oxidize readily is available in powder form because the surface (contact) area of each particle is very large in relation to its mass (Figure 17-8). The U.S. Department of Agriculture Technical Bulletin 490 lists 133 dusts according to their degree of explosiveness (Table 17-D). In addition, many synthetic resins and powders used in the plastics industry present a dust-explosion hazard comparable to that of coal (USBM Report No. 3751). The group includes phenolic, urea, vinyl, and other types of resins and a number of molding compounds, primary ingredients, and fillers.

There are two ways to prevent dust explosions: (1) prevent the formation of explosive mixtures of dust and air, and (2) prevent the ignition of such mixtures if their formation cannot be prevented.

An explosive mixture of dusts (or gases and vapors) can effectively be prevented by providing an inert atmosphere (oxygen limited to concentrations below which flame propagation does not occur). To be reliable, the concentration of oxygen must be monitored by an oxygen analyzer. This can be an effective method where enclosure of the system is possible. Inert atmospheres should not be provided where operators would have to enter the atmosphere to perform job functions.

Extreme precautions, if such are necessary, should be taken to prevent accumulations of dust that may build up to explosive proportions. Extensive use of local exhaust and regular cleaning will do much to minimize the hazard. Where possible, dusty operations should be segregated, and dust-producing equipment should be totally enclosed and exhausted to prevent leakage of dust into the general work area.

Buildings with high explosion hazards, such as grain elevators or plastics plants, are generally provided with extensive dust-collecting equipment and are constructed so that explosive pressures will push out hinged windows or blow out wall sections or panels designed and built to fail in a predetermined area, rather than cause structural collapse. (See NFPA 68, *Explosion Venting.*) Floor openings should be kept to a minimum and openings for pipes and duct work should be sealed.

Portable fire extinguishing equipment should be readily available. Fog nozzles or finely divided streams of water are more effective than are solid streams of water, which stir up dust.

Ignition of an explosive mixture of dust and air may be prevented by eliminating open flames, friction sparks, static electricity, welding, and excessive heat, by increased humidity, or by using inert gas. Every precaution should be taken to prevent overheated bearings, smoking, friction sparks, and sparks from hand tools, and from grinding or welding operations. Sources of static electricity, explained in the previous chapter, should also be controlled.

Nonferrous material should also be used for truck wheels, and bucket conveyors should be prevented from striking sparks. Pneumatic or magnetic separators can be used to remove stones, nails, and other spark-producing foreign objects from the material being processed. Conveyor belt rollers and similar moving parts should be kept properly lubricated and maintained to avoid excessive friction and heat.

As discussed under Electrical equipment earlier, only approved dust-tight wiring, fixtures, and motors should be used, and employees should be instructed to use only extension cords, lamps, and portable electric tools designed for protection against this hazard. (See NFPA *National Fire Codes,* volume 6.)

Gases and vapors. Gases and vapors that produce flammable mixtures with air or oxygen are common in industry. Some such gases are hydrogen, acetylene, propane, LP-gas, carbon monoxide, methane, natural gas, and manufactured gas.

Highly volatile liquids that emit flammable vapors include gasoline, benzene, naphtha, and methyl alcohol. Kerosene, turpentine, Stoddard solvent, and other liquids with flash points above 100 F (38 C) must generally be heated above normal room temperatures before they give off sufficient vapors to form ignitable concentrations.

When flammable liquids, including those with flash points above 100 F (38 C), must be handled and used, only minimum amounts, in safety containers, should be allowed in work areas.

The flash point of a liquid is defined as the lowest temperature at which it gives off sufficient vapor to form an ignitable mixture with air and produce a flame when an ignition source is brought near the surface of the liquid. Flash point temperatures are not always exact physical constants under work environment conditions because published figures are based on carefully controlled testing conditions. For instance, the amount of vapor that will accumulate in the air above a volatile liquid depends upon its relative saturation in the air above the liquid surface. The amount of vapor given off is also directly related to the surface area of the volatile liquid. Thus, even a high-flash liquid may be dangerous if considerable surface area is exposed, such as in a mist or froth. (It might be thought of as a "liquid dust.") Conversely, limiting the surface area of a liquid reduces its flash or explosion hazard.

The autoignition temperature of a gas or vapor in air must also be considered. For example, a mixture of carbon disulfide vapor and air will ignite spontaneously at temperatures around 194 F (90 C), which is below that of an uninsulated high-pressure steam line. However, the majority of gases and vapors will not self-ignite in air until they reach temperatures of about 500 to 900 F (260 to 482 C).

FIRE DETECTION

A look at statistical data tells us that, despite good construction, cleanliness, and modern firefighting methods, a considerable number of fire losses occur. Losses would be reduced if each developing fire were detected in one way or another so it could be attacked and extinguished. This means that some detection facilities must be a part of every fire protection system. It may be a watchman; automatic sprinklers; or smoke, flame, or heat detectors; or, more likely, a combination of these.

There are two main tasks for the detection part of a fire protection system: (1) giving an early warning to enable building occupants to escape, and (2) starting up extinguishing procedures. The latter may be as simple as alerting the fire brigade or department or triggering the operation of an automatic sprinkler.

Each automatic fire-detecting system requires a sensor, which observes a physically measurable quantity. This quantity must undergo measurable variations when a fire begins in the vicinity of the detectors. Normally, air temperature, smoke density, or flame radiation is used for this purpose. The sensing element in the detector is coupled to a decision-making device, which compares the measured quantity with a predetermined value, and when a value is different, an alarm is sounded. A detector both detects and signals.

Logically, there are really *no* fire detectors, but rather heat detectors, smoke detectors, and radiation detectors.

Combined systems using two or more principles are also possible. It is not worthwhile to connect an insensitive measuring device to a sophisticated information-processing (decision-making) system—neither component will be able to perform adequately.

Human observer

A good fire-detecting system is the human observer, for several reasons:

- Several different measuring facilities are available; namely for temperature, smoke, and flame radiation. The facilities are sensitive over the range important for fire protection.
- An able and effective information processing system not only processes the data from each sensor separately, but also can add other information. Unfortunately, unless taught by experience, it needs a teacher.
- The human observer is able to take immediate action in a flexible way, for example, run away, call the fire department, put the fire out with an extinguisher.

Automatic fire-detection system

By comparing with the human system, it is possible to point out some problems that are involved in the use of automatic fire-detection systems. In general, there are two possible errors in any nonhuman fire-detection system—it may give an alarm without any fire in the room or space being monitored, and it may not detect a fire or detect it too late. The overall effectiveness or efficiency of fire-detecting systems can be uniquely described by these two parameters. The cause of false alarms may be human interference, mechanical or electrical faults, or special environmental effects. They are important considerations.

Where proper installation and planning of fire detection systems are concerned, a checklist might cover the following items.
- What is the main purpose of the system? Typical purposes include warning the occupants of a building, protection of irreplaceable goods, protection of highly valuable goods, protection against interruption of production, and protection against corrosion or radioactive contamination.
- What are the possible sources of ignition?
- What kinds of material will probably be ignited first?
- What kind of building construction is used?
- What are the environmental conditions?
- What kind of detection system has been installed and what are the reasons for choosing this system?
- How much time delay can the fire stand?

Fire detectors

Properly engineered fire-detection systems are sound investments. The service of a fire-protection consultant should be obtained in engineering the system and establishing the control procedures. Here are four steps to consider:

1. Select the proper type detector(s) for the hazard areas. For example, a computer area may require ionization or combination detectors. A warehouse may have infrared and ionization detectors. In low-risk areas, thermal detectors or combinations of detectors may be used.
2. Determine the spacing and locations of detectors to provide the earliest possible warning.
3. Select the best control system arrangement to provide fast identification of the exact source of alarm initiation.
4. Assure notification of responsible authorities who can immediately respond to the alarm and can take appropriate action. Every detection system must have an alarm signal transmitted to a point of constant supervision.

This process of being alerted to a fire involves the human response factor with assistance from a wide variety of automatic fire-detecting devices. A detector with human senses of smell, feeling, hearing, and sight, providing it were in the right place, could do the job more reliably than a human being. It seems that good design and mechanical monitoring with reliable maintenance could overcome human mental and behavioral shortcomings, but still not match the advantages of in-built senses.

There are many types of fire detectors to handle various situations and detect various stages of an incipient fire. Most manufacturers and distributors offer several or all of the commonly used types, and will engineer a "mixture" of equipment into a coordinated system to meet the particular set of performance parameters.

From a system analysis approach, a suitable detector and system must be fitted for a given hazard. Sensitivity, reliability, maintainability, and stability are the critical variables in the selection of a detection system.

Sensitivity is established by the design. How well can the device detect what is wanted?

Reliability is the ability to perform its intended function, when needed. It should not malfunction.

Maintainability means the unit should be easy to service and keep at operating efficiency. A minimum of maintenance should be needed.

Stability means the unit holds sensitivity over time.

The four stages of fire

Fire is a chemical combustion process created by the combination of fuel, oxygen, and heat, as described earlier under The Chemistry of Fire. Fire development relative to detection instruments may be considered to progress through four distinct stages.

Incipient stage. No visible smoke, flame, or significant heat is developed. However, a condition exists that generates a significant amount of combustion particles. These particles, created by chemical decomposition, have weight and mass, but are too small to be visible to the human eye. They behave according to gas laws, and quickly rise. This stage usually develops over an extended time period.

Smoldering stage. As the fire condition develops, the quantity of combustion particles increases to the point where their collective mass becomes visible. This is referred to as "smoke." There is still no flame nor significant heat.

Flame stage. As the fire condition develops further, the point of ignition occurs. Infrared energy is now given off by the flames. The level of visible smoke usually decreases and more heat is developed.

Heat stage. At this point, large amounts of heat, flame, smoke and toxic gases are produced. The transition from the third to the fourth stage develops very quickly.

Fire-detection equipment

There are many types of fire detectors, suitable for various situations, and particularly useful at various stages of a fire. Most manufacturers and distributors offer several or all of the commonly used types, and will engineer a "mixture" of equipment into a coordinated system to meet the particular set of performance parameters under consideration.

Thermal detectors.

Fixed-temperature detectors detect the heat of a fire (fourth stage). They are based on a bimetallic element, made of two metals that have different coefficients of expansion. When heated, the element will bend to close a circuit, initiating the alarm. Or a thermal detector may use a fusible, spring-loaded element (such as found in a sprinkler head), which melts at a certain temperature, releasing an arm to close a circuit.

This is a simple device, inexpensive, and requiring a very low voltage draw to keep the normally open contacts supervised through an end-of-line resistor.

Rate-compensated thermal detectors work by the expansion characteristics of a hollow tubular shell containing two curved expansion struts under compression, fitted with a pair of normally open, opposed contacts. (See Figure 17-9.) When subjected to a rapid heat rise, the shell expands and lengthens at a faster rate than the struts, permitting the contacts to close. When heated slowly, both the shell and the struts lengthen at about the same rate until the struts are fully extended, making contact at a preset temperature point.

Rate-of-rise thermal detectors use an enclosed vented

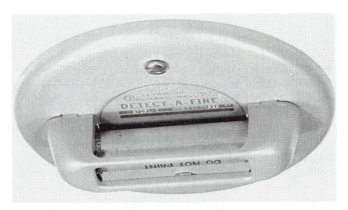

Figure 17-9. This rate-compensated thermal detector will activate a system at a predetermined danger point. It can be used as an alarm device or to actuate fire attack systems. (Reprinted with permission from Fenwal Incorporated.)

hemispherical chamber containing air at atmospheric pressure, with a small pressure-sensitive diaphragm at top. At a normal rise in temperature the excess pressure is relieved through small vents, but rapid heat rise will deflect the diaphragm faster than the vents can operate, triggering an alarm. This unit responds quickly to a fast heat rise.

Line thermal detectors use a small diameter, long run of tubing, which can be as long as 1,000 ft (305 m). When exposed to the heat of a fire, air inside the tube expands, sending a pressure wave to expand a diaphragm at the end, triggering an alarm. This is an unobtrusive, inexpensive detector. For example, it can be run along the ceiling molding, where it is nearly invisible. No maintenance is needed. It can be painted over, and, in fact, it will even work with the tubing broken, if the heat rise is fast enough.

Eutectic salt line thermal detectors consist of pliant metal tubing containing a eutectic salt in which a wire is embedded. At a preselected temperature level the salt will create a short-circuit between the internal wire and the outside tubing, thereby triggering an alarm. This detector, along with a continuous-resistance unit, has been widely applied over the past several years to guard against fires in aircraft jet engine nacelles. The pliant tubing can be wound around and shaped to the various components of the engines, to signal any increase in temperature that might be the result of a fire caused by leaking oil, hydraulic fluid, etc.

These line detectors are also used in conveyor belt systems where the bearings supporting the rubberized belt may ignite because of friction and lack of lubrication. If the system is transporting coal, for instance, the resultant fire could be expensive and hard to extinguish.

The bulb detection system is completely mechanical and especially desirable in locations where the explosive nature of the fire hazard makes it wise or essential to avoid the use of electricity. The system involves a number of bulbs containing air at atmospheric pressure. One or more of these bulbs are installed along the ceiling of the hazardous area, all connected back to a diaphragm at the control center. When the rate of heat rise striking one or more of the bulbs deflects the diaphragm, a mechanical extinguishing system can be activated. This system has been widely used over the past few years as a release mechanism for carbon dioxide fire extinguishers in marine and industrial applications.

Thermal detectors are reliable for what they do. However, they can only detect the *heat* of a fire, which usually will not build up to significant levels until the fourth stage. Many fires start slowly, with little heat generated at the beginning, and will be well under way by the time a thermal detector comes into operation. They are generally used where no life hazard is involved and some fire loss can be tolerated.

Smoke detectors respond to the particles of combustion, both visible (smoke) and invisible, which occur in the second fire phase. A conventional smoke detector operates on a light principle; it can be triggered by either less light or more light. When smoke enters a light beam, it either absorbs light so the receiver end of the circuit registers less light, or else the smoke scatters light so that a terminal, normally bypassed by the light beam, now receives part of the light.

Photoelectric detectors are line powered and usually include lamp supervision circuitry and annunciation in case of lamp failure. An incandescent light source may be used, or a high-intensity strobe lamp that generates a stronger reflection can be used so that fewer or smaller smoke particles will actuate the photocell.

The beam photoelectric detector works on the obscuration (less light) principle. A long beam is directed at a photocell. Rising smoke tends to obscure the beam, decreasing light transmission and sounding an alarm. This method is an inexpensive way to cover large spaces, such as a warehouse. It is, however, sensitive to voltage variations, dirt on lamp or lens, and also to flying insects or spiders, which sometimes accumulate near the lamp, seeking warmth.

"Reflected beam" photoelectric detectors use a beam of light in a chamber, with the photocell being normally in darkness. Should visible smoke particles enter the chamber, they scatter the light and reflect it onto the cell, causing a change in electric conductivity which results in an alarm.

The more sophisticated detectors in the smoke category are termed as being responsive to both gas or products of combustion and smoke. These products of combustion detectors are capable of detecting an incipient fire long before there is visible smoke or fire.

Products of combustion (or ionization) detectors sense both visible and invisible products of combustion suspended in the air. The system consists of a chamber with positive and negative plates and a minute amount of radioactive material that ionizes air in the chamber. The potential between the two plates causes ions to move across the chamber, setting up a small current. When aerosols from incipient fires enter the chamber, they cling to masses of moving ions. This slows the ion movement and increases the voltage necessary for the ions to make contact. This voltage imbalance, amplified by an electric circuitry, triggers an alarm.

The single-chamber ionization detector is most economical. The chamber is open to the atmosphere. Current flows between two poles and gets an increased voltage from combustion aerosols. This closes the contact and sends an alarm through the relay.

The dual-chamber ionization detector has two identical sources of radiation—one is a sealed chamber, the other is open to the atmosphere. The inner ionization chamber monitors ambient conditions and compensates for the effect on the ionization rate of barometric pressure, temperature, and relative humidity. This construction accepts a wider range of atmospheric variations without giving false alarms.

The low-voltage ionization detector is relatively new. While the conventional type needs 220 volts, a low-voltage detector needs only 24 volts (Figure 17-10). Theoretically, an installation would cost less, because low-energy nonarmored cable is less expensive and easier to install, with essentially no danger of short-circuit or electrical shock. Most large cities, however, require armored conduit for low-voltage detectors, so cost savings may be less than expected. Nevertheless, most systems specified today are low-voltage, as the low-profile detector heads are less obtrusive, while being equally sensitive and reliable.

Ionization detectors sense fire at the earliest practical detection stage. They are the best method for detecting slow, incipient fires in commercial occupancies: a cigaret in a wastepaper basket, for example, which might be in a pre-smoldering condition for 30 minutes or more. Because the ionization detector

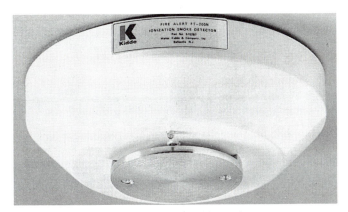

Figure 17-10. Ionization smoke detector is used with compatable fire alarm panels. Detector circuitry is solid-state and maximum reliability is provided by the fact that three samples must be omitted before an alarm can be initiated.

operates in the fail-safe mode, a further feature is provided. If excessive dust should be present, however, the device will give an alarm.

Flame detectors respond to the optical radiant energy of combustibles. A flame fire detector senses light from flames. Sometimes it works at the ultraviolet end, but more often at the infrared end of the visible spectrum. To avoid false alarms from ambient light sources, it is set to detect the typical flicker of a flame, perhaps at 5 to 30 Hz. Or, there may be a few seconds delay before the alarm to eliminate false alarms from transient flickering light sources, such as flashlights, or headlights.

Flame detectors have some very important applications, as in large aircraft hangars, and guarding against fires in fuel and lubricant drips. In general, however, by the time flame is visible, a fire has a good foothold. Either an infrared or ultraviolet detector can be used to sense the flame.

Infrared (IR) detectors sense a portion of the radiant IR energy of flames. It is often used in operations requiring extremely fast response (for example, where flammable liquids are stored or used).

Ultraviolet (UV) detectors react only to actual flame; they do not respond to glowing embers or incandescent radiation. This unit also is insensitive to heat, infrared radiation, and ordinary illumination.

Other detectors.

The combustion gas detector is the closest to a general-purpose detector and it also dramatically reduces the time of alarm. The combustion gas detector does not rely on heat; it, in effect, "smells" a potential fire by measuring the percent of gas present and does not wait until the dangerous conditions of flame occur. Most fires detected by combustion gas detectors can be extinguished by workers on the site.

The instrument can usually be set to automatically sound an alarm or set off extinguishing equipment. Some advance conditions may require periodic maintenance and calibration. In work areas, spacing can vary up to 40 ft (12 m) in semienclosed application.

There are areas, however, where a combustion gas detector cannot be used. For example, in an area where a certain level of combustion gases may be tolerated at times because a particular process emits them, flash fires—for example, from nearby flammable liquids—may be anticipated and detection must be almost instantaneous if it is to be at all effective. In such an area, a flame detector is best used. In another location where chemicals and some plastics could generate great volumes of smoke with negligible combustion, a smoke detector would be preferable.

Extinguishing system attachments. An important type of automatic detection equipment is often not even thought to be detection. These are devices that fall under the heading "extinguishing system attachments." These, nonetheless, are truly detectors. The primary function of alerting people that there is a fire may be handled by waterflow indicators in a sprinkler system. These may operate on the principle of a sudden increase or decrease in pressure or by detection by a vane inside the piping of the flow of water. These are designed to detect a flow of water that exceeds so many gallons per minute.

A second type of device indicates that the fire extinguishing system is jeopardized. A few of these are:

- Water level devices. These warn when a self-contained water storage supply is low, or, in the case of a low differential dry pipe valve, when the level of water is too high.
- Water temperature switch. This warns when storage water supply is approaching the freezing point.
- Water supply valve position switches signal if someone accidentally or purposely starts to turn off the water supply. These should give the alarm before the handle is closed two revolutions or $\frac{1}{5}$ the distance from the open position but, in practice, are often set to signal an alarm within $\frac{1}{2}$ a revolution.
- Closed circuit TV. Often used for security system.

Monitoring systems. In smokestacks, storage tanks, or other areas where several variables are at work simultaneously, use of a linear sensor system, alone, or in combination with other systems for greater accuracy in detection of fire, is often recommended.

The common heat sensors simply provide averages, and an average obscures many sins. The linear sensor systems give continuous point-by-point readings; they pick up and monitor trouble spots instantly over an entire area.

The concept of totally integrated fire detection protection combats fragmentation and brings together the entire establishment's firefighting capacity, thereby saving time and, in turn, costly loss of lives and property. This concept also demands the continuing close cooperation of all persons involved in design and construction, as well as safety engineering and fire protection.

A modern fire alarm system should no longer be a thermal system—or a combustion detection system, or flame, or smoke—but a combination of the various types in one integrated system, depending on the activities and processes being carried out (Table 17-E).

ALARM SYSTEMS

Regardless of its type, each individual detector must be part of an overall system if it is to be effective. These alarm systems are discussed in the chapter Planning for Emergencies, in the *Administration and Programs* volume of this Manual. The section below presents general principles for building or plant alarm systems, followed by a discussion of the most commonly used types.

Table 17-E. Relative Sensitivity of Fire Detectors When Means of Detection Is Matched to the Class of Fire

Type of Detector	Class of Fire		
	A	B	C
Fixed temperature	L	H	L
Rate of rise	M	H	L
Rate compensator	M	H	L
Particulate matter	H	H	M
Visible smoke	H	L	M
Ultraviolet	L	H	H
Infrared	L	H	L

George J. Grabowski, "Fire Detection and Activation Device for Halon Extinguishing System," National Academy of Science, 1972.

A plant or building should be equipped with fire alarm signal systems that are clearly communicated to all persons in the building, plant, or laboratory whenever the alarm is sounded in any portion of the building or area, with certain exceptions discussed later. All employees should be made aware of the sound and meaning of every type of alarm system used in the facility. Every facility fire alarm system, whether it is currently in place, to be newly installed, or to be revised, should meet the following criteria.

- When alarms are audible, the alarm sound should be clearly and immediately distinguishable from other signals which might be in use at a given facility.
- Audible alarm devices should be strategically located so they are clearly audible to all facility personnel, with the latter trained to recognize the signal and to respond in accordance with that location's specific disaster control procedure.
- The fire alarm system should be comprised of the type equipment that conforms to the NFPA standards and bears certification by Underwriters Laboratories Inc.
- The alarm system should be maintained in good working order with tests at intervals sufficiently frequent to provide assurance of its proper working order. It is recommended that the interval between tests of the alarm system not exceed one month.
- The location of and means of contacting external fire protection sources should be communicated to all facility personnel, especially those trained in fire protection. This information should also be conspicuously posted in strategic areas. It is very important that all employees know the proper procedures for turning in a fire alarm if they are the ones who detect the fire.

Alarm systems (as differentiated from fire detectors) can be divided into four groups: Local, auxiliary, central station, and proprietary.

Local alarms

A local alarm simply consists of bells, horns, sirens, or other warning devices right in the building. Local alarms are generally used for life protection; that is, to evacuate everyone and thereby limit loss of life or injury from the fire. A local alarm can be tied in with another system to summon the fire department.

In a large building, local alarm signals should be coded in a recognizable pattern of sound and silence to tell where the fire is. Otherwise the alarm would simply indicate a fire without showing where it is located, which would give the fire a chance to get underway.

A presignal system may be especially advantageous to prevent panic. A presignal system rings specific alarm stations before a general alarm is given. In a multi-building operation, the manager, plant superintendent, and security chief might be alerted ahead of the general alarm. And alarms may be given by floors or building units: with a fire in one building the immediately adjacent buildings might also be alerted and evacuated, but the more remote buildings would remain undisturbed.

Local systems are inexpensive, available from a wide range of suppliers, and are easy and inexpensive to install. By themselves, however, they do not buy much protection. Essentially they alert personnel; they do not summon the fire department.

Auxiliary alarms

An auxiliary system is even less expensive than a local alarm. Such a system simply ties a fire detector to a nearby fire-call box. In effect, it becomes a relay station triggered by fire detectors inside the building.

Central station service

Central station service is available in most major cities around the country. Manned by trained personnel, a central station continually monitors a number of establishments and, in case of an alarm, calls a nearby fire station and alerts the building personnel. In some cases central station personnel also go out and check on an alarm.

Central station devices are virtually always leased; the central station company sprinkles fire detectors throughout the building, then ties them to an alarm board back at the central station, commonly using leased phone lines. In rare cases the central station may be connected directly to the fire department; generally, however, when the central station attendant gets an alarm signal from a subscriber, he or she picks up a telephone and dials the fire department.

Central station personnel are competent and expensive. Costs for leasing the equipment alone are usually set high enough to amortize the central station's entire investment in three years or less; payment for professional manning of the station is high enough to cover actual cost and return a profit besides.

One argument used to sell central station protection is that some of the cost will be offset by a reduced insurance premium. However, a company can get as much or more reduction with its own security force (almost always required by insurance regulations when the total policy covers $10 million or more).

Proprietary systems

Proprietary systems feed the alarms to the building's own watchman or maintenance force, and, optionally, to the fire department as well. One reason for their acceptance is that insurance regulations generally require guards, as discussed in the previous paragraph.

Proprietary systems are especially effective where new plants or buildings have sprung up around the periphery of a metropolitan area. Most central station headquarters are downtown, close by the established mercantile districts. As cities have grown, many of these older districts have been superseded by outward metropolitan expansion, miles and miles away from

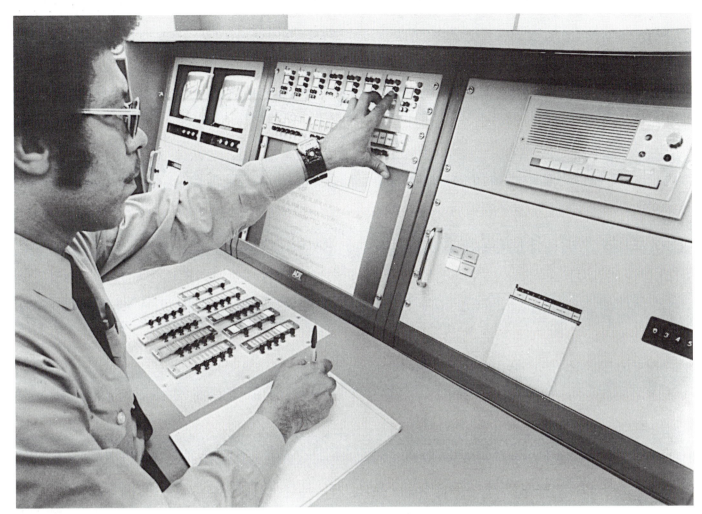

Figure 17-11. This console is part of a comprehensive proprietary system capable of detecting fire, burglary, and any unwanted fluctuations in industrial process conditions. The system incorporates closed circuit television, portable radio communication equipment, and other instruments to handle indoor and outdoor security needs of a building or a group of buildings. (Reprinted with permission from ADT— American District Telegraph Company.)

what was once the heart of the city. "Guarding" a suburban establishment from a central station downtown is impractical and uneconomic.

While a proprietary system can mean substantial reductions in insurance premiums, it can mean more dramatic savings by cutting guard requirements. Unless guards are supplemented by electronic devices, insurance companies usually demand that guards tour the building every hour. Figuring two shifts, weekends, vacations, etc., this requires at least 13 persons. Installing an approved proprietary fire-detection system can cut this requirement to one patrol every four hours, freeing at least three employees on each tour of duty (for a total of nine) for more important duties elsewhere.

Proprietary system consoles can include much more than fire alarms. As they will be monitored constantly, such command consoles can include intrusion detection, security protection, equipment surveillance, and dozens of monitoring functions usually ignored or haphazardly scattered throughout a building. (See Figure 17-11.)

Some engineers recommend an electronic central processor unit (CPU), through which all remote sensors report. Located in a plant's main control room or emergency control center, the processor receives signals not only from supervisory fire-detection equipment, but from detectors of other plant process variables not necessarily associated with fire conditions. The CPU compares these parameters in ways not normally done in routine plant operations—that is, for conditions that may lead to a fire. Therefore, if operating vessel pressure, for example, were to exceed a certain limit, the CPU might arm an existing suppression system.

A more complex situation might be a precipitator failure combined with low wind velocity at the chimney output. These conditions might cause a stack fire. Signals monitored by CPU should either sound an alert or call for a shutdown.

It is possible to assemble one's own proprietary system by buying components from a number of sources, but this may have a number of drawbacks. First, the services of a competent fire-protection engineer should be used to make sure the protection is adequate. For another, components would have to carry UL or FM ratings and then the entire system would have to be rated.

Though proprietary systems are especially advantageous for large operations, they offer significant advantages to smaller establishments as well. They can be purchased outright or leased. Costs can be slashed further by installing proprietary systems in a number of buildings in an area—factories, stores, warehouses, office buildings. This would permit hiring professionals to patrol the entire complex, bringing costs down.

Multiplant operations can install proprietary systems in several small plants in the same area for cost efficiency. The systems can then be linked to a central control console.

System design considerations

The spacing of detectors. Because a fire-resistive structure can be expected to withstand the burnout of its contents for a longer period of time without structural collapse than a structure constructed of combustible materials, it follows that early warning is of greater importance in the latter structure. This point should be kept in mind by the system designer, who should cover combustible buildings with a heavier concentration of detectors than would normally be used. This, coupled with a tie-in to the local fire department or central station, will go a long way in minimizing building damage.

To determine the number of detectors required for a given area, consideration must be given to a number of factors. In general, the more detectors that are installed, the greater the coverage that is provided. If the number of detectors in a given area were doubled, the distance and the time that combustion products would have to travel from the farthest point in the room to a detector would be proportionately reduced.

The exact area to be covered by an ionization detector, for example, depends upon total area, building construction, area contents, air movement, value of building and contents, ceiling obstructions, and cost of equipment downtime.

It would be well at this point to quote from the NFPA standards on this subject.

Detector spacings shall not exceed the linear maximum indicated by tests of Underwriters Laboratories Inc. and Factory Mutual Laboratories Inc. for the particular device used. Closer spacing may be required due to structural characteristics of the protected area and possible drafts or other conditions affecting detector operation. Factors such as air velocity, ceiling shape and height, ceiling material, and configuration of building structure influence the proper spacing.

Location of smoke detectors. Smoke detectors shall be located and adjusted to operate reliably in case of smoke at any part of the protected area. The location of detectors should be based upon an engineering survey of the application of this form of protection to the area under consideration. These features include air velocity, number of detectors to provide adequate coverage of cross-sectional area of the space with respect to travel, diffusion, or stratification of smoke; location of detectors with respect to return, supply or circulating blowers, air conditioning facilities, temperature variations and the like. Such conditions vary with different installations and should be dealt with on the basis of experience.

Special consideration shall be given to the storage of contents of a protected space to provide unobstructed openings for the travel of smoke to the smoke detector.

Where air conditioning or ventilating equipment serves the space to be protected by a smoke detector, particular attention shall be given to the supply, return, and circulation of smoke under any condition of operation of the equipment to ensure prompt detection.

There is a temptation to try to protect many thousands of square feet of building space by spacing smoke detectors in the air-handling system. Since the products of combustion become diluted by air in their travel toward the detectors in the air-handling system, it follows that in order to detect smoke with the detectors located in the vicinity of the fan, a very heavy smoke concentration must exist in the occupied area of the building.

Detector maintenance. Every fire protection system needs periodic maintenance and must be installed, regularly inspected, and periodically maintained or tested by a knowledgeable, responsible person. Without a periodic inspection and testing of each component, no system can be considered reliable.

This work may be done by an organization whose specialty is installing and servicing the type of equipment selected. It is also possible for user personnel to become expert at routine inspection procedures.

Proper functioning of the system should be checked at regular intervals. This can be done by applying heat to a thermal detector or introducing smoke directly into a smoke or ionization detector. A spot check of one or two detectors each month can be made part of a regular fire prevention inspection. Different detectors should be actuated each time so that all components of the system will have been tested in the course of a year.

Every six months, each detector head screen in the system should be inspected for dust accumulation and cleaned if necessary. In a dusty location, more frequent cleaning may be necessary. Then a detector should be activated and control unit indications checked: supervisory circuits are checked via the "reset" switch.

On a yearly basis, each detector on the circuit should be checked for operation and sensitivity. Alarm relay contacts should be checked for proper operation.

PORTABLE FIRE EXTINGUISHERS

Equipment used to extinguish and control fires is of two types: fixed and portable. Fixed systems include water equipment (automatic sprinklers, hydrants, and standpipe hoses) and special pipe systems for dry chemical, carbon dioxide, Halon, and foam. Special pipe systems are applicable to areas of high fire potential where water may not be effective, such as tanks for storage of flammable liquids and on electrical equipment. Fixed systems are discussed later in this chapter.

Fixed systems, however, must be supplemented by portable fire extinguishers. These often can preclude the action of sprinkler systems because they can prevent a small fire from spreading as well as provide rapid extinguishment in the early stages of a fire.

Principles of use

Even though the plant may be equipped with automatic sprinklers or other means of fire protection, portable fire extinguishers should also be available and ready for emergency. "Portable" is applied to manual equipment used on small incipient fires or in the interim between discovery of fire and the functioning of automatic equipment or the arrival of professional firefighters.

Figure 17-12. Picture-symbol use labels promulgated by the National Association of Fire Equipment Distributors. The symbols illustrated are for use on a Class A extinguisher (for extinguishing fires in trash, wood, or paper). Symbol at left is in blue. Since extinguisher is not recommended for use on Class B or C fires, these two illustrations are in black, with diagonal red line through them. For use on a Class A/B extinguisher, first two illustrations would be in blue, third would be in black with red diagonal. For use on Class B/C, last two would be in blue; on Class A/B/C, all three would be in blue.

To be effective, portable extinguishers must be:
1. Approved by a recognized testing laboratory.
2. The right type for each class of fire that may occur in the area.
3. In sufficient quantity and size to protect against the expected exposure in the area.
4. Located where they are readily accessible for immediate use and the location is kept accessible and clearly identified.
5. Maintained in operating condition, inspected frequently, checked against tampering, and recharged as required.
6. Operable by area personnel who can find them and who are trained to use them effectively and promptly.

Classification of fire extinguishers. Portable extinguishers are classified to indicate their ability to handle specific classes and sizes of fires. This classification is necessary because of the constant development of improved and new extinguishing agents and devices, and because of the availability of larger portable extinguishers. Labels on extinguishers indicate the class and relative size of fire that they can be expected to handle.

The following paragraphs are a guide to the selection of portable fire extinguishers for given exposures, in accordance with classifications set forth in NFPA 10, *Portable Fire Extinguishers.* (Be sure to review Classification of Fires earlier in this chapter.) It is essential that plant protection and insurance recommendations be observed, based upon fire protection requirements of the authority having jurisdiction.

CLASS A EXTINGUISHERS for ordinary combustibles, such as wood, paper, some plastics, and textiles, where a quenching-cooling effect is required.

CLASS B EXTINGUISHERS for flammable liquid and gas fires, such as oil, gasoline, paint, and grease, where oxygen exclusion or flame-interrupting effect is essential.

CLASS C EXTINGUISHERS for fires involving electrical wiring and equipment where the dielectric nonconductivity of the extinguishing agent is of first importance. (These units are not classified by a numeral because Class C fires are essentially either Class A or Class B, but also involve energized electrical wiring and equipment.) Therefore, the coverage of the extinguisher must be chosen for the burning fuel.

CLASS D EXTINGUISHERS for fires in combustible metals, such as magnesium potassium, powdered aluminum, zinc, sodium, titanium, zirconium, and lithium. Persons working in areas where Class D fire hazards exist must be aware of the dangers in using Class A, B, or C extinguishers on a Class D fire, as well as the correct way to extinguish Class D fires. These units are not classified by a numerical system and are intended for a special hazard protection only.

The following recommendations are given in NFPA Standard No. 10 as a guide in marking extinguishers, and/or extinguisher locations, to indicate the suitability of the extinguisher for a particular class of fire. Extinguishers suitable for more than one class of fire may be identified by multiple symbols and are described at the end of this subsection.

Markings should be applied by decalcomanias, painting, or similar methods having at least equivalent legibility and durability.

1. Extinguishers suitable for Class A fires should be identified by a triangle containing the letter "A." If colored, the triangle should be colored green.

2. Extinguishers suitable for Class B fires should be identified by a square containing the letter "B." If colored, the square should be colored red.

3. Extinguishers suitable for Class C fires should be identified by a circle containing the letter "C." If colored, the circle should be colored blue.

4. Extinguishers suitable for fires involving metals should be identified by a five-pointed star containing the letter "D." If colored, the star should be colored yellow.

Figure 17-13a. Fire equipment should be conspicuously located, appropriately marked, and inspected regularly.

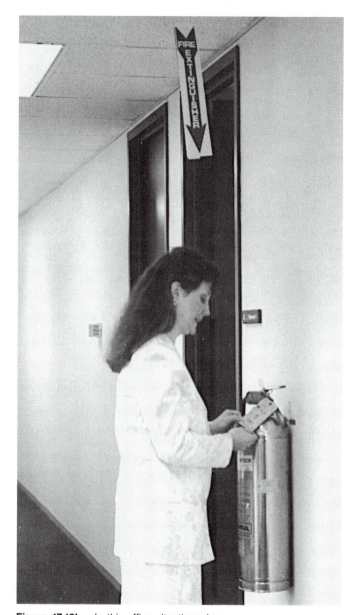

Figure 17-13b. In this office situation where marking a wall would be incongruous with the decor, a three-sided, red and white FIRE EXTINGUISHER sign is suspended from the ceiling.

Markings should be applied to the extinguisher, on the front, and be of a size and form to give easy legibility at a distance of three feet. An easily visible picture-symbol label has been devised by the National Association of Fire Equipment Dealers, NAFED (see Figure 17-12).

Where markings are applied to walls and panels in the vicinity of extinguishers, they should be of a size and form to give easy legibility at a distance of 15 ft (4.6 m).

Extinguishers listed by Underwriters Laboratories Inc. are rated after physical testing. These ratings, which are indicated by a numeral and a letter, define the extinguishing potential of an extinguisher because they specify the type and size or number that should be installed in a specific area.

The numeral signifies the relative extinguishing potential, and the letter(s) signifies the class(es) of fire on which the particular extinguisher is most effective for extinguishment. For example, a 4A:10B:C rating signifies that the extinguisher is (1) adequate for a small Class A fire, (2) better for extinguishing a Class B fire, and (3) safe to use in or near electrical equipment.

Location of units. Extinguishers should be located close to the likely hazards, but not so close that they would be damaged or cut off by the fire. They should be located along the normal path of egress from the building, preferably at the exits. Where highly combustible material is stored in small rooms or enclosed spaces, the extinguishers should be located outside the door, rather than inside. This requires potential users to exit the room and then make a conscious decision to re-enter and fight the fire.

The location of the extinguisher should be made conspicuous (Figure 17-13). For example, if it is hung on a large column or post, a distinguishing red band can be painted around the post. Also, large signs can be posted to direct attention to extin-

guishers. The extinguisher should be kept clean and should not be painted in any way that will camoflage it or obscure labels and markings on it.

If the extinguisher itself is not already marked plainly to indicate the classifications of fire or types of material for which it is intended, signs or cards should be placed on the wall close to where it hangs. Special labels are available from manufacturers of the extinguishers, insurance companies, the National Safety Council, NAFED, and NFPA (see References for addresses); markings indicating special uses can be stenciled on the extinguisher or on an adjacent wall.

Fire extinguishers must not be blocked or hidden by stock, finished material, or machines. They should be hung in accordance with NFPA Standard No. 10, where they will not be damaged by trucks, cranes, and other operations, or corroded

Table 17-F. Extinguishers Suitable for Class A Fires

Basic Minimum Extinguisher Rating for Area Specified	Maximum Travel Distances to Extinguishers	Areas to Be Protected per Extinguisher		
		Light Hazard Occupancy	Ordinary Hazard Occupancy	Extra Hazard Occupancy
1A	75 ft	3,000 sq ft	*	*
2A	75 ft	6,000 sq ft	3,000 sq ft	*
3A	75 ft	9,000 sq ft	4,500 sq ft	3,000 sq ft
4A	75 ft	11,250 sq ft	6,000 sq ft	6,000 sq ft
6A	75 ft	11,250 sq ft	9,000 sq ft	6,000 sq ft
10A	75 ft	11,250 sq ft†	11,250 sq ft†	9,000 sq ft
20A	75 ft	11,250 sq ft†	11,250 sq ft†	11,250 sq ft†
40A	75 ft	11,250 sq ft†	11,250 sq ft†	11,250 sq ft†

*The protection requirements specified in this table may be fulfilled by several extinguishers of lower ratings for ordinary or extra hazard occupancies, subject to the approval of the authority having jurisdiction. Consideration should be given to the number of persons available to operate the extinguishers, the degree of training provided, and the physical ability of employees likely to handle extinguishers.

‡11,250 sq ft is considered a practical limit, which is 75 ft (22.8 m) from an extinguisher in a circle.

From NFPA No. 10, Table 3-2.1

by chemical processes, and where they will not obstruct aisles or injure passersby. If installed out of doors, the extinguisher should be protected from the elements.

Plant and warehouse aisles should (1) be wide enough that mobile fire-protection units can be brought close to a fire and (2) be kept free of obstructions. Floor spaces may be marked to allow access to fire extinguishing equipment, and units may be protected with bumpers or guardrails.

Extinguishers weighing over 40 lb (18 kg) should not be more than 3½ ft (1 m) above the floor and a clearance of at least 4 in. (10 cm) should be maintained between the bottom of the extinguisher and the floor.

Distribution of extinguishers. The minimum required number and type of portable extinguishers to be installed is determined for each floor or area by the relative hazard of the occupancy, the nature of any anticipated fires, and protection for special hazards.

- Extinguishers suitable for Class A fire hazards are installed according to the classification of occupancy: light, ordinary, or extra hazard. These are summarized in Table 17-F.

Light hazard occupancies include office buildings, schools (exclusive of trade schools and shops), churches, and public buildings, where because of the relatively small amount of combustibles, incipient fires of minimum severity may be anticipated.

Ordinary hazard occupancies include department stores, warehouses, and manufacturing buildings of average hazard where incipient fires of average severity in combustibles may be anticipated.

Extra hazard occupancies include some warehouses, woodworking shops, textile mills, and paper mills, where because of the character or quantity of combustibles, extra severe incipient fires may be anticipated.

- Extinguisher requirements for Class B protection (of a "special hazard" area, such as a laboratory or kitchen) are in addition to the requirements of extinguishers for Class A protection

Table 17-G. Extinguishers Suitable for Class B Fires, for Fires in Flammable Liquids ¼ Inch and Under in Depth

Type of Hazard	Basic Minimum Extinguisher Rating	Maximum Travel Distance to Extinguishers
For Extinguishers Labeled Between 1955 and 1969		
Light	4-B	50 ft (15 m)
Ordinary	8-B	50 ft
Extra	12-B	50 ft
For Extinguishers Labeled After June 1, 1969		
Light	5-B	30 ft (9 m)
	10-B	50 ft (15 m)
Ordinary	10-B	30 ft
	20-B	30 ft
Extra	20-B	50 ft
	40-B	50 ft

Note. For flammable liquid hazards of depth greater than ¼ in. (6 mm), Class B fire extinguishers shall be provided on the basis of one numerical unit of Class B extinguishing potential per square foot of flammable liquid surface of the largest tank hazard within the area.

From NFPA Standard No. 10, *Portable Fire Extinguishers,* Table 3-3.1.1.

except where the total area under consideration presents wholly Class B hazards. Extinguishers are installed according to the nature of anticipated fires and protection for special hazards.

The requirements for fire extinguisher size and placement for Class B fires other than those in flammable liquids over ¼ in. (6 mm) deep are shown in Table 17-G.

Two or more extinguishers of lower rating, except for foam extinguishers, shall not be used to fulfill the protection requirements of Table 17-G. Up to three foam extinguishers of 2½ gallons (9.5 liters) approximately may be used to fulfill light hazard requirements and up to three AFFF (aqueous film-forming foam) solution extinguishers (2½ gallons approximately) may be used to fulfill ordinary or extra-hazard requirements.

The protection requirements may be fulfilled with extinguishers of higher ratings provided the travel distance to such larger extinguishers does not exceed 75 ft (23 m).

For flammable liquids of appreciable depth (over ¼ in.), such as those in dip or quench tanks, the following recommendations apply.

1. Class B fire extinguishers should be provided on the basis of two numerical units of Class B extinguishing potential per square foot (0.09 m²) of flammable liquid surface of the largest tank to be protected within the area.
2. Two or more extinguishers of lower ratings, except for foam extinguishers, shall not be used instead of the extinguisher required for the largest tank, but up to three AFFF (aqueous film-forming foam) extinguishers may be used to fulfill these requirements.
3. Portable fire extinguishers shall not be installed as the sole protection for flammable light hazard of appreciable depth (¼ in.) whose surface exceeds 10 sq ft (0.9 m²). Where personnel are trained in the extinguishment of such fires, the maximum surface area shall not exceed 20 sq ft (1.9 m²). Fixed fire protection systems should be considered for protecting tanks in excess of 20 sq ft; the portable extinguishers would then be used for putting out burning liquid spills outside the range of fire equipment or putting out fires that originate outside the tank.
4. Where approved automatic fire protection devices or systems have been installed for a flammable liquid, additional portable Class B fire extinguishers, as required in paragraph 1, may be waived. When so waived, Class B extinguishers should be provided to protect areas near such protected hazards.
5. Travel distances should be given consideration with reference to special hazards and the availability of the extinguisher for such protection. Scattered or widely separated hazards shall be individually protected if the specified travel distances in Table 17-G are exceeded. Likewise, extinguishers in the proximity of a hazard shall be carefully located so as to be accessible in the presence of a fire without undue danger to the operator.

▪ Extinguishers with Class C ratings are required where energized electrical equipment may be encountered which would require a nonconducting extinguishing media. This will include fire either directly involving or surrounding electrical equipment. Since the fire itself is a Class A and/or Class B, the extinguishers are sized and located on the basis of the anticipated Class A or B hazard.

Whenever possible, electrical equipment should be deenergized before attacking a Class C fire.

▪ Extinguishers used for Class D protection and other special fires are installed according to the size and type of the special hazard. The type of combustible material, the quantity, and its physical form are all determining factors that must be considered when selecting the proper type of agent and the method of application.

Selection of extinguishers

Operating characteristics that make one type of portable fire extinguisher suitable for certain fire hazards may make the same type dangerous for others.

The purchaser and user of extinguishers should secure on-the-job advice from fire inspection bureaus, fire insurance carriers, and fire protection engineers.

Choosing the right extinguisher is extremely important. All too frequently cost is given more consideration than adequate protection. Remember, good extinguishers are worth their cost because of the protection they give. Obviously, only extinguishers listed by nationally recognized agencies such as Underwriters Laboratories Inc. or Factory Mutual Engineering Corporation should be purchased.

Table 17-H and Figure 17-14 give an overview of common types of extinguishers and their operating characteristics.

Although suitable types and sizes are installed in conformance with NFPA 10, there is a certain flexibility that can be used to the purchaser's advantage. For example, if a certain condition calls for Class B extinguishers, the required units could be obtained by using various size dry-chemical, carbon dioxide, AFFF, or Halon extinguishers. The relative advantages and disadvantages of each of these units should be considered with respect to all the conditions in the area. It is also wise to investigate the relative merits of any particular extinguisher available on the market, because the UL rating can vary widely for extinguishers of the same size but of different manufacture.

The design and operating features, ease of maintenance, and the availability of repair service should be considered too. If possible, actually operating and testing the extinguisher may make a great deal of difference in the final selection.

The following discussion of portable extinguishers is divided into (1) water solution extinguishers, (2) dry-chemical, carbon dioxide, and Halon extinguishers, and (3) dry-powder extinguishers. (Some vaporizing liquid extinguishers, such as those containing carbon tetrachloride or chlorobromomethane, are excluded as they are not recommended because of their toxic properties.)

Water solution extinguishers

Fire extinguishers that use water or water solutions include: pump tank, cartridge actuated,* stored pressure, loaded stream, and foam.** These extinguishers are effective against Class A fires because of the quenching and cooling effect of water. These units cannot be used on fires in or near electrical equipment since they can produce a shock hazard to the operator.

---

*Underwriters Laboratories Inc., *Fire Protection Equipment List;* and Factory Mutual Engineering Corporation, *Approval Guide—Equipment, Materials, Services for Conservation of Property.*

**National Fire Protection Association, *Installation of Portable Fire Extinguishers,* NFPA 10.

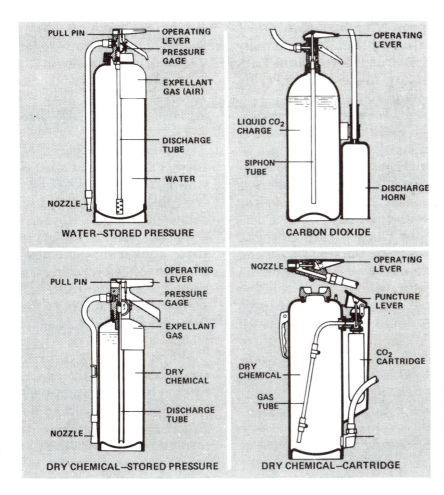

PULL PIN — OPERATING LEVER — PRESSURE GAGE — EXPELLANT GAS (AIR) — DISCHARGE TUBE — WATER — NOZZLE

WATER–STORED PRESSURE

OPERATING LEVER — LIQUID CO₂ CHARGE — SIPHON TUBE — DISCHARGE HORN

CARBON DIOXIDE

PULL PIN — OPERATING LEVER — PRESSURE GAGE — EXPELLANT GAS — DRY CHEMICAL — DISCHARGE TUBE — NOZZLE

DRY CHEMICAL–STORED PRESSURE

NOZZLE — OPERATING LEVER — PUNCTURE LEVER — CO₂ CARTRIDGE — DRY CHEMICAL — GAS TUBE

DRY CHEMICAL–CARTRIDGE

Figure 17-14. Among the major types of portable fire extinguishers are the stored-pressure water unit, the carbon dioxide unit, the stored-pressure dry-chemical unit, and the cartridge-operated dry-chemical unit. The internal parts of each type are shown and labeled. (Reprinted with permission from *Plant Engineering* Magazine.)

One important maintenance item on all water extinguishers is to inspect the nozzle frequently for foreign particles that may prevent discharge, such as dirt, match sticks, and paper. The pressure-relief hole in the cap of gas cartridge and foam units must also be free from obstruction as it is designed to release any residual pressure that may injure a worker when removing the cap before recharging.

In locations exposed to freezing temperatures, extinguishers must be installed in heated cabinets, or they must be charged with a nonfreezing solution. However, antifreeze or salt solutions should not be used unless the equipment has been designed for such use, because these chemicals can cause rapid corrosion. Be sure to check manufacturer's recommendations. Calcium chloride and other salt solutions are good conductors of electricity and, therefore, are especially dangerous if applied to live electrical apparatus.

When charging any extinguisher with an antifreeze solution, the chemical should be thoroughly dissolved in warm water in a separate container and then poured into the extinguisher through a fine strainer to remove any foreign particles that may clog the unit. Greater care must be given in maintaining extinguishers containing antifreeze solutions, because they may corrode more easily than those filled with plain water. In recharging, all parts including the hose and nozzle should be thoroughly flushed with plain water. Be sure that all the plain water is drained off to prevent freezing and clogging.

Dry-chemical, carbon dioxide, and liquefied-gas extinguishers

Dry-chemical extinguishers. The dry-chemical extinguisher is one of the most versatile units available. It extinguishes by interrupting the chemical flame chain reaction. It is not to be confused with dry-powder extinguishers discussed in the next section.

- Four types of base agents are used—sodium bicarbonate, potassium bicarbonate, urea potassium bicarbonate, potassium chloride, or monoammonium phosphate base (multipurpose). When recharging, it is important to use only the dry chemical agent and cartridge that is recommended by the extinguisher manufacturers.

Sodium bicarbonate-base dry chemical, the most common agent, is available in ordinary and foam form. The ordinary type can be used simultaneously with foam without causing the foam blanket to break down.

Potassium bicarbonate-base dry chemical, "purple K," is similar in extinguishing properties to sodium bicarbonate-base dry chemical, but it has twice the effective firefighting capacity on a pound-to-pound basis. This agent has good moisture repellency and is compatible with the simultaneous use of water or foam. It does not allow fuel to reflash as easily or as rapidly as sodium bicarbonate. The absence of momentary flare-up also permits the user to approach fires more closely. Also, the potassium compound extinguishes the

Table 17-H. Fire Extinguisher Selection Chart

| | Class A | | | | | | Class A/B | Class B/C | | | |
| | Water Types | | Multipurpose Dry Chemical | | AFFF Foam | Halon 1211 | AFFF Foam | Carbon Dioxide | Dry Chemical Types | | Halon 1211 |
	Stored Pressure*	Pump Tank*	Stored Pressure	Cartridge Operated	Stored Pressure*	Stored Pressure	Stored Pressure*	Self Expelling	Stored Pressure	Cartridge Operated	Stored Pressure
Sizes Available	2½ gal	2½ and 5-gal	2½-30 lb *ALSO* Wheeled 50-350 lb	5-30 lb *ALSO* Wheeled 50-350 lb	2½ gal	9 to 22 lb	2½ gal	5-20 lb *ALSO* Wheeled 50-100 lb	2½-30 lb *ALSO* Wheeled 150-350 lb	4-30 lb *ALSO* Wheeled 50-350 lb	2-22 lb
Horizontal Range (Approx.)	30 to 40 ft	30 to 40 ft	10-15 ft (Wheeled-15-45 ft)	10-20 ft (Wheeled-15-45 ft)	20 to 25 ft	14 to 16 ft	20 to 25 ft	3-8 ft (Wheeled-10 ft)	10-15 ft (Wheeled-15-45 ft)	10-20 ft (Wheeled-15-45 ft)	10-16 ft
Discharge Time (Approx.)	1 min.	1 to 3 min.	8-25 s (Wheeled-30-60 s)	8-25 seconds (Wheeled-20-60 s)	50 seconds	10 to 18 seconds	50 seconds	8-15 seconds (Wheeled-8-30 s)	8-25 seconds (Wheeled-30-60 s)	8-25 seconds (Wheeled-20-60 s)	8 to 18 seconds

| | Class A/B/C | | | Class D |
| | Multipurpose Dry Chemical | | Halon 1211 | Dry Powder |
	Stored Pressure	Cartridge Operated	Stored Pressure	Cartridge Operated
Sizes Available	2½-20 lb *ALSO* Wheeled 150-350 lb	5-30 lb *ALSO* Wheeled 50-350 lb	9 to 22 lb	30 lb *ALSO* Wheeled 150-350 lb
Horizontal Range (Approx.)	10-15 ft Wheeled-15-45 ft	10-20 ft (Wheeled-15-45 ft)	14 to 16 ft	5 ft (Wheeled-15 ft)
Discharge Time (Approx.)	8-25 seconds (Wheeled-30-60 s)	8-25 seconds (Wheeled-20-60 s)	10 to 18 seconds	20 seconds (Wheeled-150 lb 70 s, 350 lb 1¾ min.)

*Must be protected from freezing.
Reprinted with permission from the National Association of Fire Equipment Distributors.

leading edges of contained flammable liquid fires more easily than the sodium compound. This agent is also marketed as potassium bicarbonate/urea.

Ammonium phosphate-base dry chemical (multipurpose) has operating characteristics on Class B and C fires similar to sodium- and potassium-base dry-chemical extinguishers. When discharged into a Class A fire, chemical reaction will destroy the flames and a coating, formed when the extinguishing agent softens, adheres to the burning surface, thereby retarding further combustion. To obtain complete extinguishment on Class A materials, all burning areas must be thoroughly exposed to the extinguishing agent. Because any small burning ember may be a source of reignition, the importance of proper application on Class A fires is more critical than with water extinghisher. In the presence of moisture, multipurpose dry chemicals may cause corrosion when discharged on metals. It is therefore very important to clean up the multipurpose agent immediately after the fire is extinguished. It is also important never to mix the ammonium phosphate-base agent with the potassium or sodium bicarbonate-base agent because dangerous pressure can be developed by even a trace of moisture.

■ There are two basic styles of dry chemical extinguishers as defined by the propellant technique: gas cartridge or stored pressure. In the gas cartridge type, pressure is supplied by a gas stored in a separate cylinder; whereas in the stored-pressure type, the entire container is pressurized. The extinguisher may be used intermittently or continuously, depending on the nature of the fire.

Many models have a high-velocity discharge; therefore, care should be taken not to aim the initial discharge directly into the burning area since it may cause the fire to spread. For best results when attacking a Class B fire, use a "fanning" action—rapidly move the nozzle from side-to-side so that the agent can thoroughly intermix with the flames. Start well in advance of the burning edge and go beyond the burning edge on each side to avoid leaving any burning pockets behind.

To minimize the possibility of reflash, the operator should continue to discharge the chemical after the fire has gone out.

Carbon dioxide extinguisher. The CO_2 extinguisher puts out a fire by diluting the amount of available oxygen. It does not leave a residue.

Liquefied-gas extinguisher. The liquefied-gas extinguisher puts out a fire by interrupting the flame chain reaction. Because of toxicity considerations, do not use this extinguisher in confined spaces or small, unventilated rooms.

Halogenated compounds

There are two halogenated compounds that are used as fire extinguishing agents. One is Halon 1211, bromochlorodifluoromethane. Although it is a liquid when discharged, it quickly vaporizes; this characteristic provides a desirable compromise between reach and dispersion.

The other is Halon 1301, bromotrifluoromethane, a liquefied gas that provides for immediate dispersion in the fire area. It is about twice as effective as CO_2 on a pound-for-pound basis. Halon 1301 is a clean agent and in its undecomposed state has been determined to be safe to humans (in concentrations up to about 10 percent by volume in air for relatively short exposures of up to 20 minutes). However, when it is exposed to temperatures of about 900 F (480 C), it becomes unstable and breaks down to give off hydrogen fluoride, hydrogen bromide, and free bromine.

Dry-powder extinguishers for metal fires

The use of combustible metals has increased until protection is now needed for: sodium, titanium, uranium, zirconium, lithium, magnesium, sodium-potassium alloys (NaK), and other less common materials. There are several powdered agents approved for use on metal fires, the oldest being the G-1 type, which is a graphite-organic phosphate compound. When it is applied with a scoop or shovel to a metal fire, the phosphate material generates vapors that blanket and smother the flames, and the graphite, being a good conductor of heat, cools the metal below its ignition temperature.

Care should be taken to assure that the depth of the cover is adequate to provide a smothering blanket. If hot spots should occur, they should be covered by additional powder. The burning metal should be allowed to cool before disposal is attempted.

Another material is "Met-L-X," composed of a sodium chloride base with additives to make it free flowing, increase water repellency, and create the property of heat caking. This material is dispensed from a 30-pound dry-powder extinguisher which is similar in appearance and physical features to the cartridge-operated dry-chemical extinguisher, or from larger wheeled or stationary units.

The technique used to extinguish a metal fire with Met-L-X is to open the nozzle of the extinguisher fully and apply a thin layer of Met-L-X over the burning mass from a safe distance, until control is established. Then throttle the nozzle to produce a soft heavy stream and completely cover with a heavy layer from close range. The heat of the fire causes the Met-L-X to cake, forming a crust that excludes air.

"Lith-X" is also a dry powder. It is composed of a special graphite base with additives to render it free flowing so it can be discharged from an extinguisher. Lith-X was developed mainly for use on lithium fires, but is also effective on other combustible metals. Lith-X does not cake or crust when applied over a burning metal. It excludes air and conducts heat away from the burning mass, extinguishing the fire.

The extinguishing effects of dry-powder agents on a given metal depend on the physical form and quantity of the metal involved and on ambient conditions. Although dry powders are very effective on combustible metal fires, they all have certain limitations. The type, quantity, and form of the metal and the existing physical conditions must be considered when selecting the proper type of dry powder and the method of application.

A problem recently developed in firefighting involves pyrophoric liquids, such as triethylaluminum. These liquids ignite spontaneously and the resulting fires cannot be easily extinguished by dry powder or other commonly used agents. A special material ("Met-L-Kyl") has been developed, consisting of a bicarbonate base dry chemical and an activated absorbent.

The principle of extinguishment involves the combination effect of dry chemical, which extinguishes the flames, and the absorbent which absorbs the remaining fuel and prevents reignition. This extinguishing agent has been designed so that it

can be discharged from an extinguisher similar to the standard cartridge-operated dry chemical model.

Miscellaneous equipment

Wheeled equipment. Large portable units on wheels are commercially available and include 17-gal and 30-gal (64- and 115-liter) soda-acid, loaded stream, calcium chloride, and foam types; 50-, 75-, and 100-lb (23-, 34-, and 45-kg) carbon dioxide types; and 150- and 350-lb (68- and 160-kg) dry-chemical types.

Wheeled "twinned" extinguishers. With the development of water-soluble fluorocarbon surface-active agents, foaming agents are available that give water the property of floating in thin layers on liquid fuel surfaces ("light water"). This characteristic provides excellent protection against reflash on liquid hydrocarbon fires with only one-fourth the volume as compared to protein–air foam.

A combination wheeled extinguisher with "purple K" dry chemical and "light water" fluorocarbon foam provides a synergistic extinguishing system that has both rapid knockdown of flames and complete protection against reflash. The two agents are simultaneously applied through dual pistol-grip nozzles.

Vehicle-mounted equipment. Water, foam, carbon dioxide, and dry chemical extinghishing agents are available in units that are mounted on vehicles. They range in size from in-plant fire vehicles capable of turning in warehouse aisles (Figure 17-15), to large trucks.

Fire blankets. In some cases, fire blankets can be used to smother a small fire. Their major purpose is to extinguish burning clothing, but they are useful, too, for smothering flammable liquid fires in small open containers.

Flame-retardant blankets are available. The most common size is 66 by 80 in. (1.7 × 2 m). Blankets are usually stored in containers mounted on a wall or column, so arranged that they can be readily pulled out.

Miscellaneous hand equipment. Five gallon or 2½ gallon water or antifreeze backpack tanks are available with hand pumps built into the hose nozzle handles. This type of unit is carried on the back, and the slide-action pump is operated with both hands. The backpack unit is frequently used for combating brush fires.

Training of employees

Extinguishers are effective only when fires are in the first stages. Extinguishers must be immediately accessible and promptly used by trained personnel. Extinguishers are only as good as the operators using them. It pays to thoroughly train key workers on each shift.

Instruction of employees in the use of extinguishers can best be given by demonstration, preferably at the time when extinguishers are scheduled for recharging. At the demonstrations, fire conditions should be simulated and an instructor should explain the fundamentals of firefighting and the use of the equipment, and employees should then be allowed to get the "feel" of the extinguisher. In small organizations, everyone in the plant should attend and participate in the demonstrations. In larger plants, it is advisable to train a suitable number of employees, so that there is a good distribution of trained personnel.

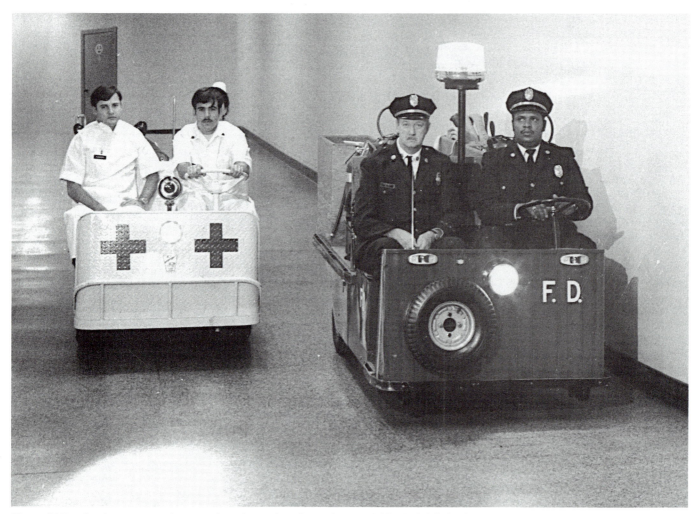

Figure 17-15. Serving a community many times larger than those often served by their full-sized counterparts, a diminutive fire truck and ambulance negotiate a ramp between floors in the Pentagon. Manned around the clock, the units provide emergency service to some 30 thousand employees. (Official Department of Defense Photograph.)

One of the most difficult decisions faced by anyone is the decision of whether to fight the fire or get out safely. If the following conditions are met, an employee could decide to try to fight the fire with an extinguisher: (1) there is a clear exit, (2) the fire brigade or department is already called or being called, (3) the fire is small, like in a waste basket or small tool housing, (4) the employee knows how to use the extinguisher, and (5) the extinguisher is in working order.

An employee should *not* fight the fire if it is clearly spreading beyond the point of origin, if the fire could block the exit, or if he or she is unsure how to use the extinguisher.

The education of employees should be continued with demonstrations, practice drills, and lectures at yearly intervals, or more often if there is a special fire hazard. It may be advisable to put printed instructions regarding the use of fire extinguishers into the hands of the employee. Mimeographed or printed sheets, leaflets, or cards can give both general and detailed instructions regarding the use of the extinguishers.

Fire extinguisher manufacturers, insurance companies, fire departments, and the National Safety Council have films, posters, and cut-away displays useful in explaining the construction, maintenance, and operation of portable extinguishers.

NFPA 10, *Portable Fire Extinguishers,* gives particularly valuable information for the training of employees in practices related to portable fire extinguishers.

In addition, instructions in how to use extinguishers should be permanently posted near or on the extinguishers themselves.

Fire extinguisher training is intended to teach the employees how to stop small fires from spreading out of control. Underwriters Laboratories Inc. considers a fire extinguisher to be only 40 percent of its firefighting capacity when it is used by an inexperienced operator, which would include most people because 9 out of 10 people have never used a fire extinguisher. This means that any fire extinguisher in the hands of a trained operator has 2½ times the firefighting capacity it has when used by a novice. This underscores the critical importance of training.

Maintenance and inspection

One person in the plant or establishment should be given the responsibility for the maintenance and inspection of fire extinguishing equipment. This employee may have one or more assistants to make routine inspections weekly or monthly and to do testing and repair work in accordance with procedures in NFPA

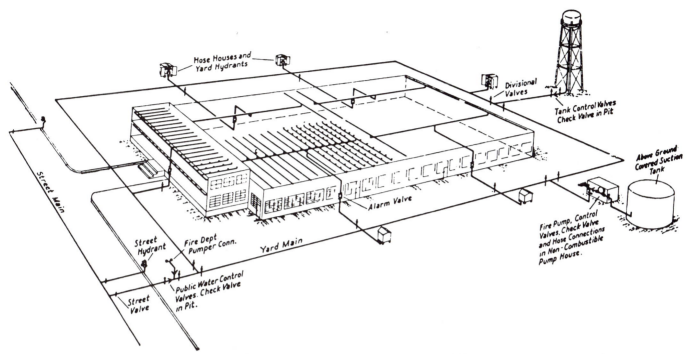

Figure 17-16. Typical automatic sprinkler installation for a modern industrial facility. Note sprinkler systems, yard mains, and water supplies. (Reprinted with permission from Factory Mutual Engineering Corp.)

10. There should be a record system and an organized plan for checking and repairing various types of extinguishers.

The inspection and maintenance records should at least consist of durable tags fastened to the extinguishers (Figure 17-13b) showing dates of inspection (at least monthly), and examination for recharge and other maintenance work (at least annually). The tag or label must indicate if the unit was recharged during maintenance.

A duplicate record should also be kept in office files. Other records may be kept that list extinguishers by type, by location, and by recharge periods. The office record system should give the history of each extinguisher, the type and quantity of refills on hand, and other pertinent information.

Efficient in-plant maintenance requires that supplies of spare parts and refills be kept on hand. In larger plants, specially trained personnel can test and refill extinguishers on a full-time basis. In other plants, the maintenance service can be under contract with a reliable service organization or the factory representative.

Generally, it is desirable to seal hand extinguishers or to install them in cabinets to discourage tampering and to facilitate inspection. However, extinguishers should not be locked in cabinets nor stored in any way that would prevent immediate use in an emergency.

One of the most important tasks involved in keeping fire extinguishers in top condition is periodic hydrostatic pressure testing, a mandatory requirement set forth in NFPA 10. The occasional reports of injuries and fatalities due to extinguishers rupturing during operation should be proof enough of the importance of this requirement. Be sure to check this standard for details of testing and marking the extinguishers.

Whether the testing is done by an outside agency or within the company, there should be no mass removal of extinguishers from an occupancy. A testing program should be established so only a few scattered units are removed at one time. They should be tested and returned to service promptly. Temporary replacement units or additional protection may be needed when extinguishers are being removed for testing.

WATER SYSTEMS

Water supply and storage

Sprinkler systems need a reliable water supply of ample capacity and pressure for efficient fire extinguishment (Figure 17-16). The water supply should be engineered with the sprinkler protection to provide a hydraulically balanced system at the least cost. For example, to supply all sprinklers likely to open in addition to hose streams, a volume of 500 to 3,000 gpm (31.5–190 liters/second), and sometimes more, may be needed. The precise need depends on the maximum water flow through all openings. The pressure requirement will vary; however, it should be high enough to maintain a residual pressure of 15 psig (103 kPa) (pressure while required water volume is flowing) in topstory sprinklers.

Water may be supplied from the following:
1. Underground supply mains from public water works.
2. Automatically or manually controlled pumps drawing water from lakes, ponds, rivers, surface storage tanks, underground reservoirs, or similar adequate sources.
3. Pressure tanks containing water in a quantity determined by the formula in NFPA 13, *Installation of Sprinkler Systems,* and compressed air for expelling the water into the piping supply system. The smallest tank for light-hazard occupancy is 2,000 gallons (7,570 liters).

4. Elevated tanks or reservoirs that depend on gravity to force water through the system.

Usually consideration should be given to providing at least two of these independent water sources. In case of fire, the main source furnishes water to the system immediately; it is reinforced by the other source that also supplies emergency protection if the primary source is out of service. The preferred source is a connection from a reliable public water works system. Connection should be made to two different mains to provide greater volume and give flexibility in case of failure of one water main.

A fire pump that can deliver water at high pressure over an extended time can be a second source. This pump should be located where it will not be put out of service by a fire, and it should have a reliable power supply, independent of the plant system and routed so that a fire anywhere in the protected areas will not expose the pumps' power supply.

Care must be taken that the water connections do not pollute drinking water, especially where emergency supplies are taken from a river or other nonpotable source.

Stored water for private fire protection should not be used for other purposes. Everyday use of tank-stored water necessitates constant refilling, hence the danger of accumulated sediment circulating into the hydrant and sprinkler system; also the varying water level may shorten the life of a wooden tank or require frequent painting of a steel tank.

Construction and installation of water tanks are beyond the scope of this chapter; however, one basic consideration is mentioned—in freezing temperatures, the tanks should be heated. In support of this consideration, the heating system should be inspected daily during freezing weather; the control valves, weekly; and the entire system (tank, supporting tower, piping, valves, heating system, and all components and accessories), annually. (See NFPA 22, *Water Tanks for Private Fire Protection;* also see NFPA *Fire Protection Handbook.*)

To facilitate inspection, wood or steel gravity tanks supported on steel towers should be ringed by a platform that is protected by a substantial steel rail having a vertical height (nominal), and 6-in. (15-cm) toeboards. They should have substantial steel ladders equipped with approved steel cages or basket guards.

Water tanks that are accessible to the public should have their ladders protected by locked gates or fences to discourage unauthorized persons from climbing them.

Automatic sprinklers

Automatic sprinklers are the most extensively used and most effective installations of fixed fire-extinguishing systems. These systems are basic and have proved so effective that most fire-protection engineers consider them the most important firefighting tool. Nationwide figures from NFPA indicate that sprinklers have a very high efficiency rating for satisfactory extinguishment, usually over 95 percent.

Although many other types of protection have been developed to deal effectively with special hazards, sprinklers are often recommended as "backup" protection.

The cost of automatic sprinkler protection is relatively small compared with the total plant investment; it generally averages about 2 percent. Experience shows that a sprinkler system often can pay for itself in ten years (and sometimes less). The saving in insurance cost is usually 40 to 90 percent lower than buildings without a sprinkler system. The cost of a sprinkler system is much less if it is installed when the building is built rather than later.

In addition to the economic factors, automatic sprinklers have an impressive life-saving record. Loss of life by fire is rare where properly designed and maintained sprinkler systems have been installed.

The surprising fact is that in the instances where sprinklers fail to operate (perhaps 3 to 4 percent of the cases), the failure is due to some readily preventable condition. Over one-third of all failures can be attributed to closed water supply valves.

Although the primary function of the sprinkler system is to deliver water automatically to a fire, the system can also serve as a fire alarm. This is done by installing an electrical water-

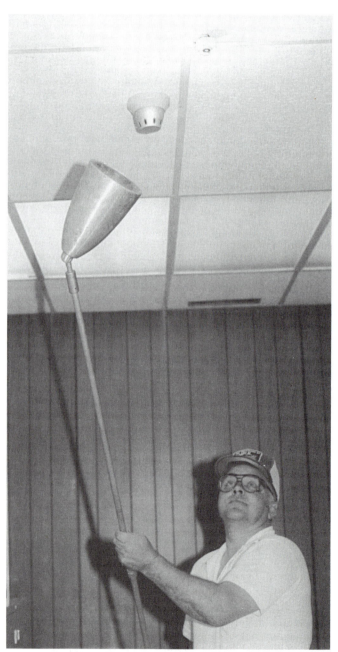

Figure 17-17. Safety Supervisor demonstrates method of checking detector with heat lamp. (Reprinted with permission from Ansul Fire Protection.)

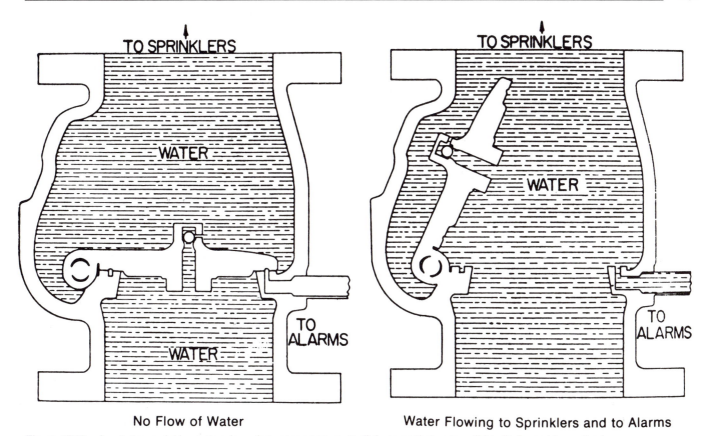

TO SPRINKLERS

WATER

WATER

TO ALARMS

No Flow of Water

TO SPRINKLERS

WATER

TO ALARMS

Water Flowing to Sprinklers and to Alarms

Figure 17-18. A wet-pipe sprinkler system is under water pressure at all times so that water will be discharged immediately when an automatic sprinkler operates. The automatic alarm valve triggers a warning signal when water flows through the sprinkler piping.

flow alarm switch in each main riser pipe. When a fire occurs and the first sprinkler opens, the water rushing though the pipe sets off an alarm which alerts the control system or firefighters.

Dependable sprinkler protection requires a systematic maintenance and inspection program that includes periodic inspection of water supply valves, water supply tests, physical inspection of system piping for obstructions to distribution, and similar items. (See Figure 17-17). See NFPA 13A, *Inspection, Testing, and Maintenance of Sprinkler Systems,* for additional maintenance requirements.

There are six basic types of automatic sprinkler systems: wet-pipe, dry-pipe, pre-action, deluge, combined dry-pipe and pre-action, and sprinklers for limited water supply systems. (See NFPA 13, *Installation of Sprinkler Systems.*) The combination dry-pipe and pre-action system is used on installations that are larger than can be accommodated by one dry-pipe valve. The limited water supply system is used for installations that do not have access to a continual or large supply of water. The other four types of sprinkler systems are discussed in detail in the following paragraphs. Also included are (1) a description of the simultaneous tripping of alarms, (2) an explanation of temperature rating, and (3) a list of reasons for sprinkler failure with remedies for avoiding them.

Wet pipe. In the wet-pipe system, which represents the greatest percentage of sprinkler installations, all parts of the system piping are filled up to the sprinkler heads with water under pressure (Figure 17-18). Then, when heat actuates the sprinkler, water is immediately sprayed over the area below. If a portion

of the wet-pipe system is subjected to freezing temperatures, as on a loading dock, the exposed sections should be filled with an antifreeze solution, or these sprinklers should be connected to a dry-pipe system. (A good rule-of-thumb is to use a dry-pipe system when more than 20 sprinklers are involved.) Shutting off and draining a wet-pipe system is hazardous in that it removes the protection when it may be needed.

The antifreeze should be water-soluble liquid that is proportioned to give low-temperature protection without producing a combustible mixture, as specified in NFPA 13 and 13A, which cover the installation, care, and maintenance of sprinkler systems. When the system is supplied from public water connections, antifreeze should be chemically pure glycerine (U.S. Pharmacopoeia 96.5 percent grade) or propylene glycol, and then added only in accordance with local health regulations.

A disadvantage of wet-pipe sprinkler systems is that, once the fusible link melts, water flow continues until the main water-supply valve is manually closed. Thus, the sprinkler may operate considerably longer than required to extinguish the fire, causing severe water damage.

To eliminate this problem, use automatic on-off (cycling) sprinkler heads. A bimetallic (snap disk) element in the sprinkler head expands when temperatures exceed 165 F (74 C), allowing water to flow from the head. As the heat subsides and the snap disk cools below 100 F (38 C), it contracts and water flow stops. The on-off cycle repeats as required. The main water-supply valve does not have to be shut off manually.

The dry-pipe system generally substitutes for a wet-pipe

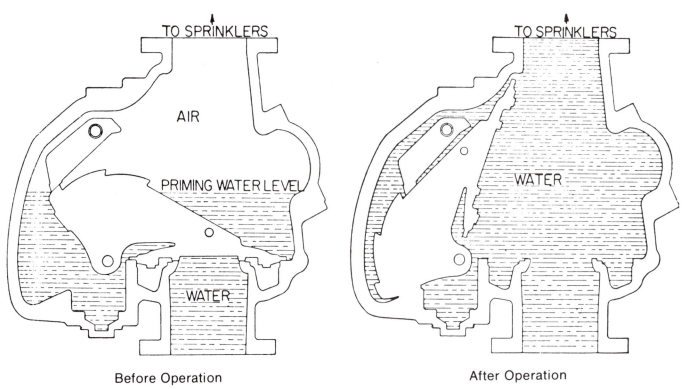

TO SPRINKLERS

AIR

PRIMING WATER LEVEL

WATER

Before Operation

TO SPRINKLERS

WATER

After Operation

Figure 17-19. The principle of a dry-pipe system is illustrated by these simplified drawings of a dry-pipe valve. Compressed air in the sprinkler system holds the dry valve closed, preventing water from entering the sprinkler piping until the air pressure has dropped below a predetermined point. (Reprinted with permission from National Fire Protection Association.)

system in areas where piping is exposed to freezing temperatures. It is essential, however, that the dry-pipe valve and water-supply line be located in a heated enclosure.

In the dry-pipe system, the piping contains compressed air that holds back the water by means of a dry-pipe valve (Figure 17-19). When a sprinkler opens, the air is released, the pressure drops, and the dry-pipe valve opens to admit water into the risers and branch lines. These sequential actions delay the actual wetting when compared to a wet-pipe system. Because of this delay, extra-hazard occupancies are difficult to protect with a dry-pipe system. In general, more water damage may result with the dry-pipe system because more sprinklers open than with the wet-pipe system; that is, the fire progresses further, hence more sprinklers are tripped before the extinguishing action of the water takes effect. To reduce this delay, quick-opening devices, such as exhausters or accelerators, can be added to dry-pipe systems to expel the air more rapidly.

The system air pressure should be kept 15 to 20 psi (110–140 kPa) above the normal tripping pressure; higher pressure delays the action of the dry-pipe valve. The system should not lose more than 10 psi (70 kPa) air pressure per week, which would require repressurizing more often than once a week; equipment that leaks more is not acceptable. (See NFPA 13 for details of testing dry-pipe systems.)

It is essential that all parts of a dry-pipe system be installed so that they can be thoroughly drained. Therefore, where it is necessary to use inverted (pendant) sprinklers or "drop piping," a special type of pendant sprinkler can be used or the piping can be filled with an antifreeze solution.

Pre-action systems are similar to dry-pipe systems, but react faster and, hence, minimize water damage in case of fire or

mechanical damage to sprinklers or piping. Supervision of the system against mechanical damage can be accomplished by connecting to the piping an automatic, low-pressure air supply that compensates for minor system leakage. A rapid reduction in pressure resulting from, say, an accidental breakage of the piping, sends a trouble signal without tripping the water-control valve.

The pre-action valve, which controls the water supply to the system piping, is actuated by a separate automatic detection system, which is located in the same area as the sprinkler, and not by the operation of a sprinkler. Because the detection system is more heat sensitive than the sprinklers, the water-supply valve opens sooner than in a dry-pipe system. The water-supply valve can also be operated manually.

Usually an alarm is sounded when the valve opens and starts filling the system with water. There may then be time to put out the fire with portable equipment before the sprinklers go into action and drench the area with water. This system is especially effective where valuable merchandise is handled or stored.

The deluge system wets down an entire area by admitting water to sprinklers that are open at all times. Deluge valves that control the water supply to the system are actuated by an automatic detection system located in the same area as the sprinklers. The water supply valves can also be operated manually.

This type of system is primarily designed for extra-hazard occupancies where great quantities of water may have to be applied immediately over large areas. Deluge systems are ordinarily used to best advantage where rapidly spreading or flash fires may be anticipated, such as in explosives plants, plants handling or processing nitrocellulose materials, lacquer plants, and buildings that contain large quantities of flammable materials.

Another application of a deluge system is an open system of

Table 17-I. Standard Temperature Ratings of Automatic Sprinklers

Rating	Operating Temperature (F)	Color	Maximum Ceiling Temperature (F)	(C)
Ordinary	135-150-160-165	Uncolored*	100	38
Intermediate	175-212	White*	150	66
High	250-280-286	Blue	225	107
Extra high	325-340-350-360	Red	300	150
Very extra high	400-415	Green	375	190
Very extra high	450	Orange	425	218
Very extra high	500	Orange	475	246

*The 135 F sprinklers of some manufacturers are half black and half uncolored. The 175 F sprinklers of the same manufacturers are yellow.
Reprinted with permission from NFPA 13, *Installation of Sprinkler Systems.*

outside sprinklers for distributing water over the roof, exterior of a building, and at windows and cornices to protect the building against fire from adjoining property. This type is usually manually operated and is used where construction is inadequately protected by design or by distance from adjacent fire hazards.

In special applications, open sprinklers and closed sprinklers may be combined in a single system where deluge protection is not needed over the entire area. However, it must be remembered that separate automatic detectors are also required in the area covered by the closed sprinklers, that operation of a closed sprinkler will not activate the entire system, and that a fire in the area of the closed sprinkler will also cause water to discharge from all of the open sprinklers.

Automatic alarms, operated by the flow of water through the system, should be a part of every standard sprinkler installation. Such an alarm may be connected to a central-station fire alarm service or to the municipal fire department, or may be a local alarm signal. Its purpose is to give prompt notice that the sprinkler system is operating. It also signals water leakage or discharge from causes other than fire.

The automatic alarm system must be tested and inspected frequently and be maintained by persons thoroughly familiar with it.

Temperature rating of sprinklers. Sprinklers should be selected on the basis of temperature rating and occupancy. Sprinklers are built with heat-actuated elements of solder that melt, or with special devices in which chemicals melt or expand to open them.

Table 17-I shows the ratings and distinguishing colors of sprinklers.

Quick operation of sprinklers, an advantage when over a fire, may be a disadvantage elsewhere be wasting water and wetting down materials that might otherwise be unaffected. Tests conducted by the Factory Mutual Engineering Corporation resulted in the ratings recommended in Table 17-J.

Causes of failure of sprinkler systems. Sprinklers seldom fail to control fires, but when they do, failure is usually due to (1) not keeping all supply valves open, and (2) shutting off the supply valves prematurely during fire.

Other causes of sprinkler failure, with corresponding remedies (in italics), are:
1. Freezing of wet system sprinkler pipes. *Heat the building or convert to a dry system.*
2. Defective dry-pipe valve, or slow operation of dry system because of its excessive size. *Check valves at frequent intervals, "trip test" in accordance with insurance company recommendations or subdivide system.*
3. Foreign material obstructing the system. *Flush out system on a regular schedule and provide debris-clear water at intake through use of filter screens.*
4. Improper drainage through faulty installation. *Check pitch of pipes and eliminate low spots.*
5. Sprinklers obstructed by stock piled too high, sprinklers isolated by temporary partitions or shelving, and sprinklers shielded from heat. *Improve housekeeping and maintain a minimum of 18 in. (46 cm) clearance between the top of*

Table 17-J. Selection of Sprinkler Ratings

Rate of Heat Release	Maximum Temperature (F) at Sprinkler Level Under Other Than Fire Conditions					
	100	150	225	300	365	465
Low rate of heat release from fire. (Light occupancies such as offices, schools, hotels, hospitals, apartments)	135, 160, 165	175, 212	250, 280, 286	325, 350, 360	400	500
Moderate rate of heat release from fire. (Ordinary industrial occupancies)	175, 212	175, 212	250, 280, 286	325, 350, 360	400	500
High rate of heat release from fire. (Flammable liquids, rubber tires, rubber and plastic foams, high-piled combustible storage, and similar locations.)	250, 280, 286	250, 280, 286	250, 280, 286	325, 350, 360	400	500

Reprinted with permission from Factory Mutual Engineering Corporation.

Figure 17-20. A water spray system for oil-filled electric power transformers. A thick layer of crushed stone and subsurface drainage is provided around the base of the transformer installation to prevent buring oil from flowing beyond the area protected by the spray. (Reprinted with permission from National Fire Protection Association.)

material storage and the deflector. It is a good practice to increase the clearance up to 36 in. (0.9 m) over large, closely packed piles of combustible cases, bales, cartons, or similar stock.

6. Corrosion of sprinklers in such locations as bleacheries, dye houses, or chemical operations. *Use sprinklers specially protected for such locations.*

7. Inadequate supply of water because of faulty design or poor maintenance. *Verify that water conditions have not changed since the original plant installation. It is important that the system be sturdily installed and anchored because explosions can jolt the piping and render the system ineffective.*

Water spray systems

Water spray is effective on all types of fires where there is no hazardous chemical reaction between the water and the material that is burning. Although these systems are independent of, and supplemental to, other forms of protection, they are not a replacement for automatic sprinklers.

Fixed water spray systems are similar to the standard deluge system except that the open sprinklers are replaced with spray nozzles. The water supply to the system can be controlled automatically or manually. They are generally used to protect flammable liquid and gas tankage, piping and equipment, cooling towers, and electrical equipment, such as transformers, oil switches, and motors (Figure 17-20). Because of its low electrical conductivity, water spray applied through fixed piping systems on electrical equipment with voltages as high as 345,000 volts

has proved practical. When applied on some types of electrical equipment, water spray may cause short circuits by forming a continuous path of water between energized parts. In such cases, means should be provided for cutting off the electrical current before the water spray is applied. (See NFPA 15, *Water Spray Fixed Systems for Fire Protection.*)

The type of water spray required depends upon the nature of the hazard and the purpose for which the protection is provided. The basic principle of water spraying is to give a complete surface wetting with a preselected water density, taking into consideration nozzle types, sizes, spacing, and water supply.

Water spray systems can be designed effectively for any one or any combination of the following:

1. Extinguishment of fire
2. Control of fire where extinguishment is not desirable, such as gas leaks
3. Exposure protection; that is, absorb heat transferred from equipment by the spray
4. Prevention of fire; water spray is used to dissolve, dilute, disperse, or cool flammable materials.

Because the passages in a water spray nozzle are small in comparison with those in the ordinary sprinkler, they can easily be clogged by foreign matter in water. Therefore, strainers are ordinarily required in the supply lines of fixed-piping spray systems. The strainer basket should have holes small enough to protect the smallest orifices of the nozzles used. In cases where the nozzles have extremely small water passages, they may have their own internal strainer in addition to the supply-line strainer.

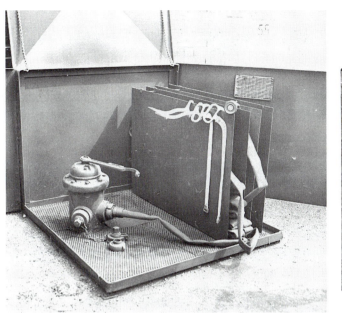

Figure 17-21. Hose houses may be made of metal (*left*) or wood (*right*). (*Right:* Reprinted courtesy of National Fire Protection Association.)

Hydrants, hoses, and nozzles

Fire hydrants. In larger plants where parts of the plant are a considerable distance from public fire hydrants or where no public hydrants are available, hydrants should be installed at convenient locations in the plant yard. The number needed depends on the fire exposure, and the hose-laying distance to the built-up plant areas. (See NFPA 24, *Installation of Private Fire Service Mains.*

Exterior fire department connections serving sprinkler or hose systems should be kept accessible and unobstructed. The discharge ports should be at least 18 in. (46 cm) above the ground or floor level. Vegetation, snow, and stored materials should be kept away from hydrants or hydrant houses. A hydrant must also be protected from mechanical injury, but this protection cannot interfere with efficient use.

Before cold weather sets in, hydrants should be drained or pumped out, if they are not the type that normally drain. Drainage must be checked whenever hydrants are used during freezing weather. Frozen hydrants may be discovered by sounding (striking the hand over the open outlet), by a partial turn of the hydrant stem (if the hydrant is frozen, the stem will not turn), or by lowering a weight on a string into the hydrant barrel.

Frozen hydrants can best be thawed with steam introduced through the outlet by means of a steam hose that is pushed slowly down the barrel, thawing as it goes. Corrosive chemicals, such as calcium chloride, caustics, or salts, should not be used.

Control valves should be tested frequently and be well maintained, and a number of persons, including plant fire brigade members, should know the location of valves and the sections of the pipe controlled by them.

Connections should be checked with the municipal fire department to be sure that they are of a size and thread that will fit its equipment. If special adapters are required, they should be supplied to municipal firefighters and also be available on the premises.

Fire hoses. Like other fire-extinguishing equipment, hose lines should be available for immediate use, and should not be obstructed nor inaccessible. Space around hose lines and control valves should be clear. The equipment should be visible and conspicuously indicated, and employees should know its location and understand its operation. Aisles and doorways should be kept clear and should be wide enough to allow rapid use of hose reel carts or other mobile equipment.

Fire hose must be rugged and dependable, capable of carrying water under substantial pressures, yet flexible and sufficiently easy to handle. Therefore, the fire hose should be of suitable quality, properly cared for when used, and carefully maintained in storage. Yard hose should be stored in standard hose houses for protection from weather (Figure 17-21).

Woven-jacket, lined hose with an outer rubber or plastic cover is chiefly used in industries where the hose jacket must be protected against chemicals and abrasions. Lined hose with a rubber or plastic cover is available in ¾- and 1-in. sizes, and is generally used as a booster hose, or as a hose on chemical engines, wheeled extinguishers, and wall-mounted or vehicle-mounted pressurized hose reels.

For inside use by building occupants (as opposed to standpipe systems designed for the fire department), 1½-in. unlined linen hose was often used. Since June 1976, however, only lined 1½-in. hose has been specified by NFPA. Unlined hose on existing system may be used provided it is maintained in good condition. However, with the introduction of lightweight, lined fire hose, which has twice the hydraulic efficiency of unlined hose, many firms are replacing the unlined hose on their industrial standpipe installations. The unlined hose has its place in such locations as office buildings, where it would be discarded and replaced if ever used. The hose should be kept on a swinging rack or reel of approved type, approximately 5 ft (1.5 m) above the floor or high enough so that it will not be a hazard to passers-by or be damaged by trucking operations. Hose stations intended for employee use should be located at the exits. See NFPA 14, *Standpipe and Hose Systems.*

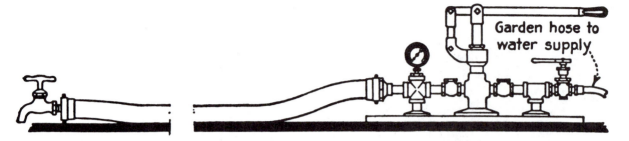

Figure 17-22. Hand-pump assembly for hydrostatic testing of hose.

Figure 17-23. Hydrant monitor nozzle used for cooling oil tanks or extinguishing fires. (Reprinted with permission from U.S. Steel.)

The hose should be so arranged that it will not kink or tangle when pulled out. One end should be kept connected to the standpipe, and the other end should be equipped with a ⅜- or ½-in. nozzle tip or a combination spray-solid stream nozzle. To prevent kinking in use, not more than 150 ft (46 m) of hose should be placed at a standpipe outlet.

Except for unlined linen hose, all fire hose should be hydrostatically tested annually, thoroughly inspected, dried, and returned to service (Figure 17-22).

Monitor nozzles. Permanent-mounted monitor nozzles are frequently used to protect pulpwood storage piles at paper mills, in lumber yards, in stock yards, in railway car storage yards, and near oil storage tanks (Figure 17-23). Nozzles are often elevated to clear obstructions so that the operator can stand on a shielded platform and direct a high-pressure stream of water over a wide area. Such systems are especially useful in large congested areas where it is impractical to lay hose lines in an emergency.

Hose nozzles. Effective streams for firefighting are controlled by the size and type of nozzle. The nozzle, in turn, must be supplied with the correct quantity of water at the discharge pressure for which it is designed. Nozzles are designed for solid streams, spray streams (frequently referred to as fog), or combination streams. And nozzles for special extinguishing agents, such as foam and dry chemical, are also available. Solid stream nozzles are designated by the diameter of the nozzle tip; whereas spray nozzles are designated according to the amount of water they discharge at 100 psi (690 kPa) nozzle pressure.

Spray nozzles are widely used in both public and private fire protection and make the application of water more effective under many conditions. They are of three general types:
1. Open nozzles of fixed (nonadjustable) spray pattern, usually attached to shut-off valves. Some nonadjustable nozzles can be equipped with an applicator (a long pipe extension, curved at the end and fitted with a fixed spray nozzle) for fighting fires where extended reach is necessary.
2. Adjustable nozzles which provide variable discharges and patterns from shutoff to solid stream and from narrow- to wide-angle spray.
3. Combination nozzles in which a solid stream, a fixed or adjustable spray, and shutoff are selected usually by a two- or three-way control valve.

Maintenance of fire hose

Hose for outdoor use. Inspect and test woven-jacket lined hose periodically to make sure that it will be in good condition when emergencies arise. Water should be run through the hose at least twice a year. Fire hose should be reserved for fighting fires; if hose is needed for other uses, separate hose should be provided.

Mildew may attack untreated hose fabric containing cotton if the hose is stored in a damp location or if it is not thoroughly dried after wetting. Fire hose is available with chemically treated fabric. The treatment is primarily for protection against mildew and rot. Treated jackets also absorb less water and therefore dry more quickly. The resistance to dampness and mildew is not one hundred percent effective even when the treatment is new, and it deteriorates with age.

Jackets made entirely with synthetic warp and filler are impervious to mildew and rot. Drying of such hose is not imperative. Washing after use and before storing is, however, recommended.

It is as important to carefully dry hose with jackets made from a combination of cotton and synthetic yarns as it is to dry hose with all-cotton jackets. Drying of both jacket constructions takes about the same time.

For plant yards containing rough surfaces that will cause heavy wear or where working pressures are above 150 psi (1060 kPa), double-jacket lined hose is advised. If hose may be subjected to acids, acid gases, and other corrosive materials, such as found in chemical plants, rubber-covered woven-jacket lined hose is advised. For such conditions, it can also be obtained with a neoprene-impregnated all-synthetic jacket.

Hose for indoor use. Maintain and test unlined linen hose and woven-jacket lined hose as follows: (it only can be wet twice; then it rots):

1. Reserve the hose for firefighting.
2. Keep hose valves tight; leakage will rot linen hose.
3. Examine hose visually each year for mildew, rot, damage by chemicals, vermin, and abrasions. If the hose is in doubtful condition, give it a hydrostatic pressure test. Damaged hose should be replaced.
4. Give hose a pressure test after the 5th and 8th year of service, and repeat the test every second year afterwards. (Unlined linen hose cannot be pressure tested.)
5. Keep hose clean. Wash woven-jacket lined hose with laundry soap if necessary.
6. Dry hose jacket thoroughly after use and keep them dry.

The local fire department will often pressure test hose.

SPECIAL SYSTEMS AND AGENTS

Special hazards may require methods of extinguishment or control other than water. Each of the several methods available offers certain advantages and limitations that must be considered in making a selection. These systems are usually installed to supplement, rather than replace, the automatic sprinkler system. They should be engineered to fit the circumstances of the particular hazard. It is common practice to install them in such a manner that their operation will shut down other processes (such as pumps and conveyors) that might intensify the fire.

The following special agents and methods are currently in use. Specific details are found in the following NFPA publications and also in the NFPA *Fire Prevention Handbook.*

Carbon dioxide—NFPA 12
Foam—NFPA 11 and No. 16
Dry chemical—NFPA 17
Water spray and sprinkler—NFPA 15, 13, and 231C
Steam smothering—NFPA 86 (ovens and furnaces)
Halon 1301—NFPA 12A
Halon 1211—NFPA 12B
Wetting agent—NFPA 18
Inerting—NFPA 69
Explosion venting—NFPA 68

Foam systems

Foam systems are often used to protect dip tanks, oil and paint storage rooms, and asphalt coating tanks. Foam systems also have been developed to put out tank fires by subsurface injection of foam. Foam can also be used to extinguish fires in laboratories.

Firefighting foam is an aggregate of gas-filled bubbles formed in aqueous solution. It is lower in density than most flammable liquids and as a result forms a continuous floating blanket of material on flammable liquids. It extinguishes fire in these liquids by smothering and cooling the fuel, thus halting the production of vapors, excluding the availability of oxygen to the fire, and physically separating the flames from the uninvolved fuel surface.

There are two types of foam-generation methods—(1) chemical and (2) mechanical (air generated), which includes the protein and synthetic types. Chemical foam is formed by a chemical reaction in which masses of bubbles of carbon dioxide gas and a foaming agent produce an expanded froth. Mechanical foam consists of bubbles of air produced when air and water are mechanically agitated with a foam-making agent.

Firefighting foams are of two major types—low-expansion foam used mainly for Class B (flammable liquid) fires and high-expansion foam used principally for Class A (ordinary combustibles) fires.

Low-Expansion foam is available in four types:
Chemical foam
Mechanical or air-generated foam
Protein foam
Synthetic (fluorinated surface-active agent) foam

Chemical foam is formed by the reaction in water of a mixture of aluminum sulfate with sodium bicarbonate, a foaming agent, and a stabilizer. As stated earlier, carbon dioxide that evolves in the reaction produces bubbles and is trapped in the foam. Because of the high solids content, its chief advantage is that it is very resistant to flame and mechanical disruption. It also has an ability to "set up" and assist surface flow. Chemical foam gains much of its consistency from the hydrated aluminum hydroxide in the reacted foam mass. Very little, if any, fire extinguishing capability is provided by the carbon dioxide-containing bubbles.

Some special types of chemical foam powders are available for fighting fires in polar solvents, such as alcohols, ethers, ketones, and acetates. These contain heavy-metal soaps to render the foam bubbles less susceptable to breakdown.

When reacted at solution temperatures of 60 to 85 F (16 to 29 C), thick foam with volume expansions of 7:1 to 16:1 can be produced. Severe limitations of these foams are that above and below these temperatures, chemical foam-making compounds give poorer quality foams or low expansion and rapid breakdown. Dry powders or solutions of the acid and basic components deteriorate with storage (especially if the temperature drops below 50 F or exceeds 100 F [10–38 C]). Chemical foam is also not capable of being transported through long pipes or hose lines, or being applied with devices requiring moderate pressures because the foam breaks down.

There are four general types of equipment for producing chemical foam: self-contained units, closed generators, hopper generators, and stored-solution systems.

In the self-contained unit, two solutions that produce foam on contact are stored independently in a single vessel and are caused to mix either manually or automatically. The amount of foam produced is determined by the quantity of foam-producing materials within the vessel.

In the closed-type generator system, chemical foam powder is stored in large hoppers permanently fixed to foam generators that mix the powder with water and then pump or use the water pressure to force the foam to special outlets. This kind of installation may be either the one- or two-powder type and is mainly

Figure 17-24. Proportioner permits operator to vary percentage of foam-producing concentrate.

used on flammable liquid storage tank farms, where a single foam-producing installation can service a number of tanks.

Hopper generators can be either permanent installations or portable. The advantage of portable generators is that they are not limited in foam production by a fixed storage of foam powder, but can be continuously refilled. Generators employing either one or two foam powders to produce foam are available. On large flammable liquid storage tanks, permanent lines affixed to foam chambers at the top of the tank are sometimes provided. Portable foam towers are used for foam application to burning oil storage tanks. Generator and water connections can be made at a calculated safe distance from the tank at the time of a fire.

Stored-solution systems have large permanently installed tanks that contain two foam-producing solutions stored separately. At the time of a fire, either duplex or twin pumps force the solutions to outlets where they mix and discharge foam. Foam production is limited by the size of the solution storage facilties.

Because of their limitations, chemical foams, the oldest of the three types of low-expansion foams, are gradually being replaced by mechanical or air-generated foams, which include the protein and synthetic types.

Mechanical or air-generated foam is produced by the mechanical action of adding the proper amounts of a liquid concentrate into a water stream via a proportioner, and then introducing and mixing air into the water concentrate solution.

There are four basic methods of producing mechanical (air) foam—nozzle aspirating systems, in-line foam pump systems, in-line aspirating systems, and in-line compressed air systems. The names of these systems indicate where and how air is injected into the water-concentrate solution to produce mechanical foam.

Each one of these systems uses a proportioner to introduce the foam concentrate into the water stream. There are a number of types of proportioners that can either be located at the main pump or in between the main pump and the foam maker.

Protein concentrates are available in two strengths; one is used with the proportioner at a three percent by volume ratio, and the other, at a six percent by volume ratio. It is imporant to note that both the three and six percent foam liquids are satisfactory only on hydrocarbons, such as benzene, toluene, xylene, gasoline, naphtha, and kerosene.

Water-miscible solvents, such as esters, ethers, alcohols, and ketones, destroy the regular three and six percent foam liquids as rapidly as they are applied. Therefore, an alcohol-resistant foam should be used in combating fires in these later materials. Where such uses are contemplated, a sample of the material should be submitted for evaluation by a supplier of alcohol-resistant foam. Because the stabilizer in this foam is chemically different from other stabilizers, the system in which the foam is to be used must be carefully designed with due regard to its limitations.

Because these foams must be used in 6 percent or greater concentrations, they are more costly and require greater quantities to be on hand. A 6 percent solution of protein foam permits vertical surfaces to be covered with an insulating blanket to aid in confining the fire. The high burnback resistance of protein foams proves especially valuable when the foam must seal to the hot sides of a metal tank or vessel.

Protein concentrates consist primarily of high-molecular weight digestion products obtained by the chemical hydrolysis of vegetable or animal protein materials. Metallic salts are included to maintain bubble strength in the presence of heat and mechanical action. Organic solvents are also added to improve the foaming characteristics and to control the freezing point. Additional additives are used to prevent corrosion, resist bacterial decomposition, and control viscosity.

Fluorinated surfactant foams, sometimes referred to as "light water" or aqueous film-forming foams (AFFF), consist of a fluorinated surfactant (surface-active agent) and a foam stabilizer as part of the formulation. AFFF is applied in 3 or 6 percent solutions in either fresh or sea water. Unlike protein foam, which must be applied through conventional foam nozzles, AFFF may be applied with fog or water spray nozzles, although with very reduced resistance to flashover and burnback. Special AFFF formulations also are available to overcome these limitations.

Conventional AFFF is limited to petroleum product (nonpolar) spill fires where rapid knockdown is desired and where cost is secondary. It is useful in preventing flashover (reignition) of fuel vapors exposed to lingering open flames or heated surfaces. The surfactant solution floats on the surface of the fuel, forming a barely visible film that reduces the release of vapors and the subsequent flashover hazard.

Because air foams possess low viscosity, they have fast spreading and leveling characteristics. They exclude air and halt fuel vaporization. Foam-generating devices or water-air spray devices can usually generate the easily formed foam.

A proportioner unit and siphon hose can be installed on a standard fire hose to draw the concentrate from 55-gal (210-liter) drums, 5-gal pails, or smaller portable containers. Some proportioner units can vary the percentage of concentrate from a plain water stream to a foam stream (Figure 17-24). A plain water stream can also be quickly obtained by removing the siphon tube from the concentrate.

A major limitation is that AFFF is susceptible to breakdown, loss of burnback resistance, and failure to seal against the wall of tanks or vessels in cases where prolonged freeburning has occurred prior to foam application. Therefore the utility of AFFF on flammable liquids contained in metal tanks, the hot sides of which require a relatively high water retention capability, is limited. As mentioned earlier, protein foams do not have this weakness.

Foam-water sprinkler and spray systems use mechanical (air) foam equipment with a deluge sprinkler system. These systems are generally used to protect flammable liquid hazard areas and can discharge foam or plain water selectively through aspirating devices (water sprinklers or foam-water spray nozzles).

Wet-water foam is generated by using an aspirating nozzle or through the injection of air under pressure into water containing a wetting agent, a chemical compound that reduces the surface tension of the water and increases its penetrating, spreading, and emulsifying properties. Wetting agents that are listed by Underwriters Laboratories Inc. are siphoned through a proportioner at rates not exceeding two percent by volume.

The reflective white opaque surface formed by the air bubbles and wet water particles makes this agent a good medium for protection from exposure fires. The cellular structure of the foam also retards heat conduction, thus affording insulation. As the foam blanket absorbs the heat and breaks down, the wet water released from the air bubbles carries the heat away from the protected surface. Mixing wetting agents with other wetting agents or with mechanical or chemical foam is not recommended since it may neutralize the effect of the agents and thus destroy the firefighting properties.

Wet water. The wet water principle can also be used without generating foam. In liquid form, wet water has the same general extinguishing properties as plain water. However, its cooling ability is increased, and there is a greater penetration of porous surfaces because of its reduced surface tension.

High-expansion foam is particularly suited for control and extinguishment of Class A and B fires in confined spaces, such as basements, buildings, shafts, and sewers. High-expansion foam is a blend of surface-active agents and a synthetic detergent foaming agent. High-expansion foam is made by introducing a small amount (usually 1½ percent) of foam liquid into a foam generator where water and large quantities of air are mixed. Firefighting foams of expansions from 100:1 up to 1,000:1 can be produced. In general, one 5-gal can of high-expansion foam can produce approximately one-third of a million gallons of foam, enough to cover a football field 12 in. (0.3 m) deep. Such foams provide a unique agent for transporting water to inaccessible places, for total flooding of confined spaces, and for displacement of air, vapors, heat, and smoke.

In addition to being a superior foaming agent, high-expansion foams have an emulsifying ability to clean up petroleum product spills, as well as a wetting ability to increase the penetrating effect of water on deep seated Class A fires. When accumulated in depth, high-expansion foams can provide an insulating barrier for the protection of exposed materials or structures not involved in a fire and thus prevent fire spread.

The temperature of the water and the quality of the air can affect the properties of the high-expansion foam. Water should be less than 90 F (32 C), and air, for foam production, must be taken where smoke and the pyrolysis products of fire cannot be drawn into the foam generator. These products can decrease the amount of foam produced, increase the drainage rate by chemical interaction with the foaming agent, and trap the gases in the bubble aggregate to create a toxic foam.

Ventilation is an important operation in using high-expansion foam for firefighting. Ventilation must be provided at the side opposite to foam application. Foam will not flow into a confined space unless provision is made for venting the displaced air and gases.

Carbon dioxide extinguishing systems

Fixed (local or flood) CO_2 (carbon dioxide) systems are often installed for the protection of rooms that contain electrical equipment, flammable liquid or gas processes, dry cleaning machinery, and other exposures where fire can be extinguished by diluting the oxygen content of the air or where water must not be used because of electrical hazard or the nature of the product. Details on installation and storage are given in NFPA 12, *Carbon Dioxide Extinguishing Systems.*

In the high-pressure system, CO_2 is stored in compressed gas cylinders at normal temperatures. It is released by manual operation or by automatic devices through nozzles close to the expected source of fire. Carbon dioxide has definite advantages, unlike water or other chemical extinguishing materials, in that it generally does not damage stock or equipment.

In the low-pressure fixed installation, carbon dioxide is stored in an insulated pressure vessel and maintained at 0 F (−18 C) by mechanical refrigeration. At this temperature the pressure is approximately 300 psi (2,070 kPa). At such low pressure, 500

Figure 17-25. This 16- by 24-ft flammable liquids storage building is protected by a total-flooding dry-chemical system. On actuation of the system by a heat-sensitive device, nitrogen is discharged into the 150-lb storage container, and dry chemical is thereby expelled to eight nozzles beneath the roof. (Reprinted with permission from National Fire Protection Association.)

pounds (226 kg) to more than 125 tons of CO_2 can be stored more economically than at a higher pressure.

Relief valves are provided in case of refrigeration failure. Liquid CO_2 is delivered through pipelines to nozzles that may have delivery capacities as high as 2,500 lb/min (15 kg/sec). As in the high-pressure CO_2 system, release can be either manual or automatic. Fixed local application systems provide for extinguishment of the fire at its source. Total flood systems may be used in small buildings, compartments, or rooms where wall or other openings can be automatically shut when the gas is released. Warning alarms must be provided to alert persons working in areas protected by this type of system. Sufficient time must be allowed to evacuate the area.

In confined locations, it is important to ventilate the area thoroughly after a fire is extinguished, because available oxygen may not be sufficient to sustain life.

Hand-hoseline systems combine fixed tanks with hose reel attachments, which permit a limited range of firefighting. Range is predicated on the length of the hose plus the effective range of CO_2.

Dry-chemical systems

Dry-chemical piped systems have been developed for situations where quick extinguishment is needed, either in a confined area or for localized application, and where reignition is unlikely. They are adaptable to flammable liquid and electrical hazards and are available for either manual or automatic operation, activated at the system or by remote control (Figure 17-25). A rate-of-rise heat-actuated device or an electrical release controls the automatic operation.

Installations can provide for simultaneous closing of fire doors, operating valves, windows, and ventilation ducts, as well as for shutting off fans and machinery and for actuating alarms. Piped systems providing either local application or total flooding are available.

The dry-chemical agent is neither toxic nor a conductor of electricity, nor does it freeze. In piped systems and most hand-hoseline systems, it is stored in a tank that is pressurized by an inert gas cylinder when the controls are actuated. In some hand-hoseline systems, the agent is stored in a pressurized container. Extinguishing action results mainly from the interruption of the chemical flame chain reaction by the dry chemical agent (see Chemistry of Fire earlier in this chapter for details of this mechanism).

Fixed storage tanks and pressurized cylinders, similar to those used in piped systems, are available for monitor operation and for mounting on vehicles.

See NFPA 17, *Dry Chemical Extinguishing Systems*.

Halon systems

Halogenated compounds were described earlier in this chapter under Portable Fire Extinguishers. To review, Halons are colorless, odorless, nonconductive gases or vaporizing liquids that apparently extinguish fires by inhibiting the chemical reaction of fuel and oxygen. The Halon agents do not work by displacing or diluting oxygen and thus do not add to the human safety concern. However, in a total flooding system, human exposure should be limited to one minute if the concentration is as high as 7 to 10 percent. People should get out immediately.

Most applications require a concentration in the range 4 to 6 percent by volume. The gas is relatively safe to human exposure at less than 7 percent concentration. A feeling of lightheadedness increases at about 10 percent. At 15 percent symptoms can be severe and near 20 percent, there is the possibility of a cardiac arrhythmia.

Halon 1301 has three major firefighting advantages for many situations, which tend to mitigate its higher cost. First is a reduction in weight and space compared to a CO_2 system, because firefighting ability is nearly 5 times better than CO_2. Second, Halon 1301 fire system can discharge in 10 seconds or less.

Normally a CO_2 system takes 1 to 2 minutes. Third, Halon 1301 is relatively safe to people, as described earlier.

Halons are, however, ineffective for fire extinguishment where the fuel is either self-oxidizing, a reactive metal, a metal hydride, or a material that undergoes autothermal decomposition.

Halon 1301 is intended to suppress a fire in the building contents, not the building itself, where generally, a water system would be a better agent. Halon 1301 will not completely extinguish deep-seated Class A fires when used in a low concentration (less than 7 percent).

Halon 1301 may not be suitable for the protection of ovens and furnaces where there may be high-temperature surfaces. The Factory Insurance Association will not approve Halon 1301 for use in ovens and furnaces operating over 500 F (260 C).

"Automatic total flooding" Halon 1301 systems are being widely used to extinguish fires in enclosed areas where waterbase or dry chemical systems would severely damage the contents, and where the use of CO_2 would create a danger to personnel within the room. These systems are used in computer rooms, magnetic tape storage vaults, electronic and electrical control rooms, storage areas containing highly valued documents or materials, aircraft and ship cargo areas, processing and storage areas for paints, solvents, and other flammable liquids and gases.

The extinguishing system consists of a supply of Halon 1301; detection and activation devices responding to smoke, flame, and/or heat; and one or more discharge nozzles. Automatic shutdown of the ventilation system, closing of doors, provision for remote and local alarms and other auxiliary features can be provided to tailor the capability of the system to the needs of the user.

"Local application" systems are designed to surround a burning object with a high concentration of Halon. Examples of installations where a local application system might be desirable are printing presses, dip and quench tanks, spray booths, oil-filled electric transformers, flammable vapor vents, and other operations where it is either unnecessary or impractical to employ a total flooding system.

Halon 1211, because it is a liquid with low volatility and as such can be conveyed as a stream, is ideally suited for local application, although the toxicity of this material is greater than Halon 1301. Supply, detection, activation, and discharge of 1211 equipment, although similar in concept to that used in a 1301 flooding system, is mechanically different. Refer to NFPA 12A and 12B, *Halon 1301 and Halon 1211 Fire Extinguishing Agent Systems,* for further information.

Steam systems

Automatic or manually controlled steam jet systems have been used to smother fires in closed containers or in small rooms, such as heaters, drying kilns, smoke ovens, asphalt mixing tanks, and dry cleaning tumbler dryers. However, the system is practical only where a large supply of steam is continuously available; unfortunately, most plants do not have this steam-generating capacity.

Steam has not been found effective on deep-seated fires that may form glowing coals, or in enclosures where the normal operating temperature is not considerably higher than air temperatures.

The possibility of a personal injury hazard from burns should be considered in any steam-extinguishing installation. See NFPA 86, *Ovens and Furnaces.*

Inerting systems

Inert gas systems can prevent fires and explosions by replacing the oxygen in the air with an inert gas, such as carbon dioxide, nitrogen, flue gas, or other noncombustible gas, until it reaches a level (or percentage) where combustion will not take place. (See the discussion earlier in this chapter and in NFPA 69, *Explosion Prevention Systems.*)

To be effective, the inert gas must reduce the amount of oxygen in the system, from the normal 21 percent to between 2 and 16 percent, depending upon the type of hazard involved and the type of inerting gas. For instance, an inert gas such as CO_2 must reduce the oxygen in air to 6 percent to prevent fire or explosion of carbon monoxide (CO), 14 percent for gasoline, and 15 percent for cotton dust.

Inert gases, such as flue gas from power plant stacks, have been used extensively to prevent explosion. Flame producers operating on either fuel oil or gas, which yield products of combustion with a high percentage of CO_2 and nitrogen, are often used in fixed installations where large quantities of the inert gas are required, as in the purging of storage tanks, in pipelines, and in manufacturing processes with high explosion hazards.

Use of an inert gas in a confined space can result in an oxygen-deficient atmosphere, and even carbon monoxide (if flue gas is used); therefore, before anyone is allowed to enter a confined space into which an inert gas has been introduced, the space should be thoroughly ventilated and tested with instruments that indicate if the enclosed atmosphere will support life. Otherwise, the individual should wear approved respiratory protective equipment and harness with lifeline for such entry. Furthermore, a watcher similarly equipped should stand by to observe the person in the confined space and to rescue in case of an emergency.

There are three ways of applying inert gas to assure the formation of a noncombustible atmosphere within an enclosed tank or space. The methods are fixed volume, fixed rate, and variable rate.

The fixed volume method introduces the inerting gas into the equipment chamber by either reducing the pressure within the chamber and allowing the inert gas to flow in until the pressure is equalized, or by pressurizing the chamber with the inert gas and then letting off the overpressure to the atmosphere after mixing has taken place. Several pressurizing cycles may be necessary to reduce the oxygen content sufficiently.

The fixed rate method adds a continuous supply of inerting gas in amounts sufficient to accommodate peak requirements. The quantity required is based on the maximum inbreathing rate which may result by sudden cooling, such as that caused by rain or a sudden drop in temperature, plus maximum product withdrawal. Although this method is relatively simple, it has the disadvantage of wasting inert gas and increasing the rate of the evaporation of the product.

The variable rate method supplies the inert gas to the system on a demand basis. The inerting gas is continuously released to a low-volume supply line to compensate for minor pressure changes. When rapid changes take place, such as caused by product withdrawal, a means is provided which opens a large supply line until the pressure equalizes. This type of system is extremely efficient and has the advantage of reducing product vapor losses by maintaining a slight positive pressure within the chamber.

Inert gases are sometimes used to prevent explosions of gases, vapors, and dusts. Their function is to keep the concentration of oxygen below the point at which it will support combustion. High fire-hazard commercial processes, such as lacquer manufacturing, are sometimes flooded with carbon dioxide, nitrogen, or other inert gases. This method should not be solely relied upon to provide a "safe" atmosphere, unless it is continuously analyzed for oxygen content.

Inert gases can also be used as a means of transferring flammable liquids, for inerting the atmosphere of storage tanks of volatile flammable liquids, and for purging gas holders or pipelines. Inert gases, since they dilute the oxygen in the air, also have widespread use for standby emergency fire extinguishing.

Preventing explosions

Preventing the development of explosive mixtures is the best defense against explosions. Equipment for handling and storing of flammable gases should be designed, constructed, inspected, and maintained so that the danger of leakage and explosive mixture formation is reduced to a minimum. Equipment should be inspected at regular intervals by qualified individuals either from the regular plant staff or from an outside source.

Ventilation will prevent excessive accumulations of gases and vapors under certain conditions. The best method of ventilation necessarily varies with the nature of the gas or vapor to be removed and depends upon whether it is heavier or lighter than air. Inasmuch as heating or cooling of the gas or vapor can change its density, the ventilation or exhaust system, or both, should be designed for operating conditions and should not be based only on the published density figures. (See NFPA 91, *Blower and Exhaust Systems for Dust, Stock, and Vapor Removal or Conveying*.)

Natural draft ventilation may be through openings near the floor, near the ceiling, or both. However, the best method is a positive local exhaust system, using explosion-proof electrical equipment and taking suction as close to the source of a vapor or gas as possible. A nonexplosion-proof exhaust fan motor may be used if it is properly installed outside the duct work and outside the hazardous area.

In general, flammable industrial gases, such as acetylene, carbon monoxide, hydrogen, and natural gas, are lighter than air. The vapors of flammable liquids, however, are generally heavier than air. Examples are alcohol, naphtha, gasoline, benzene, amyl acetate, and carbon disulfide. This is logical because gas or vapor density is proportional to molecular weight, and those compounds that are normally liquids at room temperature have higher molecular weights than those that are gases.

Unburned gases or flammable vapors in the combustion chambers of unit heaters, boilers, furnaces, enameling, drying, and bakery ovens may form an explosive mixture with air. Interruption of the gas feed pressure or extinguishment of the flame or pilot light may cause an accumulation of unburned fuel.

A number of safety devices have been developed to overcome this hazard. Most of them operate automatically to provide ventilation and to control the interlocking of gas and air supplies to safeguard against explosive mixtures.

Means should be provided on gas- and oil-fired equipment to ventilate or purge the combustion zone thoroughly in case of flame failure. Those who are in charge of firing these devices should know the ventilating or purging time required in the event of flame failure. In the event of flame failure, the program controller should take over; this forces the operator to go through the interlocked startup procedure which includes a timed preignition purge cycle.

It may be advisable to check the atmospheres inside large industrial equipment with a combustible gas indicator before relighting.

Gas-fired equipment, including its controls, should be inspected and tested at regular intervals and be kept in good repair in accordance with the manufacturer's recommendations. Only trained personnel should be permitted to operate it.

Gas valves should be inspected frequently for leaks. If gas is present, ventilation is needed immediately and the condition must be corrected before the equipment is used. The recommendations of the manufacturer of the equipment and of the public utility supplying the gas should be followed.

Suppressing explosions. Under certain conditions, an explosion-suppression system can be used to reduce the destructive pressure of an explosion. These systems are designed to detect an explosion as it is starting and actuate devices that suppress, vent, or take other action to prevent the full explosive force.

These systems require split-second timing. The mechanism for dispersal of the suppression agent must operate at extremely high speed to fill the enclosure completely within milliseconds after detection (Figure 17-26). The suppression agent must be dispersed from the suppressors in the form of a very fine mist at a rapid speed, normally through the use of a small secondary explosive force. The suppression agent is normally a noncombustible liquid compatible with the combustion process to be suppressed.

IDENTIFICATION OF HAZARDOUS MATERIALS

Fires and other emergency situations often involve chemicals that have varying degrees of toxicity, flammability, and reactivity (or stability). Information on these relative hazards must be readily available to those confronted with such emergencies if life safety, fire prevention, and effective fire extinguishment are to be achieved.

A system for the quick identification of hazardous properties of chemicals has been developed by the National Fire Protection Association. (See NFPA 704, *Identification of the Fire Hazards of Materials*.) For uniformity, this system recommends the use of a diamond-shape symbol and numerals indicating the degree of hazard. (See Figures 17-27 and 17-28.)

The three categories of hazards are identified for each material: health, flammability, and reactivity (stability). The order of severity (under fire conditions) in each category is indicated numerically by five divisions ranging from 4, which indicates a severe hazard, to 0, which indicates that no special hazard is involved. Colors may be used to better identify each hazard category—blue for health, red for flammability, and yellow for reactivity.

An explanation of the degrees of hazard follows.

NFPA health hazards. In general, the health hazard in firefighting is that of a single exposure which may vary from a few seconds up to an hour. The physical exertion caused by firefighting or other emergency may intensify the effects of any exposure.

Health hazards arise from two sources: (1) the inherent properties of the material, and (2) the toxic products of combustion or

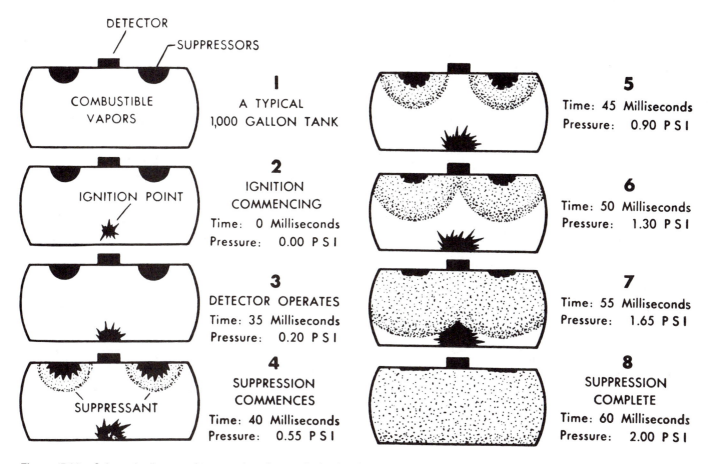

DETECTOR

SUPPRESSORS

1

**A TYPICAL
1,000 GALLON TANK**

COMBUSTIBLE
VAPORS

2

IGNITION
COMMENCING

Time: 0 Milliseconds

Pressure: 0.00 P S I

IGNITION POINT

3

DETECTOR OPERATES

Time: 35 Milliseconds

Pressure: 0.20 P S I

4

SUPPRESSION
COMMENCES

Time: 40 Milliseconds

Pressure: 0.55 P S I

SUPPRESSANT

5

Time: 45 Milliseconds

Pressure: 0.90 P S I

6

Time: 50 Milliseconds

Pressure: 1.30 P S I

7

Time: 55 Milliseconds

Pressure: 1.65 P S I

8

SUPPRESSION
COMPLETE

Time: 60 Milliseconds

Pressure: 2.00 P S I

Figure 17-26. Schematic diagram of suppression of an explosion in a large cylindrical tank. Reprinted with permission from National Fire Protection Association.

decomposition of the material. (Common hazards from burning of ordinary combustible materials are not included.)

The degree of hazard should indicate (1) that people can work safely only with specialized protective equipment, (2) that they can work safely with suitable respiratory protective equipment, or (3) that they can work safely in the area with ordinary clothing.

A health hazard, as defined by the NFPA, is any property of a material which either directly or indirectly can cause injury or incapacitation, either temporary or permanent, for exposure by contact, inhalation, or ingestion.

The degrees of hazard under fire conditions are ranked according to the probable severity of hazard to personnel, as shown in Figure 17-27.

NFPA flammability hazards deal with the degree of susceptibility of materials to burning, even though some materials that burn under one set of conditions will not burn under others. The form or condition of material, as well as its properties, affects the hazard.

The degrees of hazard are ranked according to the susceptibility of materials to burning; again, see Figure 17-27.

NFPA reactivity (instability) hazards deal with the degree or susceptibility of materials to release energy. Some materials are capable of rapid release of energy by themselves (as by self-reaction or polymerization), or they can undergo violent eruptive or explosive reaction if contacted with water or other extinguishing agents or with certain other materials.

The violence of reaction or decomposition of materials may be increased by heat or pressure, by mixture with certain other materials to form fuel-oxidizer combinations, or by contact with incompatible substances, sensitizing contaminants, or catalysts.

Because of the wide variations of accidental combinations possible in fire emergencies, these extraneous hazard factors (except for the effect of water) cannot be applied in a general numerical scaling of hazards. Such extraneous factors must be considered individually in order to establish appropriate safety factors such as separation or segregation. Such individual consideration is particularly important where significant amounts of materials are to be stored or handled. Guidance for this consideration is provided in NFPA 49, *Hazardous Chemicals Data*.

The degree of hazard should indicate to firefighting personnel that the area should be evacuated, that the fire may be fought from a protected location, that caution must be used in approaching the fire and applying extinguishing agents, or that the fire may be fought using normal procedures.

The relative reactivity of a material is defined as follows.

Reactive materials are those which can enter into a chemical reaction with other stable or unstable materials. For purposes of this guide, the other material to be considered is water and only if its reaction releases energy. While it is recognized that reactions with common materials, other than water, may release energy violently, and that such reactions must be considered in individual cases, inclusion of these reactions is beyond the scope of this identification system.

Unstable materials are those which in the pure state or as

BLUE	RED	YELLOW
IDENTIFICATION OF HEALTH HAZARD	**IDENTIFICATION OF FLAMMABILITY**	**IDENTIFICATION OF REACTIVITY**
Type of Possible Injury	**Susceptibility to Burning**	**Susceptibility to Release of Energy**
Signal	**Signal**	**Signal**
4 Materials which on very short exposure could cause death or major residual injury even though prompt medical treatment were given.	**4** Materials which will rapidly or completely vaporize at atmospheric pressure and normal ambient temperature, and which will burn.	**4** Materials which are readily capable of detonation or of explosive decomposition or reaction at normal temperatures and pressures.
3 Materials which on short exposure could cause serious temporary or residual injury even though prompt medical treatment were given.	**3** Liquids and solids that can be ignited under almost all ambient temperature conditions.	**3** Materials that are capable of detonation or explosive reaction but require a strong initiating source, or that must be heated under confinement before initiation, or react explosively with water.
2 Materials which on intense or continued exposure could cause temporary incapacitation or possible residual injury unless prompt medical treatment is given.	**2** Materials that must be moderately heated or exposed to relatively high ambient temperatures before ignition can occur.	**2** Materials that are normally unstable and readily undergo violent chemical changes but do not detonate; also materials that may react with water violently, or that may form potentially explosive mixtures with water.
1 Materials which on exposure would cause irritation but only minor residual injury even if no treatment is given.	**1** Materials that must be preheated before ignition can occur.	**1** Materials that are normally stable, but that can become unstable at elevated temperatures and pressures, or that may react with water with some release of energy, but *not* violently.
0 Materials which on exposure under fire conditions would offer no hazard beyond that of ordinary combustibles.	**0** Materials that will not burn.	**0** Materials that are normally stable even under fire explosive conditions, and that are not reactive with water.

FIRE
HEALTH   SAFETY

Figure 17-27. Degree of hazard can be quickly identified with this system. (Adopted from National Fire Protection Association.)

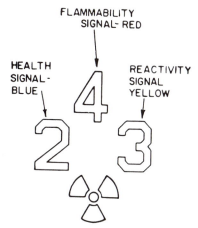

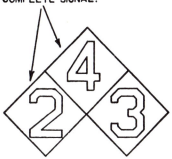

Fig. 1. For Use Where White Background is Not Necessary.

Fig. 2. For Use Where White Background is Used With Numerals Made From Adhesive-Backed Plastic

Fig. 3. For Use Where White Background is Used With Painted Numerals, or, For Use When Signal is in the Form of Sign or Placard

ARRANGEMENT AND ORDER OF SIGNALS — OPTIONAL FORM OF APPLICATION

Distance at Which Signals Must be Legible	Size of Signals Required
50 feet	1″
75 feet	2″
100 feet	3″
200 feet	4″
300 feet	6″

NOTE:
This shows the correct arrangement and order of signals used for identification of materials by hazard

Figure 17-28. NFPA hazard signal arrangement for in-plant use only. See NFPA Standard No. 704 for dimensional and other details. Meaning of number system is explained in Figure 17-27. (Reprinted with permission from National Fire Protection Association.)

commercially produced will vigorously polymerize, decompose or condense, or become self-reactive and undergo other violent chemical changes.

Stable materials are those that normally have the capacity to resist changes in their chemical composition, despite exposure to air, water, and heat as encountered in fire emergencies.

The degrees of hazard are ranked according to ease, rate, and quantity of energy release, as described in Figure 17-27.

At the bottom of the three-part diagram is an open space. This may be used to indicate additional information such as radioactivity hazards, proper fire extinguishing agent, skin hazard, pressurized containers, protective equipment required, or unusual reactivity with water. The recommended signal to indicate this unusual reactivity with water and to alert the firefighting personnel not to use water is the letter "W" with a long line through the center as shown in Figure 17-28.

Shipping regulations

Identification of hazardous materials during shipment is specified by U.S. Government regulations and differs from the NFPA identification system just described. Hazardous materials carried by rail, aircraft, vessel, and public highway are regulated by *Title 49—Transportation, U.S. Code of Federal Regulations,* Parts 170–180. "An Index to the Hazardous Materials Regulations" is published by the Department of Transportation. (See References.)

REFERENCES

American Gas Association, 1515 Wilson Blvd., Arlington, Va. 22209. (Lists of approved gas appliances and accessories.)

American Insurance Association, 85 John St., New York, N.Y. 10038. (Publications catalog available.)

Alliance of American Insurers, 1501 Woodfield Rd., Schaumburg, Ill. 60195.
"Judging the Fire Risk," LC-PM-05-678.
"Tested Activities for Fire Prevention Committees," LC-PM-02-678.

American Petroleum Institute, 1220 L St., NW., Washington, D.C. 20005. (Publications and materials list available.)

American National Standards Institute, 1430 Broadway, New York, N.Y. 10018.
Color Code for Marking Physical Hazards, Z53.1.
Practices for Respiratory Protection for the Fire Service, Z88.5.
Safety in Welding and Cutting, Z49.1.

American Welding Society, P.O. Box 351040, Miami, Fla. 33135. (Publications catalog available.)

Building Officials & Code Administrators International, Inc., 4051 W. Flossmoor Rd., Country Club Hills, Ill. 60477. *Basic Fire Prevention Code* (issued every three years).
The BOCA Basic Building Code.

Dillon, James C. "Smoke, the Silent Killer." *Fire Command!* April 1976.

Factory Mutual Engineering Corp., 1151 Boston-Providence Turnpike, Norwood, Mass. 02062

FM *System Approval Guide—Equipment, Materials, Services for Conservation of Property.* (Loss prevention data and books)

Fire Equipment Manufacturers' Association, Inc., c/o Thomas Associates, Inc., 1230 Keith Bldg., Cleveland, Oh. 44115. *Inspection, Recharging and Maintaining Portable Fire Extinguishers.*

International Association of Fire Fighters, 1750 New York Ave., NW., Washington, D.C. 20006. "Firefighter Mortality Report," 1976.

National Association of Fire Equipment Distributors, Inc., c/o Smith, Bucklin and Associates, Mgrs., 111 E. Wacker Dr., Chicago, Ill 60601. "Portable Fire Extinguisher Selection Guide," G-117.

National Fire Protection Association, Batterymarch Park, Quincy, Mass. 02269.

Fire Protection Handbook, current edition.

Fire Brigade Training Manual.

Inspection Manual.

Principles of Fire Protection.

"Publications Catalog."

National Safety Council, 444 N. Michigan Ave., Chicago, Ill. 60611

Industrial Data Sheets

Cleaning Small Containers of Combustibles, 432.

Fire Prevention and Control on Construction Sites, 491.

Fire Prevention in Stores, 549.

Fire Protection for Combustible Metals, 567.

Flammable and Combustible Liquids in Small Containers, 532.

Society of Fire Protection Engineers, 60 Batterymarch St., Boston, Mass. 02110.

Underwriters Laboratories Inc., 333 Pfingsten Road, Northbrook, Ill. 600062 (Catalog available yearly)

Building Construction and Materials List.

Electrical Equipment List.

Fire Protection Equipment List.

Gas and Oil Equipment List.

Hazardous Location Electrical Equipment List.

U.S. Department of Commerce, National Bureau of Standards, Washington, D.C. 20234.

U.S. Department of Commerce, National Fire Prevention and Control Administration, Washington, D.C.

U.S. Department of Transportation, Office of Hazardous Materials, Washington, D.C. 20590. "An Index to the Hazardous Materials Regulations." Available from Superintendent of Documents, U.S. Government Printing Office, Washington, D.C. 20402.

18

Boilers and Unfired Pressure Vessels

BOILERS (FIRED PRESSURE VESSELS) AND UNFIRED PRESSURE VESSELS have many potential hazards in common, as well as having those unique to a specific operation. These vessels hold gases, vapors, liquids, and solids at various temperatures and pressures, ranging from almost a full vacuum to pressures of thousands of pounds per square inch. In some applications, extreme pressure and temperature changes may occur in a system in a rapid succession, imposing special strains.

Design, fabrication, testing, and installation of boilers and unfired pressure vessels should be in compliance with the applicable sections of the American Society of Mechanical Engineers' *Boiler and Pressure Vessel Code* (hereafter referred to as the ASME Code), and any federal, state or provincial, or local governing codes.

A Synopsis of Boiler and Pressure Vessel Laws, Rules and Regulations, by States, Cities, Counties and Provinces, in the United States and Canada is available from the Uniform Boiler and Pressure Vessel Laws Society. This document details which governing bodies have made the ASME Code a legal requirement in their jurisdictions and what other compliances are required. However, if there is any question, the owner should check directly with the jurisdictional authority.

Compliance with the ASME Code is determined by authorized inspectors commissioned by the National Board of Boiler and Pressure Vessel Inspectors. (See Figure 18-1.)

The ASME Code contains eleven sections:

I   Power Boilers
II   Material Specifications
III Nuclear Power Plant Components
IV   Heating Boilers
V   Nondestructive Examination
VI   Recommended Rules for Care and Operation of Heating Boilers
VII Recommended Rules for Care of Power Boilers
VIII Pressure Vessels
      Division 1
      Division 2—Alternate Rules
IX   Welding and Brazing Qualifications
X   Fiberglass-Reinforced Plastic Pressure Vessels
XI   Rules for In-Service Inspection of Nuclear Power Plant Components.

Any or all of these sections may be purchased from ASME (see References for address).

The minimum requirements for the installation of high-pressure boilers are covered in the "Boiler-Furnace Standards," National Fire Protection Association Standards Nos. 85A, 85B, 85D, and 85E. When pressure vessels are to be installed, it is usually advisable to secure the services of a competent boiler or unfired pressure vessel engineering consultant. Such a professional can survey the plant or operation to determine the requirements, design a system that will satisfy them, and supervise installation and testing.

If the consultant arranges for the purchase of secondhand boilers or pressure vessels, they should be inspected by authorized inspectors who will report if repairs are necessary before purchase. Arrangements for inspectors usually can be made through the user's insurance carrier. The government inspection department that has jurisdiction can also provide this service. (See *Synopsis* mentioned above for listings.)

Because the ASME Code covers only safety of design, fabrication, and inspection during construction of boilers and

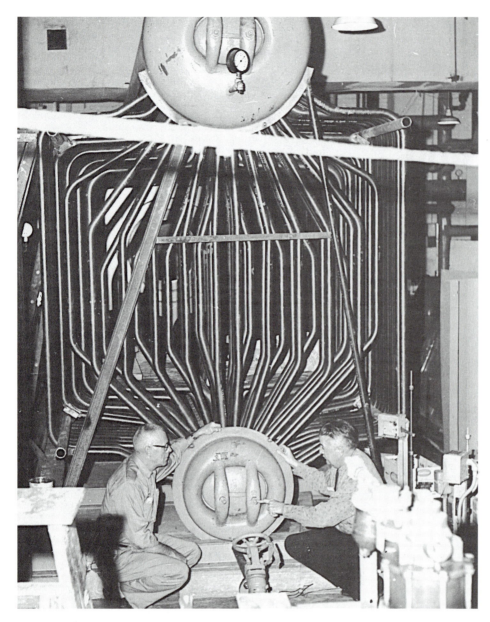

Figure 18-1. Boilers and pressure vessels subject to ASME Code regulations must be checked during construction at all points by authorized inspectors who can certify compliance with the Code. Inspectors are commissioned by the National Board of Boiler and Pressure Vessel Inspectors. (Reprinted with permission from Lutheran General Hospital, Park Ridge, Ill.)

pressure vessels, the National Board of Boiler and Pressure Vessel Inspectors has published its own *Inspection Code,* (hereafter referred to as the NB Code), designed to provide rules and guidelines for inspection *after* installation, repair, alteration, and rerating. It is strongly recommended, therefore, that the NB Code be referred to for guidance when repairing and altering boilers and pressure vessels. The NB Code should also be used to supplement and expand upon the specific safety and inspection procedures discussed in the remainder of this chapter.

In general, installation and maintenance should be in accordance with manufacturers' instructions. Further, operating personnel should be trained, not only to operate equipment properly, but also to make routine safety checks and to know when to call in qualified maintenance personnel.

Some common causes of failure in pressure vessels which should be anticipated and avoided insofar as possible are:
1. Errors in design, construction, and installation
2. Improper operation, human failure, and improper training of operators
3. Corrosion or erosion of the metal
4. Mechanical breakdown, failure, or blocking of safety devices, and failure or blocking of automatic control devices
5. Failure to inspect thoroughly, properly, and frequently
6. Improper application of equipment
7. Lack of planned preventive maintenance.

In addition to hazards presented by the possibility of explosion, boilers may also present fire hazards. They are a significant factor in hotel, store, apartment house, and religious institution fires. Oil-fired equipment is usually at fault, although losses from explosion of gas-fired equipment have occasionally been catastophic.

The majority of boilers in use are automatically or semi-automatically fired, and may operate unattended for long periods. Many are not maintained and checked regularly, leaving them in less than perfect condition. Because such boilers are unattended, when fires do start they can gain considerable headway before being detected unless adequate precautions have been installed.

The means for controlling and containing fires from boilers include:

1. Provide a fully enclosed boiler room of noncombustible construction (5/8 in. [1.6 cm] gypsum wallboard or better). Be sure to leave sufficient space for maintenance, including pulling of tubes. (Many jurisdictions have clearance requirements that must be met.)

2. All boiler rooms and/or buildings should be equipped with large door openings in order to allow easy access to all boiler room equipment. These doors should be sized to allow for easy installation and removal of all equipment. Entryways leading to the boiler room should be equipped with 1½ hour fire-resistance rated doors and door frames.

3. Provide a noncombustible ceiling over the boiler and automatic sprinkler protection over the firing end of the boiler and in areas where gas and oil pipelines are located. If the boilers are coal fired, automatic sprinkler protection should be provided over the coal augers, feeders, chutes, and indoor coal piles. (Note: other protection features may be required for coal piles, conveyor belts, etc.)

4. Provide proper clearance around exteriors of boiler room walls so that materials are not stored against the walls. Boiler rooms should not be used to store miscellaneous combustibles. Only materials and items pertinent to the boiler room operation should be kept in the room. If there is a need for combustible items in the boiler room, they should be stored in noncombustible approved cabinets. It is important not to expose the boiler room equipment to unnecessary, unimportant combustibles because of the possibility of seriously damaging, destroying, and/or interrupting boiler room equipment.

In order to minimize low-pressure boiler fires and explosions caused by faulty controls and safety devices:

1. Establish a test and servicing program whereby operating controls, safety controls, and safety and relief valves are tested and maintained at regular intervals.

2. Make sure that safety and relief valves are always tested with pressure on the boiler to prevent damage to the valve seats.

3. Have repairs made *immediately* upon any indication of malfunction or leakage of operating controls, safety controls or safety and relief valves. *Never operate with a malfunctioning safety or relief valve.*

4. Have a reliable service organization check and service the boiler during the heating season as well as during the normal out-of-season checking and cleaning.

5. Enforce the keeping of a boiler log to make sure that necessary tests, maintenance, and services are performed, and that records are available at all times.

BOILERS

In its simplest definition, a boiler is a closed vessel in which water is heated by combustion of fuel or heat from other sources to form steam, hot water, or high temperature water (HTW) under pressure.

Design and construction

Good standard references covering the design and construction of boilers are the *Standard Handbook for Mechanical Engi-*

neers, Theodore Baumeister, editor, and *Combustion Engineering,* G. R. Fryling, editor.

Instruments. Subsection C6 of Section VII of the ASME Code states that, in general, a boiler unit should include a meter-and-control board located on the operating floor so that the operator can see either the furnace door or the lighting ports of the burners and the water column of the boiler without leaving the control board. If it is not possible to see these directly, then reliable remote-indicating equipment should be installed, or someone should make a visual check during lighting off.

Economizers, usually an integral part of the heat exchanger system, are the last step in utilizing as much heat as possible. The exhausted flue gas from combustion of the fuel is used to heat incoming cold makeup feedwater—the hot flue gas being passed over tubes conveying the makeup water.

Cast iron and steel tube economizers should be equipped with at least one safety valve (two are preferrable).

Superheaters. After the heat transfer medium (water, steam, or other fluid) leaves the boiler unit itself, its temperature can be raised even more by passing through a superheater.

Detailed operational procedures are given in Section VII of the ASME Code.

Air preheaters. Fires may occur in an air preheater immediately after lighting a boiler and during periods of low-load operation. This condition can be detected by a sudden rise in air heater temperature.

To prevent fires, maintain proper combustion and use soot-blowers properly. Do not use the soot-blower when it is suspected that there is a fire in the gas passages because this environment could cause a serious explosion.

Chimneys, whether made of brick, concrete, or steel, should be equipped with grounded lightning arresters, and if not self-supporting, should be stayed to solid building structures. Any ladders added to a chimney should be of permanent construction, securely fastened to the chimney, and protected with hoop enclosures.

Ash disposal equipment. Hoistways, driving machinery, conveyors, wormgears, ash sluices, and reciprocating pumps should be properly guarded. An alarm bell can be hooked up to the driving machinery to warn that doors are about to be opened.

Exercise special care to prevent injury to operating personnel from steam or hot water that may be present when ash gates are opened. When excess carbon is present in ash pits and it is not properly wetted down, a gas explosion can result when gates are opened.

Ashes should never be stored against boilers or combustible materials. Ashes contain sulfur compounds which, on contact with water, form highly corrosive acids.

Water treatment. There are many professional consultants on feedwater treatment whose services are worthwhile. Generally, treatment that will remove dissolved oxygen and carbon dioxide and maintain a maximum pH of 11 in the boiler water will effectively minimize corrosion. Operators have sustained injuries while introducing boiling compounds into feedwater, and therefore should be adequately protected against scalds and caustic burns. Often, automatic feeding or softening equipment is used.

Blowdown pipes and valves. Blowdown piping is used to remove sludge and other impurities in boiler water which, if not removed, would seriously impede the efficiency and safety of the boiler. All piping, operating, and discharge valves should conform to the ASME Code, Section I or Section IV.

Blowdown piping and boiler drains should be conducted to a discharge point which will not present a hazard to operators or other personnel.

Safety valves and fusible plugs. Selection, fabrication, installation, testing, and replacement of safety valves and fusible plugs should be in accordance with the ASME Code. From the viewpoint of safety professionals, safety valves and fusible plugs are important, even vital devices requiring attention. Insurance company boiler inspectors and other specialists can advise on specific procedures for checking each type of equipment.

Safety valves, when properly applied and installed, will relieve excess pressure or vacuum (depending on design) that would otherwise damage equipment or result in injury to personnel. Safety valves must be kept in good operating order at all times and should be checked by qualified personnel in accordance with insurance company recommendations and jurisdictional regulations. Safety valves that are set by screwing down the body should be avoided because installers usually jam them by screwing them down tight.

Safety valves for water heaters differ from those used for boilers in that they must sense excessive temperatures as well as over-pressures, as specified in ANSI standard Z21.22.

Fusible plugs are designed to relieve pressure and to indicate certain conditions that contribute to low water. When these plugs are used, they must be manufactured, installed, inspected, repaired, or replaced according to the ASME Code.

Boiler operators should be trained to check the safety controls, preferably once a week, but at least once a month. Often a checklist is used, and a form is filled out and sent to management or owner.

If a safety valve opens, fails to reseat correctly, and cannot be freed by use of the hand-lifting lever, then the boiler should be taken out of service and the safety valve repaired.

Valves should be tested whenever a boiler is returned to service. To test safety valves by hand, the valve should be held wide open for a sufficient period to blow out possible accumulation of dirt and chips. The steam pressure should be at least 75 percent of the safety valve set pressure when opening with the hand-lifting gear. A more meaningful test is to raise the boiler pressure to the safety valve set pressure and let the valve open.

If boilers are kept in continuous operation for several months it may be desirable, depending on boiler conditions, to repeat the hand-lifting or pressure raising at intervals during operation. Small chains or wires attached to the levers of pop safety valves and extended over pulleys to other parts of the boiler room may be used. A counterspring or weight prevents the weight of the chain or wire from pulling the valves partly open.

These testing procedures, however, should not be used on valves with a set pressure over 400 psi (2,760 kPa). These should be periodically checked according to manufacturer's recommendations.

Discharge pipes, individual escape pipes designed to carry discharge away from each safety valve, should be supported so as to prevent any stress upon the safety valve and must not be rigidly connected to the valve. Clearance must be allowed for boiler expansion.

Drain pipes from a drip pan and valve body should be carried clear of the boiler setting and discharged into an open funnel providing a clear view of the drip. They should be installed to avoid freezing at any point.

Steam and water indicators. Steam gages indicate the pressure of the steam generated. All gages should be graduated to approximately double the pressure at which the safety valve is set, but in no case less than 1½ times that pressure. Pressure gages installed on a multiple boiler setup should be of the same type and graduated alike.

Good piping practice

A number of accidents can be prevented by installing steam lines with enough forethought to reduce the necessity for maintenance work on them. The ANSI standard B31 Series, "Pressure Piping," (1) prescribes minimum requirements for design, materials, fabrication, erection, test, and inspection of various piping systems, and (2) discusses the expansion, flexibility, and supporting of lines.

Another consideration in providing safety through good piping practice is that valves and other operating controls of boilers should be easy to reach. Many operators and maintenance personnel have been hurt when they fell from ladders or inadequate work stands while trying to operate an infrequently used and inaccessible valve.

If it is necessary to open lines, maintenance personnel should always assume that the lines are loaded and under pressure. A supervisor who is completely familiar with the system should then certify that the proper line is being opened and that all steps possible have been taken to drain and vent the line. Also, safe work stands must be provided. Setting up a system of "line breaking permits" to make sure that precautions are carried out is an excellent and proven idea.

Placing boilers in and out of service

This Manual cannot cover all details of placing boilers in and out of service but it emphasizes that the ASME Code and manufacturers' recommendations should be followed.

Cleaning and maintenance

Whenever a boiler is taken out of service for a prolonged period, it should be cleaned promptly and inspected for defects by the plant engineer. Authorized boiler inspectors can also view the boiler at this time.

Cleaning. Prompt cleaning is important. Soot gathers moisture rapidly and this contributes to deterioration of the metal surfaces. Soot and fly ash should be removed as soon as the boiler has cooled. Ashes may remain hot for days, presenting a hazard to anyone entering the combustion chamber. They are, therefore, usually wetted down with a hose.

As in spraying ashes for disposal, the operator should wet down from the outside toward the center and stay clear of any steam and dust that will come up. A jet of water driven into the center of a hot ash pile can literally explode it. When removing ashes, exercise care to prevent injury to personnel from steam or hot water that may be present when ash gates are opened.

For boilers in continuous service, planned and scheduled boiler shutdowns for preventive maintenance are far safer than risking an extensive shutdown caused by boiler failure. At least once a year, or more often, the boiler, the flame safeguard supervisory system, and other safety controls should be inspected during a scheduled shutdown by an authorized inspector who is accompanied by the plant inspector. Defective parts should be repaired or replaced.

Scheduled outages for maintenance should be carefully planned so that a minimum interruption to production results. Be sure that by the day the inspector is scheduled, the boilers will have been properly prepared. They must be cool enough so that the inspector will not have to rush the work and clean enough that metal parts can be examined thoroughly for corrosion, pitting, cracking, and other defects. Internal parts should be readily accessible for a close and thorough examination. Handholes and manholes should be open and the boiler should be ventilated. Adequate lighting and protective equipment for work in the boiler should be provided. (See entry precautions next.)

Boiler and furnace entry precautions. General precautions for entering boilers include having proper ventilation, proper equipment, and proper protection. To make certain that no flammable or toxic gases are present, ventilate boiler settings thoroughly and then check the atmosphere with a testing instrument before permitting anyone to enter. This is especially important when more than one boiler is connected to one breeching or chimney, because under certain circumstances flue gases can come back into the boiler from other boilers.

One cannot overemphasize the need for caution when employees clean or do maintenance work on boilers. A great number of injuries have occurred in this work. Good vessel entry procedures and good lockout procedures to prevent steam, hot or high-pressure water, or hot gases from reaching employees must be followed. (See the discussion of zero mechanical state (ZMS) procedures in Chapter 8.)

In the case of fuel gases, closing and leak testing may not be sufficient. Positive blanking or block and vent valving is advisable. All valves that are closed must be checked for leakage. Lines that are interconnected between boilers must be positively sealed off at both ends and locked out. Also consider providing workstands and protection from overhead hazards of ash deposits falling on employees entering boilers.

Ventilation can be provided by portable power-driven blowers operated outside the boiler setting and having canvas tubes leading in through access doors. Draft fans can be operated for short periods of time to provide ventilation.

As a precaution against electric shock, many firms only permit 6- or 12-volt lights and tools inside a boiler. They are connected to small portable power tranformers outside the boiler. Battery-powered lights are an even safer alternative. All electrically operated tools used inside a boiler should be properly grounded, and along with extension cords, be thoroughly inspected before use. (See Chapter 15.)

When cleaning a boiler, employees should wear hats, goggles, dust respirators, and heavy, leather-palm gloves. Wearing safety shoes is also recommended.

Personnel working in confined areas should wear a lifeline, if necessary, and be kept under constant observation.

Boiler rooms

Floors. Boiler room floors can become very slippery and dirty, so a surface that can be easily cleaned should be provided. Consideration should be given to drainage and protection against flooding.

Lighting. In addition to being well illuminated, the boiler room should have a source of emergency lighting. Gages and controls should be especially well lit so they can be read easily. Well-maintained flashlights should be provided for personal use in case of power failure, or other emergency. Exits, too, should be well lit and identified.

Exits. Each boiler room should have two or more exits, remotely located from one another. If a boiler extends more than one story above ground level, the room should have an exit at the boiler runway or floor level for each story. This exit should lead to a fire escape on the outside of the building. If the boiler room is in a basement (or subbasement), exits should lead to outside stairways and runways, and should have landings leading to the exit doors.

Stairs, ladders, and runways. Some state and local boiler codes require the installation of stairs, ladders, and runways around boilers which extend 10 ft (3.1 m) or more above floor level. Even if not required by law, such access is desirable so that personnel can operate and service the boiler safely and without having to step on hot steam or water lines or stand on valve stems or handles.

Stairs, ladders, and runways shall have standard guardrails, handrails, and toeboards. Runways should be fabricated of steel grating to provide a slip-resistant surface and at the same time permit circulation of air. Walkways should *not* be near water glasses or safety valve discharge areas where an operator might accidentally be scalded.

Boiler room emergencies

So many boiler accidents have been investigated thoroughly that necessary preventive action is generally known. Safety professionals should know enough about boiler room procedures that they can be sure their units are protected.

Rules for both routine and emergency boiler operation should be posted permanently and legibly in the boiler room (Figure 18-2). Manufacturers can supply rules applicable to their equipment. In addition, all operators and substitute operators should be furnished copies for their guidance. Supervisory personnel should make sure that boiler operators know the rules and are capable of performing the necessary operations under emergency conditions.

Many plants have only one boiler room operator. Should this employee become sick or injured, the boilers may be left unattended and an accident could occur. In plants with isolated boiler rooms operated by one person, it may be advisable to have this person call a central location at half hour intervals to make sure that everything is operating on schedule in the boiler room. Plants having a plant protection patrol may have patrol officers check the boiler room. An intercom system could also be of value.

For this reason, the safety professional should recommend and train someone (a supervisor, a night shift employee, or some other substitute) who can take over the boiler operation in case of emergency.

SAFETY OF HIGH TEMPERATURE WATER (HTW)

Because of certain economic advantages, high temperature water (HTW) is sometimes used instead of steam for transferring heat in both manufacturing and district heating uses. The water is

Boiler Operating Instructions

DANGER

OVERHEATED BOILER—
Do NOT add water—STOP FIRE
Call service company representative or supervisor for assistance

FLAME FAILURE—
Do NOT restart until thoroughly vented
Call service company representative or supervisor for assistance

STARTING and DAILY CHECK
1. Be sure water is at proper level.
2. Do not start fire until after furnace has been thoroughly vented.
3. Use small fire during warm-up.
4. Check boiler frequently while in normal operation.

WEEKLY CHECK
Test low-water cut-out control—Record Test on tag.

MONTHLY CHECK
Test safety valve and record on tag.

YEARLY
Replace or disassemble and overhaul low-water cut-out control during annual boiler cleanup and repair period. The combustion safeguards, such as fuel pressure switches, limit switches, motor starter interlocks, shutoff valves, prepurge cycle, etc., on the firing end of the boiler should be tested at least once annually by a recognized vendor. All test results should be documented. (Note: Combustion protection safeguards on a boiler should not be jumped, tampered with, and/or bypassed.)

W.S. 171 NEW 11-63 PRINTED IN U.S.A. SAFETY ENGINEERING SERVICE THE TRAVELERS INSURANCE COMPANIES • HARTFORD, CONNECTICUT

Figure 18-2. Emergency procedure and checklist poster, similar to this one, should be posted permanently in the boiler room. (Reprinted with permission from The Travelers Insurance Companies.)

kept in a closed system under high pressures so that it will remain in liquid form instead of turning into steam. Conditions such as 400 F (200 C) and 247 psi (1700 kPa) pressure often exist, and if one considers the potential volume increase when liquid under this pressure and temperature expands to steam at atmospheric pressure, it can be quite upsetting.

However, HTW is very different from steam or cold water when it discharges through a break in a pipe or equipment: it has a very high rate of increase of volume and very low energy release during expansion. Energy liberated in the expansion is spent in accelerating the particles of water and vapor, and in pushing air out of the way so the steam-water mixture that is being formed can occupy the vacated space. Practically no energy is left over and available for rupturing equipment and imparting kinetic energy to fragments.

When steam escapes, approximately 16 times the energy is available from its expansion than from HTW expansion, hence considerable energy is left over to provide the explosive effect. Although fragments of fracturing cast iron valves on steam service have been known to penetrate a 10-in. (25-cm) thick brick wall, no case has been observed where parts of fractured valves on HTW service have been projected any distance at all.

Because the volume increase of escaping HTW continues after leaving the pipe, the mixture does not form a long jet as does steam or cold water issuing from an orifice, but spreads out practically at right angles from the centerline of the jet to form a wet fog. The rate of flow has been found to be less than half that of cold water issuing from a similar hole and over the same pressure range.

Nevertheless, such considerations must not cause any feeling of false security or negligence on the part of the design engineers and operating personnel. Although HTW is safer than steam and accidents are rare, such accidents have happened. Even 180 F (80 C) water can fatally burn if sufficient body area is immersed in it.

When failure of equipment or piping does occur in HTW systems, it is usually caused *not* by the inherent thermodynamic forces, but by mechanical forces, such as water hammer, thermal expansion, thermal shock, and faulty material. It is, therefore, imperative that only experienced engineers should be allowed to design HTW systems. These engineers must be willing and capable of minutely analyzing the entire design and equipment selection for the presence of possible dangers.

A good design is neat and simple but it does not overlook the essentials. Overloading of systems with automatic controls should be avoided since, in case of malfunction, these can *introduce* more hazards into the system than they avoid. They furthermore tend to turn the operator into an attendant who, in an emergency, is incapable of operating the plant.

But disruptive forces may also be caused by faulty operation, and it is the plant management's responsibility to select and train qualified operators.

UNFIRED PRESSURE VESSELS

Unfired pressure vessels are compressed air tanks, steam-jacketed kettles, digesters, vulcanizers, and other vessels that can be subjected to internal pressure or vacuum, but that do not have the direct fire of burning fuel impinging on them. If heat is generated in the vessel, it is by chemical action within the vessel or by application of electric heat, steam, hot oil, or other heating medium to the contents of the vessel.

Design

Unfired pressure vessels are covered in the ASME Code, Section VIII, Divisions 1 and 2. Certain classes of vessels, however, are exempt from its scope. Some of these are:

1. Vessels subject to federal control
2. Vessels with a nominal capacity of 120 gal (450 liters) or less of water under pressure, in which any trapped air serves only as a cushion
3. Vessels having an internal or external operating pressure not exceeding 15 psi (103 kPa), with no limitation on size
4. Vessels with an inside diameter not exceeding 6 in. (15 cm), and no limitation on pressure
5. Hot water storage tanks heated by steam or other indirect means—heat input to be 200,000 Btuh (59,000 joules/second) or less, and water temperature to be 200 F (93 C) or less, and nominal capacity of 120 gal (450 liters) or less water.

The ASME Code provides that vessels designed for pressures over 3,000 psi (20,700 kPa) may be code stamped. This is covered under Section VIII of the ASME Code, Divisions 1 and 2, described next.

Division 1 normally covers vessels with ratings of 3,000 psi or less (with the exemptions listed earlier). Vessels may be constructed for pressures above 3,000 psi; however, design principles and construction practices in addition to the minimum ASME Code requirements must be considered.

Vessel designs under Division 1 rules are calculated according to the principal stress theory; and a design factor of 4 is provided on tensile strength. Vessels built to these specifications may be used anywhere the pressure and temperature do not exceed the ratings allowed by the Code.

Before the pressure vessel is designed under Division 1 rules, some questions to consider are:

1. Will the material used in construction of the vessel damage or chemically change the material in process?
2. Will the material in process affect or damage the metal in the vessel?
3. Will the filled vessel carry the weight of its contents (plus internal pressure)?
4. Will it resist both the pressure introduced into it and any additional pressure that may be caused by chemical reaction during processing?
5. Will the vessel withstand any vacuum that may be created intentionally or accidentally without collapsing?

Specifications for construction of Division 1 pressure vessel should include, in addition to general requirements, the working pressure range, working temperature range, data as to whether or not the pressure and/or temperature range is cyclic, a description of what contents are to be, and all other information of a specific nature that may affect fabrication and installation of the vessel, such as stress relieving, radiography, welding, and other requirements.

Division 2 is entitled "Alternate Rules for Pressure Vessels." Vessel design under these rules is based on a detailed stress analysis using Tresca's Maximum Shear Theory; and a design factor of 3 is provided on tensile strength. Design calculations are more complex than those in Division 1, but they allow thinner wall sections and may provide for vessels used at pressures in excess of 3,000 psi.

The "alternate rules" apply only to vessels installed in a fixed location and subjected to a specific service. To obtain a vessel with an ASME stamp under these rules, a prospective purchaser must prepare a "user's design specification" and have it certified by a registered professional engineer experienced in pressure vessel design.

Other codes. Although the ASME Code has been adopted by many governing bodies, and therefore has the force of law, other codes, such as the American Petroleum Institute and state and local codes, with size or service limitations which are different or more restrictive may be required by the legal jurisdiction in which the vessel is located. It is best to check.

Second-hand vessels. Prospective purchasers of second-hand vessels must comply with the jurisdictional requirements for second-hand vessels. Usually one of the requirements will be to have the equipment inspected by an authorized inspector. Before the vessel is purchased, a written report that the equipment meets the requirements of the jurisdiction where it is to be installed should be obtained. A great deal of trouble has arisen when second-hand equipment has been purchased and reinstalled before inspection.

Inspection and entry

Pressure vessels should be inspected regularly by persons who are qualified and trained for this work. Inspectors should be instructed to be conservative in approving borderline cases. Be sure to check whether state or local governing bodies or insurance companies require their own people to make inspections. Often these individuals can make suggestions for refinements which can contribute to lower insurance rates by contributing to greater safety.

A large company or plant may find it advantageous to employ a full-time inspection staff to administer a regular inspection program for all its pressure vessels. Such a program, coupled with good preventative maintenance, prolongs vessel life and prevents accidents.

A log of the history of each vessel should be kept by the inspector or the maintenance department. Included in it should be blueprints, manufacturer's data report and instructions, design data (including location of dimensional check points), installation information, records of process changes, and records of all repairs and conditions found on inspection. This log will prove valuable in operating the equipment, and in design, installation, and operation of new equipment.

When corrosive, poisonous, or toxic materials are used in the plant, management should so advise the operating personnel and the inspectors. When the inspection is carried out by the state, city, or insurance inspector, a plant chemical engineer or other competent person should accompany the inspector to describe the processes in detail so the vessels' conditions will be known.

When new processes are developed, the inspectors and the operators should be advised in detail what these processes are and how they may affect the pressure vessels.

Entry. A safe tank-entry procedure is absolutely necessary (and sometimes required by law) in order to eliminate the sizable number of fatalities that come from entering dangerous vessels and confined spaces. The hazard arises when workers or inspectors cannot get out of a vessel without help; difficulty in communication compounds the problem. (Review the procedures described in Chapter 16 under Proper Procedures and Cleaning Storage Tanks.)

Hazards to those in confined spaces include:

1. Exposure to toxic materials already in the confined space or introduced later
2. Lack of sufficient oxygen

3. Heat—a fire might start, hot gases or liquids might enter, or the vessel might be heated inadvertently
4. Agitators might be started or the vessel itself might be set in motion.

Before being entered, a vessel must be properly prepared. It must be drained, ventilated, and cleaned. All connecting pipelines should be disconnected and blanked, or valves on the line should be closed, locked out, and tagged (Figure 18-3). All power-driven devices (such as agitators) must be positively disconnected, locked out, and tagged. See the discussion of zero mechanical state (ZMS) procedures in Chapter 2, Construction and Maintenance of Plant Facilities, and Chapter 8, Principles of Guarding.

When purging a tank, the vent should discharge outside into an area where no hazard will be created for persons. In some instances, the vessels may be purged with an inert gas such as CO_2 (carbon dioxide) or nitrogen. It must be remembered that this inert gas will not support life, so persons entering the vessel must wear air-supplied respirators or self-contained or self-generating breathing equipment.

Using forced ventilation for confined spaces is usually safer than requiring employees to wear respiratory protection (see the Council's *Fundamentals of Industrial Hygiene*, 3rd edition, Chapter 23, Respiratory Protective Equipment). Air should be blown in until tests of the exhaust and of the interior of the enclosed vessel show that the space is safe for entry. All areas of the vessel should be tested for flammable and toxic gases, and for inadequate oxygen. These tests must be repeated at intervals to make sure that conditions remain safe while employees are in the vessel. Air should be introduced to make sure there are several air changes per minute in the vessel.

After all preparations for entry are completed, the supervisor of the area should check that the vessel is safe, that all lines are closed off, that power sources are locked out, that ventilation and personal protective equipment are adequate, and that safe work procedures are planned. A "vessel entry permit" that certifies all precautions have been carried out can then be issued.

Straight ladders or rope or chain ladders with rigid wood rungs should be provided. Also, employees should not be required to go into an opening that they must squeeze through. In an emergency they cannot exit or be removed quickly.

Another necessary precaution is to have employees don safety harnesses attached to lifelines before entering any vessel. An observer equipped with similar respiratory protection and a harness with a lifeline should be stationed outside the vessel. This person should also have some device for signaling for more help.

Depending on the previous content of the vessel, the person entering the vessel should be equipped with a vapor-proof flashlight or vapor-proof low-voltage extension light. At times, a chemical protective suit is advisable. All inspection equipment (including tools) should be made of nonsparking materials.

The method of cleaning depends on the use of the vessel. If it has contained petroleum or chemical products, the vessel may be filled with water, a caustic solution, or a neutralizing agent to remove sludge and adhered materials. Vessels used for flammable liquids should be washed, steamed, and/or ventilated until a test with an approved explosion meter shows the level is safe. (See Chapter 16, Flammable and Combustible Liquids.)

Figure 18-3. Valves on lines leading to pressure vessels should be locked out when workers are inside. If contents of the line are very hazardous, the line should be disconnected and blanked off.

Hydrostatic tests

If a pressure vessel is so constructed that an internal inspection cannot be made periodically (length of intervals depending on the corrosivity of the contents), it should be subjected to a hydrostatic test if the weight of water will not in turn set up damaging stresses. In this latter case, a pneumatic test can be applied. See the ASME Code, Section VIII, for new construction, and see the NB Code for existing vessels.

Compressed gas or air should never be used to test an unfired pressure vessel above its safe working pressure, although it can be used to test for leaks at pressures below the working pressure. Great care must be used because a vessel may fail under test and shatter. Testing should follow procedures in the ASME and NB Codes and be conducted under the supervision of qualified personnel.

The required pressure for a standard hydrostatic test is normally no greater than 1½ times the maximum allowable working pressure of the vessel being tested. Division 2 of the ASME Code, Section VIII, provides for the establishment of upper limits by the design engineer, in terms of stress-intensity limits relative to the yield strength or tensile strength, or creep rupture strength, at test temperature. Inspection of Division 2 ves-

sel is made at a pressure equal to the greater of the design pressures, or ¾ of the test pressure.

Test areas should be isolated as far as possible from other operations and suitable barricades should be provided for protection of personnel and valuable equipment. This is especially important when conducting proof tests and tests to destruction.

All personnel should keep clear of the vessel under full test pressure and no one should be allowed to approach it until the pressure has been reduced to, or is very close to, the maximum allowable working pressure.

Detecting cracks and measuring thickness

Pressure vessels used to process gases or oily materials may have very small leaks which will not show under hydrostatic tests. To detect them, a small amount of ammonia is released inside the vessel and compressed air is then applied until a maximum pressure of 50 percent of the working pressure is attained. A swab soaked in hydrochloric (muriatic) acid is passed over all seams and other suspect areas. Leakage will be indicated by a white vapor, (ammonium chloride) formed by contact of escaping ammonia and the acid. Using a burning sulfur stick is also effective—a change in the flame indicating the presence of ammonia.

Radiography or ultrasonic examination is especially good for finding cracks. For example, ammonia tanks used for agriculture are subject to corrosion and to stress cracking. If such tanks are not routinely checked for cracks, accidents could occur.

There are also a number of applications where it is vital to check the thickness of an unfired pressure vessel without damaging it. A number of manufacturers make instruments for doing this. These instruments measure with ultrasonic and electronically produced rays. Using these instruments, a qualified operator can determine the thickness of metal to within two or three percent. Some will disclose cracks which extend to or are slightly below the surface. The use of a radiograph will then be necessary to determine how deep the cracks are.

Another method of detecting microscopic hairline cracks is the lacquer method. When the head of a pressure vessel is suspected, it is cleaned and given a coat of clear lacquer. After the lacquer hardens, a hydro test is applied. The weak spots, hairline cracks, or fatigue stress cracks will show up on the head, expand and crack the lacquer. The more modern dye penetrant tests, when made by competent inspectors according to manufacturers' instructions, are considered very reliable.

At each inspection of such pressure vessels as vulcanizers, digesters, and autoclaves which have removable cover plates, heads, or doors, the holding bolts, cover plate bolts, slots, and retaining rings should be checked for wear, and hammer-tested for soundness. Because these parts are badly abused in service and receive considerable wear, it is advisable to replace them periodically. They are comparatively inexpensive. The width of the slot should be gaged and a careful check should be made of the retaining rings and cover plates, as cracks are set up by the stress of improper adjustment or closing of the door plate.

If cracks cannot be satisfactorily repaired, the vessel should be condemned.

Operator training and supervision

It is important that employees working with pressure vessels, particularly with those that are used in chemical processes, be thoroughly trained both in routine duty and emergency procedure. Supervision, too, should be competent and alert. A new employee being trained as an operator or helper should have the entire process explained, including the hazards involved and just how this operation affects the entire process.

In some plants a checklist is used to make certain that no step has been overlooked in the processing cycle. The operator or helper records on a card the information obtained from recording apparatus and thermometers, and the time and frequency at which the valves to each pressure vessel are operated.

After each complete processing cycle, the operator initials a card checklist.

If the contents of a pressure vessel are being discharged to a vessel that the operator cannot see, or one run by another operator, a whistle, bell, or light signal system should be installed. The time that these signals are given and action is taken should be noted on the checklist so that the wrong valve will not be opened or closed.

The plant operations supervisor should instruct operators of vessels with cover plates or removable doors how to tighten the bolts or quick-closing lugs without damaging them or the retain-ing rings. Operators likewise should be instructed to open cover plates or removable doors only after the vessel has been relieved of all pressure. Improper vessel opening can be avoided more simply, however, by the use of a safety interlock system (discussed below under Autoclaves and interlocks).

Operators should know where to look for wear on holding bolts, quick-opening lugs, and lug openings on the cover plate or removable door, and when a bolt has worn to the point at which the supervisor should be notified. A torque wrench used to tighten bolts will assure uniform tightness and reduce wear and damage.

So operators will not open or close the wrong valves, valves and pipelines should be tagged and marked as described in ANSI standard A13.1, *Scheme for the Identification of Piping Systems*. (Chapter 1, Industrial Buildings and Plant Layout, discusses this subject.)

Safety devices

Because pressure vessels are used to process such a great variety of materials, each vessel should be equipped with safety devices designed for the type of vessel and for the work it is to do.

Safety valves for each vessel should be ASME/NB rated and stamped "safety valves." The vessel should be provided with safety devices that will adequately protect it against overpressure, chemical reaction, or other abnormal conditions.

Safety valves. ASME/NB rated and stamped safety valves of the spring-loaded type are the most commonly used safety devices for pressure vessels. They are used on vessels containing air, steam, gases, and liquids that will not solidify as they pass out through the safety valve discharge.

Valves on pressure vessels containing air or steam should be large enough to discharge the contents at a rate to prevent pressure buildup as prescribed in the ASME Code. On vessels that contain liquids, the safety valve seat should be so made that it will not collapse nor the contents plug the discharge opening. The safety valve discharge line should lead to a point where it is safe to discharge. When practical, valves should be equipped with test levers. Frequent testing prevents sticking of the valves. For vessels with dangerous contents (toxic, flammable, etc.), the safety valve should not have a test lever. (See details under previous section on Boilers—Design and construction.)

If liquid contents are heated, the safety valve should be designed to operate if the vessel is overpressured as the liquid expands. For pressure vessels containing hot water, or in which water is heated, the valve should be sized to relieve the contents on the basis of the total number of Btu's that can be applied to the vessel.

The old ball-and-lever safety valve has been condemned for use by all Code states, because its setting is easily tampered with and it can be accidentally reset.

Rupture disks. A frangible disk may not clog as easily as a spring-loaded safety valve and is easily and inexpensively replaced. A rupture disk may clog or become coated with material in such processes as the manufacture of varnish and other resins. This coating at times becomes thick enough to affect the rupturing pressure of the disk so that it is necessary either to replace it or clean it with a solvent.

The condition of these disks should be checked not less than once annually to see that they are clear. It must be remembered that when the disk ruptures, all pressure is relieved from the vessel. This could result in complete loss of product or spoiled in-process material.

A rupture disk must function within ±5 percent of its specified bursting pressure at a specific temperature. Disks may be installed between a spring-loaded safety or relief valve and the pressure vessel in order to prevent unnecessary corrosion of the valve and to prevent it from becoming plugged by the contents of the vessel. These multiple installations must be in accordance with the ASME Code, Section VIII.

Various designs of rupture disks are available.

Vacuum breakers. It is just as important to protect a pressure vessel from collapsing under a vacuum as it is to protect it against bursting from overpressure. Several safety devices provide such protection. One, the mechanical vacuum breaker, similar to a spring-loaded safety valve, has a spring set at a predetermined vacuum. Another, the weight balanced vacuum breaker, using a weight suspended from a fulcrum attached to the gate, is generally used on pressure vessels working intermittently on pressure and vacuum. If the vacuum exceeds the setting of the weight on the fulcrum, the breaker then opens. Some vessels that ordinarily work under pressure but in which a vacuum may occur because of rapid cooling (as when steam condenses), may install a check valve with a flap or valve disk facing into the vessel. Whenever a vacuum occurs, the check valve disk opens automatically.

Water seal. A water seal is used on pressure vessels that operate on low pressure or slight vacuum, such as alcohol stills and gas holders. A water seal is a U-pipe filled with water, with one end connected to the pressure side of the vessel and other vented to the atmosphere.

Since the vessel operates under a pressure of only a few pounds, the degree of pressure can be regulated by the height of the water in the vent pipe. If the pressure rises above the set limit, the water is forced out of the pipe, thus relieving the pressure.

Vents. In many processes, pressure must be relieved before the pressure vessel can be opened. An easy means of relieving this pressure is to vent it to the atmosphere. Condensate tanks, which operate under very low pressure or no pressure at all, but in which excessive pressure can build up, should also be equipped with vent pipes and safety valves.

Vent pipes should be large enough in diameter to relieve the contents of the vessel before excess pressure can build up. A vent pipe should preferably be installed with a U-bend at the atmospheric discharge to prevent dirt from clogging the pipe. Be careful to direct the flow away from the vessel in case of fire, so it will not impinge on the metal.

Vent pipes must also be protected in cold weather—vapor may freeze as it leaves, rendering vents inoperative as safety devices. If a vent pipe is so placed that it may freeze or become clogged by dirt, a relief valve should be installed on the pipe as added protection.

Regulating or reducing valves. Some vessels are operated under steam pressure much lower than that obtained from the boiler or steam transmission line. A regulating or reducing valve reduces high-pressure steam to the pressure required for a specific operation.

There should be a safety valve on the low-pressure side of the reducing valve. The relieving capacity of the safety valve should be sufficient to assure that the pressure on the vessel that is being fed the steam will not exceed the vessel's safe working pressure in the event the reducing valve fails.

To provide protection for all pressure vessels in a battery of the same type, one reducing valve and one safety valve may be installed in the main steam line. This is the usual method in the case of steam-jacketed kettles in which ordinary pressures do not exceed 10 to 25 psi (70 to 170 kPa). Safety valves should be so connected that there is no stop valve between them and any vessel they protect.

Autoclaves and interlocks. Autoclaves, vulcanizers, retorts, digesters, and all pressure vessels that may contain large volumes of steam during operation and that must be opened for charging should be equipped with interlocks. An interlock will prevent the opening of the charging door until all pressure has been relieved, or prevent the pressurizing of the vessel until the door is in the fully closed position.

The most hazardous part of these vessels is their closure, although they should be inspected for cracks like any other pressure vessel. Opening an autoclave with pressure in it will cause the door to be flung open with explosive violence. The contents may be fired out like projectiles, and the reaction to the blowout may cause it to move back an impressive distance. No matter whether a bolted door, a rotary lug door, a shearing door, a clamp door, or a screwed-on door is used, the sealing mechanism must be maintained in good shape (as described under Detecting cracks and Operator training), and an interlock system must be provided.

Steam-jacketed vessels and evaporating pans

Steam-jacketed vessels are used to heat liquid mixtures to a moderate degree. Steam is circulated between the outer and inner shells of the vessel at pressures which are usually 10 to 30 psi. Occasionally the process may require that they be operated at pressures up to 100 psi, (690 kPa). Heat is transmitted through the inner shell to the contents.

Such vessels are used principally in commercial preparation of food, in candy manufacture, and for cooking starch in laundries and textile mills. They are also used in the chemical industry for low-temperature "cooking." On a steam-jacketed vessel which has a tight cover, a separate safety valve must be provided for the inner kettle.

If a steam-jacketed vessel can be completely valved off, the vessel should be protected with a vacuum-breaker to keep it from collapsing.

Precautions to be followed in the operation of steam-jacketed kettles include:

1. The steam space should be thoroughly drained before steam is admitted to the jacket. It is advisable to open drain lines even though traps are installed, because water in the steam space may cause serious damage.
2. Steam should be admitted to cold vessels slowly, in order to allow ample time for uniform heating and expansion of all parts. This becomes more important as vessel size and steam temperature are increased.
3. Unless automatic protection is provided, vents should be opened when the steam supply is shut off to prevent damage or even collapse of the kettle upon condensation of the steam.

Figure 18-4. Two methods of securing a high-pressure line: line was secured on either side of the fitting *(lower right arrow)* and at the bends *(upper arrow)*. As a secondary protective measure, channel iron was placed over the lines.

4. Where the agitators are used, paddles must not strike the kettle. Even a slight deformity in the inside may mean extensive repairs. Hand stirrers should also be used with care.
5. Kettle edges should be sufficiently high, or guardrails should be provided, so that employees will not accidentally fall in.
6. Kettles should be filled only to a point where undue splashing will not occur when the contents are heated or agitated. Splash guards or loose covers may be used.

Evaporating pans ordinarily are shallow pans containing steam coils which, when the pans are in operation, are immersed in the material being treated. If the coils become exposed, the material may be overheated or ignited, and a fire or explosion may result. To prevent accidents:

1. Pans should be continuously attended as long as they are in operation.
2. After each use, the pans and coils should be thoroughly cleaned.
3. After steam is shut off in the coils, the coils should be drained to prevent the product from being drawn into lines when steam condenses and creates a vacuum. (Installing a vacuum breaker would also prevent this.)

A general precaution to follow in maintenance work is "never allow employees to climb up over large open vats or kettles filled with hot, corrosive, or viscous fluids." Vats should be drained or covered, safe work stands should be provided, and employees should wear lifelines if necessary. All boards in the workstands should be fastened in place, not left loose. Every other precaution should be taken to make sure that people and objects do not fall into these liquids.

HIGH-PRESSURE SYSTEMS

The hazards of high-pressure systems largely arise from failures caused by leaks, pulsation, vibration, and overpressure. Besides the damage that can be expected from the release of high-pressure gases if a vessel or pipe ruptures, fatal injuries can result from the blowout of high-pressure gages or from whiplash of broken high-pressure pipe, tubing, or hose. The potential of injury and damage from pressure system accidents is very high. In the research laboratories of the rocket and missile industry, for instance, it is exceeded only by propellant explosions.

Because reciprocating pumps and compressors are normally needed to generate high pressures, the inner fibers of the piping and vessels of the system are subjected to pulsating pressures. Pipe and vessels must, therefore, be absolutely free of internal notches or severe scratches. These defects are stress raisers that will surely lead to fatigue failure. Also, the pipe must avoid stress concentrations arising from ill-planned holes or cross bores. The stress concentration factor from a radial entry to a pipe or cylinder wall will reduce the pressure endurance limit by almost one-half. Because of the pulsating pressure condition, nothing can be labeled "safe" and then be forgotten. Constant surveillance is essential.

Leaks in pressurized systems can also be hazardous. Liquids expelled can easily penetrate clothing and skin. A sudden leak might instantly fill an enclosure with an explosive gas mixture. Because most leaks occur at joints, the number of joints should be kept at a minimum.

In piping, vibration should be limited by pulse dampening, if possible. Designers use various means to do this in hydraulic and pressurized gas systems. Because pulse dampening is an ideal never fully achieved, many rugged pipe supports must be used in high-pressure systems. The supports must be strong enough to resist deflection from any direction; they must be securely anchored to the structure of the building to prevent whiplash (Figure 18-4).

Bourdon tubes in pressure gages will almost all eventually fail from fatigue caused by the constant pulsing of the pressure. They can fail after many cycles, or even when the gage is new. Gages on large vessels are not as subject to large pressure oscillations as those in lines where pressure is maintained by a compressor.

Figure 18-5. A reference gage has been installed for each of the four separate pressure stages of this compressor. Note the plastic shield *(arrow)* over the gages.

Pressure gages used at a 1,800 psi (12,400 kPa) or more (except Underwriters Laboratories listed gas regulator gages) should have full-size blow-out backs, integral sides and front designed to withstand internal explosion, and either multi-ply plastic or double-laminated safety glass gage-face cover (Figure 18-5). Tests at 3,000 psi (20,700 kPa) show that gages not constructed in this manner will have a hole blown in the face of the gage. Providing a ½- to 1 in. (1.3 to 2.5 cm) hole in the back of the gage does not give enough vent area for safe clearance of gases.

A substantial shield should be provided in front of high-pressure gages. This shield should be made of acrylic plastic (at least ⅝-in. [1.6-cm] thick, meeting MIL P5225B-Finish A Specification), and be free of scratches, gripper marks, tool chatter marks, and other stress raisers.

When mounted, a gage should have at least ½ in. (1.3 cm) clearance between itself and the item to which it is attached. If the gage is mounted flush to a backing plate, a diameter hole at least equal to the gage should be cut through the plate, leaving, of course, sufficient area to mount the face flange.

Shields mounted behind the gages should be substantial and a clear area no less than ½ inch (1.3 cm) wide should be left behind the gage for vent gases to escape between the shield and the back of the gage.

Areas where high gas-pressure systems are operating should be restricted to all but necessary personnel. Often reactors, pressure vessels, and heat exchangers having great hazard potential should be located behind barricades and have remote control and monitoring. Particularly, vessels and systems being tested should be placed behind barricades.

REFERENCES

American Society of Heating, Refrigerating and Air-Conditioning Engineers, 1791 Tullie Circle, N.E., Atlanta, Ga. 30329.
Applications.
Handbook of Fundamentals.
Systems and Equipment.
American Society of Mechanical Engineers, 345 E. 47th St., New York, N.Y. 10017. *ASME Boiler and Pressure Vessel Code.*
American National Standards Institute, 1430 Broadway, New York, N.Y. 10018.
Safety Color Code for Marking Physical Hazards, Z53.1.
Scheme for the Identification of Piping Systems, A13.1.
Baumeister, Theodore, ed. *Mark's Standard Handbook for*

Mechanical Engineers. New York, N.Y. McGraw-Hill Book Co., 1978.

Combustion Institute, 5001 Baum Blvd., Pittsburgh, Pa. 15213. (General.)

Compressed Gas Association, 1235 Jefferson Davis Hwy., Arlington, Va. 22202. *Cylinder Service Life: Seamless, High-Pressure Cylinders,* Pamphlet C-5.

Fryling, G. R., *Combustion Engineering,* New York, N.Y., Combustion Engineering, Inc. 1966.

National Board of Boiler and Pressure Vessel Inspectors, 1055 Crupper Ave., Columbus, Ohio 43229.
  Inspection Code.

National Fire Protection Association, Batterymarch Park, Quincy, Mass. 02269.
  Fire Protection Handbook.
  Life Safety Code, NFPA 101.
  "Boiler-Furnace Standards," NFPA 85A, 85B, 85D, and 85E.

National Safety Council, 444 N. Michigan, Chicago, Ill. 60611.
  Industrial Data Sheets
    Cleaning with Hot Water and Steam, 238.
    Maintenance of High-Pressure Gate and Plug Valves, 440.

Uniform Boiler and Pressure Vessel Laws Society, Long Beach Rd., P.O. Box 512, Oceanside, N.Y. 11572. *Synopsis of Boiler and Pressure Vessel Laws, Rules, and Regulations by States, Cities, Counties, and Provinces, in the United States and Canada.*

19

Safety Engineering Tables

THE HEALTH AND SAFETY PROFESSIONAL requires information on many subjects usually not available in any one source, but fortunately it is possible to gather the more generally pertinent information into one chapter such as this. However, other more specialized material must be sought out as required when the health and safety professional must deal with a specific subject in great detail and for a specific location. This latter information includes (1) applicable federal, state or provincial, and local code requirements which are beyond the scope of this Manual and which must be checked locally, and (2) specific scientific and engineering information which generally is available in handbooks devoted to specific subject fields. For the reader's convenience in developing this additional information, a number of these handbooks are listed under References at the end of this chapter.

FACTORS OF SAFETY

In applying scientific data to actual work situations, the "ideal conditions" under which the experimental data were gathered often do not exist, and so a safety factor is often applied to compensate for this. The magnitude of this factor depends on how great the cost of failure will be in terms of life or damage. While pertinent safety factors are suggested in the various chapters of this Manual, many are summarized here.

By definition, this factor of safety is the ratio of the ultimate (breaking) strength of a member or piece of material to the actual working stress or to the maximum permissible (safe load) stress when in use. This term is basic to the whole safety program, as the following few examples will indicate.

Stairs and landings should sustain a live load of not less than 100 psf (480 kg/m²) with a factor of safety of 4—that is, the maximum live load should never exceed ¼ of the breaking strength. OSHA requires standard railings to withstand at least 200 lb (890 newtons) pressure applied in any direction at any point on the rail—§1910.23e(2)(v).

Every scaffold and its supporting members should be so designed to support a given load with a safety factor of 4.

Boilers should have a safety factor of about 5, which figure is sometimes raised to 5.5 with used boilers, and to 6 when the boiler is 10 or more years old. Unfired pressure vessels should have a safety factor of approximately 5.

For each section of a refrigerating system, the factor of safety is commonly 5 times the safe working pressure, or 5 times the pressure at which the safety valve on the section is set to relieve.

Hydraulic piping and hose should have a safety factor of 8.

In the case of cranes, the hook load should have a safety factor of between 4 and 5; hoisting ropes, gears, and other parts subject to wear, a factor of not less than 8; and all other parts, including structural steel, a factor not less than 5. On hot metal cranes, however, all these safety factors should be higher, even as high as 10 throughout.

For general hoisting purposes, it is not advisable for the working load of a rope to exceed 1/5 of its breaking strength. This means a safety factor of 5, but factors that are larger than this (varying up to 8 or more) are often required for safe and economical operation.

Cast iron flywheels should have a safety factor of 10, but those made of wood should have the factor of safety raised to 20.

In addition to data on construction materials, information

Table 19-A. General Factors of Safety for Common Construction Materials

Material	Steady Load	Load Varying from Zero to Maximum in one Direction	Load Varying from Zero to Maximum in both Directions	Suddenly Varying Loads and Shocks
Cast iron	6	10	15	20
Wrought iron	4	6	8	12
Steel	5	6	8	12
Wood	8	10	15	20
Brick	15	20	25	30
Stone	15	20	25	30

Reprinted with permission from *Machinery's Handbook,* 16th ed., The Industrial Press.

is supplied on lighting and ventilation, chemicals and compounds, mathematical functions, and abbreviations used in this Manual.

- The illumination levels listed in Table 19-G provide a guide for efficient visual performance rather than for safety alone and they should not be considered as regulatory minimum illumination standards.

These sample levels are taken from the American National Standard *Practice for Industrial Lighting* ANSI/IES RP7–1983. They are based upon research conducted on young adults with normal and better than 20/30 corrected vision. More illumination might be provided for older workers to compensate for degeneration of vision due to age.

Levels of 200 foot-candles or more were obtained with a combination of general lighting plus specialized supplementary lighting. Consideration should be given to glare, objectionable shadows, color, contrast and eye protection.

Minimum illumination for safety of personnel, at any time and at any place where safety is related to seeing conditions, is as follows:

Hazards requiring visual detection	Slight		High	
Normal activity level	Low	High	Low	High
Illumination levels, foot-candles	0.5	1.0	2.0	5.0
levels, dekalux	0.54	1.1	2.2	5.4

Special conditions may require different levels of illumination. In some cases, higher levels may be required—for example, where security is a factor. In some other cases, greatly reduced levels of illumination, including total darkness, may be necessary (notably in connection with photographic products). In these situations, alternate methods of ensuring safe operation must be relied upon.

See specific industry reports of the Industrial Lighting Committee of the Illuminating Engineering Society for guidelines. ANSI/IES RP7–1983 has more details.

- There is an appendix on conversion between metric and English units at the end of this chapter. This follows, insofar as possible, the recommendations of the *Standard Metric Practice Guide,* E 380, adopted by the American Society for Testing and Materials.

Table 19-B. Basic Stresses For Laminated Structural Members

	Extreme fiber in bending or tension parallel to grain (psi)	Maximum longi-tudinal shear (psi)	Compression perpen-dicular to grain (psi)	Compression parallel to grain (psi)	Modulus of elasticity in bending (1,000 psi)
SOFTWOODS					
Baldcypress (southern cypress)...	2,400	170	330	2,000	1,300
Cedars:					
Redcedar, western...	1,600	135	220	1,300	1,100
White-cedar, Atlantic (southern white cedar) and northern...	1,400	115	195	1,050	900
White-cedar, Port-Orford...	2,000	150	275	1,650	1,600
Yellow-cedar, Alaska (Alaska-cedar)...	2,000	150	275	1,450	1,300
Douglas-fir, coast type...	2,750	150	350	2,000	1,800
Douglas-fir, coast type, close-grained...	2,950	150	375	2,150	1,800
Douglas-fir, Rocky Mountain type	2,000	135	310	1,450	1,300
Douglas-fir, all regions, dense...	3,200	150	410	2,350	1,800
Fir, balsam...	1,600	115	165	1,300	1,100
Fir: California red, grand, noble, and white...	2,000	115	330	1,300	1,200
Hemlock, eastern...	2,000	115	330	1,300	1,200
Hemlock, western (west coast hemlock)...	2,400	125	330	1,650	1,500
Larch, western...	2,750	150	350	2,000	1,800
Pine, eastern white (northern (white), ponderosa, sugar, and western white (Idaho white)	1,600	135	275	1,400	1,100
Pine, jack...	2,000	135	240	1,450	1,200
Pine, lodgepole...	1,600	100	240	1,300	1,100
Pine, red (Norway pine)...	2,000	135	240	1,450	1,300
Pine, southern yellow...	2,750	180	350	2,000	1,800
Pine, southern yellow, dense...	3,200	180	410	2,350	1,800
Redwood...	2,200	115	275	1,850	1,300
Redwood, close-grained...	2,400	115	295	2,000	1,300
Spruce, Engelmann...	1,400	115	195	1,100	1,100
Spruce, red, white, and Sitka...	2,000	135	275	1,450	1,300
Tamarack...	2,200	160	330	1,850	1,400
HARDWOODS					
Ash, black...	1,800	150	330	1,150	1,200
Ash, white...	2,550	210	550	2,000	1,600
Beech, American...	2,750	210	550	2,200	1,800
Birch, sweet and yellow...	2,750	210	550	2,200	1,800
Cottonwood, eastern...	1,400	100	165	1,100	1,100
Elm, American and slippery (white or soft elm)...	2,000	170	275	1,450	1,300
Elm, rock...	2,750	210	550	2,200	1,400
Hickory and pecan...	3,500	235	660	2,750	2,000
Maple, black and sugar (hard maple)...	2,750	210	550	2,200	1,800
Oak, red and white...	2,550	210	550	1,850	1,600
Sweetgum (redgum or sapgum)...	2,000	170	330	1,450	1,300
Tupelo, black (blackgum)...	2,000	170	330	1,450	1,300
Tupelo, water...	2,000	170	330	1,450	1,300
Yellow-poplar...	1,800	150	240	1,450	1,300

Reprinted with permission from Forest Products Laboratory. 1 psi = 6.895 kilopascals.
Note: This table is for clear material under dry conditions as in most covered structures.

Table 19-C. Allowable Unit Stresses, Structural Timber (Pounds per square inch)

Material	$K^\S$ (columns)	Bending and tension	Compression°	Shear†	Bearing‡	Modulus of elasticity
Douglas-fir:						
Parallel to grain.........	23.4	1,500	1,200	100	1,500	1,600,000
Across the grain........	......	......	......	......	350	......
Oak:						
Parallel to grain.........	23.7	1,300	1,100	110	1,500	1,500,000
Across the grain........	......	......	......	......	500	......
Spruce:						
Parallel to grain.........	21.2	1,200	1,100	100	1,200	1,200,000
Across the grain........	......	......	......	......	250	......
Longleaf yellow pine:						
Parallel to grain.........	23.4	1,400	1,200	120	1,400	1,600,000
Across the grain........	......	......	......	......	300	......

Reprinted with permission from O'Rourke, *General Engineering Handbook,* McGraw-Hill Book Co.

° On lengths not greater than eleven times the least dimension. Otherwise, use column formulas.

† In timber-joint designs, these values may be increased 50 percent.

‡ These values are for joint design, and for bearings less than 10 in. long. § See following notes. 1 psi = 6.895 kilopascals.

TIMBER, STRUCTURAL GRADE (DRY). Allowable unit stresses in bending, compression, shear, and bearing, as specified by the N.L.M.A., are given in Table 19-C. Allowable unit stresses for timber columns are as follows:

1. Short Columns. The safe load, in pounds per square inch of net cross-sectional area, for columns and other members stressed in compression parallel to the grain, with a ratio of unsupported length to least dimension (l/d) not exceeding eleven (short columns) shall not exceed the allowable unit compression stress (c) parallel to grain for short columns, *i.e.*,

$$\frac{P}{A} = c$$

2. Intermediate Columns. For columns with a ratio of unsupported length to least dimension greater than eleven but less than K (intermediate columns), the following formula shall be used until the reduction in allowable stress equals one-third the stress permitted for short columns:

$$\frac{P}{A} = c\left[1 - \frac{1}{3}\left(\frac{l}{Kd}\right)^4\right]$$

3. Long Columns. For columns with a ratio of unsupported length to least dimension greater than K (long columns), the safe load shall be determined by the following formula:

$$\frac{P}{A} = \frac{\pi^2 E}{36(l/d)^2} = \frac{0.274E}{(l/d)^2}$$

4. Notation.
    P = total load, in pounds.
    A = area, in square inches of net cross section.
  P/A = the working stress or maximum load per square inch.
    c = allowable unit stress in compression parallel to grain for short columns.
    l = unsupported length of column, in inches.
    d = least dimension of column, in inches.
    E = modulus of elasticity.
    $K = \frac{\pi}{2}\sqrt{\frac{E}{6c}}$; at which $\frac{P}{A} = \frac{2c}{3}$

5. The safe load on a column of round cross section shall not exceed that permitted for a square column of the same cross-sectional area.

6. Columns shall be limited in maximum length to $l/d = 50$.

Table 19-D1. Capacities of Horizontal Cylindrical Tanks When Filled To Various Depths
Tanks with flat ends. Contents given in U.S. gallons per one foot of length

Diam. of Tank, In.	Full Tank	3	6	9	12	15	18	21	24	27	30	33	36	39	42	45	48	51	54	57	60
12	5.88	1.15	2.94																		
18	13.22	1.45	3.86	6.61																	
24	23.50	1.70	4.60	8.05	11.75																
30	36.72	1.91	5.23	9.27	13.72	18.36															
36	52.88	2.12	5.79	10.34	15.43	20.85	26.44														
42	71.97	2.28	6.31	11.31	16.97	23.07	29.47	35.99													
48	94.01	2.45	6.78	12.20	18.38	25.10	32.20	39.54	47.00												
54	118.98	2.60	7.22	13.04	19.68	26.97	34.72	42.80	51.08	59.49											
60	146.89	2.75	7.64	13.82	20.91	28.72	37.06	45.82	54.87	64.11	73.44										
66	177.73	2.89	8.04	14.56	22.07	30.37	39.28	48.65	58.39	68.41	78.59	88.68									
72	211.52	3.02	8.42	15.26	23.17	31.92	41.36	51.32	61.71	72.45	83.41	94.54	105.76								
78	248.24	3.15	8.78	15.94	24.21	33.41	43.34	53.86	64.87	76.27	87.97	99.90	111.97	124.13							
84	287.90	3.26	9.12	16.57	25.24	34.85	45.24	56.29	67.87	79.91	92.30	104.98	117.85	130.87	143.95						
90	330.49	3.43	9.46	17.20	26.20	36.21	47.05	58.61	70.75	83.39	96.43	109.81	123.45	137.28	151.23	165.25					
96	376.02	3.50	9.79	17.80	27.13	37.52	48.81	60.84	73.52	86.73	100.39	114.44	128.79	143.40	158.17	173.06	188.01				
102	424.50	3.61	10.10	18.37	28.01	39.00	50.49	62.99	76.18	89.94	104.20	118.89	133.92	149.25	164.81	180.53	196.37	212.25			
108	476.10	3.71	10.39	18.94	28.90	40.03	52.14	65.09	78.74	93.04	107.87	123.17	138.87	154.89	171.19	187.71	204.37	221.14	238.05		
114	530.25	3.78	10.74	19.49	29.75	41.22	53.73	67.10	81.24	96.05	111.43	127.31	143.63	160.33	177.33	194.60	212.05	229.65	247.37	265.13	
120	587.54	3.91	10.98	20.02	30.57	42.39	55.26	69.06	83.65	98.95	114.87	131.32	148.25	165.58	183.27	201.24	219.46	237.87	256.43	275.08	293.77

To ascertain the contents of a tank over one-half full: Let *h* = depth of unfilled portion. Find from the table the quantity corresponding to a depth *h*. Subtract this quantity from the contents of a full tank.

1 in. = 2.54 cm. 1 gal = 3.785 liters. Reprinted with permission from *Kent's Mechanical Engineers' Handbook*. John Wiley & Sons, Inc.

Note: This table may be used in conjunction with Table 19-D2 for finding total volume of liquid in double-dished head tanks. Example: A tank 8 ft long, 3 ft diam, 15 in. liquid inside. Solution: 20.85 × 8 (this table plus 3.92 × 2) for dished heads (Table 19-D2) = 174.64 gal.

Table 19-D2. Capacities of Standard Dished Heads When Filled To Various Depths
Contents given in U.S. gallons for one head only

Diam of Head (in.)	Full Head	3	6	9	12	15	18	21	24	27	30	33	36	39	42	45	48	51	54	57	60
12	0.40	0.05	0.20																		
18	1.36	.07	.32	0.68																	
24	3.22	.08	.41	.95	1.61																
30	6.30	.10	.49	1.18	2.10	3.15															
36	10.88	.11	.56	1.39	2.54	3.92	5.44														
42	17.28	.12	.63	1.59	2.94	4.64	6.57	8.64													
48	25.79	.13	.68	1.75	3.31	5.29	7.62	10.19	12.89												
54	36.72	.14	.74	1.90	3.64	5.90	8.60	11.65	14.95	18.36											
60	50.37	.14	.82	2.07	3.98	6.49	9.54	13.03	16.87	20.96	25.18										
66	67.04	.15	.83	2.19	4.25	6.98	10.35	14.30	18.68	23.43	28.42	33.52									
72	87.04	.16	.88	2.32	4.52	7.47	11.15	15.48	20.38	25.74	31.46	37.43	43.52								
78	110.66	.17	.93	2.44	4.79	7.97	11.94	16.65	22.02	27.97	34.39	41.16	48.20	55.33							
84	138.22	.18	.98	2.59	5.07	8.44	12.69	17.78	23.60	30.11	37.19	44.75	52.67	60.83	69.11						
90	170.01	.18	1.00	2.68	5.33	8.91	13.44	18.86	25.12	32.18	39.90	48.22	56.99	66.14	75.52	85.00					
96	206.32	.20	1.07	2.83	5.59	9.36	14.14	19.90	26.60	34.17	42.52	51.53	61.13	71.22	81.66	92.34	103.16				
102	247.48	.22	1.14	3.01	5.89	9.87	14.92	21.01	28.11	36.18	45.19	54.91	65.31	76.29	87.73	99.56	111.59	123.74			
108	293.77	.20	1.13	3.03	6.04	10.21	15.50	21.93	29.47	38.03	47.56	57.97	69.14	81.05	93.53	106.47	119.76	133.26	146.88		
114	345.51	.21	1.16	3.12	6.25	10.55	16.06	22.80	30.70	39.73	49.81	60.88	72.85	85.61	99.05	113.07	127.56	142.41	157.51	172.75	
120	402.27	.21	1.19	3.23	6.47	10.93	16.68	23.70	31.96	41.43	52.04	63.73	76.40	89.95	104.32	119.39	135.04	151.15	167.62	184.32	201.13

Radius = Diameter

To ascertain the contents of a head over one-half full: Let *h* = depth of unfilled portion. Find from the table the quantity corresponding to a depth *h*. Subtract this quantity from the contents of a full head.

1 in. = 2.54 cm. 1 gal. = 3.785 liters. Reprinted with permission from Lukens Steel Company.

Table 19-E. Capacities and Liquid Surface Area of Cylindrical Vessels, Tanks, Cisterns, etc., Set in Upright (Vertical) Position
Diameter in feet and inches, area in square feet, and U.S. gallons capacity for one foot in depth

$$1 \text{ gallon} = 231 \text{ cubic inches} = \frac{1 \text{ cubic foot}}{7.4805} = 0.13368 \text{ cubic foot} = 3.785 \text{ liters.}$$

Diam Ft	In.	Area Sq Ft	Gals 1 foot depth	Diam Ft	In.	Area Sq Ft	Gals 1 foot depth	Diam Ft	In.	Area Sq Ft	Gals 1 foot depth
1		0.785	5.87	5	8	25.22	188.66	19		283.53	2,120.9
1	1	0.922	6.89	5	9	25.97	194.25	19	3	291.04	2,177.1
1	2	1.069	8.00	5	10	26.73	199.92	19	6	298.65	2,234.0
1	3	1.227	9.18	5	11	27.49	205.67	19	9	306.35	2,291.7
1	4	1.396	10.44	6		28.27	211.51	20		314.16	2,350.1
1	5	1.576	11.79	6	3	30.68	229.50	20	3	322.06	2,409.2
1	6	1.767	13.22	6	6	33.18	248.23	20	6	330.06	2,469.1
1	7	1.969	14.73	6	9	35.78	267.69	20	9	338.16	2,529.6
1	8	2.182	16.32	7		38.48	287.88	21		346.36	2,591.0
1	9	2.405	17.99	7	3	41.28	308.81	21	3	354.66	2,653.0
1	10	2.640	19.75	7	6	44.18	330.48	21	6	363.05	2,715.8
1	11	2.885	21.58	7	9	47.17	352.88	21	9	371.54	2,779.3
2		3.142	23.50	8		50.27	376.01	22		380.13	2,843.6
2	1	3.409	25.50	8	3	53.46	399.88	22	3	388.82	2,908.6
2	2	3.687	27.58	8	6	56.75	424.48	22	6	397.61	2,974.3
2	3	3.976	29.74	8	9	60.13	449.82	22	9	406.49	3,040.8
2	4	4.276	31.99	9		63.62	475.89	23		415.48	3,108.0
2	5	4.587	34.31	9	3	67.20	502.70	23	3	424.56	3,175.9
2	6	4.909	36.72	9	6	70.88	530.24	23	6	433.74	3,244.6
2	7	5.241	39.21	9	9	74.66	558.51	23	9	443.01	3,314.0
2	8	5.585	41.78	10		78.54	587.52	24		452.39	3,384.1
2	9	5.940	44.43	10	3	82.52	617.26	24	3	461.86	3,455.0
2	10	6.305	47.16	10	6	86.59	647.74	24	6	471.44	3,526.6
2	11	6.581	49.98	10	9	90.76	678.95	24	9	481.11	3,598.9
3		7.069	52.88	11		95.03	710.90	25		490.87	3,672.0
3	1	7.467	55.86	11	3	99.40	743.58	25	3	500.74	3,745.8
3	2	7.876	58.92	11	6	103.87	776.99	25	6	510.71	3,820.3
3	3	8.296	62.06	11	9	108.43	811.14	25	9	520.77	3,895.6
3	4	8.727	65.28	12		113.10	846.03	26		530.93	3,971.6
3	5	9.168	68.58	12	3	117.86	881.65	26	3	541.19	4,048.4
3	6	9.621	71.97	12	6	122.72	918.00	26	6	551.55	4,125.9
3	7	10.085	75.44	12	9	127.68	955.09	26	9	562.00	4,204.1
3	8	10.559	78.99	13		132.73	992.91	27		572.56	4,283.0
3	9	11.045	82.62	13	3	137.89	1,031.5	27	3	583.21	4,362.7
3	10	11.541	86.33	13	6	143.14	1,070.8	27	6	593.96	4,443.1
3	11	12.048	90.13	13	9	148.49	1,110.8	27	9	604.81	4,524.3
4		12.566	94.00	14		153.94	1,151.5	28		615.75	4,606.2
4	1	13.095	97.96	14	3	159.48	1,193.0	28	3	626.80	4,688.8
4	2	13.635	102.00	14	6	165.13	1,235.3	28	6	637.94	4,772.1
4	3	14.186	106.12	14	9	170.87	1,278.2	28	9	649.18	4,856.2
4	4	14.748	110.32	15		176.71	1,321.9	29		660.52	4,941.0
4	5	15.321	114.61	15	3	182.65	1,366.4	29	3	671.96	5,026.6
4	6	15.90	118.97	15	6	188.69	1,411.5	29	6	683.49	5,112.9
4	7	16.50	123.42	15	9	194.83	1,457.4	29	9	695.13	5,199.9
4	8	17.10	127.95	16		201.06	1,504.1	30		706.86	5,287.7
4	9	17.72	132.56	16	3	207.39	1,551.4	30	3	718.69	5,376.2
4	10	18.35	137.25	16	6	213.82	1,599.5	30	6	730.62	5,465.4
4	11	18.99	142.02	16	9	220.35	1,648.4	30	9	742.64	5,555.4
5		19.63	146.88	17		226.98	1,697.9	31		754.77	5,646.1
5	1	20.29	151.82	17	3	233.71	1,748.2	31	3	766.99	5,737.5
5	2	20.97	156.83	17	6	240.53	1,799.3	31	6	779.31	5,829.7
5	3	21.65	161.93	17	9	247.45	1,851.1	31	9	791.73	5,922.6
5	4	22.34	167.12	18		254.47	1,903.6	32		804.25	6,016.2
5	5	23.04	172.38	18	3	261.59	1,956.8	32	3	816.86	6,110.6
5	6	23.76	177.72	18	6	268.80	2,010.8	32	6	829.58	6,205.7
5	7	24.48	183.15	18	9	276.12	2,065.5	32	9	842.39	6,301.5

Reprinted with permission from *Kent's Mechanical Engineer's Handbook,* John Wiley & Sons, Inc.

1 ft = 0.305 m

Table 19-F. Maximum Luminance Ratios

	Environmental Classification		
	A	B	C
Between task and adjacent darker surroundings	3 to 1	3 to 1	5 to 1
Between tasks and adjacent lighter surroundings	1 to 3	1 to 3	1 to 5
Between tasks and more remote darker surfaces	10 to 1	20 to 1	*
Between tasks and more remote lighter surfaces	1 to 10	1 to 20	*
Between luminaires (or windows, skylights, etc.) and surfaces adjacent to them	20 to 1	*	*
Anywhere within normal field of view	40 to 1	*	*

* Brightness Ratio control not practical.

A Interior Areas where reflectances of entire space can be controlled in line with recommendations for optimum seeing conditions.

B Areas where reflectances of immediate work area can be controlled, but control of remote surroundings is limited.

C Areas (indoor or outdoor) where it is completely impractical to control reflectances and difficult to alter environmental conditions.

Note: From the normal point of view, brightness ratios of areas of appreciable size in industrial areas should not exceed those in the above table.

Reprinted with permission from ANSI/IES Standard RP7–1983.

Table 19-G. Levels of Illumination Recommended for Sample Occupational Tasks

Area	Foot-candles	Area	Foot-candles
Assembly-rough, easy seeing	30	Loading platforms	20
Assembly—medium	100	Machine shops—medium work	100
Building construction—general	10	Materials—loading, trucking	20
Corridors	20	Offices—general areas*	100
Drafting rooms—detailed*	200	Paint dipping, spraying	50
Electrical equipment, testing	100	Service spaces—wash rooms, etc.	30
Elevators	20	Sheet metal—presses, shears	50
Garages—repair areas	100	Storage rooms—inactive	5
Garages—traffic areas	20	Storage rooms—active, medium	20
Inspection, ordinary	50	Welding—general	50
Inspection, highly difficult	200	Woodworking—rough sawing	30

° Reprinted with permission from *Practice for Office Lighting*, ANSI/IES RP1-1982. Others from ANSI/IES RP7-1983. See explanation in the introduction to this chapter.

1 footcandle = 10.76 lux.

Table 19-H. Rpm Speed of Saws to Give 10,000 Feet per Minute Rim Speed

Diameter (in.)	Circumference (ft)	RPM	Diameter (in.)	Circumference (ft)	RPM
6	1.57	6,366	56	14.66	680
8	2.09	4,775	58	15.18	660
10	2.62	3,820	60	15.71	640
12	3.14	3,180	64	16.78	600
14	3.67	2,730	66	17.28	580
16	4.19	2,390	68	17.80	560
18	4.71	2,125	70	18.33	545
			72	18.85	530
20	5.24	1,910	74	19.37	520
22	5.76	1,740	76	19.90	500
24	6.28	1,590	78	20.42	490
26	6.81	1,470	80	20.94	480
28	7.33	1,365	82	21.47	470
30	7.85	1,275	84	22.00	455
32	8.38	1,195	86	22.52	445
34	8.90	1,125	90	23.56	425
36	9.43	1,060	96	25.13	400
38	9.95	1,005	100	26.18	380
40	10.47	955	104	27.22	370
42	11.00	910	106	27.75	364
			108	28.27	354
44	11.52	870	110	28.80	348
46	12.04	830	112	29.32	341
48	12.57	800	120	31.42	318
50	13.09	765			
52	13.61	740			
54	14.14	710			

Reprinted with permission from H.K. Porter Co., Inc., Disston Division.

Note: Saws are usually tensioned to a standard speed of 10,000 peripheral feet per minute and, with the exception of large saws, can be operated at slower speeds with good performance and safety. The above table gives the rpm corresponding to this standard speed for various diameter saws.

All saws required to operate in excess of 10,000 peripheral feet per minute are tensioned for the speed specified. This speed is etched on the saw blade. These special saws should not be operated at other than the specified speed without consultation with the saw manufacturer.

1 in. = 2.54 cm
1 ft = 0.305 m
10,000 fpm = 50.8 m/s.

Table 19-I. Strength of Ice

Load	Ice Thickness Inches (min.)
One person on skis	1³/₄
One person on foot	2
Group, single file or Snowmobile	3
Passenger car, 2 ton gross	7½
Light truck, 2½ ton gross	8
Medium truck, 3½ ton gross	10
Heavy truck, 7-8 ton gross	12

Reprinted with permission from Snow, Ice, and Permafrost Research Establishment, Corps of Engineers, U.S. Army. Snowmobile data from "A Snowmobile Accident Study," R. W. McLay and S. Chism, 1969.

Notes:
1) This table is for clear sound ice.
2) Thickness must be increased and extreme caution used after spring melt starts.
3) Snow ice—use 50% of thickness of snow ice as effective for table given.
   Example: 10 in. snow ice plus 6 in. clear ice gives 11 in. thickness. Medium truck all right, but not heavy truck.
4) Table does not apply to parked loads.
5) Key to safety on ice is to disperse the weight over a large area; do not run vehicles within 50 ft of each other.

1 in. = 2.54 cm.

Table 19-J. Design Wind Pressures for Various Height Zones of Buildings or Other Structures
(From *Mimimum Design Loads in Buildings and Other Structures,* ANSI A58.1)

Height Zone, feet	Wind Pressure pounds per sq ft
Less than 50	20
50 to 99	24
100 to 199	28
200 to 299	30
300 to 399	32
400 to 499	33
500 to 599	34
600 to 799	35
800 to 999	36
1,000 to 1,199	37
1,200 to 1,399	38
1,400 to 1,599	39
1,600 and over	40

Reprinted with permission from Kent's Mechanical Engineers' Handbook, John Wiley & Sons.

WIND LOADS. A minimum pressure in pounds per square foot is specified by all building codes as an allowance for the effect of wind pressure against exposed surfaces of buildings. The figures given are recommended in this report as minimums. These requirements do not provide for tornadoes because they are based on a design wind velocity of 75 mph, corresponding roughly to a 5-min. average of 50 mph (indicated 63 by a 4-cup anemometer) at 30 ft from the ground.

Special provision for wind bracing is required for narrow buildings of even medium height and for all high buildings. In towerlike structures wind bracing should be planned in the early stages of design. The overturning moment due to wind pressure should not exceed two-thirds of the moment of stability, disregarding live loads, except where foundations are securely anchored. The possibility of a structure sliding on its foundation bed should also be considered. In computing the wind load no allowance should be made for the shielding effect of other buildings. All roofs should be designed to resist wind suction as well as wind pressure.

1 ft = 0.305 m.   1 psf = 0.0479 kPa

Table 19-K. Safe Bearing Loads on Soils
(Values approximate pressures allowed in major city building codes)

Nature of Soil	Safe Bearing Capacity (Tons per sq ft)
Solid ledge of hard rock, such as granite, trap, etc.	25–100
Sound shale or other medium rock, requiring blasting for removal	10–15
Hardpan, cemented sand and gravel, difficult to remove by picking	8–10
Soft rock, disintegrated ledge; in natural ledge, difficult to remove by picking	5–10
Compact sand and gravel, requiring picking for removal	4–6
Hard clay, requiring picking for removal	4–5
Gravel, coarse sand, in natural thick beds	4–5
Loose, medium, and coarse sand; fine compact sand	1.5–4
Medium clay, stiff but capable of being spaded	2–4
Fine loose sand	1–2
Soft clay	1

Reprinted with permission from *Standard Handbook for Mechanical Engineers,* rev. 9th ed. Edited by T. Baumeister, III and E. A. Avallone. (Copyright 1987, McGraw-Hill Book Co.)

Table 19-L. Specific Gravity of Gases and Liquids
(Gases at 32 F: air = 1,000; liquids: water = 1,000)

Gas	Sp Gr	Gas	Sp Gr	Gas	Sp Gr
Air	1.000	Ether vapor	2.586	Methane	0.554
Acetylene	0.920	Ethylene	0.967	Nitrogen	0.971
Ethyl alcohol vapor	1.601	Helium	0.138	Nitric oxide	1.039
Ammonia	0.592	Hydrofluoric acid	2.370	Nitrous oxide	1.527
Carbon dioxide	1.520	Hydrochloric acid	1.261	Oxygen	1.106
Carbon monoxide	0.967	Hydrogen	0.069	Propane	1.554
Chlorine	2.423	Mercury vapor	6.940	Sulfur dioxide	2.250
Ethane	1.049	Marsh gas	0.555	Water vapor	0.623

1 cu ft of air at 32 F and atmospheric pressure weighs 0.0807 pounds.

Liquid	Sp Gr	Liquid	Sp Gr	Liquid	Sp Gr
Acetic acid	1.06	Fluoric acid	1.50	Palm oil	0.97
Alcohol, commercial	0.83	Gasoline	0.70	Petroleum oil	0.82
Ammonia	0.77	Glycerin	1.26	Phosphoric acid	1.78
Benzene	0.88	Kerosene	0.80	Rape oil	0.92
Bromine	2.97	Linseed oil	0.94	Vinegar	1.08
Carbolic acid	0.96	Mineral oil	0.92	Water	1.00
Carbon disulfide	1.26	Naphtha	0.76	Whale oil	0.92
Cotton-seed oil	0.93	Olive oil	0.92		

Reprinted with permission from *Machinery's Handbook,* 20th ed., The Industrial Press.

1 cu ft water at 39 F weighs 62.43 pounds
1 ml water at 4 C weighs 1 g

Table 19-M. Approximate Specific Gravities and Densities
(At room temperature with reference to water at 39 F)

Substance	Specific gravity	Avg density, lb per cu ft	Substance	Specific gravity	Avg density, lb per cu ft
Metals, Alloys, Ores			**Timber, air-dry**		
Aluminum, cast-hammered	2.55–2.80	165	Apple	0.66–0.74	44
Aluminum, bronze	7.7	481	Ash, black	0.55	34
Brass, cast-rolled	8.4–8.7	534	Ash, white	0.64–0.71	42
Bronze, 7.9 to 14% Sn	7.4–8.9	509	Birch, sweet, yellow	0.71–0.72	44
Bronze, phosphor	8.88	554	Cedar, white, red	0.35	22
Copper, cast-rolled	8.8–8.95	556	Cherry, wild red	0.43	27
Copper ore, pyrites	4.1–4.3	262	Chestnut	0.48	30
German silver	8.58	536	Cypress	0.45–0.48	29
Gold, cast-hammered	19.25–19.35	1205	Fir, Douglas	0.48–0.55	32
Gold coin (U.S.)	17.18–17.2	1073	Fir, balsam	0.40	25
Iridium	21.78–22.42	1383	Elm, white	0.56	35
Iron, gray cast	7.03–7.13	442	Hemlock	0.45–0.50	29
Iron, cast, pig	7.2	450	Hickory	0.74–0.80	48
Iron, wrought	7.6–7.9	485	Locust	0.67–0.77	45
Iron, spiegel-eisen	7.5	468	Mahogany	0.56–0.85	44
Iron, ferrosilicon	6.7–7.3	437	Maple, sugar	0.68	43
Iron ore, hematite	5.2	325	Maple, white	0.53	33
Iron ore, limonite	3.6–4.0	237	Oak, chestnut	0.74	46
Iron ore, magnetite	4.9–5.2	315	Oak, live	0.87	54
Iron slag	2.5–3.0	172	Oak, red, black	0.64–0.71	42
Lead	11.34	710	Oak, white	0.77	48
Lead ore, galena	7.3–7.6	465	Pine, Oregon	0.51	32
Manganese	7.42	475	Pine, red	0.48	30
Manganese ore, pyrolusite	3.7–4.6	259	Pine, white	0.43	27
Mercury	13.546	847	Pine, Southern	0.61–0.67	38–42
Monel metal, rolled	8.97	555	Pine, Norway	0.55	34
Nickel	8.9	537	Poplar	0.43	27
Platinum, cast-hammered	21.5	1330	Redwood, California	0.42	26
Silver, cast-hammered	10.4–10.6	656	Spruce, white, red	0.45	28
Steel, cold-drawn	7.83	489	Teak, African	0.99	62
Steel, machine	7.80	487	Teak, Indian	0.66–0.88	48
Steel, tool	7.70–7.73	481	Walnut, black	0.59	37
Tin, cast-hammered	7.2–7.5	459	Willow	0.42–0.50	28
Tin ore, cassiterite	6.4–7.0	418			
Tungsten	19.22	1200	**Various Liquids**		
Zinc, cast-rolled	6.9–7.2	440			
Zinc, ore, blende	3.9–4.2	253	Alcohol, ethyl (100%)	0.789	49
			Alcohol, methyl (100%)	0.796	50
Various Solids			Acid, muriatic (HCl), 40%	1.20	75
			Acid, nitric, 91%	1.50	94
Cereals, oats, bulk	0.41	26	Acid, sulfuric, 87%	1.80	112
Cereals, barley, bulk	0.62	39	Chloroform	1.500	95
Cereals, corn, rye, bulk	0.73	45	Ether	0.736	46
Cereals, wheat, bulk	0.77	48	Lye, soda, 66%	1.70	106
Cork	0.22–0.26	15	Oils, vegetable	0.91–0.94	58
Cotton, flax, hemp	1.47–1.50	93	Oils, mineral, lubricants	0.88–0.94	57
Fats	0.90–0.97	58	Turpentine	0.861–0.867	54
Flour, loose	0.40–0.50	28	Water, 4 C, max. density	1.0	62.428
Flour, pressed	0.70–0.80	47	Water, 100 C	0.9584	59.830
Glass, common	2.40–2.80	162	Water, ice	0.88–0.92	56
Glass, plate or crown	2.45–2.72	161	Water, snow, fresh fallen	0.125	8
Glass, crystal	2.90–3.00	184	Water, sea water	1.02–1.03	64
Glass, flint	3.2–4.7	247			
Hay and straw, bales	0.32	20	**Ashlar Masonry**		
Leather	0.86–1.02	59	Granite, syenite, gneiss	2.4–2.7	159
Paper	0.70–1.15	58	Limestone	2.1–2.8	153
Potatoes, piled	0.67	44			
Rubber, Caoutchouc	0.92–0.96	59	Marble	2.4–2.8	162
Rubber goods	1.0–2.0	94	Sandstone	2.0–2.6	143
Salt, granulated, piled	0.77	48	Bluestone	2.3–2.6	153
Saltpeter	2.11	132			
Starch	1.53	96	**Rubble Masonry**		
Sulfur	1.93–2.07	125	Granite, syenite, gneiss	2.3–2.6	153
Wool	1.32	82			

Substance	Specific gravity	Avg density, lb per cu ft	Substance	Specific gravity	Avg density, lb per cu ft
Rubble Masonry			**Minerals**		
Limestone	2.0–2.7	147	Asbestos	2.1–2.8	153
Sandstone	1.9–2.5	137	Barytes	4.50	281
Bluestone	2.2–2.5	147	Basalt	2.7–3.2	184
Marble	2.3–2.7	156	Bauxite	2.55	159
			Bluestone	2.5–2.6	159
			Borax	1.7–1.8	109
Dry Rubble Masonry			Chalk	1.8–2.8	143
			Clay, marl	1.8–2.6	137
Granite, syenite, gneiss	1.9–2.3	130	Dolomite	2.9	181
Limestone, marble	1.9–2.1	125	Feldspar, orthoclase	2.5–2.7	162
Sandstone, bluestone	1.8–1.9	110	Gneiss	2.7–2.9	175
			Granite	2.6–2.7	165
			Greenstone, trap	2.8–3.2	187
Brick Masonry			Gypsum, alabaster	2.3–2.8	159
			Hornblende	3.0	187
Hard brick	1.8–2.3	128	Limestone	2.1–2.86	155
Medium brick	1.6–2.0	112	Marble	2.6–2.86	170
Soft brick	1.4–1.9	103	Magnesite	3.0	187
Sand-lime brick	1.4–2.2	112	Phosphate rock, apatite	3.2	200
			Porphyry	2.6–2.9	172
			Pumice, natural	0.37–0.90	40
Concrete Masonry			Quartz, flint	2.5–2.8	165
Cement, stone, sand	2.2–2.4	144	Sandstone	2.0–2.6	143
Cement, slag, etc.	1.9–2.3	130	Serpentine	2.7–2.8	171
Cement, cinder, etc.	1.5–1.7	100	Shale, slate	2.6–2.9	172
			Soapstone, talc	2.6–2.8	169
Various Building Mat'ls			Syenite	2.6–2.7	165
Ashes, cinders	0.64–0.72	40–45			
Cement, portland, loose	1.5	94			
Portland cement	3.1–3.2	196	**Stone, Quarried, Piled**		
Lime, gypsum, loose	0.85–1.00	53–64	Basalt, granite, gneiss	1.5	96
Mortar, lime, set	1.4–1.9	103	Limestone, marble quartz	1.5	95
		94	Sandstone	1.3	82
Mortar, portland cement	2.08–2.25	135	Shale	1.5	92
Slags, bank slag	1.1–1.2	67–72	Greenstone, hornblende	1.7	107
Slags, bank screenings	1.5–1.9	98–117			
Slags, machine slag	1.5	96			
Slags, slag sand	0.8–0.9	49–55	**Bituminous Substances**		
			Asphaltum	1.1–1.5	81
Earth, etc., Excavated			Coal, anthracite	1.4–1.8	97
Clay, dry	1.0	63	Coal, bituminous	1.2–1.5	84
Clay, damp, plastic	1.76	110	Coal, lignite	1.1–1.4	78
Clay and gravel, dry	1.6	100	Coal, peat, turf, dry	0.65–0.85	47
Earth, dry, loose	1.2	76	Coal, charcoal, pine	0.28–0.44	23
Earth, dry, packed	1.5	95	Coal, charcoal, oak	0.47–0.57	33
Earth, moist, loose	1.3	78	Coal, coke	1.0–1.4	75
Earth, moist, packed	1.6	96	Graphite	1.64–2.7	135
Earth, mud, flowing	1.7	108	Paraffin	0.87–0.91	56
Earth, mud, packed	1.8	115	Petroleum	0.87	54
Riprap, limestone	1.3–1.4	80–85	Petroleum, refined (kerosene)	0.78–0.82	50
Riprap, sandstone	1.4	90	Petroleum, benzene	0.73–0.75	46
Riprap, shale	1.7	105	Petroleum, gasoline	0.70–0.75	45
Sand, gravel, dry, loose	1.4–1.7	90–105	Pitch	1.07–1.15	69
Sand, gravel, dry, packed	1.6–1.9	100–120	Tar, bituminous	1.20	75
Sand, gravel, wet	1.89–2.16	126			
Excavations in Water			**Coal and Coke, Piled**		
Sand or gravel	0.96	60	Coal, anthracite	0.75–0.93	47–58
Sand or gravel and clay	1.00	65	Coal, bituminous, lignite	0.64–0.87	40–54
Clay	1.28	80	Coal, peat, turf	0.32–0.42	20–26
River mud	1.44	90	Coal, charcoal	0.16–0.23	10–14
Soil	1.12	70	Coal, coke	0.37–0.51	23–32
Stone riprap	1.00	65			

1 lb per cu ft = 16.02 kg/m³

Table 19-N. Natural Trigonometric Functions

Angle	Function	Value	Angle	Function	Value	Angle	Function	Value	Angle	Function	Value
0	sin	0.0000	6	sin	0.1045	12	sin	0.2079	18	sin	0.3090
	cos	1.0000		cos	0.9945		cos	0.9781		cos	0.9511
	tan	0.0000		tan	0.1051		tan	0.2126		tan	0.3249
1	sin	0.0175	7	sin	0.1219	13	sin	0.2250	19	sin	0.3256
	cos	0.9998		cos	0.9925		cos	0.9744		cos	0.9455
	tan	0.0175		tan	0.1228		tan	0.2309		tan	0.3443
2	sin	0.0349	8	sin	0.1392	14	sin	0.2419	20	sin	0.3420
	cos	0.9994		cos	0.9903		cos	0.9703		cos	0.9397
	tan	0.0349		tan	0.1405		tan	0.2493		tan	0.3640
3	sin	0.0523	9	sin	0.1564	15	sin	0.2588	21	sin	0.3584
	cos	0.9986		cos	0.9877		cos	0.9659		cos	0.9336
	tan	0.0524		tan	0.1584		tan	0.2679		tan	0.3839
4	sin	0.0698	10	sin	0.1736	16	sin	0.2756	22	sin	0.3746
	cos	0.9976		cos	0.9848		cos	0.9613		cos	0.9272
	tan	0.0699		tan	0.1763		tan	0.2867		tan	0.4040
5	sin	0.0872	11	sin	0.1908	17	sin	0.2924	23	sin	0.3907
	cos	0.9962		cos	0.9816		cos	0.9563		cos	0.9205
	tan	0.0875		tan	0.1944		tan	0.3057		tan	0.4245

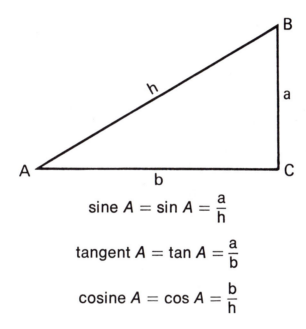

$$\text{sine } A = \sin A = \frac{a}{h}$$

$$\text{tangent } A = \tan A = \frac{a}{b}$$

$$\text{cosine } A = \cos A = \frac{b}{h}$$

SIGNS AND LIMITS OF VALUE
ASSUMED BY THE FUNCTIONS

Function	Quadrant I		Quadrant II		Quadrant III		Quadrant IV	
	Sign	Value	Sign	Value	Sign	Value	Sign	Value
sin	+	0 to 1	+	1 to 0	−	0 to 1	−	1 to 0
cos	+	1 to 0	−	0 to 1	−	1 to 0	+	0 to 1
tan	+	0 to ∞	−	∞ to 0	+	0 to ∞	−	∞ to 0

Angle	Function	Value	Angle	Function	Value	Angle	Function	Value	Angle	Function	Value
24	sin	0.4067	40	sin	0.6428	57	sin	0.8387	73	sin	0.9563
	cos	0.9135		cos	0.7660		cos	0.5446		cos	0.2924
	tan	0.4452		tan	0.8391		tan	1.5399		tan	3.2709
25	sin	0.4226	41	sin	0.6561	58	sin	0.8480	74	sin	0.9613
	cos	0.9063		cos	0.7547		cos	0.5299		cos	0.2756
	tan	0.4663		tan	0.8693		tan	1.6003		tan	3.4874
26	sin	0.4384	42	sin	0.6691	59	sin	0.8572	75	sin	0.9659
	cos	0.8988		cos	0.7431		cos	0.5150		cos	0.2588
	tan	0.4877		tan	0.9004		tan	1.6643		tan	3.7321
27	sin	0.4540	43	sin	0.6820	60	sin	0.8660	76	sin	0.9703
	cos	0.8910		cos	0.7314		cos	0.5000		cos	0.2419
	tan	0.5095		tan	0.9325		tan	1.7321		tan	4.0108
28	sin	0.4695	44	sin	0.6947	61	sin	0.8746	77	sin	0.9744
	cos	0.8829		cos	0.7193		cos	0.4848		cos	0.2250
	tan	0.5317		tan	0.9657		tan	1.8040		tan	4.3315
29	sin	0.4848	45	sin	0.7071	62	sin	0.8829	78	sin	0.9781
	cos	0.8746		cos	0.7071		cos	0.4695		cos	0.2079
	tan	0.5543		tan	1.0000		tan	1.8807		tan	4.7046
30	sin	0.5000	46	sin	0.7193	63	sin	0.8910	79	sin	0.9816
	cos	0.8660		cos	0.6947		cos	0.4540		cos	0.1908
	tan	0.5774		tan	1.0355		tan	1.9626		tan	5.1446
31	sin	0.5150	47	sin	0.7314	64	sin	0.8988	80	sin	0.9848
	cos	0.8572		cos	0.6820		cos	0.4384		cos	0.1736
	tan	0.6009		tan	1.0724		tan	2.0503		tan	5.6713
32	sin	0.5299	48	sin	0.7431	65	sin	0.9063	81	sin	0.9877
	cos	0.8480		cos	0.6691		cos	0.4226		cos	0.1564
	tan	0.6249		tan	1.1106		tan	2.1445		tan	6.3138
33	sin	0.5446	49	sin	0.7547	66	sin	0.9135	82	sin	0.9903
	cos	0.8387		cos	0.6561		cos	0.4067		cos	0.1392
	tan	0.6494		tan	1.1504		tan	2.2460		tan	7.1154
34	sin	0.5592	50	sin	0.7660	67	sin	0.9205	83	sin	0.9925
	cos	0.8290		cos	0.6428		cos	0.3907		cos	0.1219
	tan	0.6745		tan	1.1918		tan	2.3559		tan	8.1443
35	sin	0.5736	51	sin	0.7771	68	sin	0.9272	84	sin	0.9945
	cos	0.8192		cos	0.6293		cos	0.3746		cos	0.1045
	tan	0.7002		tan	1.2349		tan	2.4751		tan	9.5144
36	sin	0.5878	52	sin	0.7880	69	sin	0.9336	85	sin	0.9962
	cos	0.8090		cos	0.6157		cos	0.3584		cos	0.0872
	tan	0.7265		tan	1.2799		tan	2.6051		tan	11.43
37	sin	0.6018	53	sin	0.7986	70	sin	0.9397	86	sin	0.9976
	cos	0.7986		cos	0.6018		cos	0.3420		cos	0.0698
	tan	0.7536		tan	1.3270		tan	2.7475		tan	14.30
38	sin	0.6157	54	sin	0.8090	71	sin	0.9455	87	sin	0.9986
	cos	0.7880		cos	0.5878		cos	0.3256		cos	0.0523
	tan	0.7813		tan	1.3764		tan	2.9042		tan	19.08
39	sin	0.6293	55	sin	0.8192	72	sin	0.9511	88	sin	0.9994
	cos	0.7771		cos	0.5736		cos	0.3090		cos	0.0349
	tan	0.8098		tan	1.4281		tan	3.0777		tan	28.64
			56	sin	0.8290				89	sin	0.9996
				cos	0.5592					cos	0.0175
				tan	1.4826					tan	57.29

NOTE: Values of the sine, cosine, and tangent of each degree from 0 to 90 also permit determination of values for angles from 90 to 360 degrees if appropriate signs for the functions are used depending on the quadrant in which the angle measure belongs. For example, the sine of 4 degrees is also the sign of 176 degrees, and with a negative sign, the sine of 184 and 356 degrees. Note that if an angle is given in radians, multiply the number of radians by 57.296 to obtain the number of degrees.

SURFACE AREA AND VOLUME FORMULAS

R = radius of circle

d = diameter of circle

$\pi = 3.1416$

Circumference of circle $= \pi d = 2\pi R$

Area of circle $= \pi d^2/4 = \pi R^2$

Surface area of sphere $= \pi d^2 = 4\pi R^2$

Volume of sphere $= \pi d^3/6 = 4/3\pi R^3$

Area of a triangle $= ab/2$

b = base

a = altitude

TABLE OF UNIT PREFIXES

Multiples and submultiples	*Prefixes*	*Symbols*
$1{,}000{,}000{,}000{,}000 = 10^{12}$	tera-	T
$1{,}000{,}000{,}000 = 10^{9}$	giga-	G
$1{,}000{,}000 = 10^{6}$	mega-	M
$1{,}000 = 10^{3}$	kilo-	k
$100 = 10^{2}$	hecto-	h
$10 = 10$	deka-	D
$0.1 = 10^{-1}$	deci-	d
$.01 = 10^{-2}$	centi-	c
$.001 = 10^{-3}$	milli-	m
$.000001 = 10^{-6}$	micro-	μ
$.000000001 = 10^{-9}$	nano-	n
$.000000000001 = 10^{-12}$	pico	p

Data from National Bureau of Standards.

SIGNS AND SYMBOLS

$+$	plus, addition, positive	$\sqrt{}$	square root
$-$	minus, subtraction, negative	$\sqrt[n]{}$	nth root
$\pm$	plus or minus, positive or negative	a^n	nth power of a
$\mp$	minus or plus, negative or positive	log, $\log_{10}$	common logarithm
$\div, /, \text{———}$	division	ln, $\log_e$	natural logarithm
$\times, \cdot, ()()$	multiplication	e or ϵ	base of natural logs, 2.718
$()[]$	collection	π	pi, 3.1416
$=$	is equal to	$\angle$	angle
$\neq$	is not equal to	$\perp$	perpendicular to
$\equiv$	is identical to	$\parallel$	parallel to
$\cong$	equals approximately, congruent	n	any number
		$\lvert n \rvert$	absolute value of n
$>$	greater than	$\bar{n}$	average value of n
$\ngtr$	not greater than	a^{-n}	reciprocal of nth power of a,
$\geqq$	greater than or equal to		of a, or $\left\{\dfrac{1}{a^n}\right\}$
$<$	less than		
$\nless$	not less than	$n°$	n degrees (angle)
$\leqq$	less than or equal to	n'	n minutes, n feet
$::$	proportional to	n''	n seconds, n inches
$:$	ratio	$f(x)$	function of x
$\sim$	similar to	Δx	increment of x
$\propto$	varies as, proportional to	dx	differential of x
$\rightarrow$	approaches	Σ	summation of
∞	infinity	sin	sine
$\therefore$	therefore	cos	cosine
		tan	tangent

SELECTED COMMON ABBREVIATIONS

Å	Angstrom unit of length		liq	liquid
abs	absolute		L	liter and lambert(s)
amb	ambient		LP-gas	liquefied petroleum gas
amp	ampere		log	logarithm (common)
app mol wt	apparent molecular weight		ln	logarithm (natural)
atm	atmospheric		m, M	meter
at wt	atomic weight		ma	milliampere
Bé	degrees Baumé		MAC	maximum allowable concentration
bp	boiling point		max	maximum
bbl	barrel		mp	melting point
Btu	British thermal unit		μ	micron
Btuh	Btu per hour		mks system	meter-kilogram-second system
c	cycles per second (see Hz)		mph	miles per hour
cal	calorie		mg	milligram
cfh	cubic feet per hour		ml	milliliter
cfm	cubic feet per minute		mm	millimeter
cfs	cubic feet per second		mm (Hg)	mm of mercury
cg	centigram		mμ	millimicron
cm	centimeter		mppcf	million particles per cu ft
cgs system	centimeter-gram-second system		mr	millirem
conc	concentrated, concentration		mR	1/1000 Roentgen
cc, cm^3	cubic centimeter		min	minute or minimum
cu ft, ft^3	cubic foot		mol wt, MW	molecular weight
cu in.	cubic inch		N	newton
° or deg	degree		OD	outside diameter
C	degree Centigrade, degree Celsius		oz	ounce
F	degree Fahrenheit		ppb	parts per billion
K	degree Kelvin		pphm	parts per hundred million
R	degree Reaumur, degree Rankine		ppm	parts per million
dB	decibel		lb	pound
ET	effective temperature		psf	pounds per square foot
ft	foot		psi	pounds per square inch
ft-c	foot-candle		psia	pounds per square inch absolute
ft lb	foot pound			
fpm	feet per minute		psig	pounds per square inch gage
fps	feet per second		Rem	Roentgen equivalent man
fps system	foot-pound-second system		rpm	revolution per minute
fp	freezing point		sec	second
gal	gallon		sp gr	specific gravity
gr	grain		sp ht	specific heat
g	gram		sp wt	specific weight
gpm	gallons per minute		sq	square
Hz	hertz (cycles per second)		scf	standard cubic foot
hp	horsepower		STP	standard temperature and pressure
hr	hour			
ID	inside diameter		temp	temperature
in.	inch		TLV	threshold limit valve
kcal	kilocalorie		v	volt
kg	kilogram		W	watt
km	kilometer		wt	weight

Note. Symbols are always written in singular form. Unabbreviated units form plurals in the usual manner.

REFERENCES TO USEFUL HANDBOOKS OF ENGINEERING TABLES AND FORMULAS

Note: Because engineering and scientific handbooks are revised often, as frequently as once a year in some cases, no attempt has been made to list the "latest" edition in the following bibliography. In ordering, however, the most recent edition should be requested.

Alexander, J. M., and R. C. Brewer, *Manufacturing Properties of Materials*. Princeton, N.J. 08540. D. VanNostrand Co., Inc., 1963.

Allegheny Ludlum Steel Corp., Pittsburgh, Pa. 15222. *Tool Steel Handbook*, 1951.

American Conference of Governmental Industrial Hygienists, Committee on Industrial Ventilation, PO Box 1937, Cincinnati, Ohio 45201.
Air Sampling Instruments Manual.
Industrial Ventilation—A Manual of Recommended Practice.

American Industrial Hygiene Assn., 475 Wolf Ledges Parkway, Akron, Ohio 44311. *Industrial Noise Manual, 3rd Edition*, 1975.

American Institute of Steel Construction, Inc., 101 Park Ave., New York, N.Y. 10017. *Manual of Steel Construction.*

American National Metric Council, 1625 Massachusetts Ave., NW., Washington, D.C. 20036. "Metrication for the Manager," 1977.

American Mutual Insurance Alliance, 20 North Wacker Dr., Chicago, Ill. 60606. "Handbook of Organic Industrial Solvents," Technical Guide No. 6, 1980.

American Society for Testing and Materials, 1916 Race St., Philadelphia, Pa. 19103. "Standard Metric Practice Guide," E 380 (ANSI Z210.1).

American Society of Heating, Refrigerating and Air-Conditioning Engineers, 1791 Tullie Circle, N.E., Atlanta, Ga. 30329. *Guide and Data Books: Applications, Handbook of Fundamentals, Systems and Equipment.*

American Welding Society, 2501 NW. 7th St., Miami, Fla. 33125. *Welding Handbook.*

Baumeister, Theodore, III and E. A. Avallone, eds. *Standard Handbook for Mechanical Engineers*, New York, N.Y. 10036, McGraw-Hill Book Co.

Bennett, H., ed. *Concise Chemical and Technical Dictionary.* New York, N.Y. 10003, Chemical Publishing Co., Inc.

A. M. Best Company, Oldwick, N.J. 08858. *Best's Safety Directory*, (2 vols).

Bishop, Jerry C. *The Wall Street Journal Guide to the Metric System.* Princeton, N.J. 08540, Dow Jones Books (PO Box 300), 1977.

Bolz, Harold A., ed. *Materials Handling Handbook.* New York, N.Y. 10016. The Ronald Press Co.

Carmichael, Colin, and J. K. Salisbury, *Kent's Mechanical Engineers' Handbook.* New York, N.Y. 10016, John Wiley & Sons, Inc.

Car, Clifford C. *Croft's American Electricians' Handbook.* New York, N.Y. 10036, McGraw-Hill Book Co.

Carson, Gordon B. *Production Handbook.* New York, N.Y. 10016. The Ronald Press, 1972.

Compressed Air Magazine Co., Phillipsburg, N.J. 08865. *Compressed Air Data.*

Crane Co., Industrial Products Group, Engineering Division, 4100 S. Kedzie Ave., Chicago, Ill 60632. *Flow of Fluids Through Valves, Fittings, and Pipe,* Technical Paper 410. 1957.

Damon A., H. W. Stoudt, and R. A. McFarland. *The Human Body in Equipment Design.* Cambridge, Mass. 02138, Harvard University Press, 1966.

DeGarmo, E. Paul, and R. A. Kohser. *Materials and Processes in Manufacturing,* 6th ed. New York, N.Y. 10022. The Macmillan Co., 1984.

Elonka, Stephen Michael, and Joseph Frederick Robinson. *Standard Plant Operator's Questions and Answers.* 2nd ed. New York, N.Y. 10036, McGraw-Hill Book Co., 1981.

Eshbach, Ovid W. *Handbook of Engineering Fundamentals,* New York, N.Y. 10016. John Wiley & Sons.

Factory Mutual Engineering Corp., Norwood, Mass. 02062. *Loss Prevention Data,* 1973.

Gardner, William, Edward I. Cooke, and Richard W. I. Cooke. *Handbook of Chemical Synonyms and Trade Names.* Cleveland, Ohio 44128, CRC Press, 1978.

Gosselin, R. E. *Clinical Toxicology of Commerical Products: Acute Poisoning.* 5th ed. Baltimore, Md. 21202, The Williams & Wilkins Co., 1984.

Green, Marvin H. *International and Metric Units of Measurement.* New York, N.Y. 10003, Chemical Publishing Co., Inc., 1973.

Gunther, Raymond C. *Refrigeration, Air Conditioning, and Cold Storage,* 2nd ed. Philadelphia, Pa. 19106, Chilton Co. Inc., 1969.

Harris, Cyril M. *Handbook of Noise Control.* 2nd ed. New York, N.Y. 10036, McGraw-Hill Book Co., 1979.

Hodgman, Charles D., *et al.,* eds. *Handbook of Chemistry and Physics.* Cleveland, Ohio 44128, The Chemical Rubber Co.

Hosey, A. D., and Powell, C. H., eds. *Industrial Noise—A Guide to Its Evaluation and Control.* Washington, D.C., U.S. Dept. of Health, Education, and Welfare, 1967. (Available through U.S. Government Printing Office, Washington, D.C. 20402.)

Hudson, Ralph G. *The Engineer's Manual.* New York, N.Y. 10016, John Wiley & Sons, Inc.

Illuminating Engineering Society, 345 E. 47th St., New York, N.Y. 10017. *IES Lighting Handbook (The Standard Lighting Guide).*

Ireson, W. G., and E. L. Grant. *Handbook of Industrial Engineering and Management,* 2nd ed. Englewood Cliffs, N.J. 07632, Prentice-Hall Publishing Co., Inc., 1970.

Kidder, Frank E., and Harry Parker. *Kidder-Parker Architects' and Builders' Handbook.* New York, N.Y. 10016, John Wiley & Sons, Inc.

Knowlton, A. E., ed. *Standard Handbook for Electrical Engineers.* New York, N.Y. 10036, McGraw-Hill Book Co.

Kurtz, Edwin B. and T. M. Shoemaker. *The Linemans' and Cablemans' Handbook.* 7th ed. New York, N.Y. 10036, McGraw-Hill Book Co., 1985.

LaLonde, William S., Jr., and Milo F. Janes. *Concrete Engineers Handbook.* New York, N.Y. 10036, McGraw-Hill Book Co.

LeGrand, Rupert, ed. *The New American Machinists' Handbook.* New York, N.Y. 10036, McGraw-Hill Book Co.

Liebers, Arthur. *The Engineer's Handbook Illustrated.* Los Angeles, Calif. 90047, Key Publishing Co., 1968.

Lindsey, Forrest R. *Pipefitters Handbook.* New York, N.Y. 10016, The Industrial Press.

Mantell, Charles L., ed. *Engineering Materials Handbook*. New York, N.Y. 10036, McGraw-Hill Book Co.

Maynard, H. B. *Industrial Engineering Handbook*. 3rd ed. New York, N.Y. 10036, McGraw-Hill Book Co., 1971.

Miner, Douglas F., and John B. Seastone, eds. *Handbook of Engineering Materials*. New York, N.Y. 10016, John Wiley & Sons, Inc.

Morris, I. E. *Handbook of Structural Design*. New York, N.Y. 10022, Reinhold Publishing Corp.

Morrow, L. C., ed. *Maintenance Engineering Handbook*. New York, N.Y. 10036, McGraw-Hill Book Co.

National Association of Home Builders. *Construction Dictionary*, 1985.

National Fire Protection Assn., Batterymarch Park, Quincy, Mass. 02269. *Fire Protection Handbook*.

Oberg, Erik, and F. D. Jones. *Machinery's Handbook*. New York, N.Y. 10016, The Industrial Press.

Pender, Harold, *et al. Electrical Engineers' Handbook*. New York, N.Y. 10016, John Wiley & Sons, Inc.

Perry, John H., and Robert H. Perry. *Engineering Manual*. New York, N.Y. 10036, McGraw-Hill Book Co.

Perry, Robert H., *et al.*, eds. *Chemical Engineers' Handbook*. New York, N.Y. 10036, McGraw-Hill Book Co.

Peterson, Arnold P. G., and Ervin E. Gross. *Handbook of Noise Measurement*. West Concord, Mass. 01781, General Radio Co., 1972.

Robb, Dean A., and Harry M. Philo. *Lawyers Desk Reference: A Source Guide to Safety Information, What to Find, How to Find It*. Rochester, N.Y. 14603, The Lawyers Co-operative Publishing Co.

Rose, Arthur, and Elizabeth Rose. *The Condensed Chemical Dictionary*. New York, N.Y. 10022, Reinhold Publishing Co.

Rossnagel, W. E. *Handbook of Rigging*. New York, N.Y. 10036, McGraw-Hill Book Co.

Society of Automotive Engineers, 400 Commonwealth Dr., Warrendale, Penn. 15096. *SAE Handbook*.

Stanier, William. *Mechanical Power Transmission Handbook*. New York, N.Y. 10036, McGraw-Hill Book Co.

——. *Plant Engineering Handbook*. New York, N.Y. 10036, McGraw-Hill Book Co.

Steere, Norman V., ed. *Handbook of Laboratory Safety*. 2nd ed. Cleveland, Ohio 44128, The Chemical Rubber Co., 1971.

United States Steel Corp., National Tube Division, Pittsburgh, Pa. 15230. *Lubrication Engineers Manual*. 1963.

Urquhart, Leonard Church, ed. *Civil Engineering Handbook*. New York, N.Y. 10036, McGraw-Hill Book Co.

Wilson, Frank W., and Philip D. Harvey, eds. *Tool Engineers Handbook*. New York, N.Y. 10036, McGraw-Hill Book Co.

Conversion
of Units

ALL PHYSICAL UNITS OF MEASUREMENT can be reduced to three basic dimensions—mass, length, and time. Not only does reducing units to these basic dimensions simplify the solution of problems, but standardization of units makes comparison between operations (and between operations and standards) easier.

For example, air flows are usually measured in liters per minute, cubic meters per second, or cubic feet per minute. The total volume of air sampled can be easily converted to cubic meters or cubic feet. In another situation, the results of atmospheric pollution studies and stack sampling surveys are often reported as grains per cubic foot, grams per cubic foot, or pounds per cubic foot. The degree of contamination is usually reported as parts of contaminant per million parts of air.

If physical measurements are made or reported in different units, they must be converted to the standard units if any comparisons are to be meaningful.

To save time and space in reporting data, many units have standard abbreviations. Because the metric system (SI) is becoming more frequently used, conversion factors are given for the standard units of measurement.

FUNDAMENTAL UNITS

Conversion factors for various measurement units are listed in the tables in this section. To use a table to find the numerical value of the quantity desired, locate the unit to be converted in the first column. Then multiply this value by the number appearing at the intersection of the row and the column containing the desired unit. The answer will be the numerical value in the desired unit.

Various English systems and metric system units are given for the reader's convenience. The new system of measurement, however, is the International System of Units (SI). The official conversion factors and an explanation of the system are given to 6- or 7-place accuracy in ASTM Standard E 380-76 (ANSI Z210.1-1976).

Briefly, the SI System being adopted throughout the world is a modern version of the MKSA (meter, kilogram, second, ampere) system. Its details are published and controlled by an international treaty organization, the International Bureau of Weights and Measures (BIPM), set up by the Metre Convention signed in Paris, France, on May 20, 1875. The United States and Canada are member states of this Convention, as implemented by the Metric Conversion Act of 1975 (Public Law 94-168).

HELPFUL ORGANIZATIONS

The following four groups in the U.S. and Canada are deeply involved in planning and implementing metric conversion:

American National Metric Council
5410 Grosvenor Lane
Bethesday, MD. 20814

Metric Commission Canada
240 Sparks Street
Ottawa K1A 0H5, Canada

U.S. Metric Association, Inc.
Boulder, Colo. 80302

Office of Metric Programs
U.S. Dept. of Commerce
Washington, D.C. 20230

Conversion of Units

FAHRENHEIT-CELSIUS CONVERSION TABLE

Fahrenheit-Celsius Conversion.—A simple way to convert a Fahrenheit temperature reading into a Celsius temperature reading or vice versa is to enter the accompanying table in the center or boldface column of figures. These figures refer to the temperature in either Fahrenheit or Celsius degrees. If it is desired to convert from Fahrenheit to Celsius degrees, consider the center column as a table of Fahrenheit temperatures and read the corresponding Celsius temperature in the column at the left. If it is desired to convert from Celsius to Fahrenheit degrees, consider the center column as a table of Celsius values, and read the corresponding Fahrenheit temperature on the right.

To convert from "degrees Fahrenheit" to "degrees Celsius" (formerly called "degrees centigrade"), use the formula:

$$t_c = \frac{(t_f - 32)}{1.8} \text{ or } \frac{5}{9}(t_f - 32)$$

Conversely,
$$t_f = 1.8\, t_c + 32 \text{ or } \frac{9}{5}\, t_c + 32$$

Example, convert the boiling point of water in F to C:

$$212\ \text{F} - 32 = 180$$

$$\frac{5}{9}(180) = 100\ \text{C}$$

Fahrenheit — Celsius Conversion Table

Deg C		Deg F	Deg C		Deg F	Deg C		Deg F	Deg C		Deg F
−273	−459.4	...	−129	−200	−328	−13.9	7	44.6	1.1	34	93.2
−268	−450	...	−123	−190	−310	−13.3	8	46.4	1.7	35	95.0
−262	−440	...	−118	−180	−292	−12.8	9	48.2	2.2	36	96.8
−257	−430	...	−112	−170	−274	−12.2	10	50.0	2.7	37	98.6
−251	−420	...	−107	−160	−256	−11.7	11	51.8	3.3	38	100.4
−246	−410	...	−101	−150	−238	−11.1	12	53.6	3.9	39	102.2
−240	−400	...	− 96	−140	−220	−10.6	13	55.4	4.4	40	104.0
−234	−390	...	− 90	−130	−202	−10.0	14	57.2	5.0	41	105.8
−229	−380	...	− 84	−120	−184	− 9.4	15	59.0	5.6	42	107.6
−223	−370	...	− 79	−110	−166	− 8.9	16	60.8	6.1	43	109.4
−218	−360	...	− 73	−100	−148	− 8.3	17	62.6	6.7	44	111.2
−212	−350	...	− 68	− 90	−130	− 7.8	18	64.4	7.2	45	113.0
−207	−340	...	− 62	− 80	−112	− 7.2	19	66.2	7.8	46	114.8
−201	−330	...	− 57	− 70	− 94	− 6.7	20	68.0	8.3	47	116.6
−196	−320	...	− 51	− 60	− 76	− 6.1	21	69.8	8.9	48	118.4
−190	−310	...	− 46	− 50	− 58	− 5.6	22	71.6	9.4	49	120.2
−184	−300	...	− 40	− 40	− 40	− 5.0	23	73.4	10.0	50	122.0
−179	−290	...	− 34	− 30	− 22	− 4.4	24	75.2	10.6	51	123.8
−173	−280	...	− 29	− 20	− 4	− 3.9	25	77.0	11.1	52	125.6
−169	−273	−459.4	− 23	− 10	14	− 3.3	26	78.8	11.7	53	127.4
−168	−270	−454	−17.8	0	32−	− 2.8	27	80.6	12.2	54	129.2
−162	−260	−436	−17.2	1	33.8	− 2.2	28	82.4	12.8	55	131.0
−157	−250	−418	−16.7	2	35.6	− 1.7	29	84.2	13.3	56	132.8
−151	−240	−400	−16.1	3	37.4	− 1.1	30	86.0	13.9	57	134.6
−146	−230	−382	−15.6	4	39.2	− 0.6	31	87.8	14.4	58	136.4
−140	−220	−364	−15.0	5	41.0	0−	32	89.6	15.0	59	138.2
−134	−210	−346	−14.4	6	42.8	0.6	33	91.4	15.6	60	140.0

Deg C		Deg F	Deg C		Deg F	Deg C		Deg F	Deg C		Deg F
16.1	**61**	141.8	50.0	**122**	251.6	83.9	**183**	361.4	276.7	**530**	986
16.7	**62**	143.6	50.6	**123**	253.4	84.4	**184**	363.2	282.2	**540**	1004
17.2	**63**	145.4	51.1	**124**	255.2	85.0	**185**	365.0	287.8	**550**	1022
17.8	**64**	147.2	51.7	**125**	257.0	85.6	**186**	366.8	293.3	**560**	1040
18.3	**65**	149.0	52.2	**126**	258.8	86.1	**187**	368.6	298.9	**570**	1058
18.9	**66**	150.8	52.8	**127**	260.6	86.7	**188**	370.4	304.4	**580**	1076
19.4	**67**	152.6	53.3	**128**	262.4	87.2	**189**	372.2	310.0	**590**	1094
20.0	**68**	154.4	53.9	**129**	264.2	87.8	**190**	374.0	315.6	**600**	1112
20.6	**69**	156.2	54.4	**130**	266.0	88.3	**191**	375.8	321.1	**610**	1130
21.1	**70**	158.0	55.0	**131**	267.8	88.9	**192**	377.6	326.7	**620**	1148
21.7	**71**	159.8	55.6	**132**	269.6	89.4	**193**	379.4	332.2	**630**	1166
22.2	**72**	161.6	56.1	**133**	271.4	90.0	**194**	381.2	337.8	**640**	1184
22.8	**73**	163.4	56.7	**134**	273.2	90.6	**195**	383.0	343.3	**650**	1202
23.3	**74**	165.2	57.2	**135**	275.0	91.1	**196**	384.8	348.9	**660**	1220
23.9	**75**	167.0	57.8	**136**	276.8	91.7	**197**	386.6	354.4	**670**	1238
24.4	**76**	168.8	58.3	**137**	278.6	92.2	**198**	388.4	360.0	**680**	1256
25.0	**77**	170.6	58.9	**138**	280.4	92.8	**199**	390.2	365.6	**690**	1274
25.6	**78**	172.4	59.4	**139**	282.2	93.3	**200**	392.0	371.1	**700**	1292
26.1	**79**	174.2	60.0	**140**	284.0	93.9	**201**	393.8	376.7	**710**	1310
26.7	**80**	176.0	60.6	**141**	285.8	94.4	**202**	395.6	382.2	**720**	1328
27.2	**81**	177.8	61.1	**142**	287.6	95.0	**203**	397.4	387.8	**730**	1346
27.8	**82**	179.6	61.7	**143**	289.4	95.6	**204**	399.2	393.3	**740**	1364
28.3	**83**	181.4	62.2	**144**	291.2	96.1	**205**	401.0	398.9	**750**	1382
28.9	**84**	183.2	62.8	**145**	293.0	96.7	**206**	402.8	404.4	**760**	1400
29.4	**85**	185.0	63.3	**146**	294.8	97.2	**207**	404.6	410.0	**770**	1418
30.0	**86**	186.8	63.9	**147**	296.6	97.8	**208**	406.4	415.6	**780**	1436
30.6	**87**	188.6	64.4	**148**	298.4	98.3	**209**	408.2	421.1	**790**	1454
31.1	**88**	190.4	65.0	**149**	300.2	98.9	**210**	410.0	426.7	**800**	1472
31.7	**89**	192.2	65.6	**150**	302.0	99.4	**211**	411.8	432.2	**810**	1490
32.2	**90**	194.0	66.1	**151**	303.8	100.0	**212**	413.6	437.8	**820**	1508
32.8	**91**	195.8	66.7	**152**	305.6	104.4	**220**	428.0	443.3	**830**	1526
33.3	**92**	197.6	67.2	**153**	307.4	110.0	**230**	446.0	448.9	**840**	1544
33.9	**93**	199.4	67.8	**154**	309.2	115.6	**240**	464.0	454.4	**850**	1562
34.4	**94**	201.2	68.3	**155**	311.0	121.1	**250**	482.0	460.0	**860**	1580
35.0	**95**	203.0	68.9	**156**	312.8	126.7	**260**	500.0	465.6	**870**	1598
35.6	**96**	204.8	69.4	**157**	314.6	132.2	**270**	518.0	471.1	**880**	1616
36.1	**97**	206.6	70.0	**158**	316.4	137.8	**280**	536.0	476.7	**890**	1634
36.7	**98**	208.4	70.6	**159**	318.2	143.3	**290**	554.0	482.2	**900**	1652
37.2	**99**	210.2	71.1	**160**	320.0	148.9	**300**	572.0	487.8	**910**	1670
37.8	**100**	212.0	71.7	**161**	321.8	154.4	**310**	590.0	493.3	**920**	1688
38.3	**101**	213.8	72.2	**162**	323.6	160.0	**320**	608.0	498.9	**930**	1706
38.9	**102**	215.6	72.8	**163**	325.4	165.6	**330**	626.0	504.4	**940**	1724
39.4	**103**	217.4	73.3	**164**	327.2	171.1	**340**	644.0	510.0	**950**	1742
40.0	**104**	219.2	73.9	**165**	329.0	176.7	**350**	662.0	515.6	**960**	1760
40.6	**105**	221.0	74.4	**166**	330.8	182.2	**360**	680.0	521.1	**970**	1778
41.1	**106**	222.8	75.0	**167**	332.6	187.8	**370**	698.0	526.7	**980**	1796
41.7	**107**	224.6	75.6	**168**	334.4	193.3	**380**	716.0	532.2	**990**	1814
42.2	**108**	226.4	76.1	**169**	336.2	198.9	**390**	734.0	537.8	**1000**	1832
42.8	**109**	228.2	76.7	**170**	338.0	204.4	**400**	752.0	565.6	**1050**	1922
43.3	**110**	230.0	77.2	**171**	339.8	210	**410**	770.0	593.3	**1100**	2012
43.9	**111**	231.8	71.8	**172**	341.6	215.6	**420**	788	621.1	**1150**	2102
44.4	**112**	233.6	78.3	**173**	343.4	221.1	**430**	806	648.9	**1200**	2192
45.0	**113**	235.4	78.9	**174**	345.2	226.7	**440**	824	676.7	**1250**	2282
45.6	**114**	237.2	79.4	**175**	347.0	232.2	**450**	842	704.4	**1300**	2372
46.1	**115**	239.0	80.0	**176**	348.8	237.8	**460**	860	732.2	**1350**	2462
46.7	**116**	240.8	80.6	**177**	350.6	243.3	**470**	878	760.0	**1400**	2552
47.2	**117**	242.6	81.1	**178**	352.4	248.9	**480**	896	787.8	**1450**	2642
47.8	**118**	244.4	81.7	**179**	354.2	254.4	**490**	914	815.6	**1500**	2732
48.3	**119**	246.2	82.2	**180**	356.0	260.0	**500**	932	1093.9	**2000**	3632
48.9	**120**	248.0	82.8	**181**	357.8	265.6	**510**	950	1648.9	**3000**	5432
49.4	**121**	249.8	83.3	**182**	359.6	271.1	**520**	968	2760.0	**5000**	9032

From Machinery's Handbook, *18th ed.* (*The Industrial Press*)

Above 1000 in the center column, the table increases in increments of 50. To convert 1462 degrees F to Celsius, for instance, add to the Celsius equivalent of 1400 degrees F ⅝ths of 62 or 34 degrees, which equals 794 C.

LENGTH

Multiply Number of ↓ by ↘ \ To Obtain →	meter (m)	centimeter (cm)	millimeter (mm)	micron (μ) or micrometer (μm)	angstrom unit, (A)	inch (in.)	foot (ft)
meter	1	100	1000	10^6	10^{10}	39.37	3.28
centimeter	0.01	1	10	10^4	10^8	0.394	0.0328
millimeter	0.001	0.1	1	10^3	10^7	0.0394	0.00328
micron	10^{-6}	10^{-4}	10^{-3}	1	10^4	3.94×10^{-5}	3.28×10^{-6}
angstrom	10^{-10}	10^{-8}	10^{-7}	10^{-4}	1	3.94×10^{-9}	3.28×10^{-10}
inch	0.0254	2.540	25.40	2.54×10^4	2.54×10^8	1	0.0833
foot	0.305	30.48	304.8	304,800	3.048×10^9	12	1

AREA

Multiply Number of ↓ By ↘ \ To Obtain →	square meter (m^2)	square inch (sq in.)	square foot (sq ft)	square centimeter (cm^2)	square millimeter (mm^2)
square meter	1	1,550	10.76	10,000	10^6
square inch	6.452×10^{-3}	1	6.94×10^{-3}	6.452	645.2
square foot	0.0929	144	1	929.0	92,903
square centimeter	0.0001	0.155	0.001	1	100
square millimeter	10^{-6}	0.00155	0.00001	0.01	1

DENSITY

Multiply Number of ↓ By ↘ \ To Obtain →	gm/cm^3	lb/cu ft	lb/gal
gram/cubic centimeter	1	62.43	8.345
pound/cubic foot	0.01602	1	0.1337
pound/gallon (U.S.)	0.1198	7.481	1

1 grain/cu ft = 2.28 mg/m^3

FORCE

To Obtain → *Multiply Number of* *By* ↘	*dyne*	*newton (N)*	*kilogram-force*	*pound-force (lbf)*
dyne	1	1.0×10^{-5}	1.02×10^{4}	2.248×10^{4}
newton	1.0×10^{5}	1	0.1020	0.2248
kilogram-force	9.807×10^{-5}	9.807	1	2.205
pound-force	4.448×10^{-5}	4.448	0.4536	1

MASS

To Obtain → *Multiply Number of* *By* ↘	*gram (gm)*	*kilogram (kg)*	*grains (gr)*	*ounce (avoir) (oz)*	*pound (avoir) (lb)*
gram	1	0.001	15.432	0.03527	0.00220
kilogram	1,000	1	15,432	35.27	2.205
grain	0.0648	6.480×10^{-5}	1	2.286×10^{-3}	1.429×10^{-4}
ounce	28.35	0.02835	437.5	1	0.0625
pound	453.59	0.4536	7,000	16	1

VOLUME

To Obtain → *Multiply Number of* *By* ↘	*cu ft*	*gallon (U.S. liquid)*	*liters*	*cm³*	*m³*
cubic foot	1	7.481	28.32	28,320	0.0283
gallon (U.S. liquid)	0.1337	1	3.785	3,785	3.79×10^{-3}
liter	0.03531	0.2642	1	1,000	1×10^{-3}
cubic centimeters	3.531×10^{-5}	2.64×10^{-4}	0.001	1	10^{-6}
cubic meters	35.31	264.2	1,000	10^{6}	1

VELOCITY

To Obtain → Multiply Number of By ↘	cm/s	m/s	km/hr	ft/s	ft/min	mph
centimeter/second	1	0.01	0.036	0.0328	1.968	0.02237
meter/second	100	1	3.6	3.281	196.85	2.237
kilometer/hour	27.78	0.2778	1	0.9113	54.68	0.6214
foot/second	30.48	0.3048	18.29	1	60	0.6818
foot/minute	0.5080	0.00508	0.0183	0.0166	1	0.01136
mile per hour	44.70	0.4470	1.609	1.467	88	1

PRESSURE

To Obtain → Multiply Number of By ↓	lb/sq in. (psi)	atm	in. (Hg) 32 F 0 C	mm (Hg) 32 F 0 C	kPa (kN/m²)	ft (H₂O) 60 F 15 C	in. (H₂O)	lb/sq ft
pound/square inch	1	0.068	2.036	51.71	6.895	2.309	27.71	144
atmospheres	14.696	1	29.92	760.0	101.32	33.93	407.2	2,116
inch (Hg)	0.4912	0.033	1	25.40	3.386	1.134	13.61	70.73
millimeter (Hg)	0.01934	0.0013	0.039	1	0.1333	0.04464	0.5357	2.785
kilopascals	0.1450	9.87×10^{-3}	0.2953	7.502	1	0.3460*	4.019	20.89
foot (H₂O)(15 C)	0.4332	0.0294	0.8819	22.40	2.989*	1	12.00	62.37
inch (H₂O)	0.03609	0.0024	0.073	1.867	0.2488	0.0833	1	5.197
pound/square foot	0.0069	4.72×10^{-4}	0.014	0.359	0.04788	0.016	0.193	1

*at 4C

Index